W0112173

GUIDE TO GAS CHROMATOGRAPHY LITERATURE

GUIDE TO
GAS CHROMATOGRAPHY
LITERATURE

by

Austin V. Signeur

Canisius College
Buffalo, New York

PLENUM PRESS
NEW YORK
1964

ISBN-13: 978-1-4684-6194-7   e-ISBN-13: 978-1-4684-6192-3
DOI: 10.1007/978-1-4684-6192-3

Library of Congress Catalog Card Number 64-20743

©1964 Plenum Press
Sofcover reprint of the hardcover 1st edition 1964
A Division of Consultants Bureau Enterprises, Inc.
227 West 17th Street
New York, N.Y. 10011

All rights reserved

No part of this publication may be reproduced in any
form without written permission from the publisher

PREFACE

The bibliography which follows represents an effort to provide the active or potential worker in the field of gas chromatography with references to the theory, methodology, and applications of this phase of chemistry. A review of the cited references will afford background for proposed applications, suggest possible solution of a problem, furnish an acquaintance with trends and current work being conducted, and furnish a realization of the possibilities and potentialities of a technique for the separation, identification, and—more recently—preparation of materials.

To augment the numerous literature references, titles of papers presented at various scientific meetings are given. Some of these papers have not been published, but they represent a part of the literature of this technique since they indicate the progress and thinking of workers in this field, and provide the opportunity for those with mutual interests to communicate with each other for further details.

To afford ready referral for additional information, references are given, when available, to Chemical Abstracts, or to the abstract in the program of the meeting. To accommodate those who may desire microfilm or photostatic copies of the published works, complete pagination is given rather than initial page references.

Austin V. Signeur

CONTENTS

1. Abe, H., "A Few of the New Stationary Liquids for Gas Chromatography," Kagaku to Sosa, 12, 527-535 (1959); CA 54, 15064e

2. Abe, Y., "Analysis of Trichlorosilane, Silicon Tetrachloride, Phosphorus Trichloride and Boron Trichloride by Gas Chromatography," Bunseki Kagaku, 9, 795-798 (1960); CA 55, 18421g

3. Abegg, H., "A Simple Program Control Apparatus for Gas Chromatography," J. Chromatog., 9, 405-410 (1962)

4. Abegg, H., "A Device for the Introduction of Small Sized Samples in Gas Chromatography Columns," J. Chromatog., 9, 519-521 (1962)

5. Abel, E. W., Nickless, G., and Pollard, F. H., "Vapor-Phase Chromatography of Organometallic Compounds: Alkyl Derivatives of the Group IVB Metals," Proc. Chem. Soc. (London), 1960, 288; CA 55, 2466c

6. Abernethy, R. F., and Christos, T., "Solid and Gaseous Fuels," Anal. Chem., 35, No. 5, 78R-88R (1963)

7. Abraham, M. H., Davies, A. G., Llewellyn, D. R., and Thain, E. M., "The Chromatographic Analysis of Organic Peroxides," Anal. Chim. Acta, 17, 499-503 (1957); CA 53, 120b

8. Abraham, N. A., "Stereochemistry of the Bornylic and Fenchylic Rearrangements in the Pinene Series," Ann. chim. (Paris), 5, 961-997 (1960); CA 55, 10494i

9. Abraham, S., Matthes, K. J., and Chaikoff, I. L., "Factors Involved in Synthesis of Fatty Acids from Acetate by a Soluble Fraction Obtained from Lactating Rat Mammary Gland," Biochim. et Biophys. Acta, 49, No. 2, 268-285 (1961); CA 55, 27474a

10. Abramovitch, R. A., Notation, A. D., and Seng, G. C., "Orientation in the Reaction of Phenyllithium with 3-Substituted Pyridines," Tetrahedron Letters, 1959, No. 8, 1-3; CA 54, 3411e

11. Abramovitch, R. A., Pepper, J. M., and Saha, J. G., "Terminal Olefins in Low-Temperature Lignite Tar," Chem. & Ind. (London), 1961, 368; CA 55, 23992a

12. Ache, H. J., Thiemann, A., and Herr, W., "Radio Gas-Chromatographic Analysis of Tritium-Labeled Aromatic Nitro and Halogen Compounds at High Temperatures," Z. Anal. Chem., 181, 551-560 (1961); CA 56, 2892h

13. Ackman, R. G., "Structure and Retention Time in the Gas-Liquid Chromatography of Unsaturated Fatty Acids on Polyester Substrates," Nature, 194, 970-971 (1962); CA 57, 6045i

14. Ackman, R. G., "Revision of the End Carbon Chain Concept in the Gas-Liquid Chromatography of Methyl Esters of Unsaturated Fatty Acids," Nature, 195, 1198 (1962); CA 57, 14926b

15. Ackman, R. G., "Correlation of Polyester Gas-Liquid Chromatography Retention Times with the Structures of Unsaturated Fatty Acid Methyl Esters," 36th Fall Meeting, American Oil Chemists' Society, Toronto, Ontario, October 1962

16. Ackman, R. G., "Influence of Column Temperature in the Gas-Liquid Chromatographic Separation of Methyl Esters of Fatty Acids on Polyester Substrates," J. Gas Chromatog., 1, No. 6, 11-16 (1963)

17. Ackman, R. G., Bannerman, M. A., Retson, M. E., and Vandenheuvel, F. A., "Application of the Varrentrapp Reaction to the Preparation of Long-Chain Dicarboxylic Acids," Can. J. Chem., 39, 1730-1732 (1961); CA 55, 27055c

18. Ackman, R. G., Bannerman, M. A., and Vandenheuvel, F. A., "Decomposition of Short-Chain Dicarboxylic Acid Esters during Separation on Polyester Gas Chromatography Media," Anal. Chem., 32, 1209 (1960); CA 54, 24131f

19. Ackman, R. G., Bannerman, M. A., and Vandenheuvel, F. A., "A Simple Apparatus for Preparative Scale Gas-Liquid Chromatography," J. Chromatog., 8, 44-51 (1962); CA 58, 1651h

20. Ackman, R. G., and Burgher, R. D., "Absorption of Alcohols in Gas-Liquid Chromatography Applied to the Determination of Non-Alcoholic Impurities," J. Chromatog., 6, 541-543 (1961); CA 56, 13527h

21. Ackman, R. G., and Burgher, R. D., "The Effect of Detector Block Fouling on Observed Separations in Gas-Liquid Chromatography," Anal. Chem., 35, 413-414 (1963); CA 58, 11933b

22. Ackman, R. G., and Burgher, R. D., "Quantitative Gas-Liquid Chromatographic Estimation of Volatile Fatty Acids in Aqueous Media," Anal. Chem., 35, 647-652 (1963)

23. Ackman, R. G., Burgher, R. D., Sipos, J. C., and Odense, P. H., "Carbonaceous Fouling of Catherometer Filaments," J. Chromatog., 9, 531-534 (1962)

24. Ackman, R. G., Retson, M. E., Gallay, L. R., and Vandenheuvel, F. A., "Ozonolysis of Unsaturated Fatty Acids. I. Ozonolysis of Oleic Acid," Can. J. Chem., 39, 1956-1963 (1961); CA 56, 4885h

25. Ackman, R. G., and Vandenheuvel, F. A., "Ozonolysis of Ethylenic Fatty Acids with Gas Chromatographic Analysis of the Products," 44th CIC Meeting, Montreal, Canada, August 1961

26. Acree, Jr., F., and Beroza, M., "Quantitative Gas Chromatography of Insect Repellent Mixture M-1960," J. Econ. Entomol., 55, 469-471 (1962); CA 57, 17131i

27. Acree, Jr., F., and Beroza, M., "Quantitative Gas Chromatography of Isomers of Insect Repellent N, N-Diethyltoluamide," J. Econ. Entomol., 55, 619-622 (1962); CA 58, 7311c

28. Acrivos, A., "Combined Effect of Longitudinal Diffusion and Mass Transfer Resistance in Fixed Bed Operation, " Chem. Eng. Sci., 13, 1-6 (1960); CA 55, 5055d

29. Adams, A. W., Hawkins, E. G. E., Oldham, G. F., and Thompson, R. D., "The Reaction of 1, 4-Dichloro-2, 3-epoxybutane with Sodium Methoxide, " J. Chem. Soc., 1959, 559-568; CA 53, 15963g

30. Adams, D. F., "Silica Gel Catches Sulfur Smells," Chem. & Eng. News, 37, No. 39, 106 (Sept. 28, 1959)

31. Adams, D. F., and Koppe, R. K., "Application of Instrumentation to Pulp Mill Atmospheric Discharges, " Tappi, 41, 366-372 (1958); CA 52, 17703d

32. Adams, D. F., and Koppe, R. K., "Gas Chromatographic Analysis of Hydrogen Sulfide, Sulfur Dioxide, Mercaptans, and Alkyl Sulfides and Disulfides, " Tappi, 42, 601-605 (1959); CA 53, 21361c

33. Adams, D. F., Koppe, R. K., and Jungroth, D. M., "Adsorption Sampling and Gas Chromatographic Analysis of Sulfur Compounds in Waste Process Gases," Tappi, 43, 602-608 (1960); CA 54, 25422b

34. Adams, G. A., and Bishop, C. T., "Constitution of an Arabinogalactan from Maple Sap, " Can. J. Chem., 38, 2380-2386 (1960); CA 55, 15361g

35. Adams, V. H., and Fraade, D. J., "New Developments in Stream Analysis, " IRE Trans. on Ind. Electronics, PGIE-11, 1-9 (December 1959)

36. Addison, L. M., "Design of a High-Performance Straight-Wire Thermal Conductivity Detector, " 8th Natl. Instrumental Analysis Symposium, ISA, Charleston, W. Va., April-May 1962

37. Addison, L. M., and Lane, L. H., "Gas-Chromatographic Integrator and Digitizer for Routine Laboratory Use, " Z. Anal. Chem., 189, 80-90 (1962); CA 57, 14412c

38. Addison, W. E., and Barrer, R. M., "Sorption and Reactivity of Nitrous Oxide and Nitric Oxide in Crystalline and Amorphous Siliceous Sorbents, " J. Chem. Soc., 1955, 757-769; CA 49, 8657f

39. Adlard, E. R., "An Evaluation of Some Polyglycols Used as Stationary Phases for Gas-Liquid Partition Chromatography, " in "Vapor Phase Chromatography, " edited by D. H. Desty, Academic Press, Inc., New York, 1957, pp. 98-114

40. Adlard, E. R., "Values and Partition Coefficients for Compounds on PEG 400, PEG 350, and PPG 425 (Chromatographic Data Table XLV), " J. Chromatog., 1, xxx (1958); cf. "Gas Chromatography," edited by D. H. Desty, Butterworths, London, 1957, p. 105

41. Adlard, E. R., and Hill, D. W., "Analysis of Anesthetic Mixtures by Gas Chromatography, " Nature, 186, 1045 (1960); CA 54, 25590d

42. Adlard, E. R., Khan, M. A., and Whitman, B. T., "Determination and Use of the Specific Retention Volumes of Benzene and Cyclohexane in Dinonyl Phthalate, " in "Gas Chromatography 1960, " edited by R. P. W. Scott, Butterworths, London, 1960, pp. 251-272

43. Adlard, E. R., Khan, M. A., and Whitham, B. T., "The Application of Capillary Columns in the Study of the Thermodynamic Behaviour of Ethanol and Carbon Tetrachloride in Dinonyl Phthalate," 4th Intern. Gas Chromatography Symp., Hamburg, Germany, June 1962; pub. in "Gas Chromatography 1962," ed. by M. van Swaay, Butterworth & Co., London, 1962

44. Adlard, E. R., and Whitham, B. T., "Applications of High Temperature Gas-Liquid Chromatography in the Petroleum Industry, " in "Gas Chromatography 1958, " edited by D. H. Desty, Academic Press, Inc., New York, 1958, pp. 351-368

45. Adlard, E. R., and Whitham, B. T., "Behavior of a 5A Molecular Sieve in Subtractive Gas Chromatography, " Nature, 192, 966 (1961); CA 56, 5422b

46. Adlard, E. R., and Whitham, B. T., "The Analysis of Petroleum Fractions by Subtractive Gas Chromatography, " 3rd Intern. Symp. on Gas Chromatography, ISA, East Lansing, Mich., June 1961; ISA Proc., 3, 243-252 (1961)

47. Adlard, E. R., and Whitham, B. T., "The Analysis of Petroleum Fractions by Subtractive Gas Chromatography, " in "Gas Chromatography," edited by N. Brenner, J. E. Callen, and M. D. Weiss, Academic Press, Inc., New York, 1962, pp. 371-390

48. Adlard, E. R., and Whitham, B. T., "Analysis of Petroleum Fractions by Subtractive Gas Chromatography, " Gas Chromatog., Intern. Symp., 1961, 3, 371-390 (Pub. 1962); CA 58, 4354e

49. Adloff, J. P., "Techniques and Applications of Gas-Phase Radiochromatography, " J. Chromatog., 6, 373-380 (1961); CA 56, 12287b

50. Adloff, J. P., "Chromatographic Analysis of Radiolytic and Electrical Decomposition Products with a Flame Ionization Detector," Journees Intern. Etude Methodes Separation Immediate Chromatog., Paris, 1961, 299-305 (Pub. 1962); CA 59, 1213h

51. Adloff, J. P., "Gas Chromatography of Radioactive Substances. Techniques and Applications," in "Chromatographic Reviews," Vol. 4, edited by M. Lederer, American Elsevier Publishing Co., New York, 1962

52. Aepli, O. T., and Collier, H. G., "Determination of Lower Aliphatic Alcohols by Gas Chromatography," 8th Detroit Anachem Conf., Detroit, Mich., October 1960

53. Aeschbacher, R., "Gas Chromatography in Stearin [Fatty Acids] Manufacture, " Mitt. Lebensm. Hyg., 51, 525-532 (1960); CA 57, 7402e

54. Aeschbacher, R., "Gas Chromatography in the Manufacture of Stearic Acid, " Olearia, 16, 65-69 (1962); CA 58, 11207g

55. Agre, C. L., and Cason, J., "Complexity of the Mixture of Fatty Acids from Tubercle Bacillus. Acids with Less Than Twenty Carbon Atoms," J. Biol. Chem., 234, 2555-2559 (1959); CA 54, 2494g

56. Ahrens, E. H., Insull, Jr., W., Hirsch, J., Stoffel, W., Peterson, M. L., Farquhar, J. W., Miller, T., and Thomasson, H. J., "The Effect on Human Serum-Lipids of a Dietary Fat, Highly Unsaturated, But Poor in Essential Fatty Acids," Lancet, 1, 115-119 (1959); CA 53, 7349g

57. Ahrens, R. W., Sauer, Jr., M. C., and Willard, J. E., "Hydrogen Labelling of Hydrocarbons Using Ionizing Radiation," J. Am. Chem. Soc., 79, 3285-3286 (1957); CA 51, 14428g

58. Aikman, A. R., "Process Control by Analytical Instrumentation," Chem. Eng., 64, No. 6, 266-273 (1957); CA 51, 11775a

59. Aivazov, B. V., "Separation of Hydrocarbon Mixtures and Organic Sulfur-Containing Compounds by Vapor-Phase Chromatography," Khromatog., ee Teoriya i Primenenie, Akad. Nauk S.S.S.R., Trudy Vsesoyuz. Soveshchaniya, Moscow, 1958, 438-440 (Pub. 1960); CA 55, 18450i

60. Aivazov, B. V., and Vyakhirev, D. A., "Separation of a Mixture of Simplest Hydrocarbons by the Method of Chromathermography," Zhur. Priklad., Khim., 26, 505-511 (1953); CA 48, 6371h

61. Akamatsu, H., "Adsorption at the Solid-Liquid Interface. I. Adsorption of Some Fatty Acids by Glass," Bull. Chem. Soc. Japan, 17, 141-146 (1942); CA 41, 4349g

62. Akamatsu, H., "Adsorption at the Solid-Liquid Interface. II. The Thickness of the Adsorption Layer," Bull. Chem. Soc. Japan, 17, 161-165 (1942); CA 41, 4349h

63. Akamatsu, H., "Adsorption at the Solid-Liquid Interface. III. Polar and Nonpolar Adsorption," Bull. Chem. Soc. Japan, 17, 260-267 (1942); CA 41, 4350b

64. Akawie, R. I., "Synthesis of Deuterated Biphenyls. II.," J. Org. Chem., 26, 243-245 (1961); CA 55, 18658d

65. Akawie, R. I., Scarborough, J. M., and Burr, J. G., "Synthesis of Deuterated Biphenyls," J. Org. Chem., 24, 946-949 (1959); CA 54, 22481f

66. Akiyoshi, S., Matsuda, T., and Akune, K., "Gas Chromatography. I. Decomposition of Polyethylene Glycol on Firebrick Support for Gas Chromatography," Bunseki Kagaku, 10, 960-965 (1961); CA 58, 6165e

67. Akoyunoglon, G. A., and Calvin, M., "Some Reactions of Diazomethane with Carbon Dioxide and Ammonia in Aqueous Solution," J. Org. Chem., 28, 1484-1487 (1963); CA 59, 1481h

68. Aksenov, M. Z., Fel'dman, Y. S., and Vidavskii, N. N., "KhPA-2 Chromatograph," Novosti Neft. i Gaz. Tekhn., Neft. Obrud. i Sredstva Avtomatiz., 1962, No. 1, 28-31; CA 58, 6624c

69. Albrecht, H., "Hydrogenation of Monovinylacetylene," Monatsber. Deut. Akad. Wiss. Berlin, 3, 714-718 (1961); CA 58, 2354a

70. Albrecht, W. L., "I. Exchange of Isotopic Hydrogen Chloride and Aromatic Hydrocarbons. II. Tritium Labeling Caused by Ionizing Radiation in Gaseous Mixtures of Tritium and Aromatic Hydrocarbons," Univ. Microfilms (Ann Arbor, Mich.), L. C. Card No. Mic 60-2418, 139 pp.; Dissertation Abstr., 21, 67-68 (1960); CA 55, 2253i

71. Alekseeva, A. V., and Gol'bert, K. A., "Determination of Trace Impurities in Pure Ethylene Intended for Production of Polyethylene," Zavodskaya Lab., 27, 972-975 (1961); CA 56, 6666f

72. Alekseeva, K. V., Zhukhovitskii, A. A., and Turkel'taub, N. M., "The Effect of Different Parameters in Preparative Chromatography," Khim. i Tekhnol. Topliv i Masel, 7, No. 4, 60-66 (1962); CA 57, 2009i

73. Alexander, D., and Marsh, R. F., "High-Temperature Gas Chromatograph," Gas Chromatog., Intern. Symposium, 2nd, East Lansing, Mich., 1959, 163-169 (Pub. 1961); CA 55, 18204g

74. Alexander, D., and Marsh, R. F., "A High-Temperature Gas Chromatography," in "Gas Chromatography," edited by H. J. Noebels, R. F. Wall, and N. Brenner, Academic Press, Inc., New York, 1961, pp. 163-169

75. Alexander, K. M., and Lusena, C. F., "Fractionation of the Lipoproteins of the Fat Globule Membrane from Cream," J. Dairy Sci., 44, 1414-1419 (1961); CA 55, 27686d

76. Alford, J. A., Elliott, L. E., and Blankenship, L. C., "Action of Bacterial Lipases on Natural Fats as Determined by Gas Chromatography," 20th Annual Mtg., Institute of Food Technologists, San Francisco, Calif., May 1960; Abstr., Food Technol., 14, No. 4, 25 (1960)

77. Alford, J. A., Elliott, L. E., Horstein, I., and Crowe, P. F., "Lipolytic Activity of Microorganisms at Low and Intermediate Temperatures. II. Fatty Acids Released as Determined by Gas Chromatography," J. Food Sci., 26, 234-238 (1961); CA 55, 26292h

78. Aliev, Y. Y., and Aminova, R. I., "Application of Chromatographic Methods for Analysis of Separator and Natural Gas," Issled. Mineral'n. i Rast. Syr'ya Uzbekistana, Akad. Nauk Uz. SSR, Inst. Khim., 1962, 106-115; CA 12341e

79. Aliprandi, B., and Cacace, F., "Labeling of Organic Compounds by Exchange with Tritiated Water," Ann. chim. (Rome), 49, 2011-2016 (1959); CA 54, 18395h

80. Aliprandi, B., and Cacace, F., "Labeling of Organic Compounds by Substitution with Tritium Oxide," Ann. chim. (Rome), 50, 931-934 (1960); CA 55, 4338d

81. Aliprandi, B., and Cacace, F., "Radioactive Carboxylic Acids Obtained from a Gaseous Mixture of Pentane and Carbon-14 Dioxide by Beta-Radiation," Gazz. chim. ital., <u>89</u>, 2268-2274 (1959); CA <u>55</u>, 6364i

82. Aliprandi, B., and Cacace, F., "Labeling Aromatic Compounds by Exchange with Tritiated Trifluoro Acetic Acid. III," Ann. chim. (Rome), <u>51</u>, 397-401 (1961); CA <u>55</u>, 27152h

83. Aliprandi, B., Cacace, F., and Cieri, L., "Effect of Conditions of Irradiation on the Distribution of Radioactive Atoms in the Toluene Molecule Labeled by Exchange with Gaseous Tritium," Ricerca sci., <u>30</u>, 90-96 (1960); CA <u>54</u>, 23652e

84. Allegrini, F., Cieri, L., and Ciranni, E., "The Nature of Fats Produced by Rhodotorula Gracilis," Ricerca sci., Rend. Sez. B, <u>2</u>, 258-263 (1962); CA <u>59</u>, 1978f

85. Allen, P. W., Everett, D. H., and Penny, M. F., "The Thermodynamics of Hydrocarbon Solutions. I. Technique of Vapor-Pressure Measurements; The Vapor Pressure of Benzene," Proc. Roy. Soc. (London), <u>A212</u>, 149-163 (1952); CA <u>46</u>, 10836c

86. Allen, R. H., and Yats, L. D., "Kinetics of Three-Compound Equilibrations. II. The Isomerization of Xylene," J. Am. Chem. Soc., <u>81</u>, 5289-5292 (1959); CA <u>54</u>, 7591i

87. Allen, R. H., Yats, L. D., and Erley, D. S., "Kinetics of Three-Compound Equilibrations. III. The Isomerization of Ethyltoluene," J. Am. Chem. Soc., <u>82</u>, 4853-4856 (1960); CA <u>55</u>, 9310a

88. Allen, R. R., "Chromatography," J. Am. Oil Chemists' Soc., <u>32</u>, 638-640 (1955); CA <u>50</u>, 1338h

89. Allinger, N. L., and Coke, J. L., "The Relative Stabilities of cis and trans Isomers. VI. The Decalins," J. Am. Chem. Soc., <u>81</u>, 4080-4082 (1959); CA <u>54</u>, 3269c

90. Allinger, N. L., and Coke, J. L., "The Relative Stabilities of cis and trans Isomers. VII. The Hydrindans," J. Am. Chem. Soc., <u>82</u>, 2553-2556 (1960); CA <u>54</u>, 22522b

91. Allinger, N. L., and Coke, J. L., "Relative Stabilities of cis and trans Isomers. XI. The 9-Methyldecahydronaphthalenes," J. Org. Chem., <u>26</u>, 2096-2099 (1961); CA <u>55</u>, 27220h

92. Allinger, N. L., Conia, J. M., Ripoll, J. L., Tushaus, L. A., and Neumann, C. L., "Conformational Analysis. XXXVI. Some Consequences of the Nonplanarity of Cyclobutane and Cyclobutanone Rings," J. Am. Chem. Soc., <u>84</u>, 4982-4983 (1962); CA <u>58</u>, 10059a

93. Allinger, N. L., and Curby, R. J., "The Relative Stabilities of cis and trans Isomers. X. The 1,3-Cyclohexanedicarboxylate Esters," J. Org. Chem., <u>26</u>, 933-935 (1961); CA <u>55</u>, 27135i

94. Allinger, N. L., Nakazaki, M., and Zalkow, V., "The Relative Stabilities of cis and trans Isomers. V. The Bicycle "5.2.0" Nonanes. An Extension of the Conformational Rule," J. Am. Chem. Soc., <u>81</u>, 4074-4080 (1959); CA <u>54</u>, 3267g

95. Allison, G. M., Gibson, S. P., and Atherley, J. F., "A Continuous Stripper for the Determination of Dissolved Gases and Fission-Product Gases in High-Temperature and Pressure Water Systems," Analytical Chem. in Nuclear Reactor Techn., U. S. At. Energy Comm., <u>TID-7606</u>, 330-343 (1960); CA <u>55</u>, 23097g

96. Allred, E. L., Sonnenberg, J., and Winstein, S., "Preparation of Homobenzyl and Homoallyl Alcohols by the Hydroboration Method," J. Org. Chem., <u>25</u>, 26-29 (1960); CA <u>54</u>, 15287h

97. Alm, J., Driscoll, J. L., Smith, W. R., and Gudzinowicz, B. J., "Application of Gas Chromatography to the Analysis of Thermal Degradation Products of Hexadecane," 139th Natl. ACS Mtg., St. Louis, Mo., March 1961, Program Abstr., p. 7Q; Am. Chem. Soc., Div. Petrol. Chem., Symp., <u>6</u>, No. 2B, 41-53 (1961); CA <u>58</u>, 2352h

98. Almond, J., "A Simple Gas Inlet for Use with Vacuum Systems," J. Sci. Instr., <u>35</u>, No. 2, 70 (1958)

99. Altenau, A. G., and Rogers, L. B., "New Adsorbents for Gas Chromatography," 14th Pittsburgh Conf. on Anal. Chem. & Appl. Spectroscopy, Pittsburgh, Pa., March 1963, Program Abstr., p. 54

100. Altenau, A. G., and Rogers, L. B., "Characterization of New Adsorbents for Gas Chromatography," 145th Natl. ACS Mtg., New York, N. Y., September 1963

101. Alton, A. B., "Chromatography Has Become a Valuable Tool as a Guide to Fractionation Operation," Oil Gas J., <u>56</u>, No. 16, 122 (1958); CA <u>52</u>, 15883f

102. Alton, A. B., "Chromatography as a Guide to Fractionator Operations," 37th Annual Mtg., Natural Gasoline Assoc. of America, Dallas, Texas, April 1958; Tech. Papers, <u>37</u>, 13-14 (1958); CA <u>54</u>, 1834g

103. Altschuller, L. W., and Shreve, O. D., "Techniques and Procedures in the Application of Gas Chromatography to Paint Solvent Analysis," 3rd Delaware Valley Regional Mtg., ACS, Symposium on Vapor Phase Chromatography, Philadelphia, Pa., February 1960

104. Altshuller, A. P., "Air Pollution," Anal. Chem., <u>35</u>, No. 5, 3R-10R (1963)

105. Altshuller, A. P., and Bellar, T. A., "Gas Chromatographic Analysis of Hydrocarbons in the Los Angeles Atmosphere," J. Air Pollution Control Assoc., <u>13</u>, No. 2, 81-87 (1963); CA <u>58</u>, 11888f

106. Altshuller, A. P., Bellar, T. A., and Clemons, C. A., "Concentration of Hydrocarbons on Silica Gel Prior to Gas Chromatographic Analysis," Am. Ind. Hyg. Assoc. J., <u>23</u>, No. 2, 164-166 (1962); CA <u>57</u>, 6271g

107. Altshuller, A. P., and Clemons, C. A., "Gas Chromatographic Analysis of Aromatic Hydrocarbons at Atmospheric Concentrations Using Flame Ionization Detection," Anal. Chem., <u>34</u>, 466-472

(1962); CA <u>57</u>, 1224i; Correction, ibid., p. 747; 140th Natl. ACS Mtg., Chicago, Ill., September 1961, Program Abstr., pp. 37B, 12W

108. Altshuller, A. P., and Cohen, I. R., "Application of Diffusion Cells to the Production of Known Concentrations in Gaseous Hydrocarbons," Anal. Chem., <u>32</u>, 802-810 (1960); CA <u>54</u>, 17980d

109. Altshuller, A. P., and Cohen, I. R., "The Gas Phase Reactions of Nitrogen Oxides with Olefins," Intern. J. Air and Water Pollution, <u>4</u>, Nos. 1/2, 55 (1961); CA <u>55</u>, 20916e

110. Altshuller, A. P., and Sieva, S. F., "Vapor Phase Determination of Olefins by a Coulometric Method," Anal. Chem., <u>34</u>, 418-422 (1962); CA <u>57</u>, 5305d

111. Alvarado, G., and Manjarrez, A., "Separation and Determination of Fragrant Compounds by Gas Chromatography," Bol. Inst. Quim. Univ. Nal. Auton. Mex., <u>13</u>, 6-13 (1961); CA <u>57</u>, 2354c

112. Alvazov, B. V., "Separation of Hydrocarbon Mixtures and Organic Sulfur-Containing Compounds by Vapor-Phase Chromatography," Khromatog., ee Teoriya i Primenen., Akad. Nauk S.S.S.R., Trudy Vsesoyuz Soveshchaniya, Moscow, <u>1958</u>, 438-440 (Pub. 1960); CA <u>55</u>, 18450i

113. Amaya, K., and Sasaki, K., "The Mutual Interaction Between Solutes in Gas Chromatography," Bull. Chem. Soc. Japan, <u>35</u>, 1507-1510 (1962); CA <u>57</u>, 14456e

114. Amberg, C. H., "Gas-Liquid Partition Chromatography of Organic Sulfur Compounds," Can. J. Chem., <u>36</u>, 590-592 (1958); CA <u>52</u>, 8826c

115. Amberg, C. H., "Organic Sulfur Compounds in Alberta Cracked Distillate," J. Inst. Petrol., <u>45</u>, 1-8 (1959); CA <u>53</u>, 5650c

116. Amberg, C. H., Echigoya, E., and Kulanovic, D., "Quantitative Gas Chromatography of Reaction Products from the Catalytic Oxidation of Ethylene," Can J. Chem., <u>37</u>, 708-713 (1959); CA <u>53</u>, 16796b

117. Ambrose, D., "Gas Chromatography," Nature, <u>186</u>, 943-944 (1960) Report on Meeting of Gas Chromatography Discussion Group, Inst. Petroleum Research, London, April 1960

118. Ambrose, D., "Gas Chromatography Discussion Group," Nature, <u>194</u>, 824-825 (1962) Report on meeting of Gas Discussion Group, Inst. Petroleum Research, March 1962

119. Ambrose, D., and Ambrose, B. A., "Gas Chromatography," George Newnes, Ltd., London, 1961; D. Van Nostrand Co., Inc., New York, 1961

120. Ambrose, D., and Collerson, R. R., "A Thermal Conductivity Gauge for Use in Gas-Liquid Partition Chromatography," J. Sci. Instr., <u>32</u>, 323 (1955)

121. Ambrose, D., and Collerson, R. R., "Use of Gas-Liquid Partition Chromatography as a Preparative Method," Nature, <u>177</u>, 84 (1956); CA <u>50</u>, 8262h

122. Ambrose, D., James, A. T., Keulemans, A. I. M., Kovats, E., Rock, H., Rouit, C., and Stross, F. H., "Preliminary Recommendations on Nomenclature and Presentation of Data in Gas Chromatography," in "Gas Chromatography," edited by R. P. W. Scott, Butterworths, London, 1960, pp. 423-432

123. Ambrose, D., (Chairman), et al., "Terminology for Gas Chromatography Recommended by IUPAC (International Union of Pure and Applied Chemistry) Committee," F & M Facts and Methods for Scientific Research, <u>1</u>, No. 1, 1-4 (Spring 1960)

124. Ambrose, D., (Chairman), James, A. T., Keulemans, A. I. M., Kovats, E., Rock, H., Rouit, C., and Stross, F. H., (Committee, Section of Anal. Chem., IUPAC), "Preliminary Recommendations on Nomenclature and Presentation of Data in Gas Chromatography," Preprints, 3rd Symp. on Gas Chromatography, Edinburgh, Scotland, June 1960, pp. AA283-AA292

125. Ambrose, D., Keulemans, A. I. M., and Purnell, J. H., "Presentation of Gas-Liquid Chromatography Retention Data," Anal. Chem., <u>30</u>, 1582-1586 (1958), CA <u>53</u>, 6868b; 132nd Natl. ACS Mtg., New York, N. Y., September 1957, Program Abstr., pp. 39B-40B; Preprints, Div. Petrol. Chem., <u>2</u>, No. 4, D143

126. Ambrose, D., and Purnell, J. H., "Presentation of Gas-Liquid Chromatographic Retention Data. II," in "Gas Chromatography 1958," edited by D. H. Desty, Academic Press, Inc., New York, 1958, pp. 369-372

127. Amell, A. R., Lamprey, P. S., and Schiek, R. C., "Gas Chromatographic Separation of Simple Aliphatic Amines," Anal. Chem., <u>33</u>, 1805-1806 (1961); CA <u>56</u>, 7973g

128. Amenomiya, Y., "Utilization of Gas Chromatography in a Laboratory," Kagaku no Ryoiki, <u>12</u>, No. 8, 557 (1958); CA <u>52</u>, 19338i

129. Amenomiya, Y., "Podbielniak's Chromacon 9475-3A," Kagaku no Ryoiki, <u>12</u>, 606-607 (1958); CA <u>52</u>, 19339a

130. American Society for Testing Materials, "Tentative Method for Analysis of Certain Light Hydrocarbons by Gas Chromatography," ASTM Designation: D1717-61T, Part 7, "1961 Book of ASTM Standards - Petroleum Products and Lubricants," pp. 900-904

131. Amos, R., and Hurrell, R. A., "The Design of High-Efficiency Packed Columns for Use with Katharometer Detectors," 4th Intern. Gas Chromatography Symp., Hamburg, Germany, June 1962, pub. in "Gas Chromatography 1962," ed. by M. Van Swaay, Butterworth & Co., London, 1962

132. Amy, J., "Preparative Gas Chromatography in Academic Research," 4th Annual Eastern Analytical Symp., New York, N. Y., November 1962, Program Abstr., p. 25

133. Amy, J. W., "Gas Chromatography Features in Instruments for Education," Perkin-Elmer Instrument News, 13, No. 3, 1, 6-7 (Spring, 1962)

134. Amy, J. W., "Gas Chromatography Sample Injection - Practices and Problems," 4th Annual Gas Chromatography Institute, Canisius College, Buffalo, N. Y., April 1962

135. Amy, J. W., and Baitinger, W. E., "Gas Chromatography Sample Injection," in "Lectures on Gas Chromatography 1962," edited by H. A. Szymanski, Plenum Press, Inc., New York, 1963, pp. 19-31; CA 58, 10701g

136. Amy, J. W., Brand, L., and Baitinger, W. E., "The Effect of Restrictions in Large Columns Used for Gas Chromatography," 12th Pittsburgh Conf. on Anal. Chem. & Appl. Spectroscopy, Pittsburgh, Pa., February-March 1961, Program Abstr., p. 53

137. Amy, J. W., Brand, L., and Baitinger, W. E., "The Effect of Restrictions in Large Columns Used for Gas Chromatography," in "Progress in Industrial Gas Chromatography," Vol. 1, edited by H. A. Szymanski, Plenum Press, Inc., New York, 1961, pp. 147-161; CA 56, 7967d

138. Amy, J. W., and Dimick, K. P., "Multichannel Gas Chromatography," 14th Pittsburgh Conf. on Anal. Chem. & Appl. Spectroscopy, Pittsburgh, Pa., March 1963, Program Abstr., p. 65

139. Analytical Methods Committee, Essential Oils Sub-Committee of the Society for Analytical Chemistry, Great Britain, "Application of Gas-Liquid Chromatography to Essential Oil Analysis. Interim Report of Determination of Citronellol in Admixture with Geranoil," Analyst, 84, 690-691 (1959)

140. Anders, M. W., and Mannering, G. J., "Gas Chromatography of Some Pharmacologically Active Phenothiazines," J. Chromatog., 7, 258-260 (1962); CA 57, 10131b

141. Anders, M. W., and Mannering, G. J., "New Peak-Shift Technique for Gas-Liquid Chromatography. Preparation of Derivatives on the Column," Anal. Chem., 34, 730-733 (1962); CA 57, 2825a

142. Anders, V. R., Frolovskii, P. A., Remnev, V. F., and Slobodkin, M. S., "Automatic Chromatograph for the Analysis of Hydrocarbon Gases in Flow," Khim. i Tekhnol. Topliv i Masel, 4, No. 3, 25-29 (1959); CA 53, 10856b

143. Anderson, B. C., "Separation of Some Mixtures of Isomeric Organic Compounds by Gas Chromatography," Gas Chromatog. Intern. Symposium, 2nd, East Lansing, Mich., 1959, 327-332 (Pub. 1960); CA 55, 23309d

144. Anderson, B. C., "Separation of Some Mixtures of Isomeric Organic Compounds by Gas Chromatography," in "Gas Chromatography," edited by H. J. Noebels, R. F. Wall, and N. Brenner, Academic Press, Inc., New York, 1961, pp. 327-332

145. Anderson, C. B., and Winstein, S., "Oxidation of Cyclohexene by Thallic and Other Metal Acetates," J. Org. Chem., 28, 605-606 (1963); CA 58, 10060d

146. Anderson, D. M. W., "Application of Infrared Spectroscopy. The Identification and Determination of Gas Chromatographic Fractions," Analyst, 84, 50-55 (1959); CA 53, 13867a

147. Anderson, D. M. W., "An Application of G.L.C. - Infrared Technique," Analyst, 85, 163 (1960)

148. Anderson, D. M. W., "Some Aspects of Quantitative Vapor-Phase Infrared Spectroscopy," Chem. & Ind. (London), 1960, 647 Summary of papers presented at meeting of Edinburgh and East of Scotland Section, Society of Chemical Industry, Edinburgh, March 1960

149. Anderson, D. M. W., "Applications of Infrared Spectroscopy. VI Recent Developments in Techniques of Increased Sensitivity for the Analysis of GLC [Gas-Liquid Chromatography] Fractions and Other Small Scale Samples," Talanta, 8, 832-835 (1961); CA 56, 7968b

150. Anderson, D. M. W., and Duncan, J. L., "The Identification and Quantitative Estimation of Gas Chromatography Fractions by Vapor-Phase Infrared Spectroscopy," Chem. & Ind. (London), 1958, 1662; CA 53, 8913b

151. Anderson, J. C., and Reese, C. B., "The Photochemical Fries Reaction," J. Chem. Soc., 1963, 1781-1784; CA 58, 7983d

152. Anderson, J. R., "Partition Coefficients from Gas-Liquid Partition Chromatography," J. Am. Chem. Soc., 78, 5692-5693 (1956); CA 51, 1690g

153. Anderson, J. R., and Baker, B. G., "Hydrocracking of Neopentane and Neohexane Over Evaporated Metal Films," Nature, 187, 937-938 (1960); CA 55, 2071e

154. Anderson, J. R., and Napier, K. H., "Vapor-Phase Chromatographic Separation of Aromatics from Saturated and Olefinic Hydrocarbons," Australian J. Chem., 9, 541-543 (1956); CA 51, 5717d

155. Anderson, J. R., and Napier, K. H., "Thermodynamic Data from Gas-Liquid Chromatography," Australian J. Chem., 10, 250-255 (1957); CA 51, 17329i

156. Anderson, J. R., Stein, K. C., Feenan, J. J., and Hofer, L. J. E., "Catalytic Oxidation of Methane," Ind. Eng. Chem., 53, 809-812 (1961)

157. Anderson, R. J., "High-Temperature Gas Chromatography," in "Progress in Industrial Gas Chromatography," Vol. 1, edited by H. A. Szymanski, Plenum Press, Inc., New York, 1961, pp. 73-95; CA 56, 7967c

158. Anderson, R. J., "Recent Developments in Chromatography," in "Progress in Industrial Gas Chromatography," Vol. 1, edited by H. A. Szymanski, Plenum Press, Inc., New York, 1961, pp. 3-17;

CA 56, 7967b

159. Anderson, W. R., and Flipse, R. J., "Origin and Fatty Acid Composition of the Glycerides in the Semen of Dairy Bulls," 58th Annual Mtg., Am. Dairy Sci. Assoc., Lafayette, Ind., June 1963; Abstr., J. Dairy Sci., 46, 619 (1963)

160. Andreatch, A. J., "Industrial Health Foundation. XIII. Flame Ionization Detector. Application to Industrial Hygiene and Air Pollution Studies," Arch. Environmental Health, 4, 317-319 (1962); CA 57, 17012h

161. Andreatch, A. J., and Feinland, R., "Continuous Trace Hydrocarbon Analysis by Flame Ionization," Anal. Chem., 32, 1021-1024 (1960); CA 54, 18200h

162. Andreen, B. H., and Kniebes, D. V., "Analysis of Natural Gas for Organic Sulfur Compounds by Two-Stage Gas Chromatography," 13th Pittsburgh Conf. on Anal. Chem. & Appl. Spectroscopy, Pittsburgh, Pa., March 1962, Program Abstr., p. 54

163. Andrew, T. D., Phillips, C. S. G., and Semlyen, J. A., "Some Applications of an R. F. Oscillator in Gas Chromatography," J. Gas Chromatog., 1, No. 1, 27-30 (1963)

164. Andronikashvili, T. G., "The Effect of Certain Factors on the Separation of the C_5-C_7 Hydrocarbons by the Chromatographic Method," Soobscheniya Akad. Nauk Gruzin, S.S.S.R., 19, No. 3, 273-278 (1957); CA 53, 21429f

165. Andronikashvili, T. G., and Kuz'mina, L. P., "Chromatographic Analysis of Saturated Hydrocarbons C_3-C_7 on Natural Sorbents," Zavodskaya Lab., 22, 1403-1406 (1956); CA 51, 17606a

166. Andronikashvili, T. G., and Sabelashvili, S. D., "Gas Adsorption Chromatography on Synthetic Zeolites," Sintetich, Tseolity, Poluchenie, Issled. i Primenenie, Akad. Nauk SSSR, Otd. Khim. Nauk, 1962, 65-67; CA 58, 10759h

167. Andronikashvili, T. G., Sabelashvili, S. D., and Tsitsishvili, G. V., "Gas-Chromatographic Investigation of the Separating Properties of X-Type Sodium and Silver Forms of Molecular Sieves," Neftekhimiya, 2, 248-252 (1962); CA 58, 5425h

168. Andrychuk, D., Edds, D. L., and Knapp, R. E., "The Gas Chromatographic Analysis of Chloral," 8th Pittsburgh Conf. on Anal. Chem. & Appl. Spectroscopy, Pittsburgh, Pa., March 1957; Program Abstr., p. 34

169. Anet, F. A. L., and Hartman, J. S., "Ring Inversion in Cyclooctane," 144th Natl. ACS Mtg., Los Angeles, Calif., March-April 1963, Program Abstr., p. 25M

170. Angele, H. P., (Ed.), "Gas Chromatographie 1958," Akademie Verlag, Berlin, 1959 (in German) First Symp. on Gas Chromatography, Leipzig, East Germany, October 1958, 338 pp.

171. Angele, H. P., and Hofmann, M., "Survey of the Gas Chromatographic Literature in 1959," in "Gas Chromatographie 1959," edited by R. E. Kaiser and H. G. Struppe, Akademie Verlag, Berlin, 1959, pp. 18-41

172. Angyal, S. J., Gorin, P. A. J., and Pitman, M., "A Stereospecific Epimerization of Cyclitols," Proc. Chem. Soc. (London), 1962, 337-338; CA 58, 8029b

173. Annino, R., "Introduction to the Theory and Practice of Gas Chromatography," 4th Annual Gas Chromatography Institute, Canisius College, Buffalo, N. Y., April 1962

174. Annino, R., "Basic Theory," 5th Annual Gas Chromatography Institute, Canisius College, Buffalo, N. Y., April 1963

175. Annino, R., "Introduction to the Theory and Practice of Gas Chromatography," in "Lectures on Gas Chromatography 1962," edited by H. A. Szymanski, Plenum Press, Inc., New York, 1963, pp. 1-17; CA 58, 10701f

176. Annison, E. F., "Volatile Fatty Acids in the Sheep Rumen," Biochem. J., 57, 400-405 (1954); CA 48, 10883h

177. Annison, E. F., "Volatile Fatty Acids of Sheep Blood with Special Reference to Formic Acid," Biochem. J., 58, 670-680 (1954); CA 49, 4830e

178. Annison, E. F., Hill, K. J., and Lewis, D., "Portal Blood of Sheep. II. Absorption of Volatile Fatty Acids from the Rumen of the Sheep," Biochem. J., 66, 592-599 (1957); CA 51, 18183e

179. Annison, E. F., and Pennington, R. J., "The Metabolism of Short Chain Fatty Acids in the Sheep. III. Formic, Valeric and Some Branched Chain Acids," Biochem. J., 57, 685-692 (1954); CA 48, 12993b

180. Anonymous, "Collection of Samples," Aerograph Research Notes (Wilkens Instrument and Research, Inc.), No. 1 (1957)

181. Anon., "Comparison of Polyester Packings - Butanediol Succinate (BDS) and Diethylene Glycol Succinate (DEGS)," Aerograph Research Notes (Wilkens Instrument & Research, Inc.), Fall 1959

182. Anon., "GLC Packings and Conditions," Aerograph Research Notes (Wilkens Instrument & Research, Inc.), Fall 1961

183. Anon., "Pesticide Residue Analysis," Aerograph Research Notes (Wilkens Instrument & Research, Inc.), Summer 1962, pp. 1-3

184. Anon., "Uses of Pyrolysis," Aerograph Previews and Reviews (Wilkens Instrument & Research, Inc.), November 1962, pp. 1-3

185. Anon., "The Aerograph Steam Generator - Model A-675," Aerograph Previews and Reviews (Wilkens Instrument & Research, Inc.), May 1962

186. Anon., "Pyrolysis at 1100°C. Opens New Areas of Gas Chromatographic Research," Bulletin 2382, American Instrument Co., Inc., Silver Spring, Md., April 1962

187. Anon., "New Stabilized Polyesters - Stable to 250-270°C. for Gas Chromatography," Analabs (Analytical Engineering Laboratories, Inc.), 2, No. 1, 1-3 (Winter 1963)

188. Anon., "New Chromatography Instruments Feature Versatility," Anal. Chem., 28, No. 10, 76A (1956)

189. Anon., "Developing Gas Analyzers: New Mass Spectrometers and Gas Chromatographic Apparatus Do Continuous and Special Analysis," Anal. Chem., 29, No. 2, 23A, 25A, 26A, 28A (1957)

190. Anon., "Physiochemical Research on Flavor," Anal. Chem., 30, No. 2, 17A-21A (1958)

191. Anon., "Analytical Chemistry in Europe," Anal. Chem., 30, No. 12, 39A-40A (1958)

192. Anon., "Report for Analytical Chemists. Instrumentation of Law Enforcement," Anal. Chem., 31, No. 2, 21A-36A (1959)

193. Anon., "New Methods and Techniques Described at Trace Analysis Symposium," Anal. Chem., 32, No. 9, 51A-54A (1960)

194. Anon., "Old and New Analytical Methods Reviewed at LSU Symposium," Anal. Chem., 33, No. 3, 67A-71A (1961)

195. Anon., "Symposium on Gas Chromatography," Analyst, 81, 52-58 (1958)

196. Anon., "Application of Gas-Liquid Chromatography to Essential Oil Analysis. Interim Report on the Determination of Citronellol in Admixture with Geranoiol," Analyst, 84, 690-691 (1959)

197. Anon., "Preparative Gas Chromatography at CONOCO," Analyzer, 2, No. 4, 7-10 (October 1961)

198. Anon., "Chromatography in Aromatic Refining," British Chem. Eng., 3, 371 (1958)

199. Anon., "Gas Chromatography. II.," Bunseki Kagaku, 9, No. 1, 59-66 (1960); "III.," ibid., 9, No. 6, 524-532 (1960); "IVa.," ibid., 9, No. 11, 979-985 (1960); "IVb.," ibid., 9, No. 12, 1080-1091 (1960)

200. Anon., "Revealing Facts About Temperature Control in Gas and Vapor Chromatography," Bulletin 838 (Burrell Corp.), 8 pp., (1958)

201. Anon., "The Progress in Gas Chromatography," Can. Chem. Processing, 43, No. 4, 115-116, 120 (1959)

202. Anon., "Vapor-Phase Analysis," Chem. Age, 72, 1196 (1955)

203. Anon., "Novel Use of High Temperature Gas Chromatography Units," Chem. Age, 78, 563 (1957)

204. Anon., "Gas-Liquid-Phase Chromatography Catching on Fast," Chem. Eng., 63, No. 6, 116, 118, 120 (1956)

205. Anon., "Cyclic Analysis No Bar to Control," Chem. Eng., 66, No. 22, 34 (1959)

206. Anon., "Chromatographs Get Programmed Controls," Chem. Eng., 67, No. 17, 84, 88 (1960)

207. Anon., "Vapor-Phase Chromatography," Chem. & Eng. News, 33, No. 15, 1510-1511 (April 11, 1955)

208. Anon., "Gas Chromatography Growing," Chem. & Eng. News, 34, No. 15, 1692-1696 (April 9, 1956)

209. Anon., "Bridge to the Future? Density Balance, Similar to Wheatstone Bridge, Improves Detection of Gas Chromatography Eluates," Chem. & Eng. News, 34, No. 49, 5992-5995 (December 3, 1956)

210. Anon., "Mass vs. Vapor. Mass Spectrometry and Vapor Phase Chromatography Grow More Competitive in Process Applications," Chem. & Eng. News, 35, No. 18, 70 (May 6, 1957)

211. Anon., "Chromatography Reaches the Plant," Chem. & Eng. News, 35, No. 33, 77 (August 19, 1957)

212. Anon., "Novel Gas Chromatographic Method Finds New Acids in Cigarette Smoke - It May Find Other Uses Too," Chem. & Eng. News, 35, No. 50, 46 (December 16, 1957)

213. Anon., "Arterosclerosis Breakthrough. Gas Chromatography Advances Lead to Quick Separation of Key Fatty Acids from Complex Lipids," Chem. & Eng. News, 36, No. 5, 48 (February 3, 1958)

214. Anon., "Gas Chromatography Widens Range," Chem. & Eng. News, 36, No. 11, 59, 62, 63 (March 17, 1958)

215. Anon., "Analyzer Aims for Sensitivity," Chem. & Eng. News, 36, No. 34, 58-60 (August 25, 1958)

216. Anon., "Free Radicals Have an Aim," Chem. & Eng. News, 36, No. 38, 53-54 (September 22, 1958)

217. Anon., "Polyester Tags Fatty Acids," Chem. & Eng. News, 36, No. 45, 58 (November 10, 1958)

218. Anon., "Automation Draws Closer," Chem. & Eng. News, 37, No. 1, 41-42 (January 5, 1959)

219. Anon., "Chromatography: Full Speed Ahead," Chem. & Eng. News, 37, No. 11, 56-57 (March 14, 1959)

220. Anon., "Analyzer Fractionates, Too," Chem. & Eng. News, 37, No. 17, 68 (April 27, 1959)

221. Anon., "Computer Speeds Analysis," Chem. & Eng. News, 37, No. 23, 52 (June 8, 1959)

222. Anon., "Instrumentation Know-How Moves Forward Fast," Chem. & Eng. News, 37, No. 28, 80-90 (July 13, 1959)

223. Anon., "Gas Chromatograph Profiles Beer," Chem. & Eng. News, 37, No. 35, 44-46 (August 31, 1959)

224. Anon., "Gas Chromatography Gets Faster," Chem. & Eng. News, 37, No. 35, 52 (August 31, 1959)

225. Anon., "Tiniest GLC Yet," Chem. & Eng. News, 37, No. 38, 57 (September 21, 1959)

226. Anon., "Gas Chromatograph Hits High °C.," Chem. & Eng. News, 38, No. 5, 42-44 (February 1, 1960)

227. Anon., "Du Pont Builds Big Chromatograph," Chem. & Eng. News, 38, No. 7, 58-59 (February 15, 1960)

228. Anon., "Flavor Enzyme Process Gets Patent," Chem. & Eng. News, 38, No. 7, 66-67 (February, 15, 1960)

229. Anon., "Chromatograph Offers Versatility," Chem. & Eng. News, 38, No. 7, 77 (February 15, 1960)

230. Anon., "Improved H_2-D_2 Separation Sought," Chem. & Eng. News, 38, No. 8, 50 (February 22, 1960)

231. Anon., "Mobile Labs Gather Refinery Data," Chem. & Eng. News, 38, No. 8, 58 (February 22, 1960)

232. Anon., "FICA Find Trace Hydrocarbons," Chem. & Eng. News, 38, No. 10, 65-66 (March 7, 1960)

233. Anon., "SWRI Pegs Trace Compounds in Body Fluids. Method Uses Gas Chromatograph. May Prove Valuable in Diagnosis, Nutritional Studies," Chem. & Eng. News, 38, No. 11, 36-37 (March 14, 1960)

234. Anon., "Gas Chromatography Gains on Several Fronts," Chem. & Eng. News, 38, No. 16, 114-115 (April 18, 1960)

235. Anon., "Soviet Lags in Gas Chromatography," Chem. & Eng. News, 38, No. 16, 130 (April 18, 1960)

236. Anon., "Kraft Mill Odors Analyzed Faster: New Sampling Kit, Chromatographic Analysis Give Rapid Survey of Kraft Mill Stack Effluent," Chem. & Eng. News, 38, No. 27, 50 (July 4, 1960)

237. Anon., "GC Takes on Copolymers," Chem. & Eng. News, 38, No. 28, 37 (July 11, 1960)

238. Anon., "Fatty Acid Analysis Moves Out of the Plant: Armour Converts Fatty Acids to Esters with Boron Trifluoride Reagent, Speeds Up Chromatographic Control Analysis from Hours to Minutes," Chem. & Eng. News, 38, No. 29, 58-59 (July 18, 1960)

239. Anon., "SE-30 Aids Separation of Complex Compounds," Chem. & Eng. News, 38, No. 30, 40-42 (July 25, 1960)

240. Anon., "Chromatography Takes on Complex Analysis: Celanese Uses Custom-Built Gas Chromatograph to Analyze Alcohol Oxidation Products," Chem. & Eng. News, 38, No. 44, 58 (October 31, 1960)

241. Anon., "Research and Technology Concentrates: Gas-Liquid Chromatography Can Be Used to Analyze Aromatic Impurities in Styrene," Chem. & Eng. News, 38, No. 50, 45 (December 12, 1960)

242. Anon., "GLC Analyzes Aromatic Impurities in Styrene. Triode Cell Detector and Capillary Columns Enable Gas-Liquid Chromatograph to Separate Trace Aromatics," Chem. & Eng. News, 38, No. 51, 50 (December 19, 1960)

243. Anon., "Chromatograph Controls Multicomponent Systems," Chem. & Eng. News, 39, No. 5, 26 (January 30, 1961)

244. Anon., "Gas Chromatograph Operates at 1000°C. Sample Reactivity and Column Packing Selection Are Problems Still To Be Solved," Chem. & Eng. News, 39, No. 9, 49 (February 27, 1961)

245. Anon., "Chemists Trap Wax Odors: C_7 Aldehydes and Ketones Are Main Offenders, Sun Oil's Tests Show," Chem. & Eng. News, 39, No. 10, 41-42 (March 6, 1961)

246. Anon., "Chromatograph Monitors Blast Furnace Gases," Chem. & Eng. News, 39, No. 13, 44-45 (March 27, 1961)

247. Anon., "Studies Pinpoint Sites of Cholesterol Block," Chem. & Eng. News, 39, No. 15, 45-46 (April 10, 1961)

248. Anon., "New Method Predicts GLC Results," Chem. & Eng. News, 39, No. 15, 50 (April 10, 1961)

249. Anon., "Gas Chromatography Separates Amino Acids," Chem. & Eng. News, 39, No. 18, 40-41 (May 1, 1961)

250. Anon., "Melpar Purifies Organometallics: Gas Chromatography Technique Gives Pure Tetraethylgermanium for Semiconductors," Chem. & Eng. News, 39, No. 18, 42-43 (May 1, 1961)

251. Anon., "Polar Phase Separates Steroids Better: Neopentyl Esters of Diacids Aid Gas Chromatography; Rapid Handling Reduces Decomposition," Chem. & Eng. News, 39, No. 23, 46 (June 5, 1961)

252. Anon., "Whistle Detects Chromatograph Output," Chem. & Eng. News, 39, No. 26, 49 (June 26, 1961)

253. Anon., "Lunar Chromatograph Designed: Compact Chromatograph Will Analyze Organics on Moon Surface," Chem. & Eng. News, 39, No. 26, 49-50 (June 26, 1961)

254. Anon., "Gas Chromatography, Something for Everybody," Chem. & Eng. News, 39, No. 27, 76-83 (July 3, 1961)

255. Anon., "Hydrocarbons Adsorb at High Temperatures," Chem. & Eng. News, 39, No. 35, 34-35 (August 28, 1961)

256. Anon., "New Tests Close In on Air Pollutants: Improved Techniques, Methods Analyze Air Quickly, Easily for Hydrocarbon and Other Pollutants," Chem. & Eng. News, 39, No. 38, 82 (September 18, 1961)

257. Anon., "ISA Show Points Up Instrument Advances," Chem. & Eng. News, 39, No. 39, 54-55 (September 25, 1961)

258. Anon., "New Method Gives Components in Soil's Organic Matter: Pyrolysis and Gas Chromatography Show Carbonyl Compounds and Phenolic Materials Present," Chem. & Eng. News, 39, No. 50, 49-50 (December 11, 1961)

259. Anon., "Gas Chromatography Splits Metal Chelates," Chem. & Eng. News, 40, No. 14, 50 (April 2, 1962)

260. Anon., "Gas Chromatography Determines Cl^-, Br^-," Chem. & Eng. News, 40, No. 14, 54-55 (April 2, 1962)

261. Anon., "New Chromatograph Designed for Efficiency. Wilkens' Autoprep Uses Narrow Coiled Columns, Operates Repetitively and Automatically," Chem. & Eng. News, 40, No. 14, 73-74 (April 2, 1962)

262. Anon., "Device Generates Steam for GC. Used as a Carrier Gas in Gas Chromatography, Steam Leads to Better Resolution of Many Polar Compounds, Simplifies Column Preparation," Chem. & Eng. News, 40, No. 29, 50 (July 16, 1962)

263. Anon., "GC Moves In On Blood Gas Analysis," Chem. & Eng. News, 40, No. 33, 42-43 (August 13, 1962)

264. Anon., "Prep Chromatograph Operates Continuously. F&M Scientific Designs Completely Continuous Preparative Chromatograph for Use by Air Force in Purifying Pentaborane," Chem. & Eng. News, 40, No. 37, 74-75 (September 10, 1962)

265. Anon., "Apparatus Combination Analyzes Gasoline," Chem. & Eng. News, 40, No. 38, 61 (September 17, 1962)

266. Anon., "Unit Gives Carbon-Hydrogen-Nitrogen Analysis. Instrument, Developed by F&M Scientific, Is Based on Gas Chromatography, Analyzes for the Three Elements Simultaneously," Chem. & Eng. News, 41, No. 4, 62-63 (January 28, 1963)

267. Anon., "Process Chromatograph Gives Distillation Data. Gulf R&D's Instrument Cuts Laboratory Test Time; Results Can Be Sent Directly to a Digital Control Computer," Chem. & Eng. News, 41, No. 5, 52, 54 (February 4, 1963)

268. Anon., "Computer Interprets GC Data. Shell's Computer-GC System Avoids Major Shortcomings of Conventional Automatic Interpreting Devices," Chem. & Eng. News, 41, No. 12, 54, 56 (March 25, 1963)

269. Anon., "Analyzer Has Tapping Recorder. Gas Chromatograph Makes Trace Analyses of Chlorinated Pesticides and Herbicides," Chem. & Eng. News, 41, No. 17, 92 (April 29, 1963)

270. Anon., "Metals Analyzed by Gas Chromatography. Volatile and Stable Fluoroacetylacetonates of Some Metals Separate Cleanly Even in Submicrogram Quantities," Chem. & Eng. News, 41, No. 26, 41 (July 1, 1963)

271. Anon., "Solanesol Pyrolyzes Via C_{10} Biradical. Philip Morris Research Group Confirms New Pathway of Breakdown of Tobacco Component," Chem. & Eng. News, 41, No. 29, 38 (July 22, 1963)

272. Anon., "Cost Conscious Chromatographs," Chem. Eng. Progr., 55, No. 6, 108, 110 (1959)

273. Anon., "Process Instrumentation-Gas Chromatography," Chem. in Can., 12, No. 8, 16, 18, 52 (1960)

274. Anon., "New Control for Administering Anaesthetics," Chem. & Ind. (London), 1961, 559

275. Anon., "Vapor Phase Chromatography and Confectionery," Chem. & Ind. (London), 1958, 1136

276. Anon., "High Resolution Obtained with Column, Detector for Fractometers. Combination Expands Range of Chromatography," Chem. Processing, 22, No. 6, 106-107 (1959)

277. Anon., "High-Purity Petrochemical Products Assured by Advances in Gas Chromatography," Chem. Processing, 22, No. 8, 104-107 (1959)

278. Anon., "Aerosol Purity Proved by Gas Chromatography," Chem. Processing, 22, No. 9, 97-98(1959)

279. Anon., "Separating Fluids by Gas-Liquid Partition Chromatography," Chem. & Processing Eng., 37, No. 2, 59 (1956)

280. Anon., "Sifting Profit Out of Offbeat Organics," Chem. Week, 86, No. 6, 51-52, 54 (1960)

281. Anon., "Equipment: Gas Chromatograph," Chem. Week, 87, No. 4, 42 (1960)

282. Anon., "Phillips Petroleum Cuts Chromatographic Analysis Time to Minutes," Chem. Week, 88, No. 23, 40 (1961)

283. Anon., "Use of the 'Fractovap' Gas Chromatograph for the Identification and Determination of the Methyl Esters of Fatty Acids in Oils and Fats," Chim. anal., 43, 31-37 (1961); CA 55, 8895a

284. Anon., "Preparation of Stationary Phases for Gas Chromatography," Chim. anal., 43, No. 4, 195-196 (1961)

285. Anon., "The Ampoule Technique in Gas Chromatography," Chim. e Ind., 43, No. 7, 789 (1961)

286. Anon., "Ionization Detection System Boosts Chromatography Sensitivity and Resolution," Control Eng., 6, No. 2, 107 (1959)

287. Anon., "Automatic Flame Ionization Detector," Control Eng., 6, No. 12, 21 (1959)

288. Anon., "Thermal Conductivity Gas Analyzer of High Sensitivity," Cryogenic Engineering Laboratory, Natl. Bur. Standards, Boulder, Colorado. (USCOMM-NBS-BL, U. S. Dept. of Commerce)

289. Anon., "Technical Notes: Gas Chromatography," Erdol u. Kohle, 11, No. 10, 747 (1958)

290. Anon., "A Micro-Syringe for Chromatographic Analysis," European Tech. Digest, February 1961, pp. 7-8

291. Anon., "Use of the Gas Chromatograph in Solvent Problems. I.," Federation Paint and Varnish Production Club, Official Digest, 30, 1172, 1177 (1958)

292. Anon., "Eliminating Steroid Tailing in Coiled Columns," Gas-Chrom Newsletter (Applied Science Laboratories, Inc., State College, Pa.), 4, No. 3, 1 (June 1963)

293. Anon., "Gas Chromatographic Carrier Gases," Gas Pipe (Jarrell-Ash Co., Newtonville, Mass.), No. 1, 5-6 (January 1963); ibid., No. 2, 3 (March 1963)

294. Anon., "Gas Chromatography of Steroids," Gas Pipe (Jarrell-Ash Co., Newtonville, Mass.), No. 2, 1-2 (March 1963)

295. Anon., "Gas Chromatography in Fatty Acid Chemistry," Ideas in Develop. (Armour Chemical Division, Chicago, Ill.), 3, No. 1, 1-4 (1959)

296. Anon., "Gas-Liquid Chromatography - I.E.C. Reports," Ind. & Eng. Chem., 47, No. 10, 13A, 14A, 16A (October 1955)

297. Anon., "Gas Chromatography in the Pilot Plant," Ind. Labs., 10, No. 10, 102-104 (1959)

298. Anon., "Gas Chromatography," Intern. Perfumer, 9, No. 9, 5 (1959); CA 54, 11375d

299. Anon., "'GC' Awakens Sleeping Giant," ISA Journal, 6, No. 8, 55-56 (1959)

300. Anon., "New Chromatography Instrument," ISA Journal, 7, No. 3, 91 (1960)

301. Anon., "Chromatograph Optimizes Sulfur Recovery Process," ISA Journal, 7, No. 3, 40-43 (1960)

302. Anon., "Gas Chromatography for Copolymers," ISA Journal, 7, No. 10, 36-37 (1960)

303. Anon., "Scanning Gas Chromatography: To Capture That Aroma You've Got to Analyze It!," ISA Journal, 8, No. 4, 34 (1961)

304. Anon., "Gas Chromatography," J. Agr. Food Chem., 7, 310-321 (1959)

305. Anon., "Chromosorb - Grades and Particle Size," Bulletin FF-101, J-M Celite Division, Johns-Manville (July 1962)

306. Anon., "Diatomite Aggregates for Gas Chromatography," Bulletin FF-102, J-M Celite Division, Johns-Manville (July 1962)

307. Anon., "Deactivation of Chromosorb P & W," Bulletin FF-103, J-M Celite Division, Johns-Manville (July 1962)

308. Anon., "Elementary Gas Chromatography," Bulletin FF-104, J-M Celite Division, Johns-Manville (July 1962)

309. Anon., "Gas Analysis," Inst. Sewage Purif., J. Proc., 1962, Pt. 3, 261-267; CA 57, 13163d

310. Anon., "Quality Measurements in the Chemical Factory," Mfg. Chemist, 32, No. 8, 343-344 (1961)

311. Anon., "Sorting Molecules," Nature, 176, 1188-1190 (1955)

312. Anon., "Plasma Lipid Fatty Acids," Nutrition Revs., 18, 137-139 (1960)

313. Anon., "Gas Chromatography," Oil Gas J., 54, No. 85, 126-128, 131-134, 136, 138, 140 (1956)

314. Anon., "Chromatography Forum," Oil Gas J., 56, 114 (1958)

315. Anon., "Molecular Sieves Look Good," Oil Gas J., 57, No. 27, 58-59 (1959)

316. Anon., "New Hydrocarbon Analyzer Has Many Uses," Oil Gas J., 58, No. 10, 59 (1960)

317. Anon., "Flame Chromatograph," Perfumery Essent. Oil Record, 48, 415 (1957)

318. Anon., "Gas Chromatography," Perfumery Essent. Oil Record, 48, No. 2, 49 (1957); ibid., 49, No. 2, 56 (1958)

319. Anon., "New Pye Gas-Liquid Chromatography Equipment," Perfumery Essent. Oil Record, 49, 270-271 (1958)

320. Anon., "How to Get Clean Gas Samples for Chromatograph - Fast," Perfumery Essent. Oil Record, 50, 568 (1959)

321. Anon., "Vapor Fractometer Offers Powerful New Technique for Gas and Liquid Analyses," Perkin-Elmer Instrument News, 6, No. 4, 1, 6 (1955)

322. Anon., "VPC and IR - a Powerful Team," Perkin-Elmer Instrument News, 7, No. 4, 4-5 (1956)

323. Anon., "Gas Chromatography - Columns and Column Performance," Perkin-Elmer Instrument News, 8, No. 1, 4-5 (1956)

324. Anon., "New Printing Integrator Speeds Analysis by Gas Chromatography," Perkin-Elmer Instrument News, 9, No. 4, 1, 7-9 (1958)

325. Anon., "Determination of Water by Gas Chromatography," Perkin-Elmer Instrument News, 11, No. 1, 3 (1959)

326. Anon., "Perkin-Elmer Announces New Model 154-D Vapor Fractometer," Perkin-Elmer Instrument News, 11, No. 1, 1, 6-10 (1959)

327. Anon., "New Laboratory Hydrocarbon Detector Offers High Sensitivity," Perkin-Elmer Instrument News, 12, No. 2, 11 (1961)

328. Anon., "Linear Temperature Programmer Widens Range of Chromatography," Perkin-Elmer Instrument News, 12, No. 2, 14 (1961)

329. Anon., "Electron Capture Detection of Pesticides," Pesticide Research Bull. (Stanford Res. Inst.),

2, No. 1, 1-4 (1962)

330. Anon., "For Control of Refinery Processes . . Use of Chromatography Grows," Petrol. Week, 8, No. 19, 40 (1959)

331. Anon., "Detector Measures Trace Hydrocarbons," Petrol. Week, 10, No. 10, 34 (1960)

332. Anon., "Better, Cheaper Analyzers Needed," Petrol. Week, 11, No. 20, 80, 82 (1960)

333. Anon., "Application of Gas Chromatography to Control of Food Products," Reichstoffe u. Aromen, 10, No. 3, 90-91 (1960)

334. Anon., "Detecting Pesticides on Vegetables," Research for Ind. (Stanford Res. Inst.), 12, No. 3, 8-10 (1960)

335. Anon., "New Industrial Gas Chromatograph - Now Also for Liquid Process Streams," Seifen-Ole-Fette-Wachse, 87, No. 5, 112 (1961)

336. Anon., "The Barber-Colman Gas Chromatographic Ionization Detection System, Model 10," S. African Ind. Chemist, 13, 235 (1959)

337. Anon., "Chromatography in Medicine," What's New (Abbott Laboratories, North Chicago, Ill.), No. 228, 12-17 (Summer 1962)

338. Anselmi, S., "Chromatography in Fat Analyses," Olearia, 15, 61-64 (1961); 55, 25293a

339. Anselmi, S., Boniforti, L., and Monacelli, R., "Fatty Substance Gas Chromatography. Detection of Methanol and Ethylene Glycol in Oil," Boll. lab. chim. provinciali (Bologna), 10, 335-341 (1959); CA 54, 16873i

340. Anselmi, S., Boniforti, L., and Monacelli, R., "Chromatographic Analysis of Fatty Substances in Vapor Phase. II. Technique of Procedure and Chromatograms of Some Vegetable and Animal Oils and Fats," Boll. lab. chim. provinciali (Bologna), 11, 257-277 (1960); CA 55, 4011d

341. Anselmi, S., Boniforti, L., and Monacelli, R., "Gas-Chromatographic Analysis of Fats. III. Execution Technique for Chromatograms of Butter and Related Products," Boll. lab. chim. provinciali (Bologna), 11, 317-330 (1960); CA 55, 14746f

342. Anselmi, S., Boniforti, L., and Monacelli, R., "Vapor Phase Chromatography of Fats. III. Analysis of Butter and Allied Products," Riv. Ital. Sostanze Grasse, 38, 436-442 (1961)

343. Anson, P. C., Fredericks, P. S., and Tedder, J. M., "Free Radical Substitution in Aliphatic Compounds. I. Hydrogenation of n-Butane and iso-Butane in the Gas Phase," J. Chem. Soc., 1959, 918-922; CA 53, 14915g

344. Anson, P. C., and Tedder, J. M., "The Reactions of Gaseous Fluorine and Chlorine with Liquid n-Butane and iso-Butane," J. Chem. Soc., 1957, 4390-4392; CA 52, 5281c

345. Anthony, D. S., "Acceleration of Biochemical Research by Recent Analytical Developments," J. Chem. Educ., 36, 540-543 (1959); 135th Natl. ACS Mtg., Boston, Mass., April 1959, Program Abstr., p. 12F

346. Antonis, A., and Bersohn, I., "Influence of Diet on Serum Triglycerides in South African White and Bantu Prisoners," Lancet, 1961, No. 1, 3-9; CA 55, 7573g

347. Aoyanagi, M., and Mizuoka, T., "Application to Synthetic Organic Chemistry: Analysis of Reaction Products," Kagaku no Ryoiki Zokan, No. 44, 147-167 (1961); CA 56, 3a

348. Apostolakis, M., Grimmer, G., Glaser, A., and Voigt, K. D., "Composition of the Cholesterol Esters of the Blood and Lymph of Rats," Biochem. Z., 336, 1-9 (1962); CA 57, 10399i

349. Applequist, D. E., Fanta, G. F., and Henrikson, B. W., "Chemistry of Spiropentane. I. Improved Synthesis of Spiropentane," J. Org. Chem., 23, 1715-1716 (1958); CA 53, 21701a

350. Applequist, D. E., Fanta, G. F., and Henrikson, B. W., "Chemistry of Spiropentane. II. The Chlorination of Spiropentane," J. Am. Chem. Soc., 82, 2368-2372 (1960); CA 55, 19815a

351. Applequist, D. E., and McGreer, D. E., "The Synthesis and Some Reactions of Diazocyclobutane," J. Am. Chem. Soc., 82, 1965-1972 (1960); CA 54, 17285i

352. Applequist, D. E., and Peterson, A. H., "The Stereochemistry of the Hunsdiecker Reaction of a Cyclopropane Ring," J. Am. Chem. Soc., 82, 2372-2376 (1960); CA 55, 18617h

353. Applequist, D. E., and Peterson, A. H., "The Configurational Stability of cis- and trans-2-Methyl-cyclopropyllithium and Some Observations on the Stereochemistry of Their Reactions with Bromine and Carbon Dioxide," J. Am. Chem. Soc., 83, 862-865 (1961); CA 55, 24590f

354. April, A., and Herrmann, E., "Possibilities Offered by Spectrophotometric Methods in the Analysis of Oil Binders of Paints," Ingr. chimiste (Milan), 42, No. 218, 118-167 (1960); CA 54, 25873i

355. Arad-Talmi, Y., Levy, M., and Vofsi, D., "Gas Chromatographic Analysis of Aliphatic Nitriles in Aqueous Acidic Solution," J. Chromatog., 10, 417-420 (1963)

356. Arai, S., Sato, S., and Shida, S., "Mercury Photosensitized Decomposition of Cyclohexane at High Temperatures," Nippon Kagaku Zasshi, 81, 1790-1793 (1960); CA 55, 7007c

357. Arai, S., Sato, S., and Shida, S., "Mercury Photosensitized Decomposition of Cyclohexane at High Temperatures," J. Chem. Phys., 33, 1277-1278 (1960)

358. Araki, S., "A New Gas Sample Tube for Gas Chromatography," Bunseki Kagaku, 9, 250-251 (1960); CA 56, 9382h

359. Araki, S., and Kato, T., "Analysis of Air Pollutants by Gas Chromatography," Bunseki Kagaku, 11,

533-543 (1962); CA 57, 7562a

360. Araki, S., and Kato, T., "Peak Distortion in Gas Chromatography," Bunseki Kagaku, 12, 179-181 (1963); CA 58, 10702e

361. Araki, S., Kishimoto, K., and Yasumori, Y., "Programmed Temperature Gas Chromatography," Bunseki Kagaku, 8, 699-703 (1959)

362. Araki, S., and Mashiko, Y., "Some Comments on the Gas Chromatography Apparatus in the Present Status," Kagaku no Ryoiki, 12, 575-576 (1958); CA 52, 19338i

363. Araki, T., and Goto, R., "Gas-Liquid Partition Chromatography of m- and p-Xylenes on 1-Naphthyl-amine," Bull. Chem. Soc. Japan, 33, 115 (1960); CA 54, 22171f

364. Araki, T., Goto, R., and Munemiya, S., "Gas-Liquid Partition Chromatography. I. Effects of the Stationary Liquids for the Separation of Xylene Isomers," Nippon Kagaku Zasshi, 81, 1315-1317 (1960); CA 56, 2903d

365. Araki, T., Goto, R., and Ono, A., "Gas-Liquid Partition Chromatography. II. Separation of Xylene Isomers with Mixed Stationary Phases," Nippon Kagaku Zasshi, 81, 1318-1320 (1960); CA 56, 2903f

366. Araki, T., Goto, R., and Ono, A., "Effects of Stationary Liquids for the Separation of Xylene Isomers," Nippon Kagaku Zasshi, 82, 1081-1085 (1961); CA 59, 391f

367. Araki, Y., Ishikawa, M., Sakai, T., and Okunishi, T., "Absorption Efficiency of Benzene Scrubber by Process Gas Chromatography," Koru Taru, 14, 440-448 (1962); CA 58, 10994f

368. Aranda, V. G., Cordon, J. L. M., Lopez, G. Z., and Flaquer, J. O., "Gas Chromatography. I. The Setting Up and Operation of a Laboratory Unit. II. Application of the Laboratory Unit in the Qualitative and Quantitative Separation of Volatile Organic Compounds," Combustibles (Zaragoza), 20, No. 110-112, 79-89 (1960); CA 56, 4552b

369. Aratani, T., and Komae, H., "Analysis of Terpene Compounds and Perfumery Materials by Gas Chromatography - Solid Support," Koryo, No. 64, 45-50 (1961); CA 57, 6041a

370. Archer, D. A., and Booth, H., "The Thermal Decomposition of Quaternary Ammonium Hydroxides. I. Methohydroxides Derived from NN-Dialkylanilines and Related Compounds," J. Chem. Soc., 1963, 322-330

371. Archer, D. A., Booth, H., Crisp, P. C., and Parrick, J., "The Thermal Decomposition of Quaternary Ammonium Hydroxides. II. Methohydroxides Derived from 1,2,3,4-Tetrahydroquinolines and 1,2,3,4,4a,9,9a,10-Octahydroacridines," J. Chem. Soc., 1963, 330-338

372. Archer, E. D., Shively, J. H., and Francis, S. A., "Analysis of C_7 Olefins by a Combination of Gas Chromatography and Nuclear Magnetic Resonance Spectroscopy," 14th Pittsburgh Conf. on Anal. Chem. & Appl. Spectroscopy, Pittsburgh, Pa., March 1963, Program Abstr., p. 61

373. Archer, T. E., Bevenue, A., and Zweig, G., "Comparison of Two Partitioning Phases for Gas-Liquid Chromatography with Two Types of Detectors Utilizing Four Halogenated Hydrocarbons," J. Chromatog., 6, 457-460 (1961); CA 56, 12292i

374. Arcus, A. C., and Dunckley, G. G., "Chromatography of Some Lipids on Poly (Tetrafluoroethylene)," J. Chromatog., 5, 272-273 (1961)

375. Arcus, C. L., Cort, L. A., Howard, T. J., and Loc, L., "Olefinic Addition with Asymmetric Reactants. V. The Asymmetric Hydrogenation of (−)-3-Methyl-4-phenylbut-3-en-2-ol and of (+)-α-Pinene," J. Chem. Soc., 1960, 1195-1200; CA 54, 15428g

376. Arcus, C. L., and Howard, T. J., "Olefinic Additions with Asymmetric Reactants. VI. Disymmetric Hydrogenation of (±)-4-Phenylpent-3-en-2-ol," J. Chem. Soc., 1961, 670-674; CA 55, 12340h

377. Arcus, C. L., Page, J. M. J., and Reid, J. A., "Olefinic Additions with Asymmetric Reactants. VII. The Hydrogenation of (±)-3-Methylenepentan-2-ol," J. Chem. Soc., 1963, 1213-1216; CA 58, 8886g

378. Aris, R., "On the Dispersion of a Solute in a Fluid Flowing Through a Tube," Proc. Roy. Soc. (London), A235, 67-77 (1956)

379. Aris, R., "Dispersion of a Solute by Diffusion, Convection and Exchange Between Phases," Proc. Roy. Soc. (London), A252, 538-550 (1959); CA 54, 20360d

380. Aristov, B. G., Kiselev, A. V., Mirskii, Y. V., Pavlova, L. F., and Petrova, R. S., "Adsorption on Molecular Sieves from Vapors and Solutions," Khim. i Tekhnol. Topliv i Masel, 7, No. 8, 7-12 (1962); CA 57, 14454b

381. Aristova, V., and Karysheva, A., "Determination of Volatile Fatty Acids by the Chromatographic Method," Molochnaya Prom., 19, No. 10, 35-37 (1958); CA 53, 4016a

382. Arkell, A., and Newman, M. S., "Design and Use of a Gas Chromatographic Apparatus for the Isolation of Organic Reaction Products," PB Rept., 151,543, 23 pp.; CA 54, 11589f

383. Armitage, F., "The Analysis of Isoprene by Gas Chromatography," J. Chromatog., 2, 655-657 (1959); CA 54, 20672i

384. Armstrong, N. W., Beck, M. S., Fahey, R. L., and Bennett, C. E., "The Applications of Programmed Temperature Gas Chromatography," 13th Annual Mid-America Spectroscopy Symposium, Chicago, Ill., April-May 1962

385. Arndt, R. R., Nel, W. J., and Pretorius, V., "A Flame-Ionization Detector for Gas Chromatography. The Effect of a Number of Parameters on the Sensitivity," J. South African Chem. Inst., 12, 69-74 (1959); CA 54, 10410a

386. Arnett, E. M., "Deuterium Analysis - A Rapid, Simple Application of Gas Chromatography to an Old Method," 11th Pittsburgh Conf. on Anal. Chem. & Appl. Spectroscopy, Pittsburgh, Pa., March 1960, Program Abstr., p. 44

387. Arnett, E. M., and Duggleby, P. M., "A Practical Method of Deuterium Analysis by Gas Chromatography - a Progress Report," 14th Pittsburgh Conf. on Anal. Chem. & Appl. Spectroscopy, Pittsburgh, Pa., March 1963, Program Abstr., pp. 50-51

388. Arnett, E. M., Strem, M., Hepfinger, N., Lipowicz, J., and McGuire, D., "Deuterium Analysis - A Simple and Precise Method," Science, 131, 1680-1681 (1960); CA 54, 22812g

389. Arnett, E. M., and Wu, C. Y., "Basicity Constants for Some Saturated Ethers in Aqueous Sulfuric Acid Solutions," 139th Natl. ACS Mtg., St. Louis, Mo., March 1961, Program Abstr., p. 36-0

390. Arnett, E. M., and Wu, C. Y., "The Application of Gas Chromatography to Basicity Measurements," 12th Pittsburgh Conf. on Anal. Chem. & Appl. Spectroscopy, Pittsburgh, Pa., March 1961, Program Abstr., p. 58

391. Arnold, J. H., "Studies in Diffusion. II. A Kinetic Theory of Diffusion in Liquid Systems," J. Am. Chem. Soc., 52, 3937-3955 (1930)

392. Arnold, L. K., and Choudhury, B. R., "The Fatty Acid Composition of Cottonseed Oil at Various Stages of Solvent Extraction," J. Am. Oil Chemists' Soc., 38, 87-88 (1961); CA 55, 6890d

393. Arnold, L. K., and Milloy, A. D., "Distribution of Individual Fatty Acids in the Crystallization Fractions of Lard," J. Am. Oil Chemists' Soc., 40, 296-297 (1963)

394. Arnold, R., "A Katharometer Analyzer for Air/Fuel Ratio and Incomplete Combustion in Boiler Installations," Australian J. Instr. Technol., 12, 65-76 (1956); CA 51, 2251e

395. Arnold, R. T., and Smolinsky, G., "The Pyrolysis of β-Hydroxy-olefins," J. Am. Chem. Soc., 81, 6443-6445 (1959); CA 54, 8584a

396. Arpino, A., "The Oil of Simaruba Glauca," Riv. Ital Sostanze Grasse, 38, No. 5, 275-276 (1961); CA 56, 1544b

397. Arthur, H. R., Tam, S. W., and Ng, Y. L., "An Examination of the Rutaceae of Hong Kong. VII. The Essential Oils," J. Chem. Soc., 1961, 3551-3552; CA 56, 1539i

398. Artsybasheva, Y. P., and Favorskaya, I. A., "A Study of Structure of Vinylacetylenic Hydrocarbons by the Method of Gas-Liquid Distributive Chromatography," Zh. Obshch. Khim., 32, 2380-2381 (1962); CA 58, 8883e

399. Arumeel, E., and Eizen, O., "Gas-Chromatographic Separation of Light Products from the Thermal Treatment of Bituminous Shale by Using Different Filling Materials," Goryuch. Slantsy, Khim. i. Tekhnol., Akad. Nauk Est. SSR, Inst. Khim., 1961, No. 4, 161-165; CA 57, 15416f

400. Arvidson, G., and Olivecrona, T., "Fatty Acid Composition of Plasma and Tissue Lipids of Normal and Ethionine Rats," Acta Physiol. Scand., 55, 303-312 (1962); CA 58, 13023e

401. Asahara, T., Kondo, T., and Hamada, I, "Vapour Phase Nitration of Propane," Kogyo Kagaku Zasshi, 62, 1659-1661 (1959)

402. Asahara, T., and Takagi, Y., "Gas Chromatography," Abura Kagaku, 6, 32-36, 108-112 (1957); CA 51, 11810b

403. Asahara, T., and Yamashita, K., "Gas Chromatography of Fats, Fatty Acids, Their Esters, and Unsaponifiable Matter," Yukagaku, 8, 590-598 (1959); CA 54, 13938h

404. Asatoor, A., and Dalgleish, C. E., "The Use of Adsorption for Isolation of Aromatic Substances," J. Chem. Soc., 1958, 1498-1501; CA 52, 10686i

405. Ascoli, F., and Crescenzi, V., "The Chemistry of Terpenes by Vapor-Phase Chromatography and Ultraviolet and Infrared Spectroscopy," Chim. e Ind.(Milan), 40, 724-727 (1958); CA 53, 6286e

406. Ascoli, F., Pispisa, B., and Servello, F., "The Use of Pentabenzoyl-α-glucose as Stationary Phase in Gas Chromatography," J. Chromatog., 6, 544-546 (1961); CA 56, 14902i

407. Ashbury, G. K., Davies, A. J., and Drinkwater, J. W., "Versatile Gas-Liquid Partition Chromatography Apparatus Developed for Analytical Use," 129th Natl. ACS Mtg., Dallas, Texas, April 1956, Program Abstr., p. 16B

408. Ashbury, G. K., Davies, A. J., and Drinkwater, J. W., "Versatile Gas-Liquid Partition Chromatography Apparatus," Anal. Chem., 29, 918-925 (1957); CA 51, 12567h

409. Ashley, J. W., and Reilley, C. N., with Hurwitz, P., and Rogers, L. B., "Observations on the Effects of Temperature, Support, and Amount of Partitioning Liquid in Gas Chromatography," Anal. Chem., 34, 1537-1540 (1962); CA 58, 3h

410. Asinger, F. A., and Halcour, K., "The Dependence of the Reactivity of Functional Groups in Paraffinic Hydrocarbons on Their Position in the Molecule. XIV. The Composition of the Reaction Products Obtained by the Action of Dinitrogen Pentoxide, Colorless or Fuming Nitric Acid on n-Octane," Chem. Ber., 94, 83-93 (1961); CA 55, 9267i

411. Askins, J. W., "Gas Chromatography Is Paying Off," Oil Gas J., 57, No. 17, 111 (1959); CA 53, 18455f

412. Aspinall, G. O., "Gas-Liquid Partition Chromatography of Methylated and Partially Methylated Methyl Glycosides," J. Chem. Soc., 1963, 1676-1680; CA 58, 11459f

413. Aspinall, G. O., and Baillie, J., "Gum Tragacanth. I. Fractionation of the Gum and the Structure of Tragacanthis Acid," J. Chem. Soc., 1963, 1702-1714; CA 58, 12657d

414. Aspinall, G. O., and Baillie, J., "Gum Tragacanth. II. The Arabinogalactan," J. Chem. Soc., 1963, 1714-1721; CA 58, 12657f

415. Aspinall, G. O., Cairncross, I. M., and Ross, K. M., "A Xylan from the Roots of Perennial Rye-grass (Lolium Perenne)," J. Chem. Soc., 1963, 1721-1727

416. Aspinall, G. O., Charlson, A. J., Hirst, E. L., and Young, R., "The Location of L-Rhamnopyranose Residues in Gum Arabic," J. Chem. Soc., 1963, 1696-1702; CA 58, 11459f

417. Aspinall, G. O., and Wood, T. M., "The Structures of Two Water-Soluble Polysaccharides from Scots Pine (Pinus Sylvestris)," J. Chem. Soc., 1963, 1686-1696; CA 58, 12657h

418. Asscher, M., and Vofsi, D., "Chlorine Activation by Redox Transfer. II. The Addition of Carbon Tetrachloride to Olefins," J. Chem. Soc., 1963, 1887-1896; CA 58, 10063d

419. Asselineau, C., Asselineau, J., Ryhage, R., Stallberg-Stenhagen, S., and Stenhagen, E., "Synthesis of (−)-Methyl-2D, 4D, 6D-trimethylnonacosanoate and Identification of C_{32}-Mycocerosic Acid as a 2, 4, 6, 8-Tetramethyloctacosanoic Acid," Acta Chem. Scand., 13, 822-824 (1959); CA 55, 14293i

420. Asselineau, J., "Applications of Chromatography to Fatty Acids," Chim. anal., 39, 375-383 (1957); CA 52, 4415h

421. Asselineau, J., "Some Applications of Gas Chromatography to the Study of Bacterial Fatty Acids," Ann. inst. Pasteur, 100, 109-119 (1961); CA 55, 16654f

422. Asselineau, J., and Lederer, E., "Chemistry of Lipids," Ann. Rev. Biochem., 30, 71-92 (1961); CA 55, 27470h

423. Ast, H. J., and Van der Wal, R. J., "The Structural Components of Milk Triglycerides," J. Am. Oil Chemists' Soc., 38, 67-69 (1961); CA 55, 7690g

424. Atkinson, E. P., and Tuey, G. A. P., "An Automatic 'Preparative-Scale' Gas Chromatography Apparatus," in "Gas Chromatography 1958," edited by D. H. Desty, Academic Press, Inc., New York, 1958, pp. 270-287

425. Atkinson, E. P., and Tuey, G. A. P., "Automatic 'Preparative-Scale' Gas Chromatography Apparatus," Gas Chromatog., 1958, 270-283; CA 53, 16609c

426. Atkinson, R. d'E., "Statistical Experiments on the Motion of Electrons in Gases," Proc. Roy. Soc. (London), A119, 335-348 (1928); CA 22, 3349

427. Attaway, J. A., Wolford, R. W., Alberding, G. E., and Edwards, G. J., "Identification of Alcohols and Volatile Organic Acids from Natural Orange Essence," 144th Natl. ACS Mtg., Los Angeles, Calif., March-April 1963, Program Abstr., p. 18A

428. Attaway, J. A., and Wolford, R. W., with Edwards, G. J., "Isolation and Identification of Some Volatile Carbonyl Components from Orange Essence," J. Agr. Food Chem., 10, 102-104 (1962); CA 57, 7693e; 139th Natl. ACS Mtg., St. Louis, Mo., March 1961, Program Abstr., p. 2A

429. Attrill, J. E., Boyd, C. M., and Meyer, Jr., A. S., "Gas Chromatographic Analysis of Helium at Reduced Pressures," 144th Natl. ACS Mtg., Los Angeles, Calif., March-April 1963, Program Abstr., p. 21B

430. Atwood, J. G., "How Golay Works," 139th Natl. ACS Mtg., St. Louis, Mo., March 1961, Program Abstr., p. 4B

431. Aubeau, R., and Champeix, L., "Analysis of Impurities in Carbon Dioxide by Gas Chromatography," Ind. atomiques, 4, No. 11/12, 78-86 (1960); CA 55, 13171h

432. Aubeau, R., and Champeix, L., "Gas Chromatography Applied to the Separation of Permanent Gases," Bull. inform. sci. et tech. (Paris), No. 49, 35-45 (1961); CA 55, 18204b

433. Aubeau, R., Champeix, L., and Reiss, J., "Separation and Determination of Krypton and Xenon by Gas Chromatography. Application to the Fission of Gases," J. Chromatog., 6, 209-219 (1961); CA 56, 5396e

434. Aubeau, R., Champeix, L., and Reiss, J., "Separation and Determination of Helium and Hydrogen by Gas Phase Chromatography," J. Chromatog., 7, 447-454 (1962); CA 58, 10706a

435. Aubeau, R., Reiss, J., Champeix, L., and Ravnik, V., "Chromatographic Determination of Low Concentrations of Oxygen and Argon in Heat-Transfer Gases," J. Nucl. Mater., 6, 271-280 (1962); CA 58, 8399h

436. Audette, R. C. S., Baxter, R. M., and Walker, G. C., "A Study of the Lipid Content of Trichophyton Mentagrophytes," Can. J. Microbiol., 7, No. 4, 282-283 (1961); CA 55, 15616d

437. Audran, R., and Reutenauer, G., "Chromatographic Apparatus Permitting Concentration of Constituents," Compt. rend., 245, 168-171 (1957); CA 51, 16005g

438. Augustine, R. L., and Broom, A. D., "Catalytic Hydrogenation of α, β-Unsaturated Ketones. II. Mechanism of Hydrogenation in Acidic Medium," J. Org. Chem., 25, 802-804 (1960); CA 55, 1537i

439. Aunstrup, K., and Djurtoft, R., "Simple Set-Up for the Determination of Titrable, Volatile Com-

pounds by Gas Chromatography," Brygmesteren, 17, No. 6, 137-143 (1960); CA 54, 20355c; Radiometer News, No. 1, 4 (March 1960)

440. Ausloos, P., and Murad, E., "The Photolysis of 2-Pentanone and 2-Pentanone-1, 1, 1, 3, 3-d$_5$," J. Am. Chem. Soc., 80, 5929-5932 (1958); CA 53, 6998e

441. Ausloos, P., and Rebbert, R. E., "Intramolecular Rearrangements. III. Formation of 1-Methyl-cyclobutanol in the Photolysis of 2-Pentanone," J. Am. Chem. Soc., 83, 4897-4899 (1961); CA 56, 6822a

442. Aust, R., and Heft, K. H., "Gas Chromatographic Investigations of Compounds Capable of Polymerization and the Evaluation of the Results," in "Gas Chromatographie 1959," edited by R. E. Kaiser and H. G. Struppe, Akademie Verlag, Berlin, 1959, pp. 42-68

443. Austin, F. L., and Boruff, C. S., "Concentration of Congeners of Grain Spirits and Their Analysis by Gas Chromatography," J. Assoc. Offic. Agr. Chemists, 43, 675-679 (1960); CA 54, 25555d

444. Avdeevz, A. A., "Analysis of Combustion Products in Boiler Installations by Gas Chromatography," Teploenergetika, 6, No. 8, 16-20 (1959); CA 54, 7406h

445. Averill, W., "New Chromatographic Techniques Simplify Flavor and Odor Analysis," Perkin-Elmer Instrument News, 12, No. 1, 1, 6-7 (1960)

446. Averill, W., "Use of Gas Chromatography for Analysis of Odors, Flavors, and Air Pollution," in "Progress in Industrial Gas Chromatography," Vol. 1, edited by H. A. Szymanski, Plenum Press, Inc., New York, 1961, pp. 31-36; CA 56, 7967c

447. Averill, W., "Columns with Minimum Liquid Phase Concentration for Use in Gas-Liquid Chromatography," 3rd. Intern. Symp. on Gas Chromatography, ISA, East Lansing, Mich., June 1961; ISA Proc., 3, 1-7 (1961)

448. Averill, W., "Columns with Minimum Liquid Phase Concentration for Use in Gas-Liquid Chromatography," Gas Chromatog., Intern. Symp., 3, 1-6 (1961) (Pub. 1962); CA 58, 9607h

449. Averill, W., "Capillary Columns in Gas Chromatography," 4th Annual Gas Chromatography Institute, Canisius College, Buffalo, N. Y., April 1962

450. Averill, W., "Program Temperature Capillary Column Unit and Its Use for Analysis of Wide Boiling Range and Complex Samples," 4th Annual Gas Chromatography Institute, Canisius College, Buffalo, N. Y., April 1962

451. Averill, W., "Parameters of the Capillary Column," 2nd Symp. on Gas Chromatography, Toronto Section, Chemical Institute of Canada, Toronto, Ontario, February 1962

452. Averill, W., "Columns with Minimum Liquid Phase Concentration for Use in Gas-Liquid Chromatography," in "Gas Chromatography," edited by N. Brenner, J. E. Callen, and M. D. Weiss, Academic Press, Inc., New York, 1962, pp. 1-6

453. Averill, W., "Trace Analysis," 5th Annual Gas Chromatography Institute, Canisius College, Buffalo, N. Y., April 1963

454. Averill, W., "Capillary Columns," 5th Annual Gas Chromatography Institute, Canisius College, Buffalo, N. Y., April 1963

455. Averill, W., "Gas Chromatography of Free Fatty Acids Using Golay Columns," J. Gas Chromatog., 1, No. 1, 22 (1963)

456. Averill, W., "In-Place Coating of the Solid Support for Gas Chromatography," J. Gas Chromatog., 1, No. 1, 34-35 (1963)

457. Averill, W., and Ettre, L. S., "Gas-Chromatographic Analysis of C$_1$-C$_4$ Hydrocarbons with Open Tubular Columns," Nature, 196, 1198-1199 (1962); CA 58, 5022b

458. Averill, W., Scholly, P., and Ettre, L. S., "Trace Analysis Using Capillary and Packed Columns with the Flame Ionization Detector," 12th Pittsburgh Conf. on Anal. Chem. & Appl. Spectroscopy, Pittsburgh, Pa., February-March 1961, Program Abstr., p. 57

459. Avgul, V. T., Elovich, S. Y., Semenovskaya, T. D., and Chmutov, K. V., "Chromatographic Column for High Temperature Operation," Zhur. Fiz. Khim., 35, 946-947 (1961)

460. Avgul, V. T., and Yudilevich, M. D., "Apparatus for Thermostating a Chromatographic Column," Zavodskaya Lab., 25, 1403 (1959); cf. Ind. Lab., 25, 1469 (1959) (Pub. 1960); CA 57, 1514e

461. Ayers, B. O., "Chromatographic Analyzers in Process Instrumentation," in "Gas Chromatography," edited by V. J. Coates, H. J. Noebels, and I. S. Fagerson, Academic Press, Inc., New York, 1958, pp. 249-267

462. Ayers, B. O., Loyd, R. J., and DeFord, D. D., "Principles of High-Speed Gas Chromatography with Packed Columns," Anal. Chem., 33, 986-991 (1961); CA 56, 9377h

463. Ayers, B. O., Loyd, R. J., and Reinecke, M. E., "Chromatographic Analysis of Trace Quantities of Carbon Monoxide in Ethylene," 9th Pittsburgh Conf. on Anal. Chem. & Appl. Spectroscopy, Pittsburgh, Pa., March 1963, Program Abstr., p. 53

464. Ayers, B. O., Loyd, R. J., and Sanford, R. A., "Data Systemization for the Selection and Optimization of Chromatographic Columns," 14th Pittsburgh Conf. on Anal. Chem. & Appl. Spectroscopy, Pittsburgh, Pa., March 1963, Program Abstr., p. 53

465. Aylward, F., and Nichols, B. W., "Plant Lipids. III. Phosphatidic Acid and Phosphatidylinositol in

Cereal Grains," J. Sci. Food Agr., 13, 92-95 (1962); CA 56, 14677g

466. Aylward, F., and Showler, A. J., "Plant Lipids. IV. Glycerides and Phosphatides in Cereal Grains," J. Sci. Food Agr., 13, 492-496 (1962); CA 58, 8235d

467. Aylward, F., and Wills, P., "Isolation of Sterol Esters from Human Feces," Nature, 191, 1397 (1961)

468. Ayrapaa, T., Holmberg, J., and Sellmann-Persson, G., "Lipids and Lipid-Soluble Substances in Beer and Wort," European Brewery Conv., Proc. Congr., 1961, 286-297; CA 58, 13091c

469. Aznavourian, W., and McIntyre, E. A., "Gas Chromatographic Analysis of Aqueous Organic Solutions," 13th Pittsburgh Conf. on Anal. Chem. & Appl. Spectroscopy, Pittsburgh, Pa., March 1962, Program Abstr., p. 57

470. Aznavourian, W., and McIntyre, E. A., "Gas Chromatographic Analysis of Ammonia-Fixed Gas Streams," 14th Pittsburgh Conf. on Anal. Chem. & Appl. Spectroscopy, Pittsburgh, Pa., March 1963, Program Abstr., p. 50

471. Baba, T., and Tokumaru, S., "Identification of Rubber by Gas Chromatography. I.," Nippon Gomu Kyokaishi, 35, 162-166 (1962); CA 57, 16821d

472. Babini, B., Paolucci, G., Salvioli, Jr., G. P., Manfredi, G., and Corsini, F., "Gas-Chromatographic Determination of Plasma and Erythrocyte Fatty Acids in Thalassemia," Boll. Soc. Ital. Biol. Sper., 38, 728-729 (1962); CA 58, 6070g

473. Bachman, G. B., Logan, T. J., Hill, K. R., and Standish, N. W., "Nitration Studies. XII. Nitrohelogination of Negatively Substituted Olefine with Mixtures by Dinitrogen Tetroxide and Halogens," J. Org. Chem., 25, 1312-1322 (1960); CA 54, 24373a

474. Bachmann, L., Bechtold, E., and Cremer, E., "Gas-Chromatographic Separation of Ortho- and Parahydrogen on Molecular Sieves," J. Catalysis, 1, 113-120 (1962); CA 57, 9267g

475. Baddiel, C. B., and Cullis, C. F., "The Use of a Carbon Monoxide Flame Detector in Gas Chromatography," Chem. & Ind. (London), 1960, 1154-1155; CA 55, 8148g

476. Bade, M. L., and Clayton, R. B., "Cholesterol Esters of the Cockroach Eurycotis Floridana," Nature, 197, 77-79 (1963)

477. Badger, G. M., and Buttery, R. G., "The Formation of Aromatic Hydrocarbons at High Temperature. IV. The Pyrolysis of Styrene," J. Chem. Soc., 1958, 2458-2463; CA 52, 20096c

478. Badger, G. M., Buttery, R. G., Kimber, R. W. L., Lewis, G. E., Moritz, A. G., and Napier, I. M., "The Formation of Aromatic Hydrocarbons at High Temperatures. I. Introduction," J. Chem. Soc., 1958, 2449-2452; CA 52, 20094a

479. Badger, G. M., and Kimber, R. W. L., "The Formation of Aromatic Hydrocarbons at High Temperatures. VI. The Pyrolysis of Tetralin," J. Chem. Soc., 1960, 266-270; CA 54, 9868b

480. Badger, G. M., and Kimber, R. W. L., "The Formation of Aromatic Hydrocarbons at High Temperatures. XIV. Pyrolysis of Ethylbenzene-α-C^{14}," J. Chem. Soc., 1961, 3407-3414; CA 56, 4690f

481. Badger, G. M., Kowanko, N., and Sasse, W. H. F., "Synthetic Applications of Activated Metal Catalysts. IX. A Comparison of the Desulfurizing Abilities of Some Transition Metals," J. Chem. Soc., 1960, 1658-1662; CA 54, 18449e

482. Badger, G. M., Kowanko, N., and Sasse, W. H. F., "Synthetic Applications of Activated Metal Catalysts. X. Desulfurization of Thionaphtheno[3, 2-b]thianaphthene," J. Chem. Soc., 1960, 2969-2972; CA 54, 24637h

483. Badger, G. M., Lewis, G. E., and Napier, I. M., "The Formation of Aromatic Hydrocarbons at High Temperatures. VIII. The Pyrolysis of Acetylene," J. Chem. Soc., 1960, 2825-2827; CA 55, 493g

484. Badger, G. M., and Novotny, J., "Formation of Aromatic Hydrocarbons at High Temperatures. XII. Pyrolysis of Benzene," J. Chem. Soc., 1961, 3400-3402; CA 56, 4689h

485. Badger, G. M., and Novotny, J., "The Formation of Aromatic Hydrocarbons at High Temperatures. XIII. Pyrolysis of 3-Vinylcyclohexene," J. Chem. Soc., 1961, 3403-3407; CA 56, 4689h

486. Badger, G. M., and Spotswood, T. M., "The Formation of Aromatic Hydrocarbons at High Temperatures. V. The Pyrolysis of 1-Phenyl-1, 3-butadiene," J. Chem. Soc., 1959, 1635-1641; CA 54, 461e

487. Badings, H. T., "Fatty Acid Composition of Cow Milk Phospholipids as Determined by Gas Chromatography and Thin-Layer Chromatography," Neth. Milk Dairy J., 16, 217-225 (1962); CA 58, 8287d

488. Badings, H. T., and Koops, J., "The Component Fatty Acids of the Milk Phosphatides," Fette, Seifen, Anstrichmittel, 62, 302-303 (1960); CA 54, 19991i

489. Bagby, M. O., Smith, Jr., C. R., Miwa, T. K., Lohmar, R. L., and Wolff, I. A., "A Unique Fatty Acid from Limnanthes Douglasii Seed Oil: The C_{22} Diene," J. Org. Chem., 26, 1261-1265 (1961); CA 55, 19770c

490. Bahr, G., and Meier, G., "Urea Inclusion Products of Non-Branched Chain Mercury Dialkyls," Z. Anorg. Chem., 294, 22 (1958)

491. Bailey, S. D., Bazinet, M. L., Driscoll, J. L., and McCarthy, A. I., "The Volatile Sulfur Components of Cabbage," J. Food Sci., 26, 163-170 (1961); CA 56, 1813a
17

492. Bailey, W. J., and Baylouny, R. A., "Pyrolysis of Esters. XVII. Effect of Configuration on the Direction of Elimination," J. Am. Chem. Soc., 81, 2126-2129 (1959); CA 54, 339h

493. Bailey, W. J., and Baylouny, R. A., "Synthesis of 5-Methylene-1,3-cyclohexadiene," 138th Natl. ACS Mtg., New York, N. Y., September 1960, Program Abstr., p. 73P

494. Bailey, W. J., and Hale, W. F., "Pyrolysis of Esters. XV. Effect of Temperature on Direction of Elimination of Tertiary Esters," J. Am. Chem. Soc., 81, 647-651 (1959); CA 53, 21711b

495. Bailey, W. J., and Hale, W. F., "Pyrolysis of Esters. XVI. Effect of Ring Size on Formation of Alicyclic Olefins," J. Am. Chem. Soc., 81, 651-655 (1959); CA 53, 21711b

496. Bailey, W. J., and Marktscheffel, F., "Cleavage of Tetrahydrofuran During Reductions with Lithium Aluminum Hydride," J. Org. Chem., 25, 1797-1800 (1960); CA 55, 2598b

497. Bailly, F. H., "Gas Chromatograph Analyses Aid New Oilfield Detection," World Oil, 155, No. 7, 75-78 (1962); CA 58, 3243h

498. Baines, C. B., and Proctor, K. A., "Gas Chromatography in Routine Pharmaceutical Analysis," J. Pharm. and Pharmacol., 11, Suppl., 230T-234T (1959); CA 54, 12483i

499. Baker, A. R., and Hartwell, F. J., "Gas Chromatography for Mine-Gas Analysis," Safety Mines Research Estab., No. 189, 31 pp. (1960); CA 55, 1283d

500. Baker, B. E., Papacenstantinou, J. A., Cross, C. K., and Khan, N. A., "Protein and Lipid Constitution of Some Pakistani Pulses," J. Sci. Food Agr., 12, 205-207 (1961); CA 55, 20258f

501. Baker, C. D., and Gunstone, F. D., "Fatty Acids. X. The Synthesis of the cis-Isomers of Tetradec-8-enoic, Hexadec-10-enoic, Octadec-12-enoic, Eicos-14-enoic Acids," J. Chem. Soc., 1963, 489-491; CA 58, 10067f

502. Baker, C. D., and Gunstone, F. D., "Fatty Acids. XI. The Synthesis of 9D-Hydroxyoctadecanoic Acid," J. Chem. Soc., 1963, 759-760

503. Baker, D., "Fatty Acid Composition of Oil from Damaged Corn and Wheat," Cereal Chem., 39, 393-397 (1962); CA 58, 1652e

504. Baker, J. M., and King, B. D., "Rapid Blood Ether Determinations Using Gas Chromatography," Annual Mtg., Am. Soc. Anesthesiologists, New York, N. Y., October 1962; Abstr., Anesthesiology, 24, 112 (1963)

505. Baker, R. A., "Gas-Chromatographic Analysis of Aqueous Solutions," Mater. Res. Std., 2, 983-988 (1962); CA 58, 3861h

506. Baker, R. A., and Doerr, R. C., "Methods of Sampling and Storage of Air Containing Vapors and Gases," Intern. J. Air Pollution, 2, 142-158 (1959); CA 54, 9174f

507. Baker, R. W. R., "Ester Linked Long Chained Fatty Acids of Nervous Tissue," Biochem. J., 79, 642-648 (1961); CA 55, 21300f

508. Baker, R. W. R., and Gower, D. B., "Gas Chromatography of Androst-16-3α-ol and Related Steroids," Nature, 192, 1074-1075 (1961); CA 56, 12293e

509. Baker, W. J., "The Effect of Operational Variables on Column Efficiency," 7th Natl. Symp. on Instrumental Methods of Analysis, ISA, Houston, Texas, April 1961

510. Baker, W. J., Lee, E. H., and Wall, R. F., "Chromatographic Solid-Support Studies," Gas Chromatog. Intern. Symposium, 2nd, East Lansing, Mich., 1959, 21-32 (Pub. 1961); CA 55, 19583f

511. Baker, W. J., Norlin, H. L., Zinn, T. L., and Wall, R. F., "Design Parameters for Process Gas Chromatography," 132nd Natl. ACS Mtg., New York, N. Y., September 1957, Program Abstr., p. 32B; Preprints, Div. Petrol. Chem., 2, No. 4, D43-D49 (1957); CA 54, 20354d

512. Baker, W. J., and Wall, R. F., "Comparative Studies of Gas Chromatograph Column Parameters," Proc. of the Natl. Symposium on Instrumental Methods of Analysis, ISA, Houston, Texas, May 1958, pp. 129-140

513. Baker, W. J., and Zinn, T. L., "Preparation of Gas Calibration Samples," Perkin-Elmer Instrument News, 11, No. 3, 1 (Summer, 1960)

514. Baker, W. J., and Zinn, T. L., "Multiple Columns in Chromatography," Control Eng., 8, No. 1, 77-81 (1961)

515. Baker, W. J., Zinn, T. L., Wise, K. V., and Wall, R. F., "Observations on the Anomalous Chromatographic Behavior of Hydrogen," Proc. of the Natl. Symposium on Instrumental Methods of Analysis, ISA, Houston, Texas, May 1958, pp. 159-162

516. Balandin, A. A., Marukyan, G. M., Lavrovskaya, T. K., Seimovich, R. G., and Gryzlova, L. V., "Catalytic Dehydrogenation of Chloroethylene," Izv. Akad. Nauk SSSR, Otdel. Khim. Nauk, 1962, 2031-2036; CA 58, 8938g

517. Balasubrahmanyam, S. N., and Quin, L. D., "Pyrolytic Degradation of Nornicotine and Myosmine," Tobacco Sci., 6, 133-136 [Pub. in Tobacco, 155, No. 6, 34-37 (1962)]; CA 58, 4609d

518. Baldwin, R. A., Smitheman, K. A., and Washburn, R. M., "Dichlorophosphination," J. Org. Chem., 26, 3547-3549 (1961); CA 56, 11614e

519. Ball, D. H., and Parrish, F., "Acetylation of Carbohydrates in Aqueous Solution," J. Org. Chem., 27, 4120 (1962); CA 58, 11454h

520. Ballance, P. E., "Production of Volatile Compounds Related to the Flavor of Foods from the Strecker

Degradation of DL-Methionine," J. Sci. Food Agr., 12, 532-536 (1961); CA 56, 535f

521. Ballinger, P., De la Mare, P. B. D., and Williams, D. L. H., "The Kinetics and Mechanism of Additions to Olefinic Substances. VI. Olefin-Forming Proton-Loss Accompanying Addition of Hypochlorous Acid to 2,3-Dichloropropene," J. Chem. Soc., 1960, 2467-2472; CA 54, 20834h

522. Ballod, A. P., Galanina, N. L., Patsevich, I. V., Topchiev, A. V., and Yanyukova, A. M., "Gas-Liquid Chromatography of the Liquid Products from the Vapor Phase, Thermal Nitration of Propane and the Products Resulting from the Interaction Between Methyl Radicals and Nitrogen Dioxide," Neftekhimiya, 2, 924-927 (1962); CA 59, 1461f

523. Banerjee, S. K., Manolopoulo, M., and Pepper, J. M., "The Synthesis of Lignin Model Substances: 5-Hydroxyvanillin and 5-Hydroxyacetoguaiacone," Can. J. Chem., 40, 2175-2177 (1962); CA 58, 6727d

524. Bang, H. O., Jesting, E., and Thaysen, E. H., "Fat Analyses with the Aid of Gas Chromatography," Ugeskrift Laeger, 123, 1787-1790 (1961); CA 56, 8861f

525. Banks, R. E., Ginsberg, A. E., and Haszeldine, R. N., "Heterocyclic Polyfluoro-Compounds. I. Pentafluoropyridine," J. Chem. Soc., 1961, 1740-1743; CA 55, 17630c

526. Bannister, D. W., Phillips, C. S. G., and Williams, R. J. P., "Adsorption Chromatography and Liquid Partition of High Polymers," Anal. Chem., 26, 1451-1454 (1954); CA 49, 2256i

527. Bannister, M. H., Brewerton, H. V., and McDonald, I. R. C., "Vapor-Phase Chromatography in a Study of Hybrids of Pinus," Svensk Papperstidn., 62, 567-573 (1959); CA 53, 22925d

528. Bannister, M. H., Williams, A. L., McDonald, I. R. C., and Forde, M. B., "Variation of Turpentine Composition in Five Population Samples of Pinus Radiata," New Zealand J. Sci., 5, 486-495 (1962); CA 59, 1960f

529. Baque, C., and Champeix, L., "Chromatographic Analysis of Gaseous Mixtures Extracted from Metals by Reducing Fusion Under Vacuum," Comm. energie at. (France), Rappt. No. 1386, 10 pp. (1959); CA 54, 19262d

530. Baque, C., and Champeix, L., "Application of Chromatography to the Analysis of Gas Mixtures Extracted from Vacuum-Fused Metals and Reducing Conditions," Rev. met., 57, 919-924 (1960); CA 55, 11182i

531. Barabanov, N. L., and Mukhina, T. N., "Pyrolysis of Ethane-Propane Mixtures," Neftekhimiya, 1, No. 3, 386-391 (1961); CA 57, 1150f

532. Baraud, J., "Determination of Certain Amino Acids by Gas Chromatography," Bull. soc. chim. France, 1960, 785

533. Baraud, J., "Quantitative Study by Vapor Phase Chromatography of Alcohols and Esters of Alcoholic Fermentation," Bull. soc. chim. France, 1961, 1874-1877; CA 56, 6474d

534. Baraud, J., and Genevois, L., "Presence of 3-Methylbutanol-2(isopropylmethylcarbinol) and of Pentanol-3(diethylcarbinol) in the Products of Alcoholic Fermentation," Compt. rend., 247, 2479-2481 (1958); CA 53, 11751e

535. Baraud, J., and Genevois, L., "Composition of Fusel Oils," Bull. soc. chim. France, 1959, 779

536. Baraud, J., and Genevois, L., "Fatty Acids in Brandy and Yeast," Bull. soc. chim. France, 1960, 212

537. Baraud, J., and Maurice, A., "The Alcohols and Esters in Cane and Apple Brandy," Ind. Aliment. Agr. (Paris), 80, No. 1, 3-7 (1963); CA 59, 2130d

538. Barber, D. W., Phillips, C. S. G., Tusa, G. F., and Verdin, A., "The Chromatography of Gases and Vapors. VI. Use of Stearates of Bivalent Manganese, Cobalt, Nickel, Copper, and Zinc as Column Liquids in Gas Chromatography," J. Chem. Soc., 1959, 18-24; CA 53, 7718h

539. Barber, H. J., and Lunt, E., "Ring Enlargement During the Reduction of Nitrocycloalkanes by Lithium Aluminum Hydride," J. Chem. Soc., 1960, 1187-1194; CA 54, 16434c

540. Barbier, M., and Pain, J., "The Secretion of Mandibular Glands of Queen and Worker Bees (Apis Mellifica) Determined by Gas Phase Chromatography," Compt. rend., 250, 3740-3742 (1960); CA 54, 19979f

541. Barbour, W., and Rushneck, D. R., "A Broad Range Ionization Detector," 14th Annual Mid-America Spectroscopy Symp., Chicago, Ill., May 1963; in "Developments in Applied Spectroscopy," Vol. 3, J. E. Forrette and E. Lanterman, eds., pp 398-408, Plenum Press, New York, 1964

542. Barclay, L. R. C., Milligan, C. E., and Hall, N. D., "Ortho Diquaternary Compounds. I. The Synthesis of o-di-tert-Butylbenzene. Some Reactions of Side Chain Substituted Derivatives," Can. J. Chem., 40, 1664-1671 (1962); CA 58, 1375c

543. Bardwell, J., "Some Observations on Thermal Conductivity Cells," 2nd Alberta Gas Chromatography Discussion, Edmonton, Alberta, February 1959

544. Barefoot, R. R., and Currah, J. E., "Some Applications of Vapor-Phase Chromatography," Chem. in Can., 7, No. 11, 45-48, 50, 52 (1955); CA 50, 3143f

545. Barefoot, R. R., and Currah, J. E., "Gas Chromatography. Review of Recent Progress," Chem. in Can., 9, No. 3, 68-72 (1957); CA 51, 7802c

546. Barford, R. A., Herb, S. F., Luddy, F. E., Magidman, P., and Riemenschneider, R. W., "Alcoholysis of Vernonia Anthelmintica Seed Oil and Isolation of Methyl Epoxyleate," J. Am. Oil

Chemists' Soc., 40, 136-138 (1963); CA 59, 844g

547. Barker, E., and Sloman, K. G., "Food," Anal. Chem., 35, No. 5, 62R-77R (1963); CA 58, 14621h

548. Barker, H., "Comparison of Hydrogen Flame Ionization and Thermal Conductivity for Detection of Lower Nitriles," 13th Annual Mid-America Spectroscopy Symp., Chicago, Ill., April-May 1962

549. Barker, P. E., and Critcher, D., "The Separation of Volatile Liquid Mixtures by Continuous Gas-Liquid Chromatography," Chem. Eng. Sci., 13, 82-89 (1960); CA 55, 5050i

550. Barlow, A., Lehrle, R. S., and Robb, J. C., "Direct Examination of Polymer Degradation by Gas Chromatography. I. Applications to Polymer Analysis and Characterization," Polymer, 2, 27-40 (1961); CA 55, 17071a

551. Barlow, J.S., "Fatty Acid Characteristics of Some Insect Taxa," Nature, 197, 311 (1963); CA 58, 8257f

552. Barnard, D., Evans, M. B., Higgins, G. M. C., and Smith, J. F., "Gas-Liquid Chromatographic Separation and Polarographic Reduction of Thiosulfonates," Chem. & Ind. (London), 1961, 20-21; CA 55, 23189b

553. Barnard, J. A., "The Pyrolysis of Butanol," Trans. Faraday Soc., 53, 1423-1430 (1957); CA 52, 12528b

554. Barnard, J. A., "The Pyrolysis of tert-Butanol," Trans. Faraday Soc., 55, 947-951 (1959); CA 54, 6270d

555. Barnard, J. A., "The Pyrolysis of Isopropanol," Trans. Faraday Soc., 56, 72-79 (1960); CA 54, 12749g

556. Barnard, J. A., and Hughes, H. W. D., "Analysis of Light Hydrocarbon Gas Mixtures," Nature, 183, 250 (1959); CA 53, 12920a

557. Barnard, J. A., and Hughes, H. W. D., "The Pyrolysis of Ethanol," Trans. Faraday Soc., 56, 55-63 (1960); CA 54, 12749d

558. Barnes, C. S., "The δ-Hydroxy Acids of Wool Grease," Australian J. Chem., 13, 184-186 (1960); CA 54, 16875b

559. Baron, C., and Maume, B., "Gas Chromatographic Behavior and Stereochemical Structure of Menthol, Menthone, Borneol, Menthoglycol, and Their Stereoisomers, and Camphor," Bull. soc. chim. France, 1962, 1113-1117; CA 57, 16662g

560. Barrall, E. M., and Ballinger, P., "Gas Chromatographic Analysis of Lead Alkyls with Electron Affinity Detectors," 144th Natl. ACS Mtg., Los Angeles, Calif., March-April 1963, Program Abstr., pp. 18B-19B

561. Barrall, II, E. M., Porter, R. S., and Johnson, J. F., "Gas Chromatographic Analysis of Poly-(ethylene Ethyl Acrylate) and Poly(ethylene Vinyl Acetate) Pyrolyzates," Anal. Chem., 35, 73-76 (1963); CA 58, 10349b; 142nd Natl. ACS Mtg., Atlantic City, N. J., September 1962, Program Abstr., p. 10B

562. Barras, R. C., and Boyle, J. F., "Applications of Laboratory Gas Chromatography to Refinery Operations," 27th Mtg. of the API Division of Refining, San Francisco, Calif., May 1962

563. Barrer, R. M., "The Sorption of Polar and Non-Polar Gases by Zeolites," Proc. Roy. Soc. (London), A167, 393-420 (1938); CA 33, 4104[6]

564. Barrer, R. M., "Zeolites as Adsorbents and Molecular Sieves," Ann. Repts. Progr. Chem., 41, 31-46 (1944); CA 42, 8573d

565. Barrer, R. M., "Separation of Mixtures Using Zeolites as Molecular Sieves. I. Three Classes of Molecular Sieve Zeolite," J. Soc. Chem. Ind., 64, 130-131 (1945); CA 39, 4789[8]

566. Barrer, R. M., "Separation of Mixtures Using Zeolites as Molecular Sieves. III. The Use of Zeolites to Separate Polar Molecules from Mixtures Containing Them," J. Soc. Chem. Ind., 64, 133-134 (1945); CA 39, 4790[2]

567. Barrer, R. M., "Synthesis of a Zeolite Mineral with Chabazite-like Sorptive Properties," J. Chem. Soc., 1948, 127-132; CA 42, 4423a

568. Barrer, R. M., "Synthesis and Reactions of Mordenite," J. Chem. Soc., 1948, 2158-2163; CA 43, 5323c

569. Barrer, R. M., "Separations with Zeolite Materials," Disc. Faraday Soc., 1949, No. 7, 135-141; CA 45, 18h

570. Barrer, R. M., "Transient Flow of Gases in Sorbents Providing Uniform Capillary Networks of Molecular Dimensions," Trans. Faraday Soc., 45, 358-373 (1949); CA 43, 6488h

571. Barrer, R. M., "Aspects of Intra-Crystalline Sorption," J. chim. phys., 47, 82-94 (1950); CA 44, 6701a

572. Barrer, R. M., "The Separation of Molecules with the Aid of Crystal Sieves," Brennstoff-Chem., 35, 325-334 (1954); CA 49, 3617c

573. Barrer, R. M., "Stoichiometry of Inclusion Compounds," Nature, 176, 745-746 (1955); CA 50, 3840a

574. Barrer, R. M., "Separation of Molecular Mixtures Using Crystal Sieves," Chem. & Ind. (London), 1958, 252

575. Barrer, R. M., "New Selective Sorbents: Porous Crystals as Molecular Filters," Brit. Chem. Eng., 4, 267-279 (1959); CA 54, 979d

576. Barrer, R. M., and Baynham, J. W., "Hydrothermal Chemistry of the Silicates. VII. Synthetic Potassium Aluminosilicates," J. Chem. Soc., 1956, 2882-2891; CA 50, 16506i

577. Barrer, R. M., and Baynham, J. W., "Synthetic Chabazites: Correlation Between Isomorphous Replacements, Stability and Sorption Capacity," J. Chem. Soc., 1956, 2892-2903; CA 51, 7099h

578. Barrer, R. M., Baynham, J. W., and McCallum, N., "Hydrothermal Chemistry of Silicates. V. Compounds Structurally Related to Analcite," J. Chem. Soc., 1953, 4035-4041; CA 48, 4348h

579. Barrer, R. M., and Belchetz, L., "Separation of Mixtures Using Zeolites as Molecular Sieves. II. The Use of a Zeolite to Resolve Hydrocarbon Mixtures," J. Soc. Chem. Ind., 64, 131-132 (1945); CA 39, 4790[1]

580. Barrer, R. M., and Brook, D. W., "Sorption and Reactivity of Simple Organic Molecules in Chabazite," Trans. Faraday Soc., 49, 940-948 (1953); CA 48, 5596h

581. Barrer, R. M., and Brook, D. W., "Molecular Diffusion in Chabazite, Mordenite, and Levynite," Trans. Faraday Soc., 49, 1049-1059 (1953); CA 48, 5597a

582. Barrer, R. M., Bultitude, F. W., and Sutherland, J. W., "Structure of Faujasite and Properties of Its Inclusion Complexes with Hydrocarbons," Trans. Faraday Soc., 53, 1111-1123 (1957); CA 52, 8667h

583. Barrer, R. M., and Gabor, T., "Sorption and Diffusion of Simple Paraffins in Silica-Aluminum Cracking Catalyst," Proc. Roy. Soc. (London), A256, 267-290 (1960); CA 54, 23289b

584. Barrer, R. M., and Hampton, M. G., "Gas Chromatography and Mixture Isotherms in Alkyl Ammonium Bentonites," Trans. Faraday Soc., 53, 1462-1475 (1957); CA 52, 12511a

585. Barrer, R. M., Hinds, L., and White, E. A. D., "Hydrothermal Chemistry of Silicates. III. Reactions of Analcite and Leucite," J. Chem. Soc., 1953, 1466-1475; CA 47, 8571g

586. Barrer, R. M., and Ibbitson, D. A., "Occlusion of Hydrocarbons by Chabazite and Analcite," Trans. Faraday Soc., 40, No. 5, 195-206 (1944); CA 38, 5445[9]

587. Barrer, R. M., and Ibbitson, D. A., "Kinetics of Formation of Zeolite Solid Solutions," Trans. Faraday Soc., 40, No. 5, 206-216 (1944); CA 38, 5451[8]

588. Barrer, R. M., and Mackenzie, N., "Sorption by Attapulgite. I. Availability of Intracrystalline Channels," J. Phys. Chem., 58, 560-568 (1954); CA 48, 11870h

589. Barrer, R. M., Mackenzie, N., and MacLeod, D. M., "Sorption by Attapulgite. II. Selectivity Shown by Attapulgite, Sepiolite, and Montmorillonite for n-Paraffins," J. Phys. Chem., 58, 568-572 (1954); CA 48, 11871a

590. Barrer, R. M., and MacLeod, D. M., "Activation of Montmorillonite by Ion Exchange and Sorption Complexes of Tetra-Alkyl Ammonium Montmorillonites," Trans. Faraday Soc., 51, 1290-1300 (1955); CA 50, 4582h

591. Barrer, R. M., and McCallum, N., "Hydrothermal Chemistry of Silicates. IV. Rubidium and Cesium Aluminosilicates," J. Chem. Soc., 1953, 4029-4035; CA 48, 4348f

592. Barrer, R. M., and Meier, W. M., "Structural and Ion Sieve Properties of a Synthetic Crystalline Exchanger," Trans. Faraday Soc., 54, 1074-1085 (1958); CA 53, 8764i

593. Barrer, R. M., and Reay, J. S. S., "Interlamellar Sorption by Montmorillonite," Proc. Intern. Congr. Surface Activity, 2nd, London, 1957, 2, 79-89; CA 52, 17888f

594. Barrer, R. M., and Reay, J. S. S., "Sorption and Intercalation by Methyl-Ammonium Montmorillonites," Trans. Faraday Soc., 53, 1253-1261 (1957); CA 52, 9711e

595. Barrer, R. M., and Reay, J. S. S., "Sorption by NH_4^+ and Cs^+ Montmorillonites and Ion Fixation," J. Chem. Soc., 1958, 3824-3830; CA 53, 2733i

596. Barrer, R. M., and Rees, L. V., "Sorption of Mixtures. III. Polar Sorbates as Modifiers of Zeolite Crystals," Trans. Faraday Soc., 50, 852-863 (1954); CA 49, 2812e

597. Barrer, R. M., and Rees, L. V., "Sorption of Mixtures. IV. Molecular Diffusion in Crystals Modified by Polar Sorbates," Trans. Faraday Soc., 50, 989-999 (1954); CA 49, 4365b

598. Barrer, R. M., and Rees, L. V., "Henry's Law Adsorption Constants," Trans. Faraday Soc., 57, 999-1007 (1961); CA 56, 38b

599. Barrer, R. M., and Riley, D. W., "Sorptive and Molecular Sieve Properties of a New Zeolite Mineral," J. Chem. Soc., 1948, 133-143; CA 42, 4423c

600. Barrer, R. M., and Robins, A. B., "Sorption of Mixtures. I. Molecular Sieve Separations of Permanent and Inert Gases," Trans. Faraday Soc., 49, 807-815 (1953); CA 48, 3099h

601. Barrer, R. M., and Robins, A. B., "Sorption of Mixtures. II. Equilibria Between Binary Gas Mixtures and Some Zeolites," Trans. Faraday Soc., 49, 929-939 (1953); CA 48, 5597d

602. Barrer, R. M., and Strachan, E., "Sorption and Surface Diffusion in Microporous Carbon Cylinders," Proc. Roy. Soc. (London), A231, 52-74 (1955); CA 49, 14419d

603. Barrer, R. M., and Sutherland, J. W., "Inclusion Complexes of Faujasite with Paraffins and Permanent Gases," Proc. Roy. Soc. (London), A237, 439-463 (1956); CA 51, 15214a

604. Barrer, R. M., and White, E. A. D., "Hydrothermal Chemistry of Silicates. I. Synthetic Lithium

Aluminosilicates," J. Chem. Soc., 1951, 1267-1278; CA 45, 7460d

605. Barrer, R. M., and White, E. A. D., "Hydrothermal Chemistry of Silicates. II. Synthetic Crystalline Sodium Aluminosilicates," J. Chem. Soc., 1952, 1561-1571; CA 46, 9007i

606. Barrow, J. G., Quinlan, C. B., Edmands, R. E., Whitner, V. S., and Goodloe, M. H. R., "Comparison by Gas Chromatography of the Fatty Acid Content of Adipose Tissue with Dietary Intake of Fat in Vegetarian and Nonvegetarian Males," 33rd Sci. Sessions, Am. Heart Assoc., St. Louis, Mo., October 1960; cf. Circulation, 22, 720-721 (1960)

607. Barry, J. A., Vasishth, R. C., and Shelton, F. J., "Analysis of Chlorophenols by Gas-Liquid Chromatography," Anal. Chem., 34, 67-69 (1962); CA 56, 9389b

608. Barry, R., "Ultrasensitive Ionization Detector for Permanent Gas Analysis," Nature, 188, 578-579 (1960); CA 55, 9976f

609. Barsky, M. H., and Jordan, J., "Gas Chromatography. Significance of Elution Areas Obtained with 'Thermal Conductivity' Detectors," Central Pennsylvania Mtg. in Miniature, ACS, Erie, Pa., March 1959

610. Bartels-Keith, J. R., "Alternaric Acid. III. Structure," J. Chem. Soc., 1960, 1662-1665; CA 54, 18492h

611. Bartle, J., "GLC Automatic Analyzer Developed for Wilton Works," Chem. Age, 83, 535-536 (1960)

612. Bartlet, J. C., and Smith, D. M., "The Determination of the Areas of Resolved and Partially Resolved Chromatography Peaks," Can. J. Chem., 38, 2057-2065 (1960); CA 55, 9145f

613. Bartley, W., Getz, G. S., Notton, B. M., and Renshaw, A., "The Lipid Composition of Phosphorylating 'Digitonin Particles' and Water- and Saline-Extracted Mitochondria from Rat Liver," Biochem. J., 82, 540-553 (1962)

614. Bartok, W., and Lucchesi, P. J., "The Chain Alkylation of Acetylene with Propane Induced by Nuclear Radiation," J. Am. Chem. Soc., 81, 5918-5921 (1959); CA 54, 7300b

615. Bartok, W., and Lucchesi, P. J., "The Radiation-Induced Chain Alkylation of Ethylene with Propane," J. Am. Chem. Soc., 82, 4525-4528 (1960); CA 55, 5322g

616. Barton, D. H. R., and Vries, J. X. de, "The Constitution of Sedanolide," J. Chem. Soc., 1963, 1916-1919; CA 58, 10096c

617. Bartsch, R. C., Miller, F. D., and Trent, F. M., "Quantitative Analysis of Some C_{10} Dibasic Acids and Associated Monobasic Acids by High Temperature Gas Chromatography," Anal. Chem., 32, 1101-1103 (1960); CA 54, 22154c

618. Bashkirov, A. N., and Pal, S., "Oxidation of Olefins in Liquid Phase in the Presence of Boric Acid," Doklady Akad. Nauk S.S.S.R., 128, 1175-1178 (1959); CA 54, 7531h

619. Basmadjian, D., "Adsorption Equilibria of Hydrogen, Deuterium and Their Mixtures. I.," Can. J. Chem., 38, 141-148 (1960); CA 55, 3152h

620. Basmadjian, D., "Adsorption Equilibria of Hydrogen, Deuterium and Their Mixtures. II. "Can. J. Chem., 38, 149-156 (1960); CA 55, 3152h

621. Bassemir, R. W., and Rusterholz, W. E., "Application of Vapour-Liquid Partition Chromatography to Analysis and Control of Printing Ink Volatiles," American Ink Mkr., 35, No. 11, 44-47 (1957); J. Appl. Chem., 8, 536 (1958); CA 52, 9626c

622. Bassett, D. W., and Habgood, H. W., "A Gas Chromatographic Study of the Catalytic Isomerization of Cyclopropane," J. Phys. Chem., 64, 769-773 (1960); CA 54, 23599c; 137th Natl. ACS Mtg., Cleveland, Ohio, April 1960, Program Abstr., p. 28B

623. Bassett, D. W., and Habgood, H. W., "The Catalytic Isomerization of Cyclopropane. Effects of Residual Water and the Nature of the Exchangeable Cation," Abstr., Chem. in Can., 13, No. 5, 50 (1961); 44th Canadian Chemical Conf. and Exhibition of CIC, Montreal, Canada, August 1961

624. Bassette, R., Ozeris, S., and Whitnah, C. H., "Gas Chromatographic Analysis of Head Space Gas of Dilute Aqueous Solutions," Anal. Chem., 34, 1540-1543 (1962); CA 58, 738f

625. Bassette, R., Ozeris, S., and Whitnah, C. H., "Direct Chromatographic Analysis of Milk," 57th Annual Mtg., Am. Dairy Sci. Assoc., College Park, Md., June 1962; Abstr., J. Dairy Sci., 45, 660-661 (1962)

626. Bassette, R., and Whitnah, C. H., "Removal and Identification of Organic Compounds by Chemical Reaction, in Chromatographic Analysis," Anal. Chem., 32, 1098-1100 (1960); CA 54, 22171g

627. Bassette, R., and Whitnah, C. H., "Mass Spectrometry and Gas Chromatography Applied to the Study of Volatiles in Milk," 56th Annual Mtg., Am. Dairy Sci. Assoc., Madison, Wisc., June 1961; Abstr., J. Dairy Sci., 44, 1164 (1961)

628. Basson, R. A., DeWet, C. R., Nel, W., and Pretorius, V., "Glow-Discharge Detector for Gas Chromatography," J. S. African Chem. Inst., 12, 62-68 (1959); CA 54, 10410c

629. Bastin-Merkeman, M. J., and Dietz, H. G., "Analysis of Technical Xylenes by Using Gas Chromatography and Ultraviolet Absorption Spectrometry," Chim. anal., 42, 493-497 (1960); CA 55, 10190d

630. Bates, R. B., Gale, D. M., and Gruner, B. J., "Terpenoids. VII. The Stereoisomeric Farnesols," J. Org. Chem., 28, 1086-1089 (1963); CA 58, 12604c

631. Bates, R. B., Gale, D. M., Gruner, B. J., and Nicholas, P. P., "The Stereoisomeric Farnesols," Chem. & Ind. (London), 1961, 1907-1908; CA 3487a

632. Bates, R. B., Gale, D. M., and Nicholas, P. P., "Trans, Trans- and Cis, Trans- Farnesols," 138th Natl. ACS Mtg., New York, N. Y., September 1960, Program Abstr., p. 27P

633. Bates, R. B., and Slagel, R. C., "Conversion of Bulnesol to Patchouli Alcohol, Guaiol, and 'δ-Guaiene'," J. Am. Chem. Soc., 84, 1307-1308 (1962); CA 57, 2259e

634. Bates, R. B., and Slagel, R. C., "Terpenoids. VI. β-Guaiene, β-Patchoulene, and Guaioxide in Essential Oils," Chem. & Ind. (London), 1962, 1715-1716; CA 58, 5728b

635. Bates, T. H., and Williams, T. F., "Radiolysis of Terpene Hydrocarbons. Isomerization and Polymerization of α- and β-Pinene," Nature, 187, 665-669 (1960); CA 55, 2720f

636. Baumann, F., and Olund, S. A., "Analysis of Liquid Odorants by Gas Chromatography," J. Chromatog., 9, 431-438 (1962)

637. Baumann, F. B., Johnson, J. F., and Klaver, R. F., "Subambient Programmed Temperature Gas Chromatography," 4th Intern. Gas Chromatography Symp., Hamburg, Germany, June 1962; pub. in "Gas Chromatography 1962," ed. by M. Van Swaay, Butterworth & Co., London, 1962

638. Baumann, F. B., White, F. A., and Johnson, J. F., "Automatic Attenuator for a Hydrogen Flame Gas Chromatograph," Anal. Chem., 34, 1351-1352 (1962); CA 57, 11835g

639. Baumgarten, H. E., and Anderson, C. H., "Reactions of Amines. VII. The Reactions of α-Amino Ketones with Nitrous Acid," J. Am. Chem. Soc, 83, 399-404 (1961); CA 55, 10373i

640. Bavisotto, V. S., and Roch, L. A., "Gas Chromatography of Volatiles in Beer During Its Brewing, Fermentation, and Storage," Brewers Digest, July 1959, p. 58

641. Bavisotto, V. S., and Roch, L. A., "Gas Chromatography of Volatiles in Beer During Its Brewing, Fermentation, and Storage," Am. Soc. Brewing Chemists, Proc., 1959, 63-75; CA 54, 7062d

642. Bavisotto, V. S., and Roch, L. A., "Gas Chromatography of Volatiles in Beer During Its Brewing, Fermentation, and Storage," Am. Soc. Brewing Chemists, Proc., 1960, 101-112; CA 55, 27763f

643. Bavisotto, V. S., Roch, L. A., and Heinisch, B., "Gas Chromatography of Volatiles in Beer During Its Brewing, Fermentation, and Storage. III. Quantitative Effect of Physical, Chemical, and Physiological Variables," Am. Soc. Brewing Chemists, Proc., 1961, 16-23; CA 56, 7799g

644. Bavisotto, V. S., Roch, L. A., and Lesniewski, R. S., "Gas Chromatography of Volatiles of Fermented Milk Products," 55th Annual Mtg., Am. Dairy Sci. Assoc., Logan, Utah, June 1960; Abstr., J. Dairy Sci., 43, 849 (1960)

645. Baxter, R. A., "Analysis of Biphenyl-Terphenyl Organic Coolant Mixtures by Gas Chromatography," U. S. At. Energy Comm. NAA-SR-Memo-3793, 8 pp. (1959); CA 55, 21987f

646. Baxter, R. A., and Keen, R. T., "High Temperature Gas Chromatography of Aromatic Hydrocarbons," Anal. Chem., 31, 475-476 (1959); CA 53, 10899i

647. Baxter, R. A., and Keen, R. T., "High Temperature Gas Chromatography of Aromatic Hydrocarbons: Instrument Design and Exploratory Studies at Temperatures Up to 430°," U. S. At. Energy Comm., NAA-SR-3154, 30 pp. (1959); CA 54, 4269h

648. Baxter, R. M., Dandiya, P. C., Kandel, S. I., Okany, A., and Walker, G. C., "Separation of the Hypnotic Potentiating Principles from the Essential Oil of Acorus Calamus L. of Indian Origin by Liquid-Gas Chromatography," Nature, 185, 466-467 (1960); CA 54, 15847c

649. Baxter, R. M., Fan, M. C., and Kandel, S. I., "Cis-Trans Isomers of Asarone, Their Liquid-Gas Chromatographic Behavior, and That of Certain Other Propenylphenol Ethers," Can. J. Chem., 40, 154 (1962); CA 57, 7146h

650. Bayer, E., "Separation of Derivatives of Amino Acids Using Gas-Liquid Chromatography," in "Gas Chromatography 1958," edited by D. H. Desty, Academic Press, Inc., New York, 1958, pp. 333-342; CA 53, 16261h

651. Bayer, E., "Selectivity of the Liquid Phase in Gas Chromatography, and Choice of Support," Angew. Chem., 71, 299-302 (1959); CA 53, 17748g

652. Bayer, E., "Gas Chromatography," Elsevier Monographs, No. 10, Chemistry Section, D. Van Nostrand Co., Inc., Princeton, N. J., 1961; "Gas Chromatographie," Springer-Verlag, Berlin, 1959

653. Bayer, E., "Gas Chromatographie," 2nd ed., Springer Verlag., Berlin, 1962

654. Bayer, E., "Theoretical and Practical Considerations on Choice of Solvent and Filling of Analytical and Preparative Columns," University of Houston Intern. Symp. on Advances in Gas Chromatography, Houston, Texas, January 1963

655. Bayer, E., and Anders, F., "Biological Objects as Detectors in Gas Chromatography," Naturwissenschaften, 46, 380 (1959); CA 53, 20228e

656. Bayer, E., and Bassler, L., "Systematic Identification of Volatile Organic Substances. II. Esters in Wine Aroma," Z. Anal. Chem., 181, 418-424 (1961); CA 56, 779f

657. Bayer, E., Hupe, K. P., and Mack, H., "Filling of Analytical and Preparative Columns for Gas Chromatography," Anal. Chem., 35, 492-496 (1963)

658. Bayer, E., Hupe, K. P., and Witsch, H. G., "Preparative Gas Chromatography," Symp. on Modern Methods for Analysis of Organic Compounds, Munich, Germany, October, 1960; Abstr., Angew.

Chem., 73, 145-146 (1961)

659. Bayer, E., Hupe, K. P., and Witsch, H. G., "Preparative Gas Chromatography, IV.," Angew. Chem. 73, 525-530 (1961); CA 56, 9902g

660. Bayer, E., Kupfer, G., and Reuther, K., "Gas Chromatographic Analysis of Synthetic and Natural Aroma Materials," Z. Anal.Chem., 164, 1-10 (1958); CA 53, 6866i

661. Bayer, E., Kupfer, G., and Reuther, K. H., "Relative Retention Volumes of Terpenes and Non-Cyclic Unsaturated Alcohols and Esters (Chromatographic Data Table 15)," J. Chromatog., 2, D7 (1959); cf. Z. Anal. Chem., 164, 4 (1958) (Camphor = 1)

662. Bayer, E., Reuther, K. H., and Born, F., "Analysis of Amino Acid Mixtures by Vapor-Phase Chromatography," Angew. Chem., 69, 640 (1957); CA 53, 17749i

663. Bayer, E., and Rock, H., "Thermodynamic Characterization of Selectivity in Gas Chromatography," Angew. Chem., 71, 407 (1959); CA 53, 19517g

664. Bayer, E., Wahl, G., and Witsch, H. G., "Preparative Gas Chromatography," Z. Anal. Chem., 181, 384-390 (1961); CA 56, 4594b

665. Bayer, E., and Witsch, H. G., "Preparative Gas Chromatography," Z. Anal. Chem., 170, 278-285 (1959); CA 54, 2832a

666. Bayes, K., "The Photolysis of Carbon Suboxide," J. Am. Chem. Soc., 83, 3712-3713 (1961); CA 56, 4282f

667. Bayle, G. G., and Klinkenberg, A., "The Theory of Adsorption Chromatography for Liquid Mixtures," Rec. trav. chim., 73, 1037-1057 (1954); CA 49, 12082i

668. Baylouny, R. A., "The Pyrolysis of Esters: Synthesis of 5-Methylene-1, 3-Cyclohexadiene, and Pyrolysis of Cyclohexenyl Acetates," Univ. Microfilms (Ann Arbor, Mich.), Order No. 61-879, 77 pp.; Dissertation Abstr., 22, 734-735 (1961); CA 56, 342a

669. Bazinet, M. L., and Walsh, J. T., "Combination Gas Sampler and Fraction Collector for Gas Chromatography and Mass Spectrometer Applications," Rev. Sci. Instr., 31, 346-347 (1960); CA 55, 8956a

670. Beare, J. L., "Fatty Acid Composition of Food Fats," J. Agr. Food Chem., 10, 120-123 (1962); CA 57, 7690g

671. Beare, J. L., Craig, B. M., and Campbell, J. A., "Nutritional Studies of Partially Hydrogenated Rapeseed Oil," J. Am. Oil Chemists' Soc., 38, 310-312 (1961); CA 55, 18914d

672. Beare, J. L., Gregory, E. R. W., Smith, D. M., and Campbell, J. A., "The Effect of Rapeseed Oil on Reproduction and on the Composition of Rat Milk Fat," Can. J. Biochem. & Physiol., 39, 195-201 (1961); CA 55, 8591d

673. Beaven, G. H., James, A. T., and Johnson, E. A., "Stearic Effects in the Gas-Liquid Chromatography of Some Alkylbiphenyls," Nature, 179, 490-491 (1957); CA 51, 11003c

674. Becher, P., and Birkmeier, R. L., "The Determination of Hydrophile-Lipophile Balance of Gas-Liquid Chromatography," 37th Natl. Colloid Symp., ACS Division of Colloid and Surface Chemistry, Ottawa, Ont., June 1963

675. Beck, E., Principe, A., "Application of Gas Chromatography to Forensic Chemistry," 13th Annual Mid-America Spectroscopy Symp., Chicago, Ill., April-May 1962; see Item 1914

676. Beck, E. C., Jungermann, E., and Linfield, W. M., "Identification of Soap Stocks by Gas Chromatographic Techniques," J. Am. Oil Chemists' Soc., 39, 53-55 (1962); CA 56, 7455b

677. Beck, M. G., "Programmed Gas Chromatography," 4th Annual Gas Chromatography Institute, Canisius College, Buffalo, N. Y., April 1962

678. Beck, M. G., Faley, R. L., and Bennett, C. E., "A Review of Programmed-Temperature Gas Chromatography," in "Progress in Industrial Gas Chromatography," Vol. 1, edited by H. A. Szymanski, Plenum Press, Inc., New York, 1961, pp. 97-124; CA 56, 7967d

679. Beck, M. G., and Mikkelsen, L., "Programmed Temperature Gas Chromatography," Facts & Methods for Scientific Research (F&M Scientific Corp.), 3, No. 1, 6-7 (Spring 1962)

680. Beckman, H., and Bevenue, A., "Pesticide Residue Analysis, Gas Chromatographic Analysis of 2, 6-Dichloro-4-nitroaniline," J. Food Research, 27, 602-604 (1962)

681. Beckman, H., and Bevenue, A., "Analysis of Grapes and Cottonseed for Chlorobenzilate Residues," 144th Natl. ACS Mtg., Los Angeles, Calif., March-April 1963, Program Abstr., p. 8A

682. Beckman, H., and Bevenue, A., "Nematocide Analysis by Gas Chromatography with Electron Capture Detector," 144th Natl. ACS Mtg., Los Angeles, Calif., March-April 1963, Program Abstr., p. 8A

683. Beckman, H., and Bevenue, A., "The Effect of the Column Tubing Composition on the Recovery of Chlorinated Hydrocarbons by Gas Chromatography," J. Chromatog., 10, 231-233 (1963)

684. Beckman, H. F., and Berkenkotter, P., "Gas Chromatography of the Reduction Products of Chlorinated Organic Pesticides," Anal. Chem., 35, 242-246 (1963); CA 58, 8367d

685. Beckman Instruments, Inc., "Analysis of Aerosol-Type Propellents by Gas Chromatography," Application Data Sheet GC-69-MI; "Determination of Fatty Acid Esters by Gas-Liquid Chromatography. Accurate Analysis Important in Medical Research, Soap, and Paint Development and Qual-

ity Control," Application Data Sheet GC-89-B; "Collection of Chromatograms Shows Wide Use of Gas Chromatography in Gas and Liquid Analysis. Data Helps Chemist Choose Right Column for Particular Application," Application Data Sheet GC-90-MI; "Beckman GC-2 Chromatograph Improves Process Efficiency in Manufacture of Formaldehyde," Application Data Sheet GC-8068-C; "Paul-Lewis Laboratories Uses Beckman GC-2 Chromatograph for Analysis of Flavor Components in Beer," Application Data Sheet GC-8067-F; Beckman/Scientific Process Instrument Division, Beckman Instruments, Inc., Fullerton, Calif.

686. Beckwith, A. L. J., and Waters, W. A., "Reaction of Chlorobenzene with Methyl Radicals," J. Chem. Soc., 1957, 1665-1668; CA 51, 13822e

687. Bednas, M. E., and Russell, D. S., "Silver Nitrate Solutions in Gas Chromatography," Can. J. Chem., 36, 1272-1276 (1958); CA 53, 1979a

688. Beebe, R. A., and Emmett, P. H., "A Comparison of the Measurement of Heats of Adsorption by Calorimetric and Chromatographic Methods on the System Nitrogen-Bone Mineral," J. Phys. Chem., 65, 184-185 (1961); CA 55, 23022f

689. Beerthuis, R. K., "Gas Chromatography," Chem. Weekblad, 53, 545-547 (1957)

690. Beerthuis, R. K., Dijkstra, G., Keppler, J. G., and Recourt, J. H., "Gas-Liquid Chromatographic Analysis of Higher Fatty Acids and Fatty Acid Methyl Esters," Ann. N. Y. Acad. Sci., 72, 616-632 (1959); CA 53, 15601g

691. Beerthuis, R. K., and Keppler, J. G., "Gas-Chromatographic Analysis of Higher Fatty Acids: Up to and Including Cerotic Acid," Nature, 179, 731-732 (1957); CA 51, 10313c

692. Beerthuis, R. K., and Recourt, J. H., "Sterol Analysis by Gas Chromatography," Nature, 186, 372-374 (1960); CA 54, 19813d

693. Beerthuis, R. K., and Recourt, J. H., "Sterol Analysis by Gas Chromatography," Colloq. Intern. Centre Natl. Rech. Sci. (Paris), No. 99, 205-219 (1961); CA 58, 5751c

694. Behmann, F. W., and Kerckoff, W. G., "Analysis of Biological Gas Mixtures," Arch. tech. Messen Lfg., 263, 271-274 (1957); CA 52, 7936f

695. Behr, O. M., Eglinton, G., Galbraith, A. R., and Raphael, R. A., "Macrocyclic Acetylenic Compounds. II. 1,2:7,8-Dibenzocyclododeca-1,7-diene-3,5,9,11-tetrayne," J. Chem. Soc., 1960, 3614-3625; CA 55, 2588h

696. Behrendt, S., "Limit of Flammability Detector for Gas Chromatography," Z. Physik. Chem. (Frankfurt), 21, 240-243 (1959); CA 53, 20937f

697. Behrendt, S., "Gas-Chromatography Detection by Labeling with Radioactive Reagents," Z. Physik. Chem. (Frankfurt), 20, 367-370 (1959); CA 54, 3043h

698. Behrendt, S., "Simple Combustion Detector for Gas Chromatography," Z. Physik. Chem. (Frankfurt), 30, 357-359 (1961); CA 56, 13987i

699. Bekesy, M., "Filling of Adsorption Tubes (for Chromatography)," Z. Anal. Chem., 157, 272-274 (1957); CA 52, 1689b

700. Bekkum, H. van, Kleis, A. A. B., Medema, D., Verkade, P. E., and Wepster, B. M., "Studies on Cyclohexane Derivatives. IV. Preparation and Chemical Proof of the Configurations of the Two 4-Isopropylcyclohexanecarboxylic Acids," Rec. Trav. Chim., 81, 833-840 (1962); CA 59, 460e

701. Bell, K. M., and McDowell, C. A., "A Mercury-Photosensitized Oxidation of Hydrocarbons. II. The Mercury-Photosensitized Oxidation of Isobutane," Can. J. Chem., 39, 1424-1433 (1961); CA 55, 27132b

702. Bellar, T. A., Brown, M. F., and Sigsby, Jr., J. E., "Determination of Atmospheric Pollutants in the Part Per Billion Range by Gas Chromatography," 142nd Natl. ACS Mtg., Atlantic City, N. J., September 1962, Program Abstr., p. 19V

703. Bellar, T. A., and Sigsby, Jr., J. E., "Application of Electron Capture Detector to Gas Chromatography in Air Pollution," 144th Natl. ACS Mtg., Los Angeles, Calif., March-April 1963, Program Abstr., pp. 25R-26R

704. Bellar, T. A., Sigsby, Jr., J. E., Clemons, C. A., and Altshuller, A. P., "Direct Application of Gas Chromatography to Atmospheric Pollutants," Anal. Chem., 34, 763-765 (1962); CA 57, 3745c; 140th Natl. ACS Mtg., Chicago, Ill., September 1961, Program Abstr., p. 11W

705. Bellis, H. E., and Slowinski, Jr., E. J., "Application of Vapor Chromatography to Infrared Spectroscopy of Liquids," J. Chem. Phys., 25, 794 (1956); CA 51, 1733e

706. Belyakova, L. D., and Kiselev, A. V., "The Effect of Dehydration of Silica Gel on the Adsorption of Benzene and Hexane," Doklady Akad. Nauk S.S.S.R., 119, 298-301 (1958); CA 52, 19337g

707. Bendel, E., Kern, M., Janssen, R., and Steffan, G., "Separation by Gas Chromatography of All Double Bond Isomers of Octene," Angew. Chem., 74, 905-906 (1962); CA 58, 2827c

708. Bender, S. R., and Kroman, H. S., "Occurrence of Estrogens in Atherogenesis," 16th Annual Mtg., Council on Arteriosclerosis, Am. Soc. for the Study of Arteriosclerosis, Cleveland, Ohio, October 1962; Abstr., Circulation, 26, 645 (1962)

709. Bendoraitis, J. G., Brown, B. L., and Hepner, L. S., "Isoprenoid Hydrocarbons in Petroleum. Isolation of 2,6,10,14-Tetramethylpentadecane by High Temperature Gas-Liquid Chromatography,"

      Anal. Chem., 34, 49-53 (1962); CA 56, 11881g

710. Bendz, G., "The Alkali Isomerization Product of Marasin," Arkiv. Kemi., 14, 475-481 (1959); CA 54, 22556c

711. Benedek, P., "Separation of Gas Mixtures with Solid and Liquid Absorbers," Magyar Kem. Lapja, 15, 432-441 (1960); CA 55, 4102a

712. Benedek, P., "Continuous Gas Chromatography. VII. Separation of Mixtures with Solid and Liquid Sorbents," Acta. Chim. Acad. Sci. Hung., 34, 257-279 (1962); CA 58, 7345d

713. Benedek, P., and Szepesy, L., "Obtaining Pure Acetylene from Gas Mixtures of Low Acetylene Content by Continuous Gas Chromatography," Erdol u. Kohle, 9, 593-597 (1956); CA 51, 1506a

714. Benedek, P., and Szepesy, L., "Continuous Gas Chromatography. II. Dynamic Adsorption on Fixed and Moving Carbon Beds," Acta Chim. Acad. Sci. Hung., 14, 19-29 (1958); CA 53, 18454g

715. Benedek, P., and Szepesy, L., "Continuous Gas Chromatography. III. Adsorption Equilibrium of Acetylene-Carbon Dioxide Mixtures," Acta Chim. Acad. Sci. Hung., 14, 31-41 (1958); CA 53, 18454h

716. Benedek, P., Szepesy, L., and Nagy, Z., "Continuous Chromatography (Hypersorption) on an Experimental Assembly," Gazovaya Prom., 1958, No. 2, 30-38; CA 52, 9676h

717. Benedek, P., Szepesy, L., and Nagy, Z., "Technological Problems of Continuous Gas Chromatography (Hypersorption)," Magyar Kem. Lapja, 13, 117-123 (1958); CA 53, 2700i

718. Benedek, P., Szepesy, L., and Szepe, I., "Design of a Column for Continuous Chromatography of Gases," Gazovaya Prom., 1958, No. 9, 41-45; CA 53, 3g

719. Benedek, P., Szepesy, L., and Szepe, I., "Continuous Gas Chromatography. IV. Design of a Continuous Chromatographic Column for the Separation of Binary Mixtures," Acta Chim. Acad. Sci. Hung., 14, 339-351 (1958); CA 53, 18455a

720. Benedek, P., Szepesy, L., and Szepe, S., "Continuous Gas Chromatography. V. Design of a Three-Product Gas Chromatographic Column," Acta Chim. Acad. Hung., 14, 353-358 (1958); CA 53, 18455b

721. Benedek, P., Szepesy, L., and Szepe, S., "Continuous Gas Chromatography. VI. Design of a Two-Product Gas Chromatographic Column for Multicomponent Gaseous Feed," Acta Chim. Acad. Sci. Hung., 14, 359-367 (1958); CA 53, 18455c

722. Benedek, P., Szepesy, L., and Szepe, S., "The Calculation of the Continuous Gas Chromatographic Column," in "Gas Chromatography," edited by V. J. Coates, H. J. Noebels, and I. S. Fagerson, Academic Press, Inc., New York, 1958, pp. 225-236

723. Benediktova, V., and Janak, J., "Separation of Some Purine Alkaloids by Means of Gas-Liquid Chromatography," Gas Chromatog. Symp., Brno, Czech., June 1962; Abstr., J. Gas Chromatog., 1, No. 4, 10 (1963)

724. Ben-Efraim, D. A., and Sondheimer, F., "o-Quinodimethane as an Intermediate in the Isomerization of cis-4-Octene-1, 7-diyne," Tetrahedron Letters, 1963, 313-315; CA 59, 528f

725. Benjamin, W., Gellhorn, A., Wagner, M., and Kundel, H., "Effect of Aging on Lipid Composition and Metabolism in the Adipose Tissues of the Rat," Am. J. Physiol., 201, 540-546 (1961); CA 56, 1879a

726. Benkeser, R. A., Burrous, M. L., Nelson, L. E., and Swisher, J. V., "The Stereochemistry of the Addition of Silicochloroform to Acetylenes. A Comparison of Catalyst Systems," J. Am. Chem. Soc., 83, 4385-4389 (1961); CA 57, 7296d

727. Benkeser, R. A., Hazdra, J. J., Lambert, R. F., and Ryan, P. W., "Reduction of Organic Compounds by Lithium in Low Molecular Weight Amines. V. Mechanism of Formation of Cyclohexanes. Utility of the Reducing Medium in Effecting Stereospecific Reactions," J. Org. Chem., 24, 854-856 (1959); CA 54, 333g

728. Bennett, C. E., Dal Nogare, S., and Safranski, L. W., "Chromatography: Gas," in "Treatise on Analytical Chemistry," Part 1, Vol. 3, edited by I. M. Kolthoff and P. J. Elving, Interscience Publishers, Inc., New York, 1961, pp. 1657-1715

729. Bennett, C. E., Dal Nogare, S., Safranski, L. W., and Lewis, C. D., "Trace Analysis by Gas Chromatography," Anal. Chem., 30, 898-902 (1958); CA 52, 13516d; 131st Natl. ACS Mtg., Miami, Fla., April 1957, Program Abstr., p. 35B

730. Bennett, C. E., Martin, A. J., and Martinez, Jr., F. W., "Direct Analysis of Fatty Acids by Gas Chromatography," 138th Natl. ACS Mtg., New York, N. Y., September 1960, Program Abstr., pp. 7B-8B

731. Bennett, C. E., Martin, A. J., and Martinez, Jr., F. W., "A Linear-Programmed Temperature Gas Chromatograph Operable to 500°C.," 137th Natl. ACS Mtg., Cleveland, Ohio, April 1960, Program Abstr., p. 5B

732. Bennett, C. E., Martin, A. J., Wisniewski, J., and Faley, R., "Solving Difficult Problems by High Temperature Gas Chromatographic Analysis, 16th Southwest Regional Mtg., ACS, Oklahoma City, Okla., December 1960

733. Bens, E. M., "Some Gas-Liquid Chromatographic Solid Supports as Gas-Solid Chromatographic Ad-

sorbents," 11th Pittsburgh Conf. on Anal. Chem. & Appl. Spectroscopy, Pittsburgh, Pa., February-March 1960, Program Abstr., p. 45

734.   Bens, E. M., "Adsorption Characteristics of Some Gas-Liquid Chromatographic Supports," Anal. Chem., <u>33</u>, 178-182 (1961); CA <u>55</u>, 10014f

735.   Bens, E. M., and McBride, W. R., "Preparative Aspects of Gas-Liquid Chromatographic Separation. Quantitative Determination of Tetraalkyltetrazenes," Anal. Chem., <u>31</u>, 1379-1383 (1959); CA <u>53</u>, 19663f; 134th Natl. ACS Mtg., Chicago, Ill., September 1958, Program Abstr., p. 23B

736.   Bens, E. M., and Stewart, D. H., "Rapid Determination of Acetone in Pentaerythritol Trinitrate by Gas Chromatography," U. S. Dept. Com., Office Tech. Serv., PB Rept. <u>154,210</u>, 12 pp. (1959); CA <u>57</u>, 2856a

737.   Bens, E. M., and Stewart, D. H., "Rapid Determination of Acetone in Pentaerythritol Trinitrate by Gas Chromatography," Navord Rept. No. <u>7014</u>, AD-231, 855 (February 10, 1960)

738.   Bensadoun, A., "Direct Estimation of the Absorption of Volatile Fatty Acids from the Gastrointestinal Tract of Ruminants: Short Chain Volatile Fatty Acid Analysis in Biological Materials; Estimation of the Total Amount of Volatile Fatty Acids Absorbed from the Gastrointestinal Tract of Sheep," Univ. Microfilms (Ann Arbor, Mich.), L. C. Card No. Mic. <u>60-2259</u>, 86 pp.; Dissertation Abstr., <u>21</u>, 5 (1960); CA <u>55</u>, 1761b

739.   Benson, G. W., Cowan, C. B., and Stirling, P. H., "A Theoretical Treatment of the Sensitivity of Katharometers," 1958 Natl. ISA Symp. on Instrumental Methods of Analysis, Houston, Texas, May 1958

740.   Benson, R. E., Lindsey, Jr., R. V., "Chemistry of Allene. II. Reaction of Allene with Acetylene," J. Am. Chem. Soc., <u>81</u>, 4250-4253 (1959); CA <u>54</u>, 4418b

741.   Benson, S. W., and Anderson, K. H., "Free Radical Processes in the Decomposition of Dimethyl Ether," 144th Natl. ACS Mtg., Los Angeles, Calif., March-April 1963, Program Abstr. p. 27P

742.   Bentley, R., Sweeley, C. C., Makita, M., and Wells, W. W., "Gas Chromatography of Sugars and Other Polyhydroxy Compounds," Biochem. Biophys. Res. Commun., <u>11</u>, 14-18 (1963); CA <u>59</u>, 1729h

743.   Berces, T., and Trotman-Dickenson, A. F., "The Reactions of Methoxyl Radicals from Cyclopropane and Isobutane," J. Chem. Soc., <u>1961</u>, 348-350; CA <u>55</u>, 13294e

744.   Bercev, B., Starr, J., Hall, E., Dunlop, A., and Moir, R., "Gas Chromatography," Panel discussion at 8th Regional Conf. of Anal. Chem. Div., CIC, Sarnia, Ontario, April 1959; Chem. in Can. <u>11</u>, No. 10, 27-34 (1959)

745.   Berck, B., and Solomon, J., "Wheat is a Chromatographic Column Towards Methyl Bromide, Ethylene Dibromide, Acrylonitrile, Chloropicrin, and Carbon Tetrachloride in the Vapor Phase," 138th Natl. ACS Mtg., New York, N. Y., September 1960, Program Abstr., p. 6A

746.   Berdick, M., and Cohen, E. M., "Gas Chromatography in Food Additive Problems," 3rd Annual Eastern Analytical Symp. & Instrument Exhibit, New York, N. Y., November 1961

747.   Berezkin, V. G., "Method and Apparatus for the Chromatographic Study of the Products of Radiolysis of Hydrocarbons," Trudy Tashkent. Konf. po Mirnomu Ispol'zovan. At. Energii, Akad. Nauk, Uzbek SSR, <u>2</u>, 425-429 (1960); CA <u>56</u>, 7974c

748.   Berezkin, V. G., "A Method for Chromatogram Calculation in Gas-Liquid Chromatography," Neftekhimiya, <u>1</u>, 169-171 (1961); CA <u>57</u>, 6206f

749.   Berezkin, V. G., and Janak, J., "A Simple Method for Sampling of Solid Materials in Gas Chromatographs," Gas Chromatog. Symp., Brno, Czech., June 1962; Abstr., J. Gas Chromatog., <u>1</u>, No. 4, 8 (1963)

750.   Berezkin, V. G., Janak, J., and Hrivnac, M., "Identification of Sulfur and Nitrogen Compounds in Tar Fractions of Unknown Composition," Gas Chromatog. Symp., Brno, Czech., June 1962; Abstr., J. Gas Chromatog., <u>1</u>, No. 4, 10 (1963)

751.   Berezkin, V. G., and Krasheninnikov, S. K., "Chromatographic Diagrams and Standardized Parts for Gas-Chromatographic Apparatus," Neftekhimiya, <u>1</u>, 700-705 (1961); CA <u>57</u>, 2827d

752.   Berezkin, V. G., and Polak, L. S., "Gas-Liquid Chromatography of C_7-C_{12} Paraffins and Aromatic Hydrocarbons," Doklady Akad. Nauk S.S.S.R., <u>140</u>, 115-117 (1961); cf. ibid., <u>131</u>, 593 (1960); CA <u>56</u>, 626i

753.   Berezkin, V. G., Polak, L. S., and Shakhrai, V. A., "Mechanism of Formation of Heavy Products by Radiolysis of Hexane in Liquid and Solid Phases," Tr. 2-go [Vtorogo] Vses. Soveshch. po Radiats. Khim. Akad. Nauk SSSR, Otd. Khim. Nauk, Moscow, <u>1960</u>, 312-316 (Pub. 1962); CA <u>59</u>, 403f

754.   Berg, C., "Hypersorption: A Process for Separation of Light Gases," Gas, <u>23</u>, No. 1, 32-37 (1947); CA <u>41</u>, 1412h

755.   Berg, C., "Hypersorption Design. Modern Advancements," Chem. Eng. Progr., <u>47</u>, 585-591 (1951); CA <u>46</u>, 296e

756.   Berg, C., "Hypersorption in Modern Gas Processing Plants," Petrol. Refiner, <u>30</u>, No. 9, 241-243, 245-246 (1951); CA <u>46</u>, 243g

757.   Berg, C., and Bradley, W. E., "Hypersorption - New Fractionating Process," Petrol. Engr., <u>18</u>,

115-118 (1947); CA 41, 4912e

758. Berg, C., Fairfield, R. G., Imhoff, D. H., and Multer, H. J., "The Hypersorption Process for Separation of Gases and Vapors," Petrol. Refiner, 28, No. 11, 113-120 (1949); CA 44, 3746d

759. Berg, E. W., "Physical and Chemical Methods of Separation," McGraw-Hill Book Co., Inc., New York, 1963

760. Berge, P. C. van, Haarhoff, P. C., and Pretorius, V., "Solute-Band-Broadening in Packed Gas Chromatographic Columns: Mechanisms Residing in the Mobile Phase," Trans. Faraday Soc., 58, 2272-2281 (1962); CA 58, 10760a

761. Berger, R., "The Proton Irradiation of Methane, Ammonia and Water at 77°K," Proc. Natl. Acad. Sci. U. S., 47, 1434-1436 (1961); CA 56, 4282i

762. Bergmann, G., and Jentzsch, D., "Quantitative Multicomponent Analyses. I. Gas Chromatographic Investigations of Mixtures of Phenols," Z. Anal. Chem., 164, 10-29 (1958); Erdol u. Kohle, 11, 339 (1958); CA 53, 6910f

763. Bergmann, G., and Jentzsch, D., "Relative Retention Volumes of Phenols (Chromatographic Data Table 17)," J. Chromatog., 2, D8 (1959); cf. Z. Anal. Chem., 164, 10 (1958) (Temperature = 150°, Anisole = 1)

764. Bergmann, J. G., and Martin, R. L., "Gas-Chromatographic Method for Determining Chloride and Bromine Ions in Aqueous Solution," Anal. Chem., 34, 911-913 (1962); CA 57, 9211d; 141st Natl. ACS Mtg., Washington, D. C., March 1962, Program Abstr., p. 19B

765. Berka, I., "Determination of Noxious Chemicals in the Atmosphere and Biological Material by Means of Gas Chromatography," Pracovni Lekar., 14, 294-296 (1962); CA 58, 870c

766. Berkowitz-Mattuck, J. B., and Noguchi, T., "Pyrolysis of Untreated and APO-THPC-Treated Cotton Cellulose During One-Second Exposure to Radiant Flux Levels of 5-25 cal./sq.cm.-sec.," J. Appl. Polymer Sci., 7, 709-725 (1963); CA 58, 12715c

767. Berl, W. G., (Ed.), "Physical Methods in Chemical Analysis," "Chromatography," Vol. II, Academic Press, Inc., New York, 1951, pp. 591-617

768. Bernard, A. H., and Rost, H. E., "Analysis of Lipids and Oxidation Products by Partition Chromatography: Dimeric and Polymeric Products," J. Am. Oil Chemists' Soc., 39, 479 (1962); CA 58, 2573f

769. Bernhard, R. A., "Separation and Identification of Some Terpenes by Gas Partition Chromatographic Analysis," J. Assoc. Offic. Agr. Chemists, 40, 915-921 (1957); CA 52, 6068d

770. Bernhard, R. A., "Examination of Lemon Oil by Gas-Partition Chromatography," Food Research, 23, 213-216 (1958); CA 53, 19210e

771. Bernhard, R. A., "Effect of Flow Rate and Sample Size on Column Efficiency in Gas-Liquid Chromatography," Nature, 185, 311-312 (1960); CA 55, 8954e

772. Bernhard, R. A., "Stationary Liquid Phases for Use in Gas-Liquid Chromatography Suitable for the Separation of the Components of Essential Oils," Food Research, 25, 531-537 (1960); CA 55, 21490d

773. Bernhard, R. A., "Analysis and Composition of Oil of Lemon by Gas-Liquid Chromatography," J. Chromatog., 3, 471-476 (1960); CA 55, 7764f

774. Bernhard, R. A., "Citrus Flavor. Volatile Constituents of the Essential Oil of the Orange (Citrus Sinensis)," J. Food Sci., 26, 401-411 (1961); CA 56, 3578d

775. Bernhard, R. A., "Separation of Terpene Hydrocarbons by Gas Liquid Chromatography Utilizing Capillary Columns and Flame Ionization Detection," Anal. Chem., 34, 1576-1579 (1962); CA 58, 7g

776. Bernhard, R. A., and Marr, A. G., "The Oxidation of Terpenes. I. Mechanism and Reaction Products of D-Limonene Autoxidation," Food Research, 25, 517-530 (1960)

777. Bernhard, R. A., and Scrubis, B., "The Isolation and Examination of the Essential Oil of the Kumquat (F. Margarita (Lour.) Swingle)," J. Chromatog., 5, 137-146 (1961); CA 56, 6103b

778. Bernsohn, J., "Fatty Acid Composition of Phospholipids Isolated from Brain Mitochondria," 13th Annual Mid-America Spectroscopy Symp., Chicago, Ill., April-May 1962

779. Beroes, C. S., "A Critical Study of the Variables Affecting Retention Time and Resolution in the Gas-Liquid Chromatographic Column," Univ. Microfilms (Ann Arbor, Mich.), Publ. No. 22843, 106 pp.; Dissertation Abstr., 17, 2229 (1957); CA 52, 1725g

780. Beroza, M., "Determination of the Chemical Structure of Microgram Amounts of Organic Compounds by Gas Chromatography," Anal. Chem., 34, 1801-1811 (1962); CA 58, 13106c

781. Beroza, M., "Ultramicrodetermination of Chemical Structure of Organic Compounds by Gas Chromatography," Nature, 196, 768-769 (1962)

782. Beroza, M., "Determining Chemical Structure Using Carbon-Skeleton Chromatography - A New Technique," 145 Natl. ACS Mtg., New York, N. Y., September 1963

783. Berridge, N. J., and Watts, J. D., "Separation of Mixtures of Methyl Ketones," J. Sci. Food Agr., 5, 417-421 (1954); CA 49, 783d

784. Berry, R., "An Ultra-sensitive Ionization Detector for Permanent Gas Analysis," Nature, 188,

28

578-579 (1960); CA 55, 9976f

785. Berry, R., "Analysis of Milli-Microlitre Quantities of Permanent Gas Mixtures, " 4th Intern. Gas Chromatography Symp., Hamburg, Germany, June 1962; pub. in "Gas Chromatography 1962," ed. by M. van Swaay, Butterworth & Co., London, 1962

786. Berson, J. A., and Mueller, W. A., "Solvent Effects on the Stereoselectivity of Diels-Alder Reactions, " Tetrahedron Letters, 1961, 131-135; CA 55, 16443h

787. Berson, J. A., Olsen, C. J., and Walia, J. S., "A Free Radical Wagner-Meerwein Rearrangement, " J. Am. Chem. Soc., 82, 5000-5001 (1960); CA 55, 2721a

788. Berson, J. A., Olsen, C. J., and Walia, J. S., "Reactions of the 2-Bornyl Radical. II. A Free Radical Wagner-Meerwein Rearrangement, " J. Am. Chem. Soc., 84, 3337-3348 (1962); CA 58, 1326c

789. Berson, J. A., and Remanick, A., "The Mechanism of the Diels-Alder Reaction. The Stereochemistry of the endo-exo Isomerization of the Adducts of Cyclopentadiene with Acrylic and Methacrylic Esters, " J. Am. Chem. Soc., 83, 4947-4956 (1961)

790. Berson, J. A., Walia, J. S., Remanick, A., Suzuki, S., Reynolds-Warnhoff, P., and Willner, D., "The Absolute Configurations of Some Simple Norbornane Derivatives. A Test of the 'Conformational Asymmetry' Model, " J. Am. Chem. Soc., 83, 3986-3997 (1961)

791. Berton, A., "Galvanic Cells Sensitive to Traces of Gaseous, Liquid, or Solid Substances, " Chim. anal., 41, 351-358 (1959); CA 54, 1d

792. Berton, A., "Infrared and Ultraviolet Absorption Spectra and Gas-Phase Chromatography in Industrial Toxicology, " Publ. groupe. avance. methodes spectrog., 1960, 339-354; CA 55, 26823i

793. Bestougeff, M. A., "Chromatographic Methods in the Study of the Number of Isomers of the Constituents of Petroleum, " Journees Intern. Etude Methodes Separation Immediate Chromatog., Paris, 1961, 123-131 (Pub. 1962); CA 59, 1417e

794. Bethea, R. M., "Scale-Up of Gas Chromatography Columns, " J. Chromatog., 9, 21-27 (1962); CA 58, 6468b

795. Bethea, R. M., and Adams, Jr., F. S., "Gas Chromatography of the C_1 to C_4 Nitroparaffins. Isothermal vs. Linear Temperature Programming, " Anal. Chem., 33, 832-839 (1961); CA 55, 20790b

796. Bethea, R. M., and Adams, Jr., F. S., "Gas Chromatography of the C_1 to C_4 Nitroparaffins: Ramp Function Temperature Programming, " J. Chromatog., 8, 532-534 (1962); CA 58, 3871c

797. Bethea, R. M., and Adams, Jr., F. S., "Vapor-Phase Butane Nitration. Product Analysis by Parallel Column Gas Chromatography, " J. Chromatog., 10, 1-8 (1963)

798. Bethea, R. M., and Smutz, M., "Gas Chromatography. Effect of Sample Size on Height of Equivalent Theoretical Plate (HETP) and Retention Volume, " Anal. Chem., 31, 1211-1214 (1959); CA 54, 13939b

799. Bethea, R. M., and Wheelock, T. D., "Gas Chromatography of the C_1 to C_4 Nitroparaffins, " Anal. Chem., 31, 1834-1836 (1959); CA 54, 24132b

800. Bethea, R. M., and Wheelock, T. D., "Effect of Flow Rate on Optimum Sample Size for Minimum Values of Retention Time and HETP [Height Equivalent to a Theoretical Plate], " Gas Chromatog. Intern. Symposium, 2nd, East Lansing, Mich., 1959, 1-10 (Pub. 1961); CA 55, 21687d

801. Bethea, R. M., and Wheelock, T. D., "Effect of Flow Rate on Optimum Sample Size for Minimum Values of Retention and HETP, " in "Gas Chromatography, " edited by H. J. Noebels, R. F. Wall, and N. Brenner, Academic Press, Inc., New York, 1961, pp. 1-10

802. Bethge, P. O., and Lingren, B. O., "Composition of Spruce Fat and Its Change During Wood Storage," Svensk Papperstid., 65, 640-646 (1962); CA 58, 1634g

803. Bethune, J. L., and Rigby, F. C., "Determination of Oxygen Content of Air in Beer by Gas-Solid Chromatography, " J. Inst. Brewing, 64, 170-175 (1958)

804. Bethune, J. L., and Rigby, F. C., "Effect of Ascorbic Acid on the Oxygen Content of Beer as Determined by Gas Chromatography, " Proc. Am. Soc. Brewing Chemists, 1958, 62-65; CA 53, 5585e

805. Beuerman, D. R., and Meloan, C. E., "Determination of Sulfur in Organic Compounds by Gas Chromatography, " Anal. Chem., 34, 319-322 (1962); CA 57, 52b; 140th Natl. ACS Mtg., Chicago, Ill., September 1961, Program Abstr., p. 27B

806. Beuerman, D. R., and Meloan, C. E., "Simultaneous Determination of Carbon and Sulfur in Organic Compounds by Gas Chromatography, " Anal. Chem., 34, 1671-1672 (1962); CA 58, 932f

807. Bevenue, A., "Gas Chromatography, " in "Analytical Methods for Pesticides, Plant Growth Regulators, and Food Additives, Vol. 1, Principles, Methods, and General Applications, " edited by G. Zweig, Academic Press, Inc., New York, 1963

808. Bevilacqua, E. M., English, E. S., and Gall, J. S., "Substrates for Gas Chromatography of Polar Compounds, " Anal. Chem., 34, 861-862 (1962); CA 57, 2827i

809. Bevilacqua, E. M., English, E. S., and Philipp, E. E., "Volatile Methyl Ketone Formed in Rubber Oxidation, " J. Org. Chem., 25, 1276-1277 (1960)

810. Bey, K., "The Application of Modern Analytical Methods to the Chemistry of Detergents, " Fette, Seifen, Anstrichmittel, 64, 900-907 (1962); CA 58, 3615b

811. Beynon, J. H., "Qualitative Analysis of Organic Compounds by Mass Spectrometry, " Nature, 174, 735-737 (1954); CA 49, 4457c

812. Beynon, J. H., "The Use of Mass Spectroscopy for Identification of Organic Compounds," Mikrochim. Acta., 1956, 437-453; CA 50, 8394i

813. Beynon, J. H., Clough, S., Crooks, D. A., and Lester, G. R., "A Theory of the Gas-Liquid Chromatographic Process," Trans. Faraday Soc., 54, 705-714 (1958); CA 53, 1890b

814. Beynon, J. H., Saunders, R. A., and Williams, A. E., "Collection of Chromatographic Fractions in a Mass Spectrometer Sample System," J. Sci. Instr., 36, 375-376 (1959)

815. Beynon, K. I., and Jackson, A., "Preparative Gas-Liquid Chromatographic Column," Chem. & Ind. (London), 1961, 1822-1824; CA 56, 9913h

816. Bhalerao, V. R., Endres, J., and Kummerow, F. A., "Fatty Acid Composition of Lipides Extracted from Rats Fed Milk Fat, Corn Oil, and Lard," J. Dairy Sci., 44, 1283-1292 (1961)

817. Bhalerao, V. R., and Kummerow, F. A., "A Summary of Methods for the Detection of Foreign Fats in Dairy Products," J. Dairy Sci., 39, 956-964 (1956)

818. Bhati, A., "Course of Bromination of 4-Chloro-3-methylanisole with N-Bromosuccinimide," J. Chem. Soc., 1963, 730-732; CA 58, 8953d

819. Bhatnagar, V. M., and Dhont, J. H., "Pyrolysis and Gas Chromatography of Benzene Clathrate," Nature, 196, 769-770 (1962)

820. Biddiscombe, D. P., Collerson, R. R., Handley, R., Herington, E. F. G., Martin, J. F., and Sprake, C. H. S., "Thermodynamic Properties of Organic Oxygen Compounds. VIII. Purification and Vapor Pressures of the Propyl and Butyl Alcohols," J. Chem. Soc., 1963, 1954-1957; CA 58, 10793g

821. Bidmead, D. S., and Welti, D., "Isolation and Identification of Volatile Fruit Flavors," Research (London), 13, No. 8, 295-299 (1960); CA 55, 838h

822. Biemann, K., Spiteller-Friedmann, M., and Spiteller, G., "Application of Mass Spectrometry to Structure Problems. X. Alkaloids of the Bark of Aspidosperma Quebracho Blanco," J. Am. Chem. Soc., 85, 631-638 (1963); CA 58, 10247h

823. Biemann, K., and Vetter, W., "Separation of Peptide Derivatives by Gas Chromatography Combined with the Mass Spectrometric Determination of the Amino Acid Sequence," Biochem. Biophys. Res. Commun., 3, 578-584 (1960); CA 55, 16649g

824. Bier, M., and Teitelbaum, P., "Gas Chromatography in Amino-Acid Analysis," Ann. N.Y. Acad. Sci., 72, Art. 13, 641-648 (1959); CA 53, 14842d

825. Biermann, W. J., and Gesser, H., "The Analysis of Metal Chelates by Gas Chromatography," Anal. Chem., 32, 1525-1526 (1960); CA 55, 1298i

826. Biernacki, W., and Urbanski, T., "Gas-Liquid Chromatography of the C_1-C_4 Nitroparaffins," Bull. Acad. Polon. Sci., Ser. Sci. Chim., 10, 601-604 (1962); CA 58, 11936e

827. Biggers, R. E., and Horton, A. D., "Mathematical Resolution of Overlapping Bands of Gas Chromatograms," Oak Ridge National Laboratory, Anal. Chem. Div., Annual Progr. Rept. for Period Ending December 31, 1959, p. 20

828. Bignardi, G., and Munari, S., "Chlorination of Cineole," Farmaco (Pavia), Ed. Sci., 17, 222-233 (1962); CA 58, 551g

829. Binder, R. G., Applewhite, T. H., Kohler, G. O., and Goldblatt, L. A., "Chromatographic Analysis of Seed Oils. Fatty Acid Composition of Castor Oil," J. Am. Oil Chemists' Soc., 39, 513-517 (1962); CA 58, 3620c

830. Bindernagel, H., "Analysis of Gases in Metallurgy," Technik (Berlin), 16, No. 1, 28-33 (1961); CA 55, 12149i

831. Biran, L. A., and Bartley, W., "Distribution of Fatty Acids in Lipids of Rat Brain, Mitochondria, and Microsomes," Biochem. J., 79, No. 1, 159-176 (1961); CA 55, 22397f

832. Birch, S. F., Cullum, T. V., Dean, R. A., and Redford, D. G., "Sulfur Compounds in the Kerosene Boiling Range of Middle East Distillates. Occurrence of a Bicyclic Thiophene and a Thionyl Sulfide," Tetrahedron, 7, 311-318 (1959); CA 54, 7678c

833. Birchall, J. M., Haszeldine, R. N., and Parkinson, A. R., "Polyfluoroarenes. III. A New Synthesis of Hexafluorobenzene," J. Chem. Soc., 1961, 2204-2206; CA 55, 21002h

834. Bird, G. R., "Infrared Gas Microcell," J. Opt. Soc. Am., 51, 579-580 (1961); CA 55, 17111e

835. Birdwell, B. F., and Crawford, G. W., "Gamma Radiolysis of Propane," J. Chem. Phys., 33, 928-929 (1960); CA 55, 7007i

836. Bischoff, C., "Determination of the Location of Double Bonds in Monoolefins," Monatsber. Deut. Akad. Wiss. Berlin, 3, 674-678 (1961); CA 58, 2353d

837. Bishop, C. T., "Separation of Carbohydrate Derivatives by Gas-Liquid Partition Chromatography," in "Methods of Biochemical Analysis," Vol. 10, edited by D. Glick, John Wiley & Sons, Inc., New York, 1962

838. Bishop, C. T., Blank, F., and Gardner, P. E., "The Cell Wall Polysaccharides of Candida Albicans: Glucan, Mannan, and Chitin," Can. J. Chem., 38, 869-881 (1960)

839. Bishop, C. T., and Copper, F. P., "Separation of Carbohydrate Derivatives by Gas-Liquid Partition Chromatography," Can. J. Chem., 38, 388-395 (1960); CA 54, 24410f; 136th Natl. ACS Mtg., At-

lantic City, N. J., September 1959, Program Abstr., p. 15E

840. Bishop, C. T., and Cooper, F. P., "Constitution of a Glucomannan from Jack Pine (Pinus Banksiana Lamb)," Can. J. Chem., 38, 793-803 (1960); CA 54, 22385h

841. Bishop, C. T., and Cooper, F. P., "Glycosidation of Sugars. I. Formation of Methyl D-Xylosides," Can. J. Chem., 40, 224-232 (1962); CA 57, 15218f

842. Bishop, D. G., and Still, J. L., "Fatty Acid Metabolism in Serratia Marcescens. II. The Occurrence of Hydroxy Acids," Biochem. Biophys. Res. Communs., 7, 337-341 (1962); CA 57, 17178d

843. Bishop, J. R., Liebmann, H., and Humphrey, M., "The Determination of Hydrogen in Head-Space Gases in Food Cans by Gas Chromatography," Chem. & Ind. (London), 1957, 360-362

844. Bissot, T. C., and Benson, K. A., "Oxidation of Butane to Maleic Anhydride," Prod. Res. & Develop., 2, No. 1, 57-60 (1963)

845. Black, L. T., and Eisenhauer, R. A., "Determination of Cyclic Fatty Acids by Gas Liquid Chromatography," J. Am. Oil Chemists' Soc., 40, 272-274 (1963); 53rd Spring Mtg., American Oil Chemists' Soc., New Orleans, La., May 1962

846. Blackie, A., "Apparatus for Taking an Average Sample of a Variable Flow of Gas," J. Soc. Chem. Ind., 58, No. 9, 293-296 (1939); CA 34, 289^7

847. Blackie, A., "Apparatus for Taking an Average Sample of a Variable Flow of Gas," J. Inst. Fuel, 13, 98-101 (1940); CA 34, 7146^7

848. Blackmore, R. L., and Fordham, W. D., "New Analysis of Essential Oils and Materials Used in Scents," Parfum. cosmet. savons, 1, No. 2, 54-56 (1958); CA 52, 13195b

849. Blackwell, J., and Hickinbottom, W. J., "Alkylation of the Aromatic Nucleus. X. Cyclohexylation of the Monoalkylbenzenes, and the Course of Thermal Alkylation," J. Chem. Soc., 1963, 524-527

850. Blades, A. T., "The Hydrogen Isotope Effect in the Pyrolysis of Cyclopropane," Can. J. Chem., 39, 1401-1407 (1961); CA 56, 2029b

851. Blair, J. W., and Amis, E. S., "Chromatography of a Mixture of Hexane, Chloroform, and Benzene on Silica Gel," Anal. Chem., 30, 329-332 (1958); CA 52, 10808e

852. Blake, A. R., and Kutschke, K. O., "The Reaction of Methyl Radicals with Formaldehyde," Can. J. Chem., 37, 1462-1468 (1959); CA 54, 14897c

853. Blake, A. R., and Kutschke, K. O., "The Oxidation of Di-tert-butyl Peroxide," Can. J. Chem., 39, 278-284 (1961); CA 55, 14285g

854. Blakemore, G., "Simple Chromatographic Gas-Analysis Apparatus," Analyst, 87, 737-742 (1962); CA 58, 1281b

855. Blakeway, J. M., and Thomas, D. B., "Gas-Liquid Chromatography of Linear Detergent Alkylates," J. Chromatog., 6, 74-79 (1961); CA 56, 8867d

856. Blanchard, Jr., E. P., and Buechi, G., "The Conversion of Glycidic Esters to Aldehydes and Ketones," J. Am. Chem. Soc., 85, 955-958 (1963); CA 59, 531g

857. Blanchard, H. S., "Mechanism of Cumene Autoxidation. Mechanism of the Interaction of tert-Peroxy Radicals," J. Am. Chem. Soc., 81, 4548-4552 (1959); CA 54, 3295h

858. Blanchard, K. R., and Schleyer, P. v. R., "Quantitative Study of the Interconversion of Hydrindan Isomers by Aluminum Bromide," J. Org. Chem., 28, 247-248 (1963); CA 58, 10061b

859. Blank, F., Shortland, F. E., and Just, G., "The Free Sterols of Dermatophytes," J. Invest. Dermatol., 39, 91-94 (1962)

860. Blankenhorn, D. H., and Chin, T. W., "A Micro-Determination of Cholestanol in Atheromatous Lesions Employing Gas-Liquid Chromatography," 46th Annual Mtg., Federation of Am. Soc. for Exper. Biol., Atlantic City, N. J., April 1962; Abstr., Federation Proc., 21, 97 (1962)

861. Blaustein, B. D., Wender, I., and Anderson, R. B., "Ethyl Branching in the Fischer-Tropsch Synthesis," Nature, 189, 224-225 (1961); CA 55, 11280i

862. Blay, N. J., Dunstan, I., and Williams, R. L., "Boron Hydride Derivatives. III. Electrophilic Substitution in Pentaborane and Decaborane," J. Chem. Soc., 1960, 430-433; CA 54, 12979b

863. Blay, N. J., Williams, J., and Williams, R. L., "Boron Hydride Derivatives. II. The Separation and Identification of Some Ethylated Pentaboranes and Decaboranes," J. Chem. Soc., 1960, 424-429; CA 54, 12978i

864. Blecharczyk, S. S., "The Efficiency of Gas-Liquid Chromatographic Micro-Bead Columns," Univ. Microfilms (Ann Arbor, Mich.), Order No. 62-990, 77 pp.; Dissertation Abstr., 22, 4169 (1962); CA 57, 10512i

865. Bloch, M. G., "Determination of C_1-Through C_7-Hydrocarbons in a Single Run by a Four-Stage Gas Chromatograph," Gas Chromatog., Intern. Symposium, 2nd, East Lansing, Mich., 1959, 133-162 (Pub. 1961); CA 55, 18204f

866. Bloch, M. G., "Determination of C_1 Through C_7 Hydrocarbons in a Single Run by a Four-Stage Gas Chromatograph," in "Gas Chromatography," edited by H. J. Noebels, R. F. Wall, and N. Brenner, Academic Press, Inc., New York, 1961, pp. 133-162

867. Blom, L., "Applications of a Direct Micro Carbon Dioxide Titration," Compt. rend. congr. intern. chim. ind., 31^e, Liege, 1958 (Pub. as Ind. chim. belge, Suppl.), 1, 193-199 (Pub. 1959); CA 54, 4242g

31

868. Blom, L., and Edelhausen, L., "Vapor-Phase Chromatographic Determination of Benzene, Naphthalene, and Other Hydrocarbons in Wash Oil. Application of a Simple and Accurate Detection Method for Vapor-Phase Chromatography," Anal. Chim. Acta, 15, 559-566 (1956); CA 51, 5637h

869. Blom, L., Edelhausen, L., and Smeets, T., "Quantitative Gas Chromatography Without Calibration," Z. Anal. Chem., 189, 91-100 (1962); CA 57, 13166f

870. Blomquist, A. T., and Connolly, D. J., "Chemistry of Small Carbon Rings. XII. Methylenecyclopropane via Thermal Decomposition of Dimethylaminomethylcyclopropane-N-oxide," J. Org. Chem., 26, 2573-2575 (1961); CA 56, 8577b

871. Blomstrand, R., "The Fatty Acid Composition of Cerebrospinal Fluid Lipids," Acta Chem. Scand., 14, 775-776 (1960)

872. Blomstrand, R., "Analysis of Human Bile Lipids by Gas-Liquid Chromatography," Acta Chem. Scand., 14, 1006-1010 (1960); CA 56, 9267f

873. Blomstrand, R., "Gas-Liquid Chromatography of Human Bile Acids," Proc. Soc. Exptl. Biol. Med., 107, 126-128 (1961); CA 55, 21218b

874. Blomstrand, R., and Christensen, S., "Fatty Acid Pattern in the Aorta Lipids of Cockerels in the Initial Stage of Cholesterol- or Stilboesterol-Induced Atherosclerosis," Nature, 189, 376-378 (1961); CA 55, 15706b

875. Blomstrand, R., and Dahlback, O., "Gas-Liquid Chromatography of Human Lymph Fatty Acids After Feeding C¹⁴-Labeled Fats," Acta Soc. Med. Upsaliensis, 64, 177-184 (1959); CA 54, 7837e

876. Blomstrand, R., and Dahlback, O., "The Fatty Acid Composition of Human Thoracic Duct Lymph Lipides," J. Clin. Invest., 39, 1185-1191 (1960); CA 55, 1861a

877. Blomstrand, R., and Ekdahl, P., "Fatty Acid Patterns of Human Bile Under Normal and Pathological Conditions," Proc. Soc. Exptl. Biol. Med., 104, 205-209 (1960); CA 54, 17649e

878. Blomstrand, R., and Gurtler, J., "Separation of Glycerol Ethers by Gas-Liquid Chromatography," Acta Chem. Scand., 13, 1466-1467 (1959); CA 56, 6283i

879. Bloomfield, D. K., "Quantitative Analysis of Complex Mixtures of Steroids and Bile Acids by Gas Chromatography," Anal. Chem., 34, 737-741 (1962); CA 57, 3727b

880. Bloomfield, D. K., "Quantitative Analysis of Fecal Steroids and Bile Acids by Gas Chromatography," 46th Annual Mtg., Federation of Am. Soc. for Exper. Biol., Atlantic City, N. J., April 1962; Abstr., Federation Proc., 21, 283 (1962)

881. Bloomfield, D. K., "The Relationship Between Solid Support, Column Efficiency, and Sterol Quantitation by Gas Chromatography," J. Chromatog., 9, 411-418 (1962)

882. Blum, H. A., "The Vortex Tube as a Gas Separator," 16th Southwest Regional Mtg., ACS, Oklahoma City, Okla., December 1960

883. Blum, M. S., Traynham, J. G., Chidester, J. B., and Boggus, J. D., "N-Tridecane and trans-2-Heptenal in Scent Gland of the Rice Stink Bug Oebalus Pugnax (F.)," Science, 132, 1480-1481 (1960); CA 55, 9694i

884. Blumenthal, J. L., Sourirajan, S., and Nobe, K., "Effect of Mean Pore Size on the Low Temperature Adsorption of Nitrogen on Alumina," Can. J. Chem., 38, 783-786 (1960); CA 54, 18018h

885. Blundell, R. V., Griffiths, S. T., and Wilson, R. R., "Analysis of the Full-Boiling Range Gasolines by Chromatographic Methods," Gas Chromatog. Proc. Symposium, 3rd, Edinburgh, 1960, 360-371; CA 56, 7580d

886. Blundell, R. V., Griffiths, S. T., and Wilson, R. R., "Analysis of Full Boiling Range Gasolines by Chromatographic Methods," in "Gas Chromatography 1960," edited by R. P. W. Scott, Butterworths, London, 1960, pp. 360-371

887. Blyholder, G., "Integrating Counter Cell for Use with Vapor Phase Chromatography," Anal. Chem., 32, 572 (1960); CA 54, 11589h

888. Blyholder, G., and Emmett, P. H., "Fischer-Tropsch Synthesis Mechanism Studies. The Addition of Radioactive Ketene to the Synthesis Gas," J. Phys. Chem., 63, 962-965 (1959); CA 54, 1051a

889. Blyholder, G., and Emmett, P. H., "Fischer-Tropsch Synthesis Mechanism Studies. II. The Addition of Radioactive Ketene to the Synthesis Gas," J. Phys. Chem., 64, 470-472 (1960); CA 54, 20825c

890. Boatman, C., Decoteau, A. E., and Hammond, E. G., "Trisaturated Glycerides of Milk Fat," J. Dairy Sci., 44, 644-661 (1961); CA 55, 15766b

891. Boch, R., Shearer, D. A., and Stone, B. C., "Identification of Isoamyl Acetate as an Active Component in the Sting Pheromone of the Honey Bee," Nature, 195, 1018-1020 (1962); CA 57, 17223h

892. Bochinski, J. H., Gardiner, K. W., and Juvet, Jr., R. S., "Potential of Gas Chromatography for Purifying Semiconductor Materials," Ultrapurif. Semicond. Mater., Proc. Conf., Boston, Mass., 1961, 239-252 (Pub. 1962); CA 57, 5757f

893. Bochinski, J. H., Johns, T., and Jones, T. E., "Determination of Blood Gases by Gas Chromatography," 18th Southwest Regional Mtg., ACS, Dallas, Texas, December 1962

894. Bochinski, J. H., Johns, T., and Porter, J. A., "Pre-Extractive Techniques as Applied to Gas Chromatography," 14th Pittsburgh Conf. on Anal. Chem. & Appl. Spectroscopy, Pittsburgh, Pa.,

March 1963, Program Abstr., p. 51

895. Bock, H., "The Thermal Conductivity of Gas Mixtures," Ann. Physik, 8, 134-155 (1950); CA 48, 11860f; Chem.-Z., 1951, I 3306

896. Boddy, P. J., and Robb, J. C., "The Hydrogenation of Olefins. I. The Hydrogenation of Ethylene and Propene," Proc. Roy. Soc. (London), A249, 518-531 (1959); CA 54, 24328f

897. Bodnar, S. J., and Mayeux, S. J., "Estimation of Trace and Major Quantities of Lower Alcohols, Ethers, and Acetone in Aqueous Solutions by Gas Liquid Partition Chromatography," Anal. Chem., 30, 1384-1387 (1958); CA 52, 16983i

898. Boehle, E., Schrade, W., Biegler, R., Larbig, D., and Karytsiotis, J., "Gas Chromatographic Studies of Serum Fatty Acids of Man. IV. The Alimentary Hyperlipemia After Various Dietary Fats," Klin. Wochschr., 39, 5-17 (1961); CA 55, 9592b

899. Boehm, E. E., Thaller, V., and Whiting, M. C., "Synthetical Studies on Terpenoids. II. The Structure of 'Tagetone'," J. Chem. Soc., 1963, 2535-2540; CA 58, 12603a

900. Boehm, E. E., and Whiting, M. C., "Stereoisomeric 2,4-Pentadienals and 3-Methyl-2,4-pentadienals," J. Chem. Soc., 1963, 2541-2543; CA 59, 430a

901. Boeke, J., "Vapor Flow Chromatography," Gas Chromatog., Proc. Symposium, 3rd, Edinburgh, 1960, 88-103; CA 55, 26823c

902. Boeke, J., "Vapour Flow Chromatography," in "Gas Chromatography 1960," edited by R. P. W. Scott, Butterworths, London, 1960, pp. 88-103

903. Boeke, J., and Parke, III, N. G., "High Speed Vapor Flow Chromatography," 3rd Intern. Symposium on Gas Chromatography, ISA, East Lansing, Mich., June 1961, 3, 391-422 (Pub. 1962); CA 58, 3867g

904. Boeke, J., and Parke, III, N. G., "High Speed Vapor Flow Chromatography," in "Gas Chromatography," edited by N. Brenner, J. E. Callen, and M. D. Weiss, Academic Press, Inc., New York, 1962, pp. 391-422

905. Boer, H., "A Comparison of Detection Methos for Gas Chromatography Including Detection by Beta Ray Ionization," in "Vapor-Phase Chromatography," edited by D. H. Desty, Academic Press, Inc., New York, 1957, pp. 169-184

906. Boer, H., "The Use of Ozonolysis in Oil Constitution Research. II. Side-Chain Analysis of Aromatics and Benzothiophenes from Various Oil Fractions," J. Inst. Petrol., 46, 234-236 (1960); CA 54, 18946a

907. Boettger, H. G., "Theory and Application of Temperature Programming of Packed and Capillary Columns at Elevated Temperatures," 4th Annual Gas Chromatography Institute, Canisius College, Buffalo, N. Y., April 1962

908. Boettger, H. G., "Evaluation of Programmed Temperature Gas Chromatography for Packed and Capillary Columns at Elevated Temperatures," in "Lectures on Gas Chromatography 1962," edited by H. A. Szymanski, Plenum Press, Inc., New York, 1963, pp. 133-152; CA 58, 10701h

909. Boggus, J. D., and Adams, N. G., "Gas Chromatography for Trace Analysis," Anal. Chem., 30, 1471-1473 (1958); CA 52, 19669b

910. Bogue, D. C., "A Note on the Theory of Solid-Phase Diffusion in Chromatography," Anal. Chem., 32, 1777-1778 (1960); CA 55, 6100i

911. Bohemen, J., Langer, S. H., Perrett, R. H., and Purnell, J. H., "A Study of the Adsorptive Properties of Firebrick in Relation to Its Use as a Solid Support in Gas-Liquid Chromatography," J. Chem. Soc., 1960, 2444-2451; CA 54, 22135h

912. Bohemen, J., and Purnell, J. H., "Katharometric Measurement of Theoretical Plate Numbers in Gas Chromatography," Chem. & Ind. (London), 1957, 815-816; CA 51, 15324g

913. Bohemen, J., and Purnell, J. H., "The Behavior of Katharometers for Gas Chromatography in Carrier Gases of Low Thermal Conductivity," J. Appl. Chem. (London), 8, 433-440 (1958); CA 52, 19269g

914. Bohemen, J., and Purnell, J. H., "Some Applications of Theory in the Attainment of High Column Efficiencies in Gas Liquid Chromatography," in "Gas Chromatography 1958," edited by D. H. Desty, Academic Press, Inc., New York, 1958, pp. 6-22

915. Bohemen, J., and Purnell, J. H., "Diffusional Band-Spreading in Gas-Chromatographic Columns. I. Elution of Unsorbed Gases," J. Chem. Soc., 1961, 360-367; CA 55, 9974e

916. Bohemen, J., and Purnell, J. H., "Diffusional Band-Spreading in Gas-Chromatographic Columns. II. The Elution of Sorbed Vapors," J. Chem. Soc., 1961, 2630-2638; CA 56, 285i

917. Bohm, Z., "Electronic Voltage Amplifier for Indirect Registration in Gas Chromatography," Chem. Listy, 52, 359-360 (1958); CA 52, 19285c

918. Bohm, Z., "Analog Integrator for Gas-Chromatographic Apparatus," J. Chromatog., 3, 265-272 (1960)

919. Bokhoven, C., and Dijkstra, A., "Effect of Carrier Gas on the Sensitivity of Thermal Conductivity Detectors," Nature, 186, 793-794 (1960); CA 54, 20359b

920. Bolland, J. L., and Melville, H. W., "On Micro Thermal Conductivity Gauges," Trans. Faraday Soc.,

        <u>33</u>, 1316-1329 (1937); CA <u>31</u>, 8264[6]

921. Bolling, J. M., Mitzner, S., and Policastro, S. G., "Gas Chromatographic Determination of Ethanol in Methanol on a Routine Basis," 8th L. H. Baekeland Award Mtg. of ACS, South Orange, N. J., January 1959

922. Bombaugh, K. J., "Gas Chromatographic Analysis of Chloronitrobenzene Isomers," Anal. Chem., <u>33</u>, 29-32 (1961); CA <u>55</u>, 10211h

923. Bombaugh, K. J., "Application of Substrate Selectivity to Analytical Problems in Industrial Research," 2nd Symp. on Gas Chromatography, Toronto Section, CIC, Toronto, Ontario, February 1962

924. Bombaugh, K. J., "Improved Efficiency in Gas Chromatography by Molecular Sieve Flour," Nature, <u>197</u>, 1102-1103 (1963); CA <u>58</u>, 10702d

925. Bombaugh, K. J., and Bull, W. C., "The Use of Gas Liquid Chromatography to Characterize Liquids from Coal Hydrogenation," 12th Pittsburgh Conf. on Anal. Chem. & Appl. Spectroscopy, Pittsburgh, Pa., February-March 1961, Program Abstr., p. 58

926. Bombaugh, K. J., and Bull, W. C., "Analysis of Formaldehyde by Gas Chromatography," 13th Pittsburgh Conf. on Anal. Chem. & Appl. Spectroscopy, Pittsburgh, Pa., March 1962, Program Abstr., p. 57

927. Bombaugh, K. J., and Bull, W. C., "Gas Chromatographic Determination of Formaldehyde in Solution and High Purity Gas," Anal. Chem., <u>34</u>, 1237-1241 (1962); CA <u>58</u>, 1912b

928. Bond, G. C., Dowden, D. A., and Mackenzie, N., "Selective Hydrogenation of Acetylene," Trans. Faraday Soc., <u>54</u>, 1537-1546 (1958); CA <u>53</u>, 15947g

929. Bond, G. C., and Newham, J., "Catalysis on Metals of Group 8. V. The Kinetics of the Hydrogenation of Cyclopropane and Methylcyclopropane," Trans. Faraday Soc., <u>56</u>, 1501-1514 (1960); CA <u>55</u>, 9014e

930. Bond, G. C., and Newham, J., "Catalysis on Metals of Group 8. VI. Hydrogenation of Methylenecyclopropane and Methylenecyclobutane Over Platinum Catalysts," Trans. Faraday Soc., <u>56</u>, 1851-1860 (1960); CA <u>55</u>, 25429b

931. Bonelli, E., Dimick, K. P., and Hartmann, H., "G. C. Retention Time and Sensitivity Data for Insecticides and Herbicides," 145th Natl. ACS Mtg., New York, N. Y., September 1963

932. Boniforti, L., "Gas Chromatographic Study of Fatty Acids in Italian Butter and in Butter from Other Countries; Application to the Investigation of Adulterations in Commercial Butter," Ann. Fals. Expert. Chim., <u>55</u>, 255-263 (1962); CA <u>58</u>, 11894a

933. Boniforti, L., DiStefano, F., and Vercillo, A., "Gas Chromatographic Determination of Sorbic Acid in Fruit Juices," Boll. Lab. Chim. Provinciali (Bologna), <u>12</u>, 505-519 (1961); CA <u>57</u>, 7692d

934. Bonner, W. A., "Deuterium Isotope Effects During the Raney Nickel Catalyzed C_1-C_2 Cleavage of 2-Phenylethanol," J. Am. Chem. Soc., <u>82</u>, 1382-1385 (1960); CA <u>54</u>, 17312f

935. Bonner, W. A., and McKay, J. B., "Hydrogen Migration Studies During the Catalytic Reduction of 3-Phenyl-1-butene with Deuterium," J. Am. Chem. Soc., <u>82</u>, 5350-5353 (1960); CA <u>55</u>, 18631h

936. Bonnet, J., "Quantitative Analysis of Benzo[a]pyrene in Vapors Coming from Melted Tar," Natl. Cancer Inst., Monograph No. 9, 221-223 (1962); CA <u>58</u>, 4963c

937. Bonnichsen, R., and Linturi, M., "Gas Chromatographic Determination of Some Volatile Compounds in Urine," Acta Chem. Scand., <u>16</u>, 1289-1290 (1962); CA <u>57</u>, 14096b

938. Bonnier, J. M., and Gaudemaris, G. de, "Thermal Stability of Hydrocarbons," Rev. Inst. Franc. Petrole Ann. Combust. Liquides, 17, 852-882 (1962); CA 57, 10103a

939. Bonnier, J. M., and Gaudemaris, G. de, "Thermal Stability of Hydrocarbons. II. Pyrolysis," Rev. Inst. Franc. Petrole Ann. Combustibles Liquides, <u>17</u>, 1016-1035 (1962); CA <u>57</u>, 14051c

940. Boord, C. E., Derfer, J. M., Menapace, H. R., Wiley, V. G., and Smith, M. L., "Precombustion Reaction of Pure Hydrocarbons," 135th Natl. ACS Mtg., Boston, Mass., April 1959, Program Abstr., p. 12Q

941. Boord, C. E., Derfer, J. M., Smith, M. L., Menapace, H. R., and Kyryacos, G., "The Behavior of Blends of Pure Hydrocarbons in Cool Flames as Related to Their Deviations from Linear Blending on the Performance-Number Scale," Proc. Am. Petrol. Inst., <u>38</u>, Sect. III, 112-116 (1959); CA <u>53</u>, 13547h

942. Booth, H., Johnson, A. W., Johnson, F., and Langdale-Smith, R. A., "Methylation of Some Pyrroles and 2-Pyrrolines," J. Chem. Soc., <u>1963</u>, 650-661

943. Boreham, G. R., and Marhoff, F. A., "Gas Chromatographic Analysis," Research Commun., GC<u>54</u>, 28 pp. (1958); CA <u>53</u>, 3973b

944. Boreham, G. R., and Marhoff, F. A., "Fuel Gas Analysis: An Apparatus Incorporating a Multi-Cell Thermal Conductivity Detector," in "Gas Chromatography 1960," edited by R. P. W. Scott, Butterworths, London, 1960, pp. 412-422

945. Boreham, G. R., and Marhoff, F. A., "Fuel Gas Analysis. An Apparatus Incorporating a Multicell Thermal Conductivity Detector," Gas Chromatog. Proc. Symposium, 3rd, Edinburgh, <u>1960</u>, 412-422; CA <u>56</u>, 7611g

946. Boren, H. G., and Kracke, F. L., "Ventilation-Perfusion Ratio Determination Using a Gas Parti-

tioner," Annual Mtg., Am. Trudeau Soc., Los Angeles, Calif., May 1960; Abstr., Am. Rev. Respiratory Diseases, 81, 944 (1960)

947. Borer, K., Littlewood, A. B., and Phillips, C. S. G., "Gas-Chromatographic Study of Diborane Pyrolysis," J. Inorg. & Nuclear Chem., 15, 316-319 (1960); CA 55, 6233c

948. Borer, K., and Phillips, C. S. G., "The Separation of Volatile Silanes and Germanes by Gas-Liquid Chromatography," Proc. Chem. Soc. (London), 1959, 189-190; CA 54, 3180i

949. Borfitz, H., "A Simple Method of Temperature Programming for Gas Chromatography," Anal. Chem., 33, 1632 (1961); CA 56, 1979d; cf. ibid., 34, 167 (1962); CA 56, 13983c

950. Borkowski, R., and Ausloos, P., "Intramolecular Rearrangements. I. sec-Butyl Acetate and sec-Butyl Formate," J. Am. Chem. Soc., 83, 1053-1056 (1961); CA 55, 16406f

951. Bornemann, P., Finke, M., and Heinze, G., "Analysis of Petroleum Products. Group Analysis of Lubricating Oil Fractions," Chem. Tech. (Berlin), 14, 292-299 (1962); CA 57, 14050b

952. Bosanquet, C. H., "The Diffusion at a Front in Gas Chromatography," in "Gas Chromatography 1958," edited by D. H. Desty, Academic Press, Inc., New York, 1958, pp. 107-115

953. Bosanquet, C. H., "Peak Dimensions in Gas Chromatography," Nature, 183, 252-253 (1959); CA 53, 12919i

954. Bosanquet, C. H., and Morgan, G. O., "The Concentration Factor in Vapour-Phase Chromatography," in "Vapour-Phase Chromatography," edited by D. H. Desty, Academic Press, Inc., New York, 1957, pp. 35-51

955. Boschan, R., "Decomposition of Acyl Nitrates. Reaction of Trifluoroacetic Anhydride with Nitric Acid," J. Org. Chem., 25, 1450-1451 (1960); CA 55, 1425a

956. Bosin, W. A., "Modified Gas Chromatography with a Microcoulometric Detector Provides Temperature Programming for Pesticide Residue Analysis," 142nd Natl. ACS Mtg., Atlantic City, N. J., September 1962, Program Abstr., p. 18A

957. Bosin, W. A., "Analysis of Pesticide Residues Using Microcoulometric Temperature-Programmed Gas Chromatography," Anal. Chem., 35, 833-837 (1963)

958. Bossart, C. J., and Heller, H., "Process Chromatography with Infinitely Flexible Programming," 7th Natl. Symp. on Instrumental Methods of Analysis, ISA, Houston, Texas, April 1961

959. Bosshard, E., Goeckner, N. A., and Keller-Schierlein, W., "Metabolic Products of Actinomycetes. XX. Metabolism of Actinomycetene; Synthesis of α,β-Dimethyllevulinaldehyde, a Degradation Product of Acetomycin," Helv. Chim. Acta, 42, 2746-2750 (1959); CA 54, 10850b

960. Bota, T., Bucur, C., Drimus, I., Stanescu, L., and Sandulescu, D., "Catalyzed Diels-Alder Reaction," Rev. Chim. (Bucharest), 12, 503 (1961); CA 56, 5848g

961. Bothe, H. K., "Investigations on the Argon-Beta-Ionization Detector," in "Gas Chromatographie 1959," edited by R. E. Kaiser and H. G. Struppe, Akademie Verlag, Berlin, 1959, pp. 69-79

962. Bothe, H. K., "Ionization Detectors Using Radioactive Emanation," Abhandl. Deut. Akad. Wiss. Berlin, Kl. Chem., Geol., Biol., 1959, No. 9, 203-214; CA 58, 5018g

963. Bothe, H. K., "Ionization Detectors with Radioactive Sources," in "Gas Chromatographie 1958," edited by H. P. Angele, Akademie Verlag, Berlin, 1959, pp. 203-214

964. Böttcher, C. J. F., "Sterols, Fatty Acids, and Atherosclerosis," Neth. Milk Dairy J., 12, 351-359 (1958); CA 53, 9448e

965. Böttcher, C. J. F., "Lipids in the Arterial Walls, with Special Reference to Atherosclerosis," Angew. Chem., 71, 435-436 (1959)

966. Böttcher, C. J. F., "Linoleic Acid in the Cholesterol Esters of the Aortic Wall," Lancet, 1960-I, 877; CA 54, 13386f

967. Böttcher, C. J. F., Boelsma-van Houte, E., Romeny-Wachter, C. C. ter H., Woodford, T. H., and Gent, C. M. van, "Lipide and Fatty Acid Composition of Coronary and Cerebral Arteries at Different Stages of Atherosclerosis," Lancet, 1960-I, 1162-1166; CA 55, 5736i

968. Böttcher, C. J. F., and Gent, C. M. van, "Changes in the Composition of Phospholipids and of Phospholipid Fatty Acids Associated with Atherosclerosis in the Human Aortic Wall," J. Atherosclerosis Research, 1, 36-46 (1961); CA 56, 5296b

969. Böttcher, C. J. F., and Meijer, W. A., "Use of an Asphalt Fraction for the Gas-Liquid Chromatography of Steroids," J. Chromatog., 6, 535-537 (1961); CA 56, 13527i

970. Böttcher, C. J. F., Romeny-Wachter, C. C. ter H., Boelsma-van Houte, E., and Gent, C. M. van, "Analysis of Lipids of the Arterial Wall," Lancet, 1958-II, 1207-1209; CA 53, 4526c

971. Böttcher, C. J. F., Woodford, F. P., Boelsma-van Houte, E., and Gent, C. M. van, "Methods of the Analysis of Lipids Extracted from Human Arteries and Other Tissues," Rec. trav. chim., 78, 794-814 (1959); CA 54, 14349a

972. Böttcher, C. J. F., Woodford, F. P., Romeny-Wachter, C. C. ter H., Boelsma-van Houte, E., and Gent, C. M. van, "Composition of Lipids Isolated from the Aorta, Coronary Arteries and Circulus Willisii of Atherosclerotic Individuals," Nature, 183, 47-48 (1959); CA 53, 14286e

973. Böttcher, C. J. F., Woodford, F. P., Romeny-Wachter, C. C. ter H., Boelsma-van Houte, E., and Gent, C. M. van, "Fatty Acid Distribution in Lipides of the Aortic Wall," Lancet, 1960-I, 1378-

1383; CA 55, 3805h

974. Botter, F., Perriere, G. de la, and Tishchenko, S., "Isotopic Analysis of H_2, HD, D_2 Mixtures and Analysis of Mixtures of Ortho- and Parahydrogen by Gas Chromatography," Comm. energie at. (France) Rappt., No. 1962, 27 pp. (1961); CA 55, 25591c

975. Boufford, C. E., and Ring, R. D., "Determination of Carbon Dioxide in Alkylene Oxide by Gas-Liquid Chromatography," 8th Detroit Anachem Conf., Detroit, Mich., October 1960

976. Boughton, B., Mackenna, R. M. B., Wheatley, V. R., and Wormall, A., "The Fatty Acid Composition of the Surface Skin Fats (Sebum) in Acne Vulgaris and Seborrheic Dermatitis," J. Invest. Dermatol., 33, 57-64 (1959); CA 54, 15631h

977. Boughton, B., and Wheatley, V. R., "Studies of Sebum. IX. Further Studies of the Composition of the Unsaponifiable Matter of Human Forearm Sebum," Biochem. J., 73, 144-149 (1959); CA 53, 22329f

978. Bouthilet, R. J., and Lowrey, W., "The Use of the Gas Chromatograph in the Determination of Fusel Oil in Grape Brandy," J. Assoc. Offic. Agr. Chemists, 42, 634-637 (1959); CA 53, 18377h

979. Bovijn, L., Pirotte, J., and Berger, A., "Determination of Hydrogen in Water by Means of Gas Chromatography," in "Gas Chromatography 1958," edited by D. H. Desty, Academic Press, Inc., New York, 1958, pp. 310-320; CA 53, 15851f

980. Bowman, R. E., and Hartley, C. B., "Determination of Impurities in Inert Gases," Welding J.(N.Y.), 29, No. 5, 258S-262S (1950); CA 44, 9297c

981. Bowman, R. L., and Karmen, A., "Micro Sample Introduction System for Gas Chromatography," Nature, 182, 1233-1234 (1958); CA 53, 4826c

982. Boyd, G. S., "Effect of Linoleate and Estogen on Cholesterol Metabolism," Federation Proc., 21, Suppl. No. 11, 86-92 (1962); CA 57, 14389h

983. Boyle, Jr., J. J., "Gas Chromatographic Analysis of Fatty Acids from Uninfected and Virus-Infected Continuously Cultured Animal Cells," Univ. Microfilms (Ann Arbor, Mich.), Order No. 62-4083, 62 pp.; Dissertation Abstr., 23, 795 (1962); CA 58, 2697b

984. Boyle, J. J., and Ludwig, E. H., "Analysis of Fatty Acids of Continuously Cultured Mammalian Cells by Gas-Liquid Chromatography," Nature, 196, 893-894 (1962); CA 58, 5972d

985. Boys, S. L., "Temperature Programmed Capillary Columns," 13th Annual Mid-America Spectroscopy Symp., Chicago, Ill., April-May 1962; pub. in "Developments in Applied Spectroscopy," Vol. 2, J. R. Ferrero and J. S. Ziomek, eds., Plenum Press, N. Y., 1962, pp. 415-425

986. Brace, R. O., "Quantitative Analysis of Phenols, Cresols, and Xylenols by Gas Chromatography," Application Data Sheet GC-96-0, Beckman Scientific and Process Instruments Division, Beckman Instruments, Inc., Fullerton, Calif., 1960

987. Bracht, G., "Advances in Orsat Analysis, with Particular Reference to Coke-Oven Gas," Brennstoff-Chem., 42, 123-129 (1961); CA 55, 20391h

988. Bradford, B. W., Harvey, D., and Chalkley, D. E., "The Chromatographic Analysis of Hydrocarbon Mixtures," J. Inst. Petrol., 41, 80-91 (1955); CA 49, 7839h

989. Bradford, B. W., and Nicholson, D. L., "Process Analytical Control: The Problems of Manpower, Productivity, and Automation," Proc. Congr. Modern Anal. Chem. in Industry, St. Andrews, Scotland, June 1957, pp. 161-167

990. Bradley, D. C., and Hill, D. A. W., "Reactions of Titanium Tetrachloride with Ethoxychlorosilanes," J. Chem. Soc., 1963, 2101-2107

991. Bradley, J. K., "Highly Selective Method for the Determination of Aldrin in N-P-K Fertilizers," Chem. & Ind. (London), 1961, 1876; CA 57, 1311d

992. Bradley, J. N., and Ledwith, A., "The Reactions of Carbene with Alkyl Halides," J. Chem. Soc., 1961, 1495-1498; CA 55, 19753e

993. Bradley, Jr., R. L., and Stine, C. M., "Simple Device for Obtaining Samples of Headspace Gas Directly from Sealed Containers for Analysis by Gas Chromatography," J. Dairy Sci., 45, 1259 (1962); CA 58, 2775d

994. Brady, A. P., Huff, H., and McBain, J. W., "Measurement of Vapor Pressures by Means of Matched Thermistors," J. Phy. & Colloid Chem., 55, 304-311 (1951); CA 45, 4530e

995. Brady, R. O., Bradley, R. M., and Trams, E. G., "Biosynthesis of Fatty Acids. I. Studies with Enzymes Obtained from Liver," J. Biol. Chem., 235, 3093-3098 (1960); CA 55, 4621i

996. Bragdon, J. H., and Karmen, A., "The Fatty Acid Composition of Chylomicrons of Chyle and Serum Following the Ingestion of Different Oils," J. Lipid Research, 1, 167-170 (1960); CA 54, 11230b

997. Branch, R. F., "New Methods of Analysis," Can. Chem. Processing, 40, No. 11, 105-106; ibid., No. 12, 80-84 (1956); CA 51, 6423g

998. Brand, J. C. D., Eglinton, G., and Morman, J. F., "The Ethynyl-Hydrogen Bond. I. Association in Ether Solution," J. Chem. Soc., 1960, 2526-2533; CA 54, 21067a

999. Brandenberger, H., and Müller, S., "A Gas Chromatographic Separation of the Volatile Fatty Acids of Black Tea," J. Chromatog., 7, 137-141 (1962); CA 57, 6041d

1000. Brandt, L. W., "Improved Gas-Sampling Tube for Mass Spectrometer Use," Chemist-Analyst, 45, 106 (1956)

1001. Brandt, W. W., "Gas Chromatography Growing Phenomenally," Anal. Chem., 32, No. 9, 56A, 64A-

68A (1960)

1002. Brandt, W. W., "Informal Gas Chromatography Symposium," Anal. Chem., 32, 339-340 (1960) General discussion on gas chromatography sponsorbed by Div. of Anal. Chem., 136th Natl. ACS Mtg., Atlantic City, N. J., September 1959

1003. Brandt, W. W., "Trace Analysis - Summer Symposium of the Analytical Division, Houston, Texas, 1960," Anal. Chem., 32, 1595-1598 (1960)

1004. Brandt, W. W., "The Column in Gas Chromatography," Anal. Chem., 33, No. 8, 23A-31A (1961); 139th Natl. ACS Mtg., St. Louis, Mo., March 1961, Program Abstr., p. 5B

1005. Brandt, W. W., "Gas Chromatography in Chemical Research," 13th Pittsburgh Conf. on Anal. Chem. & Appl. Spectroscopy, Pittsburgh, Pa., March 1962, Program Abstr., p. 59

1006. Brandt, W. W., and Heveran, J. E., "The Determination of Trace Amounts of Chromium by Gas Chromatography," 142nd Natl. ACS Mtg., Atlantic City, N. J., September 1962, Program Abstr., p. 9B

1007. Brauer, G. M., and Lehman, F., "Gas Chromatographic Analysis of Pyrolyzates of Polystyrene and Polymethylmethacrylate," 138th Natl. ACS Mtg., New York, N. Y., September 1960, Program Abstr., p. 24T

1008. Brauer, H., "Relation Between Pressure Drop and Rectifying Action in Packed Columns," Chem. Ing. Tech., 29, 520-530 (1957); CA 51, 16011i

1009. Braverman, J. B. S., and Solomiansky, L., "Separation of Terpeneless Essential Oils by the Chromatographic Method," Perfumery Essent. Oil Record, 48, 284-287 (1957); CA 51, 17106i

1010. Brealey, L., Elvidge, D. A., and Proctor, K. A., "The Determination of Chloroform in Aqueous Pharmaceutical Preparations," Analyst, 84, 221-225 (1959); CA 53, 17420g

1011. Breck, D. W., Eversole, W. G., and Milton, R. M., "New Synthetic Crystalline Zeolites," J. Am. Chem. Soc., 78, 2338-2339 (1956); CA 50, 11180f

1012. Breck, D. W., Eversole, W. G., Milton, R. M., Reed, T. B., and Thomas, T. L., "Crystalline Zeolites. I. The Properties of a New Synthetic Zeolite Type A," J. Am. Chem. Soc., 78, 5963-5971 (1956); CA 51, 5498a

1013. Bredel, H., "A Microflame Apparatus for High Temperature Gas Chromatography," Symposium on Gas Chromatography, Leipzig, Germany, October 1958; Abhandl. Deut. Akad. Wiss. Berlin, Kl. Chem., Geol., Biol., 1959, No. 9, 215-242; CA 58, 5018g

1014. Bredel, H., "Quantitative Evaluation of Results Obtained with the Flame-Ionization Detector," Chem. Tech. (Berlin), 13, 46-47 (1961); CA 55, 13162c

1015. Breed, L. W., and Haggerty, Jr., W. J., "Aryl and Allyl Chlorodialkoxysilanes," J. Org. Chem., 25, 126-128 (1960); CA 54, 15277f

1016. Brennan, D., and Kemball, C., "Gas-Phase Chromatography: A Class Experiment," J. Chem. Educ., 33, 490-492 (1956); CA 51, 2343e

1017. Brennan, D., and Kemball, C., "Chromatographic Separation," Petrol. Refiner, 37, No. 11, 255-258 (1958); CA 53, 2585h

1018. Brennan, D., and Kemball, C., "Resolution in Gas-Liquid Chromatography," J. Inst. Petrol., 44, 14-17 (1958); CA 52, 5088a

1019. Brenner, N., "Applications of Gas Chromatography to Toilet-Goods Analysis," Proc. Sci. Sect. Toilet Goods Assoc., 26, 3-8 (1956); CA 51, 3934h

1020. Brenner, N., "Special Applications of Gas Chromatography Equipment to Chemical Laboratory Problems," 8th Pittsburgh Conf. on Anal. Chem. & Appl. Spectroscopy, Pittsburgh, Pa., March 1957, Program Abstr., p. 34

1021. Brenner, N., "Modification of a Gas Chromatography Instrument for Special Laboratory Problems," 131st Natl. ACS Mtg., Miami, Fla., April 1957, Program Abstr., p. 36B

1022. Brenner, N., "Gas Chromatography," Drug & Cosmetic Ind., 80, 166-167, 261-266 (1957); CA 51, 9085i

1023. Brenner, N., "Analytical Application of a Triple Stage Gas Chromatography Instrument," 9th Pittsburgh Conf. on Anal. Chem. & Appl. Spectroscopy, Pittsburgh, Pa., March 1958, Program Abstr., p. 45

1024. Brenner, N., "New Developments in Vapor Phase Chromatography," 1st Annual Gas Chromatography Institute, Canisius College, Buffalo, N. Y., May 1959

1025. Brenner, N., "Gas Chromatography as a Microanalytical Tool," Microchem. J., 3, 155-166 (1959); CA 53, 13688g

1026. Brenner, N., "Bootlegging, Gas Chromatography and the Fourth Estate," Perkin-Elmer Instrument News, 10, No. 3, 9-10 (Spring 1959)

1027. Brenner, N., "Elementary Theory of Gas Chromatography," 2nd Gas Chromatography Institute, Canisius College, Buffalo, N. Y., April 1960

1028. Brenner, N., "I.S.A. Third International GC Symposium Reports Advances in Theory and Practice," Perkin-Elmer Instrument News, 12, No. 4, 1 (Summer 1961)

1029. Brenner, N., "Application of Instrumentation in the Paint Industry," Offic. Dig. Federation Soc.

Paint Technol., <u>33</u>, 51-61 (1961)

1030. Brenner, N., "Open Tubular or Capillary Column Gas Chromatography," Gas Chromatography Session of the 13th Annual Fisk Infrared Institute, Fisk Univ., Nashville, Tenn., August 1962

1031. Brenner, N., "Solid Supports and the Gas Phase," Gas Chromatography Session of the 13th Annual Fisk Infrared Institute, Fisk Univ., Nashville, Tenn., August 1962

1032. Brenner, N., "Chromatography with Open Tubular Columns," Gas Chromatography Session of the 14th Annual Fisk Infrared Institute, Fisk Univ., Nashville, Tenn., August 1963

1033. Brenner, N., Callen, J. E., and Weiss, M. D., (Eds.), "Gas Chromatography," Academic Press, Inc., New York, 1962. Collection of 37 papers presented at the 3rd Intern. Symposium, Analysis Instrumentation Division of the ISA, June 1961: CA <u>58</u>, 10735b

1034. Brenner, N., and Cieplinski, E., "Gas Chromatographic Analysis of Mixtures Containing Oxygen, Nitrogen, and Carbon Dioxide," Ann. N. Y. Acad. Sci., <u>72</u>, Art. 13, 705-713 (1959); CA <u>53</u>, 14829b

1035. Brenner, N., Cieplinski, E., and Coates, V. J., "The Employment of Molecular Sieves as Subtractive Substrates in Gas Chromatographic Columns," 10th Pittsburgh Conf. on Anal. Chem. & Appl. Spectroscopy, Pittsburgh, Pa., March 1959, Program Abstr., p. 57

1036. Brenner, N., Cieplinski, E., Ettre, L. S., and Coates, V. J., "Molecular Sieves as Subtractors in Gas Chromatographic Analysis. II. Selective Adsorptivity with Respect to Different Homologous Series," J. Chromatog., <u>3</u>, 230-234 (1960); CA <u>54</u>, 19086g

1037. Brenner, N., and Coates, V. J., "Molecular Sieves as Subractors in Gas Chromatographic Analysis," Nature, <u>181</u>, 1401-1402 (1958); CA <u>52</u>, 15884e

1038. Brenner, N., and Ettre, L. S., "Condensing System for Determination of Trace Impurities in Gases by Gas Chromatography," Anal. Chem., <u>31</u>, 1815-1818 (1959); CA <u>54</u>, 3063e; 135th Natl. ACS Mtg., Boston, Mass., April 1959, Program Abstr., p. 20B

1039. Brenner, N., and Ettre, L. S., "Characteristics of the Capillary Gas Chromatograph and Its Application to Quantitative Analysis," Acta Chim. Acad. Sci. Hung., <u>27</u>, 205-214 (1961); CA <u>55</u>, 23159i

1040. Brenner, N., and Hausdorff, H., "A Comprehensive Study of Instrumentation for Gas Chromatography," 1958 Natl. ISA Symp. on Instrumental Methods of Analysis, Houston, Texas, May 1958; ISA Proc., 213-221 (1958)

1041. Brenner, N., and Kirby, G., "The Analysis of Tar Acids by Gas Chromatography," Southeastern Regional Mtg., Local Sections of ACS, Durham, N. C., November 1957

1042. Brenner, N., Maier, H. J., Karpathy, O. C., and Bresky, D. R., "Automatic Analyses of Furnace Atmospheres," 11th Pittsburgh Conf. on Anal. Chem. & Appl. Spectroscopy, Pittsburgh, Pa., February-March 1960, Program Abstr., p. 41

1043. Brenner, N., Maier, H. J., Karpathy, O. C., and Bresky, D. R., "Total Analysis of Furnace Atmosphere," 12th Pittsburgh Conf. on Anal. Chem. & Appl. Spectroscopy, Pittsburgh, Pa., February-March 1961, Program Abstr., p. 52

1044. Brenner, N., Scholly, P., and O'Brien, L., "The Analysis of Fatty Acid Esters by Gas Chromatography," J. Natl. Lubricating Grease Inst., Spokesman, <u>23</u>, 137-142 (1959); CA <u>53</u>, 16829b

1045. Brenner, R. R., Thomas, M. E. de, Mercuri, O. F., and Peluffo, R. O., "Identification of Fatty Acids of Rio de la Plata Fresh Water Fish," Rev. Arg. Grasas Aceites, <u>3</u>, 65-75 (1961); CA <u>57</u>, 16775e

1046. Bresler, S. E., "The Theory of Nonequilibrium Chromatography," Doklady Akad. Nauk S.S.S.R., <u>90</u>, 205-208 (1953); CA <u>49</u>, 8724i

1047. Bresler, S. E., "Theory of Nonequilibrium Chromatography. The Formation of a Stationary Front of the Zone," Doklady Akad Nauk S.S.S.R., <u>97</u>, 699-702 (1954); CA <u>51</u>, 8508f

1048. Bresler, S. E., "Zone Diffusion in Chromatography," Kromatografiya, Leningrad, Gosudarst. Univ. im. A. A. Zhdanova, Sbornik Statei, <u>1956</u>, 106-126; CA <u>52</u>, 8675c

1049. Bresler, S. E., "The Theory of Chromatographic Separation of Isotopes," Zhur. Fiz. Khim., <u>32</u>, 628-634 (1958); CA <u>52</u>, 14282c

1050. Bresler, S. E., and Uflyand, Y. S., "The Theory of Nonequilibrium Chromatography," Zhur. Tekh. Fiz., <u>23</u>, 1443-1451 (1953); CA <u>49</u>, 3717h

1051. Breton, C., Deutschman, J. E., Ede, P., and Zwicker, J. D., "Determination of Traces of Carbon Dioxide in Liquid Chlorine," 13th Pittsburgh Conf. on Anal. Chem. & Appl. Spectroscopy, Pittsburgh, Pa., March 1962, Program Abstr., p. 54

1052. Breton, J. L., and Gonzalez, A. G., "Glucosides and Aglycones from Canary Scrophulariaceae. VI. Structure of Two New Triterpenes from Scrophularia Smithii Wydler," J. Chem. Soc., <u>1963</u>, 1401-1406

1053. Brieskorn, C. H., and Wenger, E., "Analysis of Essential Oil of Sage by Means of Gas and Thin Layer Chromatography," Arch. Pharm., <u>293</u>, 21-26 (1960); CA <u>54</u>, 15846i

1054. Brill, W. F., "Autooxidation of Liquid Allylic Chlorides," J. Org. Chem., <u>26</u>, 2969-2972 (1961); CA <u>56</u>, 300g

1055. Brill, W. F., and Lister, F., "Metal Salt Catalyzed Oxidation of Methacrolein," J. Org. Chem., <u>26</u>, 565-569 (1961); CA <u>55</u>, 17482e

1056. Brillyantov, N. A., and Fradkov, A. B., "Degree of Purification of Hydrogen and Helium by a Chromatographic Process on Activated Carbon," Zhur. Tekh. Fiz., 27, 2404-2409 (1957); Soviet Phys.-Tech. Phys., 2, 2239-2244 (1957); CA 52, 17893f

1057. Brimley, R. C., and Barrett, F. C., "Practical Chromatography," Chapman and Hall, Ltd., London, 1953

1058. Brinckman, F. E., and Stone, F. G. A., "Organoboron Halides. II. The Vinylhaloboranes, A Preliminary Study of Their Preparation and Properties," J. Am. Chem. Soc., 82, 6218-6223 (1960); CA 55, 23315i

1059. Brinton, R. K., "The High Temperature Photolysis of Acetone," J. Am. Chem. Soc., 83, 1541-1546 (1961); CA 55, 15080c

1060. Brister, T. B., "Parts Per Million Analysis Using Process Chromatographic Analyzers," 6th Instrumental Methods of Analysis Symp., ISA, Montreal, Canada, June 1960

1061. Broadbent, H. S., Campbell, G. C., Bartley, W. J., and Johnson, J. H., "Rhenium and Its Compounds as Hydrogen Catalysts. III. Rhenium Heptoxide," J. Org. Chem., 24, 1847-1854 (1959); CA 54, 9714i

1062. Brochere-Ferreol, G., and Polonsky, J., "Structure of a New Alicyclic Acid. Gascardic Acid, Isolated from Gum Lac of the Cochineal, Gascardia Madagascariensis," Bull. soc. chim. France, 1960, 963-967; CA 55, 1688f

1063. Brochmann-Hanssen, E., "Gas Chromatography and Its Application to Pharmaceutical Analysis," J. Pharm. Sci., 51, 1017-1031 (1962); CA 58, 3269g

1064. Brochmann-Hanssen, E., and Svendsen, A. B., "Gas Chromatography of Barbiturates and Related Compounds," J. Pharm. Sci., 50, 804 (1961); CA 55, 27780d

1065. Brochmann-Hanssen, E., and Svendsen, A. B., "Gas Chromatography of Sympathomimetic Amines," J. Pharm. Sci., 51, 393 (1962); CA 57, 958c

1066. Brochmann-Hanssen, E., and Svendsen, A. B., "Separation and Identification of Barbiturates and Some Related Compounds by Gas-Liquid Chromatography," J. Pharm. Sci., 51, 318-321 (1962); CA 57, 958d

1067. Brochmann-Hanssen, E., and Svendsen, A. B., "Separation and Identification of Sympathomimetic Amines by Gas-Liquid Chromatography," J. Pharm. Sci., 51, 938-941 (1962); CA 58, 2322c

1068. Brochmann-Hanssen, E., and Svendsen, A. B., "Gas Chromatography of Alkaloids, Alkaloidal Salts, and Derivatives," J. Pharm. Sci., 51, 1095-1098 (1962); CA 58, 8222b

1069. Brodel, H., "On a Microflame Apparatus for High Temperature Gas Chromatography," in "Gas Chromatographie 1958," edited by H. P. Angele, Akademie Verlag, Berlin, 1959, pp. 215-242

1070. Brodskii, A. M., Kalinenko, R. A., and Lavrovskii, K. P., "Analysis and Separation of Gaseous Hydrocarbons by the Adsorption Method," Khim. i Tekhnol Topliva, 1956, No. 8, 18-22; CA 51, 135h

1071. Brodskii, A. M., Kalinenko, R. A., and Lavrovskii, K. P., "Use of Adsorption Methods of Analysis and the Separation of Gaseous Hydrocarbons in Kinetic Studies by the Use of Labelled Atoms," Problemy Kinetiki i Kataliza, Akad. Nauk S.S.S.R., 9, 399-404 (1957); CA 53, 7714c

1072. Brodskii, A. M., Kalinenko, R. A., and Lavrovskii, K. P., "Theory of High-Temperature Ethane Cracking," J. Chem. Soc., 1960, 4443-4454; CA 55, 19213h

1073. Brodskii, A. M., Kalinenko, R. A., Lavrovskii, K. P., and Titov, V. B., "The Mechanism of the High-Temperature Cracking of Ethane," Russian J. Phys. Chem., 33, 474-479 (1959)

1074. Brodskii, A. M., Kolbanovskii, Y. A., Filatova, E. D., and Chernysheva, A. S., "Radiolysis of Heptane," Doklady Akad. Nauk S.S.S.R., 122, 1035-1038 (1958); CA 54, 23665g

1075. Brodskii, A. M., Kolbanovskii, Y. A., Filatova, E. D., and Chernysheva, A. S., "Radiolysis of Normal Heptane and Its Inhibition by Dibenzyl Sulfide and Dibenzyl Additions," Intern. J. Appl. Radiation and Isotopes, 5, No. 1, 57-62 (1959); CA 53, 12024h

1076. Brodskii, A. M., Lavrovskii, K. P., Naimushin, N. N., Titov, V. B., and Filatova, E. D., "The Chromatographic Analysis of Mixtures of Alkenes with Diolefins," Khim. i Tekhnol. Topliv i Masel, 4, No. 3, 30-32 (1959); CA 53, 11110g

1077. Brodskii, A. M., Zvonov, N. V., Lavrovskii, K. P., and Titov, V. B., "Thermal-Radiation Conversions of Petroleum Fractions," Neftekhimiya, 1, 370-381 (1961); CA 57, 1149i

1078. Brodsky, J., Macka, M., and Mikl, O., "The Analysis of Products of Chloroprene Production by Two-Step Gas Chromatography," Chem. prumysl., 10, 460-463 (1960); CA 55, 4253d

1079. Brodsky, J., and Zmitko, J., "Preparation and Properties of Kieselguhr Used as a Support in Gas Chromatography," Chem. listy, 52, 2012-2013 (1958); CA 53, 1588g

1080. Bromley, L. A., "Thermal Conductivity of Gases at Moderate Pressures," Atomic Energy Comm., Tech. Inform. Service, UCRL-1852, 31 pp. (1952); CA 48, 417g

1081. Brook, B. M., and Whitham, B. T., "A Rapid Method for the Determination of the Aromatic Content of Petroleum Fractions Boiling Above the Kerosene Range," J. Inst. Petrol., 44, 212-215 (1958); CA 52, 15037e

1082. Brook, J. H. T., and Glazebrook, R. W., "Free Radical Reactions in Hydrocarbon Mixtures," Trans. Faraday Soc., 56, 1014-1021 (1960); CA 55, 421c

1083. Brooks, C. J. W., and Hanaineh, L., "Regularities of Group Retention Factors in Gas-Liquid Chromatography of Steroids," Biochem. J., 84, 102P (1962); CA 58, 3644a

1084. Brooks, C. J. W., and Young, J. S., "Analysis of Sterols in Blood Serum by Gas-Liquid Chromatography," Biochem. J., 84, 53P (1962); CA 57, 10130d

1085. Brooks, G. T., Harrison, A., and Cox, J. T., "Significance of the Epoxidation of the Isomeric Insecticides Aldrin and Isodrin by the Adult Housefly in Vivo," Nature, 197, 311-312 (1963); CA 58, 8372c

1086. Brooks, J., Murray, W., and Williams, A. F., "Apparatus for Vapor Phase Chromatography. Analysis of Pyridine Homologs. Separation of the Isomers of Picoline," 15th Intern. Congr. of Pure and Appl. Chemistry, Lisbon, Portugal, September 1956

1087. Brooks, J., Murray, W., and Williams, A. F., "Apparatus for Vapour-Phase Chromatography with Ancillary Unit for the Determination of Isopropyl Nitrate in Heavy Oils," in "Vapour Phase Chromatography," edited by D. H. Desty, Academic Press, Inc., New York, 1957, pp. 281-290

1088. Brooks, V. T., "Gas-Liquid Chromatography. Separation of Close-Boiling Phenol Isomers," Chem. & Ind. (London), 1959, 1317-1318; CA 54, 4272e

1089. Brooks, V. T., and Collins, G. A., "Gas-Liquid Chromatography. Separation of Hydrocarbons Using Various Stationary Phases," Chem. & Ind. (London), 1956, 921; CA 51, 3234h

1090. Brooks, V. T., and Collins, G. A., "Separation of Pyridine Bases by Vapour-Phase Chromatography," Chem. & Ind. (London), 1956, 1021; CA 51, 2355h

1091. Brooks, V. T., and Collins, G. A., "Relative Retention Volumes of Hydrocarbons (Chromatographic Data Table XLIII)," J. Chromatog., 1, xxvii (1958); cf. Chem. & Ind. (London), 1956, 921 (Temperature 120°, Benzene = 1)

1092. Brooks, V. T., and Collins, G. A., "Relative Retention Volumes of Pyridine Homologues (Chromatographic Data Table LIII)," J. Chromatog., 1, xxxv (1958); cf. Chem. & Ind. (London), 1956, 1021

1093. Broome, J., Brown, B. R., Roberts, A., and White, A. M. S., "Reduction of α,β-Unsaturated Ketones with Lithium Aluminum Hydride and Aluminum Chloride," J. Chem. Soc., 1960, 1406-1408; CA 54, 14301i

1094. Broughton, D. B., and Carson, D. B., "Industry Gets Two Processes to Upgrade Gasoline by Removing Normal Paraffins. Molex - A Truly Continuous Process Using Molecular Sieves - It's UOP's Newest Octane Booster," Oil Gas J., 57, No. 15, 112-115 (1959); CA 53, 17490f; cf. Petrol. Refiner, 38, No. 4, 130-134 (1959)

1095. Broussard, L., and Shoemaker, D. P., "The Structure of Synthetic Sieves," J. Am. Chem. Soc., 82, 1041-1051 (1960); CA 54, 14862b

1096. Brown, A. L., and Buck, K. R., "Gas Chromatography of Quinoline Bases," Chem. & Ind. (London), 1961, 714; CA 55, 27315i

1097. Brown, G. M., and Satsmadjis, J., "Gas Chromatography," Coke and Gas, 23, 48-54 (1961); CA 55, 18073i

1098. Brown, G. R., Rightmire, R. A., and Strecker, H. A., "Selective Adsorbents Upgrade Gasoline by Removing Normal Paraffins," Oil Gas J., 57, No. 24, 189-192 (1959); CA 54, 872b

1099. Brown, H. C., and Korytnyk, W., "Hydroboration. IV. A Study of the Relative Reactivities of Representative Functional Groups Toward Diborane," J. Am. Chem. Soc., 82, 3866-3869 (1960); CA 55, 15398a

1100. Brown, H. C., and Marino, G., "Directive Effects in Aromatic Substitution. Rate Data and Isomer Distribution in the Acetylation and Benzoylation of Ethyl-, iso-Propyl- and tert-Butylbenzene. Partial Rate Factors for the Acylation Reaction," J. Am. Chem. Soc., 81, 5611-5615 (1959); CA 55, 15378i

1101. Brown, H. C., Marino, G., and Stock, L. M., "Directive Effects in Aromatic Substitutions. XXXIII. Relative Rate and Isomer Distribution in the Acetylation of Benzene and Toluene Under the Influence of Aluminum Chloride," J. Am. Chem. Soc., 81, 3310-3314 (1959); CA 54, 1396h

1102. Brown, H. C., and Moerikofer, A. W., "Hydroboration. X. Rates of Reaction of Bis-3-methyl-2-butylborane with Representative Cycloalkenes and Isomeric Cis-Trans Alkenes," J. Am. Chem. Soc., 83, 3417-3422 (1961); CA 56, 4783g

1103. Brown, H. C., Murray, K. J., Murray, L. J., Snover, J. A., and Zweifel, G., "Hydroboration. V. A Study of Convenient New Preparative Procedures for the Hydroboration of Olefins," J. Am. Chem. Soc., 82, 4233-4241 (1960)

1104. Brown, H. C., and Snyder, C. H., "The Reaction of Trialkylboranes With Alkaline Silver Nitrate. A New General Coupling Reaction," J. Am. Chem. Soc., 83, 1002-1003 (1961); CA 55, 16392b

1105. Brown, H. C., and Zweifel, G., "Hydroboration of Acetylenes. A Convenient Conversion of Internal Acetylene to cis Olefins of High Purity and of Terminal Acetylenes to Aldehydes," J. Am. Chem. Soc., 81, 1512 (1959); CA 53, 15947e

1106. Brown, H. C., and Zweifel, G., "Hydroboration. VII. Directive Effects in the Hydroboration of Olefins," J. Am. Chem. Soc., 82, 4708-4712 (1960); CA 55, 8281h

1107. Brown, H. C., and Zweifel, G., "Hydroboration. VIII. Bis(3-Methyl-2-butylborane) as a Selective Reagent for the Hydroboration of Alkenes and Dienes," J. Am. Chem. Soc., 83, 1241-1246 (1961);

CA 55, 16401g

1108. Brown, H. C., and Zweifel, G., "Hydroboration. IX. The Hydroboration of Cyclic and Bicyclic Olefins - Stereochemistry of the Hydroboration Reaction," J. Am. Chem. Soc., 83, 2544-2551 (1961); CA 55, 20993h

1109. Brown, H. C., and Zweifel, G., "Hydroboration. XI. The Hydroboration of Acetylenes - A Convenient Conversion of Internal Acetylenes into cis-Olefins and of Terminal Acetylene into Aldehydes," J. Am. Chem. Soc., 83, 3834-3840 (1961); CA 56, 4784g

1110. Brown, I., "Identification of Organic Compounds by Gas Chromatography," Nature, 188, 1021-1022 (1960); CA 55, 8179b

1111. Brown, I., "The Role of the Stationary Phase in Gas Chromatography," J. Chromatog., 10, 284-293 (1963)

1112. Brown, J. F., and Stanley, A. G., "Process Analytical Instrumentation," Soc. Instr. Techn. Trans., 8, No. 4, 156-164 (1956)

1113. Brown, P., Burdon, J., Smith, T. J., and Tatlow, J. C., "5,5,5-Trifluorolaevulinic Acids and Derived Compounds," Tetrahedron, 10, 164-170 (1960); CA 55, 1431c

1114. Brown, S. A., and Shyluk, J. P., "Gas Liquid Chromatography of Some Naturally Occurring Coumarins," Anal. Chem., 34, 1058-1061 (1962); CA 57, 10514f

1115. Brownell, W. B., Chadde, F. E., Theivagt, J. G., and Wimer, D. C., "Pharmaceuticals and Related Drugs," Anal. Chem., 35, No. 5, 143R-160R (1963)

1116. Browning, L. C., and Watts, J. O., "Design and Construction of a Gas-Liquid Partition Chromatographic Unit, and Its Application to the Quantitative Analysis of Liquid Solutions," PB Rept. 151,008, 39 pp.; CA 54, 10409d

1117. Browning, L. C., and Watts, J. O., "Interpretation of Areas Used for Quantitative Analysis in Gas-Liquid Partition Chromatography," Anal. Chem., 29, 24-27 (1957); CA 51, 11911f

1118. Bruenner, R. S., Haertle, W. R., Lundquist, R. T., Wnuk, R. J., and Ruby, A., "The Gas Chromatographic Separation and Infrared Identification of the 2,4- and 2,5-Isomers of Dimethyloxaloline," 14th Pittsburgh Conf. on Anal. Chem. & Appl. Spectroscopy, Pittsburgh, Pa., March 1963, Program Abstr., pp. 84-85

1119. Bruk, A. I., Vinogradova, L. M., and Vyakhirev, D, A., "A New Modified Sorbent for Use in Gas Chromatography," Tr. po Khim. i Khim. Tekhnol., 1962, No. 1, 211-212; CA 58, 7349a

1120. Brunauer, S., "Physical Adsorption of Gases and Vapours," Oxford University Press, Inc., New York, 1945

1121. Bruner, F., and Cartoni, G. P., "Gas Chromatographic Separation of Benzene and Deuterobenzene," J. Chromatog., 10, 306 (1963)

1122. Brunnée, C., Jenckel, L., and Kronenberger, K., "Continuous Mass Spectrometric Analysis of Fractions Separated by Gas Chromatography," Z. Anal. Chem., 189, 50-66 (1962); CA 57, 13166d

1123. Bruno, S., "Gas-Chromatographic Analysis of Some Essential Oils in Biological Matter," Farmaco (Pavia) Ed. Prat., 16, 481-486 (1961); CA 56, 7450b

1124. Brus, G., Legendre, P., and Niolle, G., "The Analysis of Turpentine Oils by Chromatography in the Vapor Phase," Ann. Fals. et Expert. Chem., 54, 142-150 (1961); CA 56, 1552a

1125. Bruyn, J. de, and Schogt, J. C. M., "Isolation of Volatile Constituents from Fats and Oils by Vacuum Degassing," J. Am. Oil Chemists' Soc., 38, 40-44 (1961); CA 55, 2206e

1126. Bryan, F. A., and Silis, V., "A Generator for Producing Low Concentrations of Vapor in Inhalation Chambers," Am. Ind. Hyg. Assoc. J., 21, 423-427 (1960); CA 55, 2206e

1127. Bryan, F. R., and Neerman, J. C., "Gas Chromatographic Identification of Major Constituents of Bubbles in Glass," Anal. Chem., 34, 278-280 (1962); CA 57, 35f

1128. Bryce, W. A., and Kebarle, P., "Thermal Decomposition of 1-Butene and 1-Butene-4-d_3," Trans. Faraday Soc., 54, 1660-1677 (1958); CA 53, 18848d

1129. Bryce, W..A., and Ruzicka, D. J., "Reactions of Allyl Radicals with Olefins," Can. J. Chem., 38, 835-844 (1960); CA 54, 24330i

1130. Bua, E., Manaresi, P., and Motta, L., "Determination of Propadiene Traces in Propene," Anal. Chem., 31, 1910-1911 (1959); CA 54, 4269a

1131. Buck, K. W., and Foster, A. B., "Reactions of Some Alkyl Chloroformates," J. Chem. Soc., 1963, 2217-2221

1132. Buckles, R. E., and Deeds, M. L., "Diels-Alder Reactions on 1,2-Cyclohexene Dicarboxylic Anhydride," J. Org. Chem., 23, 485-486 (1958); CA 53, 8014i

1133. Buckles, R. E., Forrester, J. L., Burham, R. L., and McGee, T. W., "Addition Reactions of Mixtures of Bromine and Chlorine," J. Org. Chem., 25, 24-26 (1960); CA 54, 15304b

1134. Bucur, R., "Chromatographic Analysis of Gaseous Hydrocarbons," Rev. chim. (Bucharest), 7, 163-165 (1956); CA 52, 9856e

1135. Bucur, R., Mercea, I., and Mercea, V., "Thermal Conductivity Cells and Their Use for Isotope Analysis," Acad. rep. populare Romine, Inst. fiz. atomica si Inst. fiz., Studii cerecetari fiz., 10, 753-770 (1959); CA 55, 4200b

1136. Buechel, K. H., and Korte, F., "Chemistry of Iridolactones," Intern. Kongr. Entomol. Verhandl., 11th, Vienna, 1960, No. 3, 60-65; CA 58, 4602f

1137. Buechi, J., Iconomou-Petrovitch, N., and Schumacher, H., "Use of Gas Chromatography in the Testing of Purity of Drugs," Pharm. Acta Helv., 37, 379-395 (1962); CA 57, 11307f

1138. Bukata, S. W., Zabrocki, L. L., and McLaughlin, M. F., "Gas Chromatography of Organic Peroxides," Anal. Chem., 35, 885-886 (1963); 142nd Natl. ACS Mtg., Atlantic City, N. J., September 1962, Program Abstr., p. 13B

1139. Bukhari, M. A., Foster, A. B., Lehmann, J., Webber, J. M., and Westwood, J. H., "Aspects of Stereochemistry. XI. Isopropylidene Derivatives of L-Arabitol and Ribitol," J. Chem. Soc., 1963, 2291-2295

1140. Bumb, F. C., and Marks, M. L., "Development of a Laboratory Vapor-Phase Chromatograph," Southwide Chem. Conf. on Instrumentation, ISA Symp. on Gas Chromatography, Memphis, Tenn., December 1956

1141. Bumgardner, C. L., "Formation of Cyclopropane Derivatives and Olefins from Quaternary Ammonium Halides," Chem. & Ind. (London), 1958, 1555-1556; CA 53, 21700f

1142. Bumgardner, C. L., "Elimination Reactions. I. Formation of Cyclopropane Derivatives from Quaternary Ammonium Halides," J. Am. Chem. Soc., 83, 4420-4423 (1961); CA 56, 4627d

1143. Bumgardner, C. L., "Elimination Reactions. II. Some Electronic and Steric Effects in γ-Elimination Reactions of Quaternary Ammonium Compounds," J. Am. Chem. Soc., 83, 4423-4427 (1961); CA 56, 4627e

1144. Buoncristiani, D., Toponeco, G., and Salvadorini, R., "Gas Chromatography of the Methyl Esters of Fatty Acids. Influence of Temperature on Column Efficiency in Separating Linolenic, Arachidic, and Eicosenoic Acids," Olearia, 16, 99-112, 122-126 (1962); CA 58, 4749g

1145. Buoncristiani, D., Toponeco, G., and Salvadorini, R., "Gas Chromatography of Fatty Acid Esters. The Influence of Temperature on Column Efficiency, Particularly in the Separation of Linolenic, Arachic, and Eicosenoic Acids," Boll. Lab. Chim. Provinciali (Bologna), No. 13, No. 2, 218-235 (1962); CA 58, 6169a

1146. Burchfield, H. P., "Research on Vanilla. A Summary of Progress," 15th Annual Conv. of the Flavoring Extract Manufacturers Assoc. of U. S., New York, May 1959

1147. Burchfield, H. P., and Prill, E. A., "Characterization of the Noncarbonyl Volatiles of Vanilla by Gas Chromatography," Boyce Thompson Inst. Contrib., 20, 217-229 (1959); CA 53, 20611g

1148. Burchfield, H. P., and Storrs, E. E., "Residue Analysis of 2, 4-D and Other Chlorine-Containing Herbicides in Milk," 140th Natl. ACS Mtg., Chicago, Ill., September 1961, Program Abstr., p. 19A

1149. Burchfield, H. P., and Storrs, E. E., "Biochemical Applications of Gas Chromatography," Academic Press, Inc., New York, 1962

1150. Burdon, J., Gilman, C. J., Patrick, C. R., Stacey, M., and Tatlow, J. C., "Pentafluoropyridine," Nature, 186, 231-232 (1960); CA 55, 529g

1151. Burford, R. R., and DeTomaso, P., "Analysis of Perfluoro Carboxylic Acids by Gas-Liquid Chromatography," 136th Natl. ACS Mtg., Atlantic City, N. J., September 1959, Program Abstr., p.23M

1152. Burg, S. P., and Burg, E. A., "Ethylene Evolution and Sub-Cellular Particles," Nature, 191, 967-969 (1961); CA 56, 5120e

1153. Burg, S. P., and Stolwijk, J. A. A., "A Highly Sensitive Katharometer and Its Application to the Measurement of Ethylene and Other Gases of Biological Importance," J. Biochem., Microbiol. Technol. Eng., 1, 245-259 (1959); CA 54, 5796d

1154. Burg, S. P., and Thimann, K. V., "The Physiology of Ethylene Formation in Apples," Proc. Natl. Acad. Sci. U. S., 45, 335-344 (1959); CA 53, 15216h

1155. Burg, S. P., and Thimann, K. V., "Ethylene Production of Apple Tissue," Plant Physiol., 35, 24-35 (1060); CA 54, 17583f

1156. Burgess, A. R., "Formation of Hydrogen Peroxide in the Gaseous Oxidation of Isopropyl Alcohol," J. Appl. Chem. (London), 11, 235-243 (1961); CA 55, 27028d

1157. Burgess, A. R., Cullis, C. F., and Newitt, E. J., "The Gaseous Oxidation of Isopropyl Alcohol. I. The Influence of Temperature, Pressure, and Mixture Composition on the Formation of Hydrogen Peroxide and Other Products," J. Chem. Soc., 1961, 1884-1893; CA 55, 20910e

1158. Burgt, M. J. van der, Dijkstra, F., Waterman, H. I., and Weerdt, W. J. van de, "Selectivity in Catalytic Conversions of Cyclohexane in a Hydrogen Atmosphere," Brennstoff-Chem., 40, 383-389 (1959); CA 54, 7003a

1159. Burk, M. C., and Karasek, F. W., "Data Converter Adapts Chromatograph to Process Control," ISA Journal, 5, No. 10, 28-31 (1958); CA 53, 8721c

1160. Burke, J., "Cleanup in the Use of Microcoulometric Gas Chromatographic Apparatus for Pesticide Residue Analysis," 140th Natl. ACS Mtg., Chicago, Ill., September 1961, Program Abstr., p. 58B

1161. Burke, J., and Johnson, L., "Investigations in the Use of the Microcoulometric Gas Chromatograph for Pesticide Residue Analysis," J. Assoc. Offic. Agr. Chemists, 45, 348-354 (1962); CA 57,

2626g; 140th Natl. ACS Mtg., Chicago, Ill., September 1961, Program Abstr., p. 62B

1162. Burkin, A. R., and Halsey, G., "Chemisorption at Solid-Liquid Interfaces," Nature, 191, 348-349 (1961); CA 56, 5424g

1163. Burks, Jr., R. E., "Study of Flavor and Chemical Changes in Foods During Sterilization by Irradiation," U. S. Govt. Research Repts., 33, No. 2, PB 143,775 (February 12, 1960)

1164. Burks, Jr., R. E., Baker, E. B., Clark, P., Esslinger, J., and Lacey, Jr., J. C., "Detection of Amines Produced on Irradiation of Beef," J. Agr. Food Chem., 7, 778-782 (1959); CA 54, 2625c

1165. Burn, A. J., Cadogan, J. I. G., and Bunyan, P. J., "The Reactivity of Organophosphorus Compounds. XV. Reactions of Diaroyl Peroxides with Triethyl Phosphite," J. Chem. Soc., 1963, 1527-1533; CA 58, 10232c

1166. Burnell, M. R., "A New Hydrogen Flame Detector," Analyzer, 1, No. 2, 3-7 (April 1960); 11th Pittsburgh Conf. on Anal. Chem. & Appl. Spectroscopy, Pittsburgh, Pa., February-March 1960, Program Abstr., p. 42

1167. Burnell, M. R., and Said, A. S., "Optimized Temperature Programming in Gas Chromatography," in "Progress in Industrial Gas Chromatography," Vol. 1, edited by H. A. Szymanski, Plenum Press, Inc., New York, 1961, pp. 19-30

1168. Burnell, M. R., and Said, A. S., "Optimized Temperature Programming in Gas Chromatography," Analyzer, 2, No. 2, 4-7 (April 1961); 12th Pittsburgh Conf. on Anal. Chem. & Appl. Spectroscopy, Pittsburgh, Pa., February-March 1961, Program Abstr., p. 58

1169. Burnell, M. R., and Sternberg, J. C., "Far-Out Chromatography," 144th Natl. ACS Mtg., Los Angeles, Calif., March-April 1963; Program Abstr., p. 18J

1170. Burnett, M. C., and Lohmar, R. L., "Lipides in Feedstuffs. Fatty Acids of Sorghum Leaf and Stem," J. Agr. Food Chem., 7, 436-437 (1959); CA 53, 20614d

1171. Burnett, M. G., and Swoboda, P. A. T., "A Simple Method for the Calibration of Sensitive Gas Chromatographic Detectors," Anal. Chem., 34, 1162-1163 (1962); CA 57, 10513b

1172. Burov, A. N., Kalmanovskii, V. I., and Yashin, Y. I., "Rapid Chromatographic Determination of Small Quantities of Gaseous Hydrocarbons," Trudy po Khim. i Khim. Tekhnol., 4, 345-350 (1961); CA 56, 4073g

1173. Burrell Corp., "Adsorption Fractionation of Gases and Vapors (Gas Chromatography)," Burrell Corp., Pittsburgh, Pa., 1955

1174. Burrows, G., "Process of Gas-Liquid Chromatography," Trans. Inst. Chem. Eng. (London), 35, 245-257 (1957)

1175. Burt, R., Ebeid, F. M., and Minkoff, G. J., "Point of Attack in Hydrocarbon Oxidation," Nature, 180, 188 (1957); CA 52, 3675c

1176. Bush, I. E., "The Chromatography of Steroids," Pergamon Press, Ltd., Oxford, 1961

1177. Bush, S. J., "Chemical Analysis in the Cosmetic Industry," J. Soc. Cosmetic Chemists, 10, 258-271 (1959)

1178. Bushong, P. A., "The Fisher Prep/Partitioner," in "Lectures on Gas Chromatography 1962," ed. by H. A. Szymanski, Plenum Press, New York, 1963, pp. 247-264; CA 58, 10702a (cf. Item 5615)

1179. Butler, J. N., and Brokaw, R. S., "Thermal Conductivity of Gas Mixtures in Chemical Equilibrium," J. Chem. Phys., 26, 1636-1643 (1957); CA 52, 17389c

1180. Butler, J. N., and Kistiakowsky, G. B., "Reactions of Methylene. IV. Propylene and Cyclopropane," J. Am. Chem. Soc., 82, 759-765 (1960); CA 54, 15266g

1181. Butler, J.N., and Kistiakowsky, G. B., "Reactions of Methylene. V. The Effect of Inert Gases on the Reaction with Cyclopropane," J. Am. Chem. Soc., 83, 1324-1326 (1961); CA 55, 19814a

1182. Butler, R. A., and Hill, D. W., "Estimation of Volatile Anaesthetics in Tissues by Gas Chromatography," Nature, 189, 488-489 (1961); CA 55, 17732i

1183. Buttery, R. G., Hendel, C. E., and Boggs, M. M., "Off-Flavors in Potato Products. Autooxidation of Potato Granules. Part I. Changes in Fatty Acids; Part II. Formation of Carbonyls and Hydrocarbons," J. Agr. Food Chem., 9, 245-248, 248-252 (1961); CA 55, 21415ef

1184. Buttery, R. G., and Stuckey, B. N., "Determination of Butylated Hydroxyanisole and Butylated Hydroxytoluene in Potato Granules by Gas-Liquid Chromatography," J. Agr. Food Chem., 9, 283-285 (1961); CA 56, 5172d

1185. Buttery, R. G., and Teranishi, R., "Gas-Liquid Chromatography of Aroma of Vegetables and Fruit. Direct Injection of Aqueous Vapors," Anal. Chem., 33, 1439-1441 (1961); CA 56, 745c

1186. Butz, W. H., and Johns, T., "Preparative Gas Chromatography: New Laboratory Technique," Research/Development, 11, No. 7, 90-92, 94-95 (1960)

1187. Butz, W. H., and Noebels, H. J., "Instrumental Methods for the Analysis of Food Additives," Interscience Publishers, Inc., New York, 1961

1188. Buzon, J., "Recent Applications of Gas Chromatography to the Analysis of Petroleum Products," Bull. Soc. Chim. France, 1963, 526-531; CA 59, 1417e

1189. Buzon, J., Chovin, P., Fanica, L., Ferrand, R., Guiochon, G., Huguet, M., Lebbe, J., Serpinet, J., and Tranchant, J., "Gas Phase Chromatography. Proposals for a Vocabulary and a System of Notation Relating to Retention Values, " Bull. Soc. Chim. France, <u>1959</u>, 1137-1140; CA <u>54</u>, 2879d

1190. Buzon, J., and Follain, G., "Analysis of Light Hydrocarbons by Gaseous Chromatography. Evaluation of the Precision of the Results, " Rev. inst. franc. petrole et Ann. Combustibles liquides, <u>16</u>, 715-735 (1961); CA <u>55</u>, 25227i

1191. Buzon, J., and Moghadame, P. E., "Vapor-Phase Chromatography. I., " Rev. inst. franc. petrole et Ann. combustibles liquides, <u>11</u>, 1616-1628 (1956); CA <u>51</u>, 17302a

1192. Cabiddu, S., Maccioni, A., and Secci, M., "Wittig Synthesis. Synthesis of Anethole and Isosafrole, " Ann. chim. (Rome), <u>52</u>, 1261-1266 (1962); CA <u>59</u>, 489g

1193. Cacace, F., "Labelled Organics in Gas Chromatography, " Nucleonics, <u>19</u>, No. 5, 45-50 (1961)

1194. Cacace, F., "Distribution of Radioactive Atoms in Aromatic Molecules Labeled by Exposure to Tritium Gas, " Comit. Nazl. Energia Nucl. RT/CHI, <u>3</u>, 133-134 (1962); CA <u>58</u>, 2340a

1195. Cacace, F., Cipollini, R., and Perez, G., "Continuous Elemental Analysis of Organic Compounds in Gas-Chromatographic Effluents, " Science, <u>132</u>, 1253-1254 (1960); CA <u>55</u>, 3275i

1196. Cacace, F., Cipollini, R., Perez, G., and Possagno, E., "Continuous Elemental Analysis of Volatile Compounds Separated by Gas Chromatography, " Gazz. Chim. Ital., <u>91</u>, 804-824 (1961); CA <u>57</u>, 1550i

1197. Cacace, F., Ciranni, E., and Ciranni, G., "Distribution of Radioactive Atoms in Organic Molecules Labeled by Exchange with Gaseous Tritium, " Atti. accad. nazl. Lincei Rend., Classe Sci. fis. mat. e nat., <u>28</u>, 865-875 (1960); CA <u>55</u>, 13342e

1198. Cacace, F., Ciranni, G., and Possagno, E., "Radioactive Substances Obtained by Radiation with Gamma-Rays from Gas Mixtures of Propane Plus $C^{14}O_2$ and Cyclopentane Plus $C^{14}O_2$, " Ann. chim. (Rome), <u>50</u>, 920-930 (1960); CA <u>55</u>, 1150a

1199. Cacace, F., Giacomello, G., and Montefinale, G., "Neutron Irradiation of Benzamide and Benzenesulfonate in the Solid State, " Gazz. chim. ital., <u>89</u>, 1829-1836 (1959); CA <u>55</u>, 3507h

1200. Cacace, F., Guarino, A., and Haq, I. U., "Gas Chromatographic Separation and Radiometric Analysis of High-Boiling Substances Marked with C^{14}, " Ann. chim. (Rome), <u>50</u>, 915-919 (1960); CA <u>55</u>, 3269e

1201. Cacace, F., Guarino, A., and Montefinale, G., "Labelling of Organic Compounds by Mercury-Photosensitized Reaction with Tritium Gas, " Nature, <u>189</u>, 54-55 (1961); CA <u>55</u>, 12010c

1202. Cacace, F., Guarino, A., Montefinale, G., and Possagno, E., "Distribution of Radioactive Atoms in Aromatic Compounds Labeled by Exposure to Tritium Gas, " Intern. J. Appl. Radiation and Isotopes, <u>8</u>, 82-89 (1960); CA <u>55</u>, 10353b

1203. Cacace, F., Guarino, A., and Possagno, E., "Chemical Effect of Ionization Radiations. Formation of Carboxylic Acids Following Irradiation of a Mixture of Pentane and $C^{14}O_2$ with X-Rays, " Gazz. chim. ital., <u>89</u>, 1837-1842 (1959); CA <u>55</u>, 3421c

1204. Cacace, F., and Haq, I. U., "Radiometric Analysis of Volatile Organic Compounds Marked with C^{14} and H^3 by Vapor Phase Chromatography, " Ricerca Sci., <u>30</u>, 501-508 (1960); CA <u>54</u>, 22174c

1205. Cacace, F., and Haq, I. U., "Radiometric Analysis of Tritiated Organic Compounds by Means of Vapor Phase Chromatography, " Science, <u>131</u>, 732-733 (1960); CA <u>54</u>, 15064c

1206. Cacace, F., Ikram, M., and Stein, M. L., "Applications of Gas Phase Chromatography in the Analysis of Distilled Alcoholic Beverages, " Ann. chim.(Rome), <u>49</u>, 1383-1390 (1959); CA <u>54</u>, 14571f

1207. Cacace, F., and Possagno, E., "Effects of Irradiation Conditions on the Exchange Reactions of Tritium Gas with Organic Compounds. II. The Tritium-Hydrogen Exchange in Toluene, " Gazz. chim. ital., <u>90</u>, 1800-1806 (1960); CA <u>57</u>, 65g

1208. Cadman, W. J., "Analysis of Some Vaporizable Materials of Interest to the Laboratory of Criminalistics, " 137th Natl. ACS Mtg., Cleveland, Ohio, April 1960, Program Abstr., p. 8G

1209. Cadman, W. J., "The Application of Gas Chromatography in Forensic Science, " in "Methods of Forensic Science, " Vol. 2, edited by F. Lundquist, Interscience Publishers, Inc., New York, 1963

1210. Cadman, W. J., and Johns, T., "Application of the Gas Chromatograph in the Laboratory of Criminalistics, " J. Forensic Sci., <u>5</u>, 369-385 (1960); CA <u>54</u>, 20613f

1211. Cadman, W. J., and Johns, T., "The Analysis of Some Vaporizable Materials of Interest to the Laboratory of Criminalistics, " Microchem. J., <u>5</u>, 573-585 (1961)

1212. Cadman, W. J., and Johns, T., "The Identification of Dangerous Drugs with Gas Chromatography and Infrared Spectrophotometry, " 1st Annual Pacific Regional Mtg. for Applied Spectroscopy and Analytical Chemistry, Pasadena, Calif., October 1962

1213. Cadogan, J. I. G., Hey, D. H., and Sanderson, W. A., "Organic Peroxides. III. Bis(9-benzyl-9-fluorenyl) Peroxide. A New Source of Benzyl Radicals, " J. Chem. Soc., <u>1960</u>, 3203-3210; CA <u>55</u>, 1545i

1214. Cadogan, J. I. G., Hey, D. H., and Sanderson, W. A., "Organic Peroxides. IV. The Decomposition of Benzophenone Peroxide, " J. Chem. Soc., <u>1960</u>, 4897-4900; CA <u>55</u>, 10380a

1215. Cady, G. H., and Siegwarth, D. P., "Fractional Codistillation in Gas Chromatography Apparatus, " Anal. Chem., <u>31</u>, 618-620 (1959); CA <u>53</u>, 11901h

1216. Cady, P., Abraham, S., and Chaikoff, I. L., "Relative Incorporation of the Various Propionate Carbons into Fatty Acids by Lactating Rat Mammary Gland," Biochim. Biophys. Acta, 70, 118-131 (1963); CA 58, 12945e

1217. Caffrey, Jr., J. M., and Allen, A. O., "Radiolysis of Pentane Adsorbed on Mineral Solids," J. Phys. Chem., 62, 33-37 (1958); CA 52, 7877f

1218. Cain, E. F. C., and Stevens, M. R., "Chromatographic Determination of Water in Hydrazine," Gas Chromatog., Intern. Symposium, 2nd, East Lansing, Mich., 1959, 343-350 (Pub. 1961); CA 55, 18426i

1219. Cain, E. F. C., and Stevens, M. R., "Chromatographic Determination of Water in Hydrazine," in "Gas Chromatography," edited by H. J. Noebels, R. F. Wall, and N. Brenner, Academic Press, Inc., New York, 1961, pp. 343-350

1220. Caldarera, C. M., Ronca, G., and Lenaz, G., "The Fatty Acids in the Lipids of Wheat Gluten Determined by Gas Chromatography," Quaderni Nutr., 20, Nos. 3-4, 100-106 (1960); CA 57, 7685f

1221. Call, F., "Microsampling Method of Determining Gases and Vapors, Particularly Halogenated Hydrocarbons in Air," J. Appl. Chem. (London), 7, 210-215 (1957); CA 51, 16206f

1222. Callear, A. B., and Cvetanovic, R. J., "The Application of Gas-Liquid Partition Chromatography to Problems of Chemical Kinetics," Chem. in Can., 7, 5 (May 1955)

1223. Callear, A. B., and Cvetanovic, R. J., "The Application of Gas-Liquid Partition Chromatography to Problems of Chemical Kinetics," Can. J. Chem., 33, 1256-1267 (1955); CA 49, 14421i

1224. Callear, A. B., and Robb, J. C., "An Experimental Method of Measuring the Thermal Conductivity of Gases," Trans. Faraday Soc., 51, 630-638 (1955); CA 49, 14390e

1225. Callisen, F. I., "Fractography (Chromatography) of Gases and Vapors," Ind. y quim (Buenos Aires), 18, 226-229 (1957); CA 52, 2639f

1226. Calvarano, I., "Chromatographic Methods for the Analysis of Essential Oils and Their Constituents," Essenze deriv. agrumari, 30, 32-54 (1960); CA 55, 18019g

1227. Calvarano, M., "Gas Chromatography Applied to Essential Oils and Their Constituents," Essenze deriv. agrumari, 27, 208-220 (1957); CA 52, 10507c

1228. Calvarano, M., "Composition of Essential Oils Distilled from Rosa Centifolia Grown in Calabria," Essenze deriv. agrumari, 28, 157-163 (1958); CA 53, 22752h

1229. Calvarano, M., "Examination of Tangerine Oil by Means of Partition Chromatography in Vapor Phase," Essenze deriv. agrumari, 28, 107-118 (1958); CA 53, 11702a

1230. Calvarano, M., "Coumarin of the Essential Oil of Bergamot," Essenze deriv. agrumari, 31, 167-174 (1961); CA 57, 2355b

1231. Calzolari, C., and Furlani, A., "Applications of Chromatographic Methods to the Analysis of Food," Pubbl. Univ. Cattolica S. Cuore, Ann. fac. agrar. Ser. 5, Atti convegno appl. tec. cromatogr. prod. agr., 53, 22-53 (1957); CA 52, 20711e

1232. Cameron, D. W., and Sutherland, M. D., "Some Constituents of Pinus Sylvestris Oil," Perfumery Essent. Oil Record, 50, 200-203 (1959); CA 53, 16476d

1233. Cameron, D. W., and Sutherland, M. D., "Chemical Constituents of Evodia Micrococca," Australian J. Chem., 14, 135-142 (1961); CA 55, 14420a

1234. Camin, D. L., King, R. W., and Shawhan, S. D., "Capillary Column Gas Chromatography Using Micro-Thermal Conductivity Cells as Detectors," 14th Pittsburgh Conf. on Anal. Chem. & Appl. Spectroscopy, Pittsburgh, Pa., March 1963, Program Abstr., p. 66

1235. Campbell, J. K., Rhoades, J. W., and Gross, A. L., "Acetonitrile as a Constituent of Cigarette Smoke," Nature, 198, 991-992 (1963)

1236. Campbell, R. H., and Gudzinowicz, B. J., "Separation of Some Fluorocarbon and Sulfur-Fluoride Compounds by Gas-Liquid Chromatography," Anal. Chem., 33, 842-845 (1961); CA 55, 18421c

1237. Cannings, F. R., Fisher, A., Ford, J. F., Holmes, P. D., and Smith, R. S., "Dehydrocyclization of 2,2-Dimethyl-4-(methyl-C14)pentane and 3-(Methyl-C14)heptane Over Chromia-Alumina," Radioisotopes Phys. Sci. Ind., Proc. Conf. Use, Copenhagen, 1960, 3, 205-216 (Pub. 1962); CA 58, 4391g

1238. Cannon, H. J., "Plant Analyzer Sampling Systems," 5th ISA Mtg., Symposium on Instrumental Methods of Analysis, Houston, Texas, May 1959

1239. Cannon, P., "Reaction of Chlorodifluoromethane with Linde Molecular Sieve 5A," J. Am. Chem. Soc., 80, 1766-1767 (1958); CA 52, 11730h

1240. Caprioli, G., Pavan, E., and De Vita, M., "Industrial Analytical Applications of Vapor-Phase Chromatography. II. Analysis of Hydrocarbon Mixtures," Ann. chim. (Rome), 49, 1120-1124 (1959); CA 55, 16282i

1241. Caran, J. G., and Caran, S. H., "What's New in Mine-Logging Techniques," Oil Gas J., 57, 116-120 (1959)

1242. Carberry, J. J., "Determination of Heats of Adsorption by Transient-Response Techniques," Nature, 189, 391-393 (1961); CA 55, 14011i

1243. Carberry, J. J., "Determination of Molecular Diffusivity by Gas-Chromatography Techniques," J.

Chem. Phys., 35, 2241-2242 (1961); CA 56, 12329c

1244. Carel, A. B., Lively, L. D., and Hamilton, W. C., "Investigation of Parameters in Preparative Gas Chromatography," 12th Pittsburgh Conf. on Anal. Chem. & Appl. Spectroscopy, Pittsburgh, Pa., February-March 1961, Program Abstr., p. 57

1245. Carew, J. R., "A Compact Easily Cleaned and Packed Column for Gas-Liquid Partition Chromatography Instruments," Anal. Chem., 33, 156-157 (1961); CA 55, 6944i

1246. Carle, D. W., "Design and Performance of the Beckman Gas Chromatograph," 7th Pittsbrugh Conf. on Anal. Chem. & Appl. Spectroscopy, Pittsburgh, Pa., February-March 1956, Program Abstr., p. 41

1247. Carle, D. W., "Gas Chromatography: Role of Column-Packing Materials, Carrier Gases - 3," Oil Gas J., 54, No. 85, 128, 131-132 (1956)

1248. Carle, D. W., "Precise Liquid Sampling in Gas Chromatography," in "Gas Chromatography," edited by V. J. Coates, H. J. Noebels, and I. S. Fagerson, Academic Press, Inc., New York, 1958, pp. 67-72

1249. Carle, D. W., and Burnell, M., "A Study in Preparative Scale Gas Chromatography," 9th Pittsburgh Conf. on Anal. Chem. & Appl. Spectroscopy, Pittsburgh, Pa., March 1958, Program Abstr., p. 45

1250. Carle, D. W., and Donner, W., "Design and Performance of a New High Temperature Gas Chromatograph," Abstr., Applied Spectroscopy, 11, 101 (1957); 8th Pittsburgh Conf. on Anal. Chem. & Appl. Spectroscopy, Pittsburgh, Pa., March 1957, Program Abstr., p. 34

1251. Carle, D. W., and Johns, T., "Design and Application of Preparative Scale Gas Chromatography," ISA Analytical Symposium, Houston, Texas, May 1958; Beckman Reprint R-6122

1252. Carlstrom, A. A., Spencer, C. F., and Johnson, J. F., "Determination of Trace Water in Butane by Gas Chromatography," Anal. Chem., 32, 1056 (1960); CA 54, 19261g

1253. Carman, P. C., "Flow of Gases Through Porous Media," Academic Press, Inc., New York, 1956

1254. Carnahan, C. L., "Gas-Chromatographic Separations of Rare Gases," U.S. Dept. Com., Office Tech. Serv., AD 268, 156, 36 pp. (1961); CA 58, 3081e

1255. Caroti, G., "Chromatography of Gas - Technique of Gas-Liquid Distribution," Riv. Combustibili, 10, 456-471 (1956); CA 50, 15324i

1256. Carpenter, F. G., "Use of Gas Chromatography for Rapid Determination of Carbonate at Low Levels," Anal. Chem., 34, 66-67 (1962); CA 56, 9417i

1257. Carr, Jr., H. F., and Wotiz, H. H., "The Measurement of Urinary Tetrahydroaldosterone by Gas-Liquid Partition Chromatography," Biochim. Biophys. Acta, 71, 178-184 (1963); CA 58, 11640a

1258. Carroll, K. K., "Quantitative Estimation of Peak Areas in Gas-Liquid Chromatography," Nature, 191, 377-378 (1961); CA 56, 5405c

1259. Carroll, K. K., "The Fatty Acids of Beef Brain and Spinal Cord Sphingolipid Preparations," J. Lipid Res., 3, 263-268 (1962); CA 57, 6464f

1260. Carroll, K. K., "Gas-Liquid Chromatography of Fat-Soluble Vitamins," 36th Fall Mtg., Am. Oil Chemists' Soc., Toronto, Ontario, October 1962

1261. Carroll, K. K., "The Mechanisms by Which Erucic Acid Affects Cholesterol Metabolism. Distribution of Erucic Acid in Adrenal and Plasma Lipids," Can. J. Biochem. & Physiol., 40, 1115-1122 (1962); CA 57, 13067g

1262. Carroll, R. B., and O'Brien, L., "Gas-Liquid Chromatography of Whiskies," 135th Natl. ACS Mtg., Boston, Mass., April 1959, Program Abstr., p. 10A

1263. Carruthers, W., and Johnstone, R. A. W., "Composition of a Paraffin Wax Fraction from Tobacco Leaf and Tobacco Smoke," Nature, 184, 1131-1132 (1959); CA 54, 9215c

1264. Carruthers, W., and Johnstone, R. A. W., "Some Phenolic Constituents of Cigarette Smoke," Nature, 185, 762-763 (1960); CA 54, 16753b

1265. Carruthers, W., Johnstone, R. A. W., and Plimmer, J. R., "Gas-Liquid Partition Chromatography of Mixtures of Aryl Methyl Ethers," Chem. & Ind. (London), 1958, 331; CA 52, 12678d

1266. Carson, J. F., Weston, W. J., and Ralls, J. W., "A Rapid Method for Qualitative Analysis of Volatile Mercaptan Mixtures," Nature, 186, 801 (1960); CA 54, 20668c

1267. Carson, J. F., and Wong, F. F., "Gas-Liquid Chromatography of the Volatile Components of Onions," 132nd Natl. ACS Mtg., New York, N. Y., September 1957, Program Abstr., p. 37B; Preprints, Div. Petrol. Chem., 2, No. 4, D115; J. Agr. Food Chem., 9, 140-143 (1961); CA 55, 17944i

1268. Carson, J. F., and Wong, F. F., "Separation of Aliphatic Disulfides and Trisulfides by Gas-Liquid Partition Chromatography," J. Org. Chem., 24, 175-179 (1959); CA 53, 16933f

1269. Carter, Jr., E. H., and Smith, H. A., "Separation of Hydrogen Isotopes by Gas Chromatography," Southeastern Regional ACS Mtg., Gatlinburg, Tenn., November 1962

1270. Carter, H. E., Hendry, R. A., Nojima, S., and Stanacev, N. Z., "Isolation and Structure of Cerebrosides from Wheat Flour," Biochim. Biophys. Acta, 45, 402-404 (1960); CA 55, 14744d

1271. Carter, H. E., Hendry, R. A., Nojima, S., Stanacev, N. Z., and Ohno, K., "Biochemistry of the Sphingolipids. XIII. Determination of the Structure of Cerebrosides from Wheat Flour," J. Biol. Chem., 236, 1912-1916 (1961); CA 56, 691f

1272. Carter, H. E., Ohno, K., Nojima, S., Tipson, C. L., and Stanacev, N. Z., "Wheat Flour Lipids. II. Isolation and Characterization of Glycolipids of Wheat Flour and Other Plant Sources," J. Lipid Res., 2, 215-222 (1961); CA 55, 23862c

1273. Carter, H. V., "Micro-Packed Columns for Gas Chromatography," Nature, 197, 684-685 (1963); CA 58, 8393b

1274. Carter, J. W., "Adsorption Processes," Brit. Chem. Eng., 6, 308-314 (1961); CA 55, 15013d

1275. Cartoni, G. P., and Liberti, A., "Gas Chromatography of Oxygen-Containing Terpenes," J. Chromatog., 3, 121-124 (1960); CA 54, 16254f

1276. Cartoni, G. P., Lowrie, R. S., Phillips, C. S. G., and Venanzi, L. M., "The Use of Some Complexes of the Transition Metals as Column Liquids in Gas Chromatography," Gas Chromatog., Proc. Symposium, 3rd, Edinburgh, 1960, 273-283; CA 56, 10942a

1277. Cartoni, G. P., Lowrie, R. S., Phillips, C. S. G., and Venanzi, L. M., "The Use of Some Complexes of the Transition Metals as Column Liquids in Gas Chromatography," in "Gas Chromatography 1960," edited by R. P. W. Scott, Butterworths, London, 1960, pp. 273-283

1278. Cartwright, P. F. S., and Wilson, D. W., "Analytical Chemistry - Methods of Separation - Chromatography," Ann. Repts. Progr. Chem. (Chem. Soc. London), 57, 422-439 (1960); CA 56, 3c

1279. Carugno, N., and Giovannozzi-Sermanni, G., "Analytical Research on Tobacco by Means of Gas Chromatography. I. Some Gas Components of Cigarett Smoke," Tabacco Il., 62, 255-264 (1958); CA 53, 7516i

1280. Carugno, N., and Giovannozzi-Sermanni, G., "Analytical Research on Tobacco by Means of Gas Chromatography. II. Effect of Fermentation on Smoke Components as Studied on Kentucky Tobacco," Tabacco Il., 62, 265-268 (1958); CA 53, 7516i

1281. Carugno, N., and Giovannozzi-Sermanni, G., "Determination of 3, 4-Benzopyrene and Other Aromatic Hydrocarbons by Gas Chromatography," Tabacco, 63, 285-292 (1959); CA 54, 5345a

1282. Casals, P. F., "Duplicative Reduction of 1-Acetylcycloolefins of the Cyclopentene and Cyclohexene Series. I. General Orientation of the Reduction. Stereospecific Synthesis of ϵ-Diketones Derived from Dicyclohexyl and from Dicyclopentyl," Bull. Soc. Chim. France, 1963, 253-264; CA 59, 471e

1283. Casanova, Jr., J., and Corey, E. J., "Resolution of (±)-Camphor by Gas-Liquid Chromatography," Chem. & Ind. (London), 1961, 1664-1665; CA 56, 7364d

1284. Case, J. R., Ray, N. H., and Roberts, H. L., "Sulfur Chloride Pentafluoride: Reaction with Unsaturated Hydrocarbons," J. Chem. Soc., 1961, 2066-2070; CA 55, 24533b

1285. Case, J. R., Ray, N. H., and Roberts, H. L., "Sulfur Chloride Pentafluoride: Reaction with Fluoro-Olefins," J. Chem. Soc., 1961, 2070-2075; CA 55, 24534a

1286. Case, L. C., "Use of 2-Methylpiperazine Diformamide in Gas Chromatography," J. Chromatog., 5, 181-182 (1961); CA 56, 5384c

1287. Case, L. C., "Separation of Aromatic Hydrocarbons by Gas Chromatography," J. Chromatog., 6, 381-384 (1961); CA 56, 12293b

1288. Caserio, M. C., Graham, W. H., and Roberts, J. D., "Small-Ring Compounds. XXIX. Reinvestigation of the Solvolysis of Cyclopropylcarbinyl Chloride in Aqueous Ethanol. Isomerization of Cyclopropylcarbinol," Tetrahedron, 11, 171-182 (1960); CA 55, 7312g

1289. Casey, K., Edgecombe, F. H. C., and Jardine, D. A., "Automatic Gas-Chromatographic Sampling from Static Systems of Sub-Atmospheric Pressure," Analyst, 87, 835-837 (1962); CA 58, 1895c

1290. Casey, P. S., "Construction of a Gas Chromatographic Apparatus," Proc. Penn. Acad. Sci., 33, 97-101 (1959); CA 54, 7a

1291. Casini, G., and Goodman, L., "Abnormal Direction of Ring-Opening of a 2, 3-Anhydrofuranoside," J. Am. Chem. Soc., 85, 233 (1963); CA 58, 10289g

1292. Cason, J., and Fessenden, R., "Instability of the Butenylcadmium Reagent," J. Org. Chem., 25, 477-478 (1960); CA 54, 20852i

1293. Cason, J., Fessenden, J. S., and Agre, C. L., "Location of a Branch in a Saturated Carbon Chain," Tetrahedron, 7, 289-298 (1959); CA 54, 7541a

1294. Cason, J., and Harris, E. R., "Utilization of Gas Phase Chromatography for Identification of Volatile Products from Alkaline Degradation of Herqueinone," J. Org. Chem., 24, 676-679 (1959); CA 54, 18342i

1295. Cason, J., and Kraus, K. W., "Ketone Synthesis by the Grignard Reaction with Acid Chlorides in Presence of Ferric Chloride," J. Org. Chem., 26, 1768-1772 (1961); CA 55, 23321i

1296. Cason, J., and Kraus, K. W., "Mechanism of the Reactions of Grignard Reagents and of Organocadmium Reagents with Acid Chlorides in the Presence of Ferric Chloride," J. Org. Chem., 26, 1772-1779 (1961); CA 55, 23338e

1297. Cason, J., and Kraus, K. W., "Influence of Stearic Hindrance in the Reaction of Acid Chlorides with Alcohols," J. Org. Chem., 26, 2624-2626 (1961); CA 55, 25748e

1298. Cason, J., Kraus, K. W., and McLeod, Jr., W. D., "The Grignard Reaction with 2(β-Cyanoethyl)-2-ethylhexanol and Further Conversions of the Reaction Products. Limitations of Anisole as Solvent for the Grignard Reaction," J. Org. Chem., 24, 392-397 (1959); CA 54, 9746f

1299. Cason, J., and Miller, W. T., "Versatility and Temperature Range of Silicone Grease as a Partitioning Agent for Gas Chromatography," J. Org. Chem., 24, 1814-1816 (1959); CA 55, 10189b

1300. Cason, J., and Schmitz, F. J., "Reaction of Triphenylacetyl Chloride with Organometallic Reagents. Preparation of Alkyl Trityl Ketones," J. Org. Chem., 25, 1293-1296 (1960); CA 54, 24538f

1301. Cason, J., and Schmitz, F. J., "Epimerization of (S)-Dimethylsuccinic Acids and Derivatives. Cadmium Reactions on the Ester Acid Chlorides," J. Org. Chem., 28, 555-561 (1963); CA 58, 7824g

1302. Cason, J., and Tavs, P., "Separation of Fatty Acids from Tubercle Bacillus by Gas Chromatography. Identification of Oleic Acid," J. Biol. Chem., 234, 1401-1405 (1959); CA 53, 16268b

1303. Cassidy, H. G., "Adsorption and Chromatography," Vol. V of "Technique of Organic Chemistry," edited by A. Weissberger, Interscience Publishers, Inc., New York, 1951

1304. Cassidy, H. G., "The Nature of Chromatography," J. Chem. Educ., 33, 482-485 (1956); 129th Natl. ACS Mtg., Dallas, Texas, April 1956, Program Abstr., p. 13B

1305. Cassidy, H. G., "Fundamentals of Chromatography," Vol. X of "Technique of Organic Chemistry," edited by A. Weissberger, Interscience Publishers, Inc., New York, 1957

1306. Cassil, C. C., "Determination of Thiodan in the Presence of DDT by the Microcoulometer-Gas Chromatograph," 140th Natl. ACS Mtg., Chicago, Ill., September 1961, Program Abstr., p. 62B

1307. Cassil, C. C., "Pesticide Residue Analysis by Microcoulometric Gas Chromatography," Residue Rev., F. Gunther, editor; Academic Press, Inc., New York, 1, 37-65 (1962); CA 58, 7294e

1308. Cassuto, H., "Vapor Chromatography Applied to the Resins and Essential Oils of Hops," Brasserie, 15, 40-45, 64-70 (1960); CA 55, 14809g

1309. Castiglioni, A., "Gas Chromatographic Analysis of Tetralin-Decalin Mixtures," Z. Anal. Chem., 161, 191-192 (1958); CA 53, 131c

1310. Castiglioni, A., "Gas Chromatography of Ethylenediamine-Piperazine Mixtures," Z. Anal. Chem., 182, 428-429 (1961); CA 56, 2873e

1311. Castiglioni, A., "Determination of Acetylpyridines by Gas Chromatography," Rass. Chim., 14, No. 4, 141 (1962); CA 58, 6195f

1312. Casu, B., and Cavallotti, L., "A Simple Device for Qualitative Functional Group Analysis of Gas Chromatographic Effluents," Anal. Chem., 34, 1514-1516 (1962); CA 57, 14412a

1313. Catalette, G., Beaufils, J. P., Gras, B., and Germain, J. E., "Construction and Experimental Study of a Preparative Gas Chromatographic Apparatus," Bull. soc. chim. France, 1961, 786-790; CA 55, 15005i

1314. Cates, V. E., and Meloan, C. E., "Separation of Sulfoxides by Gas Chromatography," Anal. Chem., 35, 658-660 (1963); CA 59, 1065h

1315. Cavagnol, J. C., "Food," Anal. Chem., 33, 50R-61R (1961); CA 55, 12137f

1316. Cavanagh, L., and Coulson, D. M., "The Design-Parameters and Application of a Microcoulometric Gas Chromatograph," 9th Detroit Anachem Conf., Detroit, Mich., October 1961, Program Abstr., p. 26

1317. Cavanagh, L. A., Coulson, D. M., and Devries, J. E., "Equipment for Microcoulometric Gas Chromatography of Pesticides," Proc. of Symposium and Workshop, Instrumental Methods for Analysis of Food Additives, East Lansing, Mich., March 1960, pp. 230-234

1318. Cavill, G. W. K., and Whitfield, F. B., "Synthesis of the Enantiomer of Natural Dolichodial," Proc. Chem. Soc., 1962, 380-381; CA 59, 458g

1319. Caws, A. C., and Foster, G. E., "Purity of Chloroform B. P.," J. Pharm. and Pharmacol., 9, 824-833 (1957); CA 52, 4927c

1320. Cazes, J., and Kauss, J. M., "The Isolation of Tertiary Butyl Biphenyls from a Peroxide Decomposition Reaction Mixture. An Application of Preparative Scale Gas Chromatography," Facts & Methods for Scientific Research (F&M Scientific Corp.), 3, No. 1, 1-2 (Spring 1962)

1321. Cecil, O. B., and Munch, R. H., "Thermal Conductivity of Some Organic Liquids," Ind. Eng. Chem., 48, 437-440 (1956); CA 50, 9130d

1322. Celades, R., and Paquot, C., "Chromatography in the Gaseous Phase and Lipid Chemistry. V. Analysis of the Polyethylene Glycols 200, 300, and 400," Rev. Franc. Corps Gras, 9, 145-149 (1962); CA 57, 3575e

1323. Chadd, G. K., and White, G. B., "New Uses for Gas Chromatography in Process and Product Quality Control," 13th Southwest Regional ACS Mtg., Tulsa, Okla., December 1957

1324. Chaikin, A. M., and Markevich, A. M., "The Determination of the Thermal Conductivities of Gases and Gaseous Mixtures," Zhur. Fiz. Khim., 32, 116-120 (1958); CA 52, 12538b

1325. Chalkley, D. E., "Vapor-Phase Chromatographic Analysis of Hydrocarbon Mixtures," Symp. on Gas Chromatography, Soc. for Anal. Chem., Stevenson, Scotland, May 1955; Abstr., Anal. Chem., 27, 1667 (1955)

1326. Challacombe, J. A., and McNulty, J. A., "Application of the Microcoulometric Titrating System as a Detector in Gas Chromatography of Pesticide Residues," 144th Natl. ACS Mtg., Los Angeles, Calif., March-April 1963, Program Abstr., p. 3A

1327. Chamberlain, J., and Thomas, G. H., "Characterization of Steroid Ketones by Gas Chromatography,"

Biochem. J., <u>86</u>, 3P (1963); CA <u>58</u>, 7095h

1328. Chambers, R. D., Goggin, P., and Musgrave, W. K. R., "The Oxidation of Aromatic Hydrocarbons and Phenols by Trifluoroperoxyacetic Acid," J. Chem. Soc., <u>1959</u>, 1804-1807; CA <u>53</u>, 17038d

1329. Chambliss, K. W., and Nouse, D. C., "Blood Oxygen Determination by Gas Chromatography," Clin. Chem., <u>8</u>, 654-659 (1962); Biol. Abstr., <u>42</u>, 566 (1963)

1330. Chandra, G. R., and Spencer, M., "Ethylene Production by Subcellular Particles from Rat Liver, Rat Intestinal Mucosa, and Penicillium Digitatum," Nature, <u>197</u>, 366-367 (1963); CA <u>58</u>, 9364f

1331. Chang, S. S., "A New Technique for the Isolation of Flavor Components from Fats and Oils," J. Am. Oil Chemists' Soc., <u>38</u>, 669-671 (1961); CA <u>56</u>, 12047b; 52nd Annual Mtg., Am. Oil Chemists' Soc., St. Louis, Mo., April-May 1961

1332. Chang, S. S., Brobst, K. M., Ireland, C. E., and Tai, H., "A Capillary Trap for the Collection of Gas Chromatographic Fractions for Infrared Spectrophotometry," Applied Spectroscopy, <u>16</u>, 106 (1962)

1333. Chang, S. S., Ireland, C. E., and Tai, H., "An Infrared Gas Cell for the Direct Collection of Gas Chromatographic Fractions," Anal. Chem., <u>33</u>, 479 (1961); CA <u>55</u>, 11950i

1334. Chang, T-C. L., and Karr, Jr., C., "Gas-Liquid Chromatographic Analysis of Aromatic Hydrocarbons Boiling Up to 218° in a Low-Temperature Coal Tar," Anal. Chim. Acta, <u>21</u>, 474-490 (1959); CA <u>54</u>, 3086a; 135th Natl. ACS Mtg., Boston, Mass., April 1959, Program Abstr., p. 2J

1335. Chang, T-C. L., and Karr, Jr., C., "Gas-Liquid Chromatographic Analysis of Aromatic Hydrocarbons Boiling Between 202° and 280° in a Low-Temperature Coal Tar," Anal. Chim. Acta, <u>24</u>, 343-356 (1961); CA <u>55</u>, 19201f; 138th Natl. ACS Mtg., New York, N. Y., September 1960, Program Abstr., p. 9K

1336. Chang, T-C. L., and Karr, Jr., C., "Gas-Liquid Chromatographic Analysis of C_{10}-C_{16} n-Paraffins, Isoparaffins and α-Olefins in a Low-Temperature Coal Tar," Anal. Chim. Acta, <u>26</u>, 410-418 (1962)

1337. Chang, T-C. L., and Sweeley, C. C., "Canine Adrenal Polyenoic Acids. Locating Double Bonds by Periodate-Permanganate Oxidation and Gas-Liquid Chromatography," J. Lipid Res., <u>3</u>, 170-176 (1962); CA <u>57</u>, 6262i

1338. Chang, T-C. L., and Sweeley, C. C., "Characterization of Lipids from Canine Adrenal Glands," Biochemistry, <u>2</u>, 592-604 (1963); CA <u>58</u>, 12810b

1339. Chao, T. H., and Cipriani, L. P., "The Chlorination of N, N-Dimethylaniline," J. Org. Chem., <u>26</u>, 1079-1081 (1961); CA <u>55</u>, 19849c

1340. Chapman, A. C., Paddock, N. L., Paine, D. H., Searle, H. T., and Smith, D. R., "Phosphonitrilic Derivatives. III. Cyclic Phosphonitrilic Fluorides," J. Chem. Soc., <u>1960</u>, 3608-3614; CA <u>55</u>, 3263d

1341. Chapman, N. B., Parker, R. E., and Smith, P. J. A., "Conformation and Reactivity. II. Kinetics of the Alkaline Hydrolysis of the Acetates of the Methylcyclohexanols and of Related Alcohols," J. Chem. Soc., <u>1960</u>, 3634-3643; CA <u>55</u>, 3642e

1342. Chapman, R. L., "An Inspection Method for Automobile Hydrocarbon Emission," J. Air Pollution Control Assoc., <u>10</u>, 463-464 (1960); CA <u>55</u>, 8716h

1343. Charles, S. W., and Whittle, E., "The Reactions of Trifluoromethyl Radicals with Aromatic Hydrocarbons. I. Benzene, Toluene, and o-Xylene," Trans. Faraday Soc., <u>56</u>, 794-801 (1960); CA <u>54</u>, 23656g

1344. Chaudet, J. H., Kagan, M. R., and Briden, F. E., "Detection of Gases Through Minority Carrier Lifetime Determinations," 142nd Natl. ACS Mtg., Atlantic City, N. J., September 1962, Program Abstr., p. 8B

1345. Chayen, J., Chayen, R., and Aves, E. K., "The Extraction of Lipid Matter by Formalin," 393rd Mtg., Biochemical Society, London, England, April 1960; Abstr., Biochem. J. Proc., <u>76</u>, 14P (1960)

1346. Chaykovsky, M., and Corey, E. J., "New Reactions of Methylsulfinyl and Methylsulfonyl Carbanion," J. Org. Chem., <u>28</u>, 254-255 (1963); CA <u>58</u>, 10105a

1347. Chemla, M., "The Separation of Isotopes by Chromatography," J. Chromatog., <u>1</u>, 2-23 (1958); CA <u>52</u>, 9791g

1348. Chemodanova, L. S., "Determination of Diethylamine and Triethylamine in the Presence of Ammonia," Gasovaya Kromatografiya, Trudy 1-oi [Pervoi] Vsesoyuz. Konf., Akad. Nauk S.S.S.R., Moscow, <u>1959</u>, 299-301 (Pub. 1960); CA <u>56</u>, 4105i

1349. Chemodanova, L. S., and Turkel'taub, N. M., "Chromatothermographic Method of Determining Benzene, Toluene, Isopentane, Hexane, and Iso-Octane," Zavodskaya Lab., <u>22</u>, 1406-1407 (1956); CA <u>51</u>, 17610i

1350. Chen, C., and Lantz, C. D., "Gas Chromatography of Some Steroid Hormones and Metabolites," Biochem. Biophys. Res. Commun., <u>3</u>, 451-452 (1960); CA <u>55</u>, 15601c

1351. Cheng, L-P., and Hsu, Y-S., "Analysis of C_5-C_6 Saturated Hydrocarbons by Gas-Liquid Chromatography," K'o Hsueh T'ung Pao, No. 3, 89-90 (1958); CA <u>52</u>, 18092i

1352. Cherdron, H., Hohr, L., and Kern, W., "Ampul Technique, A Simple Precision Method for Gas Chromatographic Research," Angew. Chem., 73, 215-218 (1961); CA 55, 19354c

1353. Chernozhukov, N. I., Kazakova, L. P., and Shchegrova, K. A., "Chromatographic Separation of Naphthenic Hydrocarbons from Aromatic Compounds in Heavy Fractions of Crude Oil," Izvest. Vysshikh Ucheb. Zavedenii, Neft i Gaz, 1960, No. 5, 93-104; CA 54, 18944i

1354. Cheshire, J. D., "Gas Chromatography for the Cosmetic and Pharmaceutical Industries," Chemist Druggist, 177, 205-208 (1962); CA 56, 14412g

1355. Cheshire, J. D., and Scott, R. P. W., "Gas-Liquid Chromatography: Effect of Suport Size and Proportion of Liquid Phase on Column Efficiency," J. Inst. Petrol., 44, 74-79 (1958); CA 52, 8681g

1356. Chesick, J. P., "The Kinetics of Thermal Isomerization of Methylcyclopropane," J. Am. Chem. Soc., 82, 3277-3284 (1960); CA 54, 21953d

1357. Chesunov, V. U., "Application of Gas Chromatography in Synthetic Leather Manufacturing. I. Survey of Methods of Analysis of Vapor-Gas Mixtures," Izvest. Vysshikh. Ucheb. Zavedenii Tekhnol. Legkoi. Prom., 1959, No. 6, 34-42; CA 54, 11546c

1358. Chesunov, V. U., "Application of Gas Chromatography in the Production of Synthetic Leather. II. Separation and Analysis of Mixtures of Solvents," Izvest. Vysshikh. Ucheb. Zavedenii Tekhnol. Legkoi. Prom., 1960, No. 2, 31-35; CA 54, 25963c

1359. Childs, C. E., and Henner, E. B., "Determination of Occluded Solvent in Organic Compounds," Chemist-Analyst, 49, 26 (1960); CA 58, 10731h

1360. Chilwell, E. D., and Hughes, D., "Detection of Traces of Some Triazine Herbicides by Gas-Liquid Chromatography," J. Sci. Food Agr., 13, 425-427 (1962); CA 57, 14224a

1361. Ching, T. M., and Ching, K. K., "Fatty Acids in Pollens of Some Coniferous Species," Science, 138, 890-891 (1962); CA 58, 3688f

1362. Chisholm, M. J., and Hopkins, C. Y., "Fatty Acids of the Seed Oil of Cardiospermum Halicacabum," Can. J. Chem., 36, 1537-1540 (1958); CA 53, 7629d

1363. Chisholm, M. J., and Hopkins, C. Y., "11-Octadecenoic Acid and Other Fatty Acids of Asclepias Syriaca Seed Oil," Can. J. Chem., 38, 805-812 (1960); CA 54, 20248b

1364. Chisholm, M. J., and Hopkins, C. Y., "Conjugated Fatty Acids from Tragopogon and Calendula Seed Oils," Can. J. Chem., 38, 2500-2507 (1960); CA 55, 8895b

1365. Chism, Jr., M. R., Scott, W. B., and Karas, R. S., "Gas Chromatograph Using Catalytic Combustion Detection Applied to Oil and Gas Exploration," 13th Annual Instrument Automation Conf., ISA, Philadelphia, Pa., September 1958

1366. Chiu, J., "Kinetic and Thermodynamic Studies of the Hydrolysis and Alcoholysis of Acetals by Gas Chromatography: Determination of Carbon in Organic Substances by an Oxygen-Flask Method," Univ. Microfilms (Ann Arbor, Mich.), L. C. Card No. Mic 61-1601, 132 pp.; Dissertation Abstr., 21, 3608-3609 (1961); CA 55, 22953a

1367. Chlouverakis, C., and Harris, P., "Composition of the Free Fatty Acid Fraction in the Plasma of Human Arterial Blood," Nature, 188, 1111-1112 (1960); CA 55, 11592b

1368. Chmielowski, J., and Isaac, P. C. G., "Gas Chromatographic Observation of the Reduction of Carbon Dioxide to Methane During Anaerobic Digestion," Nature, 183, 1120-1121 (1959); CA 53, 17230f

1369. Chmielowski, J., Simpson, J. R., and Isaac, P. C. G., "Use of Gas Chromatography in Sludge Digestion," Sewage and Ind. Wastes, 31, 1237-1258 (1959); CA 54, 4975f

1370. Chmielowski, J., Simpson, J. R., and Isaac, P. C. G., "Gas Chromatographic Investigation of Methane Production by High Rate Anaerobic Digestion on Laboratory Scale," 17th Intern. Congr. Pure Appl. Chem., Munich, W. Germany, August-September 1959; Abstr., 2, 90-91

1371. Chmutov, K. V. and Avgul, V. T., "Sampling Device for Chromatographic Analysis," Zhur. Fiz. Khim., 31, 724-725 (1957); CA 52, 4e

1372. Chmutov, K. V., and Filatova, N. V., "A Hydrodynamic Model of Sorption Columns," in "Gas Chromatography 1958," edited by D. H. Desty, Academic Press, Inc., New York, 1958, pp. 99-106

1373. Chou, T. C., and Harper, W. J., "Chemical Nature of the Characteristic Flavor of Cultured Buttermilk," 58th Annual Mtg., Am. Dairy Sci. Assoc., Lafayette, Ind., June 1963; Abstr., J. Dairy Sci., 46, 614 (1963)

1374. Chovin, P., "Chromatography in the Gaseous Phase," Bull. soc. chim. France, 1957, 83-101; CA 51, 6273i

1375. Chovin, P., "Gas Phase Chromatography. General Characteristics and Applications in the Field of Detection of Frauds," Ann. fals. et fraudes, 1958, 253-268; CA 53, 3519b

1376. Chovin, P., "Efficiency of Gas Chromatographic Columns," Bull. soc. chim. France, 1958, 905-910; CA 52, 19405h

1377. Chovin, P., "Gas-Phase Chromatography," Parfum. cosmet. savons, 1958, No. 7, 261-269; CA 52, 17623b

1378. Chovin, P., "Recent Progress in Gas Partition Chromatography," Bull. Soc. chim. France, 1960, 755-768; CA 54, 19085i

1379. Chovin, P., "Gas Chromatography. Principles and Applications," Mises au point chim. anal. pure et appl. et anal. bromatol., No. 8, 5-29 (1960); CA 55, 7987b

1380. Chovin, P., "Three New Aspects of Gas-Phase Chromatography," Bull. soc. chim. France, 1961, 875-881; CA 56, 4552a

1381. Chovin, P., "Programmed Temperature Gas-Liquid Chromatography," Ann. Chim. (Paris), 7, 727-743 (1962); CA 58, 8389g

1382. Chovin, P., and Ducros, M., "Gas-Phase Chromatography. Some Properties of Activity Coefficients at Infinite Dilution," Compt. rend., 253, 2352-2354 (1961); CA 56, 10941h

1383. Chovin, P., and Lebbe, J., "Definition and Properties of the Totally Corrected Retention Volume Per Molecule," Compt. rend., 247, 596-599 (1958); CA 53, 21039b

1384. Chovin, P., Lebbe, J., and Moureu, H., "Gas-Phase Chromatography: Unequivocal Identification of Peaks of Halogenated Derivatives in a Complex Chromatogram," J. Chromatog., 6, 363-365 (1961); CA 56, 10887f

1385. Chovin, P., Thirion, B., and Tranchant, J., "A Device for the Introduction of Microvolumes of Sample for Gas Chromatography," Journees Intern. Etude Methodes Separation Chromatog., Paris, 1961, 248-254 (Pub. 1962); CA 59, 1060c

1386. Chrapowa, E. W., Petrova, R. S., and Stscherbakowa, K. D., "The Physical-Chemical Characteristics of the Absorption Processes for Phase Limits with the Help of Gas Chromatography," 4th Intern. Gas Chromatography Symposium, Hamburg, Germany, June 1962

1387. Christakis, G., Hashim, S., Rinzler, S. H., Archer, M., and Van Itallie, T. B., "Effect of a Cholesterol-Lowering Diet on Fatty Acid Composition of Subcutaneous Fat in Man," 16th Annual Mtg., Council on Arteriosclerosis, Am. Soc. for the Study of Arteriosclerosis, Cleveland, Ohio, October, 1962; Abstr., Circulation, 26, No. 4, part 2, 648 (1962)

1388. Christie, P. E., (Ed.), "The Chromatography of Gases and Vapours. XVIII. A Bibliography," British Petroleum Co., Ltd., Research Station, Sunbury-on-Thames, 1958

1389. Christol, H., Mousseron, M., and Plenat, F., "Synthesis of 2,2-Trimethylene Cyclanones. Study of Alkylation," Bull. soc. chim. France, 1959, 543-553; CA 54, 8671g

1390. Christol, H., Plenat, F., and Vedel, M., "Alkylations of Dibromo-1,3-propane. Use of Gas Chromatography," Bull. soc. chim. France, 1958, 920

1391. Christophe, J., and Popjak, G., "Studies on the Biosynthesis of Cholesterol. XIV. The Origin of Prenoic Acids from Allyl Pyrophosphates in Liver Enzyme Systems," J. Lipid Res., 2, 244-257 (1961)

1392. Chu, A., "High-Sensitivity Detectors in Gas Chromatography," Hua Hsueh Tung Pao, 1962, 21-28; CA 58, 5015b

1393. Chu, J., "Quantitative Determination of Volatiles in Photographic Film by Means of Chromatographic Analyses," 14th Annual Instrument Automation Conf. and Exhibit, ISA, Chicago, Ill., September 1959; Preprint No. 82-59

1394. Chumachenko, M. N., and Tverdyukova, L. B., "Microdetermination of Active Hydrogen by Gas Chromatography," Dokl. Akad. Nauk S.S.S.R., 142, 612-614 (1962); CA 57, 50i

1395. Chundela, B., and Janak, J., "Determination of Ethanol in Blood by Means of Gas Chromatography," Casopis lekaru ceskych., 99, 90-95 (1960); CA 54, 25008g

1396. Chundela, B., and Janak, J., "Quantitative Determination of Ethanol and Other Volatile Substances in Blood and Other Body Liquids, by Gas Chromatography," J. Forensic Med., 7, 153-161 (1960); CA 56, 2672f

1397. Chundela, B., Janak, J., Nikolicova, L., and Kacl, K., "Comparison of the Determination of Alcohol in Blood by the Widmark Method and by Gas Chromatography," Acta Univ. Carolinae, Med. Suppl. No. 14, 303-308 (1961); CA 57, 16984e

1398. Chundela, B., and Slechtova, R., "A Gas Chromatographic Proof of Trichlorobutanol in Case of Poisoning by Navyton Spofa," Gas Chromatog. Symp., Brno, Czech., June 1962; Abstr., J. Gas Chromatog., 1, No. 4, 10 (1963)

1399. Cieplinski, E. W., "Prevention of Peak Tailing in the Direct Gas Chromatographic Analysis of Barbiturates," Anal. Chem., 35, 256-257 (1963); CA 58, 7784f

1400. Cieplinski, E. W., and Averill, W., "The Direct Analysis of Atmospheric Pollutants Using the Gas Chromatographic Method," 140th Natl. ACS Mtg., Chicago, Ill., September 1961, Program Abstr., pp. 36B, 10W

1401. Cieplinski, E. W., and Averill, W., "The Gas Chromatographic Analysis of Essential Oils Using Golay Columns and a Flame Ionization Detector," Inst. of Food Technologists Mtg., Miami Beach, Fla., June 1962; Application Bulletin No. GC-AP-002, Perkin-Elmer Corp., Norwalk, Conn.

1402. Cieplinski, E. W., Averill, W., and Ettre, L. S., "Simplified Analysis of Light Gas Mixtures with Gas Chromatography," J. Chromatog., 8, 550-554 (1962); CA 58, 3869b

1403. Cieplinski, E. W., and Ettre, L. S., "Gas-Chromatographic Analysis of Volatile Components in the Presence of Excess Nonvolatiles," J. Chromatog., 4, 169-171 (1960); CA 55, 3269c

1404. Cieplinski, E. W., Kabot, F. J., and Ettre, L. S., "Application of Dual Column Gas Chromatographic Systems for the Solution of Complex Problems," 14th Pittsburgh Conf. on Anal. Chem. & Appl. Spectroscopy, Pittsburgh, Pa., March 1963, Program Abstr., p. 54

1405. Cincotta, J. J., and Feinland, R., "Determination of Polyfunctional Amines, Guanidines, Amidino-guanidines, and Melamines by Gas-Liquid Chromatography," Anal. Chem., 34, 774-776 (1962); Correction, ibid., p. 1222

1406. Ciola, R., "Vapor-Phase Chromatography. II. Study of Some Variables in the Preparation of High-Efficiency Chromatographic Columns," Anais assoc. brasil. quim., 18, 191-210 (1959); CA 54, 23599g

1407. Ciola, R., "Vapor-Phase Chromatography. III. Quantitative Analysis of Refinery Gases by Vapor-Phase Chromatography," Anais assoc. brasil. quim., 19, 35-40 (1960); CA 55, 15898f

1408. Ciola, R., "Vapor-Phase Chromatography. IV. Macro Gas-Chromatographic Separation of α, β-Unsaturated Aldehydes and Preparation of Their 2,4-Dinitrophenylhydrazones," Anais assoc. brasil. quim., 20, 33-43 (1961); CA 58, 3870c

1409. Cirillo, V. A., Skahan, D. J., Hollis, B., and Morgan, H., "Rapid Mass Spectrometric-Gas Chromatographic Analysis of Nonolefinic Naphthas," Anal. Chem., 34, 1353-1354 (1962); CA 57, 15413b

1410. Claessen, J. H., "Development of a Column Assembly for the Analysis of Impurities in Ethylene by Means of Gas Chromatography," 2nd Alberta Gas Chromatography Discussion, Edmonton, Alberta, February 1959

1411. Claesson, S., "Theory for Frontal Analysis," Arkiv. Kemi, Mineral. Geol., A20, No. 3, 14 pp. (1945); CA 41, 1169a

1412. Claesson, S., "Studies on Adsorption and Adsorption Analysis with Special Reference to Homologous Series," Arkiv. Kemi, Mineral. Geol., A23, No. 1, 1-133 (1946); CA 40, 3665⁶

1413. Claesson, S., "Mathematical Characteristics of Adsorption Isotherms in Frontal Analysis of Several Solutes," Arkiv. Kemi, Mineral. Geol., A24, 7 pp. (1946); CA 41, 6834f

1414. Claesson, S., "Frontal Analysis and Displacement Development in Chromatography," Ann. N. Y. Acad. Sci., 49, 183-203 (1948); CA 42, 5300d

1415. Claesson, S., "Theory of Frontal Analysis and Displacement Development," Disc. Faraday Soc., 7, 34-38 (1949); CA 44, 9769g

1416. Claesson, S., "Chromatographic Analysis. High Molecular Polymer Separation," Disc. Faraday Soc., 7, 321-325 (1949); CA 44, 7620b

1417. Clark, I. J., "Analysis of Essential Oils. I. Injection System for a Gas-Liquid Chromatography Apparatus," J. Chromatog., 7, 433-437 (1962); CA 58, 1298h

1418. Clark, J. R., and Bernhard, R. A., "Examination of Lemon Oil by Gas-Liquid Chromatography. II. The Hydrocarbon Fraction," Food Research, 25, 389-394 (1960); CA 56, 3867g

1419. Clark, J.R., and Bernhard, R. A., "Examination of Lemon Oil by Gas-Liquid Chromatography. III. The Oxygenated Fraction," Food Research, 25, 731-738 (1960); CA 56, 3867g

1420. Clark, S. J., "Ionization Detectors," Publication No. 26-750A, Jarrell-Ash Co., Newtonville, Mass.

1421. Clark, S. J., "Appraisal of Some Problems in Quantitative Gas Chromatography," 8th Detroit Anachem. Conf., Detroit, Mich., October 1960

1422. Clark, S. J., "Gas Chromatographic Analysis of Pesticide Residues Using the Electron Affinity Detector," 140th Natl. ACS Mtg., Chicago, Ill., September 1961, Program Abstr., p. 63B

1423. Clark, S. J., "The Cross-Section Ionization Detector," Gas Pipe (Jarrell Ash Co., Newtonville, Mass.), No. 1, 1-3 (January 1963)

1424. Clarke, D. R., "A Quantitative Gas Stream Splitting Injection System Suitable for Use with Capillary Columns," Nature, 198, 681-682 (1963)

1425. Clarke, J. R. P., "Apparatus for Packing Gas-Liquid Chromatography Columns," Chem. & Ind. (London), 1962, 1830-1831; CA 58, 246h

1426. Clarke, R. L., "The Preparation of trans-anti-trans-Perhydroanthracene," J. Am. Chem. Soc., 83, 965-968 (1961); CA 55, 13394c

1427. Claudy, H. N., "Economics and Philosophy of the Application of Gas Chromatography to Process Monitoring and Control," 135th Natl. ACS Mtg., Boston, Mass., April 1959, Program Abstr., p. 19B

1428. Claudy, H. N., "Chromatographic Backflushing for Heavy End Analysis," 5th Natl. ISA Mtg., Symposium on Instrumental Methods of Analysis," Houston, Texas, May 1959

1429. Claudy, H. N., and Ettre, L. S., "Design and Performance of a Novel Hydrocarbon Detector," Summer Instrument Automation Conf., ISA, San Francisco, Calif., May 1960; Preprint No. 20-SF60

1430. Claudy, H. N., Helms, C. C., Scholly, P. R., and Bresky, D. R., "Adapting Gas Chromatography to Automatic Process Stream Analysis," Ann. N. Y. Acad. Sci., 72, 779-785 (1959)

1431. Claudy, H. N., Watson, E. S., Coates, V. J., Kaye, M., and Davis, J. J., "Design and Performance of a Printing Integrator for Gas Chromatography," 10th Pittsburgh Conf. on Anal. Chem. & Appl. Spectroscopy, Pittsburgh, Pa., March 1959, Program Abstr., p. 54

1432. Claxton, G., "Detector for Liquid-Solid Chromatography," J. Chromatog., 2, 136-139 (1959); CA 53, 14600b

1433. Clayton, R. A., and Strong, F. M., "Partition Chromatography of a Homologous Series of Volatile Primary Amines," Anal. Chem., 26, 579-580 (1954); CA 48, 7480f

1434. Clayton, R. B., "Gas-Liquid Chromatography of Sterol Methyl Ethers," Nature, 190, 1071-1072 (1961); CA 56, 4107h

1435. Clayton, R. B., "Group Retention Factors in the Gas-Liquid Chromatography of Steroids," Nature, 192, 524-526 (1961); CA 57, 12b

1436. Clayton, R. B., "Gas-Liquid Chromatography of Sterol Methyl Esters and Some Correlations Between Molecular Structure and Retention Data," Biochemistry, 1, 357-366 (1962); CA 57, 10130b

1437. Clemens, G. F. G., and Gent, C. M. van, "Response of the Beta-Ray Ionization Detector to Unesterified Lower Fatty Acids in Gas-Liquid Chromatography," J. Chromatog., 3, 582-584 (1960); CA 55, 5225d

1438. Clement, G., and Bezard, J., "Determination of [the Composition of] a Mixture of Fatty Acids from Butanoic to Docosanoic by Means of Gas Chromatography," Compt. rend., 253, 564-566 (1961); CA 56, 10307b

1439. Clements, R. L., "Low-Temperature Chromatography as a Means for Separating Terpene Hydrocarbons," Science, 128, 899-900 (1958); CA 53, 5980b

1440. Clemo, G. R., "Chemistry of Cigarette Smoke. I.," Tetrahedron, 3, 168-173 (1958); CA 53, 1642i

1441. Clemo, G. R., "Chemistry of Cigarette Smoke. II.," Tetrahedron, 11, 11-14 (1960); CA 55, 3930g

1442. Clerck, J. de, "The Influence of Essential Oils in Hops on the Flavor of Beer," Brauwissenschaft, 14, 48-50 (1961); CA 55, 15825f

1443. Clifford, J., "Preparative Gas-Liquid Chromatography," Chem. & Ind. (London), 1960, 1595-1596

1444. Clifford, J., "The Determination of Trimethylene Glycol in Glycerol by Gas Chromatography," Analyst, 85, 475-478 (1960); CA 55, 2360i

1445. Clift, T. L., "Considerations in Establishing and Maintaining Automatic Analysis of Process Systems," Ann. N. Y. Acad. Sci., 91, 825-837 (1961); CA 55, 20532e

1446. Closs, G. L., and Schwartz, G. M., "Carbenes from Alkyl Halides and Organolithium Compounds. II. The Reactivity of Chlorocarbene in its Addition to Olefins," J. Am. Chem. Soc., 82, 5729-5731 (1960); CA 55, 6397e

1447. Clough, K. H., "Analysis of Gaseous Mixtures with a New Unit," Petrol. Engr., 27, No. 10, C26-C31 (1955); CA 49, 14574d

1448. Clough, K. H., "High Temperature Chromatography," 13th Southwest Mtg., ACS, Tulsa, Okla., December 1957

1449. Clough, K. H., "Operating Your Gas Chromatograph Unit," 21st Mtg., Gulf Coast Spectroscopic Group, Baton Rouge, La., September 1956; Abstr., Anal. Chem., 29, 166 (1957)

1450. Clough, K. H., "The Foundations and Uses of Gas Chromatography," Rev. soc. quim. Mex., 2, 81-91 (1958); CA 53, 8913g

1451. Clough, K. H., "Gas Chromatography and Refinery Operations," Western Petrol. Refiners Assoc., Tech. 58-1, 4 pp. (1958); CA 54, 6102e

1452. Clough, K. H., "Control of Operating Parameters in Gas Chromatography at Variable Temperatures," Southeastern Regional Mtg., ACS, Gainesville, Fla., December 1958

1453. Coates, J. I., and Glueckauf, E., "Theory of Chromatography. III. Experimental Separation of Two Solutes and Comparison with Theory," J. Chem. Soc., 1947, 1308-1314; CA 42, 2490d

1454. Coates, V.J., and Brenner, N., "Fuel-Gas Analysis by Chromatography?," Petrol. Refiner, 35, No. 11, 197-201 (1956); CA 51, 1587a

1455. Coates, V. J., Brenner, N., O'Brien, L., and Marcus, N., Multiple-Stage Gas Chromatography Instrumentation for the Analysis of Complex Mixtures," 133rd Natl. ACS Mtg., San Francisco, Calif., April 1958, Program Abstr., p. 49B

1456. Coates, V. J., Noebels, N. H., and Fagerson, I. S., (Eds.), "Gas Chromatography," Academic Press, Inc., New York, 1958. A compilation of 28 papers presented at the Intern. Symposium on Gas Chromatography, ISA, East Lansing, Mich., August 1957

1457. Cobb, W. Y., and Patton, S., "Gas Chromatographic Analysis of Ethyl Ether-Extractable Flavor and Odor Compounds from Evaporated Milk," 57th Annual Mtg., Am. Dairy Sci. Assoc., College Park, Md., June 1962; Abstr., J. Dairy Sci., 45, 659 (1962)

1458. Cobler, J. G., and Samsel, E. P., "Determination of Alkyl Cellulose Ethers by Gas Chromatography," 140th Natl. ACS Mtg., Chicago, Ill., September 1961, Program Abstr., p. 51B

1459. Cobler, J. G., and Samsel, E. P., "Gas Chromatography - A New Tool for the Analysis of Plastics," S.P.E. (Soc. Plastics Engrs.) Trans., 2, 145-151 (1962); CA 57, 3607e

1460. Cobler, J. G., Samsel, E. P., and Beaver, G. H., "Determination of Alkyl Cellulose Ethers by Gas Chromatography," Talanta, 9, 473-481 (1962); CA 57, 5307a

1461. Cochrane, C. C., and Harwood, H. J., "The Decomposition of 11-Cyano-12-tricosanone," J. Org. Chem., 26, 2601 (1961); CA 56, 315h

1462. Cocker, W., Dahl, T., and McMurry, T. B. H., "Fatty Acids of the Lugworm, Arenicola Marina," J. Chem. Soc., 1963, 1654-1659; CA 58, 10415a

1463. Cocker, W., Lipman, C., McMurry, T. B. H., and Wheeler, B. M., "The Volatile Oils of Artemisia Brevifolia and Kurramensis," J. Sci. Food & Agr., 9, 828-835 (1958); CA 53, 7515d

1464. Cocker, W., and Shaw, S. J., "Extractives from Woods. III. Extractives from Manilkara Bidentata," J. Chem. Soc., 1963, 677-680

1465. Codegone, C., "Thermal Conductivity of Gases and Vapors," Atti accad. sci. Torino Classe sci. fis. mat. e nat., 86, 288-290 (1951-2); CA 48, 8606h

1466. Codegone, C., "Thermal Conductivity and Thermodynamic Properties of Gases and Vapors," Termotecnica (Milan), 6, 507-511 (1952); CA 48, 8637h

1467. Coe, F. R., and Jenkins, N., "An Improved Carrier-Gas Technique for the Determination of Hydrogen in Steel," Iron Steel Inst. (London), Spec. Report No. 68, 229-235 (1960); CA 55, 10195h

1468. Coe, P. L., Patrick, C. R., and Tatlow, J. C., "Aromatic Polyfluoro-Compounds. V. Preparation of Highly Fluorinated Benzenes by Defluorination of Polyfluorocyclohexanes, -hexenes, and -hexadienes," Tetrahedron, 9, 240-245 (1960); CA 54, 20920i

1469. Coetzee, J. F., Cunningham, G. P., McGuire, D. K., and Padmanabhan, G. R., "Purification of Acetonitrile as a Solvent for Exact Measurements," Anal. Chem., 34, 1139-1143 (1962); CA 57, 10513h

1470. Coffman, J. R., and Schwecke, W. M., "Gas-Liquid Chromatography of Vanillin and Related Substances," in "Gas Chromatography," edited by N. Brenner, J. E. Callen, and M. D. Weiss, Academic Press, Inc., New York, 1962, pp. 471-474

1471. Coffman, J. R., and Schwecke, W. M., "Gas-Liquid Chromatography of Vanillin and Related Substances," Gas Chromatog. Intern. Symp., 1961, 3, 471-474 (Pub. 1962); CA 58, 11896e

1472. Coffman, J. R., and Smith, D. E., "Analysis of Volatile Food Flavors by Gas Liquid Chromatography. III. Separation and Identification of 2,3-Butanediols from Bread Preferment Neutrals," 137th Natl. ACS Mtg., Cleveland, Ohio, April 1960, Program Abstr., p. 33B

1473. Coffman, J. R., Smith, D. E., and Andrews, J. S., "Analysis of Volatile Food Flavors by Gas-Liquid Chromatography. I. The Volatile Components from Dry Blue Cheese and Dry Romano Cheese," Food Research, 25, 663-669 (1960); CA 56, 2740i

1474. Coggeshall, N. D., and Hubis, W., "The Combination of Methods in the Analysis of Complex Hydrocarbon System," 3rd Pacific Area Natl. ASTM Mtg., San Francisco, Calif., October 1959; 7th Detroit Anachem Conf., Detroit, Mich., October 1959

1475. Colard, P., Elphimoff-Felkin, I., and Verrier, M., "Relative Stability of Some Isomeric α-Hydroxyketones, Containing a Phenyl Conjugated or Not Conjugated with the Carbonyl Group. Factors Stabilizing Non-Conjugated Hydroxyketones," Bull. Soc. chim. France, 1961, 516-520; CA 56, 7199f

1476. Colburn, C. B., "The Use of Vapor Phase Chromatography in Kinetic Investigations. The Thermal Decomposition of Neopentyl Nitrate," 129th Natl. ACS Mtg., Dallas, Texas, April 1956, Program Abstr., p. 20B

1477. Cole, B. T., "Modification of an Ionization Chamber Gas-Liquid Chromatography Instrument," Anal. Chem., 33, 317-318 (1961); CA 55, 20530d

1478. Cole, D. D., Harper, W. J., and Hankinson, C. L., "Observations on Ammonia and Volatile Amines in Milk," J. Dairy Sci., 44, 171-172 (1961); CA 55, 9710b

1479. Coleman, H. J., Thompson, C. J., Hopkins, R. L., Foster, N. G., Whisman, M. L., and Richardson, D. M., "Identification of Benzo[b]thiophene and the 2- and 3-Methyl Homologs in Wasson, Texas, Crude Oil," J. Chem. Eng. Data, 6, 464-468 (1961); CA 56, 3717d

1480. Coleman, H. J., Thompson, C. J., Hopkins, R. L., and Rall, H. T., "Identifying Individual Sulfur Compounds by Using a Series of Gas-Liquid Chromatographic Stationary Liquids," Regional Mtg., ACS Southeast and Southwest Sections, New Orleans, La., December 1961

1481. Coleman, H. J., Thompson, C. J., Ward, C. C., and Rall, H. T., "Identification of Low-Boiling Sulfur Compounds in Agha Jari Crude Oil by Gas-Liquid Chromatography," Anal. Chem., 30, 1592-1594 (1958); CA 53, 7569c; 133rd Natl. ACS Mtg., San Francisco, Calif., April 1958, Program Abstr., p. 14P

1482. Coleman, M. H., and Fulton, W. C., "Structural Investigation of Natural Fats by the Partial Hydrolysis Technique," Proc. Intern. Conf. Biochem. Lipids, 6th, Marseilles, 1960, 127-137 (Pub. 1961); CA 58, 12776g

1483. Collins, C. H., and Hammond, G. S., "The Steric Course of Hydration of 1,2-Dimethylcyclohexene," J. Org. Chem., 25, 911-913 (1960); CA 54, 20917b

1484. Collins, R. P., and Morgan, M. E., "Esters Produced by Chalaropsis Thielavioides," Science, 131, 933-934 (1960); CA 54, 15517g

1485. Colson, E. R., "A Partition Sampler for Vapor Analysis by Gas Chromatography," Anal. Chem., 35, 1111-1112 (1963)

1486. Comings, E. W., Lee, W. B., and Kramer, F. R., "A Cylindrical Thermal Conductivity Cell for Gases at Pressures to 3000 Atmosphere," Proc. Conf. Thermodynamic and Transport Properties Fluids, London, 1957, 188-192; CA 53, 4828c

1487. Condon, R. D., "Wide Range Analysis with the Golay Capillary Chromatograph," 135th Natl. ACS Mtg., Boston, Mass., April 1959, Program Abstr., p. 20B

1488. Condon, R. D., "Design Considerations of a Gas Chromatography System Employing High Efficiency Golay Columns," Anal. Chem., 31, 1717-1722 (1959); CA 54, 2831h; 10th Pittsburgh Conf. on Anal. Chem. & Appl. Spectroscopy, Pittsburgh, Pa., March 1959, Program Abstr., p. 58

1489. Condon, R. D., "Recent Advances in Golay Column Technology," 1960 Eastern Analytical Symposium, New York, N. Y., November 1960

1490. Condon, R. D., "Capillary Columns," 2nd Gas Chromatography Institute, Canisius College, Buffalo, N. Y., April 1960

1491. Condon, R. D., "Dual Column Chromatography Utilizing Ionization Detection," 13th Pittsburgh Conf. on Anal. Chem. & Appl. Spectroscopy, Pittsburgh, Pa., March 1962, Program Abstr., p. 46

1492. Condon, R. D., Claudy, H. N., and Scholly, P. R., "Quantitative Aspects of the Flame Ionization Detector," 11th Pittsburgh Conf. on Anal. Chem. & Appl. Spectroscopy, Pittsburgh, Pa., February-March 1960, Program Abstr., p. 42

1493. Condon, R. D., Scholly, P. R., and Averill, W., "Comparative Data on Two Ionization Detectors," Gas Chromatog. Proc. Symposium, 3rd, Edinburgh, 1960, 30-45; CA 56, 15312f

1494. Condon, R. D., Scholly, P. R., and Averill, W., "Comparative Data on Two Ionization Detectors," in "Gas Chromatography 1960," edited by R. P. W. Scott, Butterworths, London, 1960, pp. 30-45

1495. Conia, J. M., and LeCraz, A., "Alkylation of Ketones by Means of Sodium tert-Amylate. IX. 6. Orientation in Alkylation of 3-Unsubstituted-2-cyclohexenones," Bull. soc. chim. France, 1960, 1934-1937; CA 57, 3310f

1496. Conia, J. M., and Rouessac, F., "Cycloalkylation. I. New Method of Synthesis of Decalones," Tetrahedron, 16, 45-58 (1961); CA 57, 694h

1497. Conner, A. Z., "Gas Chromatography," Hercules Chemist, 36, 6-10 (1959)

1498. Cook, C. D., Elgood, E. J., Shaw, G. C., and Solomon, D. H., "Gas Chromatographic Analysis of High Boiling Point Plasticizers Using a Short Column," Anal. Chem., 34, 1177-1178 (1962); CA 57, 12710h

1499. Cook, J. G. H., Riley, C., Nunn, R. F., and Budgen, D. E., "Gas Chromatography of Methyl Derivatives of Some Barbiturates," J. Chromatog., 6, 182-185 (1961); CA 56, 7429h

1500. Cook, J. W., "Assay of Insecticides and Herbicides in Fats and Oils," J. Am. Oil Chemists' Soc., 40, 313-318 (1963)

1501. Cooke, N. J., and Hansen, R. P., "Tall Oil. Isolation and Identification of n-Heptadecanoic Acid," Chem. & Ind. (London), 1959, 1516-1517; CA 54, 25804i

1502. Cooke, W. D., "The Use of Glass Microbead Supports for Separation of Higher Molecular Weight Compounds at Lowered Temperatures," 1960 Eastern Analytical Symposium, New York, N. Y., November 1960

1503. Cooke, W. D., "Gas Chromatography of Solid Organic Compounds," Joint Mtg. of Am. Assoc. of Clin. Chemists and Am. Assoc. Adv. of Science, December 1960; Abstr., Clin. Chem., 7, 307 (1961)

1504. Cooke, W. D., "Use of Lightly Loaded Columns in Gas Chromatography," 2nd Symposium on Gas Chromatography, Toronto Section, CIC, Toronto, Ont., February 1962

1505. Cooke, W. D., "Recent Developments in Column Technology," 4th Annual Gas Chromatography Institute, Canisius College, Buffalo, N. Y., April 1962

1506. Cooper, G. D., "Preparation and Thermal Rearrangement of Poly(trimethylsilyl)phenols," J. Org. Chem., 26, 925-929 (1961); CA 55, 24619g

1507. Cooper, J. A., Abbott, J. P., Rosengreen, B. K., and Claggett, W. R., "Gas Chromatography of Urinary Steroids. I. A Preliminary Report on the Demonstration and Identification of Pregnanediol in Pregnancy Urine by Means of Gas Chromatography," Am. J. Clin. Pathol., 38, 388-389 (1962); CA 58, 4800a

1508. Cooper, J. A., Canter, R., Estes, F. L., and Gast, J. H., "Apparatus for the Equilibration of Columns Prior to Use in Vapor-Phase Chromatography," J. Chromatog., 3, 87-90 (1960); CA 54, 16038a

1509. Cooper, J. A., and Creech, B. G., "Urinary 17-Ketosteroids," Joint Mtg. of Am. Assoc. of Clin. Chemists and Am. Assoc. Adv. of Science, December 1960; Abstr., Clin. Chem., 7, 306 (1961)

1510. Cooper, J. A., and Creech, B. G., "The Application of Gas-Liquid Chromatography to the Analysis of 17-Keto Steroids," Anal. Biochem., 2, 502-506 (1961); CA 56, 6284i

1511. Cope, A. C., and Acton, E. M., "Amine Oxides. V. Olefins from N,N-Dimethylmenthylamine and N,N-Dimethylneomenthylamine Oxides," J. Am. Chem. Soc., 80, 355-359 (1958); CA 52, 10006g

1512. Cope, A. C., Ambros, D., Ciganek, E., Howell, C. F., and Jacura, Z., "Acid-catalyzed Equilibrium of Endocyclic and Exocyclic Olefins," J. Am. Chem. Soc., 82, 1750-1753 (1960); CA 54, 22412g

1513. Cope, A. C., Berchtold, G. A., and Ross, D. L., "1-Phenylcyclohexene from trans-2-Phenylcyclohexyltrimethylammonium Hydroxide by cis Elimination," J. Am. Chem. Soc., 83, 3859-3861 (1961); CA 56, 4643i

1514. Cope, A. C., Bly, R. K., Burrows, E. P., Ceder, O. J., Ciganek, E., Gillis, B. T., Porter, R. F., and Johnson, H. E., "Fungichromin: Complete Structure and Absolute Configuration at C_{26} and C_{27},"

J. Am. Chem. Soc., 84, 2170-2178 (1962); CA 57, 7099e

1515. Cope, A. C., Bumgardner, C. L., and Schweizer, E. E., "Amine Oxides. IV. Alicyclic Olefins from Amine Oxides and Quaternary Ammonium Hydroxides," J. Am. Chem. Soc., 79, 4729-4733 (1957); CA 52, 2771c

1516. Cope, A. C., and Burton, P. E., "Proximity Effects. XX. Search for Transannular Reactions in Carbon-Substituted Cyclooctane Derivatives," J. Am. Chem. Soc., 82, 5439-5445 (1960); CA 55, 7317e

1517. Cope, A. C., Ciganek, E., Howell, C. F., and Schweizer, E. E., "Amine Oxides. VIII. Medium Sized Cyclic Olefins from Amine Oxides and Quaternary Ammonium Hydroxides," J. Am. Chem. Soc., 82, 4663-4669 (1960); CA 55, 7432a

1518. Cope, A. C., Ciganek, E., and Lazar, J., "Amine Oxides. X. Thermal Decomposition of the N-Oxides and Methohydroxides of cis- and trans-N, N-Dimethyl-2-aminocyclohexanol and cis- and trans-N, N-Dimethyl-2-aminocyclooctanol," J. Am. Chem. Soc., 84, 2591-2596 (1962); CA 57, 7121a

1519. Cope, A. C., Ciganek, E., and LeBel, N. A., "Amine Oxides. VI. The Formation of 2, 5-Norbornadiene and 2-Norbornene from 5-Dimethylamino-2-norbornene and 2-Dimethylaminonorbornane," J. Am. Chem. Soc., 81, 2799-2804 (1959); CA 54, 6581h

1520. Cope, A. C., LeBel, N. A., Lee, H. H., and Moore, W. R., "Amine Oxides. III. Selective Formation of Olefins from Unsymmetrical Amine Oxides and Quaternary Ammonium Hydroxides," J. Am. Chem. Soc., 79, 4720-4729 (1957); CA 52, 2769h

1521. Cope, A. C., LeBel, N. A., Moore, P. T., and Moore, W. R., "Mechanism of the Hofmann Elimination Reaction: Evidence That an Ylide Intermediate Is Not Involved in Simple Compounds," J. Am. Chem. Soc., 83, 3861-3865 (1961); CA 56, 4589h

1522. Cope, A. C., Moore, P. T., and Moore, W. R., "Equilibration of cis- and trans-Cycloalkenes," J. Am. Chem. Soc., 82, 1744-1749 (1960); CA 54, 14886a

1523. Cope, A. C., and Ross, D. L., "Amine Oxides. IX. Comparison of the Elimination Reactions Forming Olefins from the N-Oxides and Methylhydroxides of Amines of the Type $R_2CHCH_2N(CH_3)_2$," J. Am. Chem. Soc., 83, 3854-3858 (1961); CA 56, 7109f

1524. Coppens, L., and Bricteux, J., "Gas Chromatographic Determination of Carbon Monoxide," Inst. natl. ind. charbonniere, Bull. tech. Houille et derives, 1959, No. 15, 487-506; CA 54, 3044d

1525. Coppens, L., and Bricteux, J., "The Fractional Simultaneous Determination of Carbon Dioxide and Carbon Monoxide," Inst. natl. ind. charbonniere, Bull. tech. Houille et derives, 1959, No. 17, 524-543; CA 54, 7423b

1526. Coppens, L., Bricteux, J., and Venter, J., "Fractometric Determination of Methane," Bull. tech.-houille (Liege), 1958, No. 13, 426-446; CA 54, 24131h

1527. Coppens, L., Neuray, M., and Bricteux, J., "Low-Temperature Tars," Inst. natl. ind. charbonniere, Bull. tech. Houille et derives, 1960, No. 21, 591-606; CA 55, 15888h

1528. Coppens, L., Venter, J., and Bricteux, J., "Notes on Gas Chromatography," Inst. natl. ind. charbonniere, Bull. tech. Houille et derives, 1956, 311-334; CA 51, 5386g

1529. Coppock, J. B. M., and Daniels, N. W. R., "Lipids in Flour and Their Participation in Oxidative Reactions During Dough Preparation," Brot Gebaeck, 16, No. 6, 117-119 (1962); CA 58, 3824g

1530. Coppock, J. B. M., Daniels, N. W. R., Blount, W. P., and Fox, S., "'Essential' Fatty Acid Content of Hen's Eggs," Lancet, 1961-I, 117-118; CA 55, 7573h

1531. Coppock, J. B. M., Daniels, N. W. R., and Eggitt, P. W. R., "Essential Fatty Acid Retention in Flour Treatment," Chem. & Ind. (London), 1960, 17-18; CA 54, 16683b

1532. Coppock, J. B. M., Fisher, N., and Ritchie, M. L., "The Role of Lipides in Baking. V. Chromatographic and Other Studies," J. Sci. Food Agr., 9, 498-505 (1958); CA 52, 20717d

1533. Corbin, J. R., "Developments in Chromatography," 31st Annual Mtg., California Natural Gas Assoc., Los Angeles, Calif., November 1956, Proc., 3-5; cf. Oil Gas J., 54, 127-128 (1956)

1534. Corbin, J. R., "Gas Chromatography: New Developments - 2," Oil Gas J., 54, No. 85, 127-128 (1956)

1535. Corbin, J. R., "Application of Gas Chromatography to Process Monitoring," Western Petrol. Refiners Assoc. Tech. 58-4, 4 pp. (1958); CA 54, 6102e

1536. Corbin, J. R., and Coates, V. J., "The Qualitative and Quantitative Analysis of Multicomponent Mixtures by Vapor Fractometry," 7th Pittsburgh Conf. on Anal. Chem. & Appl. Spectroscopy, Pittsburgh, Pa., February-March 1956, Program Abstr., p. 41

1537. Cordon, J. L. M., and Lopez, G. Z., "Separation of Volatile Organic Compounds by Vapor-Phase Chromatography," Combustibles, 16, 65-75 (1956)

1538. Cornforth, J. W., and James, A. T., "Structure of a Naturally Occurring Antagonist of Dihydrostreptomycin," Biochem. J., 63, 124-130 (1956); CA 50, 12169d

1539. Corse, J., and Dimick, K. P., "The Volatile Flavors of Strawberry," in "Flavor Research and Food Acceptance," Reinhold Publishing Corp., New York, 1958, pp. 302-314

1540. Corse, J., and Teranishi, R., "Stabilization of Polyester Stationary Phases for Gas-Liquid Chromatography," J. Lipid Res., 1, 191-192 (1960); CA 54, 16989g

1541. Cotabish, H. N., McConhaughey, P. W., and Mosser, H. C., "Making Known Concentrations for Instrument Calibration," Am. Ind. Hyg. Assoc. J., 22, 392-403 (1961); CA 55, 26553e

1542. Cotter, J. L., and Maley, L. E., "Selection of Components for Gas Analyzer Sampling Systems," Chem. Eng. Progr., 55, No. 5, 122-124 (1959); CA 53, 14602g

1543. Cotter, R. J., Sauers, C. K., and Whelan, J. M., "The Synthesis of B-Substituted Isomaleimides," J. Org. Chem., 26, 10-15 (1961); CA 55, 19782i

1544. Coull, J., Engel, H. C., and Miller, J., "A New Technique for Adsorption Studies," Ind. Eng. Chem., Anal. Ed., 14, 459-462 (1942); CA 31, 4747²

1545. Coulson, D. M., "Pesticide Residues on Fresh Vegetables," Stanford Research Inst., Rept. No. 3, Tech. Rept. No. 1, May 1958

1546. Coulson, D. M., "Instrumentation for Gas Chromatography of Pesticides," 18th Intern. Congr. of Pure and Applied Chemistry, Montreal, Canada, August 1961

1547. Coulson, D. M., "Theory and Equipment for Microcoulometric Gas Chromatography," 140th Natl. ACS Mtg., Chicago, Ill., September 1961, Program Abstr., p. 62B

1548. Coulson, D. M., "Gas Chromatography of Pesticides," in "Advances in Pest Control Research," edited by R. L. Metcalf, Interscience Publishers, Inc., New York, 5, 153-190 (1962); CA 58, 10671b

1549. Coulson, D. M., and Cavanagh, L. A., "Microcoulometric Detection in Gas Chromatography," 12th Pittsburgh Conf. on Anal. Chem. & Appl. Spectroscopy, Pittsburgh, Pa., February-March 1961, Program Abstr., p. 55

1550. Coulson, D. M., Cavanagh, L. A., DeVries, J. E., and Walther, B., "Microcoulometric Gas Chromatography of Pesticides," Proc. of Symposium and Workshop, Instrumental Methods for the Analysis of Food Additives, East Lansing, Mich., March 1960, pp. 219-229

1551. Coulson, D. M., Cavanagh, L. A., DeVries, J. E., and Walther, B., "Microcoulometric Gas Chromatography of Pesticides," J. Agr. Food Chem., 8, 399-402 (1960); CA 55, 17997f

1552. Coulson, D. M., Cavanagh, L. A., McCarthy, E. M., Salas, L. J., and Wilton, V. B., "Improvements in Microcoulometric Gas Chromatography," 13th Pittsburgh Conf. on Anal. Chem. & Appl. Spectroscopy, Pittsburgh, Pa., March 1962, Program Abstr., p. 46

1553. Coulson, D. M., Cavanagh, L. A., and Stuart, J., "Gas Chromatography of Pesticides," J. Agr. Food Chem., 7, 250-251 (1959); CA 53, 19280d

1554. Courtenay, S. G. P., "Gas Chromatography," in "Physical Methods of Chemical Analysis," Vol. III, edited by W. G. Berl, Academic Press, Inc., New York, 1956, pp. 1-28

1555. Cowan, C. B., and Stirling, P. H., "The Selection and Operation of Thermistors for Katharometers," in "Gas Chromatography," edited by V. J. Coates, H. J. Noebels, and I. S. Fagerson, Academic Press, Inc., New York, 1958, pp. 165-190

1556. Cowan, C. T., and Hartwell, J. M., "An Organo-Clay Complex for the Separation of Isomeric Dichlorobenzenes Using Gas Chromatography," Nature, 190, 712 (1961); CA 9g

1557. Cowan, P. J., and Sugihara, J. M., "A Gas Chromatography Demonstration Apparatus," J. Chem. Educ., 36, 246-247 (1959); CA 53, 19475h

1558. Cowley, B. R., Norman, R. O. C., and Waters, W. A., "A Quantitative Study of Homolytic Methylation of Some Monosubstituted Benzenes," J. Chem. Soc., 1959, 1799-1803; CA 53, 21729b

1559. Cox, J. S. G., High, I. B., and Jones, E. R. H., "The Determination of Steroid Side-Chains," Proc. Chem. Soc. (London), 1958, 234-235; CA 53, 6292d

1560. Coxon, R. V., Bainster, P. G., and Kay, R. H., "The Katharometer in Gas Analysis," Rec. trav. chim., 74, 513-517 (1955); CA 49, 16034d

1561. Craats, F. van de, "Application of Vapor Phase Chromatography in the Gas-Analytical Field," Anal. Chim. Acta, 14, 136-149 (1956); CA 51, 949a

1562. Craats, F. van de, "Some Quantitative Aspects of the Chromatographic Analysis of Gas Mixtures, Using Thermal Conductivity as Detection Method," in "Gas Chromatography 1958," edited by D. H. Desty, Academic Press, Inc., New York, 1958, pp. 248-264

1563. Craig, B. M., "Varietal and Environmental Effects on Rapeseed. III. Fatty Acid Composition of 1958 Varietal Tests," Can. J. Plant Sci., 41, 204-210 (1961); cf. ibid., 39, 437-443 (1959); CA 58, 8152f

1564. Craig, B. M., "Use of Ortho-phthalic-ethylene Glycol Polyester in Gas Liquid-Chromatographic Analysis of Fatty Acid Esters," Chem. & Ind. (London), 1960, 1442; CA 55, 22110b

1565. Craig, B. M., "Some Applications of Gas-Liquid Chromatography in Research on Fats and Oils," Can. Food Inds., 31, No. 7, 41-44 (1960); CA 54, 23373d

1566. Craig, B. M., "A Study of Solid Supports, Liquid Phases and Their Interaction," 3rd Intern. Symposium on Gas Chromatography, ISA, East Lansing, Mich., June 1961; ISA Proc. 3, 27-39 (1961)

1567. Craig, B. M., "A Study of Solid Supports, Liquid Phases and Their Interaction," in "Gas Chromatography," edited by N. Brenner, J. E. Callen, and M. D. Weiss, Academic Press, Inc., New York, 1962, pp. 37-56

1568. Craig, B. M., "Study of Solid Supports, Liquid Phases, and Their Interaction," Gas Chromatog., Intern. Symp., 3, 37-56 (1961) (Pub. 1962); CA 58, 8394g

1569. Craig, B. M., Mallard, T. M., and Hoffman, L. L., "Collection Unit for Gas-Liquid Chromatography Under Reduced Pressure," Anal. Chem., 31, 319-320 (1959); CA 53, 10856c

1570. Craig, B. M., and Murty, N. L., "The Separation of Saturated and Unsaturated Fatty Acid Esters by Gas-Liquid Chromatography," Can. J. Chem., 36, 1297-1301 (1958); CA 53, 2647f

1571. Craig, B. M., and Murty, N. L., "Quantitative Fatty Acid Analysis of Vegetable Oils by Gas-Liquid Chromatography," J. Am. Oil Chemists' Soc., 36, 549-552 (1959); CA 54, 927g

1572. Craig, B. M., Tulloch, A. P., and Murty, N. L., "Quantitative Determination of Short Chain (C_3-C_9) Fatty Acids by Gas-Liquid Chromatography," 33rd Fall Mtg., Am. Oil Chemists Soc., Los Angeles, Calif., September 1959; J. Am. Oil Chemists' Soc., 40, 61-63 (1963); CA 58, 8349f

1573. Craig, B. M., Youngs, C. G., Beare, J. L., and Campbell, J. A., "Fatty Acid Composition and Glyceride Structure in Rats Fed Rapesed Oil or Corn Oil," Can. J. Biochem. Physiol., 41, 43-49 (1963); CA 58, 9461d

1574. Craig, D., and Fowler, R. B., "Deutero-1,3-butadienes Derived by Reductive Dechlorination," J. Org. Chem., 26, 713-716 (1961); CA 55, 22089g

1575. Craig, D., Regenass, F. A., and Fowler, R. B., " Isoprene-d_8 and Other Intermediates for the Synthesis of Deuterio-SN Rubber," J. Org. Chem., 24, 240-244 (1959); CA 53, 16930a

1576. Craig, L. C., "Identification of Small Amounts of Organic Compounds by Distribution Studies. II. Separation by Countercurrent Distribution," J. Biol. Chem., 155, 519-534 (1944); CA 39, 1372[5]

1577. Craig, L. C., and Craig, D., "Laboratory Extraction and Countercurrent Distribution," in "Technique of Organic Chemistry," Vol. III, Part I - Separation and Purification, 2nd ed., edited by A. Weissberger, Interscience Publishers, Inc., New York, 1956

1578. Craig, N. C., and Enteman, E. A., "Thermodynamics of cis-trans Isomerizations. The 1,2-Difluoroethylenes," J. Am. Chem. Soc., 83, 3047-3050 (1961); CA 56, 1003b

1579. Cram, D. J., and Rickborn, B., "Electrophilic Substitution at Saturated Carbon. IX. Stereochemistry at Second Carbon," J. Am. Chem. Soc., 83, 2178-2183 (1961); CA 55, 24669g

1580. Cram, W. W., Abrahamovitch, R. A., and Pepper, J. M., "Separation of the Components of Lignite Tar," 2nd Alberta Gas Chromatography Discussion, Edmonton, Alberta, February 1959

1581. Craven, E. C., and Ward, W. R., "Phorone and Isomeric Forms," J. Appl. Chem. (London), 10, 18-23 (1960); CA 54, 11979f

1582. Crawford, H. M., "Testing Stream-Analyzer Applications,,' Control Eng., 6, No. 9, 200 (1959)

1583. Crawford, R. V., "Gas Chromatography for the Routine Analysis of Fatty Oils," Chem. & Ind. (London), 1960, 68-69

1584. Creech, W., "Gas Chromatographic Clinical Analysis of Steroids," Biochemical Gas Chromatography Seminar, Wilkens Instrument & Research, Inc., New Orleans, La., July 1963

1585. Cremer, E., "Outlines of Gas Chromatography," in "Gas Chromatographie 1958," edited by H. P. Angele, Akademie Verlag, Berlin, 1959, pp. 1-14

1586. Cremer, E., "Fundamental Aspects of Gas Chromatography," Abhandl. Deut. Akad. Wiss. Berlin, Kl. Chem., Geol., Biol., 1959, No. 9, 1-14; CA 58, 5018a

1587. Cremer, E., "Determination of Physical Constants from Chromatograms," in "Gas Chromatographie 1959," edited by R. E. Kaiser and H. G. Struppe, Akademie Verlag, Berlin, 1959, pp. 80-84

1588. Cremer, E., "Physicochemical Measurements by Gas Chromatography," Z. Anal. Chem., 170, 219-232 (1959); CA 54, 2879e

1589. Cremer, E., "Gas Adsorption Chromatography," Arch. Biochem. Biophys., 83, 345-349 (1959); CA 53, 21359a

1590. Cremer, E., "New Developments in the Field of Gas Chromatography," Angew. Chem., 71, 457-458 (1959)

1591. Cremer, E., "Testing of Adsorbents and Catalysts by Means of Gas Chromatography," Angew. Chem., 71, 512-514 (1959); CA 54, 5226c

1592. Cremer, E., "Physico-Chemical Measurement With Help of Gas Chromatography," German Chemical Society Mtg., Div. of Anal. Chem., Freiberg, West Germany, April 1959

1593. Cremer, E., "Microdetermination of Adsorption Isotherms by Gas Chromatography," Monats, 92, 112-115 (1961); CA 55, 16073h

1594. Cremer, E., and Bechtold, E., "Characterization of Band Asymmetry in Gas Chromatography," Z. Anal. Chem., 189, 78-80 (1962); CA 57, 13166d

1595. Cremer, E., and Haupt, R., "Gas Chromatography of Very Small Quantities (Microgram Range)," Angew. Chem., 70, 310-311 (1958); CA 52, 16115g

1596. Cremer, E., and Hausdorff, H. H., "Chromatographic Separation and Determination of Gases," Chem.-Ing.-Tech., 28, 805 (1956)

1597. Cremer, E., and Huber, H. F., "Measurement of Adsorption Isotherms at High Temperature with the Help of Gas-Solid Elution Chromatography," Angew. Chem., 73, 461-465 (1961); CA 56, 39d

1598. Cremer, E., and Huber, H. F., "Measurement of Adsorption Isotherms by Means of High-Temperature Elution Gas Chromatography," Gas Chromatog. Intern. Symp., 3, 169-182 (1961) (Pub.1962); CA 58, 6219g

1599. Cremer, E., and Huber, H. F., "Measurement of Adsorption Isotherms by Means of High Temperature Elution Gas Chromatography," in "Gas Chromatography," edited by N. Brenner, J. E.
Callen, and M. D. Weiss, Academic Press, Inc., New York, 1962, pp. 169-182

1600. Cremer, E., and Muller, R., "Separation and Determination of Substances by Chromatography in the
Gas Phase," Z. Elektrochem., 55, 217-220 (1951); CA 45, 9335a

1601. Cremer, E., and Muller, R., "Separation and Determination of Small Quantities of Gases by Chromatography," Mikrochemie ver. Mikrochim. Acta, 36/37, 553-560 (1951); CA 45, 5057h

1602. Cremer, E., and Prior, F., "Application of Chromatographic Methods of the Separation of Gases and
Determination of Adsorption Energies," Z. Elektrochem., 55, 66-70 (1951); CA 45, 9334h

1603. Cremer, E., and Roselius, L., "Gas Chromatography," Angew. Chem., 70, 42-50 (1958)

1604. Crespi, V., and Cevolani, F., "Gas-Chromatographic Analysis of Hydrocarbons from C_1 to C_4 by
Means of Magnesium Silicate," Chim. e ind. (Milan), 41, 215-217 (1959); CA 53, 14840e

1605. Crippen, R. C., "Use of the Gas Chromatograph in Detection of Volatile Substances in Body Fluids:
I. Oxygenated Compounds," 137th Natl. ACS Mtg., Cleveland, Ohio, April 1960, Program Abstr.,
p. 36B

1606. Crippen, R. C., and Emmerling, J., "Solvent Studies. II. Gas Phase Chromatographic Examination
of Chlorinated Solvents," Offic. Dig. Federation Soc. Paint Technol., 32, No. 430, 1517-1521
(1960); cf. ibid., 30, 1172 (1958); CA 57, 4790i; 38th Annual Mtg., Soc. Paint Tech., Chicago, Ill.,
October-November 1960

1607. Crippen, R. C., and Freimuth, H., "The Determination of Acetylsalicylic Acid (Aspirin) by Gas
Chromatography," 142nd Natl. ACS Mtg., Atlantic City, N. J., September 1962, Program Abstr.,
p. 10B

1608. Crisler, R. O., and Benford, C. L., "Analysis of Ionones and Methylionones by Gas-Liquid Partition Chromatography," Anal. Chem., 31, 1516-1518 (1959); CA 53, 21358i

1609. Croitoru, P. P., and Freedman, R. W., "Quantitative Gas Liquid Chromatography of Thiocresols and
Thioxylenols as Acetate Esters," Anal. Chem., 34, 1536-1537 (1962); CA 58, 2827e

1610. Crombie, L., and Griffin, B. F., "Lipids. VII. Synthesis of 8-Hydroxyoctadec-cis-11- and -trans-
11-en-9-ynoic Acid. The Status of Natural 8-Hydroxyximenynic Acid," J. Chem. Soc., 1958, 4435-
4444; CA 53, 10022d

1611. Cronan, C. S., "Analyzers Unlock Process Control," Chem. Eng., 63, No. 12, 252, 254, 256, 258
(1956)

1612. Cropper, F. R., and Heywood, A., "Analytical Separation of the Methyl Esters of the C_{12}-C_{22} Fatty
Acids by Vapour-Phase Chromatography," Nature, 172, 1101-1102 (1953); CA 48, 3203c

1613. Cropper, F. R., and Heywood, A., "Improvements in Vapour-Phase Chromatography at Relatively
High Temperature," Nature, 174, 1063-1064 (1954); CA 49, 5197f

1614. Cropper, F. R., and Heywood, A., "The Analysis of Fatty Acids and Fatty Alcohols by Vapour-Phase
Chromatography," in "Vapour-Phase Chromatography," edited by D. H. Desty, Academic Press,
Inc., New York, 1957, pp. 316-331

1615. Cropper, F. R., and Kaminsky, S., "Determination of Toxic Organic Compounds in Admixture in the
Atmosphere by Gas Chromatography," Anal. Chem., 35, 735-743 (1963); CA 59, 2094a

1616. Crouse, R. H., Garner, J. W., and O'Neill, H. J., "Determination of Phenolic Constituents of Cigarette Smoke by Gas Chromatography," J. Gas Chromatog., 1, No. 6, 18-22 (1963)

1617. Crowther, S., Fulton, J. D., and Joyner, L. P., "The Metabolism of Leishmania Donovani in Cul-
ture," Biochem. J., 56, 182-185 (1954); CA 48, 5926h

1618. Crum, W. M., "Design and Application of a Small Volume Plug Valve for Gas Chromatography," 7th
Natl. Symp. on Instrumental Methods of Analysis, ISA, Houston, Texas, April 1961

1619. Crum, W. M., "Some Applications of Multiple Column Configurations in Process Gas Chromatography," 8th Natl. Analysis Instrumentation Symp., ISA, Charleston, W. Va., April-May 1962

1620. Crumb, J. W., "The cis-trans Isomerization of Some Simple Ethylene Derivatives," J. Org. Chem.,
28, 953-956 (1963)

1621. Crump, G. B., "The Analysis of Edible Oils Contaminated with Synthetic Ester Lubricants," Analyst,
88, 456-463 (1963)

1622. Csicsery, S. M., "Methylcyclopentadiene Isomers," J. Org. Chem., 25, 518-521 (1960); CA 54,
18385c

1623. Csicsery, S. M., and Pines, H., "Relative Retention Times of C_1-C_4 Hydrocarbons Over Different
Columns and at Different Temperatures," J. Chromatog., 9, 34-43 (1962); CA 58, 5021h

1624. Culberson, C. F., and Wilder, Jr., P., "The Synthesis of 2-aza-1,2-Dihydrodicyclopentadienes,"
J. Org. Chem., 25, 1358-1362 (1960); CA 54, 24728i

1625. Cull, N. L., and Brenner, H. H., "Applying Nonlinear Regression to Kinetics of Hexane Isomerization," Ind. Eng. Chem., 53, 833-836 (1961); CA 56, 3332h

1626. Cullis, C. F., Fish, A., Hardy, F. R. F., and Warwicker, E. A., "Estimation of Combustion Products by Gas Chromatography," Chem. & Ind. (London), 1961, 1158-1159; CA 56, 3718g

1627. Cullis, C. F., Fish, A., and Turner, D. W., "The Gaseous Oxidation of 2-Methyl-2-butene. I. Kinetic and Analytical Studies," Proc. Roy. Soc. (London), A262, 318-327 (1961); CA 55, 24524f

1628. Cullis, C. F., Hardy, F. R. F., and Turner, D. W., "Point of Oxygen Attack in the Combustion of Hydrocarbons. II. Formation and Origin of Ketones," Proc. Roy. Soc. (London), A251, 265-273 (1959); CA 54, 17011f

1629. Cundall, R. B., and Palmer, T. F., "The Photosensitized Isomerization of 2-Butene," Trans. Faraday Soc., 56, 1211-1224 (1960); CA 55, 5097d

1630. Curren, W. J., and McIntyre, E. A., "The Practical Aspects of Column Design," 13th Pittsburgh Conf. on Anal. Chem. & Appl. Spectroscopy, Pittsburgh, Pa., March 1962, Program Abstr., p. 46

1631. Curry, A. S., Hurst, G., Kent, N. R., and Powell, H., "Rapid Screening of Blood Samples for Volatile Poisons by Gas Chromatography," Nature, 195, 603-604 (1962); CA 57, 12811a

1632. Curtin, D. Y., and Fraser, R. R., "Synthesis of Cyclohexadienones," Chem. & Ind. (London), 1957, 1358; CA 52, 7165f

1633. Curugno, N., "Determination of Paraffin Waxes of Tobacco and Tobacco Smoke by Gas-Liquid Chromatography," Natl. Cancer Inst. Monograph No. 9, 171-181 (1962); CA 58, 2662e

1634. Cuthbertson, F., and Musgrave, W. K. R., "1 : 1 Dihalocyclohexanes," J. Appl. Chem. (London), 7, 99-104 (1957); CA 51, 12834f

1635. Cvejanovich, G. J., "Neopentane and Cyclobutane in Western Venezuelan Crude Oils," J. Chem. Eng. Data, 4, 170-173 (1959); CA 54, 872i

1636. Cvejanovich, G. J., "Separation and Analysis of Mixtures of Hydrocarbon and Inorganic Gases by Gas Chromatography," Anal. Chem., 34, 654-657 (1962); CA 57, 1534f

1637. Cvetanovic, R. J., "Molecular Rearrangements in the Reactions of Oxygen Atoms with Olefins," Can. J. Chem., 36, 623-634 (1958); CA 52, 15415e

1638. Cvetanovic, R. J., and Doyle, L. C., "Reaction of Oxygen Atoms with Butadiene," Can.J. Chem., 38, 2187-2195 (1960)

1639. Cvetanovic, R. J., and Kutschke, K. O., "Micro-Vapour-Phase Chromatography. Effect of Column Temperature," in "Vapour Phase Chromatography," edited by D. H. Desty, Academic Press, Inc., New York, 1957, pp. 87-97

1640. Cvrkal, H., "Biochemical Diagnosis of Pine Trees in Smoky Areas," Sbornik Ceskoslov. akad. zemedel ved. Lesnictvi, 5, 1033-1048 (1959); CA 54, 9213h

1641. Cvrkal, H., and Janak, J., "Identification of Some Terpenes in Essential Oils of Conifers by Gas Chromatography," Collection Czechoslov. Chem. Communs., 24, 1967-1974 (1959); CA 53, 15489i

1642. Cymerman-Craig, C., and Horning, E. C., "Preparation of Esters by Hemiacetal Oxidation," J. Org. Chem., 25, 2098-2102 (1960); CA 55, 13289d

1643. Dabney, III, W. T., "Fatty Acid Composition of Adipose Tissue in Patients with Carcinoma of the Breast," J. Nat. Cancer Inst., 27, 25-28 (1961); CA 55, 23783d

1644. Dahmen, E. A. M. F., "Methods for the Analysis of Substances Containing Acidic Groups: Identification of Homologous Members of a Certain Group of Acids," Chimie Anal., 40, 430-434 (1958); CA 53, 5963g

1645. Dahmen, E. A. M. F., and Van der Larse, J. D., "Analysis of Products Containing Cyclopentadiene, with Special Reference to Gas Chromatography," Z. Anal. Chem., 164, 37-48 (1958); CA 53, 6914d

1646. Dailey, R. E., Swell, L., Field, Jr., H., and Treadwell, C. R., "Adrenal Cholesterol Ester Fatty Acid Composition of Different Species," Proc. Soc. Exptl. Biol. Med., 105, 4-6 (1960); CA 55, 2836i

1647. Dal, V. I., and Nabivach, V. M., "Analysis and Separation of Hydrocarbons of the Benzene Series by Gas-Liquid Chromatography," Uspekhi Khim., 29, 1353-1361 (1960); CA 55, 6262e

1648. Dal, V. I., and Nabivach, V. M., "The Use of Benzoic Anhydride as the Stationary Phase in Gas-Liquid Chromatography," Khim. i Tekhnol, Topliv i Masel, 6, No. 10, 51-54 (1961); CA 56,10938i

1649. Dal, V. I., Raskina, L. S., Nabivach, V. M., and Martsinkevich, L. E., "The Composition of Gas Condensate from the Shebelinsk Formation and Its Catalytic Cracking Products," Izvest. Vysshikh Ucheb. Zavedenii Neft i Gaz, 1961, No. 10, 59-63; CA 56, 7582b

1650. Dal Nogare, S., "Gas Chromatography," Anal. Chem., 32, 19R-25R (April 1960); CA 54, 11804f

1651. Dal Nogare, S., "Introduction to Trace Analysis by Gas Chromatography," 13th Summer Symp. on Anal. Chem., University of Houston, Houston, Texas, June 1960; cf. Anal. Chem., 32, No. 9, 51A-54A (August 1960)

1652. Dal Nogare, S., "Resolution and Efficiency in Gas Chromatography," 3rd Annual Eastern Analytical Symposium & Instrument Exhibit, New York, N. Y., November 1961

1653. Dal Nogare, S., "Programmed Temperature Gas Chromatography," 14th Annual Fisk University Infrared Spectroscopy Institute, Gas Chromatography Session, Nashville, Tenn., August 1963

1654. Dal Nogare, S., and Bennett, C. E., "Programmed Temperature Gas Chromatography,"Anal. Chem., 30, 1157-1158 (1958); CA 52, 16972i

1655. Dal Nogare, S., Bennett, C. E., and Harden, J. C., "A Simple Electromechanical Integrator," in "Gas Chromatography," edited by V. J. Coates, H. J. Noebels, and I. S. Fagerson, Academic Press, Inc., New York, 1958, pp. 117-129

1656. Dal Nogare, S., and Chiu, J., "Study of Packed Column Efficiency," 2nd Symp. on Gas Chromatography, Toronto Section, CIC, Toronto, Ontario, February 1962

1657. Dal Nogare, S., and Chiu, J., "A Study of the Performance of Packed Gas Chromatography Columns," Anal. Chem., 34, 890-896 (1962); CA 58, 921b

1658. Dal Nogare, S., and Harden, J. C., "Programmed Temperature Gas Chromatography Apparatus," Anal. Chem., 31, 1829-1832 (1959); CA 54, 23460e

1659. Dal Nogare, S., and Juvet, Jr., R. S., "Gas-Liquid Chromatography. Theory and Practice," Interscience Publishers, Inc., New York, 1962

1660. Dal Nogare, S., and Juvet, Jr., R. S., "Gas Chromatography," Anal. Chem., 34, No. 5, 35R-47R (April 1962); CA 56, 13520e

1661. Dal Nogare, S., and Langlois, W. E., "Programmed Temperature Gas Chromatography," Anal. Chem., 32, 767-770 (1960); CA 54, 19087a

1662. Dal Nogare, S., and Safranski, L. W., "Gas Chromatography to 350°C.," Anal. Chem., 29, No. 3, 23A, 26A, 28A (1957)

1663. Dal Nogare, S., and Safranski, L. W., "High Temperature Gas Chromatography Apparatus," Anal. Chem., 30, 894-898 (1958); CA 52, 13326g

1664. Dal Nogare, S., and Safranski, L. W., "Analytical Separations," J. Chem. Educ., 35, 14-17 (1958); CA 52, 4381c

1665. Dal Nogare, S., and Safranski, L. W., "Gas Chromatography," in "Organic Analysis," Vol. 4, edited by J. Mitchell, Jr., I. M. Kolthoff, E. S. Proskauer, and A. Weissberger, Interscience Publishers, Inc., New York, 1960, pp. 91-227

1666. Daly, J. W., Green, F. C., and Eastman, R. H., "Subinene Hydrate. A Constituent of American Peppermint Oil," J. Am. Chem. Soc., 80, 6330-6336 (1958); CA 53, 16195b

1667. Dan, T., and Oshima, S., "Analysis of Petroleum Hydrocarbons by Mass Spectroscopy," Bull. Japan Petrol. Inst., 2, 25-32 (1960); CA 55, 957h

1668. Danby, C. J., and Freeman, G. R., "The Thermal Decomposition of Diethyl Ether. II. Analytical Survey of the Reaction Products as a Function of Reaction Conditions," Proc. Roy. Soc. (London), A245, 40-48 (1958); CA 53, 191c

1669. Daneman, H. L., "Interpretation of Gas-Liquid Chromatograms from the Standpoint of the Recorder," 4th Delaware Valley Regional Mtg., ACS, Philadelphia, Pa., January 1962

1670. Daneman, H. L., and Ross, D. E., "The Significance of Recorder Characteristics to Gas Chromatography," 2nd Annual Gas Chromatography Institute, Canisius College, Buffalo, N. Y., April 1960

1671. Daneman, H. L., and Talbot, G. S., "The Application of Recorders to Gas Chromatography," 12th Pittsburgh Conf. on Anal. Chem. & Appl. Spectroscopy, Pittsburgh, Pa., February-March 1961, Program Abstr., p. 52

1672. Daniels, N. W. R., and Richmond, J. W., "Gas-Liquid Chromatography of Conjugated Fatty Acids," Nature, 187, 55-56 (1960); CA 55, 8895e

1673. Daniels, N. W. R., and Richmond, J. W., "Improvements of Column Efficiency in Gas-Liquid Chromatography," Chem. & Ind. (London), 1961, 1441-1442; CA 56, 2867d

1674. Darby, P. W., and Kemball, C., "Investigation of Reactions Along the Catalyst Bed in Flow Systems by Vapour-Phase Chromatography. I. Decomposition of Methanol on a Cobalt Fisher-Tropsch Catalyst," Trans. Faraday Soc., 53, 832-840 (1957); CA 52, 7121f

1675. Darby, P. W., and Kemball, C., "Observations on the Fischer-Tropsch Synthesis Over a Cobalt Catalyst at Low Pressures Using Gas Chromatography," Trans. Faraday Soc., 55, 833-841 (1959); CA 54, 6508g

1676. Darley, E. F., Kettner, K. A., and Stephens, E. R., "Analysis of Peroxyacyl Nitrates by Gas Chromatography with Electron Capture Detection," Anal. Chem., 35, 589-591 (1963)

1677. Darling, D. J., Miller, F. D., Bartsch, R. C., and Trent, F. M., "Automatic Range Changer for Beckman GC-2 Gas Chromatographs," Anal. Chem., 32, 144 (1960); CA 54, 23461a

1678. Dart, M. C., and Henbest, H. B., "Aspects of Stereochemistry. XV. Catalytic Hydrogenation of Cyclic Allylic Alcohols in the Presence of Sodium Nitrite," J. Chem. Soc., 1960, 3563-3570; CA 55, 2733i

1679. Datskevich, A. A., Zhukhovitskii, A. A., and Turkel'taub, N. M., "Sorption-Thermal Instruments for the Analysis of Gas Mixtures," Ind. Lab., 25, 222-224 (1959); Zavodskaya Lab., 25, 210 (1959); CA 54, 19034g

1680. Datta, P. R., and Susi, H., "Gas Chromatographic Separation of Oxygen-Containing Terpene Compounds on Low Temperature Columns," Anal. Chem., 34, 1028-1029 (1962); CA 57, 5306i

1681. Datta, P. R., Susi, H., Higman, H. C., and Filipic, V. J., "Use of Gas Chromatography to Identify Geographical Origin of Some Spices," Food Technol., 16, 116-119 (1962); CA 58, 1299b

1682. Dauben, W. G., and Bozak, R. E., "Lithium Aluminum Hydride Reduction of Methylcyclohexanones," J. Org. Chem., 24, 1596-1597 (1959); CA 54, 6581c

1683. Dauben, W. G., and Cargill, R. L., "Photochemical Transformations. VI. Isomerization of Cycloheptadiene and Cycloheptatriene," Tetrahedron, 12, 186-189 (1961); CA 55, 15372h

1684. Dauchy, S., and Asselineau, J., "Fatty Acids in the Lipids of Escherichia Coli. Existence of a $C_{17}H_{32}O_2$ Acid Containing a Cyclopropane Ring," Compt. rend., 250, 2635-2640 (1960); CA 55, 16670d

1685. Davenport, J. B., "Studies in the Natural Coating of Apples. V. Unsaturated and Minor Saturated Acids of the Cuticle Oil," Australian J. Chem., 13, 411-415 (1960); CA 55, 12559h

1686. Davidson, J. M., and Music, J. F., "Experimental Thermal Conductivities of Gases and Gaseous Mixtures at Zero Degrees Centigrade," U. S. At. Energy Comm., HW-29201, 7-30 (1953); CA 48, 5581a

1687. Davidson, L. V., Eanes, R. D., Eynon, J. U., and Callahan, J. A., "An Electronic Integrator for Use in Gas Chromatography Measurements," 12th Pittsburgh Conf. on Anal. Chem. & Appl. Spectroscopy, Pittsburgh, Pa., February-March 1961, Program Abstr., p. 52

1688. Davies, A. J., and Johnson, J. K., "A Thermal Conductivity Detector for Use at High Temperatures in Vapour-Phase Chromatography," in "Vapour Phase Chromatography," edited by D. H. Desty, Academic Press, Inc., New York, 1957, pp. 185-193

1689. Davies, D. I., "The Reaction of Lead Tetraacetate with Chlorobenzene and Ethyl Benzoate," J. Chem. Soc., 1963, 2351-2354

1690. Davies, G. R., "Modern Analytical Chemistry in Industry," Nature, 180, 366-368 (1957)

1691. Davies, V., and Boltz, D. F., "Gas Chromatographic Investigation of Certain Chlorinated Hydrocarbon Mixtures," 10th Detroit Anachem Conf., Detroit, Mich., October 1962, Program Abstr., p. 33

1692. Davis, A., "Introduction to Gas Chromatography," 2nd Annual Gas Chromatography Institute, Canisius College, Buffalo, N. Y., April 1960

1693. Davis, A., "Basic Units of a Gas Chromatograph, Simple Theory of Columns, Detectors, and Sampling," 4th Annual Gas Chromatography Institute, Canisius College, Buffalo, N. Y., April 1962

1694. Davis, A., "Introduction to Practical Gas Chromatography," 5th Annual Gas Chromatography Institute, Canisius College, Buffalo, N. Y., April 1963

1695. Davis, A., Roaldi, A., Michalovic, J. G., and Joseph, H. M., "Applications of Gas Chromatography to Phosphorus Containing Compounds," 14th Annual Mid-America Spectroscopy Symposium, Chicago, Ill, May 1963; pub. in "Developments in Applied Spectroscopy," Vol. 3, Plenum Press, New York, 1964, pp. 386-391

1696. Davis, A. D., and Howard, G. A., "The Use of Thermistors in Gas Chromatography," Chem. & Ind. Brit. Inds. Fair Rev., April 1956, R25-R26; CA 50, 15135f

1697. Davis, A. D., and Howard, G. A., "Thermistor Detectors in Gas Chromatography," J. Appl. Chem. (London), 8, 183-186 (1958); CA 52, 12464b

1698. Davis, C. E., and Riggs, W. A., "An Electronic Integrator-Digitizer for Gas Chromatography," 13th Pittsburgh Conf. on Anal. Chem. & Appl. Spectroscopy, Pittsburgh, Pa., March 1962, Program Abstr., p. 47

1699. Davis, D. S., "Nomogram for Gas Chromatography," Chem. & Process Eng., 41, 418 (1960); CA 54, 23598i

1700. Davis, J. J., "Compatible Readout System Design for Use with a Chromatographic Instrument Employing Golay Columns and Ionization Detectors," Gas Chromatog. Intern. Symposium, 2nd, East Lansing, Mich., 1959, 85-90 (Pub. 1960); CA 55, 20530g

1701. Davis, J. J., "Compatible Readout System Design for Use with a Chromatographic Instrument Employing Golay Columns and Ionization Detectors" in "Gas Chromatography," edited by H. J. Noebels, R. F. Wall, and N. Brenner, Academic Press, Inc., New York, 1961, pp. 85-90

1702. Davis, R. E., and McCrea, J. M., "Liquid Sample Inlet System for Gas Chromatographs," Anal. Chem., 29, 1114-1115 (1957); CA 51, 13473i

1703. Davis, R. E., and Schreiber, R. A., "Double-Column Gas Chromatography: Analysis of Noncondensable and Light Hydrocarbon Gases by a Combined Gas-Liquid, Gas-Solid Chromatograph," 132nd Natl. ACS Mtg., New York, N. Y., September 1957, Program Abstr., p. 34B; Preprints, Div. Petrol. Chem., 2, No. 4, D91-D95 (1957)

1704. Davis, T. W. M., and Farmilo, C., "The Assay of Marihuana by Gas and Paper Chromatography and Its Use in Determination of Origin of Narcotic Seizures," 141st Natl. ACS Mtg., Washington, D.C., March 1962, Program Abstr., p. 11B

1705. Davis, T. W. M., Farmilo, C. G., Martin, L., and Lane, R., "Gas Chromatography of Some High-Boiling Compounds with Particular Reference to the Fatty Acids and Opium," Proc. Can. Soc. Forensic Sci., 1, Paper 17, 8 pp. (1962); CA 58, 4374b

1706. Davis, T. W. M., and Farmilo, C. G., with Osadchuk, M., "Identification and Origin Determination of Cannabis by Gas and Paper Chromatography," Anal. Chem., 35, 751-755 (1963); CA 59, 1956b

1707. Davison, W. H. T., Slaney, S., and Wragg, A. L., "A Novel Method of Identification of Polymers," Chem. & Ind. (London), 1954, 1356; CA 49, 3736h

1708. Dawson, Jr., H. J., "Determination of Lead Alkyls by Electron Capture," Univ. of Houston Intern. Symp. on Advances in Gas Chromatography, Houston, Texas, January 1963

1709. Dawson, Jr., H. J., "Determination of Methyl-Ethyl Lead Alkyls in Gasoline by Gas Chromatography with an Electron Capture Detector," Anal. Chem., 35, 542-545 (1963)

1710. Dawson, Jr., H. J., and Schmauch, L. J., "Application of Vapor-Phase Partition Chromatography to the Analysis of Catalytic Reformates," 129th Natl ACS Mtg., Dallas, Texas, April 1956, Program Abstr., p. 17B

1711. Day, E. A., Forss, D. A., and Patton, S., "Flavor and Odor Defects of Gamma-Irradiated Skim Milk. I. Preliminary Observations and the Role of Volatile Carbonyl Compounds," J. Dairy Sci., 40, 922-931 (1957); CA 51, 18366f

1712. Day, E. A., Forss, D. A., and Patton, S., "Flavor and Odor Defects of Gamma-Irradiated Skim Milk. II. Identification of Volatile Components by Gas Chromatography and Mass Spectrometry," J. Dairy Sci., 40, 932-941 (1957); CA 51, 18366f

1713. Day, E. A., Larsen, P. B., Lindsay, R. C., and Elliker, P. R., "Some Observations on the Volatile Flavor Compounds of Ripened Cream Butter," 57th Annual Mtg., Am. Dairy Sci. Assoc., College Park, Md., June 1962; Abstr., J. Dairy Sci., 45, 660 (1962)

1714. Day, E. A., and Lindsay, R. C., "Methyl Sulfide and the Flavor of Butter," 58th Annual Mtg., Am. Dairy Sci. Assoc., Lafayette, Ind., June 1963; Abstr., J. Dairy Sci., 46, 615-616 (1963)

1715. Day, E. A., and Miller, P. H., "Decomposition of Oxygenated Terpenes in the Injection Heaters of Gas Chromatographs," Anal. Chem., 34, 869-870 (1962); CA 57, 6039i

1716. Day, E. A., and Papaionnou, S. E., "Radiation-Induced Changes in Milk Fat," 58th Annual Mtg., Am. Dairy Sci. Assoc., Lafayette, Ind., June 1963; Abstr., J. Dairy Sci., 46, 595 (1963)

1717. Day, P., "Gas Chromatography," Research (London), 11, No. 1, 39-43 (1958)

1718. Daynes, H. A., "Gas Analysis by Measurement of Thermal Conductivity," Cambridge University Press, London, 1933

1719. Dayton, S., Hashimoto, S., and Jessamy, J., "Cholesterol Kinetics in the Normal Rat Aorta and the Influence of Different Types of Dietary Fat," J. Atherosclerosis Res., 1, 444-460 (1961); CA 57, 3852a

1720. Deady, L. W., Topsom, R. D., and Vaughan, J., "Preparation of Some Chromans from 1,3-Diaryloxypropanes," J. Chem. Soc., 1963, 2094-2095

1721. Deal, C. H., "Gas-Liquid Partition Chromatography," Symp. on Recent Developments in Research Methods and Instrumentation, Bethesda, Md., May 1956; Abstr., Anal. Chem., 28, 1058 (1956)

1722. Deal, C. H., Otvos, J. W., Smith, V. N., and Zucco, P. S., "A Radiological Detector for Gas Chromatography," Anal. Chem., 28, 1958-1964 (1956); CA 51, 2335b; 129th Natl. ACS Mtg., Dallas, Texas, April 1956, Program Abstr., p. 16B

1723. DeAngelis, G., Ippoliti, P., and Spina, N., "Analysis of Polymers by Vapor Phase Chromatography," Ricerca sci., 28, 1444-1450 (1958); CA 53, 4803d

1724. Deavours, M. F., "Gas Chromatography and Interpretation of Chromatograms," 37th Annual Mtg., Natural Gasoline Assoc. of Am., Dallas, Texas, April 1958, Tech. Papers, 37, 22-24 (1958); CA 54, 1834b

1725. Debbrecht, F. J., "Factors Affecting Substrate Bleeding in Isothermal and Programmed Temperature Gas Chromatography," 12th Pittsburgh Conf. on Anal. Chem. & Appl. Spectroscopy, Pittsburgh, Pa., February-March 1961, Program Abstr., p. 59

1726. DeBoer, F. E., "Purification of Metals by Gas Chromatography," Nature, 185, 915 (1960); CA 55, 14163a

1727. Debuch, H., "Fatty Acids from Chloroplasts," Z. Naturforsch., 16b, 246-248 (1961); CA 56, 718e

1728. Decora, A. W., and Dinneen, G. U., "Gas-Liquid Chromatography of Pyridines Using a New Solid Support. Selectivity of Eleven Liquid Substrates," 134th Natl. ACS Mtg., Chicago, Ill., September 1958, Program Abstr., p. 24B

1729. Decora, A. W., and Dinneen, G. U., "Gas-Liquid Chromatography of Pyridines Using a New Solid Support," Anal. Chem., 32, 164-169 (1960); CA 54, 7405h

1730. Decora, A. W., and Dinneen, G. U., "A Solid Support for the Gas-Liquid Chromatography of Strongly Basic Nitrogen Compounds," Gas Chromatog. Intern. Symposium, 2nd, East Lansing, Mich., 1959, 33-38 (Pub. 1960); CA 55, 20530e

1731. Decora, A. W., and Dinneen, G. U., "A Solid Support for the Gas-Liquid Chromatography of Strongly Basic Nitrogen Compounds," in "Gas Chromatography," edited by H. J. Noebels, R. F. Wall, and N. Brenner, Academic Press, Inc., New York, 1961, pp. 33-38

1732. Decora, A. W., and Dinneen, G. U., "Gas-Liquid Chromatography of Basic Nitrogen Compounds," U. S. Bur. Mines Rept. Invest. No. 5768, 23 pp. (1961); CA 55, 18085e

1733. Deemter, J. J. van, "Theory and Experiment," in "Gas Chromatography 1958," edited by D. H. Desty, Academic Press, Inc., New York, 1958, pp. 3-5

1734. Deemter, J. J. van, Zuiderweg, F. J., and Klinkenberg, A., "Longitudianl Diffusion and Resistance to Mass Transfer as a Cause of Non-Ideality in Chromatography," Chem. Eng. Sci., 5, 271-289 (1956); CA 51, 801g

1735. DeFord, D. D., "Recent Advances in Column Theory," 14th Annual Mid-America Spectroscopy Symp.,

Chicago, Ill., May 1963

1736. DeFord, D. D., "Studies on the Efficiency of Packed Gas Chromatographic Columns," Univ. of Houston Intern. Symp. on Advances in Gas Chromatography, Houston, Texas, January 1963

1737. DeFord, D. D., Ayers, B. O., and Loyd, R. J., "Principles of High-Speed Gas Chromatography," 137th Natl. ACS Mtg., Cleveland, Ohio, April 1960, Program Abstr., p. 23B

1738. DeFord, D. D., with Ayers, B. O., and Loyd, R. J., "Minimization of Time in Gas Chromatographic Separations," Anal. Chem., 32, 1711-1712 (1960)

1739. DeFord, D. D., Loyd, R. J., and Ayers, B. O., "Studies on the Efficiencies of Packed Gas Chromatographic Columns," Anal. Chem., 35, 426-429 (1963)

1740. DeFrancesco, F., and Avancini, D., "Quantitative Gas Chromatography of Methyl Esters of Fatty Acids. II.," Riv. ital. sostanze grasse, 38, 128-131 (1961); cf. Olii minerali, grasse e saponi, colori e vernici, 37, 479-486 (1960); CA 55, 29293f

1741. DeFrancesco, F., and Avancini, D., "Fraudulent Addition to Butter of Transesterified Fats," Boll. Lab. Chim. Provinciali, 12, 422-444 (1961); CA 57, 7686d

1742. DeFrancesco, F., and Avancini, D., "Quantitative Gas Chromatography. V. Determination of Fatty Acids," Boll. Lab. Chim. Provinciali (Bologna), 13, 447-455 (1962); CA 59, 1096g

1743. DeFrancesco, F., Avancini, D., Maglitto, C., and Gandini, C., "Quantitative Gas Chromatography. III. Composition of the Acids in Butterfat," Riv. ital. sostanze grasse, 38, 307-314 (1961); CA 56, 5170g

1744. Deinema, M. H., and Landheer, C. A. "Extracellular Lipid Production by a Strain of Rhodotorula Graminis," Biochim. et Biophys. Acta, 37, 178-179 (1960); CA 54, 11146g

1745. Deisler, P. F., McHenry, Jr., K. W., and Wilhelm, R. H., "Rapid Gas Analyzer Using Ionization by Alpha Particles," Anal. Chem., 27, 1366-1374 (1955); CA 50, 4i

1746. De La Mere, H. E., and Rust, F. F., "Intramolecular Radical Reactions. Decomposition of Pure Bis(2-methyl-2-hexyl)peroxide in the Liquid Phase, J. Am. Chem. Soc., 81, 2691-2694 (1959); CA 54, 3172g

1747. DeMan, J. M., "Gas Chromatography of Short-Chain Fatty Acids of Milk Fats," Milchwissenschaft., 16, 245-247 (1961); CA 57, 10305f

1748. Dement'eva, M. I., Dobychin, D. P., and Shefter, V. E., "Use of Large-Pore Glass for Gas-Liquid Chromatography," Zh. Fiz. Khim., 36, 228-229 (1962); CA 58, 7571h

1749. Demole, E., "Uses of Adsorption Microchromatography on Thin Layers," J. Chromatog., 1, 24-34 (1958); CA 52, 9714d

1750. Denekas, M. O., Dunton, M. L., and Daniel, N. R., "Use of Preparative Gas Chromatography. Modified FIA Apparatus, and a Unique Chemical Reaction for Analysis of Individual Hydrocarbons in Crude Oils," Oklahoma Tetrasectional ACS Mtg., Tulsa, Okla., March 1963

1751. Denis, J., and Parc, G., "A Bibliographic Survey of Analytical Methods for Hydrocarbons in Atmospheric Pollution," Rev. Inst. Franc. Petrole Ann. Cimbust. Liquides, 17, 1473-1507 (1962); CA 58, 11889a

1752. Denney, D. B., and Boskin, M. J., "Mechanism of the Reaction of Trisubstituted Phosphines with Episulfides," J. Am. Chem. Soc., 82, 4736-4738 (1960); CA 55, 6458d

1753. Denney, D. B., and DiLeone, R., "Racemization During Chromatography of Optically Active Halides," J. Org. Chem., 26, 984 (1961); CA 55, 24527i

1754. Denning, Jr., G. S., "Thermal Decomposition of Some Medium Sized 1-Methylcycloalkyl Acetates," Univ. Microfilms (Ann Arbor, Mich.), L. C. Card No. Mic 60-6504, 132 pp.; Dissertation Abstr., 21, 1731 (1961); CA 55, 15321h

1755. Denny, B., "The Adaptation of the Chromatography Stream Analyzer to the Snyder Gasoline Plant Operation," 37th Annual Mtg., Natural Gasoline Assoc. of Am., Dallas, Texas, April 1958, Tech. Papers, 37, 19-20 (1958); CA 54, 1835a

1756. Denny, B., "Chromatography Has Quick Pay Out at Snyder Plant," Oil Gas J., 59, No. 16, 117-118 (1958); CA 52, 15883d

1757. Denny, B., "Better Absorber Control, Higher Recovery and More Profits," Petrol. Eng., 30, C17-C18 (1958)

1758. Densham, A. B., and Beale, P. A. A., "The Analysis of Oils for Gas Manufacture," Gas Council. Research Commun., No. GC79, 10 pp. (1961); CA 56, 8991g

1759. Densham, A. B., and Gough, G., "The Application of Physical Methods of Analysis in the Gas Industry," Gas World, 146, No. 3805, 118-121 (1957); CA 52, 3309f

1760. Dent, L. S., and Smith, J. V., "Crystal Structure of Chabazite, A Molecular Sieve," Nature, 181, 1794-1796 (1958); CA 52, 19735i

1761. DePuy, C. H., Bishop, C. A., and Goeders, C. N., "Pyrolytic cis Eliminations: The Pyrolysis of sec-Butyl Derivatives," J. Am. Chem. Soc., 83, 2151-2153 (1961); CA 55, 22109h

1762. DePuy, C. H., and Froemsdorf, D. H., "Electronic Effects in Elimination Reactions. III. Sulfonylhydrazone Eliminations," J. Am. Chem. Soc., 82, 634-636 (1960); CA 54, 9728c

1763. DePuy, C. H., and Goeders, C. N., "The Pyrolysis of 3-Acetoxytetrahydrofuran," J. Org. Chem.,

28, 1147-1148 (1963)

1764. DePuy, C. H., King, R. W., and Froemsdorf, D. H., "Pyrolytic Elimination of Acetates. Isotope Effect, Relative Reactivity and Mechanism," Tetrahedron, 7, 123-129 (1959); CA 54, 5501g

1765. DePuy, C. H., Mahoney, L. R., and Eilers, K. L., "Synthesis of Cyclopropanols," J. Org. Chem., 26, 3616-3617 (1961); CA 57, 3297f

1766. DePuy, C. H., Ogawa, I. A., and McDaniel, J. C., "The Solvolysis of exo- and endo-7-Isopropylidenedehydronorbornyl Tosylates," J. Am. Chem. Soc., 83, 1668-1671 (1961); CA 55, 25801h

1767. DePuy, C. H., and Story, P. R., "Gas Chromatographic Evidence for Intramolecular Hydrogen Bonding with Double Bonds," Tetrahedron Letters, 1959, No. 6, 20-21

1768. DePuy, C. H., and Story, P. R., "The Synthesis of 2,7-Disubstituted Norbornanes," J. Am. Chem. Soc., 82, 627-631 (1960); CA 54, 9786g

1769. Derby, J. V., and LaMont, B. D., "Determination of Surface Areas of Powders by Gas Chromatography," 11th Pittsburgh Conf. on Anal. Chem. & Appl. Spectroscopy, Pittsburgh, Pa., February-March 1960, Program Abstr., p. 45

1770. DeRose, A., Gerrard, W., and Mooney, E. F., "Application of Gas Chromatography to the Investigation of Ester Interchange in Dialkyl Hydrogen Phosphites," Chem. & Ind. (London), 1961, 1449-1450; CA 56, 5821d

1771. Despa, S., Iftimescu, C., and Rhoe, A., "Separation, by Chromatographic Adsorption, of Higher Alkane-Aromatic Hydrocarbon Mixtures," Lucrarile inst. petrol si gaze, Bucuresti, 3, 219-238 (1957); CA 52, 21023a

1772. Desty, D. H., "The Newest Status of Gas Chromatography," Proc. Gas-Kolloquim, Hamburg, West Germany, November 1956, pp. 13-14

1773. Desty, D. H., "Gas Chromatography," Nature, 179, 241-242 (1957)

1774. Desty, D. H., (Ed.), "Vapour Phase Chromatography," Academic Press, Inc., New York, 1957. A compilation of 34 papers presented at the Symposium on Vapor Phase Chromatography, sponsored by the Institute of Petroleum, London, May-June 1956

1775. Desty, D. H., "Vapour Detectors for Gas Chromatography," Nature, 180, 22-23 (1957)

1776. Desty, D. H., (Ed.), "Gas Chromatography 1958," Academic Press, Inc., New York, 1958. A compilation of 32 papers presented at the second symposium held under the auspices of the Hydrocarbon Research Group, Institute of Petroleum, and the Royal Chemical Society of the Netherlands, Amsterdam, May 1958

1777. Desty, D. H., "Column Packings for Gas Chromatography," Nature, 181, 604 (1958)

1778. Desty, D. H., "Gas Chromatography," Nature, 184, 327-328 (1959)

1779. Desty, D. H., "Coated Capillary Columns," Abhandl. Deut. Akad. Wiss. Berlin, Kl. Chem., Geol., Biol., 1959, No. 9, 176-184; CA 58, 5018f

1780. Desty, D. H., "Coated Capillary Columns," in "Gas Chromatographie 1958," edited by H. P. Angele, Akademie Verlag, Berlin, 1959, pp. 176-184

1781. Desty, D. H., "Recent Advances in Gas Chromatography," ISI (Indian Standards Inst.) Bull., 12, 74 (1960); CA 55, 2208e

1782. Desty, D. H., "Gas Chromatography. Recent Literature," Nature, 194, 822-823 (1962)

1783. Desty, D. H., "Recent Advances in Gas Chromatography," Riv. Combust., 16, No. 3, 115-127 (1962); CA 57, 13163e

1784. Desty, D. H., Geach, C. J., and Goldup, A., "An Examination of the Flame Ionization Detector Using a Diffusion Dilution Apparatus," Gas Chromatog., Proc. Symposium, 3rd, Edinburgh, 1960, 46-64; CA 56, 3306d

1785. Desty, D. H., Geach, C. J., and Goldup, A., "An Examination of the Flame Ionization Detector Using a Diffusion Dilution Apparatus," in "Gas Chromatography 1960," edited by R. P. W. Scott, Butterworths, London, 1960, pp. 46-64

1786. Desty, D. H., Godfrey, F. M., and Harbourn, C. L. A., "Operating Data on Two Stationary Phase Supports," in Gas Chromatography 1958," edited by D. H. Desty, Academic Press, Inc., New York, 1958, pp. 200-215

1787. Desty, D. H., Godfrey, F. M., and Harbourn, C. L. A., "Operating Data on Two Stationary Phase Supports," Gas Chromatog., Proc. Symposium, Amsterdam, 1958, 200-211; CA 53, 21038e

1788. Desty, D. H., and Goldup, A., "Coated Capillary Columns - An Investigation of Operating Conditions," Gas Chromatog., Proc. Symposium, 3rd, Edinburgh, 1960, 162-183; CA 55, 24135i

1789. Desty, D. H., and Goldup, A., "Coated Capillary Columns - An Investigation of Operating Conditions," in "Gas Chromatography 1960," edited by R. P. W. Scott, Butterworths, London, 1960, pp. 162-183

1790. Desty, D. H., and Goldup, A., "Performance of Coated Capillary Columns," 3rd Intern. Symposium on Gas Chromatography, ISA, East Lansing, Mich., June 1961; ISA Proc., 3, 83-99 (1961)

1791. Desty, D. H., and Goldup, A., "Chromatography of Hydrocarbons," in "Chromatography," edited by E. Heftmann, Reinhold Publishing Corp., New York, 1961, Chapter 27

1792. Desty, D. H., Goldup, A., Luckhurst, G. R., and Swanton, W. T., "The Effect of Carrier Gas and

Column Pressure on Solute Retention," 4th Intern. Gas Chromatography Symposium, Hamburg, Germany, June 1962; pub. in "Gas Chromatography 1962," ed. by M. van Swaay, Butterworth & Co., London, 1962

1793. Desty, D. H., Goldup, A., and Swanton, W. T., "Separation of m-Xylene and p-Xylene by Gas Chromatography," Nature, 183, 107-108 (1959); CA 53, 12096h

1794. Desty, D. H., Goldup, A., and Swanton, W. T., "Performance of Coated Capillary Columns," in "Gas Chromatography," edited by N. Brenner, J. E. Callen, and M. D. Weiss, Academic Press, Inc., New York, 1962, pp. 105-138

1795. Desty, D. H., Goldup, A., and Whyman, B. H. F., "The Potentialities of Coated Capillary Columns for Gas Chromatography in the Petroleum Industry," J. Inst. Petrol., 45, 287-298 (1959); CA 54, 1833c

1796. Desty, D. H., and Harbourn, C. L. A., "Evaluation of a Commercial Alkyl Aryl Sulfonate Detergent as a Column Packing for Gas Chromatography," 132nd Natl. ACS Mtg., New York, N. Y., September 1957, Program Abstr., p. 40B; Preprint, Div. Petrol. Chem., 2, No. 4, D157 (1957)

1797. Desty, D. H., and Harbourn, C. L. A., "Evaluation of a Commercial Alkyl Aryl Sulfonate Detergent as a Column Packing for Gas Chromatography," Anal. Chem., 31, 1965-1970 (1959); CA 54, 7004i

1798. Desty, D. H., Haresnape, J. N., and Whyman, B. H., "Construction of Long Lengths of Coiled Glass Capillary," Anal. Chem., 32, 302-304 (1960); CA 54, 10405b

1799. Desty, D. H., and Swanton, W. T., "Gas-Liquid Chromatography - Some Selective Stationary Phases for Hydrocarbon Separations," J. Phys. Chem., 65, 766-774 (1961); CA 56, 11884d; 137th Natl. ACS Mtg., Cleveland, Ohio, April 1960, Program Abstr., p. 26B

1800. Desty, D. H., Warham, T. J., and Whyman, B. H. F., "The Application of Vapour-Phase Chromatography to the Examination of Samples Taken from Internal-Combustion Engines by an Open-Hole Technique," in "Vapour Phase Chromatography," edited by D. H. Desty, Academic Press, Inc., New York, 1957, pp. 346-358

1801. Desty, D. H., and Whyman, B. H. F., "Application of Vapor-Phase Chromatography to the Analysis of Liquid Petroleum Fractions," 21st Midyear Mtg., Div. of Refining, Am. Petrol. Inst.,Montreal, Canada, May 1956; Abstr., Anal. Chem., 28, 919 (1956)

1802. Desty, D. H., and Whyman, B. H. F., "Application of Gas-Liquid Chromatography to Analysis of Liquid Petroleum Fractions," Anal. Chem., 29, 320-329 (1957); CA 51, 9135c

1803. Desty, D. H., and Whyman, B. H. F., "Relative Retention Volumes of Hydrocarbons and Sulfur pounds (Chromatographic Data Table XLIV)," J. Chromatog., 1, xxviii-xxix (1958); cf. Anal. Chem., 29, 320 (1957) (Temperature 78.5°, n-Pentane = 1)

1804. DeTar, D. F., and Wells, D. V., "The Reactivity of the 1-Hexyl Radical in Abstracting Hydrogen and Halogen Atoms," J. Am. Chem. Soc., 82, 5839-5846 (1960); CA 55, 18571c

1805. Deuel, H., "Reactions of Silicates with Organic Compounds," Makromol. Chem., 34, 206-215 (1959); CA 54, 2878c

1806. Deutsch, I., "Advanced Technology Applied to Leak Detection," Gas, 35, 66-69, 72-73 (1959); CA 54, 1938e

1807. DeVault, D., "The Theory of Chromatography," J. Am. Chem. Soc., 65, 532-540 (1943); CA 37, 3316⁶

1808. Devienne, F. M., "Thermal Conductivity in Rarefied Gases; Accommodation Coefficient," Men. sci. phys. acad. sci. Paris, No. 56, 1-71 (1953); CA 48, 4910c

1809. DeVita, M., Caprioli, G., and Pavan, E., "Analytical Applications of Gas Chromatography. I. Analysis of Impurities of Methane," Chim. e ind. (Milan), 41, 292-294 (1959); CA 53, 17749a

1810. DeVries, J. E., Coulson, D. M., and Cavanagh, L. A., "Microcoulometry in Gas Chromatography," in "Microchem. J. Symp. Ser. - Submicrogram Experimentation," edited by N. D. Cheronis, Interscience Publishers, Inc., New York, 1961, pp. 218-226; CA 56, 911h

1811. DeVries, J. E., Coulson, D. M., and Cavanagh, L. A., "Microcoulometry in Gas Chromatography," Microchem. J., Symp. Ser., 1, 219-226 (1961); CA 58, 8367f

1812. DeVries, L., "Preparation of 1,2,3,4,5-Pentamethylcyclopentadiene, 1,2,3,4,5,5-Hexamethylcyclopentadiene, and 1,2,3,4,5-Pentamethylcyclopentadienylcarbinol," J. Org. Chem., 25, 1838 (1960); CA 55, 2511e

1813. DeVries, M., and Mecke, R., "Gas Chromatographic Analysis of Alcoholic Beverages," German Chemical Society Mtg., Div. of Anal. Chem., Freiburg, West Germany, April 1959, Program Abstr., p. 21

1814. Devyatykh, G. G., Zorin, A. D., and Ezheleva, A. E., "Analysis of a Mixture of Butadiene, the Butane Isomers, and Butene by the Method of Gas-Liquid Distribution Chromatography," Nauch. Doklady Vysshei Shkoly Khim. i Khim. Tekhnol., 1958, 724-726; CA 53, 5979h

1815. Devyatykh, G. G., Zorin, A. D., and Ezheleva, A. E., "Simple Method for the Determination of Column Length for Sectional [Gas-Liquid Chromatography] Columns," Trudy po Khim. i Khim. Tekhnol., 3, No. 1, 33-35 (1960); CA 55, 25385e

1816. Dewar, M. J. S., Dietz, R., and Narayanaswami, K., "Reactions of Diazooxides," U.S. Dept. Com., Office Tech. Serv., AD 256,405, 35 pp. (1961); CA 59, 1509g

1817. Dewar, R. A., "The Flame Ionization Detector. A Theoretical Approach," J. Chromatog., 6, 312-

323 (1961); CA 56, 12290d

1818. Dewhurst, H. A., "Radiation Chemistry of Hexane and Cyclohexane Liquids," J. Chem. Phys., 24, 1254-1255 (1956); CA 50, 11120d

1819. Dewhurst, H. A., "Radiation Chemistry of Hydrocarbons: n-Alkane Liquids," 131st Natl. ACS Mtg., Miami, Fla., April 1957, Program Abstr., p. 7R

1820. Dewhurst, H. A., "Radiation Chemistry of Organic Compounds. I. n-Alkane Liquids," J. Phys. Chem., 61, 1466-1471 (1957); CA 52, 3536b

1821. Dewhurst, H. A., "Radiation Chemistry of Organic Compounds. II. n-Hexane," J. Phys. Chem., 62, 15-20 (1958); CA 52, 7877i

1822. Dewhurst, H. A., "Radiation Chemistry of Organic Compounds. III. Branched Chain Alkanes," J. Am. Chem. Soc., 80, 5607-5610 (1958); CA 53, 10999h

1823. Dewhurst, H. A., "Radiolysis of Organic Compounds. V. n-Hexane Vapor," J. Am. Chem. Soc., 83, 1050-1052 (1961); CA 55, 15081f

1824. Dewhurst, H. A., and St. Pierre, L. E., "Radiation Chemistry of Hexamethyldisiloxane, a Poly-dimethylsiloxane Model," J. Phys. Chem., 64, 1063-1065 (1960); CA 55, 4119i

1825. Dewhurst, H. A., and Winslow, E. H., "Electron and Gamma-Ray Radiolysis of n-Hexane," J. Chem. Phys., 26, 969-970 (1957); CA 51, 12670e

1826. Dhont, J. H., "Pyrolysis and Gas Chromatography for the Detection of the Benzene Ring in Organic Compounds," Nature, 192, 747-748 (1961); CA 56, 7977a

1827. Dhont, J. H., "Pyrolysis as an Additional Tool for the Identification of Organic Compounds," Chem. Weekblad., 58, 440-441 (1962); CA 57, 14410d

1828. Dhont, J. H., and Weurman, C., "The Isolation and Determination of Volatile Compounds by Adsorption on Charcoal," Analyst, 85, 419-422 (1960); CA 55, 1272a

1829. Dhopeshwarkar, G. A., and Blomstrand, R., "Occurrence of Methyl Esters in Lymph," Acta Chem. Scand., 16, 2058-2059 (1962); CA 58, 8286d

1830. Dhopeshwarkar, G. A., and Mead, J. F., "Role of Oleic Acid in the Metabolism of Essential Fatty Acids," J. Am. Oil Chemists' Soc., 38, 297-301 (1961); CA 55, 18913i

1831. Dhopeshwarkar, G. A., and Mead, J. F., "Evidence for Occurrence of Methyl Esters in Body and Blood Lipids," Proc. Soc. Exptl. Biol. Med., 109, 425-429 (1962); CA 57, 3852h

1832. Diaper, D. G. M., "Preparation of Undecenol and Undecenyl Bromide," Can. J. Chem., 39, 1723-1727 (1961); CA 55, 27010i

1833. DiCenzo, R. J., "An Improved Exhaust System for the Perkin-Elmer Vapor Fractometer," Anal. Chem., 34, 874 (1962); CA 57, 2827b

1834. Dick, A. T., Dann, A. T., Bull, L. B., and Culvenor, C. C. J., "Vitamin B_{12} and the Detoxification of Hepatotoxic Pyrrolizidine Alkaloids in Rumen Liquor," Nature, 197, 207-208 (1963); CA 58, 8299b

1835. Dickinson, J. D., and Eaborn, C., "Purification of Liquids and Low Melting Solids by Progressive Freezing," Chem. & Ind. (London), 1956, 959; CA 51, 3220a

1836. Dickman, J. T., "The Nature of the Fatty Acids of Human Depot Fat," Univ. Microfilms (Ann Arbor, Mich.), L. C. Card No. Mic 60-3630, 157 pp.; Dissertation Abstr., 21, 745-746 (1960); CA 55, 6638i

1837. Diemair, W., and Schams, E., "Gas Chromatography in Food Analysis. I. Determination of the Lower Volatile Fatty Acids in Food," Z. Lebensm.-Untersuch. u.-Forsch., 112, 457-463 (1960); CA 54, 25347g

1838. Diemair, W., and Schams, E., "Use of Physical Methods in the Analysis of Aroma and Taste Substances in Foods. I. Determination of Traces of Neutral Carbonyl Compounds," Z. Anal. Chem., 189, 149-160 (1962); CA 57, 14239f

1839. Dietrich, P., and Mercier, D., "Identification of Pyrazine Bases by Chromatography," J. Chromatog., 1, 67-69 (1958); CA 52, 9869a

1840. Dietz, W. A., "Analysis of Petroleum Fractions by Gas Chromatography," 12th Annual Mtg., Soc. for Appl. Spectroscopy, New York, N. Y., November 1957; Abstr., Appl. Spectroscopy, 12, 20 (1958)

1841. Dietz, W. A., "Analysis of Light Ends on Saturate Naphthas Using Gas Chromatography," in "Gas Chromatography," edited by V. J. Coates, H. J. Noebels, and I. S. Fagerson, Academic Press, Inc., New York, 1958, pp. 87-91

1842. Dietz, W. A., "The Use of Gas Chromatography in a Petroleum Process Laboratory," Perkin-Elmer Instrument News, 9, No. 3, 1, 4-6 (Spring 1958)

1843. Dietz, W. A., and Dudenbostel, Jr., B. F., "Applications of Gas Chromatography to Petroleum Processes," 132nd Natl ACS Mtg., New York, N. Y., September 1957, Program Abstr., p. 41B; Preprints, Div. Petrol. Chem., 2, No. 4, D171-D176; CA 54, 21722g

1844. Dietze, S., "The Modern Method of the Qualitative and Quantitative Analysis of Essential Oils," Parfüm. u. Kosmetik, 42, 43-47 (1961); CA 56, 3578f

1845. Di Giacomo, A., Rispoli, G., and Crupi, F., "Gas Chromatography of the Terpene Fraction from

Sicilian Lemon Essential Oils," Essenze deriv. agrumari, 32, No. 2, 126-134 (1962); CA 58, 1298d

1846. Dijkstra, A., "Quantitative Estimation of Peak Areas in Gas-Liquid Chromatography," Nature, 192, 965 (1961); CA 56, 5418h

1847. Dijkstra, G., and De Goey, J., "The Use of Coated Capillaries as Columns for Gas Chromatography," in "Gas Chromatography 1958," edited by D. H. Desty, Academic Press, Inc., New York, 1958, pp. 56-68

1848. Dijkstra, G., Keppler, J. G., and Schols, J. A., "Gas-Liquid Partition Chromatography," Rec. trav. chim., 74, 805-812 (1955); CA 50, 1528f

1849. Dijkstra, R., and Dahmen, E. A. M. F., "Analysis of Alkyl Aluminum Compounds by Gas Chromatography," Z. Anal. Chem., 181, 399-406 (1961); CA 56, 24h

1850. Dille, R. M., and Chapman, R. W., "Application of Process Chromatography to Synthetic Gas Analysis," 14th Annual Instrument Automation Conf. and Exhibit, ISA, Chicago, Ill., September 1959

1851. Dillon, M., and Stanton, R. E., "A Theoretical Consideration in Vapor Phase Chromatography," 1st Annual Gas Chromatography Institute, Canisius College, Buffalo, N. Y., May 1959

1852. Dimbat, M., "Factors Which Influence Efficiency in Gas-Liquid Partition Chromatography," Southwide Chemical Conference on Instrumentation, ISA, Symposium on Gas Chromatography, Memphis, Tenn., December 1956

1853. Dimbat, M., Porter, P. E., and Stross, F. H., "Gas Chromatography. Apparatus Requirements for Quantitative Application of Gas-Liquid Partition Chromatography," Anal. Chem., 28, 290-297 (1956); CA 50, 9204b

1854. Dimick, K. P., "Recent Advances in Preparative Chromatography," 14th Annual Mid-America Spectroscopy Symp., Chicago, Ill., May 1963

1855. Dimick, K. P., and Chu, T-Z., "Comparison of GLPC Polyester Packings," 33rd Fall Mtg., Am. Oil Chemists' Soc., Los Angeles, Calif., September 1959

1856. Dimick, K. P., and Chu, T-Z., "Quantitative Analysis of Fatty Alcohols by Gas Chromatography," 33rd Fall Mtg., Am. Oil Chemists' Soc., Los Angeles, Calif., September 1959

1857. Dimick, K. P., and Corse, J., "Gas Chromatography. A New Method for the Separation and Identification of Volatile Materials in Foods," Food Technol., 10, 360-364 (1956); CA 50, 15979c

1858. Dimick, K. P., and Corse, J., "Volatile Flavor of Strawberries. Minor Constituent Analysis by Gas Chromatography and Mass Spectrometry," 131st Natl. ACS Mtg., Miami, Fla., April 1957, Program Abstr., p. 2A

1859. Dimick, K. P., and Corse, J., "The Volatile Flavors of Strawberry," Quatermaster Food and Container Inst., Surveys Progr. Military Subsistence Problems, Ser. I, No. 9, 123-132 (1957); CA 52, 7564d

1860. Dimick, K. P., and Corse, J., "The Volatile Flavors of Strawberry," Am. Perfumer Aromat., 71, No. 2, 45, 48, 53 (1958); CA 52, 6661f

1861. Dimick, K. P., and Corse, J., "Vapor-Phase Chromatography - A New Method for the Separation and Identification of Volatile Materials in Foods," Proc. Conf. Appl. Phys. Sci. Food Research, Process., and Preserv., 1st San Antonio, Texas, 1959, 59-79; CA 52, 593g

1862. Dimick, K. P., and Makower, B. B., "Volatile Flavor of Strawberry Essence. I. Identification of the Carbonyls and Certain Low-Boiling Substances," Food Technol., 10, 73-75 (1956); CA 50, 12341a

1863. Dimick, K. P., Stitt, F., and Corse, J., "Volatile Flavor of Strawberries. II. Application of Gas-Liquid Partition Chromatography," 129th Natl. ACS Mg., Dallas, Texas, April 1956, Program Abstr., p. 19B

1864. Dimick, K. P., and Taft, E. M., "A Multiple Process Used in Automatic Preparative Gas Chromatography," J. Gas Chromatog., 1, No. 3, 7 (1963)

1865. Dimick, K. P., and Walker, J. Q., "Analysis of Biochemical and Low Boiling Organic Compounds," Biochemical Gas Chromatography Seminar, Wilkens Instrument & Research, Inc., New Orleans, La., July 1963

1866. Dinelli, D., Polezzo, S., and Taramasso, M., "Rotating Unit for Preparative-Scale Gas Chromatography," J. Chromatog., 7, 477-484 (1962); CA 58, 1137h

1867. Dinerstein, R. A., "Gas Density Detector," 3rd Annual Eastern Analytical Symposium & Instrument Exhibit, New York, N. Y., November 1961

1868. Dintenfass, H. T., "Selective Polar Adsorption," Chem. & Ind. (London), 1957, 560; Kolloid Z., 151, 154 (1957); CA 51, 14368c

1869. Di Prima, A., and Storto, T., "Gas Chromatography of Perfume Constituents," Riv. ital. essenze profumi, piante offic., olii vegetali, saponi, 42, 283-291 (1960); CA 54, 25567e (see item 2233)

1870. Dixon, H. B. F., "The Resolving Power of Chromatograms," J. Chromatog., 7, 467-476 (1962); CA 58, 920h

1871. Dixon, W. S., "Old and New Techniques of Gas Sampling Systems," 6th Instrumental Methods of Analysis Symp., ISA, Montreal, Canada, June 1960

1872. Djerassi, C., Antonaccio, L. D., Budzikiewicz, H., Wilson, J. M., and Gilbert, B., "Mass Spectrometry in Structural and Stereochemical Problems. XVI. Structures of the Aspidosperma Alkaloid, Aspidoalbine," Tetrahedron Letters, 1962, 1001-1009; CA 58, 10248h

1873. Djerassi, C., Eisenbraun, E. J., Finnegan, R. A., and Gilbert, B., "Naturally Occurring Oxygen Heterocycles. VII. Structure of Mammein," J. Org. Chem., 25, 2164-2169 (1960); CA 55, 10427a

1874. Djerassi, C., Osiecki, J., and Eisenbraun, E. J., "Terpenoids. XLIX. Preparation of Optically Active Polyalkylcyclohexanones from (+)-Pulegone, (−)-Menthone and (−)-Carvone," J. Am. Chem. Soc., 83, 4433-4439 (1961); CA 56, 6005f

1875. Djerassi, C., Warawa, E. J., Wolff, R. E., and Eisenbraun, E. J., "Optical Rotatory Dispersion Studies. XXXIII. α-Haloketones. trans-2-Bromo-5-tert-butylcyclohexanone," J. Org. Chem., 25, 917-921 (1960); CA 54, 20904c

1876. Djerassi, C., Wilson, J. M., Budzikiewicz, H., and Chamberlin, J. W., "Mass Spectrometry in Structural and Stereochemical Problems. XIV. Steroids with One or Two Aromatic Rings," J. Am. Chem. Soc., 84, 4544-4552 (1962); CA 58, 6880g

1877. Dobbs, H. E., "The Detection of Tritium Labelled Compounds in Vapour Phase Chromatography," J. Chromatog., 5, 32-37 (1961); CA 56, 4552a

1878. Dobiasova, M., Liebster, J., and Ekl, J., "Application of Gas Chromatography to the Preparation of C^{14} Labelled Higher Fatty Acids," Gas Chromatog. Symp., Brno, Czech., June 1962; Abstr., J. Gas Chromatog., 1, No. 4, 10 (1963)

1879. Dobychin, D. P., Burkat, T. M., and Kiseleva, N. N., "Porous Glasses as Adsorbents of the Molecular Sieve Type," Sintetich. Tseolity, Poluchenie, Issled. i Primenenie, Akad. Nauk SSSR, Otd. Khim. Nauk, 1962, 75-85; CA 58, 8743a

1880. Doering, C. E., and Hauthal, H. G., "Gas Chromatography of C_6 Hydrocarbons," J. Prakt. Chem., 19, No. 1-2, 17-32 (1962); CA 58, 7348f

1881. Doering, W. v. E., Buttery, R. G., Laughlin, R. G., and Chaudhuri, N., "Indiscriminate Reaction of Methylene with the Carbon-Hydrogen Bond," J. Am. Chem. Soc., 78, 3224 (1956)

1882. Doering, W. v. E., and Kirmse, W., "The Absolute Configuration of trans-1,2-Dimethylcyclopropane," Tetrahedron, 11, No. 4, 272-275 (1960); CA 55, 10340d

1883. Doering, W. v. E., and Knox, L. H., "Comparative Reactivity of Methylene, Carbomethoxycarbene, and bis-Carboethoxycarbene Toward Saturated Carbon-Hydrogen Bond," J. Am. Chem. Soc., 83, 1989-1992 (1961); CA 55, 22082i

1884. Doering, W. v. E., Knox, L. H., and Jones, Jr., M., "Reaction of Methylene with Diethyl Ether and Tetrahydrofuran," J. Org. Chem., 24, 136-137 (1959); CA 54, 6678c

1885. Doering, W. v. E., and LaFlamme, P. M., "A Two-Step Synthesis of Allenes from Olefins," Tetrahedron, 2, 75-79 (1958); CA 52, 11729i

1886. Doering, W. v. E., and Mole, T., "Cyclopropenes from Carbenes and Acetylenes. Stereoselectivity in the Reaction of Carbomethoxycarbene with cis-Butene," Tetrahedron, 10, 65-70 (1960); CA 55, 3460g

1887. Doering, W. v. E., and Prinzbach, H., "Mechanism of Methylene with the Carbon-Hydrogen Bond. Evidence for Direct Insertion," Tetrahedron, 6, 24-30 (1959); CA 53, 17880g

1888. Doering, W. v. E., and Roth, W. R., "The Overlap of Two Allyl Radicals or a Four-Centered Transition State in the Cope Rearrangement," Tetrahedron, 18, 67-74 (1962); CA 57, 2044g

1889. Doerrscheidt, W., and Friedrich, K., "Separation of Odorous Substances of Honey by Means of Gas Chromatography," J. Chromatog., 7, 13-18 (1962); CA 57, 7692i

1890. Dole, V. P., James, A. T., Webb, J. P. W., Rizack, M. A., and Sturman, M. F., "The Fatty Acid Patterns of Plasma Lipids During Alimentary Lipemia," J. Clin. Invest., 38, 1544-1554 (1959); CA 53, 22368c

1891. Dolejsek, Z., Grubner, O., Hanus, V., Kossler, I., Matyska, B., and Vodehnal, J., "The Purification and Analysis of Isoprene. I. The Analytical Control of Isoprene Rectification," Chem. pyrumysl., 10, 571-576 (1960); CA 55, 6021i

1892. Dolphin, J. L., and Stanley, T. W., "Vapor Phase Chromatography in Air Pollution Studies. Column Evaluation," 131st Natl. ACS Mtg., Miami, Fla., April 1957, Program Abstr., p. 36B

1893. Domange, L., and Longuevalle, S., "Gas Chromatographic Analysis of Schleich's Mixture for Anesthesia," Ann. pharm. franc., 15, 448-454 (1957); Z. Anal. Chem., 164, 371 (1958); CA 52, 5745e

1894. Domange, L., and Longuevalle, S., "Determination of Eucalyptole in Eucalyptus Oil and Cajuput Oil by Gas Chromatography," Ann. pharm. franc., 16, 557-561 (1958); CA 53, 7515d

1895. Domange, L., and Longuevalle, S., "Use of Gas-Liquid Partition Chromatography. Analysis of Oil of Eucalyptus, Menthol, and Various Medicinal Mixtures," Compt. rend., 247, 209-211 (1958); CA 53, 2542g

1896. Domange, L., and Longuevalle, S., "Use of Gas Chromatography in Drug Analysis," Mises au point chim. anal. pure et appl. et anal. bromatol., No. 9, 7-35 (1961); CA 55, 27779a

1897. Dominguez, A. M., Christensen, H. E., Goldbaum, L. R., and Stembridge, V. A., "A Sensitive Procedure for Determining Carbon Dioxide in Blood or Tissue Utilizing Gas-Solid Chromatography," Toxicol. and Appl. Pharmacol., 1, 135-143 (1959); CA 53, 14208g

1898. Donegan, L., Godin, P. J., and Thain, E. M., "The Separation and Estimation of the Insecticidal Constituents of Pyrethrum Extract by Gas Chromatography," Chem. & Ind. (London), 1962, 1420; CA 57, 15557d

1899. Donner, W., "Electronics Controls Gas Chromatography," Electronics, 30, No. 11, 164-166 (1957); CA 52, 1692b

1900. Donner, W., "What is the Moon Made Of? Lunar Chromatograph May Show Origin of the Moon, The Solar System, or Life Itself," Analyzer, 4, No. 3, 19-21 (July 1963)

1901. Donner, W., Johns, T., and Gallaway, W. S., "Use of a Mass Spectrometer as a Gas Chromatograph Detector," ASTM Committee E-14 Mtg. on Mass Spectrometry, New York, N. Y., May 1957; Abstr., Anal. Chem., 29, 1378 (1957)

1902. Doolen, O. K., "Improved Recording System for Gas Chromatography," Gas Chromatog., Intern. Symposium, 2nd, East Lansing, Mich., 1959, 111-117 (Pub. 1961); CA 55, 18204e

1903. Doolen, O. K., "An Improved Recording System for Gas Chromatography," in "Gas Chromatography," edited by H. J. Noebels, R. F. Wall, and N. Brenner, Academic Press, Inc., New York, 1961, pp. 111-117

1904. Dora, R. A., "High Accuracy Solid State Temperature Controller with One Hundred Per Cent Proportional Band for Process Instrument Applications," 8th Natl. Analysis Instrumentation Symp., ISA, April-May 1962

1905. Dora, R. A., "A New Process Gas Chromatograph - The Beckman 520-D," Analyzer, 3, No. 4, 8-10 (October 1962)

1906. Dorfman, L. M., and Wilzbach, K. E., "Tritium Labeling of Organic Compounds by Means of Electric Charge," J. Phys. Chem., 63, 799-801 (1959); CA 53, 21282g

1907. Dorfner, K., "Gas-Chromatographic Determination of the Purity of Diethyl Malonate," J. Chromatog., 4, 502-503 (1960); CA 55, 12157b

1908. Dorfner, K., "A New Technique for Introducing Solid Substances in Gas Chromatography," Brennstoff-Chem., 43, 110-111 (1962); CA 57, 6582g

1909. Dorsey, J. A., "Rapid Scanning Mass Spectrometry: The Continuous Analysis of Fractions from Capillary Gas Chromatography," Univ. of Houston Intern. Symp. on Advances in Gas Chromatography, Houston, Texas, January 1963

1910. Dorsey, J. A., Hunt, R. H., and O'Neal, M. J., "Rapid-Scanning Mass Spectrometry. Continuous Analysis of Fractions from Capillary Gas Chromatography," Anal. Chem., 35, 511-515 (1963); CA 59, 1060g

1911. Downing, D. T., "The α-Hydroxy Acids of Sheep Brain," Australian J. Chem., 14, 150-154 (1961); CA 55, 17488f

1912. Downing, D. T., Kranz, Z. H., Lamberton, J. A., Murray, K. E., and Redcliffe, A. H., "Studies in Waxes. XVIII. Beeswax: A Spectroscopic and Gas Chromatographic Examination," Australian J. Chem., 14, 253 (1961); CA 55, 27923f

1913. Downing, D. T., Kranz, Z. H., and Murray, K. E., "Studies in Waxes. XIV. An Investigation of Aliphatic Constituents of Hydrolyzed Wool Wax by Gas Chromatography," Australian J. Chem., 13, 80-94 (1960); CA 54, 14729d; and "Studies in Waxes. XX. The Quantitative Analysis of Hydrolyzed Carnauba Wax by Gas Chromatography," Australian J. Chem., 14, 619-627 (1961); CA 56, 13035b

1914. Dragel, D. T., Beck, E., and Principe, A. H., "Some Applications of Gas Chromatography to Forensic Chemistry," in "Developments in Applied Spectroscopy," Vol. 2, edited by J. R. Ferraro and J. S. Ziomek, Plenum Press, New York (1963)

1915. Drake, B., "Theory of Gradient Elution Analysis," Arkiv. Kemi., 8, 1-21 (1955); CA 49, 13815i

1916. Drawert, F., "Use of Gas Chromatography in the Determination of Quality in Wines and Musts," Vitis, 2, 172-178 (1960); CA 54, 17788a

1917. Drawert, F., "Components of Musts and Wines. II. Gas Chromatographic Methods for the Analysis of Aromatic Substances, Especially Alcohols," Vitis, 3, 104-114 (1962); cf. ibid., 2, 288-304 (1961); CA 57, 13015g

1918. Drawert, F., "Reaction Gas Chromatography," 4th Intern. Gas Chromatography Symp., Hamburg, Germany, June 1962; pub. in "Gas Chromatography 1962," ed. by M. van Swaay, Butterworth & Co., London, 1962

1919. Drawert, F., "Gas Chromatographic Examination of Aromatic Concentrates of Apples and Pears," Vitis, 3, 115-116 (1962); CA 57, 12971f

1920. Drawert, F., Felgenhauer, R., and Kupfer, G., "Reaction Gas Chromatography for the Analysis of Alcohols and the Determination of the Alcohol Content of Blood," Angew. Chem., 72, 385 (1960)

1921. Drawert, F., Felgenhauer, R., and Kupfer, G., "Reaction-Gas Chromatography," Angew. Chem., 72, 555-559 (1960); CA 55, 60a

1922. Drawert, F., Kuhn, H.-J., and Rapp, A., "Reaction Gas Chromatography. III. Gas Chromatographic Determination of Lower Fatty Acids in the Stomach of Leaf-Eating Monkeys (Colobinae)," Z. Physiol. Chem., 329, No. 1-2, 84-89 (1962); CA 58, 4802a

1923. Drawert, F., and Kupfer, G., "Gas-Chromatographic Analysis of Alcohols as Nitrites," Angew. Chem., 72, 33-34 (1960); CA 54, 12903a

1924. Drawert, F., and Kupfer, G., "Reaction Gas Chromatography. V. Analysis of Blood Alcohol," Z. Physiol. Chem., 329, No. 1-2, 90-96 (1962); CA 58, 4802b

1925. Drawert, F., Rapp, A., and Bachmann, O., "Gas-Chromatographic Analysis of Aroma Compounds and Alcohols of Fruits," Intern. Fruchtsaft-Union, Ber. Wiss.-Tech. Komm., No. 4, 235-242 (1962); CA 58, 8234f

1926. Drawert, F., and Reuther, K. H., "Reaction-Gas Chromatography. II. Reaction Products of the Thermal Cleavage of 2-Alkoxy-2-mercaptothiazolidines," Chem. Ber., 93, 3066-3070 (1960); CA 55, 5466e

1927. Drawert, F., Reuther, K. H., and Born, F., "Xanthogenates from Alcohols," Chem. Ber., 93, 3056-3065 (1960); CA 55, 5328i

1928. Drekopf. K., and Winzen, W., "Gas-Chromatographic Process for the Determination of Hydrogen in Underground Mixtures," Gluckauf, 93, 1222-1225 (1957); CA 54, 6087g

1929. Dresdner, R. D., Reed, T. M., Taylor, T. E., and Young, J. A., "Six and Twelve Carbon Fluorocarbon Derivatives of Sulfur Hexafluoride," J. Org. Chem., 25, 1464-1466 (1960); CA 55, 415e

1930. Dresdner, R. D., and Young, J. A., "Some New Sulfur-Bearing Fluorocarbon Derivatives," J. Am. Chem. Soc., 81, 574-577 (1959); CA 53, 13988d

1931. Dresdner, R. D., and Young, J. A., "Pyrolysis of Perfluorothioxane Tetrafluoride," J. Org. Chem., 24, 566-567 (1959); CA 53, 21967i

1932. Dressler, D. P., Mastio, G. J., and Allbritten, Jr., F. F., "The Clinical Application of Gas Chromatography to the Analysis of Respiratory Gases," J. Lab. Clin. Med., 55, 144-148 (1960); CA 54, 6853a

1933. Drew, C. M., "Temperature Control," in "Principles and Practice of Gas Chromatography," edited by R. L. Pecsok, John Wiley & Sons, Inc., New York, 1959, pp. 97-115

1934. Drew, C. M., "Detectors," in "Principles and Practice of Gas Chromatography," edited by R. L. Pecsok, John Wiley & Sons, Inc., New York, 1959, pp. 116-134

1935. Drew, C. M., and Johnson, J. H., "Gas Chromatography Fraction Collector and Transfer System," J. Chromatog., 9, 264-266 (1962)

1936. Drew, C. M., and McNesby, J. R., "The Application of Vapor-Phase Chromatography to the Study of Gas Phase Reactions," 129th Natl. ACS Mtg., Dallas, Texas, April 1956, Program Abstr., p. 20B

1937. Drew, C. M., and McNesby, J. R., "Some Problems Encountered with the Application of Vapour Phase Chromatography to Kinetic Studies," in "Vapour Phase Chromatography," edited by D. H. Desty, Academic Press, Inc., New York, 1957, pp. 213-221

1938. Drew, C. M., McNesby, J. R., Gordon, A. S., and Smith, S. R., "Recovery of Vapor-Phase Chromatography Fractions and Their Analysis by Mass Spectrometry," 4th Annual Mtg., ASTM Committee E-14 on Mass Spectrometry, Cincinnati, Ohio, May 1956

1939. Drew, C. M., McNesby, J. R., Smith, S. R., and Gordon, A. S., "Application of Vapor Phase Chromatography to Mass Spectrometer Analysis," Anal. Chem., 28, 979-983 (1956); CA 50, 11157i

1940. Drews, B., Specht, H., and Offer, G., "Separation of Isomeric Aliphatic Alcohols by Gas Chromatography," Z. Anal. Chem., 189, 325-330 (1962); CA 57, 14438c

1941. Drews, H., Meyerson, S., and Fields, E. K., "Transulfonation in Preparing Aromatic Sulfones," J. Am. Chem. Soc., 83, 3871-3874 (1961); CA 55, 21023i

1942. Dreyer, H., and Nehring, D., "Gas Chromatographic Investigation of Decomposition Products of Mixed Salts," Naturwissenschaften, 47, 132-133 (1960); CA 55, 1141f

1943. Drienovsky, P., "Gas Chromatography in the Radiation Chemistry of Acetone," Gas Chromatog. Symp., Brno, Czech., June 1962; Abstr., J. Gas Chromatog., 1, No. 4, 9 (1963)

1944. Drysdale, J. J., and Coffman, D. D., "Syntheses by Free-Radical Reactions. XII. Reactions on Fluoroacyl Radicals," J. Am. Chem. Soc., 82, 5111-5115 (1960); CA 55, 9410g

1945. Dubinin, M. M., "The Potential Theory of Adsorption of Gases and Vapors for Adsorbents with Energetically Nonuniform Surfaces," Chem. Rev., 60, 235-241 (1960); CA 54, 9429i

1946. Dubinin, M. M., "Theory of Physical Adsorption and Practical Use of Sorbents," Izv. Akad. Nauk Arm. SSR, Khim. Nauki, 13, 377-385 (1960); CA 57, 86e

1947. Dubinin, M. M., "Adsorption Properties and Secondary Pore Structure of Adsorbents That Exhibit Molecular Sieve Action. II. Comparison of Calculated and Experimental Limiting Values for Adsorption and Adsorption Volumes for Type A Synthetic Zeolites," Izv. Akad. Nauk SSSR, Otd. Khim. Nauk, 1961, 1183-1191; CA 58, 947a

1948. Dubinin, M. M., Vishnyakova, M. M., Zaverina, E. D., Zhukovskaya, E. G., Leont'ev, E. A., Luk'yanovich, V. M., and Sarakhov, A. I., "Adsorption Properties and Secondary Pore Structure of Adsorbents That Exhibit Molecular-Sieve Action. I. Commerical Samples of Synthetic Zeolites," Izv. Akad. Nauk SSSR, Otd. Khim. Nauk, 1961, 396-406; CA 58, 946g

1949. Dubinin, M. M., Vishnyakova, M. M., Zaverina, E. D., Zhukovskaya, E. G., and Sarakhov, A. I., "Adsorption Properties and Secondary Pore Structure of Adsorbents That Exhibit Molecular-Sieve Action. IV. Granulated Synthetic Type A Zeolites," Izv. Akad. Nauk SSSR, Otd. Khim. Nauk, 1961, 1387-1395; CA 58, 947c

1950. Dubinin, M. M., Zaverina, E. D., Luk'yanovich, V. M., and Kharlamov, N. P., "Adsorption Properties and Secondary Pore Structure of Adsorbents That Exhibit Molecular-Sieve Action. III. Granule Components for Type A Synthetic Zeolites," Izv. Akad. Nauk SSSR, Otd. Khim. Nauk, 1961, 1380-1387; CA 58, 947b

1951. Dubinin, M. M., Zhukovaskaya, E. G., and Murdmaa, K. O., "Adsorption Properties and Secondary Pore Structure of Adsorbents Having Molecular-Sieve Action. V. Limiting Adsorption Volumes of Dehydrated Crystals of Type X Synthetic Zeolite," Izv. Akad. Nauk SSSR, Otd. Khim. Nauk, 1962, 760-769; CA 57, 11884d

1952. Dubinin, M. M., Zhukovaskaya, E. G., and Murdmaa, K. O., "Adsorption Properties and the Secondary Pore Structure of Adsorbents Having a Molecular-Sieve Effect. VI. Adsorption of Nitrogen and Water Vapors on the Zeolites, Type X, and the Potential Theory of Adsorption," Izv. Akad. Nauk SSSR, Otd. Khim. Nauk, 1962, 900-968; CA 58, 42b

1953. Dubois, L., Corkery, A., and Monkman, J. L., "Chromatography of Polycyclic Hydrocarbons," 136th Natl. ACS Mtg., Atlantic City, N. J., September 1959, Program Abstr., p. 17U

1954. Dubois, L., and Monkman, J. L., "Confirmatory Tests in Gas Chromatography," in "Gas Chromatography," edited by H. J. Noebels, R. F. Wall, and N. Brenner, Academic Press, Inc., New York, 1961, pp. 237-246

1955. Dubosc, J-P., Coste, J., and Thiebaut, R., "Relation Between the Structure and the Stereospecificity of Organometallic Polymerization Catalysts. II. Organometallic Catalysts," Bull. soc. chim. France, 1961, 473-477; CA 55, 15078d

1956. Dubsky, H. E., "High-Temperature Thermostat for Gas Chromatography," Chem. listy, 54, 1183-1187 (1960); CA 55, 9975a

1957. Dubsky, H. E., and Janak, J., "Sampling Method for Solid Substances in High-Temperature Gas Chromatography Up to 500°," J. Chromatog., 4, 1-5 (1960); CA 55, 5225a

1958. Dubsky, H. E., and Sokolicek, J., "Electromechanical Integrator for Gas Chromatography," Chem. listy, 54, 724-729 (1960); CA 54, 19031e

1959. Dudenbostel, Jr., B. F., and Priestley, Jr., W., "Gas Chromatography for Process Control," Ind. Eng. Chem., 48, No. 9, 55A-56A (1956)

1960. Dudenbostel, Jr., B. F., and Priestley, Jr., W., "Application of Process Control Analysis to Petroleum Refining," Ind. Eng. Chem., 48, No. 11, 49A-50A (1956)

1961. Dudenbostel, Jr., B. F., and Priestley, Jr., W., "Developments in Process Control Analysis in 1956," Ind. Eng. Chem., 49, No. 1, 99A-100A (1957)

1962. Dudenbostel, Jr., B. F., and Skarstrom, C. W., "Gas Chromatography for Plant Stream Analysis," 132nd Natl. ACS Mtg., New York, N. Y., September 1957, Program Abstr., p. 41B; Preprints, Div. Petrol. Chem., 2, No. 4, D177 (1957); CA 54, 20354h

1963. Duffey, J. G., "A Study of the Purification of Brewery Fermentation Carbon Dioxide," Am. Soc. Brewing Chemists, Proc. 1960, 5-11; CA 55, 25149b

1964. Duffield, J. J., and Rogers, L. B., "Gas Chromatography Using Chemically Active Solids," 138th Natl. ACS Mtg., New York, N. Y., September 1960, Program Abstr., p. 6B

1965. Duffield, J. J., and Rogers, L. B., "Theoretical Plates in Gas Chromatography. Effects of Distribution Ratio, Viscosity, and Amount of Liquid Phase," Anal. Chem., 32, 340-348 (1960); CA 54, 21878c

1966. Duffield, J. J., and Rogers, L. B., "Chemically Reactive Solids as Column Packings for Gas Chromatography," Anal. Chem., 34, 1193-1195 (1962); CA 57, 14409i

1967. Duke, J. R. C., "A Liquid-Flow Regulator and Its Application to a Constant-Rate Gas Sampler," J. Soc. Chem. Ind., 58, 231-232 (1939); CA 33, 7153[9]

1968. DuMay, H., "Chromatographic Examination of Industrial Gases by Means of Semiautomatic Apparatus," J. usines gaz, 82, 107-114 (1958); CA 52, 13230f

1969. Dumazert, C., and Ghiglione, C., "Chromatography by Steam Distillation," Bull. soc. chim. France, 1959, 615-617; CA 53, 16644h

1970. Dumazert, C., and Ghiglione, C., "Chromatography by Vapor Entrainment. II. Apparatus. Application to Common Phenols," Bull. soc. chim. France, 1960, 1770-1773; CA 55, 7933f

1971. Duncan, W. R. H., and Garton, G. A., "The C_{18} Fatty Acids of Ox Plasma Lipids," J. Lipid Research, 3, 53-55 (1962)

1972. Dunn, F. J., Mann, J. B., and Mosley, J. R., "Self-Balancing System for Continuous Control of Current or Voltage," Anal. Chem., 27, 167-168 (1955); CA 49, 5170g

1973. Dunstan, I., Williams, R. L., and Blay, N. J., "Boron Hydride Derivatives. V. Nucleophilic Substitution in Decaborane," J. Chem. Soc., 1960, 5012-5015; CA 55, 16396f

1974. Dunton, M. L., "Analysis of Traces of Hydrocarbons in Water by Gas Chromatography," 141st Natl. ACS Mtg., Washington, D. C., March 1962, Program Abstr., p. 20B

1975. Dupaigne, P., "Fruit Flavors and Their Recovery," Perfumery Essential Oil Record, 50, 719-722 (1959); CA 53, 16477b

1976. Dupire, F., "Gas Chromatography of Tar," Compt. rend. congr. intern. chim. ind. 31[e], Liege, 1958;

Ind. chim. belge Suppl. 1, 159-164 (1959); CA 54, 10283e

1977. Dupire, F., "Gas Chromatography at High Temperatures. Application to Coal Tars and Their Derivatives," Z. Anal. Chem., 170, 317-326 (1959); CA 54, 2831g

1978. Dupire, F., "Analysis of Coal Tar Products by High-Temperature Gas Chromatography," Natl.
Cancer Inst., Monograph No. 9, 183-191 (1962); CA 58, 2295b

1979. Dupire, F., and Botquin, G., "Qualitative and Quantitative Analysis of Heavy Tar Oils by Gas Chromatography," Anal. Chim. Acta, 18, 282-290 (1958); J. Appl. Chem. (London), 1958, 455; CA 54,
14649c

1980. Durrell, W., Lovelace, A. M., and Adamczak, R. L., "Synthesis of Halo Olefins. Addition of Di-
bromodifluoromethane and Bromotrichloromethane to Vinylidene Fluoride," J. Org. Chem., 25,
1661-1662 (1960); CA 55, 11281i

1981. Durrett, L. R., "Determination of Solvent Impurities in Waxes and Lubricating Oil Stocks by Gas-
Liquid Chromatography," Anal. Chem., 31, 1824-1825 (1959); CA 54, 5063b

1982. Durrett, L. R., "Applications of Carbowax 400 in Gas Chromatography for Extreme Aromatic Selectivity," Anal. Chem., 32, 1393-1396 (1960); CA 55, 8829a; 137th Natl. ACS Mtg., Cleveland, Ohio,
April 1960, Program Abstr., p. 34B

1983. Durrett, L. R., Simmons, M. C., and Dvoretzky, I., "Quantitative Aspects of Capillary Gas Chromatography of Hydrocarbons," 139th Natl. ACS Mtg., St. Louis, Mo., March 1961, Program Abstr., p. 32B

1984. Durrett, L. R., Taylor, L. M., Wantland, C. F., and Dvoretzky, I., " Component Analysis of Iso-
paraffin-Olefin Alkylate by Capillary Gas Chromatography," 142nd Natl. ACS Mtg., Atlantic City,
N. J., September 1962, Program Abstr., p. 1S; Preprints, Div. Petrol. Chem. Symposium, 6,
No. 2B, 63-77; CA 57, 15408e; Correction, Anal. Chem., 36, 871 (1964)

1985. Durrett, L. R., Taylor, L. M., Wantland, C. F., and Dvoretzky, I., "Component Analysis of Iso-
paraffin-Olefin Alkylate by Capillary Gas Chromatography," Anal. Chem., 35, 637-641 (1963);
CA 59, 1094f

1986. Duskova, L., and Matyska, B., "Chromatographic Separation of Isoprene and Trimethyl Ethylene
on Some Polar Stationary Phases," Gas Chromatog., Symp., Brno, Czech., June 1962; Abstr.,
J. Gas Chromatog., 1, No. 4, 9 (1963)

1987. Duswalt, Jr., A. A., "Analytical Applications of Gas Chromatography," Univ. Microfilms (Ann Arbor,
Mich.), L. C. Card, No. 59-1615, 103 pp.; Dissertation Abstr., 20, 52 (1959); CA 53, 17749f

1988. Duswalt, Jr., A. A., and Brandt, W. W., "Carbon-Hydrogen Determination by Gas Chromatography,"
Anal. Chem., 32, 272-274 (1960); CA 54, 8419a

1989. Dutch, P. H., "Rapid Deuterium Determinations in the Presence of Air," Anal. Chem., 32, 1532
(1960); CA 55, 3271i

1990. Dutton, H. J., "Research Methods of Analysis of Drying Oils," J. Am. Oil Chemists' Soc., 36, 513-
518 (1959); CA 53, 22989f

1991. Dutton, H. J., "Monitoring Gas Chromatography for H^3- and C^{14}-Labeled Compounds by Liquid Scintillation Counting," 12th Pittsburgh Conf. on Anal. Chem. & Appl. Spectroscopy, Pittsburgh, Pa.,
February-March 1961, Program Abstr., p. 52

1992. Dutton, H. J., "Kinetics of Linolenate Hydrogenation," J. Am. Oil Chemists' Soc., 40, 35-39 (1963)

1993. Dutton, H. J., Jones, E. P., Mason, L. H., and Nystrom, R. F., "The Labelling of Fatty Acids by
Exposure to Tritium Gas," Chem. & Ind. (London), 1958, 1176-1177; CA 53, 11203h

1994. Dutton, H. J., Jones, E. P., Scholfield, C. R., Chorney, W., and Scully, N. J., "Countercurrent
Distribution of Soyabean Fatty Acid Methyl Esters Biosynthetically Labelled with H^3 and C^{14}," J.
Lipid Res., 2, 63-67 (1961); CA 55, 13878e

1995. Dutton, H. J., Scholfield, C. R., Jones, E. P., Pryde, E. H., and Cowan, J. C., "Hydrazine-Re-
duced Linolenic Acids as a Source of C_9, C_{12}, and C_{15} Dibasic Acids," J. Am. Oil Chemists' Soc.,
40, 175-179 (1963)

1996. Dutton, H. J., Scholfield, C. R., and Mounts, T. L., "Glyceride Structure of Vegetable Oils by
Countercurrent Distribution. V. Comparison of Natural, Interesterified, and Synthetic Cocoa Butter," J. Am. Oil Chemists' Soc., 38, 96-101 (1961)

1997. Duuren, B. L., van, and Kosak, A. I., "Isolation and Identification of Some Components of Cigarette
Smoke Condensate," J. Org. Chem., 23, 473 (1958); CA 53, 7515g

1998. DuVall, A. H., and Tully, W. F., "Gas-Liquid Partition Chromatography of Mixtures Containing
Phenol and Five of Its t-Butyl Derivatives," 8th Natl. Analysis Instrumentation Symposium, ISA,
Charleston, W. Va., April-May 1962

1999. Dvoretzky, I., "Applications of the Methylene Insertion Reaction to Component Analysis of Hydrocarbons," Univ. of Houston Intern. Symp. on Advances in Gas Chromatography, Houston, Texas,
January 1963

2000. Dvoretzky, I., Richardson, D. B., and Durrett, L. R., "Applications of the Methylene Insertion Reaction to Component Analysis of Hydrocarbons," Anal. Chem., 35, 545-549 (1963)

2001. Dwyer, R., "Application of Molecular Sieve Materials in Chromatographic Columns," 1st Annual

Gas Chromatography Institute, Canisius College, Buffalo, N. Y., May 1959

2002. Dyer, E., and Read, R. E., "Thermal Degradation of O-1-Hexadecyl N-1-naphthylcarbamates and Related Compounds," J. Org. Chem., 26, 4388-4394 (1961)

2003. Dykstra, S., and Mosher, H. S., "Organic Peroxides. VI. Allyl Hydroperoxide," J. Am. Chem. Soc., 79, 3474-3475 (1957); CA 51, 16276i

2004. Dzantiev, B. G., and Barkalov, I. M., "Preparation of Sulfur-Labeled Compounds by the Reaction of 'Hot' Sulfur-35 Atoms with Cyclic Hydrocarbons," Radioisotopes Phys. Sci. Ind., Proc. Conf. Use, Copenhagen, 1960, 27-40 (Pub. 1962); CA 58, 6781h

2005. Eanes, R. D., "Operating Characteristics of the Leeds & Northrup Chromatograph Process Stream Analyzer," Intern. Symp. on Gas Chromatography, ISA, East Lansing, Mich., August 1957

2006. Eastham, J. F., and Gibson, G. W., "Solvent Effects in Organometallic Reactions. I.," J. Org. Chem., 28, 280 (1963); CA 58, 10223c

2007. Eastman, R. H., "Semi-Quantitative Gas Chromatography," J. Am. Chem. Soc., 79, 4243 (1957); CA 51, 17330a

2008. Eastman, R. H., and Winn, A. V., "The Isomerization of Thujone," J. Am. Chem. Soc., 82, 5908-5914 (1960); CA 55, 23578i

2009. Easton, C. B., and Martin, A. J., "The Development of a High Temperature (1000°C.) Gas Chromatograph," F&M Scientific Corp. for Aeronautical Systems Division, U. S. Air Force, AD 268,700, 21 pp. (1961); Abstr., Anal. Chem., 34, No. 7, 69A (1962)

2010. Ebeid, F. M., "Gas Phase Chromatography," League Arab States, Arab. Petrol. Congr., 1st, Cairo, 1959, Collection Papers, 3, 128-138; CA 55, 23984h

2011. Ebeid, F. M., and Minkoff, G. J., "Sensitivity of Vapor-Phase Chromatography Detectors," Research Correspondence, 9, S24 (1956); CA 52, 1690c

2012. Eberly, P. E., "High Temperature Adsorption Studies on 13X Molecular Sieve and Other Porous Solids by Pulse Flow Techniques," J. Phys. Chem., 65, 68-72 (1961); CA 55, 21749f

2013. Eberly, P. E., and Spencer, E. H., "Mathematics of Adsorption of Pulse Flow Through Packed Columns," Trans. Faraday Soc., 57, 288-300 (1961); CA 55, 15013b

2014. Ebert, Jr., A. A., "Improved Sampling and Recording Systems in Gas Chromatography-Time-of-Flight Mass Spectrometry," Anal. Chem., 33, 1865-1870 (1961); CA 56, 7972c

2015. Eckerson, B. A., "Applications of Gas Chromatography to Natural Gasoline Plant Testing," 33rd Annual Mtg., California Natural Gasoline Assoc., October 1958

2016. Eckhardt, F., and Heinze, H. O., "New Methods for the Analysis of Hydrocarbons with Special Reference to Gas Chromatography and Infrared Spectroscopy," Erdol u. Kohle, 12, 83-87 (1959); CA 53, 12645h

2017. Eckhardt, F., and Heinze, H. O., "Application of Infrared Spectroscopy and Gas Chromatography to the Chemistry and Structure of Cumarone Resin," Z. Anal. Chem., 170, 166-176 (1959); CA 54, 4512a; Abstr., Angew. Chem., 71, 460 (1959)

2018. Eckhardt, F., and Heinze, H. O., "Comparison of Conventional Methods of Determining Useful Carbon Compounds with Gas Chromatographic and Spectroscopic Analyses," Brennstoff-Chem., 42, 136-140 (1961); CA 55, 22777d

2019. Ede, P., and Zwicker, J. D., "Determination of Traces of Carbon Dioxide in Liquid Chlorine," 13th Pittsburgh Conf. on Anal. Chem. & Appl. Spectroscopy, Pittsburgh, Pa., March 1962, Program Abstr., p. 54

2020. Eden, M., Karmen, A., and Stephenson, J. L., "Use of Katharometers in Gas Chromatography," Nature, 183, 1322 (1959); CA 53, 20926c

2021. Edgecombe, F. H. C., "Isomerization of 1-Butene on TiCl₃; An Example of Kinetically Controlled Isomerization," Tetrahedron Letters, 1962, 1161-1163; CA 58, 11184a

2022. Edgecombe, F. H. C., "The Vapor Phase Polymerization of Ethylene on an Organotitanium Catalytic Surface," Can. J. Chem., 41, 1265-1275 (1963)

2023. Edgecombe, F. H. C., "Copolymerization by Vapor Phase Chromatography," Nature, 198, 1085-1086 (1963)

2024. Edwards, Jr., H. M., and Marion, J. E., "A Simple Method of Calibrating a GLC Column for Quantitative Fatty Acid Analysis," J. Am. Oil Chemists' Soc., 40, 299-300 (1963)

2025. Edwards, J. A., "Automatic Control with High-Speed Chromatography," Petroleum (London), 23, 213-215 (1960); CA 54, 21878e

2026. Edwards, W. G. H., Clarke, J. G., and Williamson, A. G., "Detection of Trace Quantities of Acenaphthene by Gas Chromatography," Nature, 190, 531 (1961); CA 55, 20792c

2027. Effenberger, M., "Applications of Gas Chromatography to the Analysis of Sewage Gases," Gas Chromatog. Symp., Brno, Czech., June 1962; Abstr., J. Gas Chromatog., 1, No. 4, 10 (1963)

2028. Egger, K., "Differentiation of Types of Glycosides of Flavanols by Thin-Layer Chromatography on Polyamides," Z. Anal. Chem., 182, 161-166 (1961); CA 55, 23187i

2029. Eggertsen, F. T., and Groennings, S., "Determination of Five- to Seven-Carbon Saturates by Gas Chromatography," Anal. Chem., 30, 20-25 (1958); CA 52, 8519f

2030. Eggertsen, F. T., and Groennings, S., "Determination of Small Amounts of n-Paraffins by Molecular Sieve - Gas Chromatography," Anal. Chem., 33, 1147-1150 (1961); CA 56, 1678c; 139th Natl. ACS Mtg., St. Louis, Mo., March, 1961, Program Abstr., p. 6Q

2031. Eggertsen, F. T., Gorennings, S., and Holst, J. J., "Analytical Distillation by Gas Chromatography. Programmed Temperature Operation," Anal. Chem., 32, 904-909 (1960); CA 54, 25725g

2032. Eggersten, F. T., and Knight, H. S., "Gas Chromatography. Effect of Type and Amount of Solvent on Analysis of Saturated Hydrocarbons," Anal. Chem., 30, 15-20 (1958); CA 52, 8518h

2033. Eggertsen, F. T., Knight, H. S., and Groennings, S., "Gas Chromatography. Use of Liquid-Modified Solid Adsorbent to Resolve C_5 and C_6 Saturates," Anal. Chem., 28, 303-306 (1956); CA 50, 8391i

2034. Eggertsen, F. T., and Nelsen, F. M., "Gas Chromatographic Analysis of Engine Exhaust and Atmosphere. Determination of C_2 to C_5 Hydrocarbons," Anal. Chem., 30, 1040-1043 (1958); CA 52, 14416d

2035. Eglinton, G., Hamilton, R. J., Hodges, R., and Raphael, R. A., "Gas-Liquid Chromatography of Natural Products and Their Derivatives," Chem. & Ind. (London), 1959, 955-957; CA 53, 21362h

2036. Eglinton, G., and McCrae, W., "Reactive Acetylenic Intermediates: The Synthesis of 1-Bromoacetylenes and Mercury Acetylides," J. Chem. Soc., 1963, 2295-2299

2037. Eglinton, G., and Rodger, M. N., "Pyrolysis of Acetate," Chem. & Ind. (London), 1959 256; CA 54, 342g

2038. Egorov, V. E., "Basic Components of a Gas Chromatograph for the Analysis of Combustion Products," Elektr. St., 33, No. 12, 22-26 (1962); CA 58, 8393c

2039. Eidelman, M., "Determination of Micro-Quantitites of Chlorinated Organic Pesticides in Butter," J. Assoc. Offic. Agr. Chemists, 45, 672-679 (1962); CA 57, 14241a

2040. Eisenmann, J. L., Yamartino, R. L., and Howard, J. F., "Preparation of Methyl-β-hydroxybutyrate from Propylene Oxide, Carbon Monoxide, Methanol, and Dicobalt Octacarbonyl," J. Org. Chem., 26, 2102-2104 (1961); CA 55, 24556f

2041. Eisner, J., Wong, N. P., Firestone, D., and Bond, J., "Gas Chromatography of Unsaponifiable Matter. I. Butter and Margarine Sterols," J. Assoc. Offic. Agr. Chemists, 45, 337-343 (1962); CA 57, 2634e

2042. Eizen, O. G., Arumeel, E., and Joonson, V., "Use of Gas Chromatography for the Determination of the Chemical Composition of Low-Boiling Products from the Thermal Processing of Oil Shale," Izvest. Akad. Nauk Eston. S.S.R., Ser. Fiz.-Mat. i Tekh. Nauk, 9, No. 2, 113-120 (1960); CA 55, 13821e

2043. Elderfield, R. C., and McClenachan, E. C., "Pyrolysis of the Products of the Reaction of o-Aminobenzenethiols with Ketones," J. Am. Chem. Soc., 82, 1982-1988 (1960); CA 54, 17371a

2044. El-Din, S. M. Badr, and Mattick, J. F., "Isolation and Identification of Monocarbonyl Compounds from Cheddar Cheese," 58th Annual Mtg., Am. Dairy Sci. Assoc. Lafayette, Ind., June 1963; Abstr., J. Dairy Sci., 46, 602 (1963)

2045. Eliel, E. L., Haubenstock, H., and Acharya, R. V., "Conformational Analysis. VIII. The Conformational Equilibrium Constant of the Carbethoxyl Group," J. Am. Chem. Soc., 83, 2351-2354 (1961); CA 56, 13609a

2046. Eliel, E. L., and Rerick, M. N., "Reductions with Metal Hydrides. VII. Reduction of Epoxides with Lithium Aluminum Hydride-Aluminum Chloride," J. Am. Chem. Soc., 82, 1362-1367 (1960); CA 54, 15290c

2047. Eliel, E. L., and Rerick, M. N., "Reductions with Metal Hydrides. VIII. Reductions of Ketones and Epimerization of Alcohols with Lithium Aluminum Hydride-Aluminum Chloride," J. Am. Chem. Soc., 82, 1367-1372 (1960); CA 54, 15291g

2048. Ellin, R. I., Mendeloff, A. I., and Turner, D. A., "Determination of 3, 7, 12-Trioxocholanic Acid in Biological Fluids by Gas-Liquid Chromatography," Anal. Biochem., 4, 198-203 (1962); CA 58, 5968h

2049. Ellis, C. P., "A Method for the Introduction of Unstable, Solid Metallic Chlorides onto Gas Chromatography Columns," Anal. Chem., 35, 1327-1328 (1963)

2050. Ellis, J. F., and Forrest, C. W., "Analysis of Permanent Gases by Gas-Solid Chromatography Using an Ionization Method for Detection," Anal. Chim. Acta, 24, 329-333 (1961); CA 55, 18435c

2051. Ellis, J. F., Forrest, C. W., and Allen, P. L., "The Quantitative Analysis of Mixtures of Corrosive Halogen Gases by Gas-Liquid Chromatography," Anal. Chim. Acta, 22, 27-33 (1960); CA 54, 6400d

2052. Ellis, J. F., Forrest, C. W., and Howe, D. D., "Estimation of Surface Areas by a Gas Chromatographic Method," U. K. At. Energy Authority, DEG Report No. 229, 8 pp. (1960); CA 55, 6101c

2053. Ellis, J. F., and Iveson, G., "The Application of Gas-Liquid Chromatography to the Analysis of Volatile Halogen and Inter-Halogen Compounds," in "Gas Chromatography 1958," edited by D. H. Desty, Academic Press, Inc., New York, 1958, pp. 300-309; CA 53, 19475a

2054. Ellison, T., Thomson, W. A. B., and Strong, F. M., "Volatile Fatty Acids from Axenic Ascaris Lumbricoides," Arch. Biochem. Biophys., 91, No. 2, 247-254 (1960); CA 55, 7680f

2055. El-Negoumy, A. M., and Hammond, E. G., "The Characterization of the Oxidized Flavor in Butteroil," 55th Annual Mtg., Am. Dairy Sci. Assoc., Logan, Utah, June 1960; Abstr., J. Dairy Sci., 43, 840 (1960)

2056. El-Negoumy, A. M., Miles, D. M., and Hammond, E. G., "Partial Characterization of the Flavors of Oxidized Butteroil," J. Dairy Sci., 44, 1047-1056 (1961); CA 55, 20253a

2057. El-Negoumy, A. M., Puchal, M. S. de, and Hammond, E. G., "Relation of Linoleate and Linolenate to the Flavors of Autooxidized Milk Fat," J. Dairy Sci., 45, 311-316 (1962); CA 57, 6371d

2058. Elsbach, P., "Composition and Synthesis of Lipids in Resting and Phagocytizing Leukocytes," J. Exptl. Med., 110, 969-980 (1959); CA 54, 10092f

2059. Elsey, P. G., "Gas Chromatographic Determination of Dissolved Oxygen in Lubricating Oil," Anal. Chem., 31, 869-870 (1959); CA 53, 14818e

2060. Elsey, P. G., "Gas Chromatographic Separations with Dual Reversible Flow Columns," 8th Detroit Anachem Conf., Detroit, Mich., October 1960

2061. Elsey, P. G., and Rye, T., "Water-Cooled Sample Injection Port for High Temperature Gas Chromatography," J. Chromatog., 5, 88-89 (1961); CA 56, 4552d

2062. Elsey, P. G., and Wagner, W., "The Necessity for Signal Response Values in Gas Chromatographic Analysis," 135th Natl. ACS Mtg., Boston, Mass., April 1959, Program Abstr., p. 24B

2063. El-Shazly, K., "Degradation of Protein in the Rumen of Sheep. I. Volatile Fatty Acids, Including Branched-Chain Isomers, Found in Vivo," Biochem. J., 51, 642-647 (1952); CA 46, 11270e

2064. El-Shazly, K., "Degradation of Protein in the Rumen of Sheep. II. The Action of Rumen Microorganisms on Amino Acids," Biochem. J., 51, 647-653 (1952); CA 46, 11270e

2065. Elvidge, D. A., and Proctor, K. A., "The Use of Gas Chromatography for Determining Water in Pharmaceutical Preparations," Analyst, 84, 461-463 (1959); CA 54, 827i

2066. Elzakker, A. H. M. van, and Zutphen, H. J. van, "Gas-Chromatography Study of Cocoa Aroma," Z. Lebensm.-Untersuch. u. -Forsch, 115, 222-226 (1961); CA 55, 27693i

2067. Embden, I. C. M. van, "The Properties of Commerical Tetrapropylene Used as Raw Material in the Manufacture of Tetrapropylenebenzenesulfonate," Fette, Seifen, Anstrichmittel, 63, 456-460 (1961); CA 55, 18141d

2068. Emerson, W. W., "Organoclay Complexes," Nature, 180, 48-49 (1957); CA 51, 17329f

2069. Emery, E. M., and Koerner, W. E., "Gas Chromatographic Determination of Trace Amounts of the Lower Fatty Acids in Water," Anal. Chem., 33, 146-147 (1961); CA 55, 10212i

2070. Emery, E. M., and Koerner, W. E., "Double-Column Programmed Temperature Gas Chromatography," Anal. Chem., 33, 523-527 (1961); CA 55, 16027g; 139th Natl. ACS Mtg., St. Louis, Mo., March 1961, Program Abstr., p. 6Q

2071. Emery, E. M., and Koerner, W. E., "Double-Column Programmed Temperature Gas Chromatography. Volatile Polar Column Packings and Quantitative Aspects," Anal. Chem., 34, 1196-1198 (1962); CA 58, 5017e

2072. Emmerling, J., "Use of the Gas-Phase Chromatograph in Solvent Problems," Offic. Dig., Federation Paint & Varnish Production Clubs, 30, 1172-1177 (1958); CA 55, 2135g

2073. Emmett, P. H., "Applications of Gas Chromatography," in "Gas Chromatography 1958," edited by D. H. Desty, Academic Press, Inc., New York, 1958, pp. 267-269

2074. Emmett, P. H., "The Use of Tracers and Gas Chromatography in Studying Catalytic Reactions," 135th Natl. ACS Mtg., Boston, Mass., April 1959, Program Abstr., p. 3Q; Preprints, Div. Petrol. Chem., 4, No. 2, C79-C87

2075. Emmett, P. H., "A Decade of Progress in Heterogeneous Catalysis," 140th Natl. ACS Mtg., Chicago, Ill., September 1961, Program Abstr., p. 17S

2076. Ency. Chem. Tech., "Chromatography," Vol. 3, Interscience Ency., Inc., New York, 1949, pp. 928-935

2077. Endow, N., Doyle, G. J., and Schuck, E. A., "Products and Considerations on the Mechanism of Photo-Oxidation of Isobutene-Nitrogen Dioxide Mixtures in Air," 138th Natl. ACS Mtg., New York, N. Y., September 1960, Program Abstr., p. 5V

2078. Enebo, L., "Volatile Minor Compounds in Beers," Proc. European Brewing Convention, Copenhagen, 1957, 370-376; J. Sci. Food Agr., 9, 281 (1958); CA 52, 19011i

2079. Ent. W. L., "Air Products Company Wide Uniform Quality Program," Analyzer, 4, No. 3, 15-18 (July 1963)

2080. Erb, E., "Ortho-Para-Hydrogen Analysis by Gas Chromatography," Gas Chromatog., Intern. Symposium, 2nd, East Lansing, Mich., 1959, 357-362 (Pub. 1961); CA 55, 18421f

2081. Erb, E., "Ortho-Para-Hydrogen Analysis by Gas Chromatography," in "Gas Chromatography," edited by H. J. Noebels, R. F. Wall, and N. Brenner, Academic Press, Inc., New York, 1961, pp. 357-362

2082. Ergun, S., "Precision Measurements of Gas Flow Rates. Mathematical Procedure in the Calibration of Flowmeters," Anal. Chem., 25, 790-792 (1953)

2083. Erickson, E. L., "Application of Chromatography to Natural-Gas Analysis," Instr. Control Systems, 33, 1362-1365 (1960); CA 57, 10103b

2084. Erickson, L. C., and Hield, H. Z., "Determination of 2,4-Dichlorophenoxyacetic Acid in Citrus Fruit," J. Agr. Food Chem., 10, 204-207 (1962); CA 57, 10299b

2085. Erni, M., "A Perfumer's View on the Synthetic Rose Alcohols and Linalool from Pinene," Perfumery Essent. Oil Record, 51, 541-544 (1960); CA 55, 3010h

2086. Erofeev, B. V., et al., "Conjugate Decarboxylation During Autoxidation of Isopropylbenzene in a Mixture with Butyric Acid-1-C14," Dokl. Akad. Nauk Belorussk. SSR, 4, No. 4, 160-163 (1960); CA 58, 4399f

2087. Errede, L. A., and Cassidy, J. P., "Chemistry of Xylylenes. IV. Stabilization of Benzyl Radicals in Solution," J. Org. Chem., 24, 1890-1892 (1959); CA 54, 10904f

2088. Errede, L. A., and Cassidy, J. P., "The Chemistry of Xylylenes. V. The Formation of Anthracenes Via Fast Flow Pyrolysis of Toluenes and Related Compounds," J. Am. Chem. Soc., 82, 3653-3658 (1960); CA 55, 13390d

2089. Erwin, E. S., Marco, G. J., and Emery, E. M., "Volatile Fatty Acid Analyses of Blood and Rumen Fluid by Gas Chromatography," J. Dairy Sci., 44, 1768-1771 (1961); CA 56, 3755a

2090. Esayan, L., and Esayan, M., "Chromatography of Gases. I. Separation of Acetylene-Hydrogen-Methane Mixtures," Rev. chim. (Bucharest), 8, 447-452 (1957); CA 52, 4397c

2091. Esayan, L., Gherman, M., Stefan, V., and Istrate, E., "Chromatography of Gases. II.," Rev. chim. (Bucharest), 9, 125-128 (1958); CA 53, 5962b

2092. Esayan, M., "Calculation of Hypersorption Columns," Rev. chim. (Bucharest), 7, 198-205 (1956); CA 52, 19283g

2093. Escher, E. E., "In-Process Gas Chromatography," Chem. Eng., 66, No. 15, 113-118 (1959); CA 53, 19475f

2094. Escher, E. E., "Process Chromatograph Sample Handling System," 2nd Alberta Gas Chromatography Discussion, Edmonton, Alberta, February 1959

2095. Espe, W., and Kuhn, A., "Gas and Vapor Filling of Ionization Detectors," Vakuum-Tech., 10, 16-20 (1961); CA 55, 12947f

2096. Espe, W., and Kuhn, A., "Gas and Vapor Fillings of Ionization Detectors. III.," Vakuum-Tech., 10, 84-86 (1961); cf. ibid., 46-54; CA 55, 22934d

2097. Esposito, G. G., "The Application of Temperature Programmed Gas Chromatography to the Analysis of Lacquer Solvents and Thinners," U. S. Dept. Com., Office Tech. Serv., PB Rept. 171,033, 10 pp. (1960); CA 57, 4791b

2098. Esposito, G. G., "The Identification of Polyols in Synthetic Coating Resins by Programmed Temperature Gas Chromatography," U. S. Dept. Com., Office Tech. Serv., PB Rept. 181,354, 11 pp. (1962)

2099. Esposito, G. G., "Identification of Carboxylic Acids in Alkyd and Polyester Coating Resins by Programmed Temperature Chromatography," U. S. Dept. Com., Office Tech. Serv., PB Rept. 181,310, 19 pp. (1962); CA 58, 4735f

2100. Esposito, G. G., "Identification of Polyhydric Alcohols in Synthetic Resins by Programmed Temperature Gas Chromatography," Anal. Chem., 34, 1173 (1962); CA 57, 12689c

2101. Esposito, G. G., and Swann, M. H., "Direct Analysis of Solvents in Lacquers by Programmed-Temperature Gas Chromatography," Offic. Dig., Federation Soc. Paint Technol., 33, 1122-1131 (1961); Correction, ibid., p. 1460; CA 56, 4891h

2102. Esposito, G. G., and Swann, M. H., "Identification of Polyhydric Alcohols in Synthetic Resins by Programmed Temperature Gas Chromatography," Anal. Chem., 33, 1854-1858 (1961); CA 56, 11741i

2103. Esposito, G. G., and Swann, M. H., "Identification of Carboxylic Acids in Alkyd and Polyester Coating Resins by Programmed Temperature Gas Chromatography," Anal. Chem., 34, 1048-1052 (1962); CA 57, 9981d

2104. Esselborn, W., and Krebs, K. G., "Simultaneous Gas-Chromatographic Determination of α-, β-, γ-, δ-, and ε-Isomers of Hexachlorocyclohexane," Pharm. Ztg., Ver. Apotheker-Ztg., 107, 464-465 (1962); CA 57, 3827b

2105. Etemad-Moghadame, P., "Partition Chromatography Between Gas and Liquid Phases," Chim. anal., 40, 149-155 (1958); CA 52, 13516a

2106. Ettre, K., and Varadi, P. F., "Pyrolysis - Gas Chromatography Technique for Direct Analysis of Thermal Degradation Products of Polymers," Anal. Chem., 34, 752-757 (1962); CA 57, 3607f; 141st Natl. ACS Mtg., Washington, D. C., March 1962, Program Abstr., p. 20B

2107. Ettre, K., and Varadi, P. F., "Pyrolysis-Gas Chromatographic Technique. Effect of Temperature on Thermal Degradation of Polymers," Anal. Chem., 35, 69-73 (1963); CA 58, 9243b; 142nd Natl. ACS Mtg., Atlantic City, N. J., September 1962, Program Abstr., p. 9B

2108. Ettre, L. S., "Effect of the Surface Area on the Separation in Gas-Liquid Partition Chromatography," J. Chromatog., 4, 166-169 (1960); CA 55, 5049h

2109. Ettre, L. S., "Comparative Surface Measurements of Adsorbents and Support Materials for Gas Chromatography and the Effect of the Surface Area on the Chromatographic Separation," 11th

Pittsburgh Conf. on Anal. Chem. & Appl. Spectroscopy, Pittsburgh, Pa., February-March 1960, Program Abstr., p. 41

2110. Ettre, L. S., "Quantitative Reliability of Hydrocarbon Analyses with a Capillary Column-Hydrogen Flame Ionization Detector System," 64th Annual Mtg., Am. Soc. Testing Materials, Symposium on Capillary Gas Chromatography, R. D. IV, ASTM D-2, Atlantic City, N. J., June 1961

2111. Ettre, L. S., "Gas Chromatography: New Developments and Their Relative Merits," 1st Annual Conf. on Pharmaceutical Analysis, Land O'Lakes, Wisc., September 1961

2112. Ettre, L. S., "The Application of Golay Column-Flame Ionization Detector Systems for Quantitative Analysis," Regional Mtg. - ACS Southeast and Southwest Sections, New Orleans, La., December 1961

2113. Ettre, L. S., "Practical Aspects of Golay-Column Gas Chromatography," Perkin-Elmer Instrument News, 13, No. 1a, 1 (Fall, 1961)

2114. Ettre, L. S., "Bibliography on Golay Columns and Their Applications," Perkin-Elmer Instrument News, 13, No. 1a, 1-7 (Fall, 1961)

2115. Ettre, L. S., "Application of Gas Chromatographic Methods for Air Pollution Studies," J. Air Pollution Control Assoc., 11, No. 1, 34-43 (1961); CA 55, 8716b

2116. Ettre, L. S., "Bibliography on Gas Chromatography. II.," in "Gas Chromatography," edited by H. J. Noebels, R. F., Wall, and N. Brenner, Academic Press, Inc., New York, 1961, pp. 375-455

2117. Ettre, L. S., "Wide Range Application of Golay Columns," 13th Pittsburgh Conf. on Anal. Chem. & Appl. Spectroscopy, Pittsburgh, Pa., March 1962, Program Abstr., p. 51

2118. Ettre, L. S., "Relative Response of the Flame Ionization Detector," J. Chromatog., 8, 525-530 (1962); CA 58, 3861h

2119. Ettre, L. S., "Relative Molar Response of Hydrocarbons on the Ionization Detectors," Gas Chromatog., Intern. Symposium, 1961; CA 58, 3869c; pub. in "Gas Chromatography," edited by N. Brenner, J. E. Callen, and M. D. Weiss, Academic Press, Inc., New York, 1962, pp. 307-328

2120. Ettre, L. S., "Practical Aspects of Golay Columns in Gas-Chromatographic Analysis," Res. Develop. Ind., No. 15, 42-47 (1962); CA 58, 4175h

2121. Ettre, L. S., "Possibilities of Investigating and Expressing Column Efficiencies," J. Gas Chromatog., 1, No. 2, 36-47 (1963)

2122. Ettre, L. S., "The Capillary Column in VPC," 1963 Eastern Analytical Symposium, New York, N. Y., November 1963

2123. Ettre, L. S., "Open Tubular Columns in Gas Chromatography," Plenum Press, New York, in press November 1963

2124. Ettre, L. S., and Averill, W., "Investigation of the Linearity of a Stream Splitter for Capillary Gas Chromatography," 139th Natl. ACS Mtg., St. Louis, Mo., March 1961, Program Abstr., p. 33B; Preprints Div. Petrol Chem., 6, No. 2, B79-B89 (1961); Anal. Chem., 33, 680-684 (1961); CA 56, 10882a

2125. Ettre, L. S., and Averill, W., "Newest Developments in Gas Chromatography with Respect to the Analysis of Very Small Quantities," Microchem. J., Symp. Ser., 2, 715-732 (1962); CA 58, 5015a; Intern. Symp. on Microchemical Techniques, University Park, Pa., August 1961

2126. Ettre, L. S., and Averill, W., "Application of Golay Columns with Different Diameters at Both Isothermal and Programmed Temperature Conditions," 10th Detroit Anachem. Conf., Detroit, Mich., October 1962, Program Abstr., p. 31

2127. Ettre, L. S., Averill, W., and Kabot, F. J., "Gas Chromatographic Analysis of Fatty Acids," Application Bulletin No. GC-AP-001, Perkin-Elmer Corp., Norwalk, Conn.

2128. Ettre, L. S., and Brenner, N., "The Microcatlytic-Chromatographic Technique and Its Application to Commerical Gas Chromatographs," 137th Natl. ACS Mtg., Cleveland, Ohio, April 1960, Program Abstr., p. 27B

2129. Ettre, L. S., and Brenner, N., "Molecular Sieves as Subtractors in Gas Chromatographic Analysis. III. The Secondary Effect of the Molecular Trap Column," J. Chromatog., 3, 235-238 (1960); CA 54, 19086h

2130. Ettre, L. S., and Brenner, N., "Simple Accessory to a Commercial Gas Chromatograph for Microcatalytic Studies," J. Chromatog., 3, 524-530 (1960); CA 55, 5049i

2131. Ettre, L. S., Brenner, N., and Cieplinski, E. W., "Surface Determination in Constant Gas Flow. Apparatus Development and Comparative Investigations," Z. Physik. Chem. (Leipzig), 219, 17-35 (1962); CA 57, 1583e

2132. Ettre, L. S., Cieplinski, E. W., and Averill, W., "Application of Open Tubular (Golay) Columns with Large Diameter," J. Gas Chromatog., 1, No. 2, 7-16 (1963); Correction, ibid., 1, No. 4, 44 (1963)

2133. Ettre, L. S., Cieplinski, E. W., and Brenner, N., "Quantitative Aspects of Capillary Gas Chromatography Using Flame Ionization Detector," 30th Mtg., Gulf Coast Spectroscopic Group, Houston, Texas, March 1961

2134. Ettre, L. S., Cieplinski, E. W., and Brenner, N., "Quantitative Aspects of Capillary Gas Chromatography," Fall Instrument Automation Conf. and Exhibit, ISA, Los Angeles, Calif., September 1961

2135. Ettre, L. S., Cieplinski, E. W., and Brenner, N., "Quantitative Aspects of Gas Chromatography with Open Tubular (Golay) Columns," ISA Trans., 2, No. 2, 134-140 (1963); CA 59, 1064b

2136. Ettre, L. S., and Claudy, H. N., "Recent Advances in Gas Chromatographic Instrumentation. Hydrogen Flame Ionization Detector," Chem. in Can., 12, No. 9, 34-36 (1960); Symp. on Gas Chromatography, CIC, Toronto, Ontario, February 1960

2137. Ettre, L. S., Coates, V. J., and Cieplinski, E. W., "New Developments in Gas Chromatographic Instrumentation and Their Application to Industrial Analysis," Achema-European Convention of Chemical Engineering, Frankfurt on Main, West Germany, June 1961

2138. Ettre, L. S., Coates, V. J., and Cieplinski, E. W., "New Developments in Gas-Chromatographic Instrumentation and Their Application to Quantitative Analysis," Dechema Monograph., 43, 241-255 (1962); CA 58, 10703g

2139. Ettre, L. S., Golay, M. J. E., and Norem, S. D., "A Nomographic Approach to Some Problems in Linearly Programmed Temperature Gas Chromatography," 4th Intern. Gas Chromatography Symp., Hamburg, Germany, June 1962; pub. in "Gas Chromatography 1962," ed. by M. van Swaay, Butterworth & Co., London, 1962

2140. Ettre, L. S., and Kabot, F. J., "Quantitative Reproducibility of a Programmed Temperature Gas Chromatographic System with Constant Pressure Drop Using Packed and Golay Columns," Anal. Chem., 34, 1431-1434 (1962); CA 58, 4b

2141. Eucken, A., and Knick, H., "Automatic Procedure for the Microanalytical Separation of Low-Boiling Hydrocarbons by Desorption," Brennstoff-Chem., 17, 241-244 (1936); CA 31, 1727[9]

2142. Euston, C. B., "Preparative Scale Gas Chromatography," 14th Annual Fisk Univ. Infrared Spectroscopy Institute, Gas Chromatography Session, Nashville, Tenn., August 1963

2143. Euston, C. B., and Martin, A. J., "A Linear Programmed Temperature Gas Chromatograph Operable to 1000°C. for Organic or Inorganic Separations," 12th Pittsburgh Conf. on Anal. Chem. & Appl. Spectroscopy, Pittsburgh, Pa., February-March 1961, Program Abstr., p. 52

2144. Euston, C. B., and Martin, A. J., "The Development of a High Temperature (1000°) Gas Chromatograph," U. S. Dept. Com., Office Tech. Serv., AD 268, 700, 21 pp. (1961); CA 58, 7347b

2145. Evans, C. D., Cooney, P. M., and Panek, E. J., "Graphic Aid for Interpreting Gas Chromatograms," J. Am. Oil Chemists' Soc., 39, 210-213 (1962); CA 57, 7a

2146. Evans, D. E. M., Massingham, W. E., Stacey, M., and Tatlow, J. C., "Preparative-Scale Gas Chromatography," Nature, 182, 591-592 (1958); CA 53, 4827h

2147. Evans, D. E. M., and Tatlow, J. C., "The Reactions of Highly Fluorinated Organic Compounds. VIII. The Gas-Chromatographic Separation on a Preparative Scale, and Some Reactions of 3H- and 4H-Nonafluorocyclohexene," J. Chem. Soc., 1955, 1184-1188; CA 50, 1617b

2148. Evans, D. E. M., and Tatlow, J. C., "The Application of Gas Chromatography to Organic Fluorine Chemistry," in "Vapour Phase Chromatography," edited by D. H. Desty, Academic Press, Inc., New York, 1957, pp. 256-265

2149. Evans, E. A., and Stanford, F. G., "Decomposition of Tritium-Labelled Organic Compounds," Nature, 197, 551-555 (1963)

2150. Evans, E. D., "The Isolation and Estimation of Heavy Normal Paraffin Content of Chromatographic Fractions of Petroleum by Urea Adduction and Gas Chromatography," 18th Southwest Regional Mtg., ACS, Dallas, Texas, December 1962

2151. Evans, E. S., and Wing, Jr., F. E., "Design and Performance of a New Ionization Detector System for Gas-Liquid Chromatography," 10th Pittsburgh Conf. on Anal. Chem. & Appl. Spectroscopy, Pittsburgh, Pa., March 1959, Program Abstr., p. 54

2152. Evans, J. B., Quinlan, J. E., and Willard, J. E., "Making Labelled Compounds. Chemical Effects of Nuclear Transformation," Ind. Eng. Chem., 50, 192-195 (1958); CA 52, 6953f

2153. Evans, J. B., Quinlan, J. E., and Willard, J. E., "Evidence from Gas Chromatography on the Multiplicity of Radioactive Products Formed in Organic Media by the Activation of Halogens by Nuclear Processes," 130th Natl ACS Mtg., Atlantic City, N. J., September 1956, Program Abstr., p. 27R

2154. Evans, J. B., and Willard, J. E., "Use of Gas-Phase Chromatography for the Separation of Mixtures of Carrier-Free Radioactive Substances: Products of Chemical Reactions Activated by Nuclear Processes," J. Am. Chem. Soc., 78, 2908-2909 (1956); CA 50, 11847i

2155. Evans, L., Patton, S., and McCarthy, R. D., "Fatty Acid Composition of the Lipid Fractions from Bovine Serum Lipoproteins," J. Dairy Sci., 44, 475-482 (1961); CA 55, 14625c

2156. Evans, M. B., and Smith, J. F., "Prediction of Retention Data in Gas-Liquid Chromatography from Molecular Formulas," Nature, 190, 905-906 (1961); CA 55, 23159g

2157. Evans, M. B., and Smith, J. F., "Gas-Liquid Chromatography in Qualitative Analysis. I. An Interpolation Method for the Prediction of Retention Data," J. Chromatog., 5, 300-307 (1961); CA 56, 5380f

2158. Evans, M. B., and Smith, J. F., "Gas-Liquid Chromatography in Qualitative Analysis. II. The Reproducibility of Retention Data in R_{X_9} Units," J. Chromatog., 6, 293-311 (1961); CA 56, 12289d

2159. Evans, M. B., and Smith, J. F., "Gas-Liquid Chromatography in Qualitative Analysis. III. The Constancy of ΔME Values Within Homologous Series and Relations Within the Periodic Table," J. Chromatog., 8, 303-307 (1962); CA 58, 7345g

2160. Evans, M. B., and Smith, J. F., The Conversion of Relative Retention Data into R_{X_9} Units," J. Chromatog., 8, 541-544 (1962); CA 58, 3870h

2161. Evans, M. B., and Smith, J. F., "Gas-Liquid Chromatography in Qualitative Analysis. IV. A Simple Method of Calculating R_{X_9} Values at Elevated Column Temperatures," J. Chromatog., 9, 147-153 (1962); CA 58, 7345h

2162. Evans, M. V., and Lord, R. C., "Synthesis and Vibrational Spectrum of Bicyclo[3.2.0]hepta-2, 6-diene," J. Am. Chem. Soc., 83, 3409-3413 (1961); CA 56, 3368i

2163. Evans, R. J., Davidson, J. A., and Bandemer, S. L., "Fatty Acid and Lipid Distribution in Egg Yolks from Hens and Cottonseed Oil or Sterculia Foetida Seeds," J. Nutrition, 73, 282-290 (1961)

2164. Evans, R. S., "Recent Developments in Gas-Chromatographic Detectors," Dechema Monograph., 43, 211-218 (1962); CA 58, 10703f

2165. Evans, R. S., and Scott, P. G. W., "Precision and Accuracy in Gas Chromatographic Analysis," Nature, 190, 710-712 (1961); CA 56, 14900c

2166. Evered, S., and Pollard, F. H., "The Application of Gas Chromatography to the Determination of Retention Data and Activity Coefficients of Some Alkanes, Alkyl Nitrates, Nitroalkanes, and Alcohols on Selected Stationary Phases," J. Chromatog., 4, 451-457 (1960); CA 55, 14011a

2167. Everett, D. H., and Stoddart, C. T. H., "Thermodynamics of Hydrocarbon Solutions from Gas-Liquid Chromatography Measurements. I. Solutions in Dinonylphthalate," Trans. Faraday Soc., 57, 746-754 (1961); CA 55, 23020a

2168. Evrard, E., Bosch, J. van den, De Somer, P., and Joossens, J. V., "Cholesteryl Ester Fatty Acid Patterns of Plasma, Atheromata and Livers of Cholesterol-Fed Rabbits," J. Nutrition, 76, 219-222 (1962)

2169. Evrard, E., Thevelin, M., and Joossens, J. V., "Self-sustained Discharge Detector for Chromatographic Analysis of Permanent Gases," Nature, 193, 59-60 (1962); CA 56, 12290e

2170. Fabian, L. W., Dewitt, H., and Carnes, M. A., "Laboratory and Clinical Investigation of Some Newly Synthesized Fluorocarbon Anesthetics," Anesthesia & Analgesia, 39, 456-462 (1960); CA 55, 1937h

2171. Fabriani, G., "Chromatographic Studies of Lipids and Sterols in Soft and Durum Wheat," Getreide Mehl, 12, 109-111 (1962); CA 58, 3687e

2172. Fabriani, G., "Chromatographic Studies of Fats and Sterols in Soft and Durum Wheat," Ber. Getreidchemiker-Tagung, Detmold, 1962, 26-32; CA 58, 8231h

2173. Fabrizio, F. A., King, R. W., Cerato, C. C., and Loveland, J. W., "Determination of Trace Hydrocarbon Impurities in Petroleum Benzene and Toluene by Gas Chromatography," Anal. Chem., 31, 2060-2063 (1959); CA 54, 6391b; 135th Natl. ACS Mtg., Boston, Mass., April 1959, Program Abstr., p. 20B

2174. Fagerson, I. S., "Gas Chromatography. A New Analytical Technique," Mass. Agr. Expt. Station Contrib. No. 1092 (1956)

2175. Fagerson, I. S., "Gas Chromatography. Versatile Analytical Tool for Quality Control, Research," Food Eng., 29, 62-63 (1957)

2176. Fagerson, I. S., "Gas Chromatography Strides in Food Research," Food Eng., 31, No. 7, 80-81 (1959)

2177. Fagerson, I. S., Sawyer, F. M., Brenner, N., and Cieplinski, E., "A Gas Chromatographic Comparator Technique for Estimation of Flavor Differences," 134th Natl. ACS Mtg., Chicago, Ill., September 1958, Program Abstr., p. 33A

2178. Fairchild, E. J., and Stokinger, H. E., "Toxicologic Studies on Organic Sulfur Compounds. I. Acute Toxicity of Some Aliphatic and Aromatic Thiols (Mercaptans)," Am. Ind. Hyg. J., 19, 171-189 (1958); CA 55, 20181f

2179. Falconer, J. W., and Knox, J. H., "High Temperature Oxidation of Propane," Proc. Roy. Soc. (London), A250, 493-513 (1959); CA 54, 17011i

2180. Falconer, W. E., and Cvetanovic, R. J., "Separation of Isotopically Substituted Hydrocarbons by Partition Chromatography. Thermodynamic Properties as Calculated from Retention Volumes," Anal. Chem., 34, 1064-1066 (1962); CA 57, 11837d

2181. Falconer, W. E., Hunter, T. F., and Trotman-Dickenson, A. F., "The Thermal Isomerization of Cyclopropane," J. Chem. Soc., 1961, 609-611; CA 55, 14283d

2182. Falconer, W. E., Knox, J. H., and Trotman-Dickenson, A. F., "Competitive Oxidations. II. The Lower Alkanes and Cyclopropane," J. Chem. Soc., 1961, 782-792; CA 55, 16391g

2183. Fales, H. M., "Gas Chromatography of High Molecular Weight Polyfunctional Amines," 3rd Annual Eastern Analytical Symp. & Instrument Exhibit, New York, N. Y., November 1961

2184. Fales, H. M., Haahti, E. O. A., Luukainen, T., VandenHeuvel, W. J. A., and Horning, E. C., "Mil-

80

ligram-Scale Preparative Gas Chromatography of Steroids and Alkaloids," Anal. Biochem., 4, 296-305 (1962); CA 58, 11675e

2185. Fales, H. M., and Pisano, J. J., "Gas Chromatography of Biologically Important Amines," Anal. Biochem., 3, 337-342 (1962); CA 57, 10128h

2186. Faley, R. L., and Long, J. F., "High Temperature Seal for Gas Chromatography Detectors," Anal. Chem., 32, 302 (1960); CA 54, 12674h

2187. Farenden, P. J., and Mendoza, M. P., "Some Aspects of Gas Chromatography," Bull. Brit. Coal Utilisation Research Assoc., 23, 366-376 (1959); CA 54, 23598i

2188. Farges, G., and Kergomard, A., "The Action of Anhydrous Hydrofluoric Acid on Some Epoxides," Bull. soc. chim. France, 1963, 51-57

2189. Farhi, L. E., Edwards, A. W. T., and Homma, T., "Determination of Dissolved N_2 in Blood by Gas Chromatography and (a-A)N_2 Difference," J. Appl. Physiol., 97, 97-106 (1963); Biol. Abstr., 42, 5619 (1963)

2190. Farmilio, C. G., and Davis, T. W. M., "Paper and Gas Chromatographic Analysis of Cannabis," J. Pharm. Pharmacol., 13, 767 (1961); CA 56, 13010d

2191. Farnow, H., and Porsch, F., "Geraniol and Nerol," Dragoco Report, 7, 3-10 (1960); CA 57, 2351i

2192. Farnow, H., and Porsch, F., "Constitution of Ocimene," Dragoco Report, 8, 183-193 (1961); CA 57, 2351i

2193. Farquhar, J. W., "Human Erythrocyte Phosphoglycerides. I. Quantification of Plasmalogens. Fatty Acids and Fatty Glycerides," Biochim. et Biophys. Acta, 60, 80-89 (1962); Biol. Abstr., 40, 21989

2194. Farquhar, J. W., "Identification and Gas Liquid Chromatographic Behavior of Plasmalogen Aldehydes and Their Acetal, Alcohol, and Acetylated Alcohol Derivatives," J. Lipid Research, 3, 21-30 (1962)

2195. Farquhar, J. W., Insull, Jr., W., Rosen, P., Stoffel, W., and Ahrens, Jr., E. H., "The Analysis of Fatty Acid Mixtures by Gas-Liquid Chromatography: Construction and Operation of an Ionization Chamber Instrument," Nutrition Reviews, 17, No. 8, Part II, 30 pp. (1959)

2196. Farre-Rius, F., Henniker, J., and Guiochon, G., "Wetting Phenomena in Gas Chromatography Capillary Columns," Nature, 196, 63-64 (1962); CA 57, 15829i

2197. Farrington, P. S., Pecsok, R. L., Meeker, R. L., and Olson, T. J., "Detection of Trace Constituents by Gas Chromatography. Analysis of Polluted Atmosphere," Anal. Chem., 31, 1512-1516 (1959); CA 53, 22647i

2198. Farrugia, V. J., and Jarreau, C. L., "Separation of Isopropyl Alcohol from Aliphatic Sulfides and Thiols by Gas Chromatography," Anal. Chem., 34, 271-273 (1962); CA 56, 14904c

2199. Fast, P. G., and Brown, A. W. A., "Lipids of DDT-Resistant and Susceptible Larvae of Aedes Aegypti," Ann. Entomol Soc. Am., 55, 663-672 (1962); CA 58, 10549a

2200. Favre, J. A., Hines, W. J., and Smith, D. E., "Application of Gas Chromatography to the Analysis of Natural Gas," Proc. Ann. Conv. Natl. Gasoline Assoc. Am., Tech. Papers, 37, 27-32 (1958); CA 54, 1834c

2201. Favre, J. A., Hines, W. J., and Smith, D. E., "Natural Gas Analysis: How Phillips Applies Gas Chromatography," Petrol. Refiner, 37, 251-254 (1958); CA 53, 1681a

2202. Fawcett, J. S., and Taylor, B. W., "High-Temperature Gas Chromatography Apparatus," 132nd Natl. ACS Mtg., New York, N. Y., September 1957, Program Abstr., p. 34B; Preprints, Div. Petrol. Chem., 2, No. 4, D85-D89 (1957); CA 54, 20354f

2203. Fay, H., and Voelker, M. W., "Apparatus and 'Chromatographic' Method for Studying the Adsorption of Traces of Hydrocarbons from Air," 13th Pittsburgh Conf. on Anal. Chem. & Appl. Spectroscopy, Pittsburgh, Pa., March 1962, Program Abstr., p. 54

2204. Fazakerley, H., Garratt, P. G., Hills, P. R., and Roberts, R., "Olfactory and Chemical Changes in Irradiated Essential Oils," Intern. J. Appl. Radiation and Isotopes," 11, No. 4, 174-183 (1961); CA 56, 4881b

2205. Fearns, E. C., "Gas Chromatography: Sharpness Control of Flavor Components," Food Eng., 31, No. 7, 78-80 (1959)

2206. Feichtinger, H., "Exact Micro-Rapid Analysis of Gases," Arch. Eisenhittenw., 26, 127-130 (1955); CA 49, 7447a

2207. Feichtinger, H., Baechtold, H., and Brauner, K., "Determination of Gas in Metals by Gas Chromatography," Schweiz. Arch. Angew. Wiss. u. Tech., 28, No. 3, 125-126 (1962); CA 57, 9205a

2208. Feichtinger, H., Baechtold, H., and Schuhknecht, W., "Improved Method for the Rapid and Accurate Determination of Gases in Metals," Schweiz. Arch. Angew. Wiss. u. Tech., 25, 426-439 (1959); CA 54, 11826e

2209. Feinland, R., Andreatch, A. J., and Cotrupe, D. P., "Automotive Exhaust Gas Analysis by Gas-Liquid Chromatography Using Flame Ionization Detection. Determination of C_1 to C_6 Hydrocarbons," Anal. Chem., 33, 991-994 (1961); CA 56, 10508c; 138th Natl. ACS Mtg., New York, N.Y., September 1960, Program Abstr., p. 7B

2210. Feinland, R., and Cincotta, J. J., "Determination of Organic Compounds in Dilute Aqueous Solution by Gas-Liquid Chromatography," 14th Pittsburgh Conf. on Anal. Chem. & Appl. Spectroscopy, Pittsburgh, Pa., March 1963, Program Abstr., p. 62

2211. Feinland, R., Sass, J., and Buckler, S. A., "Determination of Trialkylphosphines and Their Oxidation Products by Gas Liquid Chromatography," Anal. Chem., 35, 920-921 (1963)

2212. Fejes, P., Czaran, L., and Schay, G., "Frontal Gas Chromatography by Consideration of the Change of Flow Rate Arising as a Consequence of Sorption. I. Determination of Adsorption Isotherms from the Shape of Stationary Fronts in the Diffusion Range," Magyar Kem. Folyoirat, 68, No. 1, 11-19 (1962); CA 57, 2878b

2213. Fejes, P., and Engelhardt, J., "Quantitative Calculation of Chromatographic Peaks with Correction for Thermal Conductivity," Zhur. Fiz. Khim., 34, 2355-2362 (1960); CA 55, 12139c

2214. Fejes, P., Fromm-Czaran, E., and Schay, G., "Frontal Gas Chromatography; the Variations of Flow Rate During Sorption. I. Determination of Adsorption Isotherms on the Basis of the Shape of the Stationary Front in the Diffusion Region," Acta Chim. Acad. Sci. Hung., 33, No. 1, 87-105 (1962); CA 58, 8426g

2215. Fejes, P., and Schay, G., "On the Theory of Steady Chromatographic Gas Fronts," Acta Chim. Acad. Sci. Hung., 17, 377-388 (1958); CA 53, 17630g

2216. Fejes, P., and Schay, G., "Adsorption and Desorption of Nitrogen on Charcoal," Acta Chim. Acad. Sci. Hung., 14, 439-452 (1958); CA 52, 15187a

2217. Feldman, G., "The Analysis of Fatty Acids of Tissue Lipids," Biochemical Gas Chromatography Seminar, Wilkens Instrument & Research, Inc., New Orleans, La., July 1963

2218. Feldstein, M., "Application of Infrared Spectrophotometry and Gas-Liquid Chromatography to the Analysis of Volatile Substances," J. Forensic Sci., 5, 266-275 (1960); CA 54, 12879c

2219. Fellows, E. G., "Gas Chromatography and Its Application to Continuous Analysis," 5th Natl. ISA Symp. on Instrumental Methods of Analysis, Houston, Texas, May 1958

2220. Fellows, E. G., "Chromatography Analyzes Gas and Vapor Products in the Plant," Control Eng., 4, No. 7, 75-81 (1957)

2221. Fells, I., Howells, T. J., and Patrick, M. A., "Analysis of Fuel Gases and Combustion Products," J. Inst. Fuel, 34, 283-290 (1961); CA 55, 21539h

2222. Feltkamp, H., and Thomas, K. D., "The Separation of Stereoisomeric Compounds. IV. The Gas Chromatographic Separation of Stereoisomeric Cyclohexylamines," J. Chromatog., 10, 9-14 (1963)

2223. Felton, H. R., "A Novel High Temperature Gas Chromatography Unit," in "Gas Chromatography," edited by V. J. Coates, H. J. Noebels, and I. S. Fagerson, Academic Press, Inc., New York, 1958, pp. 131-143

2224. Felton, H. R., "Developments in Gas Chromatographic Instrumentation," 6th Instrumental Methods of Analysis Symp., ISA, Montreal, Canada, June 1960

2225. Felton, H. R., "Gas Chromatography of Fluoroalcohols and Fluoroacrylates," 3rd Intern. Symp. on Gas Chromatography, ISA, East Lansing, Mich., June 1961

2226. Felton, H. R., "Gas Chromatography of Fluoroalcohols and Fluoroacrylates," in "Gas Chromatography," edited by N. Brenner, J. E. Callen, and M. D. Weiss, Academic Press, Inc., New York, 1962, pp. 437-442

2227. Felton, H. R., "Preparative Gas Chromatography," 4th Annual Eastern Analytical Symp., New York, N. Y., November 1962, Program Abstr., p. 25

2228. Felton, H. R., "Gas Chromatography of Fluoroalcohols and Fluoroacrylates," Gas Chromatog., Intern. Symp., 1961, 3, 437-441 (Pub. 1962); CA 58, 3888f

2229. Felton, H. R., "A New Concept in Gas Chromatography," Univ of Houston Intern. Symp. on Advances in Gas Chromatography, Houston, Texas, January 1963

2230. Felton, H. R., "Elementary Theory of Preparative Chromatography," 5th Annual Gas Chromatography Institute, Canisius College, Buffalo, N. Y., April 1963

2231. Felton, H. R., "Separation by Preparative Scale Gas Chromatography," J. Gas Chromatog., 1, No. 5, 12-15 (1963)

2232. Felton, H. R., and Buehler, A. A., "High Temperature Thermal Conductivity Cell," Anal. Chem., 30, 1163 (1958); Correction, ibid., p. 1425; CA 52, 14240h

2233. Fenaroli, G., "Lavender and Lavandin. II. New Variety," Riv. ital. essenze profumi, piante offic., olii vegetali, saponi, 42, 489-494 (1960); CA 52, 9794g

2234. Feng, P. Y., Glasson, W. A., and Marshall, S. A., "The Nature of Free Radicals in Irradiated Chemical [Aliphatic] Systems," U. S. Dept. Com., Office Tech. Serv., PB Rept. 171,596, 64 pp. (1960); CA 59, 1447d

2235. Feng, P. Y., and Krotoszynski, B. K., "A Split Temperature Column System for Gas Chromatographic Analyses," Nature, 188, 311-312 (1960); CA 55, 20531d

2236. Ferrand, M. R., "Gas-Chromatographic Analysis of Aromatic Fractions, Technique of Trapping Followed by Identification by Means of Infrared Spectra," Publ. Group. Avan. Methodes Spectrog., 1961, 351-368; CA 57, 12i

2237. Ferrand, R., "Gas Phase Chromatography Using Retention Indexes for the Analysis of Tars and Their Hydrogenation Products," Journees Intern. Etude Methodes Separation Immediate Chromatog., Paris, 1961, 132-140 (Pub. 1962); CA 59, 1413b

2238. Ferrand, R., "Application of Gas Chromatography to the Study of Coal Tar," Chim. Anal. (Paris), 45, 133-134 (1963); CA 59, 1415e

2239. Ferrer, P., "Gas Chromatography: New Instruments and Technique," Afinidad, 33, 199-208 (1956)

2240. Ferrero, C., "A Study of the Volatile Fractions from Bulgarian Rose Oil," Parfums, cosmet., savons, 3, 319-322 (1960); CA 54, 25595g

2241. Ferrers, P., "The Possibilities of Gas Chromatography in the Study of Tar Oils," Ind. chim. belge, 25, 237-244 (1960); CA 54, 21712f

2242. Ferrier, R. J., "The Gas-Liquid Partition Chromatography of the Tetra-O-acetylpentopyranoses," Chem. & Ind. (London), 1961, 831-832; CA 55, 27063f

2243. Ferrier, R. J., and Singleton, M. F., "Trimethylsilylation of D-Xylose and the Nuclear Magnetic Resonance Spectra of the Major Products," Tetrahedron, 18, 1143-1148 (1962); CA 58, 6907e, 6910d

2244. Ferrin, C. R., Chase, J. O., and Hurn, R. W., "Analysis of Complex Hydrocarbon Mixtures Using an Unsaturate-Discriminating Gas Chromatograph," Gas Chromatog., Intern. Symp., East Lansing, Mich., 1961, 3, 423-429 (Pub. 1962); CA 58, 3869h

2245. Ferrin, C. R., Chase, J. O., and Hurn, R. W., "Analysis of Complex Hydrocarbon Mixtures Using an Unsaturate-Discriminating Gas Chromatograph," in "Gas Chromatography," edited by N. Brenner, J. E. Callen, and M. D. Weiss, Academic Press, Inc., New York, 1962, pp. 423-430

2246. Ferrin, C. R., Latham, D. R., and Haines, W. E., "Evaluation of a Commercial Dye as the Liquid Phase in the Gas Chromatographic Separation of Nitrogen Compounds," 139th Natl. ACS Mtg., St. Louis, Mo., March 1961, Program Abstr., p. 6Q; Div. Petrol. Chem., Symp., 6, No. 2B, 23-26 (1961); CA 57, 15411a

2247. Fessenden, R. J., "Infrared Spectra of Some Silazanes and Disilazanes," J. Org. Chem., 25, 2191-2193 (1960); CA 55, 9261d

2248. Fessenden, R. J., and Crowe, D. F., "Synthesis and Cleavage of N-Trimethylsilylpyrrole," J. Org. Chem., 25, 598-603 (1960); CA 54, 18473c

2249. Fessenden, R. J., and Freenor, F. J., "The Reaction of Aminomethyltrimethylsilane with Nitrous Acid," J. Org. Chem., 26, 1681-1682 (1961)

2250. Fessenden, R. J., and Freenor, F. J., "The Chlorination of 1,1-Dimethylsilacyclopentane and 1,1-Dimethylsilacyclohexane," J. Org. Chem., 26, 2003-2006 (1961); CA 55, 24716d

2251. Fett, E. R., "Backflush Applied to Capillary Column-Flame Ionization Detector Gas Chromatography Systems," Anal. Chem., 35, 419-420 (1963); CA 58, 10703e

2252. Fettis, G. C., Knox, J. H., and Trotman-Dickenson, A. F., "The Reactions of Fluorine Atoms with Alkanes," J. Chem. Soc., 1960, 1064-1071; CA 55, 19754c

2253. Fettis, G. C., Knox, J. H., and Trotman-Dickenson, A. F., "Reactions of Bromine Atoms with Alkanes and Methyl Halides," J. Chem. Soc., 1960, 4177-4185; CA 55, 12267b

2254. Feuerberg, H., and Weigel, H., "Analysis of Elastomers in Vulcanizers. II. Gas-Chromatographic Investigations of Elastomer Pyrolyzates," Kautschuk Gummi, 15, WT276-WT282 (1962); CA 57, 16821c

2255. Fiehman, J., "Gas Analyzer," Chem. prumysl., 7, 605-606 (1957); CA 52, 13325i

2256. Field, D. C., "Sampling Systems for Chromatographs," 5th Annual ISA Mtg., Symp. on Instrumental Methods of Analysis, Houston, Texas, May 1959

2257. Filippov, G. G., and Chizhkov, V. P., "Chromatographic Analysis of Gas Mixtures Containing Concentrated Hydrogen Sulfide," Tr. Mosk. Khim.-Tekhnol. Inst., 1961, No. 35, 147-148; CA 57, 46h

2258. Findeis, A. F., "Gas Chromatographic Detection Systems Using Geiger and Proportional Counters," 142nd Natl. ACS Mtg., Atlantic City, N. J., September 1962, Program Abstr., p. 7B

2259. Findeis, A. F., and Johnson, W. H., "The Use of Geiger and Proportional Counting Gases as Flow Gases in Gas Chromatography," Regional Mtg., ACS, Southeast and Southwest Sections, New Orleans, La., December 1961

2260. Finkelstein, M., and Petersen, R. C., "Kolbe Electrolysis in Dimethyl Formamide," J. Org. Chem., 25, 136-137 (1960); CA 54, 15305f

2261. Finkelstein, M., Petersen, R. C., and Ross, S. D., "Electrochemical Degradation of Quaternary Ammonium Salts," J. Am. Chem. Soc., 81, 2361-2364 (1959); CA 53, 14779e

2262. Finlayson, A. J., and Lee, C. C., "Rearrangement Studies with C^{14}. VII. The Acetolysis of Methyl-C^{14}-isopropylcarbinol-p-toluene Sulphonate," Can. J. Chem., 37, 940-952 (1959); CA 53, 21764e

2263. Finnegan, R. A., and McNees, R. S., "Reaction of Bicyclo[2.2.1]heptadiene with Amylsodium. A Novel Cleavage Reaction," Tetrahedron Letters, 1962, 755-757; CA 58, 1370d

2264. Finnegan, W. G., and Smith, S. R., "1-Methyl-5-(2-Methoxyethyl)-tetrazole - A Stationary Phase of Gas-Liquid Partition Chromatography," J. Chromatog., 5, 461-465 (1961); CA 56, 10883e

2265. Firestone, D., "The Determination of Polymers in Fats and Oils," J. Am. Oil Chemists' Soc., 40,

247-255 (1963)

2266. Firestone, D., Horwitz, W., Friedman, L., and Shue, G. M., "Heated Fats. I. Studies of the Effects of Heating on the Chemical Nature of Cottonseed Oil," J. Am. Oil Chemists' Soc., 38, 253-257 (1961); CA 55, 14941g

2267. Firestone, D., Horwitz, W., Friedman, L., and Shue, G. M., "The Examination of Fats and Fatty Acids for Toxic Substances," J. Am. Oil Chemists' Soc., 38, 418-422 (1961); CA 55, 21411g

2268. Fischer, J., "An Automatic Gas Chromatograph," Abhandl. Deut. Akad. Wiss. Berlin, Kl. Chem., Geol., Biol., 1959, No. 9, 185-202; CA 58, 5018f

2269. Fischer, J., "Automatic Gas Chromatograph," in "Gas Chromatographie 1958," edited by H. P. Angele, Akademie Verlag, Berlin, 1959, pp. 185-202

2270. Fischer, J., "The Regulation of Distillation Columns with Help of Gas Chromatography," in "Gas Chromatographie 1959," edited by R. E. Kaiser and H. G. Struppe, Akademie Verlag, Berlin, 1959, pp. 85-95

2271. Fischer, J., "The Closed-Loop Control of (Continuous) Distillation Columns by Gas Chromatography," Z. Messen. Steuern, Regeln, 3, 16-21 (1960); CA 55, 8954f

2272. Fisher, N., and Broughton, M. E., "Studies on the Lipids of Wheat: Fractionation on Silicic Acid," Chem. & Ind. (London), 1960, 869-870; CA 55, 830e

2273. Fisher Scientific Co., "Gas Chromatography: A Bibliography," compiled by the Development Laboratories of Fisher Scientific Co., Pittsburgh, Pa., January 1957. Covers the Literature from 1938 to 1956

2274. Fisher Scientific Co., "Gas Chromatography: Bulletin 1 - Separation of Methyl Esters of Fatty Acids by Gas Chromatography; Bulletin 2 - Chromotographic Supports (Chromosorb & Chromosorb W); Bulletin 3 - 'THEED' as a Liquid Phase in Gas Chromatography; Bulletin 4 - HMPA (Hexamethylphosoramide) as a Liquid Phase in Gas Chromatography; Bulletin 5 - Determination of Hydrogen with the Fisher Gas Partitioner; Bulletin 6 - Analysis of Blast Furnace Gases with the Fisher Gas Partitioner; Bulletin 7 - Typical Applications of the Fisher Prep/Partitioner; Bulletin 8 - Fixed Gas-Hydrogen Separation with the Fisher Gas Partitioner; Bulletin 9 - Teflon as a Chromatographic Support," Fisher Scientific Co., Pittsburgh, Pa.

2275. Fisher Scientific Co., "Chromatographic Column Materials," Information Sheet No. 61-3, Fisher Scientific Co., Pittsburgh, Pa., 1961

2276. Fisher Scientific Co., "Bibliography on Gas Chromatography 1958," Fisher Scientific Co., Pittsburgh, Pa., 1959. Supplement to "Bibliography on Gas Chromatography," by C. Zahn and S. H. Langer, U. S. Bureau of Mines Information Circular 7856

2277. Fishman, J., and Brown, J. B., "Quantitation of Urinary Estrogens by Gas Chromatography," J. Chromatog., 8, 21-24 (1962); CA 58, 11641d

2278. Fitch, G. R., Probert, M. E., and Tiley, P. F., "Moving-Bed Chromatography," J. Chem. Soc., 1962, 4875-4881; CA 58, 3864b

2279. Fitzgerald, J. S., "Gas Chromatography Applied to the Analysis of Phenols," Australian J. Appl. Sci., 10, 169-189 (1959); CA 53, 14844g

2280. Fitzgerald, J. S., "The Composition of a Lurgi Brown Coal Tar. II. The Lower-Boiling Phenols," Australian J. Appl. Sci., 10, 306-320 (1959); CA 53, 22839f

2281. Fitzgerald, J. S., "Gas Chromatography Applied to the Analysis of Pyridines and Quinolines," Australian J. Appl. Sci., 12, 51-68 (1961); CA 55, 17376f

2282. Flanders, R. L., Annesser, R. J., and Saville, D. A., "Reaction-Mix Sampling Simplifies Analysis of Cat Cracker for California Standard," Oil Gas J., 59, 84-86 (1961); CA 55, 9852c

2283. Fleetwood, C. W., "Gas Chromatography for Determining the Oxygen-Nitrogen Ratio in Isoascorbic Acid Treated Carbonated Beverages," 134th Natl. ACS Mtg., Chicago, Ill., September 1958, Program Abstr., p. 33A

2284. Fleischmann, L., "Determination of Higher Alcohols in Grape Whiskeys by Vapor-Phase Chromatography," Ricerca sci., 29, 1194-1198 (1959); CA 53, 22724f

2285. Fleischmann, L., and Daghetta, A., "Analysis and Identification of Higher Alcohols in Fermented Beverages by Gas Chromatography," Ricerca sci., 28, 2286-2290 (1958); CA 53, 16462f

2286. Flett, M. St. C., "Physical Aids to the Organic Chemist," American Elsevier Publishing Co., Inc., New York, 1962

2287. Flook, W. A., "Canadian Chemical Uses Flame Ionization Detector to Measure Hydrocarbons in Air," Analyzer, 3, No. 2, 11-12 (April 1962)

2288. Flowers, M. C., and Frey, H. M., "The Thermal Isomerization of 1, 1-Dimethylcyclopropane," J. Chem. Soc., 1959, 3953-3957; CA 54, 9782c

2289. Flowers, M. C., and Frey, H. M., "The Thermal Isomerization of 1, 2-Dimethylcyclopropane. I. Cis-trans Isomerization," Proc. Roy. Soc. (London), A257, 122-131 (1960); CA 54, 23655e

2290. Flowers, M. C., and Frey, H. M., "The Thermal Isomerization of 1, 2-Dimethylcyclopropane. II. Structural Isomerization," Proc. Roy. Soc. (London), A260, 424-432 (1961); CA 55, 14324c

2291. Flowers, M. C., and Frey, H. M., "The Thermal Unimolecular Isomerization of Vinylcyclopropane to Cyclopentene," J. Chem. Soc., 1961, 3547-3549; CA 56, 7143i

2292. Flowers, M. C., and Frey, H. M., "Thermal Decomposition of Bicyclopropyl," J. Chem. Soc., 1962, 1689-1694; CA 57, 3299d

2293. Fluck, A. A. J., Mitchell, W., and Perry, H. M., "Composition of Buchu Leaf Oil," J. Sci. Food Agr., 12, 290-292 (1961); CA 55, 21411c

2294. Flumerfelt, G. C., "Selective Indication of One Component of a Multicomponent Gas Stream by Thermal Analysis," 5th Annual ISA Mtg., Symp. on Instrumental Methods of Analysis, Houston, Texas, May 1959

2295. Follain, G., "Gas Chromatography. The Influence of the Nature and Quantity of the Solute on Retention Time," Rev. Inst. Franc. Petrole Ann. Combust. Liquides, 18, 66-71 (1963); CA 58, 11980f

2296. Folmer, Jr., O. F., "Punched Card Literature Retrieval System for Gas Chromatography," 141st Natl. ACS Mtg., Washington, D. C., March 1962, Program Abstr., p. 20B

2297. Folmer, Jr., O. F., "A New Quantitative Detector," Univ. of Houston Symp. on Advances in Gas Chromatography, Houston, Texas, January 1963

2298. Folmer, Jr., O. F., Yang, K., and Perkins, Jr., G., "Use of Catalytic Combustion Filaments for Qualitative Gas Chromatography," Anal. Chem., 35, 454-459 (1963)

2299. Fontan, C. R., Smith, W. C., and Kirk, P. L., "Gas Chromatography of the Antihistamines," Anal. Chem., 35, 591 (1963)

2300. Fontell, K., Holman, R. T., and Lambertsen, G., "Some New Methods for Separation and Analysis of Fatty Acids and Other Lipids," J. Lipid Research, 1, 391-404 (1960); CA 55, 20459g

2301. Ford, H. W., and Rogers, L. H., "Irradiation Studies of Small Concentrations of Automobile Exhaust in Air Using Gas Chromatography," 133rd Natl. ACS Mtg., San Francisco, Calif., April 1958, Program Abstr., p. 4Q

2302. Ford, J. F., Pitkethly, R. C., and Young, V. O., "Stereochemistry of the Cooxidation Products of Indene and Thiophenol," Tetrahedron, 4, 325-336 (1958); CA 53, 12253b

2303. Fore, S. P., Ward, T. L., and Dollear, F. G., "The Preparation of Lauryl Alcohol and 6-Hydroxy-caproic Acid from Petroselinic Acid," J. Am. Oil Chemists' Soc., 40, 30-33 (1963); CA 58, 7051f

2304. Forrestal, L. J., and Hamill, W. H., "Effects of Ionic and Free Radical Processes in the Radiolysis of Organic Liquid Mixtures," J. Am. Chem. Soc., 83, 1535-1541 (1961); CA 55, 15081i

2305. Forrester, J. S., "Sampling Techniques in Gas Chromatography," 7th Natl. Symp. on Instrumental Methods of Analysis, ISA, Houston, Texas, April 1961

2306. Forss, D. A., "Facts on Flavor," Australian J. Dairy Technol., 12, 120-126 (1957); CA 52, 3187f

2307. Forss, D. A., Dunstone, E. A., Horwood, J. F., and Stark, W.,"Characterization of Some Unsaturated Aldehydes in Microgram Quantities," Australian J. Chem., 15, 163-168 (1962); CA 57, 1521b

2308. Forss, D. A., Dunstone, E. A., and Stark, W., "Fishy Flavor in Dairy Products. II. The Volatile Compounds Associated with Fishy Flavor in Butterfat," J. Dairy Research, 27, 210-220 (1960); CA 55, 8689c

2309. Forss, D. A., Dunstone, E. A., and Stark, W., "Fishy Flavor in Dairy Products. III. The Volatile Compounds Associated with Fishy Flavor in Washed Cream," J. Dairy Research, 27, 373-380 (1960)

2310. Forss, D. A., Dunstone, E. A., and Stark, W., "The Volatile Compounds Associated with Tallowy and Painty Flavors in Butterfat," J. Dairy Research, 27, 381-387 (1960)

2311. Forss, D. A., and Hills, G. L., "Modern Instruments for Dairy Manufacturing Research," Australian J. Dairy Technol., 16, 152-155 (1961); CA 56, 741i

2312. Forss, D. A., Pont, E. G., and Stark, W., "The Volatile Compounds Associated with Oxidized Flavor in Skim Milk," J. Dairy Research, 22, 91-102 (1955); CA 49, 9177d

2313. Forss, D. A., Pont, E. G., and Stark, W., "Further Observations on the Volatile Compounds Associated with Oxidized Flavor in Skim Milk," J. Dairy Research, 22, 345-348 (1955); CA 50, 1225g

2314. Forss, D. A., and Stark, W., "Two Types of Gas Chromatographic Apparatus," Proc. Roy. Australian Chem. Inst., 25, 201-206 (1958); CA 57, 608b

2315. Fort, A. W., and Girard, C. A., "The Reaction of 3-Phenyl-1-butene-3-C^{14} with Formic Acid," J. Am. Chem. Soc., 83, 3449-3453 (1961); CA 57, 11062b

2316. Fortune, W. B., "Control of Fine Chemicals and Pharmaceuticals," Anal. Chem., 29, No. 1, 17A-20A, 22A, 24A, 26A, 28A, 30A (1957); CA 51, 3088c

2317. Foster, A. B., "Chemistry of the Carbohydrates," Ann. Rev. Biochem., 30, 45-70 (1961); CA 56, 1517a

2318. Foster, N. F., and Cvetanovic, R. J., "Stereoselective Catalytic Isomerization of n-Butenes," J. Am. Chem. Soc., 82, 4274-4277 (1960)

2319. Foster, R. A., "A Photoelectrically-Excited Argon Ionization Detector," 12th Pittsburgh Conf. on Anal. Chem. & Appl. Spectroscopy, Pittsburgh, Pa., February-March 1961, Program Abstr., p. 51

2320. Foster, R. A., Said, A. S., and Sternberg, J. C., "Principles of Operation of Argon Ionization Detectors for Gas Chromatography," 12th Pittsburgh Conf. on Anal. Chem. & Appl. Spectroscopy, Pittsburgh, Pa., February-March, 1961, Program Abstr., p. 56

2321. Fouassin, A., "Analysis of Spirituous Liquors by Vapor-Phase Chromatography," Rev.fermentations et inds. aliment., 14, 206-212 (1959); CA 54, 9197i

2322. Foulletier, L., and Elchardus, E., "Fluorated and Chlorofluorated Light Hydrocarbons," Chim. & Ind. (Paris), 83, 242-248 (1960); CA 54, 16370h

2323. Fourroux, M. M., "What's New in Chromatography? Monitoring and Control with Automatic Chromatographic Analyzers," Oil Gas J., 56, No. 16, 114-116 (1958); CA 52, 15883b

2324. Fourroux, M. M., Karasek, F. W., and Wightman, R. E., "Automatic Control with High-Speed Chromatography," Oil Gas J., 58, No. 12, 96-99 (1960); CA 54, 18944f

2325. Fourroux, M. M., Karasek, F. W., and Wightman, R. E., "High-Speed Chromatography in Closed-Loop Fractionator Control," ISA Journal, 7, No. 5, 76-80 (1960)

2326. Fox, F. T., and Sawicki, E., "Aliphatic and Polycyclic Hydrocarbons in Urban Atmospheres," 13th Pittsburgh Conf. on Anal. Chem. and Appl. Spectroscopy, Pittsburgh, Pa., March 1962, Program Abstr., p. 58

2327. Fox, J. E., "Gas Chromatographic Analysis of Alcohol and Certain Other Volatiles in Biological Material for Forensic Purposes," Proc. Soc. Exptl. Biol. Med., 97, 236-237 (1958); CA 52, 9866b

2328. Fraade, D. J., "Better Refinery Operation with Automatic Stream Analyzers," Oil Gas J., 55, No. 42, 93-103 (1957); CA 52, 5797e

2329. Fraade, D. J., "Continuous Stream Analysis," Mesures & Controle Ind., 24, 353-371 (1959)

2330. Fraade, D. J., "Process Automation Through Closed-Loop Chromatograph's Control of Chemical Process," 6th Instrumental Methods of Analysis Symp., ISA, Montreal, Canada, June 1960

2331. Fraade, D. J., "Automatic Analysis of Ditch Gases by Vapor-Phase Chromatography," Erdol u. Kohle, 14, 720-725 (1961); CA 56, 7575i

2332. Fraade, D. J., "Improved Process Control with Continuous Stream Analysis," Instr. Meas. Chem. Anal. Elec. Quant. Nucleonics Process Control, Proc. Intern. Conf. Stockholm, 1, 174-183 (1960) (Pub. 1961); CA 56, 6242b

2333. Fraade, D. J., and Escher, E. E., "Chromatography Techniques for Process Control," 136th Natl. ACS Mtg., Atlantic City, N. J., September 1959, Program Abstr., p. 11M

2334. Franc, J., "Correlationship of the Dipole Moments with the Retention Volumes of Isomeric Phenols," in "Gas Chromatographie 1959," edited by R. E. Kaiser and H. G. Struppe, Akademie Verlag, Berlin, 1959, pp. 96-108

2335. Franc, J., "Chromatography of Aromatic Isomers. XIII. Determination of Toluenesulfonamides and p-Sulfamoylbenzoic Acid in Saccharin," Collection Czechoslov. Chem. Communs., 24, 3881-3886 (1959); CA 56, 6669a

2336. Franc, J., "Chromatography of Aromatic Isomers. XIV., The Relation Between the Value of the Dipole Moment and the Relative Elution Volume of Isomeric Phenols," Collection Czechoslov. Chem. Communs., 25, 1573-1579 (1960); CA 54, 20411g

2337. Franc, J., "Chromatography of Organic Compounds. VI. Photodetector for Gas Chromatography," Collection Czechoslov. Chem. Communs., 25, 2225-2227 (1961); CA 55, 6945a

2338. Franc, J., "Chromatography of Isomeric Compounds. XV. Relation Between Size of Dipole Moment and Relative Elution Volumes of Aromatic Isomeric Amines and Nitro-Compounds," Collection Czechoslov. Chem. Communs., 26, 596-598 (1961)

2339. Franc, J., and Blaha, J., "Group Identification of Nonvolatile Aromatic Substances with the Aid of Gas Chromatography," J. Chromatog., 6, 396-408 (1961); CA 56, 10887c

2340. Franc, J., and Celikovska, G., "Chromatography of Organic Compounds. VIII. Separation of Aldehydes and Ketones After Condensation with Cyanoacetohydrazide," Collection Czechoslov. Chem. Communs., 26, 667-672 (1961); CA 55, 19746b

2341. Franc, J., and Jokl, J., "Spectrochromatography. I. Determination of Isomeric Xylenes by Gas-Liquid Chromatography," Chem. listy, 52, 276-282 (1958); CA 52, 8825d

2342. Franc, J., and Jokl, J., "Spectrochromatography. I. Determination of Isomeric Xylenes by Gas-Liquid Chromatography," Collection Czechoslov. Chem. Communs., 24, 144-151 (1959); CA 53, 6909e

2343. Franc, J., and Michajlova, S., "Identification of Volatile Organic Materials by Gas Chromatographic Spectra," Gas Chromatog. Symp., Brno, Czech., June 1962; Abstr., J. Gas Chromatog., 1, No. 4, 7 (1963)

2344. Franc, J., and Wurst, M., "Gas-Chromatographic Determination of Phenylchlorosilanes," Gazovya Khromatografiya, Trudy 1-oi [Pervoi] Vsesoyuz. Konf., Akad. Nauk S.S.S.R., Moscow, 1959, 289-291 (Pub. 1960); CA 56, 6670f

2345. Franc, J., and Wurst, M., "Chromatography of Organic Compounds. V. Determination of Phenylchlorosilanes by Means of Gas Chromatography," Collection Czechoslov. Chem. Communs., 25, 701-705 (1960)

2346. Franc, J., and Wurst, M., "Chromatography of Organic Compounds. VII. Estimation of Aliphatic

Amines by Means of Gas Chromatography," Collection Czechoslov. Chem. Communs., 25, 2290-2294 (1960); CA 54, 24113b

2347. Franc, J., Wurst, W., and Moudry, V., "Chromatography of Organic Compounds. IX. Separation of Organic Tin Compounds by Paper and Gas Chromatography," Collection Czechoslov. Chem. Communs., 26, 1313-1319 (1961); CA 55, 2454f

2348. Francis, S. A., and Archer, E. D., "Identification of Triisobutylene and Tetraisobutylene Isomers by a Combination of GC and NMR Techniques," 14th Pittsburgh Conf. on Anal. Chem. & Appl. Spectroscopy, Pittsburgh, Pa., March 1963, Program Abstr., pp. 60-61

2349. Frank, Yu. A., and Yanovskii, M. I., "A Microionization Detector for Capillary Gas-Liquid Chromatography with Promethium-147 as Radiation Source and Without Preliminary Gas Blowing," Kinetika i Kataliz, 2, 292-294 (1961); CA 55, 21958b

2350. Frankel, E. N., McConnell, D. G., and Evans, C. D., "Analyses of Lipids and Oxidation Products by Partition Chromatography. Hydroxy Fatty Acids and Esters," J. Am. Oil Chemists' Soc., 39, 297-301 (1962); CA 57, 6045c

2351. Frankel, E. N., Nowakowska, J., and Evans, C. D., "Formation of Methyl Azelaaldehyde on Autooxidation of Lipides," J. Am. Oil Chemists' Soc., 38, 161-162 (1961); CA 55, 10925g

2352. Franz, W. F., Christensen, E. R., May, J. E., and Hess, H. V., "Industry Gets Two Processes to Upgrade Gasoline by Removing Normal Paraffins. II. Selective Finishing Texaco Process Has Reached Commerical Stage - Another Molecular Sieve Development," Oil Gas J., 57, No. 15, 116-118, 120-121 (1959); CA 53, 17490g

2353. Franzen, V., and Fikentscher, L., "Carbenes. III. Reaction of Methylene with Ethers," Ann., 617, 1-10 (1958); CA 53, 5119e

2354. Fraser, R. R., and O'Farrell, S., "Solvolysis of exo-5-p-Toluenesulfonyloxybicyclo[2, 2, 2]oct-2-ene," Tetrahedron Letters, 1962, 1143-1146; CA 58, 11193h

2355. Frederick, D. H., "Gas-Liquid Chromatography: A Critical Evaluation of Lightly Loaded Glass-Bead Columns for the Low-Temperature Separation of High-Boiling Compounds," Univ. Microfilms (Ann Arbor, Mich.), Order No. 62-106, 141 pp.; Dissertation Abstr., 22, 2563-2564 (1962); CA 56, 15305d

2356. Frederick, D. H., and Cooke, W. D., "The Effect of Liquid Loading on the Efficiency of Gas Chromatography Columns," 3rd Intern. Symp. on Gas Chromatography, ISA, East Lansing, Mich., June 1961; ISA Proc. 1961, 21-26

2357. Frederick, D. H., Miranda, B. T., and Cooke, W. D., "The Effect of Liquid Loading on the Efficiency of Gas Chromatographic Columns," in "Gas Chromatography," edited by N. Brenner, J. E. Callen, and M. D. Weiss, Academic Press, Inc., New York, 1962, pp. 27-36

2358. Frederick, D. H., Miranda, B. T., and Cooke, W. D., "Effect of Liquid Loading on the Efficiency of Gas Chromatography Columns," Gas Chromatog., Intern. Symp., 3, 27-36 (1961) (Pub. 1962); CA 58, 8426e

2359. Frederick, D. H., Miranda, B. T., and Cooke, W. D., "The Use of Lightly Loaded Columns in Gas Chromatography," Anal. Chem., 34, 1521-1526 (1962); CA 58, 2824g

2360. Fredericks, E. M., and Brooks, F. R., "Gas Chromatography. Analysis of Gaseous Hydrocarbons by Gas-Liquid Partition Chromatography," Anal. Chem., 28, 297-303 (1956); CA 50, 8391g

2361. Fredericks, E. M., Dimbat, M., and Stross, F. H., "Carrier Gas and Sensitivity in Gas Chromatography," Nature, 184, 54 (1959); CA 54, 1975e

2362. Fredericks, P. S., and Tedder, J. M., "A Stereospecific Free Radical," Proc. Chem. Soc., 1959, 9-10; 53, 16928d

2363. Fredericks, P. S., and Tedder, J. M., "Free-Radical Substitution in Aliphatic Compounds. II. Halogenation of the n-Butyl Halides," J. Chem. Soc., 1960, 144-150; CA 54, 9717i

2364. Fredericks, P. S., and Tedder, J. M., "Free-Radical Substitution in Aliphatic Compounds. III. Halogenation of the 2-Halobutanes," J. Chem. Soc., 1961, 3520-3525; CA 56, 4600d

2365. Freeguard, G. F., and Stock, R., "Partition Isotherms and Gas-Liquid Chromatography," Nature, 192, 257-258 (1961); CA 56, 13570h

2366. Freeguard, G. F., and Stock, R., "Some Static Measurements on Gas-Liquid Chromatographic Systems Involving Dinonyl Phthalate and Squalane," 4th Intern. Gas Chromatography Symp., Hamburg, Germany, June 1962; pub. in "Gas Chromatography 1962," ed. by M. van Swaay, Butterworth & Co., London, 1962

2367. Freeman, G. R., "Radiolysis of Cyclohexane. I. Pure Liquid Cyclohexane and Cyclohexane-Benzene Solutions," J. Chem. Phys., 33, 71-78 (1960); CA 55, 85e

2368. Freeman, G. R., "Radiolysis of Cyclohexane. II. Cyclohexene Solutions and Pure Cyclohexene," Can. J. Chem., 38, 1043-1052 (1960); CA 55, 85e

2369. Freeman, S. K., "Gas-Liquid Chromatographic Separation and Analysis of Chlorinated Toluenes," Anal. Chem., 32, 1304-1306 (1960); CA 55, 249d

2370. Freeman, S. K., "Isomer Distribution of Some Chloromethylated Alkylbenzenes," J. Org. Chem., 26, 212-214 (1961); CA 55, 18632i

2371. Freiser, H., "Gas-Liquid Partition Chromatography for Metal Separations," Anal. Chem., 31, 1440

(1959); CA 53, 21041e

2372. French, R. B., "Analysis of Pecan, Peanut, and Other Oils by Gas-Liquid Chromatography and Ultra-violet Spectrophotometry," J. Am. Oil Chemists' Soc., 39, 176-178 (1962); CA 56, 15896c

2373. Frenzel, J., "Use of Gas Chromatography for the Production Control of the Distillation of Gases," 3rd Intern. Conf. on Anal. Chem., Prague, Czech., September 1959

2374. Frenzel, J., "Trace Analysis of Acetylene in Liquid Oxygen," in "Gas Chromatographie 1959," edited by R. E. Kaiser and H. G. Struppe, Akademie Verlag, Berlin, 1959, pp. 109-119

2375. Freund, M., and Bathory, J., "Separation of Hydrocarbons by Urea Adduction," Erdol u. Kohle, 9, 237-241 (1956); CA 50, 17398g

2376. Freund, M., Benedek, P., Laszlo, A., and Szepesy, L., "Continuous Gas Chromatography. I. Recovery of Pure Acetylene from the End Gas of the Partial Oxidation of Methane," Acta Chim. Acad. Sci. Hung., 14, 3-18 (1958); CA 53, 18454d

2377. Freund, M., Benedek, P., and Szepesy, L., "Chemical Engineering Design of a Unit for Continuous Gas Chromatography (Hypersorption)," in "Vapour Phase Chromatography," edited by D. H. Desty, Academic Press, Inc., New York, 1957, pp. 359-376

2378. Frey, H. M., "Formation of Cyclopropane from Methylene and Ethylene," J. Am. Chem. Soc., 79, 1259-1260 (1957); CA 51, 11259e

2379. Frey, H. M., "Analysis of Hydrocarbons," Nature, 183, 743-744 (1959); CA 53, 15537a

2380. Frey, H. M., "The Addition of Methylene to Cyclobutane and the Decomposition of Excited Methyl-cyclobutane," Trans. Faraday Soc., 56, 1201-1210 (1960); CA 55, 4381g

2381. Frey, H. M., "Addition of Methylene to Allene and the Unimolecular Decomposition of Excited Meth-ylene-Cyclopropane," Trans. Faraday Soc., 57, 951-960 (1961); CA 56, 1357d

2382. Frey, H. M., and Kistiakowsky, G. B., "Reactions of Methylene. I. Ethylene, Propane, Cyclopro-pane, and n-Butane," J. Am. Chem. Soc., 79, 6373-6379 (1957); CA 52, 6204e

2383. Friedman, H. L., "The Pyrolysis of Plastics in a High-Vacuum Arc-Image Furnace," U. S. Dept. Com., Office Tech. Serv., PB Rept. 150,288, 39 pp. (1961); CA 56, 15659c

2384. Friedman, H. L., "Gas-Chromatography Valve for Vacuum Sampling of Gases," U. S. Dept. Com., Office Tech. Serv., AD 257,310, 5 pp. (1961); CA 58, 6167b

2385. Friedman, R. L., and Raab, W. J., "Determination of Tobacco Humectants by Gas Liquid Chroma-tography," Anal. Chem., 35, 67-69 (1963); CA 58, 9420g

2386. Friedrich, J. P., and Beal, R. E., "Liquid C-18 Saturated Monocarboxylic Acids - Their Prepara-tion, Characteristics, and Potential Uses," J. Am. Oil Chemists' Soc., 39, 528-533 (1962)

2387. Friedrich, K., "Partition of Methylchlorosilanes by Gas-Liquid Chromatography," Chem. & Ind. (London), 1957, 47; CA 51, 6436a

2388. Friedrich, K., "Influence of the Packing Grade in Gas Chromatography on Silica Gel," J. Chroma-tog., 2, 664-666 (1959); CA 54, 16105a

2389. Friel, D. D., "Some Recent Developments in Analytical Process Instrumentation," 136th Natl. ACS Mtg., Atlantic City, N. J., September 1959, Program Abstr., p. 13M

2390. Friel, D. D., "Continuous Process Control," Ind. Eng. Chem., 52, 494-496 (1960); CA 54, 16946i

2391. Frisone, G. J., "Trap for Liquid Fractions Separated by Gas Chromatography," Chemist Analyst, 48, 47 (1959); CA 58, 7346h

2392. Frisone, G. J., "The Design and Construction of a Two-Inch Preparative Gas Chromatographic Col-umn," J. Chromatog., 6, 97-109 (1961); CA 56, 12692g

2393. Frisone, G. J., "Use of N, N-bis(2-cyanoethyl)formamide as a Liquid Phase for Gas Chromatography," Nature, 193, 370-371 (1962); CA 57, 8g

2394. Fritz, G., "Gas Chromatographic Analysis of Silicon and Silicon/Phosphorus Compounds," German Chemist's Society Mtg., Stuttgart, Germany, April 1960

2395. Fritz, G., and Grobe, J., "Formation of Organosilicon Compounds. XII. Reactions of $(Cl_3Si)_2CCl_2$ with Organomagnesium and Organolithium Compounds," Z. anorg. u. allgem. Chem., 309, 77-97 (1961); CA 55, 23320d

2396. Fritz, G., Grobe, J., and Ksninsik, D., "Quantitative Estimation of Groups in Methylsilanes and Siliconmethylene Compounds," Z. anorg. u. allgem. Chem., 302, 175-184 (1959); CA 54, 8469f

2397. Fritz, G., and Ksninsik, D., "Gas Chromatographic Separation and Identification of Silicon Com-pounds," Z. anorg. u. allgem. Chem., 304, 241-248 (1960); CA 54, 23599a

2398. Fritz, G., and Kummer, D., "Reactions of Phenylsilanes with Halogens and Hydrogen Halides," Z. anorg. u. allgem. Chem., 308, 105-121 (1961); CA 55, 18412a

2399. Fritz, G., and Kummer, D., "The Formation of Silanes with Different Halogen Atoms (Si_2XY, $C_6H_5SiHClBr$)," Z. anorg. u. allgem. Chem., 310, 327-337 (1961); CA 56, 6876e

2400. Fritz, G., and Thielking, H., "Synthesis of Organosilicon Compounds. XI. Formation of Silicon Meth-ylene Compounds from Dichloromethane and Silicon," Z. anorg. u. allgem. Chem., 306, 39-47 (1960); CA 55, 3417a

2401. Fritz, I. T., and Davis, R. C., "Laboratory Instrumentation - or Stream Analyzers," Oil Gas J., 60, No. 121, 139-140, 142, 144 (1962); CA 57, 6215b

2402. Froemsdorf, D. H., Collins, C. H., Hammond, G. S., and DePuy, C. H., "The Direction of Elimination in the Pyrolysis of Acetates," J. Am. Chem. Soc., 81, 643-647 (1959); CA 53, 14024h

2403. Frolovskii, P. A., "Laboratory (Gas-Liquid) Chromatograph XL-3," Khim. i Tekhnol. Topliv i Masel, 6, No. 7, 44-49 (1961); CA 55, 25385c

2404. Fry, A., Eberhardt, M., and Ookuni, I., "Acid-Catalyzed Rearrangement of Diethyl Ketone and Diisopropyl Ketone," J. Org. Chem., 25, 1252-1253 (1960); CA 55, 4347c

2405. Frye, C. G., "Equilibria in the Hydrogenation of Polycyclic Aromatics," 140th Natl. ACS Mtg., Chicago, Ill., September 1961, Program Abstr., p. 6S

2406. Frye, C. G., Barger, B. D., Brennan, H. M., Coley, J. R., and Gutberlet, L. C., "Hydroisomerization of Olefins," Prod. Res. & Develop., 2, No. 1, 40-42 (1963)

2407. Fryer, F. H., Ormand, W. L., and Crump, G. B., "Triglyceride Elution by Gas Chromatography," J. Am. Oil Chemists' Soc., 37, 589-590 (1960); CA 55, 4013e

2408. Fryer, J. F., and Habgood, H. W., with Harris, W. E., "Resolution in Programmed Temperature Gas Chromatography," Anal. Chem., 33, 1515-1520 (1961); CA 56, 1979b; 139th Natl. ACS Mtg., St. Louis, Mo., March 1961, Program Abstr., p. 18B

2409. Fuchs, W., "New Advances in the Chemical Investigation of Coals," Brennstoff-Chem., 39, Sonderausgabe, 1-6 (1958); CA 52, 15872i

2410. Fuchs, W., and Nettesheim, G., "Analysis and Identification of Asphaltic Petroleum Products," Erdol u. Kohle, 10, 15-20 (1957); CA 51, 18561a

2411. Fujishima, I., and Takeuchi, T., "Hydrogen Flame Ionization Detector for Gas Chromatography," Kogyo Kagaku Zasshi, 65, 835-837 (1962); CA 58, 1895a

2412. Fujita, E., Saori, H., and Kikuchi, Y., "Analysis of the Composition of Petroleum Light Fractions : Analysis of Light Hydrocarbons in Seria, Kuwait, and Mitsuke Crude Oils," J. Japan Petrol. Inst., 4, No. 4, 209-305 (1961)

2413. Fujita, K., and Kwan, T., "Determination of Para- and Orthohydrogen, Hydrogen Deuteride, and Deuterium by Gas Chromatography," Bunseki Kagaku, 12, 15-20 (1963); CA 58, 10716d

2414. Fujita, M., "Chromatographic Investigation of Petroleum Constituents," Shoseki Giho, 1, 71-86 (1957); CA 53, 2584c

2415. Fujiwara, F., and Sugimoto, C., "Sodium Alkylbenzenesulfonate as a Stationary Phase for Gas Chromatography," Kagaku to Kogyo (Osaka), 35, 258-263 (1961); CA 55, 23164c

2416. Fujiwara, F., and Sugimoto, C., "Analysis of Phenols with Gas Chromatography. I. The Relative Retention Value and Its Column Temperature Dependency," Kagaku to Kogyo (Osaka), 36, 283-291 (1962); CA 57, 11837h

2417. Fuks, N. A., "Gas-Liquid Chromatography," Uspekhi Khim., 25, 845-858 (1956); CA 50, 15324f

2418. Fukuda, T., "Vapor Phase Chromatography. I. The Adsorbents Required for Adsorption Displacement Chromatography," Bunseki Kagaku, 6, 359-362 (1957); CA 52, 15321g

2419. Fukuda, T., "Vapor Phase Chromatography," Yuki Gosei Kagaku Kyokai Shi, 15, 577-583 (1957); CA 52, 1725g

2420. Fukuda, T., "Construction of the Apparatus for Chromatography," Kagaku no Ryoiki, 12, 585-589 (1958); CA 52, 19338i

2421. Fukuda, T., "Vapor-Phase Chromatography. III. Effect of Support and of Stationary Liquid on Relative Retention Value," Bunseki Kagaku, 8, 627-630 (1959); CA 55, 17343d

2422. Fukuda, T., "Competitive Hydrogenation of Acetylenic Compounds," Bull. Chem. Soc. Japan, 32, 1299-1302 (1959); CA 54, 19470h

2423. Fukuda, T., "A Review on Gas Chromatography," Yuki Gosei Kagaku Kyokai Shi, 19, 136-141 (1961); CA 55, 10189a

2424. Fukuda, T., and Omori, H., "Vapor-Phase Chromatography. IV. Relative Retention Value of Some Organic Compounds," Bunseki Kagaku, 8, 630-633 (1959); CA 55, 17343f

2425. Fukuda, T., Omori, H., and Kusama, T., "Vapor Chromatography. II. Quantitative Analysis with a Chromatographic Apparatus," Bunseki Kagaku, 6, 647-650 (1957); CA 52, 15321h

2426. Fukui, K., Nagatomi, H., and Murata, S., "Direct Determination of Low-Molecular-Weight Fatty Acids by Gas-Liquid Chromatography," Bunseki Kagaku, 11, 432-434 (1962); CA 57, 978b

2427. Fukunaga, S., and Koro, S., "Bromine Number of Benzene Hydrorefined Under Pressure," Koru Taru, 14, 337-343 (1962); CA 58, 3241h

2428. Fukushima, M., "Reaction of Higher Fatty Acid Esters with Metallic Sodium. XII. Gas Chromatography of Higher Fatty Alcohols," Yukagaku, 11, 128-133 (1962); CA 58, 6169c

2429. Fulco, A. J., and Mead, J. F., "Metabolism of Essential Fatty Acids. IX. The Biosynthesis of the Octadecadienoic Acids of the Rat," J. Biol. Chem., 235, 3379-3384 (1960); CA 55, 6660b

2430. Fuller, D. H., "Gas Chromatography in Plant Streams," ISA Journal, 3, 440-444 (1956)

2431. Fuller, G., and Tatlow, J. C., "Some Isomeric Hexafluorocyclobutanes and Pentafluorocyclobutenes," J. Chem. Soc., 1961, 3198-3203; CA 56, 7144a

2432. Funasaka, W., "Chromatography in General and Gas Chromatography," Kagaku (Kyoto), 14, 26-34 (1959); CA 53, 21366a

2433. Funasaka, W., and Kojima, T., "Gas-Chromatographic Analysis of High-Boiling Collidine Fractions," Bunseki Kagaku, 9, 741-747 (1960); CA 56, 9393g

2434. Funasaka, W., and Kojima, T., "Gas-Chromatographic Analysis of the Quinoline Fraction of Coal Tar," Kogyo Kagaku Zasshi, 64, 769-772 (1961); CA 57, 3717b

2435. Funk, J. E., and Houghton, G., "A Mathematical Model for Gas-Liquid Partition Chromatography," Nature, 188, 389-391 (1960); CA 55, 6990b

2436. Funk, J. E., and Houghton, G., "A Lumped-Film Model for Gas-Liquid Partition Chromatography. I. Numerical Methods of Solution," J. Chromatog., 6, 193-208 (1961); CA 56, 5380c

2437. Funk, J. E., and Houghton, G., "A Lumped-Film Model for Gas-Liquid Partition Chromatography. II. Experimental Evaluation of Analytical Solutions," J. Chromatog., 6, 281-292 (1961); CA 56, 13523b

2438. Furlani, A. D., "Constituents of Coffee. V. Separation and Determination of Furfural and Furfuryl Alcohol by Gas-Liquid Partition Chromatography," Univ. studi Trieste fac. econ. e com. ist merceol., 1959, No. 12, 13 pp.; CA 54, 17744e

2439. Fürst, H., and Heinzig, E., "Alternating-Current-Operated Thermal-Conductivity Cell for Gas Chromatographs," Chem. Tech. (Berlin), 14, 45-46 (1962); CA 56, 13984h

2440. Fürst, H., and Steege, H., "Reconstruction of Gas Chromatograph GC 012 for Work with Longer Columns and Rapidly Replaceable Packing," Chem. Tech. (Berlin), 12, 495-496 (1960); CA 55, 7934a

2441. Fürst, H., and Steege, H., "Gas Chromatographic Analysis of Octene Isomers," Annual Mtg. of the Chemical Society of the German Democratic Republic, Leipzig, Germany, November 1960; Abstr., Chem., Tech. (Berlin), 13, 182 (1961)

2442. Furuyama, S., and Kwan, T., "Gas Chromatography of Parahydrogen, Orthohydrogen, Hydrogen Deuteride and Deuterium," J. Phys. Chem., 65, 190-191 (1961); CA 55, 17343g

2443. Furuyama, S., and Kwan, T., "Gas Chromatography of Para-H_2, Ortho-H_2, HD, and D_2," Kagaku (Tokyo), 31, 145-146 (1961); CA 55, 17343g

2444. Futrell, J. H., and Newton, A. S., "The Radiation Chemistry of Symmetrical Dichloroethylenes," J. Am. Chem. Soc., 82, 2676-2681 (1960); CA 55, 3413f

2445. Gadsden, R. H., and McCord, W. M., "Determination of Methoxyflurane Blood Levels by Gas Chromatography," 46th Annual Mtg., Federation of Am. Soc. for Exper. Biol., Atlantic City, N. J., April 1962; Abstr., Federation Proc., 21, 327 (1962)

2446. Gadsden, R. H., McCord, W. M., Woods, E. F., and Bagwell, E. E., "Gas Chromatographic Determination of Methoxyfluorane in Blood," Anesthesiology, 23, 831-836 (1962); CA 59, 1943a

2447. Gage, J. C., "A Controlled Fluid-Feed Atomizer," J. Sci. Instr., 30, 25 (1953); CA 47, 3621f

2448. Gager, Jr., F. L., "Vapor Phase Chromatography in Tobacco Research," Southwide Chemical Conf. on Instrumentation, ISA, Symp. on Gas Chromatography, Memphis, Tenn., December 1956

2449. Gagnaire, D., "Ring Enlargement by Hydrolysis of Derivatives of Tetrahydrofurfurylic Alcohol," Compt. rend., 248, 420-423 (1959); CA 53, 16127i

2450. Galla, S. J., and Ottenstein, D. M., "Measurement of Inert Gases in Blood by Gas Chromatography," Ann. N. Y. Acad. Sci., 102, Art. 1, 4-14 (1962)

2451. Galla, S. J., Ottenstein, D. M., and Sancetta, S. M., "Analysis of Nitrous Oxide by Gas Chromatography in the Measurement of Cerebral and Coronary Blood Flow," 46th Annual Mtg., Federation of Am. Soc. for Exper. Biol., Atlantic City, N. J., April 1962; Abstr., Federation Proc., 21, 102 (1962)

2452. Gallagher, M. J., and Sutherland, M. D., "Terpenoid Chemistry. IV. The Turpentine of Araucaria Cunninghamii Ait," Australian J. Chem., 13, 367-371 (1960); CA 54, 25803f

2453. Gallaway, W. S., and Burnell, M. R., "Characteristics of the Hydrogen Flame Detector for Gas Chromatography," 11th Pittsburgh Conf. on Anal. Chem. & Appl. Spectroscopy, Pittsburgh, Pa., February-March 1960, Program Abstr., p. 42; Analyzer, 1, No. 2, 8 (April 1960)

2454. Gallaway, W. S., and Johns, T., "Investigation of the Combined Use of Mass Spectroscopy and Gas Chromatography," 8th Pittsburgh Conf. on Anal. Chem. & Appl. Spectroscopy, Pittsburgh, Pa., March 1957, Program Abstr., p. 34

2455. Gallaway, W. S., Johns, T., Tipotsch, D. G., and Aplin, R. J., "Micro Infrared-Gas Chromatography Techniques as Applied to Complex Organic Systems," 2nd Alberta Gas Chromatography Discussion, Edmonton, Alberta, February 1959

2456. Gallaway, W. S., Johns, T., Tipotsch, D. G., and Ulrich, W. F., "Micro Infrared-Gas-Chromatography Techniques as Applied to Complex Organic Systems," 9th Pittsburgh Conf. on Anal. Chem. & Appl. Spectroscopy, Pittsburgh, Pa., March 1958, Program Abstr., p. 44

2457. Gallaway, W. S., Jones, D. T. L., and Sternberg, J. C., "A Theoretical Interpretation of Hydrogen Flame Ionization Detector Response," 12th Pittsburgh Conf. on Anal. Chem. & Appl. Spectroscopy, Pittsburgh, Pa., February-March 1961, Program Abstr., p. 51

2458. Galwey, A. K., "Gas-Chromatographic Analysis of Hydrogen-Methane Mixtures Using the Radioactive Ionization Detector," Chem. & Ind. (London), 1960, 1417-1418; CA 55, 14156i

2459. Galwey, A. K., "Application of the Radioactive Ionization Detector to the Determination of Permanent

Gases by Gas Chromatography, and Some Uses in Studies of Chemical Kinetics," Talanta, 9, 1043-1052 (1962); CA 58, 5020d

2460. Galwey, A. K., "Use of the Argon Gas Chromatograph in the Determination of Carbon in Steel," Talanta, 10, 310-314 (1963); CA 58, 10714g

2461. Gamson, R. M., Kramer, D. N., and Miller, F. M., "A Study of the Physical and Chemical Properties of the Esters of Indophenols. II. Structural Studies of the Isomeric Esters," J. Org. Chem., 24, 1747-1750 (1959); CA 55, 12339d

2462. Gander, G. W., and Jensen, R. G., "Specificity of Milk Lipase Toward the Primary Ester Groups of Some Synthetic Triglycerides," J. Dairy Sci., 43, 1762-1765 (1960); CA 55, 8492i

2463. Gander, G. W., Jensen, R. G., and Sampugna, J., "Analyses of Milk Fatty Acids by Gas-Liquid Chromatography," J. Dairy Sci., 45, 323-328 (1962); Biol. Abstr., 40, 1031 (1962)

2464. Ganin, Y. V., Kotel'nikov, B. P., and Martynova, E. N., "Composition of Individual Intermediate Synthetic Fatty Acid Fractions as Determined by Gas-Liquid Chromatography," Masloboino-Zhirovaya Prom., 27, No. 3, 29-32 (1961); CA 55, 15961h

2465. Gant, P. L., and Yang, K., "Separation of Hydrogen Isotopes by Gas-Solid Chromatography," Science, 129, 1548-1549 (1959)

2466. Gardiner, K. W., Klaver, R. F., Baumann, F., and Johnson, J. F., "Gas Chromatographic Chart Integrator," 3rd Intern. Symp. on Gas Chromatography, ISA, East Lansing, Mich., June 1961; Proc., 3, 225-235

2467. Gardiner, K. W., Klaver, R. F., Baumann, F., and Johnson, J. F., "Gas Chromatographic Chart Integrator," in "Gas Chromatography," edited by N. Brenner, J. E. Callen, and M. D. Weiss, Academic Press, Inc., New York, 1962, pp. 349-362

2468. Gardner, J. N., Jones, E. R. H., Leeming, P. R., and Stephenson, J. S., "Chemistry of the Higher Fungi. X. Further Polyacetylenic Derivatives of Decane from Various Basidiomycetes," J. Chem. Soc., 1960, 691-697; CA 54, 14099h

2469. Gardner, K., and Overton, K. C., "Analysis of MCPA/TBA Herbicide Formulations. II. A Gas-Liquid Chromatographic Method for the Determination of 4-Chloro-2-methylphenoxyacetic Acid," Anal. Chim. Acta, 23, 337-345 (1960); CA 55, 882e

2470. Garilli, F., and Lombardo, G., "Gas-Chromatographic Determination of Aromatic Compounds in Virgin Naphthas and Cracked Petroleum Naphthas with 2,2'-Oxydipropionitrile as the Liquid Phase," Ann. Chim. (Rome), 52, 488-494 (1962); CA 57, 14053h

2471. Garn, P. D., and Kessler, J. E., "Repetitive Gas Chromatography of Thermal Analysis Effluents," Metropolitan Regional Mtg., New York and New Jersey Sections, ACS, New York, N. Y., January 1962

2472. Garner, A. Y., Chapin, E. C., and Scanlon, P. M., "Mechanism of the Michaelis-Arbuzov Reaction: Olefin Formation," J. Org. Chem., 24, 532-536 (1959); CA 53, 21750g

2473. Garner, F. H., Ellis, S. R. M., Steer, D. C., and Thompson, D. W., "Separation of Hydrocarbons at Their Boiling Points by Adsorption in Vapor Phase," Genie chim., 78, 141-151 (1957); CA 52, 9975h

2474. Garnett, J. L., Henderson, L., and Sollich, W. A., "Tritium-Labeled Aromatic Compounds by Platinum-Catalyzed Exchange with Tritium Oxide," Tritium Phys. Biol. Sci., Proc. Symp. Detection Use, Vienna, Austria, 1961, 2, 47-59 (Pub. 1962); CA 58, 4399b

2475. Garnova, T. G., Elotnikov, L. E., Moshinskaya, M. B., Paradzhanova, N. G., and Shvartsman, V.P., "Investigation of Chromatographic Gas Analyzers," Ind. Lab., 25, 167-168 (1959); Zavodskaya Lab., 25, 157-159 (1959); CA 54, 23460i

2476. Garoglio, P. G., and Giannardi, G. B., "Methodology and Equipment for the Gas-Chromatographic Analyses of Olive Oils," Olearia, 15, 127-149 (1961); CA 55, 26481e

2477. Garoglio, P. G., and Giannardi, G. B., "Gas-Chromatographic Analysis of Olive Oils," Olearia, 15, 189-197 (1961); CA 56, 3580g

2478. Garratt, D. C., "The Analysis of Commercial Sodium-2,2-dichloropropionate (Dalapon Sodium Salt)," Analyst, 86, 367-373 (1961); CA 55, 27781c

2479. Garratt, P. G., Fazackerley, H., Hills, P. R., and Roberts, R., "Radiation Effects in Brazilian Peppermint Oil," Perfumery Essent. Oil Record, 52, 488-491 (1961); CA 56, 1540g

2480. Gärtner, K., and Griessbach, R., "Selective Adsorption of Vapors by Silica Gels of Different Pore Structures," Kolloid-Z., 162, 25-27 (1959); CA 53, 7715e

2481. Garton, G. A., "Fatty Acid Composition of the Lipids of Pasture Grasses," Nature, 187, 511-512 (1960); CA 55, 701i

2482. Garzo, G., and Till, F., "Argon Ionization Detector for Gas-Liquid Chromatography of Organosilicon Compounds," Talanta, 10, 583-589 (1963)

2483. Garzo, G., Till, F., and Till, I., "Gas-Chromatographic Analysis of Methylchlorosilanes," Magy. Kem. Folyoirat, 68, 327-333 (1962); CA 58, 10734a

2484. Gasparic, J., Petranek, J., and Borecky, J., "Identification of Organic Compounds. XL. Chromatographic Methods for Analyzing Mixtures of Alkyl Phenols," J. Chromatog., 5, 408-417 (1961); CA

56, 9393c

2485. Gast, L. E., Schneider, W. J., Forest, C. A., and Cowan, J. C., "Composition of Methyl Esters from Heat-Bodied Linseed Oils," J. Am. Oil Chemists' Soc., 40, 287-289 (1963)

2486. Gaston, L. K., "Theoretical Aspects of the Application of Gas Chromatography to [Pesticide] Residue Problems," 144th Natl. ACS Mtg., Los Angeles, Calif., March-April 1963, Program Abstr., p. 2A

2487. Gatrell, R. L., "A Mixed-Substrate Column for Gas Chromatographic Analysis of Paint Thinners and Related Solvent Mixtures," 10th Detroit Anachem. Conf., Detroit, Mich., October 1962, Program Abstr., p. 32

2488. Gatrell, R. L., "A Mixed-Substrate Column for Gas Chromatographic Analysis of Lacquer Thinners," Anal. Chem., 35, 923-924 (1963)

2489. Gatrell, R. L., and Morgan, T. O., "Gas Chromatographic Determination of Carbon Dioxide and Water in Gaseous Samples," 8th Detroit Anachem. Conf., Detroit, Mich., October 1960

2490. Gaulin, C. A., Michaelsen, E. R., Alexander, A. B., and Sauer, R. W., "Chromatographic Analyzer for Separating Trace Hydrocarbons in Air Separation Plants," Chem. Eng. Progr., 54, No. 9, 49-52 (1958); CA 52, 19034a

2491. Gault, F. G., and Germain, J. E., "Synthesis of Polymethylcyclopentane Hydrocarbons," Bull. soc. chim. France, 1959, 1365-1371; CA 54, 9785d

2492. Gault, F. G., Germain, J. E., and Conia, J., "The Methylation of Cyclopentanones," Bull. soc. chim. France, 1957, 1064-1069; CA 52, 4509a

2493. Gaümann, T., "Radiation Chemistry of Hydrocarbons. II. Benzenecyclohexane," Helv. Chim. Acta, 44, 1337-1349 (1961); CA 56, 5857b

2494. Gaümann, T., and Schuler, R. H., "The Radiolysis of Benzene by Densely Ionizing Radiations," J. Phys. Chem., 65, 703-704 (1961); CA 55, 24667g

2495. Gaylor, V. F., and Jones, C. N., "Gas Chromatography of Hydrocarbons," 2nd Annual Conf., Cleveland Soc. of Spectroscopy, Cleveland, Ohio, May 1957; Abstr., Appl. Spectroscopy, 11, No. 3, 141 (1957)

2496. Gaziev, G. A., Oziraner, S. N., Yanovskii, M. I., and Kornyakov, V. S., "Effect of Some Parameters on the Performance of the Ionization Detector for Promethium-147," Zhur. Fiz. Khim., 35, 1150-1152 (1961); CA 55, 25526b

2497. Gaziev, G. A., Porshneva, N. V., Chang Lu, V. S., Yanovskii, M. I., Turkel'taub, N. M., and Kornyakov, V. S., "Comparative Characteristics of Some Detectors Used in Gas Chromatography," Gazovaya Khromatografiya, Trudy 1-oi [Pervoi] Veseoyuz. Konf., Akad. Nauk S.S.S.R., Moscow, 1959, 307-312 (Pub. 1960); CA 56, 5780a

2498. Gaziev, G. A., Yanovskii, M. I., and Brazhnikov, V. V., "A Simplified Chromatographic Method of Determining the Surface Area of Solid Absorbents and Catalysts," Kinetika i Kataliz, 1, 548-552 (1960); CA 56, 36e

2499. Gechele, G. B., Nenz, A., Garbuglio, C., and Pietra, S., "By-Products of the 2-Methyl-5-vinyl-pyridine Synthesis," Chim. e ind. (Milan), 42, 959-964 (1960); CA 55, 10431c

2500. Gee, M., and Walker, Jr., H. G., "Gas Chromatographic Analysis of Sucrose Monstearate," Chem. & Ind. (London), 1961, 829-830; CA 55, 27072e

2501. Gee, M., and Walker, Jr., H. G., "Gas-Liquid Chromatography of Some Methylated Mono-, Di-, and Trisaccharides," Anal. Chem., 34, 650-653 (1962); CA 57, 1518b

2502. Gehrke, C. W., and Goerlitz, D. F., "Quantitative Preparation of Methyl Esters of Fatty Acids for Gas Chromatography," Anal. Chem., 35, 76-80 (1963)

2503. Gehrke, C. W., Goerlitz, D. F., Richardson, C. O., and Johnson, H. D., "The Quantitative Determination of Fatty Acids by Gas Chromatography," 55th Annual Mtg., Am. Dairy Sci. Assoc., Logan, Utah, June 1960; Abstr., J. Dairy Sci., 43, 839 (1960)

2504. Gehrke, C. W., and Lamkin, W. M., "Determination of Steam-Volatile Fatty Acids by Gas-Liquid Chromatography," J. Agr. Food Chem., 9, 85-87 (1961); CA 55, 17935a; 137th Natl. ACS Mtg., Cleveland, Ohio, April 1960, Program Abstr., p. 15A

2505. Geldenhuis, P., Nel, W., and Pretorius, V., "Inexpensive Gas Chromatography Apparatus," S. African Ind. Chemist, 12, No. 10, 196-197 (1958); CA 53, 8724d

2506. Gellhorn, A., and Marks, P. A., "The Composition and Biosynthesis of Lipids in Human Adipose Tissue," J. Clin. Invest., 40, 925-932 (1961); CA 55, 20133b

2507. Gemmill, A. V., "Gas Chromatography. Striking New Guide to Better Foods," Food Eng., 31, No. 7, 77-78 (1959)

2508. Genas, M., and Rull, T., "Preparation of Cyclo-Carboxylic Acids by Action of Carbon Monoxide on Cyclododecane," Bull. soc. chim. France, 1962, 1837-1842

2509. Genevois, L., and Baraud, J., "By-Products from the Alcoholic Fermentation - Fusel Oil Composition," Inds. Aliment. et Agr. (Paris), 76, 837-844 (1959); CA 56, 7796c

2510. Genge, C. A., Hudy, J. A., and Reid, D. E., "Composition of Rosin Acids in Tall Oil Fatty Acids," 52nd Annual Mtg., Am. Oil Chemists' Soc., St. Louis, Mo., April-May 1961

2511. Genin, G., "Gas-Chromatographic Analysis of Electrolysis Gas," Compt. Rend., 254, 679-681 (1962);

CA 56, 14919i

2512. Genkin, A. N., "Analysis of Mixtures of Butylenes and Butadiene by Gas Chromatography," Fiz.-Khim. Metody Analiza i Issled. Produktov Proizv. Sintetich. Kauchuka, Vses. Nauchn. Issled. Inst. Sintetich. Kauchuka, 1961, 32-54; CA 57, 14415a

2513. Genkin, A. N., Ogorodnikov, S. K., Kogan, V. B., Nemtrsov, M. S., and Presman, B. I., "Effect of Polar Substances on the Relative Volatilities of C_5 Hydrocarbons," Zb. Prikl. Khim., 36, No. 1, 142-147 (1963); CA 58, 10789g

2514. Genkin, A. N., Ogorodnikov, S. K., and Nemtrsov, M. S., "Gas-Liquid Chromatographic Studies on the Interaction of Hydrocarbons with Polar Compounds," Neftekhimiya, 2, 837-844 (1962); CA 59, 1457b

2515. Gensler, W. J., and Bruno, J. J., "Synthesis of Unsaturated Fatty Acids. Positional Isomers of Linoleic Acid," J. Org. Chem., 28, 1254-1259 (1963); CA 58, 12412c

2516. Georgieff, K. K., and Richard, V., "Diacetylene. Preparation, Purification and Ultraviolet Spectrum," Can. J. Chem., 36, 1280-1283 (1958); CA 53, 6056c

2517. Gerberich, H. R., and Walters, W. D., "The Thermal Decomposition of cis-1,2-Dimethylcyclobutane," J. Am. Chem. Soc., 83, 3935-3939 (1961); CA 56, 14093e

2518. Gerberich, H. R., and Walters, W. D., "Thermal Decomposition of trans-1,2-Dimethylcyclobutane," J. Am. Chem. Soc., 83, 4884-4888 (1962); CA 57, 2083e

2519. Gerdes, W. F., "Gas Chromatography in a Chemical Plant," Proc. Ann. Conf. Automatic Control Petrol. and Chem. Inds., 3, 95-111 (1958); CA 53, 9517b

2520. Germain, J. E., Bassery, L., and Blanchard, M., "Catalytic Isomerizing and Dehydrating Action of Alumina," Bull. soc. chim. France, 1958, 958-964; CA 53, 8019i

2521. Germain, J. E., and Blanchard, M., "Isomerization of Cyclohexene on a Silicophosphoric Catalyst," Bull. soc. chim. France, 1958, 1000-1003; CA 53, 4161c

2522. Germain, J. E., and Blanchard, M., "Isomerization of Bicyclic Olefins on a Silicophosphoric Catalyst," Compt. rend., 248, 3301-3303 (1959); CA 54, 343a

2523. Germain, J. E., and Blanchard, M., "Isomerization of Bicyclic C_8H_{12} Olefins on a Silicophosphoric Catalyst," Bull. soc. chim. France, 1960, 473-481; CA 54, 24454d

2524. Germain, J. E., and Vaniscotte, C., "Cracking of Methane in a Tubular Reactor. III. Initiating Effect of Ethane," Bull. soc. chim. France, 1958, 964-967; CA 53, 1682g

2525. Germain, J. E., and Vaniscotte, C., "Cracking of Methane in a Tubular Reactor. I. Effect of the Ratio of Surface to Volume," Bull. soc. chim. France, 1957, 692-696; CA 51, 12464h

2526. Gerrard, W., Hawkes, S. J., and Mooney, E. F., "Temperature Limitations of Stationary Phases," Gas Chromatog. Proc. Symposium, 3rd, Edinburgh, 1960, 199-210; CA 56, 5380h

2527. Gerrard, W., Hawkes, S. J., and Mooney, E. F., "Temperature Limitations of Stationary Phases," in "Gas Chromatography 1960," edited by R. P. W. Scott, Butterworths, London, 1960, pp. 199-210

2528. Gerson, T., "Gas-Liquid Chromatography: The Introduction of Samples, the Preconditioning of Polyester Liquid Phases and the Measurement of R_F Values in the Analysis of Fatty Esters," J. Chromatog., 6, 178-181 (1961); CA 56, 9387g

2529. Gerson, T., Hawke, J. C., Shorland, F. B., and Melhuish, W. H., "The Role of n-Valeric Acid in the Synthesis of the Higher Saturated Straight-Chain Acids Containing an Odd Number of Carbon Atoms in Bovine Milk Fat," Biochem. J., 74, 366-368 (1960); CA 54, 13305g

2530. Gerson, T., Shorland, F. B., and Adams, Y., "The Effects of Corn Oil on the Amounts of Cholesterol and the Excretion of Sterol in the Rat," Biochem. J., 81, 584-591 (1961); CA 56, 5187i

2531. Getoff, N., and Sattler-Dornbacher, E., "Analysis of Hydrocarbons and Impurities in Liquefied Petroleum Gas," Prakt. Chem., 7, 159-162, 192-194 (1956); CA 50, 17401h

2532. Getz, G. S., and Bartley, W., "The Intracellular Distribution of Fatty Acids in Rat Liver. The Fatty Acids of Intracellular Compartments," Biochem. J., 78, 307-312 (1961); CA 55, 7577a

2533. Getzendaner, M. E., "Gas-Chromatographic Determination of Dalapon Residues," J. Assoc. Offic. Agr. Chemists, 46, 269-275 (1962); CA 58, 14627a

2534. Gevantman, L. H., Main, R. K., and Bryant, L. M., "Device for Regulating Small Flow Rates in Chromatographic Columns," Anal. Chem., 29, 170 (1957); CA 51, 4767e

2535. Ghanayem, I., and Swann, W. B., "Polyphenol Ether and Carbowax Mixture as Substrate for Gas Liquid Chromatographic Analysis of Glycol Mixtures," Anal. Chem., 34, 1847-1848 (1962); CA 58, 2826h

2536. Giammaria, J. J., and Norris, H. D., "High Temperature Inhibition of Normal Paraffin Oxidation by Alkylaromatics," 140th Natl. ACS Mtg., Chicago, Ill., September 1961, Program Abstr., p. 6S

2537. Gianetto, A., and Panetti, M., "Measurement of Vapor Pressure by Chromatography in the Vapor Phase," Ann. chim. (Rome), 50, 1713-1720 (1960); CA 55, 12969d

2538. Gibbs, D. S., Svec, H. J., and Harrington, R. E., "Purification of the Rare Gases. I. A Comparison of Active Metals in the Purification of Rare Gases," Ind. Eng. Chem., 48, 289-296 (1956); CA 50, 8978i

2539. Giddings, J. C., "Stochastic Considerations of Chromatographic Dispersion," J. Chem. Phys., 26, 169-173 (1957); CA 51, 7100g

2540. Giddings, J. C., "Kinetic Model for Chromatographic Dispersion and Electrodiffusion," J. Chem. Phys., 26, 1755-1756 (1957); CA 51, 17330d

2541. Giddings, J. C., "The Random Downstream Migration of Molecules in Chromatography," J. Chem. Educ., 35, 588-591 (1958); CA 53, 9778g

2542. Giddings, J. C., "Kinetic Relaxation-Time Model Applied to Nonequilibrium-Diffusion Effects in Chromatography," 133rd Natl. ACS Mtg., San Francisco, Calif., April 1958, Program Abstr., p. 8Q

2543. Giddings, J. C., "Nonequilibrium and Diffusion. A Common Basis for Theories of Chromatography," J. Chromatog., 2, 44-52 (1959); CA 53, 21039e

2544. Giddings, J. C., "Nonequilibrium Kinetics and Chromatography," J. Chem. Phys., 31, 1462-1467 (1959); CA 54, 12752b

2545. Giddings, J. C., "Eddy Diffusion in Chromatography," Nature, 184, 357-358 (1959); CA 54, 10454b

2546. Giddings, J. C., "Optimum Conditions for Separation in Gas Chromatography," Anal. Chem., 32, 1707-1711 (1960); CA 55, 5085e

2547. Giddings, J. C., "Kinetic Processes and Zone Diffusion in Chromatography," J. Chromatog., 3, 443-453 (1960); CA 55, 5089e

2548. Giddings, J. C., "Coiled Columns and Resolution in Gas Chromatography," J. Chromatog., 3, 520-523 (1960); CA 55, 8954g

2549. Giddings, J. C., "Retention Times in Programmed Temperature Gas Chromatography," J. Chromatog., 4, 11-20 (1960); CA 55, 4102b

2550. Giddings, J. C., "'Eddy' Diffusion in Chromatography," Nature, 187, 1023-1024 (1960)

2551. Giddings, J. C., "Theoretical Basis for Kinetic Effects in Gas-Solid Chromatography," Nature, 188, 847-848 (1960); CA 55, 8999a

2552. Giddings, J. C., "Plate Height Contributions in Gas Chromatography," Anal. Chem., 33, 962-963 (1961); CA 55, 17157a

2553. Giddings, J. C., "The Role of Lateral Diffusion as a Rate-Controlling Mechanism in Chromatography," J. Chromatog., 5, 46-60 (1961); CA 56, 14960b

2554. Giddings, J. C., "Lateral Diffusion and Local Nonequilibrium in Gas Chromatography," J. Chromatog., 5, 61-67 (1961); CA 56, 13569f

2555. Giddings, J. C., "Role of Pressure Gradient in Obtaining Minimum Time in Gas Chromatography," Nature, 191, 1291-1292 (1961); CA 56, 2867g

2556. Giddings, J. C., "Theory of Programmed Temperature Gas Chromatography. The Prediction of Optimum Parameters," 3rd Intern. Symposium on Gas Chromatography, ISA, East Lansing, Mich., June 1961; ISA Proc., 3, 41-63 (1961)

2557. Giddings, J. C., "Theory of Programmed Temperature Gas Chromatography," in "Gas Chromatography," edited by N. Brenner, J. E. Callen, and M. D. Weiss, Academic Press, Inc., New York, 1962, pp. 57-78

2558. Giddings, J. C., "Liquid Distribution on Gas Chromatographic Support: Relationship to Plate Height," 13th Pittsburgh Conf. on Anal. Chem. & Appl. Spectroscopy, Pittsburgh, Pa., March 1962, Program Abstr., p. 50

2559. Giddings, J. C., "The Science of Programmed Temperature Gas Chromatography," Facts & Methods for Scientific Research (F&M Scientific Corp.), 3, No. 2, 1-5 (Summer, 1962)

2560. Giddings, J. C., "Principles of Gas Chromatography," 1962 Gordon Research Conf., New Hampton, N. H., August 1962

2561. Giddings, J. C., "Theoretical Basis for a Continuous, Large-Capacity Gas Chromatographic Apparatus," Anal. Chem., 34, 37-39 (1962); CA 56, 5780b

2562. Giddings, J. C., "Theory of Minimum Time Operation in Gas Chromatography," Anal. Chem., 34, 314-319 (1962); CA 56, 13523d

2563. Giddings, J. C., "Liquid Distribution on Gas Chromatographic Support. Relation to Plate Height," Anal. Chem., 34, 458-465 (1962); CA 56, 14900h

2564. Giddings, J. C., "Plate Height Theory of Programmed Temperature Gas Chromatography," Anal. Chem., 34, 722-725 (1962); CA 57, 2825c

2565. Giddings, J. C., "Nature of Gas Phase Mass Transfer in Gas Chromatography," Anal. Chem., 34, 1186-1192 (1962); CA 57, 11834g

2566. Giddings, J. C., "Theoretical Investigation of Low Liquid Load and Low Temperature Operation in Gas Chromatography," Anal. Chem. Acta, 27, 207-212 (1962)

2567. Giddings, J. C., "Elementary Theory of Programmed Temperature Gas Chromatography," J. Chem. Educ., 39, 569-573 (1962)

2568. Giddings, J. C., "Advances in the Theory of Plate Heights in Gas Chromatography," Univ. of Houston Intern. Symp. on Advances in Gas Chromatography, Houston, Texas, January 1963

2569. Giddings, J. C., "Plate Height of Nonuniform Chromatographic Columns. Gas Compression Effects,

Coupled Columns, and Analogous Systems," Anal. Chem., 35, 353-356 (1963); CA 58, 10702a

2570. Giddings, J. C., "Advances in the Theory of Plate Height in Gas Chromatography," Anal. Chem., 35, 439-449 (1963)

2571. Giddings, J. C., "Principles of Column Performance in Large Scale Gas Chromatography," J. Gas Chromatog., 1, No. 1, 12-21 (1963)

2572. Giddings, J. C., "Generalized Nonequilibrium Theory of Plate Height in Large-Scale Gas Chromatography," J. Gas Chromatog., 1, No. 4, 38-42 (1963)

2573. Giddings, J. C., "Evidence on the Nature of Eddy Diffusion in Gas Chromatography from Inert (Non-sorbing) Column Data," Anal. Chem., 35, 1338-1341 (1963)

2574. Giddings, J. C., "Theory of Gas Solid Chromatography: Potential for Analytical Use and the Study of Surface Kinetics," 145th Natl. ACS Mtg., New York, N. Y., September 1963, Program Abstr., p. 9B

2575. Giddings, J. C., and Eyring, H., "A Molecular Dynamic Theory of Chromatography," J. Phys.Chem., 59, 416-421 (1955); CA 49, 11358b

2576. Giddings, J. C., and Fuller, E. N., "Particle Size Nonuniformity in Large Scale Columns," J. Chromatog., 7, 255-258 (1962); CA 57, 10510i

2577. Giddings, J. C., and Keller, R. A., "Theoretical Basis of Partition Chromatography," in "Chromatography," edited by E. Heftmann, Reinhold Publishing Corp., New York, 1961, Chapter 6

2578. Giddings, J. C., Mallik, K. L., and Eikelberger, M., "Comparison of Theoretical and Experimental Efficiencies in Glass Bead Gas Chromatography Columns," Anal. Chem., 34, 1026-1027 (1962); CA 57, 5282a

2579. Giddings, J. C., and Robison, R. A., "Failure of the Eddy Diffusion Concept of Gas Chromatography," Anal. Chem., 34, 885-890 (1962); CA 58, 5017g

2580. Giddings, J. C., and Seager, S. L., "Rapid Determination of Gaseous Diffusion Coefficients by Means of Gas Chromatography," J. Chem. Phys., 33, 1579-1580 (1960). CA 55, 13963h

2581. Giddings, J. C., and Seager, S. L., "Rapid Diffusional Analysis by Chromatographic Methods," J. Chem. Phys., 35, 2242-2243 (1961); CA 56, 12329d

2582. Giddings, J. C., and Seager, S. L., "Method for Rapid Determination of Diffusion Coefficients. Theory and Application," Ind. Eng. Chem. Fundamentals, 1, 277-283 (1962); CA 58, 3912g

2583. Giddings, J. C., Seager, S. L., Stucki, L. R., and Stewart, G. H., "Plate Height in Gas Chromatography," Anal. Chem., 32, 867-870 (1960); CA 54, 21929i

2584. Giever, P. M., and Cook, W. A., "Automatic Recording Instruments as Applied to Air Analysis," A. M. A. Arch. Ind. Health, 21, 233-249 (1960); CA 54, 12440e

2585. Gifford, A. P., "Vapor-Phase Chromatography," Instruments & Automation, 30, 2264-2265 (1957)

2586. Giladi, J., and Sideman, S., "Determination of Separation Factors from Unresolved Two-Component Chromatographic Peaks," 4th Intern. Gas Chromatography Symp., Hamburg, Germany, June 1962; pub. in "Gas Chromatography 1962," ed. by M. van Swaay, Butterworth & Co., London, 1962

2587. Gil-Av, E., and Herling, J., "Equilibrium Isomerization of Methylenecyclobutane and 1-Methyl-cyclobutene," Tetrahedron Letters, 1, 27-31 (1961); CA 55, 13331h

2588. Gil-Av, E., and Herling, J., "Determination of the Stability Constants of Complexes by Gas Chromatography," J. Phys. Chem., 66, 1208-1209 (1962); CA 58, 64f

2589. Gil-Av, E., Herling, J., and Shabtai, J., "Gas-Liquid Partition Chromatography of Mixtures of the Three Isomeric Methylcyclohexenes and Methylenecyclohexane," Chem. & Ind. (London), 1957, 1483-1484; CA 52, 4413i

2590. Gil-Av, E., Herling, J., and Shabtai, J., "Gas-Liquid Partition Chromatography of Mixtures of Methylenecyclohexane and the Isomeric Methylcyclohexenes," J. Chromatog., 1, 508-512 (1959)

2591. Gil-Av, E., and Herzberg-Minzly, Y., "Separation of Nitrogen and Oxygen by Gas-Liquid Partition Chromatography Using Blood as the Stationary Phase," J. Am. Chem. Soc., 81, 4749 (1959); CA 54, 6252g

2592. Gil-Av, E., and Herzberg-Minzly, Y., "Gas Chromatographic Study of the Rate of the Diels-Alder Additions," Proc. Chem. Soc., 1961, 316

2593. Gil-Av, E., and Nurok, D., "The Separation of Diastereoisomers by Gas-Liquid Chromatography," Proc. Chem. Soc., 1962, 146-147; CA 57, 1517i

2594. Gil-Av, E., and Shabtai, J., "Relative Stability of Isomeric Olefins Containing a Five- or Six-Membered Ring," Chem. & Ind. (London), 1959, 1630; CA 54, 8668b

2595. Gil-Av, E., Shabtai, J., and Steckel, F., "Thermal Aromatization of 1,3-Butadiene," Ind. Eng. Chem., 52, 31-32 (1960); CA 54, 8584f

2596. Gilbertson, J. R., "The Effect of Dietary Fat Upon the Serum Lipide Fractions," Univ. Microfilms (Ann Arbor, Mich.), L. C. Card No. Mic 60-3516, 98 pp.; Dissertation Abstr., 21, 746 (1960); CA 55, 6624i

2597. Gill, H. A., and Averill, W., "Design Considerations and Performance of a Linear Programmed-Temperature Gas-Liquid Chromatography for Golay Columns," 13th Pittsburgh Conf. on Anal. Chem. & Appl. Spectroscopy, Pittsburgh, Pa., March 1962, Program Abstr., p. 51

2598. Gilliland, E. R., "Diffusion Coefficients in Gaseous Systems," Ind. Eng. Chem., 26, 681-685 (1934); CA 28, 4643²

2599. Gilliland, E. R., Baddour, R. F., and Engel, H. H., "Flow of Gases Through Porous Solids Under the Influence of Temperature Gradients," A. I. Ch. E. Journal, 8, 530-536 (1962); CA 57, 16357b

2600. Gilliland, E. R., Baddour, R. F., and Russell, J. L., "Rate of Flow Through Micro-Porous Solids," A.I.Ch.E. Journal, 4, 90-96 (1958); CA 52, 8637e

2601. Gillis, B. T., and Beck, P. E., "Formation of Tetrahydrofuran Derivatives from 1,4-Diols in Dimethyl Sulfoxide," J. Org. Chem., 28, 1388-1390 (1963); CA 59, 1565g

2602. Gillis, R. G., "Cyclic Sulfites and the Bissinger Rearrangement," J. Org. Chem., 25, 651-653 (1960); CA 54, 18465c

2603. Gillis, R. G., "Isocyanide Bond Refraction," J. Org. Chem., 27, 4103 (1962); CA 58, 7827e

2604. Gilmour, H., and Snell, R. W., "Gas Chromatography in the Oil Field," Petroleum (London), 22, 65-66 (1959); CA 53, 14484e

2605. Gimblett, F. G. R., "Chromatographic Separation of Phosphonitrilic Chlorides by Vapour Phase Techniques," Chem. & Ind. (London), 1958, 365-366; CA 52, 12511d

2606. Ginsburg, L., "Determination of Mono-, Di-, and Triethylene Glycols in Mixtures by Gas-Liquid Chromatography," Anal. Chem., 31, 1822-1824 (1959); CA 54, 3042f

2607. Giovannozzi-Sermanni, G., and Carugno, N., "Analytical Research on Tobacco by Gas Chromatography. II. Effect of Fermentation in Gaseous Substances in the Smoke of Kentucky Tobacco," Acetes Congr. Sci. Intern. Tabac, 2e, Brussels, 1958, 550-552; CA 56, 10474e

2608. Girling, G. W., "Evolution of Volatile Hydrocarbons from Coal," Benzole Producers Ltd. (London) Res. Paper, 1962, No. 3, 48 pp.; CA 58, 2296f

2609. Girling, G. W., "Evolution of Volatile Hydrocarbons from Coal," J. Appl. Chem. (London), 13, 77-91 (1963); CA 58, 11136g

2610. Giroux, J. W., Hahto, M. P., and Pollak, P., "A Syringe Adapter and Gas-Sampling Apparatus for Synthetic Gas Mixture Preparation," Chemist-Analyst, 50, No. 4, 118 (1961); CA 56, 9382g

2611. Giullot, M., and Berton, A., "Analysis of Chlorinated Solvents by Gas Chromatography at Low Temperature with Electrochemical Detector," Compt. rend., 250, 1857-1858 (1960); CA 55, 12156f

2612. Gjaldbaek, J. C., "Gas Chromatography," Dansk. Tidskr. Farm., 31, 225-240 (1957)

2613. Gjaldbaek, J. C., "Gas Chromatographic Analyses of Pharmaceutical Preparations," Dansk. Tidskr. Farm., 33, 158-168 (1959); CA 54, 1803a

2614. Gjerstad, G., "Metabolic and Morphological Changes Induced by Gibberellic Acid on Mentha Piperita," Planta Med., 8, 127-128 (1960); CA 54, 18857c

2615. Gjerstad, G., "Gas Chromatography - A Potential Method in Pharmaceutical Analysis," Am. J. Pharm., 133, No. 2, 46-57 (1961); CA 55, 25162c

2616. Gjerstad, G., "Gas Chromatography," Am. Perfumer Cosmet., 77, No. 11, 19-22 (1962); CA 58, 5449e

2617. Gjertsen, P., "The Application of Chromatography in Brewing Research. I. Methodology," Brewers Digest, May 1959, 46-51; CA 54, 21630h

2618. Gjertsen, P., "The Application of Chromatography in Brewing Research. II. Some Results Obtained in Brewing Research by Means of Chromatographic Methods," Brewers Digest, June 1959, 36-40; CA 54, 21630h

2619. Glaser, A., Grimmer, G., Jantzen, E., and Oertel, H., "Method for Determination of Fatty Acids in Human Blood and Especially Those in Small Amounts," Biochem. Z., 336, 274-280 (1962); CA 58, 738h

2620. Glew, D. N., and Young, D. M., "Stopcock for Gas Chromatography," Anal. Chem., 30, 1890 (1958); CA 53, 1861i

2621. Glick, C. F., Miskalis, A. J., and Kessler, T., "Identification of Impurities in an Acid-Washed 1° Coke-Oven Benzene," Anal. Chem., 32, 1692-1695 (1960); CA 55, 7807b; 136th Natl. ACS Mtg., Atlantic City, N. J., September 1959, Program Abstr., p. 5K

2622. Gloesener, E., "Application of Gas Chromatography to the Analysis of Bonbons [ampuls] of Chloroform and Ether," J. pharm. Belg., 13, 585-588 (1958); CA 53, 15475g

2623. Gloesener, E., Lapiere, C. L., and Versie, J., "Toxicological Study of the Schleich Anaesthetic Mixture by Gas Chromatography," J. pharm. Belg., 13, 389-401 (1958); CA 53, 10520b

2624. Gloesener, R., "Analysis of the Alcoholic Solution of Camphor by Partition Chromatography Between Gas and Liquid," Farmaco (Pavia), Ed. pract., 13, 647-655 (1958); CA 53, 8540g

2625. Glogoczowski, J., "New Trends in Analysis of Gaseous Hydrocarbons," Nafta (Pland), 12, 268-273 (1956); CA 51, 16206d

2626. Glueckauf, E., "A Microanalysis of the Helium and Neon Contents of Air," Proc. Roy. Soc. (London), A185, 98-119 (1946); CA 40, 2760⁴

2627. Glueckauf, E., "Theory of Chromatography," Proc. Roy. Soc. (London), A186, 35-37 (1946); CA 40, 6932²

2628. Glueckauf, E., "Theory of Chromatography. II. Chromatography of a Single Solute," J. Chem. Soc.,

1947, 1302-1308; CA 42, 2490c

2629. Glueckauf, E., "Theory of Chromatography. III. Experimental Separation of Two Solutes and Comparison with Theory," J. Chem. Soc., 1947, 1308-1314; CA 42, 2490d

2630. Glueckauf, E., "Theory of Chromatography. V. Separation of Two Solutes Following a Freundlich Isotherm," J. Chem. Soc., 1947, 1321-1329; CA 42, 2490f

2631. Glueckauf, E., "Theory of Chromatography. VI. Precision Measurements of Adsorption and Exchange Isotherms from Column-Elution Data," J. Chem. Soc., 1949, 3280-3285; CA 44, 9771d

2632. Glueckauf, E., "Theory of Chromatography. VII. The General Theory of Two Solutes Following Non-Linear Isotherms," Disc. Faraday Soc., 7, 12-25 (1949); CA 44, 9771f

2633. Glueckauf, E., "Theory of Chromatography. IX. The 'Theoretical Plate' Concept in Column Separations," Trans. Faraday Soc., 51, 34-44 (1955); CA 49, 11326f

2634. Glueckauf, E., "Theory of Chromatography. X. Formulae for Diffusion into Spheres and Their Application to Chromatography," Trans. Faraday Soc., 51, 1540-1551 (1955); CA 50, 11079d

2635. Glueckauf, E., "Theory of Chromatography. XII. Chromatography of Highly Radioactive Gases," in "Gas Chromatography 1958," edited by D. H. Desty, Academic Press, Inc., New York, 1958, pp. 69-89

2636. Glueckauf, E., "Movement of Highly Radioactive Gases in Absorption Tubes," Ann. N. Y. Acad. Sci., 72, 562-591 (1959); CA 53, 14602i

2637. Glueckauf, E., "Isotope Separation by Reversible Chemical Processes," Endeavour, 20, 42-50 (1961); CA 55, 12095g

2638. Glueckauf, E., Barker, K. H., and Kitt, G. P., "Theory of Chromatography. VIII. The Separation of Lithium Isotopes by Ion Exchange and of Neon Isotopes by Low Temperature Adsorption Columns," Disc. Faraday Soc., 7, 199-213 (1949); CA 44, 9771f

2639. Glueckauf, E., and Coates, J. I., "Theory of Chromatography. IV. The Influence of Incomplete Equilibrium on the Front Boundary of Chromatograms and on the Effectiveness of Separation," J. Chem. Soc., 1947, 1315-1320; CA 42, 2490e

2640. Glueckauf, E., and Kitt, G. P., "Krypton and Xenon Contents of Atmospheric Air," Proc. Roy. Soc. (London), A234, 557-565 (1956); CA 50, 9072i

2641. Glueckauf, E., and Kitt, G. P., "Gas Chromatographic Separation of Hydrogen Isotopes," in "Vapour Phase Chromatography," edited by D. H. Desty, Academic Press, Inc., New York, 1957, pp. 422-427

2642. Glueckauf, E., and Kitt, G. P., "Gas Chromatographic Separation of Hydrogen Isotopes," Proc. Intern. Symposium Isotope Separation, Amsterdam, 1957; edited by J. Kistamaker, J. Bigeleisen, and A. O. C. Nier, 210-226 (Pub. 1958); CA 52, 11597a

2643. Gnauck, G., "Preparative Gas Chromatography of the Rare Gases," in "Gas Chromatographie 1959," edited by R. E. Kaiser and H. G. Struppe, Akademie Verlag, Berlin, 1959, pp. 120-136

2644. Gnauck, G., "The Use of Gas Chromatography in the Production of Rare Gases," Technik (Berlin), 15, 805-809 (1960); CA 56, 15310f

2645. Gnauck, G., "Problems of Rare Gas Analysis," Acta Chim. Acad. Sci. Hung., 27, 229-237 (1961); CA 55, 23173g

2646. Gnauck, G., "Gas-Chromatographic Determination of Traces of Permanent Gases with an Ionization Detector," Z. Anal. Chem., 189, 124-130 (1962); CA 57, 13171f

2647. Gnauck, G., and Frenzel, J., "Application of Gas Chromatography in the Analysis of Noble and Permanent Gases," Abhandl. Deut. Akad. Wiss. Berlin, Kl. Chem., Geol., Biol., 1959, No. 9, 154-175; CA 58, 5018e

2648. Gnauck, G., and Frenzel, J., "Application of Gas Chromatography for the Analysis of Rare Gases and Some Permanent Gases," in "Gas Chromatographie 1958," edited by H. P. Angele, Akademie Verlag, Berlin, 1959, pp. 154-175

2649. Goble, A. G., and Smith, R. S., "A Carbon-14 Tracer Technique for the Investigation of Catalytic Reactions," 135th Natl. ACS Mtg., Boston, Mass., April 1959, Program Abstr., p. 8Q; Preprints, Div. Petrol. Chem., 4, No. 2C, 141

2650. Godet, M., "Instrumentation in Gas Chromatography," 8th Annual Instrument Symp., National Institutes of Health, Bethesda, Md., May 1958

2651. Godsell, J. A., Stacey, M., and Tatlow, J. C., "Hexafluorobenzene," Nature, 178, 199-200 (1956); CA 51, 3472e

2652. Goeckner, N. A., "Odorous Constituents of Corn. Gas Chromatography of Racemic Modifications," Univ. Microfilms (Ann Arbor, Mich.), L. C. Card No. Mic 59-1680, 121 pp.; Dissertation Abstr., 19, 3127 (1959); CA 53, 19060c

2653. Goedkoop, W., "Hop Preservation. III. Analysis of Hop Oil by Means of Gas Chromatography," Intern. Tijdschr. Brouw. en Mout., 20, 39-55 (1960-61); CA 54, 25556a

2654. Goedkoop, W., "Hop Preservation. IV. Analysis of Hop Oil by Means of Gas Chromatography," Intern. Tijdschr. Brouw. en Mout., 21, 80-93 (1961-62); CA 58, 6158a

2655. Goering, H. L., Greiner, R. W., and Sloan, M. F., "Ionic Reactions in Bicyclic Systems. I. The

97

Preparation and Assignment of Configuration of the Isomeric Bicyclo[3.2.1]oct-3-en-2-ols and Bicyclo[3.2.1]octan-2-ols," J. Am. Chem. Soc., 83, 1391-1397 (1961); CA 55, 14329g

2656. Goering, H. L., and Larsen, D. W., "The Stereochemistry of Radical Additions. IV. The Radical Addition of Hydrogen Bromide and Deuterium Bromide to cis- and trans-2-Bromo-2-butene," J. Am. Chem. Soc., 81, 5937-5942 (1959); CA 54, 9720d

2657. Goering, H. L., and Sloan, M. F., "Ionic Reactions in Bicyclic Systems. II. Carbonium Ion Reactions in Bicyclo[2.2.2]-octane and Bicyclo[3.2.1]-octane Derivatives," J. Am. Chem. Soc., 83, 1397-1401 (1961); CA 55, 14330g

2658. Goetz, R. W., and Orchin, M., "The Isomerization of Allyl Alcohols with Cobalt Hydrocarbonyl," J. Am. Chem. Soc., 85, 1549-1550 (1963); CA 59, 1453f

2659. Gohlke, R. S., "Instrument Design for Gas-Liquid Partition Chromatography," Anal. Chem., 29, 1723-1727 (1957); CA 52, 4250f

2660. Gohlke, R. S., "Use of Time of Flight Mass Spectrometry and Vapor Phase Chromatography in the Identification of Unknown Mixtures," 132nd Natl. ACS Mtg., New York, N. Y., September 1957, Program Abstr., p. 34B; Preprints, Div. Petrol. Chem., 2, No. 4, D77 (1957)

2661. Gohlke, R. S., "Time-of-Flight Mass Spectrometry and Gas-Liquid Partition Chromatography," Anal. Chem., 31, 535-541 (1959); CA 53, 17595b; 9th Pittsburgh Conf. on Anal. Chem. & Appl. Spectroscopy, Pittsburgh, Pa., March 1958, Program Abstr., p. 49

2662. Gohlke, R. S., "Time-of-Flight Mass Spectrometry: Application to Capillary Column Gas Chromatography," Anal. Chem., 34, 1332-1333 (1962); CA 57, 14412c

2663. Gohlke, R. S., and McLafferty, F. W., "The Use of Vapor Phase Chromatography in the Identification of Unknown Mixtures," 129th Natl. ACS Mtg., Dallas, Texas, April 1956, Program Abstr., p. 18B

2664. Golay, M. J. E., "Vapor Phase Chromatography and the Telegrapher's Equation," Anal. Chem., 29, 928-932 (1957); CA 51, 12598a; 129th Natl. ACS Mtg., Dallas, Texas, April 1956, Program Abstr., p. 15B

2665. Golay, M. J. E., "A Performance Index for Gas Chromatographic Columns," Nature, 180, 435-436 (1957); CA 52, 805g

2666. Golay, M. J. E., "Gas Chromatographic Terms and Definitions," Nature, 182, 1146-1147 (1958); CA 53, 6867i

2667. Golay, M. J. E., "Theory and Practice of Gas-Liquid Partition with Coated Capillaries," in "Gas Chromatography," edited by V. J. Coates, H. J. Noebels, and I. S. Fagerson, Academic Press, Inc., New York, 1958, pp. 1-13

2668. Golay, M. J. E., "Theory of Chromatography in Open and Coated Tubular Columns with Round and Rectangular Cross Sections," in "Gas Chromatography 1958," edited by D. H. Desty, Academic Press, Inc., New York, 1958, pp. 36-55

2669. Golay, M. J. E., "Brief Report on Gas Chromatographic Theory," in "Gas Chromatography 1960," edited by R. P. W. Scott, Butterworths, London, 1960, pp. 139-143

2670. Golay, M. J. E., "Theoretical Considerations in Large Gas Chromatographic Columns," Gas Chromatog. Intern. Symposium, 2nd, East Lansing, Mich., 1959, 11-19 (Pub. 1960); CA 55, 21687c

2671. Golay, M. J. E., "Theoretical Considerations in Large Gas Chromatographic Columns," in "Gas Chromatography," edited by H. J. Noebels, R. F. Wall, and N. Brenner, Academic Press, Inc., New York, 1961, pp. 11-19

2672. Golay, M. J. E., "Our Present Understanding of the Gas Chromatographic Process," 139th Natl. ACS Mtg., St. Louis, Mo., March 1961, Program Abstr., p. 4B

2673. Golay, M. J. E., "Reflections of a Communications Engineer," Anal. Chem., 33, No. 6, 23A-31A (1961)

2674. Golay, M. J. E., "Gas Chromatography and Invention," in "Gas Chromatography," edited by N. Brenner, J. E. Callen, and M. D. Weiss, Academic Press, Inc., New York, 1962, pp. xi-xv

2675. Golay, M. J. E., "Recirculated Column for Preparative Gas Chromatography," Univ. of Houston Intern. Symp. on Advances in Gas Chromatography, Houston, Texas, January 1963

2676. Golay, M. J. E., Hill, H. I., and Norem, S. D., "Recirculated Column for Preparative Scale Gas Chromatography," Anal. Chem., 35, 488-491 (1963)

2677. Gol'bert, K. A., and Alekseeva, A. V., "Determination of Propylene in Ethylene and Ethane-Ethylene Fractions," Zavodskaya Lab., 24, 688-690 (1958); CA 54, 10674f

2678. Gol'bert, K. A., and Vigdergauz, M. S., "Gas Chromatography of C_{3-5} Hydrocarbons," Novosti Neft. i Gaz. Tekhn., Gaz. Delo, 1961, No. 12, 37-40; CA 58, 8394f

2679. Gold, H. J., "Micro-Scale Collection Tube for Gas Chromatography," Chemist-Analyst, 49, 112 (1960)

2680. Gold, H. J., "Estimation of Column Holdup Time in Gas Chromatography When Using Ionization Detectors," Anal. Chem., 34, 174-175 (1962); CA 56, 8489i

2681. Gold, H. J., and Wilson, C. W., "Techniques in Isolation of Volatile Materials from Celery and Identification of Some Compounds with Acidic Properties," Proc. Florida State Hort. Soc., 74,

291-296 (1961); CA 57, 1336h

2682. Gold, H. J., and Wilson, C. W., "Alkylidene Phthalides and Dihydrophthalides from Celery," J. Org. Chem., 28, 985-987 (1963); CA 58, 12457g

2683. Gold, V., and Satchell, R. S., "The Kinetics of Hydrogen Isotope Exchange Reactions. XI. Tritium Exchange Between Secondary Alcohols and Acidic Media," J. Chem. Soc., 1963, 1930-1937; CA 58, 10056c

2684. Gold, V., and Satchell, R. S., "The Kinetics of Hydrogen Isotope Exchange Reactions. XII. The Behavior of Primary Alcohols in Acidic Media," J. Chem. Soc., 1963, 1938-1947; CA 58, 10056d

2685. Goldberg, G., and Ross, W. A., "Separation of Optical Isomers by Gas-Liquid Chromatography," Chem. & Ind. (London), 1962, 657; CA 57, 7095b

2686. Golding, W. E., and Townsend, C. A., "Gas Chromatography of Pyridine Bases: Elimination of Peak-Tailing," Chem. & Ind. (London), 1960, 1476; CA 55, 25223c

2687. Gol'dinov, A. L., Lukhovitskii, V. I., and Svinina, N. A., "Gas-Liquid Chromatographic Analysis of Fluorochloromethanes," Zavodskaya Lab., 28, 150-151 (1962); CA 57, 2856d

2688. Goldschmidt, S., Zoebelein, H., and Seiz, W., "Reaction of β-(4-Hydroxyphenyl)propionic Acid with Hypobromite," Ann., 657, 25-38 (1962); CA 58, 1385e

2689. Goldsmith, D. J., "Cyclization of Epoxy Olefins: Reaction of Geraniolene Monoepoxide with Boron Fluoride Etherate," J. Am. Chem. Soc., 84, 3913-3918 (1962); CA 58, 5726c

2690. Goldup, A., "Gas Chromatography," Nature, 189, 26-28 (1961)

2691. Goldup, A., Luckhurst, G. R., and Swanton, W. T., "Gas-Liquid Chromatography - Variations in Partition Coefficients with Carrier Gas," Nature, 19, 333-334 (1962); CA 57, 2824i

2692. Gomi, S., Furukawa, J., Kobayashi, T., Kambayashi, M., and Hosoi, T., "Pyrolysis of 1,2-Dichloropropane," Kogyo Kagaku Zasshi, 65, 1384-1388 (1962); CA 58, 5492e

2693. Gonzalez, A. G., and Barrera, J. B., "Gas-Liquid Chromatography of Some Steroids, Sapogenins, and Triterpenes," Anales Real Soc. Espan. Fis. Quim. (Madrid), B58, 559-562 (1962); CA 59, 693e

2694. Good, C. D., and Ritter, D. M., "Alkenylboranes. II. Improved Preparation Methods and New Observations on Methylvinylboranes," J. Am. Chem. Soc., 84, 1162-1166 (1962); CA 57, 9865c

2695. Good, C. D., and Ritter, D. M., "Alkenylboranes. Characterization of Methylvinylboranes," J. Chem. Eng. Data, 7, 416-419 (1962); CA 57, 14417h

2696. Goodman, D. S., and Popjak, G., "Studies on the Biosynthesis of Cholesterol. XII. Synthesis of Allyl Pyrophosphates from Mevalonare and Their Conversion into Squalene with Liver Enzymes," J. Lipid Research, 1, 286-300 (1960); CA 55, 5603d

2697. Goodwin, E. S., Goulden, R., and Reynolds, J. G., "Gas Chromatography with Electron Capture Ionization Detection for Rapid Identification of Pesticide Residues in Crops," 18th Intern. Congr. on Pure and Applied Chemistry, Montreal, Canada, August 1961

2698. Goodwin, E. S., Goulden, R., and Reynolds, J. G., "Rapid Identification and Determination of Residues of Chlorinated Pesticides in Crops by Gas-Liquid Chromatography," Analyst, 86, 697-709 (1961); cf. ibid., 87, 169 (1962); CA 57, 10297d

2699. Goodwin, E. S., Goulden, R., Richardson, A., and Reynolds, J. G., "The Analysis of Crop Extracts for Traces of Chlorinated Pesticides by Gas-Liquid Partition Chromatography," Chem. & Ind. (London), 1960, 1220-1221; CA 55, 9705e

2700. Gordillo, A. L., and Montes, A. L., "Chromatography of Fatty Acids: A Semimicro Method for Determination of C_1-C_{10} Acids," Anales soc. cient. arg., 170, No. 3-4, 53-67 (1960); CA 55, 12157c

2701. Gordon, A. S., Smith, S. R., and McNesby, J. R., "Reactions of the Allyl Radical," J. Am. Chem. Soc., 81, 5059-5061 (1959); CA 54, 13020a

2702. Gordon, R. J., and Moore, R. J., "Identification of Aromatic Types in a Heavily Cracked Gas Oil," 8th Pittsburgh Conf. on Anal. Chem. & Appl. Spectroscopy, Pittsburgh, Pa., March 1957, Program Abstr., p. 35

2703. Gordon, S., Van Dyken, A. R., and Doumani, T. F., "Identification of Products in the Radiolysis of Liquid Benzene," J. Phys. Chem., 62, 20-24 (1958); CA 52, 7877a

2704. Gordon, S. M., Krige, G. J., Haarhoff, P. C., and Pretorius, V., "Mobile Phase Effects in Packed Liquid Chromatography Columns," Anal. Chem., 35, 1537-1539 (1963)

2705. Gordus, A. A., Quinlan, J. E., Evans, J. B., Sauer, M. C., and Willard, J. E., "Chemical Effects of Nuclear Transformations in the Gas Phase," 132nd Natl. ACS Mtg., New York, N. Y., September 1957, Program Abstr., p. 1S

2706. Gordus, A. A., Sauer, M. C., and Willard, J. E., "Evidence of Mechanisms of Halogen and Tritium Recoil Labeling Reactions," J. Am. Chem. Soc., 79, 3284-3285 (1957); CA 51, 14428e

2707. Gordus, A. A., and Willard, J. E., "Gas-Phase Reactions Activated by Nuclear Processes," J. Am. Chem. Soc., 79, 4609-4616 (1957); CA 51, 17468a

2708. Goto, R., and Araki, N., "Stationary Liquid Phase," Kagaku no Ryoiki Zokan, No. 44, 1-37 (1961); CA 56, 2i

2709. Gotz, A., "Use of Gas Chromatography to Shorten Analysis Time in Oxygen Determinations," Symp.

on Modern Methods for Analysis of Organic Compounds, German Chem. Soc., Munich, Germany, October 1960; Abstr., Angew. Chem., 73, 144 (1961)

2710. Gotz, A., and Bober, H., "Rapid Determination of Oxygen in Organic Substances," Z. Anal. Chem., 181, 92-100 (1961); CA 55, 23164b

2711. Gover, T. A., and Willard, J. E., "Reactions of Hydrogen and Hydrocarbons with Iodine Excited by 1849 A. Radiation," J. Am. Chem. Soc., 82, 3816-3821 (1960); CA 54, 23633e

2712. Gower, D. B., and Haslewood, G. A. D., "The Biosynthesis of Androst-16-en-3α-ol from Acetate by Testicular Slices," J. Endocrinol., 23, 253-260 (1961); CA 56, 16015e

2713. Graciantous, J., Vioque, E., Pilar de la Maza, M., "Quantitative Estimation of the Fatty Acids of Olive Oil," Nature, 184, 1941 (1959); CA 54, 13483b

2714. Graf, L., and Toth, J., "Adsorption Theory of Gas Chromatography," Acta Chim. Acad. Sci. Hung., 13, 403-428 (1958); CA 52, 14281e

2715. Graf, L., Toth, J., and Goncz, I., "Theory of the Chromatographic Analysis of Hydrocarbon Gases," Magyar Kem. Folyoirat, 62, 113-118 (1956); CA 52, 4413e

2716. Gramstad, T., "Gas Chromatography," Tidsskr. Kjemi, Bergvesen Met., 18, 81-91 (1958); CA 52, 13363g

2717. Grandy, G. L., and Koch, R. C., "On-Stream Radioactivity Monitor for Gas-Handling Systems," Rev. Sci. Instr., 31, 786 (1960); CA 55, 21688g

2718. Grant, D. H., and Grassie, N., "The Thermal Decomposition of Poly(tert-butyl Methacrylate)," Polymer, 1, 445-455 (1960); CA 55, 10950h

2719. Grant, D. H., Vance, E., and Bywater, S., "Thermal Depolymerization of Poly-α-methylstyrene in Solution," Trans. Faraday Soc., 56, 1697-1703 (1960); CA 55, 13010i

2720. Grant, D. W., "Emissivity Detector for Gas Chromatography," Gas Chromatog., Proc. Symposium, Amsterdam, 1958, 153-163; CA 53, 16609a

2721. Grant, D. W., "An Emissitivity Detector for Gas Chromatography," in "Gas Chromatography 1958," edited by D. H. Desty, Academic Press, Inc., New York, 1958, pp. 153-164; CA 53, 16609a

2722. Grant, D. W., "A Recording Dielectrometric Method for Column Chromatography," Chem. & Ind. (London), 1958, 136-140

2723. Grant, D. W., and Vaughan, G. A., "A Consideration of Factors Governing the Separation of Substances by Gas-Liquid Partition Chromatography," J. Appl. Chem. (London), 6, 145-153 (1956); CA 51, 801c

2724. Grant, D. W., and Vaughan, G. A., "The Use of Gas-Liquid Chromatography in the Determination of the Distribution of Aromatic Compounds in Coal Tar Naphthas," in "Vapour Phase Chromatography," edited by D. H. Desty, Academic Press, Inc., New York, 1957, pp. 413-421

2725. Grant, D. W., and Vaughan, G. A., "Relative Retention Volumes of Aliphatic Hydrocarbons (Chromatographic Data Table XXXIX)," J. Chromatog., 1, xxv (1958); cf. "Gas Chromatography," edited by D. H. Desty, Butterworths, London, 1957, p. 415 (Temperature, 100°)

2726. Grant, D. W., and Vaughan, G. A., "Relative Retention Volumes of Aromatic Hydrocarbons (Chromatographic Data Table XLII)," J. Chromatog., 1, xxvi (1958); cf. "Gas Chromatography," edited by D. H. Desty, Butterworths, London, 1957, p. 415 (Temperature, 100°)

2727. Grant, D. W., and Vaughan, G. A., "A Peak Interference Method for the Analysis of Close-Boiling Isomers by Gas-Liquid Chromatography," J. Appl. Chem. (London), 10, 181-187 (1960); CA 54, 16256b

2728. Grant, D. W., and Vaughan, G. A., "The Analysis of Complex Phenolic Mixtures by Capillary Column GLC After Silylation," 4th Intern. Gas Chromatography Symp., Hamburg, Germany, June 1962; pub. in "Gas Chromatography 1962," ed. by M. van Swaay, Butterworth & Co., London, 1962

2729. Grasselli, J. G., and Snavely, M. K., "Analysis of Organic Reaction Products by Combined Infrared-Gas Chromatography Techniques," Appl. Spectroscopy, 16, 190-194 (1952); CA 58, 9608c

2730. Grassmann, H., "Application of Gas Chromatography for the Analysis of Natural Gas," Z. angew. Geol., 5, 164-168 (1959); CA 53, 10706g

2731. Graven, W. M., "Gas Chromatograph. Ionization by Alpha-Particles for Detection of the Gaseous Components in the Effluent from a Flow Reactor," Anal. Chem., 31, 1197-1199 (1959); CA 54, 13937c

2732. Graven, W. M., "Kinetics of the Decomposition of N_2O at High Temperature," J. Am. Chem. Soc., 81, 6190-6192 (1959); CA 54, 11660f

2733. Gray, Jr., F. B., "Hydrocarbon Measurements in Drilling Muds," Proc. Natl. Conf. Instr. Methods Anal., Chicago, 1957, Paper No. A157-4-4, 4 pp.; CA 52, 14145a

2734. Gray, Jr., F. B., "New Analytical Techniques for the Measurement of Hydrocarbon Gases in Drilling Muds," 8th Pittsburgh Conf. on Anal. Chem. & Appl. Spectroscopy, Pittsburgh, Pa., March 1957, Program Abstr., p. 34

2735. Gray, Jr., F. B., "Reliability of Gas Chromatography as an Analytical Field Tool," 1958 Natl. ISA Mtg., Symposium on Instrumental Methods of Analysis, Houston, Texas, May 1958

2736. Gray, G. M., "The Phospholipids of Ox Spleen with Special Reference to the Fatty Acid and Fatty

Aldehyde Composition of the Lecithin and Kephalin Fractions," Biochem. J., 77, 82-91 (1960); CA 54, 22900d

2737. Gray, G. M., "Separation of the Long-Chain Fatty Aldehydes by Gas-Liquid Chromatography," J. Chromatog., 4, 52-59 (1960); CA 55, 3304i

2738. Gray, G. M., "The Structural Identification of Some Naturally Occurring Branched Chain Fatty Aldehydes," J. Chromatog., 6, 236-242 (1961); CA 56, 9034a

2739. Gray, G. M., "Cyclopropane-Ring Fatty Acids of Salmonella Typhimurium," Biochim. Biophys. Acta, 65, 135-141 (1962); CA 58, 6005e

2740. Graziano, V., "Determination of Fatty Acids in the Dental Pulp of the Ox by Gas Chromatography," Boll. Soc. Ital. Biol. Sper., 38, 1304-1306 (1962); CA 58, 12925b

2741. Green, G. E., "Hydrogen-Conversion Detector for Gas Chromatography," Nature, 180, 295-296 (1957); CA 52, 5050b

2742. Green, S. W., "Vapour-Phase Chromatography," Ind. Chemist, 32, 24-28 (1956); CA 50, 6137f

2743. Green, S. W., "The Quantitative Analysis of Mixtures of Chlorofluoromethanes," in "Vapour Phase Chromatography," edited by D. H. Desty, Academic Press, Inc., New York, 1957, pp. 388-394

2744. Greene, S. A., "The Application of Gas Adsorption Chromatography to the Analysis of Chemical Reactions in Flow Systems," J. Chem. Educ., 34, 194-195 (1957); CA 7802c

2745. Greene, S. A., "Calculation of the Limiting Retention Volume in Gas-Liquid Partition Chromatography," J. Phys. Chem., 61, 702 (1957); CA 51, 13512a

2746. Greene, S. A., "The Separation of the Rare Gases by Gas-Solid Chromatography," 132nd Natl. ACS Mtg., New York, N. Y., September 1957, Program Abstr., p. 35B; Preprints, Div. Petrol. Chem., 2, No. 4, D-105 (1957)

2747. Greene, S. A., "Gas-Solid Chromatographic Analysis of Fractions from Air Rectification Columns," Anal. Chem., 31, 480 (1959); CA 53, 11089f

2748. Greene, S. A., "Mobile Phase," in "Principles and Practice of Gas Chromatography," edited by R. L. Pecsok, John Wiley & Sons, Inc., New York, 1959, pp. 28-47

2749. Greene, S. A., and Cain, E. F. C.,"Sample Introduction," in "Principles and Practice of Gas Chromatography," edited by R. L. Pecsok, John Wiley & Sons, Inc., New York, 1959, pp. 86-96

2750. Greene, S. A., Moberg, M. L., and Wilson, E. M., "Separation of Gases by Gas Adsorption Chromatography," Anal. Chem., 28, 1369-1370 (1956); CA 50, 16531d

2751. Greene, S. A., and Pust, H., "Use of Silica Gel and Alumina in Gas-Adsorption Chromatography," Anal. Chem., 29, 1055 (1957); CA 51, 14371d

2752. Greene, S. A., and Pust, H., "Determination of Heats of Adsorption by Gas-Liquid Chromatography," J. Phys. Chem., 62, 55-58 (1958); CA 52, 7809g

2753. Greene, S. A., and Pust, H., "Determination of Nitrogen Dioxide by Gas-Solid Chromatography," Anal. Chem., 30, 1039-1040 (1958); CA 52, 14417h; 132nd Natl. ACS Mtg., New York, N. Y., September 1957, Program Abstr., p. 35B; Preprints, Div. Petrol. Chem., 2, No. 4, D-107 (1957)

2754. Greene, S. A., and Roy, H. E., "Effect of Different Carrier Gases on Retention Times in Gas-Adsorption Chromatography," Anal. Chem., 29, 569-570 (1957); CA 51, 10177f

2755. Greene, S. A., and Wachi, F. M., "Separation of Some Low Molecular Weight Fluorocarbons by Gas Chromatography," Anal. Chem., 35, 928-929 (1963)

2756. Greenwood, C. T., Knox, J. H., and Milne, E., "Analysis of the Thermal Decomposition Products of Carbohydrates by Gas Chromatography," Chem. & Ind. (London), 1961, 1878-1879; CA 56, 9394d

2757. Greenwood, F. L., "Pyrolysis of Allylic Acetates," J. Org. Chem., 24, 1735-1739 (1959)

2758. Greenwood, N. N., "A Mercury Cut-Off for Transferring Gas At Atmospheric Pressure into a High Vacuum," J. Sci. Instr., 33, 318-319 (1956); CA 51, 2h

2759. Gregg, S. J., and Stock, R., "The Adsorption of Hydrocarbon Vapors by Ammonium Phosphomolybdate," Trans. Faraday Soc., 53, 1355-1362 (1957); CA 52, 11516g

2760. Gregg, S. J., and Stock, R., "Sorption Isotherms and Chromatographic Behavior of Vapours," in "Gas Chromatography 1958," edited by D. H. Desty, Academic Press, Inc., New York, 1958, pp. 90-98

2761. Gregory, N. L., "Detection of Nanogram Quantities of Sulfur Hexafluoride by Electron-Capture Methods," Nature, 196, 162 (1962); CA 58, 1911b

2762. Gregory, N. L., and Lovelock, J. E., "Electron Capture Ionization Detectors," 3rd Intern. Symposium on Gas Chromatography, ISA, East Lansing, Mich., June 1961; ISA Proc., 3, 151-158 (1961)

2763. Greiner, A., and Mueller, U., "The Course of Ozonization of Olefins and the Splitting of Ozonides. V. Occurrence of Hydrocarbons in the Splitting of Ozonides," J. Prakt. Chem., 15, 313-321 (1962); CA 58, 1323c

2764. Grieco, D., "Identification of the Methyl Esters of Linolenic, Arachic, and Eicosenic Acids by Gas Chromatography," Olearia, 16, 11-16 (1962); CA 58, 4749f

2765. Grieco, D., "Separation of the Methyl Esters of Linolenic, Arachidic, and Eicosenoic Acids on Polyethylene Glycol Succinate," Olearia, 16, 122-126 (1962); CA 58, 5021g

2766. Griffiths, J., James, D., and Phillips, C., "Gas Chromatography: Adsorption and Partition Methods," Analyst, 77, 897-904 (1952); CA 47, 1530b

2767. Griffiths, J. H., and Phillips, C. S. G., "The Chromatography of Gases and Vapours. IV. Applications of the Surface Potential Detector," J. Chem. Soc., 1954, 3446-3453; CA 49, 2125i

2768. Grigoryan, K. A., Aliev, Z. E., and Kuliev, A. M., "The Adsorptivity and Selectivity of Various Adsorbents on Hydrocarbon Gases and Their Mixtures," Sb. Tr., Inst. Neftekhim. Protsessov, Akad. Nauk Azerb. SSR, No. 4, 229-250 (1959); CA 57, 86g

2769. Grob, C. A., Link, H., and Schiess, P. W., "Cyclodecapolyenes. II. Valency Isomerization of 1, 5-Cyclodecadiene," Helv. Chim. Acta, 46, 483-492 (1963); CA 59, 463b

2770. Grob, R. L., Mercer, D., Gribben, T., and Wells, J., "Thermal-Conductivity Cell Response and Its Relation to Quantitative Gas Chromatography," J. Chromatog., 3, 545-553 (1960); CA 55, 4231g

2771. Groringer, H. S., "Fish Spoilage. I. Determination of Bacterial Metabolites by Gas Chromatography," Com. Fisheries Rev., 20, No. 11, 23-26 (1958)

2772. Grossi, E., "Fatty Acids in Human Brain Tissue; Analysis of the Composition of Different Lipid Fractions," Ric. Sci. Rend. Sez. B, 2, 167-172 (1962); CA 58, 9476c

2773. Grossi, E., and Paoletti, P., "Gas Liquid Chromatography: The Estimation of the Liver and Serum Fatty Acids of Normal and of Rats Treated with Triton," Ital. J. Biochem., 8, 309-318 (1959)

2774. Grossi, E., and Paoletti, P., "Fatty Acid Composition of Brain Lipids from Various Animal Species. Comparative Analysis by Gas-Liquid Chromatography," Ric. Sci. Rend. Ber. B, 2, No. 1, 5-10 (1962); CA 57, 14312g

2775. Grosskopf, K., "Testing Tubes as Detectors for Gas Chromatography," Erdol u. Kohle, 11, 304-306 (1958); CA 52, 15321g

2776. Grossman, J. D., Deszyck, E. J., Ikeda, R. M., and Bavley, A., "A Study of Pyrolysis of Solanesol," Chem. & Ind. (London), 1962, 1950-1951; CA 58, 5529c

2777. Gruber, K., "The Analysis of Lacquer Solvents by Gas-Liquid Partition Chromatography. II. Quantitative Analysis," Holz-Forsch. n. Holz-Verwert., 9, No. 6, 104-107 (1957); CA 55, 1025b

2778. Grubner, O., "The Surface Area Determination of Active Substances and the Construction of Adsorption Isotherms in Gas Flow," 3rd Intern. Conf. on Anal. Chem., Prague, Czech., September 1959

2779. Grubner, O., "Remakrs on the Compressibility Correction of the Retention Volumes in a Gas Chromatographic Column According to James and Martin," in "Gas Chromatographie 1959," edited by R. E. Kaiser and H. G. Struppe, Akademie Verlag, Berlin, 1959, pp. 137-142

2780. Grubner, O., "Sorption Data from Chromatographic Measurements," Gas Chromatog. Symp., Brno, Czech., June 1962; Abstr., J. Gas Chromatog., 1, No. 4, 7 (1963)

2781. Grubner, O., and Kucera, E., "Gas Chromatography, Counter-Current Chromatography and Distillation," Gas Chromatog. Symp., Brno, Czech., June 1962; Abstr., J. Gas Chromatog., 1, No. 4, 7 (1963)

2782. Grubner, O., Ralek, M., and Svoboda, J., "Use of Less Known Types of Molecular Sieves in Gas Chromatography," Gas Chromatog. Symp., Brno, Czech., June 1962; Abstr., J. Gas Chromatog., 1, No. 4, 9 (1963)

2783. Grubner, O., and Smolkova, E., "Contribution to the Relation Between the Characteristic Retention Volume and the Adsorbent Surface Area in Gas-Solid Chromatography," 3rd Intern. Conf. on Anal. Chem., Prague, Czech., September 1959

2784. Grundon, M. F., Henbest, H. B., and Scott, M. D., "The Reactions of Hydrazones and Related Compounds with Strong Bases. I. A Modified Wolff-Kishner Procedure," J. Chem. Soc., 1963, 1855-1858; CA 58, 10263h

2785. Grune, W. N., "Application of Gas Chromatography to Sludge Digestion Gas Analysis," Water & Sewage Works, 107, 396-399 (1960); CA 55, 20271d

2786. Grune, W. N., Carter, Jr., J. Y., and Keenan, J. P., "Development of a Continuous Gas Chromatographic Analyzer for Sludge Digestion Studies," Sewage Ind. Wastes, 28, No. 12, 1433-1442 (1956); CA 51, 3067i

2787. Grune, W. N., and Chueh, C-F., "Application of Gas Chromatography to Sludge Digestion Gas Analysis," Air Water Pollution J., 6, 283-318 (1962); CA 58, 4293d

2788. Grune, W. N., Chueh, C-F., and Hawkins, J. M., "Gas Chromatography for Water Treatment Control," J. Water Pollution Control Federation, 32, 942-948 (1960); CA 55, 3890h

2789. Grune, W. N., Cossitt, R. E., and Philip, Jr., R. H., "Anaerobic Process Automation by O.R.P. Conductivity and Gas Chromatography," Purdue Univ. Eng. Bull. Ext. Ser. No. 94, 604-635 (1959); CA 53, 6494h

2790. Grune, W. N., Philip, Jr., R. H., and Borsch, R. J., "Applications of Gas Chromatography, Conductivity, and Potential Measurement in Sludge Digestion," in "Biological Treatment of Sewage and Industrial Wastes," edited by J. McCabe and W. W. Eckenfelder, Reinhold Publishing Co., New York, 1958, pp. 80-96

2791. Gubser, H., "Quantitative Trace Analysis and Chromatography," Chimia (Switz.), 13, 245-248 (1959); CA 54, 6385a

2792. Gudzinowicz, B. J., "A New Radioactive Gas Chromatographic Detector for the Identification of Strong Oxidants," Univ. of Houston Symp. on Advances in Gas Chromatography, Houston, Texas, January 1963

2793. Gudzinowicz, B. J., "Separations of Psychoactive Drugs and Their Metabolites by GLC," Biochemical Gas Chromatography Seminar, Wilkens Instrument & Research, Inc., New Orleans, La., July 1963

2794. Gudzinowicz, B. J., Alm, J., and Smith, W. R., "Gas Chromatographic Separation of Chlorophenylphenol Germicide Mixtures," Anal. Chem., 34, 1032 (1962)

2795. Gudzinowicz, B. J., and Campbell, R. H., "High Temperature Gas Chromatographic Separations of Aryl Phosphines and Phosphine Oxides," Anal. Chem., 33, 1510-1512 (1961); CA 56, 4076h; 140th Natl. ACS Mtg., Chicago, Ill., September 1961, Program Abstr., p. 25B

2796. Gudzinowicz, B. J., Campbell, R. H., Driscoll, J. L., and Martin, H. F., "Gas Chromatographic Analysis of Organic Carbonates, Aryl Phosphines and Phosphine Oxides and Organo- and Organobromoarsenic Compounds," 14th Pittsburgh Conf. on Anal. Chem. & Appl. Spectroscopy, Pittsburgh, Pa., March 1963, Program Abstr., pp. 57-58

2797. Gudzinowicz, B. J., and Driscoll, J. L., "Separation and Analysis of Organic Carbonates by Gas Chromatography," Anal. Chem., 33, 1508-1510 (1961); CA 56, 4076e

2798. Gudzinowicz, B. J., and Driscoll, J. L., "Separation of Alkyl/Aryl and Perfluorinated Organoarsenic Compounds by Gas Chromatography," J. Gas Chromatog., 1, No. 5, 25-27 (1963)

2799. Gudzinowicz, B. J., Driscoll, J. L., Alm, J., and Smith, W. R., "An Application of Gas Chromatography to the Separation and Analysis of Organic Carbonates, Chlorophenylphenol Germicide Mixtures, and Perfluoroorganic Arsenic Compounds," 10th Detroit Anachem Conf., Detroit, Mich., October 1962, Program Abstr., p. 33

2800. Gudzinowicz, B. J., and Martin, H. F., "Separation of Organo- and Organobromoarsenic Compounds by Gas-Liquid Chromatography," Anal. Chem., 34, 648-650 (1962); CA 57, 1519c; 141st Natl. ACS Mtg., Washington, D. C., March 1962, Program Abstr., p. 24B

2801. Gudzinowicz, B. J., and Smith, W. R., "High Temperature Gas-Liquid Chromatography. Exploratory Studies Using an Ionization Detector Chromatograph," Anal. Chem., 32, 1767-1771 (1960); CA 55, 8147i

2802. Gudzinowicz, B. J., and Smith, W. R., "Modification to an Ionization Detector Chromatograph for High Temperature Gas-Liquid Chromatography Exploratory Studies," Anal. Chem., 33, 1135-1136 (1961); CA 56, 9382f

2803. Gudzinowicz, B. J., and Smith, W. R., "New Radioactive Gas Chromatographic Detector for Identification of Strong Oxidants," Anal. Chem., 35, 465-467 (1963); 142nd Natl. ACS Mtg., Atlantic City, N. J., September 1962, Program Abstr., p. 8B

2804. Guenther, E., Kulka, K., and Rogers, J. A., "Review of Industrial Applications of Analysis, Control, and Instrumentation. Essential Oils and Related Products," Anal. Chem., 31, 679-687 (1959); CA 53, 8911i

2805. Guenther, E., Kulka, K., and Rogers, J. A., "Essential Oils and Related Products," Anal. Chem., 33, No. 5, 37R-45R (1961); CA 55, 12137e

2806. Guenther, E., Kulka, K., and Rogers, J. A., "Essential Oils and Related Products," Anal. Chem., 35, No. 5, 39R-58R (1963)

2807. Guertin, D. L., "The Impact of Instrumental Analyses in the Petroleum Industry," 141st Natl. ACS Mtg., Washington, D. C., March 1962, Program Abstr., p. 28B

2808. Guild, L. V., "The Analysis of Light Hydrocarbon Gases by Fractional Adsorption," Proc. 32nd Annual Conv. Natural Gasoline Assoc. Am., 1953, pp. 13-22

2809. Guild, L. V., "Some Practical Applications of Gas Chromatography," 7th Pittsburgh Conf. on Anal. Chem. & Appl. Spectroscopy, Pittsburgh, Pa., February-March 1956, Program Abstr., p. 41

2810. Guild, L. V., "Techniques and Instrumentation for Microanalysis by Gas Chromatography," 11th Annual Microchem. Symp., New York, N. Y., March 1956; Abstr., Anal. Chem., 28, 921 (1956)

2811. Guild, L. V., "Some Applications of the Elution and Displacement Techniques in Gas-Vapor Chromatography," 129th Natl. ACS Mtg., Dallas, Texas, April 1956, Program Abstr., p. 16B

2812. Guild, L. V., "Gas Chromatography: Operating Conditions and Technique," 31st Annual Mtg., California Nat. Gas Assoc., Los Angeles, Calif., November 1956; cf. Oil Gas J., 54, No. 85, 134, 136 (1956)

2813. Guild, L. V., and Bingham, S. A., "Gas Chromatographic Analysis of Samples with Diverse Properties," 10th Pittsburgh Conf. on Anal. Chem. & Appl. Spectroscopy, Pittsburgh, Pa., March 1959, Program Abstr., p. 57

2814. Guild, L. V., Bingham, S. A., and Aul, F., "Base-Line Control in Gas-Liquid Chromatography," Gas Chromatog., Proc. Symposium, Amsterdam, 1958, 226-241; CA 53, 21041e

2815. Guild, L. V., Bingham, S. A., and Aul, F., "Base-Line Control in Gas-Liquid Chromatography," in

"Gas Chromatography 1958," edited by D. H. Desty, Academic Press, Inc., New York, 1958, pp. 226-247

2816. Guild, L. V., Hollingsworth, C. A., McDaniel, D. H., and Wotiz, J. H., "Gas Chromatographic Analysis of Aryl Grignard Reagents," Anal. Chem., 33, 1156-1157 (1961); CA 56, 2001g

2817. Guild, L. V., and Lloyd, M. I., "Recent Developments in the Emission Ionization Detectors," 11th Pittsburgh Conf. on Anal. Chem. & Appl. Spectroscopy, Pittsburgh, Pa., February-March 1960, Program Abstr., p. 42

2818. Guild, L. V., Lloyd, M. I., and Aul, F., "Performance Data on a New Ionization Detector," Gas Chromatog., Intern. Symposium, 2nd, East Lansing, Mich., 1959, 91-101 (Pub. 1961); CA 55, 18201f

2819. Guild, L. V., Lloyd, M. L., and Aul, F., "Performance Data on a New Ionization Detector," in "Gas Chromatography," edited by H. J. Noebels, R. F. Wall, and N. Brenner, Academic Press, Inc., New York, 1961, pp. 91-101

2820. Guild, L. V., and Swartzel, J., "High Temperature Chromatography with Ionization Detection," 12th Pittsburgh Conf. on Anal. Chem. & Appl. Spectroscopy, Pittsburgh, Pa., February-March 1961, Program Abstr., p. 52

2821. Guillaumin, R., and Drouhin, N., "Determination of Polyunsaturated Fatty Acids in Fats: Study of Different Methods," Rev. Franc. Corps Gras., 9, 415-436 (1962); CA 57, 11329a

2822. Guillemin, C. L., "The Separation of Hexachlorocyclohexane Isomers by Vapour Phase Chromatography," Anal. Chim. Acta, 27, 213-218 (1962); CA 58, 889a

2823. Guillet, J. E., Wooten, Jr., W. C., and Combs, R. L., "Analysis of Polymethacrylates by Gas Chromatography," J. Appl. Polymer Sci., 3, No. 7, 61-64 (1960); CA 54, 20287g

2824. Guiochon, G., "Influence of Injection Time on the Efficiency of Gas Chromatography Columns," Anal. Chem., 35, 399-400 (1963); CA 58, 10702b

2825. Gunesch, H., Brandsch, J., Heitz, J., Boteiu, A., and Loffler, A., "Determination of Crotonaldehyde in Vinyl Acetate and Its Effect on Emulsion Polymerization," Rev. Chim. (Bucharest), 14, 36-39 (1963); CA 59, 762b

2826. Gunesch, H., and Stadtmuller, R., "Separation and Determination of Higher Acetylenes with Gas-Liquid Chromatography," Rev. Chim. (Bucharest), 9, 35-38 (1958); CA 52, 19712f

2827. Gunn, W. H., and Murie, R. A., "Gas Chromatographic Separation of Chlorine, Hydrochloric Acid and Phosgene," 13th Pittsburgh Conf. on Anal. Chem. & Appl. Spectroscopy, Pittsburgh, Pa., March 1962, Program Abstr., p. 55

2828. Gunner, S. W., Jones, J. K. N., and Perry, M. B., "Analysis of Sugar Mixtures by Gas-Liquid Partition Chromatography," Chem. & Ind. (London), 1961, 255-256; CA 56, 936i

2829. Gunner, S. W., Jones, J. K. N., and Perry, M. B., "The Gas-Liquid Partition Chromatography of Carbohydrate Derivatives. I. The Separation of Glycitols and Glycose Acetates," Can. J. Chem., 39, 1892-1896 (1961); CA 56, 9389d

2830. Gunstone, F. D., and Sykes, P. J., "Partial Oxidation as a Means of Determining the Structure of Poly-Unsaturated Acids," Chem. & Ind. (London), 1960, 1130; CA 55, 8283c

2831. Gunther, F. A., "Instrumentation in Pesticide Residue Determinations," in "Advances in Pest Control Research," edited by R. L. Metcalf, Interscience Publishers, Inc., New York, 5, 191-319 (1962); CA 58, 10671c

2832. Gunther, F. A., Blinn, R. C., and Barnes, M. M., "Identification of Dimethyl Sulfide as the Major Volatile Constituent of the Insect Attractant, 'Staley's Protein Insecticide Bait No. 7'," 133rd Natl. ACS Mtg., San Francisco, Calif., April 1958, Program Abstr., p. 24A

2833. Gunther, F. A., Blinn, R. C., and Kohn, G. K., "Labile Organo-Halogen Compounds and Their Gas Chromatographic Detection and Determination in Biological Media," Nature, 193, 573-575 (1962); CA 56, 15759g

2834. Gunther, F. A., Blinn, R. C., and Ott, D. E., "Gas Chromatography of Pesticide Residues. New Methods of Clean-Up and of Detection of Organohalogen Compounds," 139th Natl. ACS Mtg., St. Louis, Mo., March 1960, Program Abstr., p. 26A

2835. Gunther, F. A., Blinn, R. C., and Ott, D. E., "Beilstein Flame Method of Detection of Organohalogen Compounds Emerging from a Gas Chromatograph," Anal. Chem., 34, 302-303 (1962); CA 56, 9393a

2836. Gunzler, H., "Combination of Physical Methods for Chemical Analysis," Z. Anal. Chem., 164, 49-56 (1958); Erdol u. Kohle, 11, 339 (1958); CA 53, 6871a

2837. Gur'yakov, I. I., "New Chromatographs," Priborostroenie, 1961, No. 10, 27-28; CA 56, 9004i

2838. Gutsche, C. D., and Armbruster, C. W., "Photolysis of 10-Oxobicyclo[1.2.5]decane and Related Compounds," Tetrahedron Letters, 1962, 1297-1301; CA 59, 462f

2839. Gutsche, C. D., and Smith, T. D., "Ring Enlargements. VII. The Reaction of Cycloalkanones with Bisdiazoalkanes," J. Am. Chem. Soc., 82, 4067-4075 (1960); CA 55, 18627g

2840. Guyer, A., Ineichen, M., and Guyer, P., "The Preparation of Synthetic Zeolites and Their Properties of Molecular Sieves," Helv. Chim. Acta, 40, 1603-1611 (1957); CA 52, 2302c

2841. Guyot, A., Blanc, C., Daniel, J. C., and Trambouze, Y., "A New Method for Kinetic Study of High-

Pressure Reactions by Vapor-Phase Chromatography; Application to Polymerization, " Compt. rend., 1961, 1795-1797; CA 56, 4129c

2842. Guzman, A., and Manjarrez, M., "Components of Essential Oil from Tagetes Florida, " Bol. Inst. Quim. Univ. Nacl. Autonoma Mex., 14, 48-54 (1962); CA 58, 12366d

2843. Gygi, R., and Potterat, M., "Applications of Gas Chromatography," Mitt. Gebiete Lebensm. u. Hyg., 48, 497-504 (1957); CA 52, 12649i

2844. Haack, E., Gube, M., Kaiser, F., and Spingler, H., "Cardiac Glucosides. XII. The Isolation of Gluco-gitaloxin from the Leaves of Digitalis Purpurea," Chem. Ber., 91, 1758-1763 (1958); CA 53, 1160c

2845. Haag, W. O., and Pines, H., "Alumina: Catalyst and Support. III. The Kinetics and Mechanisms of Olefin Isomerization," J. Am. Chem. Soc., 82, 2488-2494 (1960); CA 54, 22405a

2846. Haahti, E., "Major Lipid Constituents of Human Skin Surface with Special Reference to Gas-Chromatographic Methods," Scand. J. Clin. Lab. Invest., 13, Suppl. 59, 100 pp. (1961); CA 56, 14767e

2847. Haahti, E., and Fales, H. M., "Continuous Infrared Functional Group Detection of Gas-Chromatographic Eluates," Chem. & Ind. (London), 1961, 507-508; CA 56, 7972a

2848. Haahti, E., and Nikkari, T., "A New Sensitive Detector for Gas Chromatography," Acta Chem. Scand., 13, 2125-2126 (1959); CA 56, 10881i

2849. Haahti, E., Nikkari, T., and Juva, K., "Fractionation of Serum and Skin Sterol Esters and Skin Waxes with Chromatography on Silica Gel Impregnated with Silver Nitrate," Acta Chem. Scand., 17, 538-540 (1963); CA 59, 1852g

2850. Haahti, E., Nikkari, T., and Koshinen, O., "Fatty Acid Composition of Human Cerumen (Earwax)," Scand. J. Clin. Lab. Invest., 12, 249-250 (1960); CA 56, 7839f

2851. Haahti, E., Nikkari, T., and Kulonen, E., "Ionization Detector for Gas Chromatography. A Modification Without Radiation Source," J. Chromatog., 3, 372-373 (1960); CA 54, 20355e

2852. Haahti, E., Nikkari, T., Salmi, A. M., and Laaksonen, A. L., "Fatty Acids of Vernix Caseosa," Scand. J. Clin. Lab. Invest., 13, 70-73 (1961); CA 56, 10742h

2853. Haahti, E. O. A., and Horning, E. C., "Separation of Human Skin Waxes by Gas Chromatography," Acta Chem. Scand., 15, 930-931 (1961); CA 56, 5242g

2854. Haahti, E. O. A., and Horning, E. C., "Isolation and Characterization of Saturated and Unsaturated Fatty Acids and Alcohols of Human Skin Surface Lipids," Scand. J. Clin. Lab. Invest., 15, 73-78 (1963); CA 59, 2018g

2855. Haahti, E. O. A., Horning, E. C., and Castren, O., "Microanalysis of Sebum and Sebum Like Materials by Temperature Programmed Gas Chromatography," Scand. J. Clin. Lab. Invest., 14, 368-372 (1962); CA 59, 953b

2856. Haahti, E. O. A., and VandenHeuvel, W. J. A., "Analysis of Body Metabolities," Joint Mtg., Am. Assoc. Clin. Chemists and Am. Assoc. Adv. Science, December 1960; Abstr., Clin. Chem., 7, 305 (1961)

2857. Haahti, E. O. A., VandenHeuvel, W. J. A., and Horning, E. C., "Separation of Urinary 17-Keto-steroids by Gas Chromatography," Anal. Biochem., 2, 182-187 (1961); CA 55, 22455g

2858. Haahti, E. O. A., VandenHeuvel, W. J. A., and Horning, E. C., "Two-Component Phases in Gas Chromatographic Separation of Steroids," Anal. Biochem., 2, 344-352 (1961); CA 56, 5044d

2859. Haahti, E. O. A., VandenHeuvel, W. J. A., and Horning, E. C., "Gas Chromatographic Separations of Steroids with Polyester Phases," J. Org. Chem., 26, 626-627 (1961); CA 55, 16599c

2860. Haarhoff, P. C., and Pretorius, V., "Gas-Liquid Chromatography in Capillary Columns. I. Theory," S. African Chem. Inst., 13, 97-115 (1960); CA 55, 17157b

2861. Haarhoff, P. C., and Pretorius, V., "Gas-Liquid Chromatography in Capillary Columns. II. Factors Affecting the Column Efficiency: The Distribution Coefficient, the Diffusivity of the Solute in the Gaseous and Liquid Phases and the Thickness of the Lipid Layer," S. African Chem. Inst., 13, 116-124 (1960); CA 55, 17157c

2862. Haarhoff, P. C., and Pretorius, V., "Efficiency Parameters in Linear Chromatography and Their Use for Analytical Purposes," S. African Chem. Inst., 14, 22-42 (1961); CA 56, 7a

2863. Haarhoff, P. C., Van Berge, P. C., and Pretorius, V., "Role of the Sample Inlet Volume in Preparative Chromatography," Trans. Faraday Soc., 57, 1838-1843 (1961); CA 57, 607i

2864. Habboush, A. E., and Norman, R. O. C., "Analysis of Mixtures of Isomeric Benzenoid Compounds by Gas-Liquid Chromatography," J. Chromatog., 7, 438-446 (1962); CA 58, 7349b

2865. Haber, H. S., and Gardiner, K. W., "A Rapid Method for the Determination of Carbon and Hydrogen in Organic Compounds," 138th Natl. ACS Mtg., New York, N. Y., September 1960, Program Abstr., p. 32B

2866. Haber, H. S., and Gardiner, K. W., "Electroanalytical Method for the Determination of Carbon and Hydrogen in Organic Compounds," Office Tech. Services. Tech. Rept. WADD 60-415, U. S. Govt. Repts., 35, 405 (1961)

2867. Habgood, H. W., "Gas Chromatography," Ann. Rev. Phys. Chem. (H. Eyring, editor; Annual Reviews, Inc.), 13, 259-280 (1962); CA 57, 15775d

2868. Habgood, H. W., and Hanlan, J. F., "A Gas Chromatographic Study of the Adsorptive Properties of a Series of Activated Charcoals," Can. J. Chem., 37, 843-855 (1959); CA 53, 18593c

2869. Habgood, H. W., and Harris, W. E., " The Prediction of Retention Volumes in Programmed-Temperature Gas Chromatography," 2nd Alberta Gas Chromatography Discussion, Edmonton, Alberta, February 1959

2870. Habgood, H. W., and Harris, W. E., "Retention Temperature and Column Efficiency in Programmed Temperature Gas Chromatography," Anal. Chem., 32, 450-453 (1960); CA 54, 11589h

2871. Habgood, H. W., and Harris, W. E., "Plate Height in Programmed Temperature Gas Chromatography," Anal. Chem., 32, 1206 (1960); CA 54, 23460d

2872. Habgood, H. W., and Harris, W. E., "Capillary Programmed Temperature Gas Chromatography. Some Theoretical Aspects," Anal. Chem., 34, 882-885 (1962); CA 57, 11885f

2873. Habgood, H. W., and Harris, W. E., "Homologous Series Relationships in Programmed Temperature Gas Chromatography," 145th Natl. ACS Mtg., New York, N. Y., September 1963, Program Abstr., p. 10B

2874. Habich, A., Barner, R., Roberts, R. M., and Schmid, H., "The Abnormal Claisen Rearrangement. XXVIII. Investigation with Carbon-14," Helv. Chim. Acta, 45, 1943-1950 (1962); CA 58, 4403d

2875. Hachenberg, H., "Preparative Gas Chromatography of Gases and Liquids," Brennstoff-Chem., 43, 225-234 (1962); CA 58, 1133g

2876. Hachenberg, H., Junghanns, M., and Reininger, H., "Automatic Wide Range Switching Device in Gas Chromatography," Brennstoff-Chem., 44, 6-8 (1963); CA 58, 7347b

2877. Haddock, L. A., Hill, R., and Jones, A. G., "Seven Methods of Determining the Active Constituent of MCPA Weed Killers. An Appraisal by ICI Analysts," Mfg. Chemist, 30, 57-60 (1959); CA 53, 9557h

2878. Haddock, L. A., and Phillips, L. G., "The Determination of α-(4-Chloro-2-Methylphenoxy)-propionic Acid in Commerical Acid of this Name," Analyst, 84, 94-101 (1959); CA 53, 16827c

2879. Hadley, E. H., Adams, R. M., and Katz, J. J., "Gamma-Irradiation of Some Organic Compounds Dissolved in Anhydrous Hydrogen Fluoride," 133rd Natl. ACS Mtg., San Francisco, Calif., April 1958, Program Abstr., p. 37Q

2880. Hagdahl, L., "Coupled Columns in Chromatography," Science Tools, 1, 21-28 (1955); CA 49, 5986b

2881. Haggerty, Jr., W. J., and Breed, L. W., "Interaction of Alkoxysilanes and Acetoxysilanes," J. Org. Chem., 26, 2464-2467 (1961); CA 56, 3504g

2882. Haines, T. H., Aaronson, S., Gellerman, J. L., and Schlenk, H., "Occurrence of Arachidonic and Related Acids in the Protozoon Ochromonas Danica," Nature, 194, 1282-1283 (1962)

2883. Hajra, A. K., and Radin, N. S., "Collection of Gas-Liquid Chromatographic Effluents," J. Lipid Research, 3, 131-134 (1962); CA 56, 12692e

2884. Hake, C. L., Waggoner, T. B., Robertson, D. N., and Rowe, V. K., "Metabolism of 1, 1, 1-Trichloroethane by the Rat," Arch. Environmental Health, 1, 101-105 (1960); CA 54, 23076b

2885. Halasz, I., "Quantitative Gas Chromatographic Analysis of Hydrocarbons with Capillary Column and Flame Ionization Detector (II)," in "Gas Chromatography," edited by N. Brenner, J. E. Callen, and M. D. Weiss, Academic Press, Inc., New York, 1962, pp. 287-306

2886. Halasz, I., "New Types of Columns in Gas Chromatography," Univ. of Houston Intern. Symp. on Advances in Gas Chromatography, Houston, Texas, January 1963

2887. Halasz, I., and Heine, E., "Separation of Low-Boiling Hydrocarbons by Gas Chromatography Using Packed Capillary Columns," Nature, 194, 971-973 (1962); CA 57, 5285h

2888. Halasz, I., and Horváth, C., "Open Tube Columns with Impregnated Thin Layer Support for Gas Chromatography," Anal. Chem., 35, 499-505 (1963); CA 58, 11932h

2889. Halasz, I., and Horváth, C., "Thin-Layer Graphitized Carbon Black as the Stationary Phase for Capillary Columns in Gas Chromatography," Nature, 197, 71-72 (1963); CA 58, 5017c

2890. Halasz, I., and Schneider, W., "New Time-Voltage Integrator and Further Automation of Gas Chromatographic Analysis," Proc. Symposium, 3rd, Edinburgh, 1960, 104-116; CA 55, 25385a

2891. Halasz, I., and Schneider, W., "A New Time-Voltage Integrator and Further Automation of Gas Chromatographic Analysis," in "Gas Chromatography 1960," edited by R. P. W. Scott, Butterworths, London, 1960, pp. 104-116

2892. Halasz, I., and Schneider, W., "Gas-Chromatographic Determination of Traces of Dissolved Oxygen in Benzene," Brennstoff-Chem., 41, 225-229 (1960); CA 54, 24114b

2893. Halasz, I., and Schneider, W., "New-Type Nonelectronic Integrator with Figure Integration," Z. Anal. Chem., 175, 94-96 (1960); CA 55, 2208f

2894. Halasz, I., and Schneider, W., "Quantitative Gas Chromatographic Analysis of Hydrocarbons with Capillary Column and Flame Ionization Detector. II.," 3rd Intern. Symp. on Gas Chromatography, ISA, East Lansing, Mich., June 1961; ISA Proc., 3, 195-207 (1961)

2895. Halasz, I., and Schneider, W., "Quantitative Gas Chromatographic Analysis of Hydrocarbons with Capillary Column and Flame Ionization Detector," Anal. Chem., 33, 978-982 (1961); CA 56, 13524a

2896. Halasz, I., and Schneider, W., "Quantitative Gas Chromatographic Analysis of Hydrocarbons with Capillary Column and Flame Ionization Detector," Gas Chromatog., Intern. Symp., 1961, 3, 287-

306 (Pub. 1962); CA 58, 3869f

2897. Halasz, I., and Schreyer, G., "Separation Efficiencies with Capillary Columns, Using Flame Ionization Detector Systems," in "Gas Chromatography," Chem.-Ing.-Tech., 32, 675-685 (1960); CA 55, 1096a

2898. Halasz, I., and Schreyer, G., "Efficiency and Plate Number in Gas Chromatographic Analyses, by Means of Capillary Columns," Z. Anal. Chem., 181, 367-382 (1961); CA 56, 7b

2899. Halasz, I., and Schreyer, G., "Construction and Operation of a Capillary Column Chromatographic Apparatus with Flame Ionization Detector and Its Application in Quantitative Analysis," Z. Anal. Chem., 181, 384 (1961)

2900. Halasz, I., and Wegner, E. E., "Gas Chromatographic Separation of Low-Boiling Hydrocarbons Using Active Alumina as Support for the Liquid Phase," Nature, 189, 570-571 (1961); CA 55, 13823e

2901. Halasz, I., and Wegner, E. E., "Liquid-Impregnated Active Aluminum Oxide as a Stationary Phase in the Gas Chromatographic Separations of Low Boiling Hydrocarbons," Brennstoff-Chem., 42, 261-267 (1961); CA 56, 1667f

2902. Halasz, I., and Wegner, E. E., "Activated Aluminum Oxide as a Carrier for the Liquid Phase in Gas Chromatography," Symp. on Modern Methods for Analysis of Organic Compounds, German Chemical Soc., Munich, Germany, October 1960; Z. Anal. Chem., 181, 382-383 (1961); Abstr., Angew. Chem., 73, 146 (1961)

2903. Halden, W., Schauenstein, E., Taufer, M., and Puchner, H., "Dietary Evaluation of Swedish Rye-Crisp Bread According to the Fatty Acid Composition of Total Lipids. II.," Nutritio Dieta, 3, 225-235 (1961); CA 56, 10643c

2904. Hall, Jr., H. K., "Synthesis of Two Atom-Bridged Tetracyclic Ketones," J. Org. Chem., 25, 43-44 (1960); CA 54, 13020h

2905. Hall, H. L., "Quantitative Gas-Solid Chromatographic Determination of Carbonyl Sulfide as a Trace Impurity in Carbon Dioxide," Anal. Chem., 34, 61-63 (1962); CA 56, 8001a

2906. Hall, K. D., Norris, F., and Downs, S., "Physical Chemistry of Halothane-Ether Mixtures," Anesthesiology, 21, 522-530 (1960); CA 56, 12364i

2907. Hall, W. K., and Emmett, P. H., "An Improved Microcatalytic Technique," J. Am. Chem. Soc., 79, 2091-2093 (1957); CA 51, 14333g

2908. Hall, W. K., MacIver, D. S., and Weber, H. P., "Semi-Automatic Reactor for Catalytic Research," Ind. Eng. Chem., 52, 421-426 (1960); CA 54, 17974e

2909. Hall, W. K., Sill, G., and Wolfe, C. L., "Integrating Device for Use with Potentiometers," Science, 126, 821-823 (1957)

2910. Hallgren, B., and Larsson, S., "Separation and Identification of Alkoxyglycerols," Acta Chem. Scand., 13, 2147-2148 (1959); CA 56, 12093h

2911. Hallgren, B., and Larsson, S., "The Glyceryl Ethers in the Liver Oils of Elasmobranch Fish," J. Lipid Research, 3, 31-38 (1962); CA 56, 13362h

2912. Hallgren, B., and Larsson, S., "Separation of Pristane from Herring," Acta Chem. Scand., 17, 543-545 (1963); CA 59, 1853a

2913. Hallgren, B., Stenhagen, S., and Ryhage, R., "Use of Gas Chromatography and Mass Spectrometry in the Analysis of the Fatty Acids Found in Butter and Margarine," Acta Chem. Scand., 12, 1351 (1958); CA 54, 770b

2914. Hallgren, B., Stenhagen, S., Svanborg, A., and Svennerholm, L., "Gas Chromatographic Analysis of the Plasma Lipides in Normal and Diabetic Subjects," J. Clin. Invest., 39, 1424-1434 (1960); CA 54, 25209d

2915. Hallgren, B., and Svanborg, A., "The Separation of Free Fatty Acids (FFA) from Human Plasma Lipids: Gas Chromatographic Analysis of the Free Fatty Acid Composition in Healthy Human Individuals," Scand. J. Clin. Lab. Invest., 14, No. 2, 179-184 (1962); Biol. Abstr., 39, 21867

2916. Halter, R. C., and Pohler, L. W., "Gas Chromatography Scores in Stream Analysis. It Monitors Four Refinery Fractionating Columns," Oil Gas J., 57, No. 41, 135-137 (1959)

2917. Hamelin, R., "Mixed Grignard Solutions. IV. Mechanism of the Action of Grignard Reaction on Ketones," Bull. soc. chim. France, 1961, 915-925; CA 55, 27026h

2918. Hamelin, R., "Mixed Grignard Solutions. V. Action of Grignard Derivatives on Aldehydes," Bull. soc. chim. France, 1961, 926-930; CA 55, 27027b

2919. Hamence, J. H., "Report for Analytical Chemists," Anal. Chem., 35, No. 1, 24A-31A (1963) Article gives the present position, in the United Kingdom, on food additives with particular reference to legislation, problems involved in their use, and the role that analytical chemistry plays in their control.

2920. Hamer, J. C., "The Analysis of Seed Oils of Six Tropical Plant Species by Gas Chromatography," Univ. Microfilms (Ann Arbor, Mich.), Order No. 62-5887, 109 pp.; Dissertation Abstr., 23, 2322 (1963); CA 58, 10413e

2921. Hamilton, C. H., and Meyer, R. A., "Microliter Syringes for Sample Introduction," 1958 Natl. ISA Mtg., Symposium on Instrumental Methods of Analysis, Houston, Texas, May 1958

2922. Hamilton, L. H., "Application of Gas Chromatography to Respiratory and Blood Gas Determinations," 126th Natl. Mtg., Am. Assoc. for Advancement of Sci., Chicago, Ill., December 1959

2923. Hamilton, L. H., "Range Selector to Improve Peak Height Measurements in Gas Chromatography," J. Appl. Physiol., 16, 571-573 (1961); CA 55, 22470g

2924. Hamilton, L. H., "Fold-Over Zero-Suppression Circuit for Use with Gas Chromatography," Anal. Chem., 34, 445-447 (1962); CA 56, 13523i

2925. Hamilton, L. H., "Gas Chromatography for Respiratory and Blood Gas Analysis," Ann. N. Y. Acad. Sci., 102, Art. 1, 15-28 (1962)

2926. Hamilton, L. H., and Kory, R. C., "Application of Gas Chromatography to Respiratory Analysis," J. Appl. Physiol., 15, 829-837 (1960); CA 55, 1778b

2927. Hamilton, P. B., "Biochemical Analysis," Anal. Chem., 34, No. 5, 3R-13R (1962); CA 56, 13520d

2928. Hamilton, W. C., "Trends in the Technique of Gas Chromatography," 14th LSU Analytical Symp., Baton Rouge, La., January 1961

2929. Hammar, C. G. B., "A Simple Apparatus for Adsorption Analysis of Gases," Svensk Kem. Tidskr., 63, 125-135 (1951); CA 45, 10551g

2930. Hammond, E. G., "The Trisaturated Glycerides of Butteroil," 55th Annual Mtg., Am. Dairy Sci. Assoc., Logan, Utah, June 1960; Abstr., J. Dairy Sci., 43, 839-840 (1960)

2931. Hammond, E. G., and Boatman, C., "Polyunsaturated Fatty Acid Content of Milk Fat," 58th Annual Mtg., Am. Dairy Sci. Assoc., Lafayette, Ind., June 1963; Abstr., J. Dairy Sci., 46, 614 (1963)

2932. Hammond, E. G., El-Negoumy, A. M., and Puchal, M. S. de, "The Relation of Linoleate and Linolenate to the Flavors of Autoxidized Butteroil," 56th Annual Mtg., Am. Dairy Sci. Assoc., Madison, Wisc., June 1961, Abstr., J. Dairy Sci., 44, 1170 (1961)

2933. Hammond, G. S., and Collins, C. H., "Stereochemistry of Hydrogen Halide Addition to 1, 2-Dimethylcyclopentene," J. Am. Chem. Soc., 82, 4323-4327 (1960)

2934. Hampton, B. L., and Leavens, D., "The Presence of C_{20} Unsaturated Fatty Acids in Tall Oil," J. Org. Chem., 24, 1174 (1959); CA 53, 22934e

2935. Hanack, M., "Organic Fluorine Compounds. I. Reactions of Compounds of the Pinene Series with Hydrogen Fluoride," Chem. Ber., 93, 844-849 (1960); CA 54, 15427c

2936. Hanack, M., "Organic Fluorine Compounds. III. Reaction of Reactive Compounds of the Camphor Series with Hydrogen Fluoride and Potassium Fluoride," Chem. Ber., 94, 1082-1088 (1961); CA 55, 21157c

2937. Hanack, M., and Kaiser, W., "Organic Fluorine Compounds. VII. Reaction of Bicycloheptadiene and Nortricyclanol with Hydrogen Fluoride," Ann., 657, 12-19 (1962); CA 58, 1369d

2938. Hanack, M., and Keberle, W., "Organic Fluorine Compounds. II. Reaction of Camphene with Hydrogen Fluoride," Chem. Ber., 94, 62-67 (1961); CA 55, 9455a

2939. Hanada, Y., and Kitajima, M., "The Gas Chromatographic Determination of Camphor," Nippon Kagaku Zasshi, 80, 1272-1274 (1959); CA 55, 6785d

2940. Hanahan, D. J., "Lipids Chemistry," John Wiley & Sons, Inc., New York, 1960

2941. Hanahan, D. J., Brockerhoff, H., and Barron, E. J., "Site of Attack of Phospholipase (Lecithinase) A on Lecithin: A Reevaluation," J. Biol. Chem., 235, 1917-1923 (1960); CA 54, 22786b

2942. Hanahan, D. J., Watts, R. M., and Pappajohn, D., "Some Chemical Characteristics of the Lipides of Human and Bovine Erythrocytes and Plasma," J. Lipid Research, 1, 421-434 (1960); CA 55, 8567f

2943. Hancart, J., and Marot, J., "Analysis of Hydrogen, Oxygen and Nitrogen in Metals by Melting in Argon Atmosphere Under Reducing Conditions," Rev. met., 57, 911-918 (1960); CA 55, 11180f

2944. Handa, K. L., Smith, D. M., and Levi, L., "Gas Chromatographic Examination of Oil of Seseli Sibiricum," Perfumery Essent. Oil Record, 53, 607-610 (1962); CA 58, 2320f

2945. Hanes, A., Gherman, I., and Sandulescu, D., "Gas-Chromatographic Analysis of Chlorinated Benzene Derivatives," Rev. Chim. (Bucharest), 13, 113 (1962); CA 57, 11837g

2946. Hanes, A., and Sandulescu, D., "Determination of Sylvan, Furfural, and Furfuryl Alcohol by Gas-Liquid Chromatography," Rev. Chim. (Bucharest), 12, 664 (1961); CA 57, 6617f

2947. Hanes, A., Sandulescu, D., and Lupu, C., "Modern Procedures and Apparatus for Trace Analysis in Chemical Products," Rev. Chim. (Bucharest), 12, 412-415 (1961); CA 56, 5379e

2948. Hankinson, C. L., Harper, W. J., and Mikolajcik, E., "A Gas-Liquid Chromatographic Method for Volatile Fatty Acids in Milk," J. Dairy Sci., 41, 1502-1509 (1958); CA 53, 3524b

2949. Hanlan, J. F., and Freeman, M. P., "Gas Adsorption Chromatography," Can. J. Chem., 37, 1575-1578 (1959); CA 54, 20412a

2950. Hanna, C., "Preparation of Tritium-Labeled Halothane," J. Am. Pharm. Assoc., 49, 502-503 (1960); CA 54, 23194h

2951. Hanneman, W. W., Spencer, C. F., and Johnson, J. F., "Molten Salt Mixtures as Liquid Phases in Gas Chromatography," Anal. Chem., 32, 1386-1388 (1960); CA 55, 6058b

2952. Hanneman, W. W., Spencer, C. F., and Johnson, J. F., "High Temperature Gas Chromatography," 137th Natl. ACS Mtg., Cleveland, Ohio, April 1960, Program Abstr., p. 32B

2953. Hanni, H., and Ritter, W., "Gas Chromatography and Its Application in the Dairy Industry. I. Principles and Possible Applications of the Method," Milchwissenschaft, <u>14</u>, 524-527 (1959); CA <u>54</u>, 23099b

2954. Hanni, H., and Ritter, W., "Gas Chromatography and Its Application in the Dairy Industry. III. Testing the Composition of Amyl Alcohol and Its Suitability for Use in the Gerber Test," Milchwissenschaft, <u>16</u>, 24-30 (1961)

2955. Hansen, N. R., "A Device for Preparing Gas Mixtures," J. Sci. Instr., <u>32</u>, 75-76 (1955); CA <u>49</u>, 8635g

2956. Hansen, R. P., "Occurrence of trans-Octadec-16-enoic Acid in Sheep and Ox Perinephric Fats," Nature, <u>198</u>, 995 (1963)

2957. Hansen, R. P., and McInnes, A. G., "Volatile Acids of Ox Perinephric Fat," Nature, <u>173</u>, 1093 (1954); CA <u>48</u>, 12272b

2958. Hansen, R. P., Shorland, F. B., and Cooke, N. J., "Isolation from Butterfat of 14-Methyl Pentadecanoic (Isopalmitic) Acid," Chem. & Ind. (London), <u>1959</u>, 124; CA <u>53</u>, 13439b

2959. Hansen, R. P., Shorland, F. B., and Cooke, N. J., "The Isolation of cis-9-Heptadecenoic Acid from Butterfat," Biochem. J., <u>77</u>, 64-66 (1960); CA <u>54</u>, 23104d

2960. Hansen, R. P., Shorland, F. B., and Cooke, N. J., "The C_{17} Fatty Acid Constituents of Butterfat," New Zealand J. Sci., <u>6</u>, No. 1, 101-106 (1963); CA <u>59</u>, 864d

2961. Hanson, D. N., and Maimoni, A., "Gas Blending Apparatus," Anal. Chem., <u>31</u>, 158-159 (1959); CA <u>54</u>, 13761a

2962. Hara, A., Tokairin, H., and Fujii, A., "Analysis of C_1-C_5 Hydrocarbons in Crude Oil by Gas Chromatography," J. Japan Petrol. Inst., <u>4</u>, 293-297 (1961)

2963. Hara, N., Shimada, H., Ishikawa, A., and Dohi, K., "Complete Analysis of Gaseous Hydrocarbons by Gas Adsorption-Partition Chromatography," Bull. Japan Petrol. Inst., <u>2</u>, 33-40 (1960); CA <u>55</u>, 957c

2964. Hara, N., Shimada, H., and Oe, M., "Gas Chromatography, Using Mixed Carrier Gas," Kogyo Kagaku Zasshi, <u>64</u>, 772-780 (1961); CA <u>57</u>, 2880d

2965. Haraldson, L., "Products of Pyrolysis in the Determination of Oxygen in Sulfur-Containing Organic Substances," Mikrochim. Acta, <u>1962</u>, 650-670; CA <u>57</u>, 6615b

2966. Haraldson, L., Olander, C. J., Sunner, S., and Varde, E., "Equilibrium Studies on the Disproportion Reaction Between Some Dialkyl Disulfides," Acta Chem. Scand., <u>14</u>, 1509-1514 (1960); CA <u>56</u>, 9943b

2967. Haraldson, L., and Thorneman, T., "Sampling Valve for Small Volumes of Gas," Mikrochim. Ichnoanal. Acta, <u>1963</u>, 14-18; CA <u>58</u>, 10703g

2968. Harborne, J. B., "Chromatography of Phenols," in "Chromatography," edited by E. Heftmann, Reinhold Publishing Corp., New York, 1961, Chapter 24

2969. Hardiman, J., "Polyol Fatty Acid Esters in Cosmetic Emulsions," Am. Perfumer Cosmet., <u>77</u>, No. 10, 45-47 (1962); CA <u>58</u>, 5450g

2970. Hardy, C. J., "Gas Phase Chromatography as an Analytical Technique," Soc. for Anal. Chem., Western Section Mtg., Cardiff, Wales, November 1954; Abstr., Anal. Chem., <u>27</u>, 470 (1955)

2971. Hardy, C. J., "Determination of Activity Coefficients at Infinite Dilution from Gas Chromatography Measurements," J. Chromatog., <u>2</u>, 490-498 (1959); CA <u>54</u>, 11634b

2972. Hardy, C. J., "Gas Chromatography," Nature, <u>198</u>, 741-743 (1963)

2973. Hardy, C. J., and Pollard, F. H., "Review of Gas-Liquid Chromatography," J. Chromatog., <u>2</u>, 1-43 (1959); CA <u>53</u>, 21362f

2974. Hargreaves, M. K., "Partition Chromatography: A New Design," Chem. & Ind. (London), <u>1957</u>, 1414-1415; CA <u>52</u>, 2464b

2975. Harley, J., Nel, W., and Pretorius, V., "Flame Ionization Detector for Gas Chromatography," Nature, <u>181</u>, 177-178 (1958); CA <u>52</u>, 6859e

2976. Harley, J., and Pretorius, V., "New Detector for Vapour-Phase Chromatography," Nature, <u>178</u>, 1244 (1956); CA <u>51</u>, 5472a

2977. Harold, F. V., Hildebrand, R. P., Morieson, A. S., and Murray, P. J., "Influence of Hop-Oil Constituents on the Flavor and Aroma of Beer," J. Inst. Brewing, <u>66</u>, 395-398 (1960); CA <u>55</u>, 2996c

2978. Harold, F. V., Hildebrand, R. P., Morieson, A. S., and Murray, P. J., "Trace Volatile Constituents of Beer," J. Inst. Brewing, <u>67</u>, 161-172 (1961); CA <u>55</u>, 14812d

2979. Harper, W.J., Gould, I. A., and Hankinson, C. L., "Observations on the Free Volatile Acids in Milk," J. Dairy Sci., <u>44</u>, 1764-1765 (1961)

2980. Harris, B. L., "Adsorption," Ind. Eng. Chem., <u>48</u>, 472-481 (1956); ibid., <u>49</u>, 460-469 (1957); CA <u>51</u>, 6238c

2981. Harris, B. L., "Adsorption: Gas Chromatography," Ind. Eng. Chem., <u>50</u>, 424-425 (1958); CA <u>52</u>, 6863c

2982. Harris, B. L., "Adsorption," Ind. Eng. Chem., <u>51</u>, 340-343 (1959); CA <u>53</u>, 7689c

2983. Harris, B. L., and Emmett, P. H., "Adsorption Studies. Physical Adsorption of Nitrogen, Toluene, Benzene, Ethyl Iodide, Hydrogen Sulfide, Water Vapor, Carbon Disulfide, and Pentane on Various Porous and Nonporous Solids," J. Phys. Colloid Chem., 53, 811-825 (1949); CA 43, 7288d

2984. Harris, Jr., J. F., and Stacey, F. W., "Free Radical Addition of Trifluoromethanethiol to Fluoro-olefins," J. Am. Chem. Soc., 83, 840-845 (1961); CA 55, 13305d

2985. Harris, J. J., "Biborazinyl," J. Org. Chem., 26, 2155-2156 (1961); CA 55, 25733a

2986. Harris, W. E., "Gas-Liquid Chromatography," Chem. in Can., 11, No. 7, 27-33 (1959); CA 53, 18592h

2987. Harris, W. E., "Chemical Effects of the (η, γ) Activation of Bromine in Alkyl Bromides: the Halo-methanes," Can. J. Chem., 39, 121-130 (1961); CA 55, 10135g

2988. Harris, W. E., and McFadden, W. H., "Selective Reactivity in Gas-Liquid Chromatography: The Analysis of the Bromobutanes," 2nd Alberta Gas Chromatography Discussion, Edmonton, Alberta, February 1959

2989. Harris, W. E., and McFadden, W. H., "Selective Reactivity in Gas-Liquid Chromatography. Deter-mination of 2-Bromobutane and 1-Bromo-2-methylpropane," Anal. Chem., 31, 114-117 (1959); CA 53, 6867e

2990. Harrison, G. F., "Vapor-Phase Chromatographic Analysis of Chlorinated Hydrocarbons and Hydro-carbon Gases," in "Vapour Phase Chromatography," edited by D. H. Desty, Academic Press, Inc., New York, 1957, pp. 332-345

2991. Harrision, G. F., "Uncorrected Retention Volumes of Chloro-, Bromo-, and Some Other Compounds (Chromatographic Data Table XLVI)," J. Chromatog., 1, xxx (1958); cf. "Gas Chromatography," edited by D. H. Desty, Butterworths, London, 1957, p. 336

2992. Harrison, G. F., Knight, P., Kelley, R. P., and Heath, M. T., "The Use of Multiple Columns and Programmed Column Heating in the Analysis of Wide-Boiling-Range Halogenated Hydrocarbon Samples," in "Gas Chromatography 1958," edited by D. H. Desty, Academic Press, Inc., New York, 1958, pp. 216-225

2993. Harrison, R. B., Palframan, J. F., and Rose, B. A., "Detection, Determination and Identification of Furfuraldehyde in Hydrocarbon Oil," Analyst, 86, 561-565 (1961); CA 57, 8798g

2994. Harrocks, J. A., "Neutral Materials in the Benzene Extract of Aspenwood," Univ. Microfilms (Ann Arbor, Mich.), L. C. Card No. Mic 60-2088, 58 pp.; Dissertation Abstr., 21, 53 (1960); CA 55, 2098b

2995. Harteck, P., and Suhr, K. A., "Separation of Hydrocarbons by Means of [Fractional] Desorption.III," Die Chemie, 56, 120-123 (1943); CA 37, 5227[8]

2996. Hartman, L., Hawke, J. C., Morice, I. M., and Shorland, F. B., "Component Fatty Acids of Sporo-desmium Bakeri Lipides," Biochem. J., 75, 274-278 (1960); CA 54, 25011f

2997. Hartman, L., Hawke, J. C., Shorland, F. B., and Menna, M. E., "The Fatty Acid Composition of Rhodotorula Graminis Fat," Arch. Biochem. Biophys., 81, 346-352 (1959); CA 53; 14228h

2998. Hartmann, H., "Quantitative Analyses of Trace Pesticides," Biochemical Gas Chromatography Se-minar, Wilkens Instrument & Research, Inc., New Orleans, La., July 1963

2999. Hartmann, H., and Dimick, K. P., "Gas Chromatography and Electron Capture for the Analysis of Pesticides," 144th Natl. ACS Mtg., Los Angeles, Calif., March-April 1963, Program Abstr., p. 2A

3000. Hartmann, H., Oaks, D., and Dimick, K. P., "Essential Oil Analysis by Two-Channel Gas Chro-matography," 145th Natl. ACS Mtg., New York, N. Y., September 1963

3001. Hartree, E. F., and Mann, T., "Phospholipids in Ram Semen. Metabolism of Plasmalogen and Fatty Acids," Biochem. J., 80, 464-476 (1961); CA 55, 26168g

3002. Hartung, G. K., and Jewell, D. M., "Identification of Nitriles in Petroleum Products. Complex For-mation as a Method of Isolation," Anal. Chim. Acta, 27, 219-232 (1962); CA 57, 15408g

3003. Hartung, G. K., Jewell, D. M., Larson, O. A., and Flinn, R. A., "Catalytic Hydrogenation of In-dole in Furnace Oil: Separation and Identification of Reaction Products," 138th Natl. ACS Mtg., New York, N. Y., September 1960, Program Abstr., p. 2R; Preprints, Div. Petrol. Chem., 5, No. 3, 27-36 (1960)

3004. Hartzler, H. D., "The Stereochemistry and Relative Rates of Addition of Dimethylvinylidene Carbene to Olefins," J. Am. Chem. Soc., 83, 4997-4999 (1961); CA 57, 2035d

3005. Haruki, T., "Gas Chromatography Under Reduced Pressure," Bunseki Kagaku, 9, 865-869 (1960); CA 55, 20755c

3006. Haruki, T., "An Explanation of Column Efficiency in Gas Chromatography," Bunseki Kagaku, 10, 1244-1248 (1961); CA 58, 6166a

3007. Haruki, T., "Process Gas Chromatograph," J. Japan Petrol. Inst., 4, 280-287 (1961)

3008. Haruki, T., "Process Gas Chromatography," Kagaku no Ryoiki Zokan, No. 44, 233-245 (1961); CA 56, 3b

3009. Haruki, T., "Column Efficiency in Gas Chromatography," Kogyo Kagaku Zasshi, 64, 814-819 (1961); CA 57, 4068g

3010. Haruki, T., "A New Design for Gas Chromatography," Kogyo Kagaku Zasshi, 64, 825-830 (1961); CA 57, 2880a

3011. Haruki, T., and Asai, K., "The Practical Design of a Process Gas Chromatograph," Kogyo Kagaku

Zasshi, <u>64</u>, 830-834 (1961); CA <u>57</u>, 2880c

3012. Haruki, T., and Itaya, M., "A Flame Ionization Detector for Gas Chromatography and Its Character-istics," Kogyo Kagaku Zasshi, <u>64</u>, 820-825 (1961); CA <u>57</u>, 2827c

3013. Harva, O., and Keltakallio, A., "Chromatographic Determination of Argon in Gas Analysis," Suomen Kemistilehti, <u>30B</u>, 223-224 (1957); CA <u>52</u>, 7016c

3014. Harva, O., Kivalo, P., and Keltakallio, A., "Reduction of Tailing in Gas-Liquid Chromatography," Suomen Kemistilehti, <u>32B</u>, 71-72 (1957); CA <u>53</u>, 21359b

3015. Harva, O., Kivalo, P., and Keltakallio, A., "Determination of the Hydrophilic-Lipophilic Character of Polyhydric Alcohol Esters by Gas Chromatography," Suomen Kemistilehti, <u>32B</u>, 52-54 (1959); CA <u>53</u>, 18592h

3016. Harvey, D., and Chalkley, D. E., "Gas-Liquid Partition Cheomatography," Fuel., <u>34</u>, 191-200 (1955), CA <u>49</u>, 7436i

3017. Harvey, D., and Morgan, G. O., "Factors Affecting Thermal Conductivity Detectors in Vapour-Phase Partition Chromatography," in "Vapour Phase Chromatography," edited by D. H. Desty, Academic Press, Inc., New York, 1957, pp. 74-86

3018. Harvey, D., Whiting, D. H., and Mylroi, M. G., "Vapor-Phase Partition Chromatography for Process Stream Analysis," Automatic Measurement Quality Process Plants, Proc. Conf. Swansea, <u>1957</u> (Academic Press, Inc., London), pp. 222-230; CA <u>53</u>, 5019g

3019. Harvey, F. H., and Baker, W. J., "The Sampling System - Vital Link in the Process Chromato-graphic Control Setup," Oil Gas J., <u>59</u>, No. 24, 147-148 (1961)

3020. Harvey, H. E., and Harvey, W. E., "Gas Chromatographic Determination of Dieldrin Residues on Pasture," New Zealand J. Sci., <u>6</u>, No. 1, 3-5 (1963); CA <u>58</u>, 11905b

3021. Harwood, H. J., Baikowitz, H., and Trommer, H. F., "The Investigation of Copolymerization by Vapor Phase Chromatography," 144th Natl. ACS Mtg., Los Angeles, Calif., March-April 1963, Program Abstr., p. 7Q

3022. Hasegawa, T., and Takeuchi, T., "Determination of Fatty Acids in Tissue Samples," Kagaku no Ryoiki Zokan, No. 44, 223-232 (1961); CA <u>56</u>, 3b

3023. Haskin, J. F., Warren, G. W., Kourey, R. E., and Yarborough, V. A., "Gas Chromatography. An-alysis of the Crude Reaction Product from the Hydroformylation of Isobutene," 132nd Natl. ACS Mtg., New York, N. Y., September 1957, Program Abstr., p. 38B; Preprints, Div. Petrol. Chem., <u>2</u>, No. 4, D-131

3024. Haskin, J. F., Warren, G. W., Priestley, Jr., L. J., and Yarborough, V. A., "Gas Chromatography. Determination of Constituents in the Study of Azeotropes," Anal. Chem., <u>30</u>, 217-219 (1958); CA <u>52</u>, 7945d; 132nd Natl. ACS Mtg., New York, N. Y., September 1957, Program Abstr., p. 39B; Preprints, Div. Petrol. Chem., <u>2</u>, No. 4, D135-D141

3025. Haskin, L. A., "Analysis for Uranium by Neutron Activation and Reactions of Energetic Recoil Tri-tium in Solvent Mixtures," Univ. Microfilms (Ann Arbor, Mich.), L. C. Card No. Mic <u>60-4327</u>, 78 pp.; Dissertation Abstr., <u>21</u>, 1077 (1960)

3026. Haslam, J., "Modern Analytical Chemistry in Relation to the Plastics Industry," Proc. Congr. Mod-ern Anal. Chem., St. Andrews, June 1957, Heffers, Cambridge, 1958, pp. 132-136

3027. Haslam, J., "Use of Gas Chromatography in the Plastic Industry," Chem. Age, <u>82</u>, 169-170 (1959); CA <u>58</u>, 3551e

3028. Haslam, J., Hamilton, J. B., and Jeffs, A. R., "The Determination of Polyethyl Esters in Methyl Methacrylate Copolymers," Analyst, <u>83</u>, 66-71 (1958); CA <u>52</u>, 9649c

3029. Haslam, J., and Jeffs, A. R., "Gas-Liquid Chromatography in a Plastics Analytical Laboratory," J. Appl. Chem., <u>7</u>, 24-32 (1957); CA <u>51</u>, 14315g

3030. Haslam, J., and Jeffs, A. R., "Application of Gas-Liquid Chromatography. The Examination of Sol-vents from Plastic Adhesives," Analyst, <u>83</u>, 455-462 (1958); CA <u>52</u>, 19720d

3031. Haslam, J., and Jeffs, A. R., "Application of Gas-Liquid Chromatography to the Work of a Plastics Analytical Department," Mtg. of the Plastics and Polymer Group, Society of Chemical Industry, London, April 1959; Chem. & Ind. (London), <u>1959</u>, 722

3032. Haslam, J., and Jeffs, A. R., "Application of Gas-Liquid Chromatography: The Examination of Ter-penes and Related Substances," Analyst, <u>87</u>, 658-663 (1962); CA <u>57</u>, 15781d

3033. Haslam, J., Jeffs, A. R., and Willis, H. A., "Applications of Gas-Liquid Chromatography. Collec-tion of Fractions from the Gas Chromatograph and Their Identification by Infrared Spectroscopy," Analyst, <u>86</u>, 44-53 (1961); CA <u>55</u>, 20771g

3034. Haslam, J., and Squirrell, D. C. M., "Analytical Chemistry. Chromatography," Ann. Repts. on Progr. Chem. (Chem. Soc. London), <u>55</u>, 407-415 (1958); CA <u>54</u>, 3056e

3035. Haslam, J., and Squirrell, D. C. M., "Analytical Chemistry - Qualitative and Quantitative Inorganic and Organic Analysis," Chem. Soc. Ann. Repts. Progress Chem., <u>56</u>, 394-403 (1959); CA <u>54</u>, 3056f

3036. Hasselstrom, T., "Fruit and Vegetable Flavors - Techniques Employed in the Study of the Chemistry of Cabbage Flavor," Quartermaster Food and Container Inst. Surveys Progr. Military Subsis-tence Problems, Ser. <u>I</u>, No. 9, 76-86 (1957); CA <u>52</u>, 2296h

Probelsm, Ser. I, No. 9, 76-86 (1957); CA 52, 2296h

3037. Hasselstrom, T., Hewitt, E. J., Konigsbacker, K. S., and Ritter, J. J., "Composition of Volatile Oil of Black Pepper," J. Agr. Food Chem., 5, 53-55 (1957); CA 51, 15036e

3038. Hatch, L. F., "The Use of Gas Chromatography in Kinetic Studies," in "Gas Chromatography," edited by V. J. Coates, H. J. Noebels, and I. S. Fagerson, Academic Press, Inc., New York, 1958, pp. 105-109

3039. Hatch, L. F., "Reduction of Allylic Halides by Lithium Aluminum Hydride," 136th Natl. ACS Mtg., Atlantic City, N. J., September 1959, Program Abstr., p. 92P

3040. Hatch, L. F., Gardner, P. D., and Gilbert, R. E., "The Mechanism of Bromine Addition to 1, 3-Butadiene," J. Am. Chem. Soc., 81, 5943-5946 (1959); CA 54, 9720i

3041. Hatch, L. F., and Gilbert, R. E., "Reduction of Allylic Halides by Lithium Aluminum Hydride," J. Org. Chem., 24, 1811-1812 (1959); CA 55, 14284c

3042. Hatch, L. F., and Payne, Jr., J. S., "The 1, 2-Dibromo-1-propenes," Ann. N. Y. Acad. Sci., 72, 698-704 (1959); CA 53, 17883f

3043. Hauptschein, M., and Braid, M., "Fluorocarbon Halosulfates and a New Route to Fluorocarbon Acids and Derivatives. II. Polyfluoroalkyl Fluorosulfates," J. Am. Chem. Soc., 83, 2505-2507 (1961); CA 55, 20940g

3044. Hauptschein, M., Braid, M., and Fainberg, A. H., "Addition of Iodine Halides to Fluorinated Olefins. I. The Direction of Addition of Iodine Monochloride to Perhaloolefins and Some Related Reactions," J. Am. Chem. Soc., 83, 2495-2500 (1961); CA 55, 20908a

3045. Hausdorff, H. H., "Applications of Gas Chromatography and Its Error Sources," Proc. Gas-Kolloquium, Hamburg, West Germany, November 1956, pp. 71-73

3046. Hausdorff, H. H., "Quantitative Methods of Gas Chromatography," Chem.-Ztg., 81, 392-396 (1957)

3047. Hausdorff, H. H., "Vapour Fractometry (Gas Chromatography). II. A Powerful New Tool in Chemical Analysis," in "Vapour Phase Chromatography," edited by D. H. Desty, Academic Press, Inc., New York, 1957, pp. 377-387

3048. Hausdorff, H. H., and Brenner, N., "Gas Chromatography. I. Powerful New Tool for Chemical Analysis," Oil Gas J., 56, 73-75 (June 30, 1958); CA 52, 19669d

3049. Hausdorff, H. H., and Brenner, N., "Gas Chromatography. II Instrumentation Techniques Play a Vital Role in Chemical Analysis by Gas Chromatography," Oil Gas J., 56, 122-124, 126 (July 7, 1958); CA 52, 19669d

3050. Hausdorff, H. H., and Brenner, N., "Gas Chromatography. III. Six Variables Must be Considered for Effective Gas Chromatography," Oil Gas J., 56, 86-88 (July 21, 1958); CA 52, 19669d

3051. Hausdorff, H. H., and Brenner, N., "Gas Chromatography. IV. Here Are the Limits in Application for Gas Chromatography," Oil Gas J., 56, 89-90, 93-96 (August 4, 1958); CA 52, 19669d

3052. Hause, J. A., Hubicki, J. A., and Hazen, G. G., "Determination of Sorbitol as Its Hexacetate by Gas Liquid Chromatography Using an Ionization Detector," Anal. Chem., 34, 1567-1570 (1962); CA 58, 5969d

3053. Hawke, J. C., "Volatile Fatty Acids in Phospholipids," Nature, 176, 882 (1955); CA 50, 4280e

3054. Hawke, J. C., "Volatile Fatty Acids of Bovine Muscle and Liver Phospholipides," Biochem. J., 64, 311-318 (1956); CA 51, 2143e

3055. Hawke, J. C., "The Fatty Acids of Butterfat and the Volatile Acids Formed on Oxidation," J. Dairy Research, 24, 366-371 (1957); CA 52, 15772f

3056. Hawke, J. C., "The Construction of an Apparatus for Studies in Vapour-Phase Chromatography at Temperatures Up to 300°C.," in "Vapour Phase Chromatography," edited by D. H. Desty, Academic Press, Inc., New York, 1957, pp. 266-276

3057. Hawke, J. C., "The Fatty Acids of Phosphatidylethanolamine and Phosphatidylcholine from Hen's Eggs," Biochem. J., 71, 588-592 (1959); CA 53, 7354b

3058. Hawke, J. C., "Distribution of Fatty Acids Between the α'- and β-Positions of Egg Phosphatidylcholine," Chem. & Ind. (London), 1962, 1761; CA 58, 7070b

3059. Hawke, J. C., Dunkley, W. L., and Hooker, C. N., "Separation of Fatty Esters and Aldehydes by Gas-Liquid Chromatography," New Zealand J. Sci. and Technol., 38, 925-938 (1957); CA 54, 13759g

3060. Hawke, J. C., Hansen, R. P., and Shorland, F. B., "Gas-Liquid Chromatography. Retention Volumes of the Methyl Esters of Fatty Acids with Special Reference to n-Odd-Numbered, iso and (+)-Anteiso Acids," J. Chromatog., 2, 547-551 (1959); CA 54, 13811b

3061. Hawkes, S. J., "Gas Chromatography," Nature, 190, 867 (1961); 3rd Annual General Mtg. of Gas Chromatography Discussion Group, Univ. of Birmingham, England, April 1961

3062. Hawkins, L. H. C., "The Application of Gas Chromatography to Work Control Routine Testing," Chem. & Ind. (London), 1957, 950

3063. Haworth, R. D., and Johnstone, R. A. W., "Cafestol. II.," J. Chem. Soc., 1957, 1492-1496; CA 51, 12937g

3064. Hay, D. G., "Gas Chromatography in Natural Gasoline Industry," Oil in Can., 12, No. 22, 32-33

(1960); CA 54, 13616d

3065. Hazeldean, G. S. F., and Scott, R. P. W., "Resistance to Mass Transfer in the Capillary Columns," 13th Pittsburgh Conf. on Anal. Chem. & Appl. Spectroscopy, Pittsburgh, Pa., March 1962, Program Abstr., p. 50

3066. Heald, E. F., "Chemical Study of Hawaiian Magmatic Gases," Univ. Microfilms (Ann Arbor, Mich.), L. C. Card Mic. 61-2671; Dissertation Abstr., 22, 407 (1961); CA 55, 24421i

3067. Hearfield, R. C., "Application of Gas Liquid Chromatography to Detection of Cocoa Butter Adulteration," Chem. & Ind. (London), 1961, 655

3068. Heaton, W. B., and Wentworth, J. T., "Exhaust Gas Analysis by Gas Chromatography Combined with Infrared Detection," Anal. Chem., 31, 349-357 (1959); CA 53, 11808h

3069. Heck, R. F., and Breslow, D. S., "The Reaction of Cobalt Hydrotetracarbonyl with Olefins," J. Am. Chem. Soc., 83, 4023-4027 (1961); CA 56, 3333d

3070. Hecke, F. van, "Polymer Formation in Irradiated p-Xylene," Nature, 186, 382-383 (1960); CA 55, 6413b

3071. Hedgley, E. J., Meresz, O., Overend, W. G., and Rennie, R., "4 - and 5-Deoxy-D-Glucose," Chem. & Ind. (London), 1960, 938-939; CA 55, 1458b

3072. Hedsted, D. M., Whyman, C., Gotsis, A., and Andrus, S. A., "Effects of Composition of Dietary Fat Upon Composition of Adipose Tissue," Am. J. Clin. Nutrition, 8, 209-213 (1960); CA 54, 13296a

3073. Hefendehl, F. W., "The Use of Gas Chromatography for the Analysis of the Essential Oil of Mentha Piperita," Planta Med., 10, 179-207 (1962); CA 58, 1299c

3074. Heft, C. H., "Factors Influencing the Relation Between Actual Concentration and the Mole- and Weight-Percent Results Obtained by Thermoconductivity," Abhandl. Deut. Akad. Wiss. Berlin, Kl. Chem., Geol., Biol., 1959, No. 9, 299-326; CA 58, 5019b

3075. Heft, C. H., "Mole- or Weight-Percentages at Quantitative Analyses (Thermal Conductivity Method)?" in "Gas Chromatographie 1958," edited by H. P. Angele, Akademie Verlag, Berlin, 1959, pp. 299-326

3076. Heftmann, E., "Simplified Mathematical Treatment of the Theory of Partition Chromatography," Chromatog. Methods, 2, No. 1, 5-8 (1957); CA 51, 10293g

3077. Heftmann, E., (Ed.), "Chromatography," Reinhold Publishing Corp., New York, 1961

3078. Heftmann, E., "Chromatography of Steroids," in "Chromatography," edited by E. Heftmann, Reinhold Publishing Corp., New York, 1961, Chapter 18

3079. Heide, R. ter, "An Introduction to Chromatography," Soap, Perfum. Cosmet., 33, 727-730 (1960)

3080. Heigl, J., and MacRitchie, A. L., "Data Handling Systems for Gas Chromatography," 15th Annual Summer Symp., Div. of Anal. Chem., ACS, College Park, Md., June 1962

3081. Heilbronner, I. E., Kovats, E., and Simon, W., "Program-Controlled Gas Chromatography for Preparative Separation of Organic Compounds," Helv. Chim. Acta, 40, 2410-2420 (1957); CA 52, 5891a

3082. Heinemann, W., "Analysis of Benzene Hydrocarbons by Gas Chromatography," 1960 Annual Mtg., German Soc. for Petroleum and Coal Chemistry, Frankfurt, Germany, October 1960; Abstr., Brennstoff-Chem., 41, 342 (1960); Abstr., Angew. Chem., 73, 68 (1961)

3083. Heinemann, W., "Analysis of Hydrocarbon Mixtures in the Gasoline Range by Gas Chromatography," Erdol u. Kohle, 14, 917-921 (1961); CA 56, 9001g

3084. Heines, Sister Virginia, Juhasz, Sister Roderick, O'Leary, Sister Mary Adeline, and Schramm, G., "Gas Chromatography of C_2 to C_{11} Fatty Acids with Simplified Apparatus," Trans. Kentucky Acad. Sci., 18, 1-7 (1957); CA 52, 1689h

3085. Heinze, H. O., "New Methods for the Study of Hydrocarbons," Brennstoff-Chem., 39, 347-348 (1958)

3086. Heinze, H. O., "Quantitative Multicomponent Analysis of Hydrocarbons by Means of Gas Chromatography and Infrared Spectroscopy. I. Pyridine Bases," Mtg., German Chemical Soc., Stuttgart, Germany, April 1960; Abstr., Angew. Chem., 72, 586-587 (1960)

3087. Heitzman, R. J., Patrick, C. R., Stephens, R., and Tatlow, J. C., "Fluorocyclopentanes. I. The 1H, 2H- and 1H, 3H-Octafluorocyclopentanes, and 1H, 3H/2H-Heptafluorocyclopentane," J. Chem. Soc., 1963, 281-289

3088. Hele, P., Popjak, G., and Lauryssens, M., "Biosynthesis of Fatty Acids in Cell Free Preparations. IV. Synthesis of Fatty Acids from Acetate by a Partially Purified Enzyme System from Rabbit Mammary Gland," Biochem. J., 65, 348-363 (1957); CA 51, 6722h

3089. Hellman, H. M., and Rosegay, A., "Oxidative Rearrangement of Ketones to Carboxylic Acids," Tetrahedron Letters, 1959, No. 13, 1-3; CA 54, 4345f

3090. Hellman, M., Alexander, R. L., and Coyle, C. F., "Separation of Isomeric Polyphenols by Adsorption Chromatography," Anal. Chem., 30, 1206-1210 (1958); CA 52, 15346a

3091. Hellmann, H., and Eberle, D., "Reactions of Thio Ethers with o-Fluorophenylmagnesium Bromide," Ann., 662, 188-201 (1963); CA 59, 492b

3092. Hellstrom, K., and Sjovall, J., "Turnover of Deoxycholic Acid in the Rabbit," J. Lipid Research, 3, 397-404 (1962)

3093. Hellyer, R. O., and Keyzer, H., "The Effect of Pressure Differentials on Gas Chromatography Separation," J. Chromatog., 10, 314-323 (1963)

3094. Helm, R. V., Latham, D. R., Ferrin, C. R., and Ball, J. S., "Identification of Carbazole in Wilmington Petroleum Through Use of Gas-Liquid Chromatography and Spectroscopy," Anal. Chem., 32, 1765-1767 (1960); CA 55, 5924c

3095. Helms, C. C., and Claudy, H. N., "The Practical Design of a Vapor Fractometer for Automatic Multicomponent Analysis of Process Streams," in "Gas Chromatography," edited by V. J. Coates, H. J. Noebels, and I. S. Fagerson, Academic Press, Inc., New York, 1958, pp. 269-279

3096. Helms, C. C., and Norem, S., "Vapor Phase Chromatography," Oil Gas J., 55, No. 17, 146-149 (1957); CA 51, 13367h

3097. Helms, C. C., and Norem, S., "Evaluation of an Experimental Process Vapor Fractometer," 8th Pittsburgh Conf. on Anal. Chem. & Appl. Spectroscopy, Pittsburgh, Pa., March 1957, Program Abstr., p. 34

3098. Helzhauser, H., and Kuhl, M., "Microflame Ionization Detector and Quantitative Analysis," in "Gas Chromatographie 1959," edited by R. E. Kaiser and H. G. Struppe, Akademie Verlag, Berlin, 1959, pp. 143-156

3099. Henbest, H. B., and Patton, R., " Amine Oxidation. VI. The Formation of Tetra-N-Substituted 1,2-Diamines from Tertiary Amines," J. Chem. Soc., 1960, 3557-3559; CA 55, 2529b

3100. Henchman, M., and Wolfgang, R., "Stereochemistry of Hot Hydrogen Displacement at sp³ Carbon-Hydrogen Bonds," J. Am. Chem. Soc., 83, 2991-2996 (1961); CA 56, 3332f

3101. Henderson, J. F., and Steacie, E. W. R., "The Photolysis of Deuterated Acetone-Hydrogen Mixtures," Can. J. Chem., 38, 2161-2170 (1960); CA 55, 4119c

3102. Henderson, J. I., and Knox, J. H., "The Microflame Detector in Gas-Liquid Partition Chromatography: Correlation of Response with Heats of Combustion," J. Chem. Soc., 1956, 2299-2302; CA 50, 15134h

3103. Hendricks, W. J., Soemantri, R. M., and Waterman, H. I., "Separation of Mixtures of Diphenyl, Cyclohexylbenzene, and Dicyclohexyl by Vapor-Phase Chromatography," J. Inst. Petrol., 43, 288-291 (1957); CA 52, 708f

3104. Henly, R. S., Rose, A., and Sweeney, R. F., "The Partition Coefficient as a Function of Concentration in Gas-Liquid Chromatography," 14th Pittsburgh Conf. on Anal. Chem. & Appl. Spectroscopy, Pittsburgh, Pa., March 1963, Program Abstr., p. 53

3105. Henneberg, D., "A Continuous Procedure for the Mass Spectrometric Analysis of Mixtures Separated by Gas Chromatography," Z. Anal. Chem., 170, 365-366 (1959); CA 54, 6387d

3106. Henneberg, D., "Combination of Gas Chromatography and Mass Spectrometry for the Analysis of Organic Mixtures," Z. Anal. Chem., 183, 12-23 (1961); CA 56, 5a

3107. Henneberg, D., Damen, H., and Koster, R., "Boron Compounds. IV. Mass Spectra of Lower Trialkylboranes and a Discussion of Their Significance," Ann., 640, 52-79 (1961); CA 55, 16397a

3108. Henneberg, D., and Schomburg, G., "Mass Spectrometric Identification in Capillary Column Gas Chromatography," 4th Intern. Gas Chromatography Symp., Hamburg, Germany, June 1962; pub. in "Gas Chromatography 1962," ed. by M. van Swaay, Butterworth & Co., London, 1962

3109. Hennion, G. F., and Boisselle, A. P., "Preparation of tert-Acetylenic Chlorides," J. Org. Chem., 26, 725-727 (1961); CA 55, 22090f

3110. Hennion, G. F., and Lynch, C. A., "Chlorides Derived from 1-Ethynylcyclohexanol," J. Org. Chem., 25, 1330-1333 (1960); CA 54, 24458h

3111. Hepler, L. G., "Gas Adsorption on Heterogeneous Surfaces," J. Chem. Phys., 23, 2110-2111 (1955); CA 50, 3839f

3112. Hepner, L. S., Bendoraitis, J. G., and Freeman, R. B., "High Temperature Gas-Liquid Phase Chromatography Applied to Petroleum Fractions," Symp. on V. P. C., 3rd Delaware Valley Regional ACS Mtg., Philadelphia, Pa., February 1960

3113. Hepp, H. J., and Drehman, L. E., "Isomerization of n-Heptane. Aluminum Chloride-Hydrocarbon Complex Catalyst," Ind. Eng. Chem., 52, 207-210 (1960); 135th Natl. ACS Mtg., Boston, Mass., April 1959; Preprints, Div. Petrol. Chem., 4, No. 2, A81-A88 (1959); CA 55, 23987e

3114. Herb, S. F., Magidman, P., and Barford, R. A., "A Satisfactory G. L. C. Column for the Determination of Epoxyoleic Acid in Seed Oils," 36th Fall Mtg., Am. Oil Chemists' Soc., Toronto, Ontario, October 1962

3115. Herb, S. F., Magidman, P., Barford, R. A., and Riemenschneider, R. W., "Fatty Acids of Lard. A. Identification by Gas-Liquid Chromatography," J. Am. Oil Chemists' Soc., 40, 83-85 (1963); CA 58, 13056a

3116. Herb, S. F., Magidman, P., Luddy, F. E., and Riemenschneider, R. W., "Fatty Acids of Cow Milk. II. Composition by Gas-Liquid Chromatography Aided by Other Methods of Fractionation," J. Am. Oil Chemists' Soc., 39, 142-146 (1962); CA 56, 15889h

3117. Herb, S. F., Magidman, P., and Riemenschneider, R. W., "Analysis of Fats and Oils by Gas-Liquid Chromatography and Ultraviolet Spectrophotometry," J. Am. Oil Chemists' Soc., 37, 127-129 (1960); 50th Mtg., Am. Oil Chemists' Soc., New Orleans, La., April 1959, Program Abstr., p. 26

114

3118. Heredy, L. A., Kostyo, A. E., and Neuworth, M. B., "Identification of Isopropyl Groups of Aromatic Structures in Bituminous Coal," Fuel, 42, 182-184 (1963); CA 58, 12329e

3119. Herington, E. F. G., "The Thermodynamics of Gas-Liquid Chromatography," in "Vapour Phase Chromatography," edited by D. H. Desty, Academic Press, Inc., New York, 1957, pp. 5-14

3120. Herk, L., Steffani, A., and Szwarc, M., "Methyl Affinities of Some Compounds Related to Acrylates and Acrylonitriles. Reactivities of Conjugated Systems Involving Atoms Other Than Carbon," J. Am. Chem. Soc., 83, 3008-3011 (1961); CA 56, 316b

3121. Herling, J., Shabtai, J., and Gil-Av, E., "Gas Chromatography with Stationary Phases Containing Silver Nitrate. III. Isomeric C_8 and C_9 Cyclohexenes and p-Menthenes," J. Chromatog., 8, 349-354 (1962); CA 58, 7375f

3122. Hernandez, Jr., R., and Axelrod, L. R., "Chromatographic Separation of the Steroids from Total Lipide Extracts," Anal. Chem., 35, 76-80 (1963)

3123. Herr, W., Schmidt, F., and Stocklin, G., "Radio-Gas Chromatography of Neutron-Irradiated Alkyl Halides and the Identification of Recoil Reaction Products," Z. Anal. Chem., 170, 301-310 (1959); CA 54, 3054h

3124. Herr, W., Schmidt, F., and Stocklin, G., "Radiogas-chromatographic Study of Szilard-Chalmers Reactions in n-Propyl Bromide," Z. Elektrochem., 63, 1006-1007 (1959)

3125. Herrin, C. B., and Lau, E., "Solid Supports for Gas Chromatography - A Reappraisal," 12th Pittsburgh Conf. on Anal. Chem. & Appl. Spectroscopy, Pittsburgh, Pa., February-March 1961, Program Abstr., p. 58

3126. Hersch, P., and Whittle, J. E., "Injecting Trace Impurities Into a Gas Stream," J. Sci. Instr., 35, 32-33 (1958)

3127. Hersh, C. K., "Molecular Sieves," Reinhold Publishing Corp., New York, 1961

3128. Heseltine, H. K., Pearson, J. D., and Wainman, H., "A Simple Thermal Conductivity Meter for Gas Analysis with Special Reference to Fumigation Problems," Chem. & Ind. (London), 1958, 1287-1288; CA 53, 6511c

3129. Hesse, G., "The Rate Factor of Adsorption Process in Chromatographic Columns," Z. Elektrochem., 55, 60-65 (1951); CA 45, 9335b

3130. Hesse, G., "Chromatographic Procedures in Organic Chemistry," Z. Anal. Chem., 181, 274-283 (1961); CA 55, 23160a

3131. Hesse, G., and Tschachotin, B., "Adsorption Analysis of Gases and Vapors," Naturwissenschaften, 30, 387-392 (1942); CA 37, 6211[3]

3132. Heuschkel, G., Wolny, J., and Skoczowski, S., "Determination of Trace Amounts of Methylacetylene and Propadiene in Cracking Gases by Means of Gas Chromatography," Erdol u. Kohle, 13, 98-99 (1960); CA 54, 14657f and CA 58, 8831b

3133. Hewett, D. R., Kipping, P. J., and Jeffery, P. G., "Separation, Identification, and Determination of the Fatty Acids of Montan Wax," Nature, 192, 65 (1961); CA 56, 4885i

3134. Hewitt, E. J., "Flavor Enhancement Review. Enzymatic Enhancement of Flavor," J. Agr. Food Chem., 11, 14-19 (1963)

3135. Hewitt, E. J., Mackay, D. A. M., and Lewin, S. E., "Physico-Chemical Approaches to the Study of Flavor," in "Flavor Research and Food Acceptance," Arthur D. Little, Inc., Cambridge, Mass., 1958, pp. 262-289

3136. Hewitt, G. C., and Witham, B. T., "The Identification of Substances of Low Volatility by Pyrolysis/Gas Chromatography," Analyst, 86, 643-652 (1961)

3137. Hey, D. H., Orman, S., and Williams, G. H., "Homolytic Aromatic Substitution. XXIII. A Redetermination of the Relative Rate of Phenylation of Nitrobenzene," J. Chem. Soc., 1961, 565-569; CA 55, 14335d

3138. Hey, D. H., Saunders, F. C., and Williams, G. H., "Homolytic Aromatic Substitution. XXI. Arylation of Benzotrihalides," J. Chem. Soc., 1961, 554-562; CA 55, 14334d

3139. Heydtmann, H., and Rinck, G., "Kinetics of the Thermal Decomposition of 2-Chlorobutane in the Gaseous Phase. II.," Z. Physik. Chem. (Frankfurt), 36, 75-81 (1963); CA 59, 397f

3140. Heyes, T. D., "Gas-Liquid Chromatography Applied to the Analysis of Oils and Fats," Chem. & Ind. (London), 1963, 660-665; CA 59, 843f

3141. Hickerson, J. F., "Use of Gas Chromatography in a Refinery Laboratory," Proc. Ann. Conv. Natl. Gasoline Assoc. Am., Tech. Papers, 37, 21-22 (1958); CA 54, 1834h

3142. Hickerson, J. F., "Chromatography is a Plant Workhorse," Oil Gas J., 56, No. 16, 119-121 (1958); CA 52, 15883e

3143. Higgins, C. E., and Baldwin, W. H., "The Thermal Decomposition of Tributyl Phosphate," J. Org. Chem., 26, 846-850 (1961); CA 55, 23313a

3144. Higson, H. G., and Butler, D., "The Determination of α(4-Chloro-2-methylphenoxy)propionic Acid in Chloromethylphenoxypropionic Acids by Gas-Liquid Chromatography with an Internal Standard," Analyst, 85, 657-663 (1960); CA 55, 21450h

3145. Hildebrand, R. P., and Sutherland, M. D., "Terpenoid Chemistry. II. Dysoxylonene and gamma-

Cadinene," Australian J. Chem., 12, 678-693 (1959); CA 54, 6974e

3146. Hildebrand, R. P., and Sutherland, M. D., "Terpenoid Chemistry. V. The Preparation of Pure Humulene," Australian J. Chem., 14, 272-275 (1961); CA 55, 22363c

3147. Hill, D. W., "The Application of Gas Chromatography to Anesthetic Research," Gas Chromatog. Proc. Symposium, 3rd, Edinburgh, 1960, 344-353; CA 56, 7421i

3148. Hill, D. W., "The Application of Gas Chromatography to Anaesthetic Research," in "Gas Chromatography 1960," edited by R. P. W. Scott, Butterworths, London, 1960, pp. 344-353

3149. Hill, D. W., "Production of Accurate Gas and Vapor Mixtures," Brit. J. Appl. Phys., 12, 410-413 (1961); CA 56, 2299e

3150. Hill, D. W., "The Application of Gas Chromatography to Forensic Science," Forensic Sci. Soc. J., 2, No. 1, 32-39 (1961); CA 56, 5046f

3151. Hill, D. W., and Hook, J. R., "Automatic Gas-Sampling Device for Gas Chromatography," J. Sci. Instr., 37, 253-255 (1960)

3152. Hill, D. W., Hook, J. R., and Mable, S. E. R., "A Compact Cathode-Ray Tube Gas Chromatograph," J. Sci. Instr., 39, 214-216 (1962)

3153. Hillen, L. W., and Thackray, M., "Determination of Neutron-Produced Gases in Beryllium by Gas Chromatography," J. Chromatog., 10, 309-313 (1963)

3154. Hindin, E., and Dunstan, G. H., "Analysis of Synthetic Organic Pesticides in Water by Chromatography," Proc. Brit. Insecticide Fungicide Conf., Brighton, Eng., 1961, 401-403 (Pub. 1962); CA 58, 8773f

3155. Hindin, E., Dunstan, G. H., McDonald, R., and May, D. S., "Analysis of Volatile Organic Acids by Gas Chromatography," 144th Natl. ACS Mtg., Los Angeles, Calif., March-April 1963, Program Abstr., pp. 1R-2R

3156. Hindin, E., Hatten, M. J., May, D. S., Skrinde, R. T., and Dunstan, G. H., "Analysis of Synthetic Organic Pesticides in Water," J. Am. Water Works Assoc., 54, 88-90 (1962); CA 56, 10625a

3157. Hines, W. L., and Smith, D. E., "Use of Reverse Flow in Gas Chromatography," 16th Southwest Regional Mtg., ACS, Oklahoma City, Okla., December 1960

3158. Hinkle, E. A., and Johnsen, S. E. J., "An Empirical Method for Converting Gas Chromatographic Elution Areas to Concentrations," in "Gas Chromatography," edited by V. J. Coates, H. J. Noebels, and I. S. Fagerson," Academic Press, Inc., New York, 1958, pp. 25-30

3159. Hinkle, E. A., Tucker, H. C., Wall, R. F., and Combs, J. F., "High-Vacuum Ionization Detector for Gas Chromatographic Analysis," Gas Chromatog., Intern. Symposium, 2nd, East Lansing, Mich., 1959, 55-63 (Pub. 1961); CA 55, 18204a

3160. Hinkle, E. A., Tucker, H. C., Wall, R. F., and Combs, J. F., "A High-Vacuum Ionization Detector for Gas Chromatographic Analysis," in "Gas Chromatography," edited by H. J. Noebeles, R. F. Wall, and N. Brenner, Academic Press, Inc., New York, 1961, pp. 55-63

3161. Hinnen, A., and Dreux, J., "Identification and Synthesis of 2,2,4,6-Tetramethylpyran," Compt. rend., 255, 1747-1749 (1962); CA 58, 3380g

3162. Hinsvark, O. N., and Beltz, P. B., "Microdetermination of Carbon and Hydrogen in Organic Materials: Gas Chromatographic Determination of Combustion Products," 142nd Natl. ACS Mtg., Atlantic City, N. J., September 1962, Program Abstr., pp. 20B-21B

3163. Hinsvark, O. N., and Theall, G., "Hydrogen Reduction and Determination of Metal Salts and Oxides: Gas Chromatographic Measurement of Reaction Products," 14th Pittsburgh Conf. on Anal. Chem. & Appl. Spectroscopy, Pittsburgh, Pa., March 1963, Program Abstr., p. 50

3164. Hirano, S., and Yamao, M., "Recent Developments in Gas Chromatography," Koru Taru, 14, 428-433 (1962); CA 58, 10699b

3165. Hirose, Y., Abu, M., and Sekiya, Y., "The Constituents of Sweet Potato Fusel Oil," Nippon Kagaku Zasshi, 82, 725-730 (1961); CA 57, 15569i

3166. Hirose, Y., Nishimura, K., and Sakai, T., "The Constituents of Essential Oils. I. Juniper Berry Oil," Nippon Kagaku Zasshi, 81, 1766-1769 (1960); CA 56, 6103f

3167. Hirose, Y., Ogawa, M., and Kusuda, Y., "Constituents of Fusel Oils Through the Fermentation of Corn, Barley, and Sweet Molasses," Agr. Biol. Chem. (Tokyo), 26, 526-531 (1962); CA 57, 14290f

3168. Hirsch, J., Farquhar, J. W., Ahrens, Jr., E. H., Peterson, M. L., and Stoffel, W., "Studies of Adipose Tissue in Man. A Microtechnique for Sampling and Analysis," Am. J. Clin. Nutrition, 8, 499-511 (1960); CA 54, 19892d

3169. Hirt, T. J., and Palmer, H. B., "Quantitative Determination of Carbon Suboxide by Gas Chromatography," Anal. Chem., 34, 164 (1962); CA 56, 8001b

3170. Hishta, C., and Bomstein, J., "Determination of Benzyl Chloroformate by Gas Chromatography of Its Amide. Extension of the Principle to Other Acid Chlorides," Anal. Chem., 35, 65-67 (1963); CA 58, 6194f

3171. Hishta, C., and Bomstein, J., "Gas Chromatography of Polar Compounds of Low Volatility," Anal. Chem., 35, 924-927 (1963)

3172. Histha, C., Messerly, J. P., and Reschke, R. F., "Gas Chromatography of High Boiling Compounds

116

on Low Temperature Columns," Anal. Chem., 32, 1730-1733 (1960); CA 55, 7137i; 137th Natl. ACS Mtg., Cleveland, Ohio, April 1960, Program Abstr., p. 29B

3173. Hishta, C., Messerly, J. P., and Reschke, R. F., with Fredericks, D. H., and Cooke, W. D., "Gas Chromatography of Solid Organic Compounds," Anal. Chem., 32, 880 (1960); CA 54, 17143h

3174. Hishta, C., and Reschke, R. F., "Determination of Micro Quantities of Chloroform in Pharmaceutical Products by Gas Chromatography," Intern. Symposium on Microchemical Techniques, University Park, Pa., August 1961

3175. Hishta, C., and Reschke, R. F., "Microdetermination of Chloroform in Pharmaceutical Products by Gas Chromatography," J. Symp. Ser., 2, 679-686 (1962); CA 58, 8851h

3176. Hissel, J., "Microanalysis of the Gases in Water Vapor," Bull. centre belge etude document. eaux (Liege), 42, 269-275 (1958); CA 53, 14827e

3177. Hiu, D. N., and Scheuer, P. J., "The Volatile Constituents of Passion Fruit Juice," J. Food Sci., 26, 557-563 (1961)

3178. Hively, R. A., "Identification of Hydrocarbons by Gas Chromatography," J. Chem. Eng. Data, 5, 237-240 (1960); CA 54, 20668h; 10th Pittsburgh Conf. on Anal. Chem. & Appl. Spectroscopy, Pittsburgh, Pa., March 1959, Program Abstr., p. 58

3179. Hively, R. A., "Assignment of Cis-Trans Configuration to Monoolefin Pairs," 11th Detroit Anachem Conf., Detroit, Mich, October 1963, Program Abstr., p. 27

3180. Hoag, L. E., and Reinke, H. G., "The Effects of Small Amounts of Oxygen on the Storage Stability of Beer," Am. Soc. Brewing Chemists, Proc. 1960, 141-145; CA 55, 27762h

3181. Hoare, M. R., Norrish, R. G. W., and Whittingham, G., "The Thermal Decomposition of Methylene Chloride," Proc. Roy. Soc.(London), A250, 180-196 (1959); CA 55, 18556c

3182. Hoare, M. R., Norrish, R. G. W., and Whittingham, G., "The Thermal Oxidation of Methylene Chloride," Proc. Roy. Soc. (London), A250, 197-211 (1959); CA 55, 23000b

3183. Hoare, M. R., and Purnell, J. H., "Temperature Effects in Gas Phase Partition Chromatography," Research Correspondence, 8, Suppl. S41-S42 (1955); CA 49, 15362d

3184. Hoare, M. R., and Purnell, J. H., "Temperature Effects in Gas-Liquid Partition Chromatography," Trans. Faraday Soc., 52, 222-229 (1956); CA 50, 14311g

3185. Hobbs, A. P., "Analysis of Natural Gas Samples by Vapor-Phase Chromatography," Proc. Ann. Conv., Natl. Gasoline Assoc. Am., Tech. Papers, 37, 33-36 (1958); CA 54, 1834e

3186. Hobden, F. W., "Gas Liquid Chromatography and Its Application to Paint and Allied Industries," J. Oil & Colour Chemists Assn., 4, 24-41 (1958); CA 53, 8657i

3187. Hobson, A., and Kay, R. J., "Two Designs for a Paramagnetic Oxygen Meter," J. Sci. Instr., 33, 176-181 (1956); CA 50, 11063b

3188. Hoek, A. van den. and Smit, W. M., "Theory of Chromatography," Rec. trav. chim., 76, 577-589 (1957); CA 51, 15216b

3189. Hoenes, Jr., H. J., Proehl, H. C., and Nagy, Z., "Gas Chromatographic Analysis of Impurities in Phosgene," 13th Pittsburgh Conf. on Anal. Chem. & Appl. Spectroscopy, Pittsburgh, Pa., March 1962, Program Abstr., p. 55

3190. Hoff, J. E., and Feit, E. D., "Functional Group Analysis in Gas Chromatography," Anal. Chem., 35, 1298-1299 (1963)

3191. Hoffman, A. J., and Mitchell, H. I., "Gas Chromatographic Analysis of Acetylsalicylic Acid, Acetophenetidin, and Caffeine Mixture in Pharmaceutical Tablet Formulations," J. Pharm. Sci., 52, 305-306 (1963); CA 58, 12368d

3192. Hoffmann, D., and Wynder, E. L., "Analytical and Biological Studies on Gasoline Engine Exhaust," Natl. Cancer Inst. Monograph No. 9, 91-116 (1962); CA 58, 1278c

3193. Hoffmann, E. G., "Analysis for Toxic Solvents in Technical Mixtures. Parallel Analyses by Infrared Spectroscopy and Gas Chromatography," Z. Anal. Chem., 164, 182-218 (1958); CA 53, 6870h

3194. Hoffmann, E. G., "Calculation of Relative Molar Response Factors of Thermal Conductivity Detectors in Gas Chromatography," Anal. Chem., 34, 1216-1222 (1962); CA 57, 14410a

3195. Hoffmann, G., "3-cis-Hexenal, the 'Green' Reversion Flavor of Soybean Oil," J. Am. Oil Chemists' Soc., 38, 1-3 (1961); CA 55, 5994d

3196. Hoffmann, G., "Isolation of Two Pairs of Isomeric 2,4-Alkadienals from Soybean Oil-Reversion Flavor Concentrate," J. Am. Oil Chemists' Soc., 38, 30-32 (1961); CA 55, 5994f

3197. Hoffmann, G., "1-Octen-3-ol and Its Relation to Other Oxidative Cleavage Products from Esters of Linoleic Acid," J. Am. Oil Chemists' Soc., 39, 439-444 (1962); CA 58, 1652h

3198. Hoffmann, G., and Keppler, J. G., "The Stereo-Configuration of 2,3-Decadienals Isolated from Oils Containing Linoleic Acid," Nature, 185, 310-311 (1960); CA 55, 14943b

3199. Hoffmann, R. W., "Base-Catalyzed Fragmentation of Azo Compounds, A Route to o-Bromophenyl Anion," Angew. Chem., 75, No. 3, 168 (1963); CA 59, 1509c

3200. Hofmann, J. E., and Schriesheim, A., "Ionic Reactions Occurring During Sulfuric Acid Catalyzed Alkylation. I. Alkylation of Isobutane with Butenes," J. Am. Chem. Soc., 84, 953-957 (1962); CA 57, 2045d

3201. Hofmann, M., and Kaiser, R., "Use of Gas Chromatography in Combination with IR-Spectroscopy and Mass Spectrometry for Determination of Impurities in Technical, Polymerizable Compounds Such as Methyl Methacrylate," Angew. Chem., 72, 141 (1960)

3202. Hofmann, M., Koennecke, H. G., and Leibnitz, E., "Separation of Aromatics in the C_6-C_8 Range and Non-Aromatic Hydrocarbons in a Packed Gas Chromatographic Column," Monatscher. Deut. Akad. Wiss. Berlin, 4, No. 1, 16-24 (1962); CA 58, 5539f

3203. Hofmann, M., and Struppe, H. G., "Gas Chromatographic Analysis of Higher Alcohols and Paraffins," Acta Chim. Acad. Sci. Hung., 27, 239-245 (1961); CA 55, 20789f

3204. Hofstader, R. A., "Evaluation and Application of Preparatory Scale Gas Chromatography," Metropolitan Regional Mtg., New York and New Jersey Sections, ACS, New York, N. Y., January 1962

3205. Hofstee, T., Kwantes, A., and Rijnders, G. W. A., "Determination of Activity Coefficients at Infinite Dilution by Gas-Liquid Chromatography," Proc. Intern. Symposium Distillation, Brighton, England, 1960, 105-109; CA 56, 4152b

3206. Hoftyzer, P. J., "Performance of Technical Apparatus for Gas Absorption," Joint Symposium Scaling-Up Chem. Plant and Processes, London, 1957, 53-57; CA 52, 1692g

3207. Hoh, G. L. K., Barlow, D. O., Chadwick, A. F., Lake, D. B., and Sheeran, S. R., "Hydrogen Peroxide Oxidation of Tertiary Amines," J. Am. Oil Chemists' Soc., 40, 268-271 (1963)

3208. Hoigne, J., and Gaeumann, T., "Radiation Chemistry of Hydrocarbons. V. Temperature Effect in Toluene," Helv. Chim. Acta, 46, 365-374 (1963); CA 58, 11141h

3209. Holaday, D. A., "Where Does Instrumentation Enter Medicine? Modern Technology Can Help Medicine Meet Growing Demands Through Instrumentation and Automation," Science, 134, 1172-1177 (1961)

3210. Hollis, O. L., "Gas-Liquid Chromatographic Analysis of Trace Impurities in Styrene Using Capillary Columns," Anal. Chem., 33, 352-355 (1961); CA 55, 11173h

3211. Hollis, O. L., and Hayes, W. V., "Gas Liquid Chromatographic Analysis of Chlorinated Hydrocarbons with Capillary Columns and Ionization Detectors," Anal. Chem., 34, 1223-1226 (1962); CA 57, 15780i

3212. Holmes, J. C., and Morrell, F. A., "Oscillographic Mass-Spectrometric Monitoring of Gas Chromatography," Appl. Spectroscopy, 11, 86-87 (1957); CA 51, 14329h

3213. Holmes, J. M., and Beebe, R. A., "Adsorption Studies on a Series of Heat-Treated Shawinigan Acetylene Carbon Blacks," Can. J. Chem., 35, 1542-1554 (1957); CA 52, 15186d

3214. Holmes, W., "The Use of Gas Chromatography on the Study of Hypocholesterolemic Agents," 4th Annual Eastern Analytical Symp., New York, N. Y., November 1962, Program Abstr., p. 24

3215. Holmes, W. L., and Stack, E., "Gas Chromatography of Squalene, Sterols, and Bile Acid Methyl Esters," Biochim. et Biophys. Acta, 56, 163-165 (1962); CA 56, 11943d

3216. Holness, D., "The Determination of Linaloöl. A Comparative Study of the Glichitch and Fiore Methods by Gas Liquid Chromatography," Analyst, 84, 3-10 (1959); CA 53, 14424a

3217. Holness, D., "Gas Chromatographic Examination of Essential Oils," Soap, Perfumery & Cosmetics, 32, 269-270 (1959)

3218. Holness, D., "Determination of Citronellol in Admixture with Geraniol - Further Studies of Formylation Reactions by Gas-Liquid Chromatography," Analyst, 86, 231-239 (1961); CA 55, 20327g

3219. Holness, D., "Gas-Liquid Chromatography and the Perfumer," J. Soc. Cosmetic Chemists, 12, 357-397 (1961); CA 56, 552h

3220. Holness, D., "Gas-Liquid Chromatography in Perfumery," Perfumery Essent. Oil Record, 52, 357-358 (1961)

3221. Holt, C., "Gases and Their Collection," School Sci. Rev., 42, 464-470 (1961); CA 56, 292c

3222. Holzhaeuser, H., "Trace Analysis in Synthesis Gas," Abhandl. Deut. Akad. Wiss. Berlin, Kl. Chem., Geol., Biol., 1959, No. 9, 271-298; CA 58, 5019a

3223. Holzhaeuser, H., "Gas Chromatographic Trace Analysis of Synthesis Gas" in "Gas Chromatographie 1958," edited by H. P. Angele, Akademie Verlag, Berlin, 1959, pp. 271-298

3224. Holzhaeuser, H., and Kuhl, M., "Microflame Ionization Detector and Quantitative Analysis," in "Gas Chromatographie 1959," edited by R. E. Kaiser and H. G. Struppe, Akademie Verlag, Berlin, 1959, pp. 143-156

3225. Honegger, C. G., and Honegger, R., "Occurrence and Quantitative Determination of 2-Dimethylaminoethanol in Animal Tissue Extracts," Nature, 184, 551-552 (1959); CA 54, 9046e

3226. Hoogschagen, J., "Diffusion in Porous Catalysts and Adsorbents," Ind. Eng. Chem., 47, No. 5, 906-913 (1955); CA 49, 10701a

3227. Hooimeijer, J., Kwantes, A., and Craats, F. van de, "The Automatization of Gas Chromatography," in "Gas Chromatography 1958," edited by D. H. Desty, Academic Press, Inc., New York, 1958, pp. 288-299

3228. Hoover, F. W., Webster, O. W., and Handy, C. T., "Trimerization of Acetylenes," J. Org. Chem., 26, 2234-2236 (1961); CA 55, 25810a

3229. Hopkins, C. Y., and Chisholm, M. J., "Composition of Zelkova Seed Oil," J. Am. Oil Chemists' Soc., 36, 210-212 (1959); CA 53, 12711e

3230. Hopkins, C. Y., and Chisholm, M. J., "Development of Oil in the Seed of Asclepias Syriaca," Can. J. Biochem. and Physiol., 39, 829-835 (1961); CA 55, 15638a

3231. Hopkins, C. Y., and Chisholm, M. J., "Occurrence of trans, trans-Octadeca-10,12-dienoic Acid in a Seed Oil," Chem. & Ind. (London), 1962, 2064; CA 58, 12412e

3232. Horak, W., and Lehmann, H., "The Utilization of Gas Chromatography for the Analysis of Alcoholic Mixtures and Distillery By-Products," Alkohol-Ind., 73, No. 14, 367-372 (1960); CA 55, 2007d

3233. Horn, O., Schwenk, U., and Hachenberg, H., "Gas Chromatography," Brennstoff-Chem., 38, 116-120 (1957); CA 51, 10177h

3234. Horn, O., Schwenk, U., and Hachenberg, H., "Gas Chromatography. II. Use for Gases in the Industrial Laboratory," Brennstoff-Chem., 39, 336-346 (1958); CA 53, 3973e

3235. Horner, L., and Schlafer, L., "The Course of Substitution. XX. Di- and Polychlorination of Various Monochloroalkanes," Ann., 635, 31-45 (1960); CA 55, 353g

3236. Horner, P. J., and Swallow, A. J., "The Gamma-Radiolysis of Solutions of Hydrogen Chloride in Cyclohexane," J. Phys. Chem., 65, 953-956 (1961); CA 55, 19431f

3237. Horning, E. C., "Separation of Steroids by Gas Chromatography," Joint Mtg., Am. Assoc. of Clin. Chemists and Am. Assoc. for Adv. of Science, December 1960; Abstr., Clin. Chem., 7, 304 (1961)

3238. Horning, E. C., "Gas Chromatographic Separations of Steroids and Related Substances," 3rd Annual Eastern Analytical Symp. & Instrument Exhibit, New York, N. Y., November 1961

3239. Horning, E. C., "Quantitative Aspects of Gas Chromatographic Separations in Biological Studies," Univ. of Houston Intern. Symp. on Advances in Gas Chromatography, Houston, Texas, January 1963

3240. Horning, E. C., Haahti, E. O. A., and VandenHeuvel, W. J. A., "A New Liquid Phase for Gas Chromatographic Separation of Steroids," J. Am. Chem. Soc., 83, 1513-1514 (1961); CA 55, 21226i

3241. Horning, E. C., Haahti, E. O. A., and VandenHeuvel, W. J. A., "Gas Chromatographic Separations of Steroids and Related Compounds," 11th Annual Instrument Symp. & Research Equipment Exhibit, National Institutes of Health, Bethesda, Md., October 1961

3242. Horning, E. C., Haahti, E. O. A., and VandenHeuvel, W. J. A., "Separation of Steroids by Gas Chromatography," J. Am. Oil Chemists' Soc., 38, 625-628 (1961); CA 56, 5047a

3243. Horning, E. C., Luukkainen, T., Haahti, E. O. A., Creech, B. G., and Vanden Heuvel, W. J. A., "Studies of Human Steroidal Hormones by Gas Chromatographic Techniques," in "Recent Progress in Hormone Research," Vol. XIX, edited by G. Pincus, Academic Press, Inc., New York, (in press)

3244. Horning, E. C., Maddock, K. C., Anthony, K. V., and VandenHeuvel, W. J. A., "Quantitative Aspects of Gas Chromatographic Separations in Biological Studies," Anal. Chem., 35, 526-532 (1963); CA 59, 895g

3245. Horning, E. C., Moscatelli, E. A., and Sweeley, C. C., "Polyester Liquid Phases in Gas-Liquid Chromatography," Chem. & Ind. (London), 1959, 751-752; CA 53, 21041h

3246. Horning, E. C., VandenHeuvel, W. J. A., and Creech, B. G., "Separation and Determination of Steroids by Gas Chromatography," in "Methods of Biochemical Analysis," Vol. 11, edited by D. Glick, Interscience Publishers, Inc., New York, 1963

3247. Horning, E. C., VandenHeuvel, W. J. A., and Haahti, E. O. A., "Separation of Steroids and Related Substances by Gas Chromatography," 3rd Intern. Symp. on Gas Chromatography, ISA, East Lansing, Mich., June 1961

3248. Horning, E. C., VandenHeuvel, W. J. A., and Haahti, E. O. A., "Separation of Steroids and Related Substances by Gas Chromatography," Gas Chromatog. Intern. Symp., 1961, 3, 507-517 (Pub. 1962); CA 58, 8221f

3249. Horning, E. C., VandenHeuvel, W. J. A., and Haahti, E. O. A., "Separation of Steroids and Related Substances by Gas Chromatography," in "Gas Chromatography," edited by N. Brenner, J. E. Callen, M. D. Weiss, Academic Press, Inc., New York, 1962, pp. 507-518

3250. Horning, M. G., Earle, M. J., and Maling, H. M., "Changes in Fatty Acid Composition of Liver Lipids Induced by Carbon Tetrachloride and Ethionine," Biochim. et Biophys. Acta, 56, 175-177 (1962)

3251. Horning, M. G., Martin, D. B., Karmen, A., and Vagelos, P. R., "Synthesis of Branched-Chain and Odd-Numbered Fatty Acids from Malonyl-Coenzyme A," Biochem. Biophys. Res. Comm., 3, 101-106 (1960); CA 55, 13508a

3252. Horning, M. G., Martin, D. B., Karmen, A., and Vagelos, P. R., "Fatty Acid Synthesis in Adipose Tissue. II. Enzymatic Synthesis of Branched Chain and Odd-Numbered Fatty Acids," J. Biol. Chem., 236, 669-672 (1961); CA 55, 13513i

3253. Horning, M. G., Williams, E. A., Maling, H. M., and Brodie, B. B., "Depot Fat as Source of Increased Liver Triglycerides After Ethanol," Biochem. Biophys. Res. Comm., 3, 635-640 (1960); CA 55, 15686e

3254. Hornstein, I., Alford, J. A., Elliott, L. E., and Crowe, P. F., "Determination of Free Fatty Acids in Fat," Anal. Chem., 32, 540-542 (1960); CA 54, 14727b; 136th Natl. ACS Mtg., Atlantic City,

N. J., September 1959, Program Abstr., p. 12A

3255. Hornstein, I., and Crowe, P. F., "Flavor Studies on Beef and Pork," J. Agr. Food Chem., 8, 494-498 (1960); CA 55, 17941c; 137th Natl. ACS Mtg., Cleveland, Ohio, April 1960, Program Abstr., p. 13A

3256. Hornstein, I., and Crowe, P. F., "Influence of Column Support on Separation of Fatty Acid Methyl Esters by Gas Chromatography," Anal. Chem., 33, 310-311 (1961); CA 55, 17375b

3257. Hornstein, I., and Crowe, P. F., "Gas Chromatography of Food Volatiles - An Improved Collection System," Anal. Chem., 34, 1354-1356 (1962); CA 57, 12961i

3258. Hornstein, I., Elliott, L. E., and Crowe, P. F., "Gas Chromatographic Separation of Long-Chain Fatty Acid Methyl Esters on Polyvinyl Acetate," Nature, 1710-1711 (1959); CA 54, 16381c

3259. Horrocks, L. A., "Determination of Glycol, Serine, Ethanolamine and Fatty Acids in Lipides by Gas-Liquid Chromatography," Univ. Microfilms (Ann Arbor, Mich.), L. C. Card No. Mic 61-914, 80 pp.; Dissertation Abstr., 21, 2876-2877 (1961); CA 55, 15594c

3260. Horrocks, L. A., and Cornwell, D. G., "The Simultaneous Determination of Glycerol and Fatty Acids in Glycerides by Gas-Liquid Chromatography," J. Lipid Research, 3, 165-169 (1962)

3261. Horrocks, L. A., Cornwell, D. G., and Brown, J. B., "Quantitative Gas-Liquid Chromatography of Fatty Acid Methyl Esters with the Thermal Conductivity Detector," J. Lipid Research, 2, 92-94 (1961); CA 55, 23187f

3262. Horton, A. D., "Gas-Solid Chromatography," Oak Ridge Natl. Lab., Anal. Chem. Div. Annual Progr. Rept. for Period Ending December 31, 1959, pp. 18-20; ORNL-3060: UC-4-Chemistry: TD-4500

3263. Horwitt, M. K., Harvey, C. C., and Century, B., "Effect of Dietary Fats on Fatty Acid Composition of Human Erythocytes and Chick Cerebella," Science, 130, 917-918 (1959); CA 54, 4804b

3264. Hougen, L. R., "Effect of Corona Discharges on Polyethylene," Nature, 188, 577-578 (1960); CA 55, 11911b

3265. Hough-Grossby, A. W., "Automatic Analysis in Petroleum Chemical Processing," Chem. Products, 23, 249-252 (1960); CA 54, 16043c

3266. Houghton, A. A., and Lund, N. A., "Application of Gas Chromatography to Fat Analysis," Rev. intern. chocolat., 15, 470-476 (1959); CA 54, 10178i

3267. Houghton, G., "Band Shapes in Nonlinear Chromatography with Axial Dispersion," J. Phys. Chem., 67, 84-88 (1963); CA 58, 2866e

3268. Houlihan, W. J., "Separation of Menthone-Menthol Stereoisomers by Gas Liquid Chromatography," Anal. Chem., 34, 1846 (1962); CA 58, 3871e

3269. Houlihan, W. J., "Synthesis of (+)-3-Methyl-β-citronellal," J. Org. Chem., 27, 4096-4098 (1962); CA 58, 7977b

3270. House, H. O., and Gilmore, W. F., "Application of the Favorskii Rearrangement to 2,3-Epoxy-cyclohexanones," J. Am. Chem. Soc., 83, 3972-3980 (1961); CA 57, 676c

3271. House, H. O., and Gilmore, W. F., "The Stereochemistry of the Favorskii Rearrangement," J. Am. Chem. Soc., 83, 3980-3985 (1961); CA 57, 677g

3272. House, H. O., Grubbs, E. J., and Gannon, W. F., "The Reaction of Ketones with Diazomethane," J. Am. Chem. Soc., 82, 4099-4106 (1960); CA 56, 15354e

3273. Houser, E. A., "Are You Having Sampling Problems?" ISA Journal, 7, No. 9, 95-99 (1960)

3274. Houser, E. A., "Sampling Systems for Liquid Injection Process Chromatographs," Analyzer, 2, No. 1, 5-7 (January 1961)

3275. Houston, R. H., "Theory for Industrial Gas-Liquid Chromatographic Columns," U. S. Atomic Energy Comm., UCRL-3817, 171 pp. (1958); CA 52, 14242h

3276. Hövermann, W., and Jentzsch, D., "A Critical Study on Golay Columns," 4th Intern. Gas Chromatography Symp., Hamburg, Germany, June 1962; pub. in "Gas Chromatography 1962," ed. by M. van Swaay, Butterworth & Co., London, 1962

3277. Howard, G. A. "New Approach to the Analysis of Hop Oil," J. Sci. Food Agr., 7, ii-31; J. Inst. Brewing, 62, No. 2, 158-159 (1956); CA 53, 18375c

3278. Howard, G. A., and Slater, C. A., "Hydrocarbons of the Essential Oil of Hops," Chem. & Ind. (London), 1957, 495-496; CA 51, 14207i

3279. Howard, G. A., and Slater, C. A., "Effect of Ripeness and Drying of Hops on Essential Oil," J. Inst. Brewing, 64, 234-237 (1958)

3280. Howard, G. A., and Stevens R., "The Occurrence of 2-Methylbutyl Esters in Hop Oil," Chem. & Ind. (London), 1959, 1518-1519; CA 54, 9213d

3281. Howard, G. A., and Stevens, R., "The Thermal Isomerization of Methyl Geranate and Dihydro-myrcene," J. Chem. Soc., 1960, 161-163; CA 54, 8886e

3282. Howard, G. A., and Tatchell, A. R., "Evaluation of Hops: New Approach to the Detailed Analysis of Hop Resins," J. Inst. Brewing, 62, 20-27 (1956); CA 54, 10234f

3283. Howard, H. E., and Ferguson, W. C., "Elimination of Preliminary Depentanization of Gasoline Prior to Hydrocarbon-Type Analysis by Mass Spectrometry," Anal. Chem., 31, 1048-1049 (1959); CA 53, 15539i

3284. Howe, W. H., Drinker, P. H., and Green, R. M., "Progress in Plant Measurements," Chem. Eng.,

68, No. 12, 199-204 (1961); CA 55, 16853i

3285. Howton, D. R., and Mead, J. F., "Metabolism of Essential Fatty Acids. X. Conversion of 8, 11, 14-Eicosatrienoic Acid to Arachidonic Acid in the Rat," J. Biol. Chem., 235, 3385-3386 (1960); CA 55, 6660d

3286. Hrapia, H., "Properties of Gas-Liquid Columns with Active Support Materials," in "Gas Chromatographie 1958," edited by H. P. Angele, Akademie Verlag, Berlin, 1959, pp. 93-117

3287. Hrapia, H., "A New Registering Integral Detector Based on Janak's Principles," in "Gas Chromatographie 1958," edited by H. P. Angele, Akademie Verlag, Berlin, 1959, pp. 243-253

3288. Hrapia, H., "Properties of Gas-Liquid Columns Utilizing Active Solid Carriers," Abhandl. Deut. Akad. Wiss. Berlin, Kl. Chem., Geol., Biol., 1959, No. 9, 93-117; CA 58, 5018c

3289. Hrapia, H., "A New Recording Integrator Based on Janak's Principle," Abhandl. Deut. Akad. Wiss. Berlin, Kl. Chem., Geol., Biol., 1959, No. 9, 243-253; CA 58, 5018h

3290. Hrapia, H., and Konnecke, H. G., "Chromatographic Gas Analysis," J. prakt. Chem., 3, 106-112 (1956); CA 53, 19663i

3291. Hrivnac, M., and Janak, J., "Use of Gas Chromatography for Analytical Control of the Production of Raw Naphthalene Oil and Its Processing into Pure Naphthalene," Chem. prumsyl., 9, 459-464 (1959); CA 54, 4239a

3292. Hrivnac, M., and Janak, J., "Chromatographic Estimation of the Content of Heterocyclic Sulfur Compounds in Commerical Benzenes and Naphthalenes," Chem. prumsyl., 10, 399-403 (1960); CA 55, 1299c

3293. Hrivnac, M., and Janak, J., "New Separations of Sulfur Heterocyclic Compounds by Gas-Liquid Chromatography," Chem. & Ind. (London), 1960, 930; CA 54, 24131g

3294. Hrutfiord, B. F., "Gas Chromatography Analysis of Lignin Degradation Products," 18th Annual Northwest Regional Mtg., ACS, Bellingham, Wash., June 1963

3295. Huber, J. F. K., and Keulemans, A. I. M., "Nonlinear Ideal Chromatography and the Possibilities of Linear Gas-Solid Chromatography," 4th Intern. Gas Chromatography Symp., Hamburg, Germany, June 1962; pub. in "Gas Chromatography 1962," M. van Swaay, ed., Butterworth & Co., London, 1962

3296. Huck, H. W., "Radio Frequency Detectors," 2nd Gas Chromatography Institute, Canisius College, Buffalo, N. Y., April 1960

3297. Hudson, Jr., B. E. King, Jr., W. H., and Brandt, W. W., "Detection of Hydrocarbons by a Thermionic Diode at Atmospheric Pressure," in "Gas Chromatography," edited by N. Brenner, J. E. Callen, and M. D. Weiss, Academic Press, Inc., New York, 1962, pp. 207-218

3298. Hudson, Jr., B. E., King, Jr., W. H., and Brandt, W. W., "Detection of Hydrocarbons by a Thermionic Diode at Atmospheric Pressure," 3rd Intern. Symp. on Gas Chromatography, ISA, East Lansing, Mich., June 1961; ISA Proc., 3, 143-150 (1961)

3299. Hudson, Jr., B. E., King, Jr., W. H., and Brandt, W. W., "Detection of Hydrocarbons by a Thermionic Diode at Atmospheric Pressure," Gas Chromatog., Intern. Symp., 1961, 3, 207-218 (Pub. 1962); CA 58, 3866c

3300. Hudson, J. R., and Stevens, R., "Beer Flavor. II. Fusel Oil Content of Some British Beers," J. Inst. Brewing, 66, 471-474 (1960); CA 55, 9777e

3301. Hudy, J. A., "Resin Acids. Gas Chromatography of Their Methyl Esters," Anal. Chem., 31, 1754-1756 (1959); CA 55, 3043a

3302. Huebner, V. R., "Preliminary Studies on the Analysis of Mono- and Di-Glycerides by GLPC," J. Am. Oil Chemists' Soc., 36, 262-263 (1959); CA 53, 14547h

3303. Huebner, V. R., "The Analysis of Mono-, Di-, and Triglycerides Derived from Cocoanut Oil by Means of High Temperature Gas-Liquid Phase Chromatography," Am. Oil Chemists' Soc. Mtg., Los Angeles, Calif., September 1959

3304. Huebner, V. R., "The Analysis of Glycerides by High Temperature Gas-Liquid Partition Chromatography," J. Am. Oil Chemists' Soc., 38, 628-631 (1961); CA 56, 5047b

3305. Huebner, V. R., "Determination of the Relative Polarity of Surface Active Agents by Gas-Liquid Chromatography," Anal. Chem., 34, 488-491 (1962); CA 57, 1582i

3306. Huebscher, G., Hawthorne, J. N., and Kemp, P., "The Analysis of Tissue Phospholipids - Hydrolysis Procedure and Results with Pig Liver," J. Lipid Research, 1, 433-438 (1960); CA 55, 8517h

3307. Hueckel, W., Egerer, W., and Moessner, F., "Constellation Analysis. VIII. Solvolysis of Toluene Sulfonic Esters. 10. β-Hydrindanols and the Solvolysis of Their Toluenesulfonates," Ann., 645, 162-176 (1961); CA 56, 349f

3308. Hueckel, W., Feltkamp, H., and Geiger, S., "Constellation Analysis. IV. Comparison of Isomers of 1, 4-Dimethyl-2-cyclohexanols and Those of the Menthols," Ann., 637, 1-19 (1960); CA 55, 6401i

3309. Hueckel, W., and Gelchsheimer, E., "Changes in the Molecular Configurations in Chemical Compounds. XII. Rearrangement of Pinene. 2. Methylnopinol and Pinene Hydrate," Ann., 625, 12-30 (1959); CA 54, 4655h

3310. Hueckel, W., and Hornung, W., "Apparatus for Chromatographing at Low Temperature," Chem. Ber., 90, 2023-2024 (1957); CA 54, 14814a

121

3311. Hueckel, W., Maucher, D., Fechting, O., Kurz, J., Heinzel, M., and Hubele, A., "Constellation Analysis. VII. Solvolysis of Toluenesulfonic Acid Esters. 9. Competitive Reactions with the Alcoholysis," Ann., 645, 115-162 (1961); CA 56, 345b

3312. Hueckel, W., Nag, D. S., and Zeisberger, R., "Alcoholysis of Toluene Sulfonic Esters. VIII. Camphenilols and Isopinocamphenol," Ann., 645, 101-114 (1961); CA 56, 508g

3313. Hueckel, W., and Rashingkar, R. B., "Constellation Analysis. V. The Alcholysis of Toluenesulfonic Acid Esters," Ann., 637, 20-32 (1960); CA 55, 6402f

3314. Hueckel, W., and Thiele, K., "Cis- and Trans-1-Isopropyl-2-cyclohexanol," Chem. Ber., 94, 96-102 (1961); CA 55, 9303g

3315. Hueckel, W., and Thomas, K. D., "The Walden Inversion. VIII. The Conversion of the cis- and trans-Isomers of 3- and 2-Methylcyclohexylamine with Nitrous Acid," Ann., 645, 177-194 (1961); CA 56, 342d

3316. Hueckel, W., and Ude, G., "The Walden Inversion. VII. Anomalous Side Reactions, cis- and trans-2-Cyclophenylcyclopentylamine and cis- and trans-2-Cyclohexylcyclohexylamine and Nitrous Acid," Chem. Ber., 94, 1026-1036 (1961); CA 55, 20990c

3317. Huelin, F. E., and Kennett, B. H., "Nature of the Olefins Produced by Apples," Nature, 184, 996 (1959); CA 54, 12277a

3318. Huestis, L. D., and Andrews, L. J., "The Effect of Geometry in the Alkyl Group on the Rate of the Claisen Rearrangement," J. Am. Chem. Soc., 83, 1963-1968 (1961); CA 55, 19846a

3319. Huffman, J. W., and Engle, J. E., "Stereochemistry of the Cycloheptane Ring. II. cis- and trans-2-Phenylcycloheptanol," J. Org. Chem., 26, 3116-3121 (1961); CA 56, 7159e

3320. Hughes, D., and Chilwell, E. D., "The Detection of Some Triazine Herbicides by Gas-Liquid Chromatography," Chem. & Ind. (London), 1962, 729

3321. Hughes, K. J., Ellis, C. F., and Hurn, R. W., "Composition of Automotive Engine Exhaust. Analytical Methods," 131st Natl. ACS Mtg., Miami, Fla., April 1957, Program Abstr., p. 11Q

3322. Hughes, K. J., and Hurn, R. W., "A Preliminary Survey of Hydrocarbon-Derived Oxygenated Material in Automobile Exhaust Gases," J. Air Pollution Control Assoc., 10, 367-373 (1960); CA 55, 1976h

3323. Hughes, K. J., and Hurn, R. W., "Analysis of Oxygenated Hydrocarbons in Automobile Combustion Products by Gas Chromatography," 16th Southwest Regional Mtg., ACS, Oklahoma City, Okla., December 1960

3324. Hughes, K. J., Hurn, R. W., and Edwards, F. G., "Separation and Identification of Oxygenated Hydrocarbons in Combustion Products from Automotive Engines," in "Gas Chromatography," edtied by H. J. Noebels, R. F. Wall, and N. Brenner, Academic Press, Inc., New York, 1961, pp. 171-182

3325. Hughes, M. A., "Composition of Ammoniacal Liquors. III. Analysis of the Organic Bases by Gas Chromatography," J. Appl. Chem. (London), 12, 450-457 (1962); CA 58, 3241c

3326. Hughes, M. A., White, D., and Roberts, A. I. L., "Separation of Meta- and Para-Isomers of the Xylenes, Cresols, and Toluidenes by Gas-Solid Chromatography," Nature, 184, 1796-1797 (1959); CA 55, 22083c

3327. Hughes, R. B., "Volatile Amines of Herring Fish," Nature, 181, 1281-1282 (1958); CA 52, 18950h

3328. Hughes, R. B., "Chemical Studies on the Herring (Clupea Harengus). I. Trimethylamine Oxide and Volatile Amines in Fresh, Spoiling and Cooked Herring Flesh," J. Sci. Food Agr., 10, 431-436 (1959); CA 54, 772e

3329. Hughes, R. B., "Chemical Studies on the Herring (Clupea Harengus). III. The Lower Fatty Acids," J. Sci. Food Agr., 11, 47-53 (1960); CA 54, 6993g

3330. Hughes, Jr. R. E., and Freed, V. H., "The Determination of Ethyl N,N-Dipropylthiolcarbamate (EPTC) in Soil by Gas Chromatography," J. Agr. Food Chem., 9, 381-382 (1961); CA 56, 6408g

3331. Hughes, R. W., "Use of Gas Chromatography to Study Volatile Substances in Unprocessed Whole Blood, with Special Reference to Volatile Compounds in the Blood of Ewes During Pregnancy Toxemia," Univ. Microfilms (Ann Arbor, Mich.), L. C. Card No. Mic 60-6111, 94 pp.; Dissertation Abstr., 21, 1739 (1961); CA 55, 9599e

3332. Huguet, M., "Kovats Retention Indexes in the Qualitative Analysis of Light Hydrocarbons by Gas Chromatography," Journees Intern. Etude Methodes Separation Immediate Chromatog., Paris, 1961, 69-84; CA 59, 1064e

3333. Huitric, A. C., and Carr, J. B., "Proton Magnetic Resonance Spectra and Stereochemistry of 2-Orthotolylcyclohexanol," J. Org. Chem., 26, 2648-2651 (1961); CA 55, 25831e

3334. Hull, W. Q., "Gas-Liquid Chromatography," Ind. Eng. Chem., 47, No. 10, 13A-14A, 16A (1955)

3335. Hull, W. Q., Keel, H., Kenney, J., and Gamson, B. W., "Diatomaceous Earth," Ind. Eng. Chem., 45, 256-269 (1953); CA 47, 7137b

3336. Hultschig, M., "Chromatography of Gases and Light Hydrocarbons," Freiberger Forschungsh., A80, 117-120 (1958); CA 52, 11388d

3337. Hummel, D., "The Analysis of Terephthalate Polyester Wire Lacquers," Farge u. lack., 65, 440-

449 (1959); CA 53, 20829i

3338. Hummel, R. W., "The Radiolysis of Liquid Methyl Acetate with Co^{60} Gamma Irradiation," Trans. Faraday Soc., 56, 234-245 (1960); CA 54, 13825f

3339. Hunsmann, W., "Analysis of Acetylene-Containing Gas Mixtures by Different Methods," Z. Anal. Chem., 164, 57-61 (1958); CA 53, 6887h

3340. Hunt, P. P., "Gas Chromatography of Hydrogen-Deuterium Mixtures," Univ. Microfilms (Ann Arbor, Mich.), L. C. Card No. Mic 60-2496, 104 pp.; Dissertation Abstr., 21, 299 (1960); CA 55, 2240a

3341. Hunt, P. P., and Smith, H. A., "The Separation of Hydrogen, Deuterium, and Hydrogen Deuteride Mixtures by Gas Chromatography," J. Phys. Chem., 65, 87-89 (1961); CA 55, 21968e

3342. Hunter, G. L. K., and Struck, R. F., "Nature of Extraneous Peaks in Gas Chromatographic Analysis of Girard-T Isolated Carbonyls," Anal. Chem., 34, 864-865 (1962)

3343. Hunter, I. R., "Extension of Flash Exchange Gas Chromatography to Ethyl Esters of Higher Organic Acids," J. Chromatog., 7, 288-292 (1962); CA 57, 11860g

3344. Hunter, I. R., Cole, E. W., and Pence, J. W., "Determination of Ethanol in Yeast Fermented Liquors by Gas Chromatography," J. Assoc. Offic. Agr. Chemists, 43, 769-771 (1960); CA 55, 4876g

3345. Hunter, I. R., Dimick, K. P., and Corse, J. W., "Determination of Amino-Acids by Ninhydrin Oxidation and Gas Chromatography. Separation of Leucine and iso-Leucine," Chem. & Ind. (London), 1956, 294-295; CA 50, 12754i

3346. Hunter, I. R., Ng, Hawkins, and Pence, J. W., "Nonaqueous Acidification in Preparation of Organic Acid Concentrates for Gas Chromatography," Anal. Chem., 32, 1757-1759 (1960); CA 55, 10731h

3347. Hunter, I. R., Ng, Hawkins, and Pence, J. W., "Volatile Organic Acids in Pre-Ferments for Bread," J. Food Sci., 26, 578-580 (1961)

3348. Hunter, I. R., Ortegren, V. H., and Pence, J. W., "Gas Chromatographic Separation of Volatile Organic Acids in Presence of Water," Anal. Chem., 32, 682-684 (1960); CA 54, 15089h

3349. Hunter, I. R., Stitt, F. B., and Kohler, G. O., "Determination of Amino Acids by Ninhydrin Oxidation and Gas Chromatography," 135th Natl. ACS Mtg., Chicago, Ill., September 1958, Program Abstr., p. 34A

3350. Hurn, R. W., "Analysis of Automobile Engine Exhaust by Gas Chromatography," 5th Detroit Anachem Conf., Detroit, Mich., October 1957

3351. Hurn, R. W., Chase, J. O., and Hughes, K. J., "Multistage Analyzer for Exhaust Gas Analysis," Ann. N. Y. Acad. Sci., 72, 675-684 (1959); CA 53, 20928h

3352. Hurn, R. W., and Davis, T. C., "Gas Chromatographic Analysis Shows Influence of Fuel on Composition of Automotive Engine Exhaust," Proc. Am. Petrol. Inst., 38, Sect. III, 353-372 (1958) (Pub. 1959); CA 53, 13547e

3353. Hurn, R. W., Hughes, K. J., and Chase, J. O., "Motor Vehicle Industry Efforts to Reduce Air Pollution from Exhaust," "Application of Gas Chromatography to Analysis of Exhaust Gas," Paper No. 11c, Automobile Mfg. Assoc., Detroit, Mich., 1958

3354. Hurn, R. W., Hughes, K. J., and Chase, J. O., "Application of Gas Chromatography to Analysis of Exhaust Gas," SAE Journal, 66, No. 5, 114 (1958); CA 54, 4239e

3355. Hurrell, R. A., and Perry, S. G., "Resolution in Gas Chromatography," Nature, 196, 571-572 (1962); CA 58, 5017h

3356. Hussey, A. S., Sauvage, J. F., and Baker, R. H., "Equilibrium Composition of Octahydronaphthalenes," J. Org. Chem., 26, 256-257 (1961); CA 55, 19872c

3357. Hutchinson, R. B., and Alexander, J. C., "The Structure of a Cyclic C_{18} Acid from Heated Linseed Oil," 144th Natl. ACS Mtg., Los Angeles, Calif., March-April 1963, Program Abstr., p. 41M

3358. Huyser, E. S., "The Free Radical Induced Rearrangement of 2-Methoxytetrahydropyran to Methyl Valerate," J. Org. Chem., 25, 1820-1822 (1960); CA 55, 2630c

3359. Huyser, E. S., "Trichloromethanesulfonyl Chloride as a Selective Chlorinating Agent," J. Am. Chem. Soc., 82, 5246-5247 (1960); CA 55, 4343d

3360. Huyser, E. S., "Addition and Abstraction Reactions of the Trichloromethyl Radical with Olefins," J. Org. Chem., 26, 3261-3264 (1961); CA 56, 11416a

3361. Huyser, E. S., and Wang, D. T., "Reactions of Ortho Esters with di-tert-Butyl Peroxide," J. Org. Chem., 27, 4696-4698 (1962); CA 58, 10068b

3362. Huyten, F. H., "Gas-Liquid Chromatography," Chem. Weekblad, 57, 209-214 (1961); CA 55, 18206a

3363. Huyten, F. H., Beersum, W. van, and Rijunders, G. W. A., "Improvements in the Efficiency of Large Diameter Gas-Liquid Chromatography Columns," in "Gas Chromatography 1960," edited by R. P. W. Scott, Butterworths, London, 1960, pp. 224-241

3364. Huyten, F. H., Beersum, W. van, and Rijunders, G. W. A., "Improvements in the Efficiency of Large-Diameter Gas-Liquid Chromatography Columns," Proc. Symposium, 3rd, Edinburgh, 1960, 224-241; CA 56, 5779h

3365. Huyten, F. H., Rijnders, G. W. A., and van Beersum, W., "Trace Analyses by Means of Gas-Solid Chromatography," 4th Intern. Gas Chromatography Symp., Hamburg, Germany, June 1962; pub. in "Gas Chromatography 1962," ed. by M. van Swaay, Butterworth & Co., London, 1962

3366. Hwa, J. C. H., Benneville, P. L. de, and Sims, H. J., "A New Preparation of 1,3,5-Hexatriene and the Separation of Its Geometrical Isomers," J. Am. Chem. Soc., 82, 2537-2540 (1960); CA 54, 18328c

3367. Hybl, C., and Lhotsky, G., "Gas Chromatography in the Plant-Control Laboratory," Chem. prumsyl, 7, 405-407 (1957); CA 52, 12650b

3368. Hyden, S., "Applications of Pyrolysis-Gas Chromatography. Decomposition of Dialkyl Peroxides," Anal. Chem., 35, 113-114 (1963); CA 59, 1471a

3369. Ibraev, G. Z., and Gorvaev, M. I., "Determination of Furfural by Gas Chromatography," Gidrolizn. i Lesokhim. Prom., 15, No. 8, 25-26 (1962); CA 58, 11936g

3370. Iconomou, N., Bechi, J., and Schumacher, H., "Gas Chromatography in the Testing of Purity of Drugs. IV.," Pharm. Acta Helv., 37, 622-638 (1962); CA 58, 4378h

3371. Iconomou-Petrovich, N., Beuchi, J., and Schumacher, H., "Use of Gas Chromatography in the Tests for Purity of Drugs," Pharm. Acta Helv., 37, 748-769 (1962); CA 58, 7784b

3372. Iden, R. B., and Kahler, E. J., "Comparison of Methods for the Quantitative Determination of Tall Oil Fatty Acids by Gas Chromatography," J. Am. Oil Chemists' Soc., 39, 171-173 (1962); CA 56, 13035a

3373. Iguchi, M., "Application to Analysis of Chemicals. Arsenic," Kagaku no Ryoiki Zokan, No. 44, 199-203 (1961); CA 56, 3b

3374. Iguchi, M., Nichiyama, A., and Nagase, Y., "Application of Gas Chromatography for Pharmaceutical Analysis. I. Arsenic Test," Yakugaku Zasshi, 80, 1408-1410 (1960); CA 55, 5871e

3375. Ikeda, R. M., Rolle, L. A., Vannier, S. H., and Stanley, W. L., "Isolation and Identification of Aldehydes in Cold-Pressed Lemon Oil," J. Agr. Food Chem., 10, 98-102 (1962); CA 57, 11618i

3376. Ikeda, R. M., Simmons, D. E., and Grossman, J. D., "Removal of Alcohol from Complex Mixtures During Gas Chromatography," 145th Natl. ACS Mtg., New York, N. Y., September 1963

3377. Ikeda, R. M., Stanley, W. L., Vannier, S. H., and Rolle, L. A., "A Deterioration of Lemon Oil, Formation of p-Cymene from Gamma-Terpinene," Food Technol., 15, 379-380 (1961); CA 56, 553b

3378. Ikegawa, N., "Analysis of Steroid Hormones by Using Gas Chromatography," Nippon Rinsho, 20, 541-547 (1962); CA 57, 17295i

3379. Ikekawa, N., "Recent Advances in Gas Chromatography, Especially of Steroids and Alkaloids," Kagaku no Ryoiki, 15, 449-459 (1961); CA 56, 6631b

3380. Ingram, W. T., and Dieringer, L. F., "A Critical Examination of Air-Sampling Instrumentation Methods," Am. Ind. Hyg. Assoc. Quart., 14, 121-132 (1953); CA 47, 10767g

3381. Innes, W. B., and Bambrick, W. E., "Rapid Hydrocarbon Analysis by Chromatography and Chemical Absorption," 145th Natl. ACS Mtg., New York, N. Y., September 1963, Program Abstr., p. 12B

3382. Innes, W. B., Bambrick, W. E., and Andreatch, A. J., "Hydrocarbon Gas Analysis Using Differential Chemical Absorption and Flame Ionization Detectors," Anal. Chem., 35, 1198-1203 (1963); 14th Pittsburgh Conf. on Anal. Chem. & Appl. Spectroscopy, Pittsburgh, Pa., March 1963, Program Abstr., p. 65

3383. Institute of Petroleum, "Analysis by Gas Chromatography - Petroleum Gases," I. P. Methods 169 (A, B & C)/59 (Tentative); Institute of Petroleum Methods for Testing Petroleum and Its Products, 18th ed., 1959

3384. Insull, Jr., W., and Ahrens, Jr., E. H., "The Fatty Acids of Human Milk from Mothers on Diets Taken Ad Libitum," Biochem., J., 72, 27-33 (1959); CA 53, 13299a

3385. Insull, Jr., W., Hirsch, J., James, A. T., and Ahrens, Jr., E. H., "The Fatty Acids of Human Milk II. Alterations Produced by Manipulation of Caloric Balance and Exchange of Dietary Fats," J. Clin. Invest., 38, 443-450 (1959); CA 53, 11573c

3386. Insull, Jr., W., and James, A. T., "The Quantitative and Qualitative Analysis of Fatty Acids in the Range C_1 to C_{20}," 132nd Natl. ACS Mtg., New York, N. Y., September 1957, Program Abstr., p. 36B; Preprints, Div. Petrol. Chem., 2, No. 4, D111-D113 (1957)

3387. Irby, R. M., and Harlow, E. S., "The Quantitative Determination of Certain Vapor Phase Constituents in Cigarette Smoke," 133rd Natl. ACS Mtg., San Francisco, Calif., April 1958, Program Abstr., p. 9A

3388. Irvine, L., and Mitchell, T. J., "Gas-Liquid Chromatography. I. Retention Volume Data for Certain Tar Acids," J. Appl. Chem. (London), 8, 3-6 (1958); CA 52, 8508e

3389. Irvine, L., and Mitchell, T. J., "Gas-Liquid Chromatography. II. Analysis of Alkali Extract of a Low-Temperature Coal Tar," J. Appl. Chem. (London), 8, 425-432 (1958); CA 52, 20997b

3390. Irving, H. M., and Williams, R. J. P., "Partition Chromatography," Science Progr., 41, 418-433 (1953); CA 47, 8460c

3391. Isbell, R. E., "Determination of Hydrogen Cyanide and Cyanogen by Gas Chromatography," Anal. Chem., 35, 255-256 (1963); CA 58, 7371a

3392. Ishi, Y., Sekiguchi, S., and Hayakawa, A., "Product Distribution of Ethylene Oxide Adducts of Various Alcohols," Bull. Chem. Soc. Japan, 35, 1624-1625 (1962); CA 58, 1421f

3393. Ishikawa, M., and Tsuchiya, T., "Gas Chromatography of Natural Essential Oils," Shika Zairyo

Kenkyusho Hokoku, 2, 401-408 (1962); CA 58, 11165g

3394. Issenberg, P., and Wick, E. L., "Banana Odor Components. Volatile Components of Bananas," J. Agr. Food Chem., 11, 2-8 (1963)

3395. Issoire, J., and Boileau, J., "Problems in the Synthesis and Separation of the Methylamines," Mem. Poudres, 43, 357-366 (1961); CA 58, 6660h

3396. Issoire, J., and Chaput, L., "Separation of Ammonia and the Methylamines by Gas-Phase Chromatography for Quantitative Determination," Chim. anal., 43, 313-320 (1961); CA 55, 21968b

3397. Ito, S., and Fukuzumi, K., "Quantitative Analysis of Fatty Acid Esters of Marine Animal Oils by Gas-Liquid Chromatography," Kogyo Kagaku Zasshi, 65, 1963-1968 (1962); CA 58, 12778e

3398. Ivanova, R. V., and Stepukhovich, A. D., "Initiation of the Cracking of Ethane," Zh. Fiz. Khim., 36, 222-224 (1962); CA 59, 419f

3399. Ives, I. G. C., "Analysis and Testing of Plastics," Fibres and Plastics, 22, 301-304 (1961); CA 56, 8905g

3400. Iveson, G., "Application of Gas-Liquid Chromatography to the Analysis of Corrosive Gases. I. Development of Apparatus and Technique for the Separation of Mixtures of Chlorine Trifluoride, Hydrogen Fluoride, and Uranium Hexafluoride," U. S. At. Energy Authority, Prod. Group, PG Repts., 82, 1-15 (1960); CA 55, 6058a

3401. Iveson, G., and Hamlin, A. G., "A Gas Chromatographic Apparatus for In-Line Analysis of Corrosive Inorganic Gases," in "Gas Chromatography 1960," edited by R. P. W. Scott, Butterworths, London, 1960, pp. 333-343

3402. Iveson, G., and Hamlin, A. G., "A Gas-Chromatographic Apparatus for In-Line Analysis of Corrosive Inorganic Gases," Gas Chromatog., Proc. Symposium, 3rd, Edinburgh, 1960, 333-343; CA 56, 6633i

3403. Iveson, G., and Hamlin, A. G., "Gas-Liquid Chromatography [in] the Analysis of Corrosive Gases, II. Design of In-Line Chromatographs," U. K. At. Energy Authority, Prod. Group. PG Rept. 354, 21 pp. (1962); CA 58, 9608f

3404. Iveson, G., Hamlin, A. G., and Phillips, T. R., " 'In-Line' Gas Chromatographic Analysis of Halogen and Interhalogen Compounds," 12th Pittsburgh Conf. on Anal. Chem. & Appl. Spectroscopy, Pittsburgh, Pa., February-March 1961, Program Abstr., p. 56

3405. Iyer, R. M., and Mittal, J. P., "Construction of an Argon Gas-Liquid Chromatograph," J. Sci. Ind. Res. (India), 21D, No. 1, 19-22 (1962); CA 57, 8d

3406. Jack, E. L., "The Fatty Acids and Glycides of Cow Milk Fat," J. Agr. Food Chem., 8, 377-380 (1960); CA 55, 19056i; 137th Natl. ACS Mtg., Cleveland, Ohio, April 1960, Program Abstr., p. 11A

3407. Jacknow, B. B., "The Reaction of t-Butyl Hypochlorite with Organic Compounds," Univ. Microfilms (Ann Arbor, Mich.), L. C. Card No. Mic 60-2018, 139 pp.; Dissertation Abstr., 21, 55 (1960); CA 55, 2463h

3408. Jackson, H. W., "Flavor Research on Cheese," Perfumery Essent. Oil Record, 49, 256 (1958)

3409. Jackson, H. W., "Flavor Research on Cheese," in "Flavor Research and Food Acceptance," Arthur D. Little, Inc., Cambridge, Mass., 1958, pp. 324-330; CA 53, 4600c

3410. Jackson, H. W., and Hussong, R. V., "Secondary Alcohols in Blue Cheese and Their Relation to Methyl Ketones," J. Dairy Sci., 41, 920-924 (1958); CA 52, 20740h

3411. Jackson, K. L., and Entenman, C., "A Two Recorder Integrator-Readout System for Gas-Liquid Chromatography," USNRDL-TR-397 (February 9, 1960), 24 pp.; Abstr., Nuclear Sci. Abstracts, 14, 1455 (1960)

3412. Jackson, K. L., and Entenman, C., "A Two-Recorder Integrator Readout System for Gas-Liquid Chromatography," J. Chromatog., 4, 435-445 (1960); CA 55, 11950h

3413. Jackson, M. W., and Heston, W. B., "Hydrocarbons in Exhaust Gas. Nondispersive Infrared Versus Gas Chromatographic Measurements," 7th Detroit Anachem. Conf., Detroit, Mich., October 1959

3414. Jackson, R. E., "Detection of the Stable Gases with the Argon Ionization Detector," 6th Instrumental Methods of Analysis Symp., ISA, Montreal, Canada, June 1960

3415. Jackson, R. G., "Gas Chromatography for Automatic Process Stream Analysis," Proc. Instr. Soc. Am., Natl. Symposium Progr. Trends Chem. Petrol. Instrumentation, Wilmington, 1, 61-68 (1958); CA 52, 17822a

3416. Jacobs, M. B., "Composition of Onion Oil," Am. Perfumer Aromat., 70, No. 5, 53-56 (1957); CA 52, 2344h

3417. Jacobs, M. B., "The Chemical Analysis of Air Pollutants," in "Chemical Analysis," Vol. X, Monagraph Series, Interscience Publishers, Inc., New York, 1960

3418. Jacobs, T. L., Petty, W. L., and Teach, E. G., "The Reaction of Propargyl Alcohols with Thionyl Chloride," J. Am. Chem. Soc., 82, 4094-4097 (1960); CA 55, 17472d

3419. Jacobson, M., Beroza, M., and Jones, W. A., "Insect Sex Attractants. I. The Isolation, Identification, and Synthesis of the Sex Attractant of the Gypsy Moth," Science, 132, 1011-1012 (1960); CA 56, 14038i

3420. Jaffe, F., Steadman, T. R., and McKinney, R. W., "Autoxidation of Decalin. I. Primary Products of Decalin Autoxidation," J. Am. Chem. Soc., 85, 351-353 (1963); CA 58, 11235g

3421. Jahnsen, V. J., "Complexity of Hop Oil," Nature, 196, 474-475 (1962); CA 58, 4371c

3422. Jain, T. C., Varma, K. R., and Bhattacharyya, S. C., "Terpenoids. XXVIII. Gas-Liquid Chromatography of Monoterpenes and Its Application to Essential Oils," Perfumery Essent. Oil Record, 53, 678-684 (1962); CA 58, 4371a

3423. James, A. T., "Gas-Liquid Partition Chromatography: The Separation of Volatile Aliphatic Amines and of the Homologues of Pyridine," Biochem. J., 52, 242-247 (1952); CA 47, 446b

3424. James, A. T., "Gas-Liquid Chromatography," Chem. Age (London), 73, 733-736 (1955)

3425. James, A. T., "Separation of Volatile Materials by Gas-Liquid Chromatography," Chem. and Proc. Engr., 36, No. 3, 95-100 (1955)

3426. James, A. T., "Separation of Volatile Materials by Gas-Liquid Chromatography," Mfg. Chemist, 26, 5-10 (1955)

3427. James, A. T., "A New Technique for the Separation of Volatile Materials," Research, 8, No. 1, 8-16 (1955); CA 49, 8028i

3428. James, A. T., "A New Kind of Chromatogram," Times Sci. Rev., 16, 8 (1955)

3429. James, A. T., "Gas-Liquid Chromatography. Separation and Microestimation of Volatile Aromatic Amines," Anal. Chem., 28, 1564-1567 (1956); CA 51, 951g

3430. James, A. T., "Recent Advances in Pharmaceutical Analysis. Gas-Liquid Chromatography," J. Pharm. and Pharmacol., 8, 232-240 (1956); CA 50, 13370g

3431. James, A. T., "Gas-Liquid Chromatography - A Method of Separation and Identification of Volatile Materials," Can. Chem. Processing, 40, 111-114 (1956)

3432. James, A. T., "The Gas-Liquid Chromatogram," Endeavour, 15, 73-78 (1956); CA 51, 42a

3433. James, A. T., "The Determination of the Structure of Unsaturated Fatty Acids on a Micro Scale with the Gas-Liquid Chromatogram," Biochem. J., 66, 515-529 (1957)

3434. James, A. T., "Separation and Identification of Saturated and Unsaturated Fatty Acids from Formic to Dodecanoic Acids by Gas Chromatography," Fette u. Seifen, 59, 73-77 (1957); CA 52, 6816g

3435. James, A. T., "Analysis in Medical Research," Proc. Congr. Modern Anal. Chem. Ind., Univ. St. Andrews, Heffer, Cambridge, England (1957), pp. 109-118; CA 53, 113e

3436. James, A. T., "Detection of Vapours in Flowing Gas Streams," in "Vapour Phase Chromatography," edited by D. H. Desty, Academic Press, Inc., New York, 1957, pp. 127-130

3437. James, A. T., "The Separation of the Long-Chain Fatty Acids by Gas-Liquid Chromatography," Am. J. Clin. Nutrition, 6, 595-600 (1958); CA 53, 3342e

3438. James, A. T., "Essential Fatty Acids," Am. J. Clin. Nutrition, 6, 650-651 (1958); CA 53, 3311g

3439. James, A. T., "Summing Up of Symposium," in "Gas Chromatography 1958," edited by D. H. Desty, Academic Press, Inc., New York, 1958

3440. James, A. T., "Retention Volumes of Amines Relative to that of Ethylamine with Non-Polar and Polar Liquid Phases in the Columns at 100° (Chromatographic Data Table L)," J. Chromatog., 1, xxxiii (1958); cf. Biochem. J., 52, 242 (1952)

3441. James, A. T., "Retention Volumes of Some Aromatic Bases Relative to Aniline at 137° in Three Types of Column (Chromatographic Data Table LI)," J. Chromatog., 1, xxxiv (1958); cf. Anal. Chem., 28, 1564 (1956)

3442. James, A. T., "Relative Retention Volumes of Pyridine Homologs (Chromatographic Data Table LII)," J. Chromatog., 1, xxxv (1958); cf. Biochem. J., 52, 242 (1952) (Temperature 137°, Pyridine = 1)

3443. James, A. T., "Determination of the Degree of Unsaturation of Long Chain Fatty Acids by Gas-Liquid Chromatography," J. Chromatog., 2, 552-561 (1959); CA 54, 16379b

3444. James, A. T., "Qualitative and Quantitative Determination of the Fatty Acids by Gas-Liquid Chromatography," in "Methods of Biochemical Analysis," Vol. 8, edited by D. Glick, Interscience Publishers, Inc., New York, 1960, pp. 1-59; CA 54, 14343b

3445. James, A. T., "Recent Developments in Gas-Liquid Chromatography," Lab. Practice, 9, 835-841 (1960)

3446. James, A. T., "The Development of an Idea," in "Gas Chromatography," edited by H. J. Noebels, R. F. Wall, and N. Brenner, Academic Press, Inc., New York, 1961, pp. 247-254

3447. James, A. T., "Applications of Gas Chromatography to Some Medical Problems," Chem. & Ind. (London), 1961, 1098

3448. James, A. T., "Studies by Gas-Phase Chromatography of the Biosynthesis of Mono- and Trienoic C_{18} Fatty Acids," Bull. Soc. Chim. Biol., 44, 951-963 (1962); CA 58, 4764a

3449. James, A. T., "A Simple Compact Gas-Radio Chromatogram." Univ. of Houston Intern. Symp. on Advances in Gas Chromatography, Houston, Texas, January 1963

3450. James, A. T., and Lovelock, J., "A Preliminary Investigation of the Fatty Acid Composition of Blood Lipids from Rabbit, Ox, Rat, and Normal Atherosclerotic Humans," Koninkl. Vlaam. Acad. Wetenschap. Lettermen Schone Kunsten Belg. Kl. Wetenschap. Intern. Colloq. Biochem.

Problem, Lipiden, Brussels, 3, 94-103 (1956); CA 52, 13073c

3451. James, A. T., Lovelock, J. E., and Webb, J. P. W., "The Lipids of Whole Blood. I. Lipid Biosynthesis in Human Blood in Vitro," Biochem. J., 73, 106-113 (1959); CA 53, 22328i

3452. James, A. T., and Martin, A. J. P., "Gas-Lipid Partition Chromatography: A Technique for the Analysis of Volatile Materials," Analyst, 77, 915-932 (1952); CA 47, 1529d

3453. James, A. T., and Martin, A. J. P., "Gas-Liquid Partition Chromatography: The Separation and Microestimation of Volatile Fatty Acids from Formic Acid to Dodecanoic Acid," Biochem. J. (London), 50, 679-690 (1952); CA 46, 4883i

3454. James, A. T., and Martin, A. J. P., "Gas-Liquid Partition Chromatography: The Separation and Microestimation of Volatile Fatty Acids and Bases," Congr. intern. biochim., Resumes communs., 2nd Congr., Paris, 1952, 159; CA 49, 12082h

3455. James, A. T., and Martin, A. J. P., "Gas-Liquid Chromatography: A Technique for the Analysis and Identification of Volatile Materials," British Medical Bull., 10, 170-176 (1954); CA 49, 4928b

3456. James, A. T., and Martin, A. J. P., "Gas-Liquid Chromatography. A Technique for the Analysis and Identification of Volatile Materials," Chim. anal. (Paris), 37, 321-326 (1955)

3457. James, A. T., and Martin, A. J. P., "Gas Chromatography," Mfg. Chemist, 26, No. 5, 229 (1955)

3458. James, A. T., and Martin, A. J. P., "Separation and Identification of Methyl Esters of Saturated and Unsaturated Fatty Acids from n-Pentanoic to n-Octadecanoic Acids," Biochem. Problems Lipids, Proc. Intern. Conf., 2nd, Ghent, 1955, pp. 42-48 (Pub. 1956); CA 52, 11662h

3459. James, A. T., and Martin, A. J. P., "The Separation and Identification of Some Volatile Paraffinic, Naphthenic, Olefinic, and Aromatic Hydrocarbons," J. Appl. Chem. (London), 6, 105-115 (1956); CA 50, 12455c

3460. James, A. T., and Martin, A. J. P., "Gas-Liquid Chromatography: The Separation of the Methyl Esters of Saturated and Unsaturated Acids from Formic to n-Octadecanoic Acid," Biochem. J., 63, 144-152 (1956); CA 50, 13660g

3461. James, A. T., and Martin, A. J. P., "Retention Volumes of Free Fatty Acids Relative to That of n-Butyric Acid on Silicone-Stearic Acid Columns (Chromatographic Data Table LIV)," J. Chromatog., 1, xxxvi (1958); cf. Biochem. J., 50, 679 (1952)

3462. James, A. T., and Martin, A. J. P., "Retention Volumes of Methyl Esters of Saturated Fatty Acids from Formic to n-Caproic Acid Relative to Methyl n-Butyrate in a Variety of Stationary Phases at Two Temperatures (Chromatographic Data Table LV)," J. Chromatog., 1, xxxvi (1958); cf. Biochem. J., 63, 144 (1956)

3463. James, A. T., and Martin, A. J. P., "Retention Volumes of Methyl Esters of Longer Chain Saturated Fatty Acids Relative to Methyl Myristate in Two Stationary Phases at 197° (Chromatographic Data Table LVI)," J. Chromatog., 1, xxxvii (1958); cf. Biochem. J., 63, 144 (1956)

3464. James, A. T., and Martin, A. J. P., "Retention Volumes of Methyl Esters of Some Unsaturated Acids Relative to Methyl n-Tetradecanoate in Two Stationary Phases at 197° (Chromatographic Data Table LVII)," J. Chromatog., 1, xxxvii (1958); cf. Biochem. J., 63, 144 (1956)

3465. James, A. T., and Martin, A. J. P., "Values of the Correction Factor j of James and Martin Used in Gas Chromatography for Values of the Ratio of Inlet to Outlet Pressure Between 1 and 3 (Chromatographic Data Table 54)," J. Chromatog., 2, D33-D45 (1959)

3466. James, A. T., Martin, A. J. P., and Smith, G. H., "Relative Retention Volumes of the Methylamines (Chromatographic Data Table XLIX)," J. Chromatog., 1, xxxiii (1958); cf. Biochem. J., 51, 323 (1952) (Ammonia = 1)

3467. James, A. T., Martin, A. J. P., and Smith, G. H., "Gas-Liquid Partition Chromatography: The Separation and Microestimation of Ammonia and the Methylamines," Biochem. J., 52, 238-242 (1952); CA 47, 446d

3468. James, A. T., Peters, G., and Lauryssens, M., "The Metabolism of Propionic Acid," Biochem. J., 64, 726-730 (1956); CA 51, 5223i

3469. James, A. T., and Piper, E. A., "Automatic Recording of the Radioactivity of Zones Eluted from the Gas-Liquid Chromatogram," J. Chromatog., 5, 265-270 (1961); CA 56, 5382h

3470. James, A. T., and Piper, E. A., "A Compact Radiochemical Gas Chromatograph," Anal. Chem., 35, 515-520 (1963); CA 59, 1060a

3471. James, A. T., and Webb, J., "Essential Fatty Acids. I. Behavior of Polyunsaturated Fatty Acids on Gas-Liquid Chromatogram," Proc. Intern. Conf. Biochem. Problems of Lipids, Oxford, 1957, 3-8 (Pub. 1958); CA 53, 17277d

3472. James, A. T., and Webb, J., "Determination of Structure of Unsaturated Fatty Acids on a Micro-Scale with the Gas-Liquid Chromatogram," Biochem. J., 66, 515-520 (1957); CA 52, 248e

3473. James, A. T., Webb, J. P. W., and Kellock, T. D., "Occurrence of Unusual Fatty Acids in Fecal Lipides from Human Beings with Normal and Abnormal Fat Absorption," Biochem. J., 78, 333-339 (1961); CA 55, 7677d

3474. James, A. T., Webb, J. P. W., Stapleton, T., and MacDonald, W. B., "Essential Fatty Acids and Idiophathic Hypercalcemia of Infancy," Lancet, 1958-I, 502-504; CA 52, 12112b

3475. James, A. T., and Wheatley, V. R., "Determination of the Component Fatty Acids of Human Forearm Sebum by Gas-Liquid Chromatography," Biochem. J., 63, 269-273 (1956); CA 50, 14907g

3476. James, D. H., and Phillips, C. S. G., "Simple Gas-Flow Control of High Efficiency," J. Sci. Instr., 29, 362-363 (1952); CA 47, 3048c

3477. James, D. H., and Phillips, C. S. G., "The Chromatography of Gases and Vapours. II.," J. Chem. Soc., 1953, 1600-1610; CA 47, 9861d

3478. James, D. H., and Phillips, C. S. G., "The Chromatography of Gases and Vapours. III. The Determination of Adsorption Isotherms," J. Chem. Soc., 1954, 1066-1070; CA 48, 7958g

3479. Jamieson, G. R., "Quantitative Analysis Using Thermal Conductivity Detection," Analyst, 84, 74-75 (1959); CA 55, 3268i

3480. Jamieson, G. R., "Relative Detector Response in Gas Chromatography," J. Chromatog., 3, 464-470 (1960); CA 55, 3268i

3481. Jamieson, G. R., "Relative Detector Response in Gas Chromatography. II. Benzene Hydrocarbons, Phenols, and Phenol Ethers," J. Chromatog., 3, 494-496 (1960); CA 55, 4253g

3482. Jamieson, G. R., "Relative Detector Response in Gas Chromatography. III. Aliphatic Esters," J. Chromatog., 4, 420-422 (1960); CA 55, 17374d

3483. Jamieson, G. R., "Relative Detector Response in Gas Chromatography. IV. Ethers and Acetals," J. Chromatog., 8, 544-546 (1962); CA 58, 5017f

3484. Jamieson, G. R., Relative Response of a Thermal Conductivity Detector in Gas Chromatography," Journees Intern. Etude Methodes, Separation Immediate Chromatog., Paris, 1961, 85-89 (Pub. 1962); CA 59, 2144h

3485. Janak, J., "Chromatographic Semimicroanalysis of Gases," Chem. listy, 47, 464-467 (1953); CA 48, 3196h

3486. Janak, J., "Chromatographic Semimicroanalysis of Gases. I. Theory and Method of Analysis," Chem. Listy, 47, 817-827 (1953); CA 48, 3197a

3487. Janak, J., "Chromatographic Semimicroanalysis of Gases. II. Analysis of Natural Gas and the Determination of Methane in Mine Gas," Chem. listy, 47, 828-836 (1953); CA 48, 3197b

3488. Janak, J., "Chromatographic Semimicroanalysis of Gases. III. Analysis of Hydrogen Gases," Chem. listy, 47, 837-841 (1953); CA 48, 3197c

3489. Janak, J., "Chromatographic Semimicroanalysis of Gases. IV. Analysis of Gaseous Paraffins," Chem. listy, 47, 1184-1189 (1953); CA 48, 3853f

3490. Janak, J., "Chromatographic Semimicroanalysis of Gases. VI. Analysis of Rare Gases," Chem. listy, 47, 1348-1353 (1953); CA 48, 3854b

3491. Janak, J., "Chromatographic Semimicroanalysis of Gases," Collection Czechoslov. Chem. Communs., 18, 798-802 (1953); CA 48, 3196h

3492. Janak, J., "Chromatographic Semimicroanalysis of Gases," Collection Czechoslov. Chem. Communs., 19, 684-699 (1954); CA 49, 6777h

3493. Janak, J., "Chromatographic Semimicroanalysis of Gases," Collection Czechoslov. Chem. Communs., 19, 700-711 (1954); CA 49, 6777h

3494. Janak, J., "Chromatographic Semimicroanalysis of Gases. VI. Analysis of Rare Gases," Collection Czechoslov. Chem. Communs., 19, 912-924 (1954); CA 6777h

3495. Janak, J., "Use of Zeolites in Gas Chromatography. Preliminary Communication," Chem. listy, 49, 1403-1405 (1955); CA 50, 104b

3496. Janak, J., "Use of Zeolites in Gas Chromatography. Preliminary Communication," Collection Czechoslov. Chem. Communs., 20, 1241-1243 (1955); CA 50, 7664h

3497. Janak, J., "Chromatography Analysis and Separation of Gases," Paliva, 35, 357-416 (1955); CA 50, 8173c

3498. Janak, J., "Gas Chromatography on Various Adsorbents and with Liquid-Wetted Supports," 15th Intern. Congr. of Pure and Applied Chemistry, Lisbon, Portugal, September 1956

3499. Janak, J., "Chromatographic Analysis of Connate Water in Oil Prospecting," Prace Ustavu pro Naftovy Vyskum Ser. E, No. 17/21, 5-24 (1956); CA 50, 16083d

3500. Janak, J., "Systematic Chromatographic Microanalysis of Gases," Mikrochim. Acta, 1956, 1038-1049; CA 50, 10377e

3501. Janak, J., "New Ways in Gas Analysis with the Help of Gas Chromatography," Proc. Gas-Kolloquium, Hamburg, West Germany, November 1956, pp. 2-13

3502. Janak, J., "New Methods in Gas Analysis. Gas Chromatography," Chem. Tech. (Berlin), 8, 125-132 (1956); CA 52, 155d

3503. Janak, J., "The Concept of the Chromatographic Spectrum of Gases and Volatile Materials," in "Vapour Phase Chromatography," edited by D. H. Desty, Academic Press, Inc., New York, 1957, pp. 235-246

3504. Janak, J., "Vapour-Phase Chromatography on Zeolites," in "Vapour Phase Chromatography," edited by D. H. Desty, Academic Press, Inc., New York, 1957, pp. 247-255

3505. Janak, J., "New Methods of Gas Analysis by Gas Chromatography," Erdol u. Kohle, 10, 442-444

(1957); CA <u>52</u>, 155d

3506. Janak, J., "New Trends in Gas Chromatography," Chemie (Prague), <u>9</u>, 20-24 (1957); CA <u>52</u>, 7924h

3507. Janak, J., "Use of Gas Chromatography in the Constitution Proofs of Unvolatile Materials," 3rd Intern. Conf. on Anal. Chem., Prague, Czech., September 1959

3508. Janak, J., "Contemporary Developments in Gas Chromatography Abroad and in Czechoslovakia," 3rd Intern. Conf. on Anal. Chem., Prague, Czech., September 1959

3509. Janak, J., "Chromatography in the Gas-Solid System," Ann. N. Y. Acad. Sci., <u>72</u>, Art. 13, 606-612 (1959); CA <u>53</u>, 14818b

3510. Janak, J., Statistic Comparison of the Methods for Quantitative Evaluation of the Elution Curves of Gas Chromatographs," in "Gas Chromatographie 1959," edited by R. E. Kaiser and H. G. Struppe, Akademie Verlag, Berlin, 1959, pp. 157-163

3511. Janak, J., "Identification by Pyrolysis of Nonvolatile Substances," Gas Chromatograpny Discussion Group, Bristol, Eng., 1959

3512. Janak, J., "Identification of the Structure of Nonvolatile Organic Substances by Gas Chromatography of Pyrolytic Products," Nature, <u>185</u>, 684-686 (1960); CA <u>54</u>, 16225d

3513. Janak, J., "Statistical Evaluation of Methods Used for Quantitative Measurement of Recorded Differential Curves Obtained in Gas Chromatography," J. Chromatog., <u>3</u>, 308-312 (1960); CA <u>54</u>, 19085i

3514. Janak, J., "Identification of Organic Substances by Means of Defined Pyrolysis in Combination with Gas Chromatography. I. Basic Principle and Technique of the Analysis," Collection Czechoslov. Chem. Communs., <u>25</u>, 1780-1789 (1960)

3515. Janak, J., "Identification of Organic Substances by the Gas Chromatographic Analysis of Their Pyrolysis Products," in "Gas Chromatography 1960," edited by R. P. W. Scott, Butterworths, London, 1960, pp. 387-400

3516. Janak, I., "Identification of Organic Substances by Gas Chromatographic Analysis of Their Pyrolysis Products," Gas Chromatog., Proc. Symposium, 3rd, Edinburgh, <u>1960</u>, 387-400; CA <u>56</u>, 8006d

3517. Janak, J., "Chromatography of Nonhydrogen Gases," in "Chromatography," edited by E. Heftmann, Reinhold Publishing Corp., New York, 1961, Chapter 26

3518. Janak, J., "Chromatographic Identification of Complex Mixtures of High-Boiling Compounds," Nature, <u>195</u>, 696-697 (1962); CA <u>57</u>, 12798f

3519. Janak, J., "Gas Chromatography Symposium, Brno, Czechoslovakia, June 1962," J. Gas Chromatog., <u>1</u>, No. 4, 6-10 (1963)

3520. Janak, J., "Progress in Gas Chromatography in the Light of the 4th International Symposium 1962 in Hamburg," Gas Chromatog. Symp., Brno, Czech., June 1962; Abstr., J. Gas Chromatog., <u>1</u>, No. 4, 7 (1963)

3521. Janak, J., Dobiasova, M., and Veres, K., "Separation of Saturated and Unsaturated C_4 and C_6 Fatty Acids by Gas Chromatography," Collection Czechoslov. Chem. Communs., <u>25</u>, 1566-1572 (1960); CA <u>56</u>, 7973i

3522. Janak, J., and Hrivnac, M., "Gas Chromatography of Nitrogen-Containing Heterocyclic Compounds. I. Separation and Analysis of Quinoline and Higher Pyridine Bases and Indoles by Gas-Liquid Chromatography," Collection Czechoslov. Chem. Communs., <u>25</u>, 1557-1565 (1960); CA <u>57</u>, 1560g

3523. Janak, J., and Hrivnac, M., "Use of the 'π'-Electron Interaction for Selective Separation of Some Quinoline Bases, and Aromatic and Heterocyclic Hydrocarbons from Coal Tar Distillation by Gas Chromatography," J. Chromatog., <u>3</u>, 297-302 (1960); CA <u>54</u>, 22640h

3524. Janak, J., and Komers, R., "Relative Retention Volumes of Phenols on Various Stationary Phases (Chromatographic Data Table 16)," J. Chromatog., <u>2</u>, D8 (1959); cf. Z. Anal. Chem., <u>164</u>, 69 (1958)

3525. Janak, J., and Komers, R., "Evaluation of Some Sugars as Stationary Phases for Separation of Phenols by Gas Chromatography," in "Gas Chromatography 1958," edited by D. H. Desty, Academic Press, Inc., New York, 1958, pp. 343-350; CA <u>53</u>, 17940i

3526. Janak, J., and Komers, R., "Separation of Mono- and Di-Functional Phenols by Means of Gas Chromatography," Z. Anal. Chem., <u>165</u>, 69-72 (1958); CA <u>53</u>, 6910b

3527. Janak, J., and Komers, R., "Separation and Analysis of Dihydric Phenols by Gas Chromatography," Collection Czechoslov. Chem. Communs., <u>24</u>, 1960-1966 (1959); CA <u>53</u>, 14816h

3528. Janak, J., Komers, R., and Sima, J., "Gas Chromatography of Monohydric Phenols," Chem. listy, <u>52</u>, 2296-2310 (1958); CA <u>53</u>, 9884e

3529. Janak, J., Komers, R., and Sima, J., "Gas Chromatography of Monohydric Phenols," Collection Czechoslov. Chem. Communs., <u>24</u>, 1492-1508 (1959); CA <u>54</u>, 149d

3530. Janak, J., and Krejci, M., "Problems of Chromatography in the System Gas/Absorbent," in "Gas Chromatographie 1958," edited by H. P. Angele, Akademie Verlag, Berlin, 1959, pp. 15-27

3531. Janak, J., and Krejci, M., "Some Problems of the System Gas-Adsorbent," Abhandl. Deut. Akad. Wiss. Berlin, K. Chem., Geol., Biol., <u>1959</u>, No. 9, 15-27; CA <u>58</u>, 5018a

3532.	Janak, J., Krejci, M., and Dubsky, H. E., "Zeolites in Gas Chromatography. I. Separation and Analysis in Mixtures of Hydrogen, Nitrogen, Carbon Monoxide, and Methane," Chem. listy, 52, 1099-1107 (1957); CA 52, 15331h

3533.	Janak, J., Krejci, M., and Dubsky, H. E., "Properties of Ca Zeolite as an Adsorbent for Gas Chromatography," Ann. N. Y. Acad. Sci., 72, 731-738 (1959); CA 54, 6252i

3534.	Janak, J., Krejci, M., and Dubsky, H. E., "Use of Zeolites in Gas Chromatography. I. Separation and Analysis of Mixtures of Hydrogen, Oxygen, Nitrogen, Carbon Monoxide, and Methane," Collection Czechoslov. Chem. Commun., 24, 1080-1090 (1959); CA 53, 11099a

3535.	Janak, J., Nedorost, M., and Bubenikova, V., "Chromatographic Semimicroanalysis of Gases. XIII. Separation of Chlorine, Bromine, and Iodine," Chem. listy, 51, 890-894 (1957)

3536.	Janak, J., Nedorost, M., and Bubenikova, V., "Chromatographic Semimicroanalysis of Gases. XIII. Separation of Chlorine, Bromine, and Iodine," Collection Czechoslov. Chem. Communs., 22, 1799-1804 (1957); CA 51, 11924e

3537.	Janak, J., and Novak, J., "Chromatographic Semimicroanalysis of Gases. XIV. Direct Determination of Individual Gaseous Paraffins and Olefins in 1,3-Butadiene," Chem. listy, 51, 1832-1837 (1957); CA 52, 1860d

3538.	Janak, J., and Novak, J., "Chromatographic Semimicroanalysis of Gases. XIV. Direct Determination of Individual Gaseous Paraffins and Olefins in 1,3-Butadiene," Collection Czechoslov. Chem. Communs., 24, 384-390 (1959); CA 53, 7859e

3539.	Janak, J., Novak, J., and Sulovsky, J., "Separation of Substituted Malonic Acid Esters by Gas Chromatography and a New Method of Identification," Collection Czechoslov. Chem. Communs., 27, 2541-2549 (1962); CA 58, 2827b

3540.	Janak, J., Novak, J., and Sulovsky, J., "Identification of Materials by Means of Two Different Detection Systems," Gas Chromatog. Symp., Brno. Czech., June 1962; Abstr., J. Gas Chromatog., 1, No. 4, 7 (1963)

3541.	Janak, J., Novak, J., and Zollner, G., "Separation of Ethylamines in the Presence of Ammonia and Water by Gas-Liquid Chromatography," Collection Czechoslov. Chem. Communs., 27, 2628-2636 (1962); CA 58, 3871a

3542.	Janak, J., Novak, J., and Zollner, G., "Separation of Ethylamines in the Presence of Ammonia and Water by Means of Gas-Liquid Chromatography," Gas Chromatog. Symp. Brno, Czech., June 1962; Abstr., J. Gas Chromatog. 1, No. 4, 9 (1963)

3543.	Janak, J., and Paralova, I., "Chromatographic Semimicroanalysis of Gases. VII. Analysis of Dissolved Gases," Chem. listy, 47, 1476-1480 (1953)

3544.	Janak, J., and Paralova, I., "Chromatographic Semimicroanalysis of Gases. VII. Analysis of Dissolved Gases," Collection Czechoslov. Chem. Communs. 20, 336-341 (1955); CA 48, 3854c

3545.	Janak, J., and Rusek, M., "Chromatographic Semimicroanalysis of Gases. V. Analysis of Unsaturated C_2 and C_3 Hydrocarbons," Chem. listy, 47, 1190-1196 (1953); CA 48, 3853g

3546.	Janak, J., and Rusek, M., "Chromatographic Semimicroanalysis of Gases. VIII. Separation and Analysis of Some Halogenated Hydrocarbons," Chem. listy 48, 207-212 (1954); CA 48, 6321e

3547.	Janak, J., and Rusek, M., "Chromatographic Semimicroanalysis of Gases. IX. Determination of Nitrous Oxide," Chem. listy, 48, 397-400 (1954); CA 48, 7489e

3548.	Janak, J., and Rusek, M., "Chromatographic Semimicroanalysis of Gases. IV. Analysis of Gaseous Paraffins; V. Analysis of Unsaturated C_2 and C_3 Hydrocarbons," Collection Czechoslov. Chem. Communs., 19, 700-711 (1954); CA 49, 6777h

3549.	Janak, J., and Rusek, M., "Chromatographic Semimicroanalysis of Gases. XI. Direct Determination of Individual Olefins in Gases," Chem. listy, 49, 191-199 (1955); CA 49, 8047g

3550.	Janak, J., and Rusek, M., "Chromatographic Semimicroanalysis of Gases. IX. Determination of Nitrous Oxide," Collection Czechoslov. Chem. Communs., 20, 343-347 (1955); CA 49, 11498a

3551.	Janak, J., and Rusek, M., "Chromatographic Semimicroanalysis of Gases. XI. Direct Determination of Individual Olefins in Gases," Collection Czechoslov. Chem. Communs., 20, 923-932 (1955); CA 50, 4715a

3552.	Janak, J., Rusek, M., and Lazarev, A., "Chromatographic Semimicroanalysis of Gases. XII. Separation and Analysis of Gaseous Cycloparaffins," Chem. listy, 49, 700-705 (1955); CA 49, 12201h

3553.	Janak, J., Rusek, M., and Lazarev, A., "Chromatographic Semimicroanalysis of Gases. XII. The Separation and Analysis of Gaseous Hydrocarbons," Collection Czechoslov. Chem. Communs., 20, 1199-1204 (1955); CA 50, 7672a

3554.	Janak, J., and Tesarik, K., "Chromatographic Semimicroanalysis of Gases X. Determination of Minute and Trace Amounts of Helium, Neon, and Hydrogen in Gases," Chem. listy, 48, 1051-1057 (1954); CA 48, 13536d

3555.	Janak, J., and Tesarik K., "Chromatographic Semimicroanalysis of Gases. X. Determination of Minute and Trace Amounts of Helium, Neon, and Hydrogen in Gases," Collection Czechoslov. Chem. Communs., 20, 348-355 (1955); CA 49, 11498b

3556.	Janak, J., and Tesarik, K., "Chromatographic Semimicroanalysis of Gases. XV. Automatization of the Measuring Unit of a Gas Chromatograph," Chem. listy, 51, 2048-2054 (1957); CA 52, 2652b

3557. Janak, J., and Tesarik, K., "Automation of Gas Chromatography for Volume Measurement," Z. Anal. Chem., 164, 62-69 (1958); CA 53, 6699e

3558. Janak, J., and Tesarik, K., "Chromatographic Semimicroanalysis of Gases. XV. Automation of the Measuring Unit of a Gas Chromatograph," Collection Czechoslov. Chem. Communs., 24, 536-544 (1959); CA 53, 7859e

3559. Janak. J., and Vojtovic, K. V., "Gas Chromatographic Study of the Composition of Benzol Fore-Fractions," Chem. prumsyl., 8, 127-131 (1958); CA 52, 19088c

3560. Janda, J., "Isochoric Gas Discharging Instrument," Chem. Zvesti, 13, 317-319 (1959); CA 53, 20939i

3561. Jangaard, P. M., Burgher, R. D., and Ackman, R. G., "A Preliminary Investigation of the Blubber Oil from the Atlantic Bottlenose Whale," J. Fisheries Res. Board Can., 20, 245-247 (1963); CA 58, 12777h

3562. Janitzki, U., "Gas Chromatography Identification of Volatile Substances in Blood After Respiratory Absorption," Deut. Z. Ges. Gerichtl. Med., 52, 22-27 (1961); CA 57, 7540c

3563. Jarboe, C. H., and Rosene, C. J., "Volatile Products of Pyrolysis of Nicotine," J. Chem. Soc., 1961, 2455-2458; CA 55, 25942e

3564. Jart, A., "The Fatty Acid Composition of Shea Butter and Olive Seed Oil," Acta Chem. Scand., 13, 1723-1724 (1959); CA 56, 9173h

3565. Jart, A., "Separation of Fatty Acid Esters by Gas Chromatography," Fette, Seifen, Anstrichmittel, 61, 541-546 (1959); CA 54, 3999h

3566. Jart, A., "The Infrared Absorption Spectra of Some Monounsaturated and Saturated Fatty Acids and Esters," Acta Chem. Scand., 14, 1867-1878 (1960); CA 56, 6805d

3567. Jart, A., "Gas-Liquid Chromatographic Retention Data for Some Isothiocyanates," Acta Chem. Scand., 15, 1223-1230 (1961); CA 56, 10860a

3568. Jart, A., Funch, J. P., and Dam, H., "The Composition of Rat Milk Fat," Acta Chem. Scand. 1910-1912 (1959); CA 56, 12133b

3569. Jaulmes, P., and Mestres, R., "Theory of Gas Chromatography," Compt. rend., 248, 2752-2754 (1959); CA 54, 39g

3570. Jaulmes, P., and Mestres, R., "The Theory of Vapor-Phase Chromatography," J. chim. phys., 56, 920-932 (1959); CA 55, 21749a

3571. Jaulmes, P., and Mestres, R., "Theory of Gas Chromatography," Bull. soc. chim. France, 1960, 789

3572. Jaureguiberry, G., and Wolff, R., "Presence of (+)-Sabinene in Oil of Nutmeg and Some Properties of This Hydrocarbon," Bull. soc. chim. France, 1962, 1985-1987; CA 58, 2319d

3573. Jaworski, M., "Chromatographic Determination of Aliphatic Hydrocarbons in Oxygen Compounds," Chem. Anal. (Warsaw), 6, 243-249 (1961); CA 55, 18451d

3574. Jaworski, M., "Use of Gas Chromatography in the Petrochemical Industry," Przemysl Chem., 41, 14-17 (1962); CA 56, 14523f

3575. Jay, B. E., and Wilson, R. H., "Adaptation of the Gas Adsorption Chromatographic Technique for Use in Respiratory Physiology," J. Appl. Physiol., 15, 298-302 (1960)

3576. Jay, B. E., Wilson, R. H., Doty, V., Pingree, H., and Hargis, B., "Gas Chromatographic Determination of Absorption Coefficients and Tensions of Gases in Solution," Anal. Chem., 34, 414-418 (1962); CA 57, 1212h

3577. Jayadevappa, E. S., and Hisatsune, I. C., "Liquid-Gas Infrared Intensities in Benzene," J. Karnatak Univ., 4, 21-25 (1960); CA 55, 9039b

3578. Jeffery, P. G., and Kipping, P. J., "The Determination of Constituents of Rocks and Minerals by Gas Chromatography. I. The Determination of Carbon Dioxide," Analyst, 87, 379-382 (1962); CA 57, 5301a, 57, 9212c

3579. Jeffery, P. G., and Kipping, P. J., "Determination of Carbon Dioxide and Nitrous Oxide in Solutions of Monoethanolamine," Analyst, 87, 594-595 (1962); CA 57, 11859b

3580. Jeffery, P. G., and Kipping, P. J., "The Determination of Constituents of Rocks and Minerals by Gas Chromatography. II. The Determination of Some Gaseous Constituents," Analyst, 88, 266-271 (1963); CA 59, 1078f

3581. Jelliffe, R.W., and Blankenhorn, D. H., "Gas Chromatography of Digitoxin, Digoxin, and Their Aglycones," 35th Scientific Session, Am. Heart Assoc., Cleveland, Ohio, October 1962; Abstr., Circulation, 26, 737 (1962)

3582. Jellinek, J. S., "Odor Quality," Drug Cosmetic Ind., 85, 38-39, 118-120 (1959); CA 54, 826i

3583. Jenard, H., "Volatile Constituents of Beer," Brewers Digest, 35, No. 4, 58-60 (1960); CA 55, 897a

3584. Jenkins, J. W., and Amburgey, J. M., "Determination of the Volatile Constitutents of Aerosols by Gas Chromatography, "Progr. Sci. Sect. Toilet Goods Assoc., No. 31, 19-21 (1959); Perfumery Essent. Oil Record, 50, 716 (1959); CA 53, 15851h

3585. Jenkins, P., "Modern Methods of Analysis. II. Gas Chromatography," Sci. Surface Coatings, 1962,

564-577; CA 58, 9327g

3586. Jenkins, P., "A Reliable System for the Introduction of Samples to Gas-Chromatographic Columns, Particularly Suited for Quantitative Capillary Column Analysis," Nature, 197, 72-73 (1963); CA 58, 6166h

3587. Jenkins, R. E., "Three Electronic Methods to Analyze Drilling Fluid Gases," Petrol. Eng., 31, No. 11, B75, 80, 82, 85, 88, 90, 96 (1959); CA 54, 869h

3588. Jennings, A. P. H., "Recording Integrator for Gas Chromatography," J. Sci. Instr., 38, 55-58 (1961); CA 56, 12692e

3589. Jennings, Jr., E. C., Curran, T. D., and Edwards, D. G., "Gas-Liquid Partition Chromatography. Determination of 2, 6-Di-tert-butyl-p-cresol on Antioxidant-Treated Paperboard," Anal. Chem., 30, 1946-1948 (1958); CA 53, 3691i; 133rd Natl. ACS Mtg., San Francisco, Calif., April 1958, Program Abstr., p. 48B

3590. Jennings, Jr., E. C., and Dimick, K. P., "Preparative Separation of Sex Hormones by Gas Chromatography," 139th Natl. ACS Mtg., St. Louis, Mo., March 1961, Program Abstr., pp. 33B and 8Q; Preprints, Div. Petrol. Chem., 6, No. 2B, 103-108 (1961); CA 57, 16984f

3591. Jennings, Jr., E. C., and Dimick, K. P., "Gas Chromatography of Pyrolytic Products of Purines and Pyrimidines," Anal. Chem., 34, 1543-1547 (1962); CA 58, 2827g; 13th Pittsburgh Conf. on Anal. Chem. & Appl. Spectroscopy, Pittsburgh, Pa., March 1962, Program Abstr., p. 58

3592. Jennings, W. G., "Application of Gas-Liquid Partition Chromatography to the Study of Volatile Flavor Compounds," J. Dairy Sci., 40, 271-279 (1957); CA 51, 10789d

3593. Jennings, W. G., "The Chemical Characterization of Flavor Volatiles from Bartlett Pears, With Some Attention to the Role of the Individual Fractions in Pear Flavor," Intern. Fruchsaft-Union, Ber. Wiss.-Tech. Komm., 4, 337-349 (1962); CA 58, 13057d

3594. Jennings, W. G., Leonard, S., and Pangborn, R. M., "Volatiles Contributing to the Flavor of Bartlett Pears," Food Technol., 14, 587-590 (1960)

3595. Jennings, W. G., Viljhalmsson, S., and Dunkley, W. L., "Direct Gas Chromatography of Milk Vapors," J. Food Sci., 27, 306-308 (1962); Biol. Abstr., 40, 9692

3596. Jennings, W. G., and Wrolstad, R. E., "Volatile Constituents of Black Pepper," J. Food Sci., 26, 499-509 (1961); CA 56, 2747b

3597. Jensen, P. W., "Application of Gas Chromatography to Polymers," and "Quantitative Analysis," 1st and 2nd Annual Gas Chromatography Institutes, Canisius College, Buffalo, N. Y., May 1959 and April 1960

3598. Jensen, R., and Leslie, S. W., "A Technique for the Identification of Volatile Flavor Components in Foods," 14th Annual Mid-America Spectr. Symp., Chicago, Ill., May 1963; pub. in "Developments in Applied Spectroscopy," Vol. 2, edited by J. R. Ferraro and J. S. Ziomek, Plenum Press, N. Y., (1963) pp. 426-430

3599. Jensen, R. G., and Gander, G. W., "Fatty Acid Composition of the Monoglycerides from Lipolyzed Milk Fat," J. Dairy Sci., 43, 1758-1761 (1960); CA 55, 9710i

3600. Jensen, R. G., Gander, G. W., and Sampugna, J., "Fatty Acid Composition of Diglycerides from Lipolyzed Milk Fat," 56th Annual Mtg., Am. Dairy Sci. Assoc., Madison, Wisc., June 1961; Abstr., J. Dairy Sci., 44, 1169 (1961)

3601. Jensen, R. G., Gander, G. W., and Sampugna, J., "Fatty Acid Composition of the Lipids from Pooled Raw Milk," J. Dairy Sci., 45, 329-331 (1962)

3602. Jensen, R. G., Gander, G. W., Sampugna, J., and Forster, T. L., "Lipolysis by a Beta-Esterase Preparation from Milk," J. Dairy Sci., 44, 943-944 (1961); CA 55, 20249g

3603. Jensen, R. G., and Sampugna, J., "Identification of Milk Fatty Acids by Gas-Liquid and Thin-Layer Chromatography," J. Dairy Sci., 45, 435-437 (1962)

3604. Jensen, R. G., and Sampugna, J., "Lipolysis of Glyceryl 1-Oleate 2, 3-Di-caproate by a Beta-Esterase Concentrate from Milk," 57th Annual Mtg., Am. Dairy Sci. Assoc., College Park, Md., June 1962; Abstr., J. Dairy Sci., 45, 646 (1962)

3605. Jensen, R. G., Sampugna, J., and Gander, G. W., "Glyceride and Fatty Acid Composition of Some Mono-Diglyceride Ice Cream Emulsifiers," J. Dairy Sci., 44, 1057-1069 (1961); CA 55, 20252g

3606. Jensen, R. G., Sampugna, J., and Gander, G. W., "Fatty Acid Composition of the Diglycerides from Lipolyzed Milk Fat," J. Dairy Sci., 44, 1983-1988 (1961)

3607. Jensen, R. G., Sampugna, J., Parry, Jr., R. M., and Forster, T. L., "Absence of Fatty Acid Specificity During Lipolysis of Some Synthetic Triglycerides by Beta-Esterase Preparations from Milk," J. Dairy Res., 45, 842-847 (1962); CA 58, 10462d

3608. Jentoft, R. E., Johnson, J. F., and Carlstrom, A. A., "Preparative Gas Chromatography," 144th Natl. ACS Mtg., Los Angeles, Calif., March-April 1963, Program Abstr., p. 17J

3609. Jentzsch, D., "New Apparatus for Gas Chromatography," Dechema Monograph., 43, 231-240 (1962); CA 58, 10703d

3610. Jentzsch, D., and Bergmann, G., "Relation Between Gas Chromatographic Retention Ratios and Vapor-Pressure Ratios. I. Aromatic Hydrocarbons," Z. Anal. Chem., 165, 401-415 (1959); CA 53, 16795f

3611. Jentzsch, D., and Bergmann, G., "Characteristics of the Separation Effect of the Stationary Phase in Gas Chromatography," Z. Anal. Chem., 170, 239-255 (1959); CA 54, 3043g

3612. Jentzsch, D., Bogen, P., and Friedrich, K., "The Use of Flame Ionization Detectors in Combination with Thermal Conductivity Cells for Gas Chromatographic Analyses," German Chemists' Society Mtg., Stuttgart, Germany, April 1960; Abstr., Angew. Chem., 72, 587 (1960)

3613. Jentzsch, D., and Friedrich, K., "Gas Chromatographic Procedures. I. Use of a Flame Ionization Detector," Z. Anal. Chem., 180, 96-109 (1961); CA 55, 17343h

3614. Jentzsch, D., and Hövermann, W., "New Applications of Golay Columns with Special Wall Treatments," 14th Pittsburgh Conf. on Anal. Chem. & Appl. Spectroscopy, Pittsburgh, Pa., March 1963, Program Abstr., p. 54

3615. Jesse, W. P., and Sadauskis, J., "Ionization by α-Particles in Mixtures of Gases," Phys. Rev., 100, 1755-1762 (1955); cf., ibid., 94, 764 (1954); CA 50, 4625f

3616. Jesting, E., and Bang, H. O., "Analysis of Unesterified Fatty Acids by Gas-Liquid Chromatography," Danish Med. Bull., 8, 166-168 (1961); CA 57, 11i

3617. Jesting, E., and Bang, H. O., "Gas-Liquid Chromatography Applied as a Control for Methylation Processes," Danish Med. Bull., 8, 169-170 (1961); CA 57, 12308g

3618. Jeung, E., and Helwig, H. L., "Methods for Analysis of C_1-C_4 Hydrocarbons in Atmospheric Samples by Gas Chromatography," 144th Natl. ACS Mtg., Los Angeles, Calif., March-April, 1963, Program Abstr., p. 26R

3619. Jewett, P., and Horrocks, B. J., "Improved Gas Chromatography of Unesterified Fatty Acids," Nature, 192, 966-967 (1961); CA 56, 9388b

3620. Jikilee, B. M., and Rowland, F. S., "Proportional Counter Assay of Radioactive Components in Gas Chromatographic Streams," 137th Natl. ACS Mtg., Cleveland, Ohio, April 1960, Program Abstr., p. 26B

3621. Johansson, G., "Gas Analysis by Use of Microwaves," Anal. Chem., 34, 914-916 (1962); CA 57, 5283f

3622. Johncock, P., and Musgrave, W. K. R., "Chlorofluorocyclohexanes from Benzene," Chem. & Ind. (London), 1959, 1314; CA 54, 5500i

3623. Johns, T., "The Behavior of the Solid Support in Gas-Liquid Partition Chromatography," in "Gas Chromatography," edited by V. J. Coates, H. J. Noebels, and I. S. Fagerson, Academic Press, Inc., New York, 1958, pp. 31-39

3624. Johns, T., "Beckman Gas Chromatography Application Manual," Bulletin 756, Beckman Instruments, Inc., Fullerton, Calif., 1959

3625. Johns, T., "'Two-in One' Gas Chromatography. Unique Valve Readily Adopts Beckman GC-1 and GC-2 Gas Chromatographs for Advantageous Two-Column Analysis." Application Data Sheet GC-82-MI, Beckman/Scientific and Process Instruments Division, Beckman Instruments, Inc., Fullerton, Calif., 1959

3626. Johns, T., "Analysis of Ethylene Oxide Sterilizing Mixtures by Gas Chromatography," Application Data Sheet GC-98-F, Beckman/Scientific and Process Instruments Division, Beckman Instruments, Inc., Fullerton, Calif., 1959

3627. Johns, T., "The Analysis of Chlorine Feed Gas by Gas Chromatography," Application Data Sheet GC-8064, Beckman/Scientific and Process Instruments Division, Beckman Instruments, Inc., Fullerton, Calif., 1960

3628. Johns, T., "The Analysis of Natural Gas Samples by Dual Stage Gas Chromatography," Application Data Sheet GC-8066-P, Beckman/Scientific and Process Instruments Division, Beckman Instruments, Inc., Fullerton, Calif., 1960

3629. Johns, T., "Peak Area and Peak Height Measurement Techniques Increase Quantitative Accuracy in Gas Chromatography," Application Data Sheet GC-86-MI, Beckman/Scientific and Process Instruments Division, Beckman Instruments, Inc., Fullerton, Calif., 1960

3630. Johns, T., "Application of Gas Chromatography to the Analysis of Food Additives," Proc. of Symposium and Workshop, Instrumental Methods for Analysis of Food Additives, East Lansing, Mich., March 1960, pp. 201-207; CA 57, 3842a

3631. Johns, T., "Purification and Identification of the Components of the Complex Organic Materials," in "Gas Chromatography 1960," edited by R. P. W. Scott, Butterworths, Washington, D. C., 1960, pp. 242-250; Beckman Reprint R-6168

3632. Johns, T., "Purification and Identification of the Components of Complex Organic Materials," Gas Chromatog. Proc. Symposium, 3rd, Edinburgh, 1960, 242-250; CA 56, 5386e

3633. Johns, T., "Programmed-Temperature Gas Chromatography - Theory, Control, and Optimization of Operating Parameters," 2nd Symp. on Gas Chromatography, Toronto Section, CIC, Toronto, Ontario, February 1962

3634. Johns, T., "Programmed Chromatography and Chromatographic Units for Collection of Samples," 4th Annual Gas Chromatography Institute, Canisius College, Buffalo, N. Y., April 1962

3635. Johns, T., "Recent Developments in Gas Chromatography," 4th Annual Gas Chromatography Institute, Canisius College, Buffalo, N. Y., April 1962

3636. Johns, T., "Purification of Compounds by Preparative Gas Chromatography," 13th Annual Mid-America Spectroscopy Symp., Chicago, Ill., April-May 1962

3637. Johns, T., "Applications of an Electrical Conductivity Detector in Gas Chromatography," 14th Pittsburgh Conf. on Anal. Chem. & Appl. Spectroscopy, Pittsburgh, Pa., March 1963, Program Abstr., p. 66

3638. Johns, T., "Selective Detection and Identification of Pesticide Residues," 144th Natl. ACS Mtg., Los Angeles, Calif., March-April 1963, Program Abstr., p. 3A

3639. Johns, T., Burnell, M. R., and Carle, D. W., "Selected Application With a New Preparative Gas Chromatograph," Gas Chromatog. Intern. Symposium, 2nd, East Lansing, Mich., 1959, 207-223 (Pub. 1961); CA 55, 19417h

3640. Johns, T., Burnell, M. R., and Carle, D. W., "Selected Applications With a New Prepatative Gas Chromatograph," in "Gas Chromatography," edited by H. J. Noebels, R. F. Wall, and N. Brenner, Academic Press, Inc., New York, 1961, pp. 207-223

3641. Johns, T., Cadman, J., and Ulrich, W. F., "The Determination of Blood Alcohol by Gas Chromatography," 9th Pittsburgh Conf. on Anal. Chem. & Appl. Spectroscopy, Pittsburgh, Pa., March 1958, Program Abstr., p. 45

3642. Johns, T., and Lawson, R. H., "Analysis of Alcoholic Beverages Utilizing Preparative Gas Chromatography," 11th Pittsburgh Conf. on Anal. Chem. & Appl. Spectroscopy, Pittsburgh, Pa., February-March 1960, Program Abstr., p. 41

3643. Johns, T., and Morris, R., "Gas Chromatographic Analysis of Insecticides with a Selective Detector, 18th Southwest Regional Mtg., ACS, Dallas, Texas, December 1962

3644. Johns, T., and Thompson, B., "Gas Chromatographic Determination of Blood Gases," Analyzer, 4, No. 2, 13-15 (1963)

3645. Johnson, D. E., and Meister, A., "Gas Chromatography of Amino Acid Derivatives," 138th Natl. ACS Mtg., New York, N. Y. September 1960, Program Abstr., p. 59C

3646. Johnson, D. E., Scott, S. J., and Meister, A., "Gas-Liquid Chromatography of Amino Acid Derivatives," Anal. Chem., 33, 669-673 (1961); CA 55, 18452f; 140th Natl. ACS Mtg., Chicago, Ill., September 1961, Program Abstr., p. 48C

3647. Johnson, D. F., Bennett, R. D., and Heftmann, E., "Cholesterol in Higher Plants," Science, 140, 198-199 (1963); CA 59, 1961e

3648. Johnson, E. A., "The Dimethyldiphenyls," J. Chem. Soc., 1957, 4155-4156; CA 52, 4566g

3649. Johnson, E. A., "The Study of Stearic Effects in Alkylbiphenyls by Vapor-Phase Chromatography," Steric Effects Conjugated Systems, Proc. Symposium, Hull, 1958, 174-181; CA 53, 14048i

3650. Johnson, E. A., Childs, D. G., and Beaven, G. H., "Some Comments on the Construction and Operation of the Gas Density Balance," J. Chromatog., 4, 429-434 (1960); CA 55, 20527i

3651. Johnson, H. S., and Jain, K., "Sulfur Dioxide Sensitized Photochemical Oxidation of Hydrocarbons," Science, 131, 1523-1524 (1960)

3652. Johnson, Jr., H. W., "Liquid Substrates and the Gas-Liquid Partition Column," 129th Natl. ACS Mtg., Dallas, Texas, April 1956, Program Abstr., p. 14B

3653. Johnson, Jr., H. W., "Storage and Complete Automatic Computation of Gas Chromatographic Data," Anal. Chem., 35, 521-526 (1963); CA 59, 1068a; Univ. of Houston Intern. Symp. on Advances in Gas Chromatography, Houston, Texas, January 1963

3654. Johnson, Jr., H. W., "Collection and Computer Analysis of Gas Chromatographic Data by Digital Techniques," 18th Annual ISA Instrument-Automation Conference and Exhibit, Chicago, Ill., September 1963

3655. Johnson, Jr., H. W., and Stross, F. H., "Terms and Units in Gas Chromatography," Anal. Chem., 30, 1586-1589 (1958); CA 53, 6867h

3656. Johnson, Jr., H. W., and Stross, F. H., "The Determination of Column Efficiency in Gas-Liquid Chromatography," 9th Pittsburgh Conf. on Anal. Chem. & Appl. Spectroscopy, Pittsburgh, Pa., March 1958, Program Abstr., p. 49

3657. Johnson, Jr., H. W., and Stross, F. H., "Gas Chromatography. Determination of Column Efficiency," Anal. Chem., 31, 357-365 (1959); CA 53, 13689e

3658. Johnson, Jr., H. W., and Stross, F. H., "Gas and Liquid Elution Chromatography. Quantitative Detector Evaluation," Anal. Chem., 31, 1206-1211 (1959); CA 53, 18727d

3659. Johnson, J. F., Klaver, R.F., Baumann, F., and Beach, J. Y., "Automation in an Analytical Gas Chromatographic Laboratory. Use of Chart Integrator, Automatic Attenuators, and Automatic Base Line Reset," Journees Intern. Etude Methodes, Separation Immediate Chromatog., Paris, 1961, 235-240 (Pub. 1962); CA 59, 2141g

3660. Johnson, R. E., "The Ionization Detector in Vapor Phase Chromatography," 1st Annual Gas Chromatography Institute, Canisius College, Buffalo, N. Y., May 1959

3661. Johnson, R. E., "Detection of Stable Gases with the Argon Ionization Detector," 1960 ISA Symp. on Instr. Methods of Analysis, Montreal, Canada, June 1960, ISA Proc., 6, C10-1-C10-6

3662. Johnson, R. E., "Gas Chromatograph Detectors," in "Progress in Industrial Gas Chromatography," Vol. 1, edited by H. A. Szymanski, Plenum Press, Inc., New York, 1961, pp. 163-199

3663. Johnson, R. E., "Observations on Effects of Parameters in Ionization Detection System Gas Chromatography," 9th Detroit Anachem Conf., Detroit, Mich., October 1961, Program Abstr., p. 26

3664. Johnson, R. E., "Theory and Comparative Performance of Argon and Hydrogen-Flame Ionization Detectors for Halogenated Hydrocarbons and Inorganic Compounds," 2nd Symp. on Gas Chromatography, Toronto Section, CIC, Toronto, Ontario, February 1962

3665. Johnson, R. E., "Elementary Theory of Detectors," 4th Annual Gas Chromatography Institute, Canisius College, Buffalo, N. Y., April 1962; pub. as Item 3668

3666. Johnson, R. E., "Detectors for Gas Chromatography," 13th Annual Fisk Univ. Infrared Institute, Gas Chromatography Session, Nashville, Tenn., August 1962

3667. Johnson, R. E., "Bibliography - Selected Medical & Biochemical Applications of Gas Chromatography," Barber-Colman Co., Rockford, Ill., 1963

3668. Johnson, R. E., "Gas Chromatography Detectors," in "Lectures on Gas Chromatography 1962," edited by H. A. Szymanski, Plenum Press, Inc., New York, 1963, pp. 65-86

3669. Johnson, R. E., "Elementary Theory of Detectors," 5th Annual Gas Chromatography Institute, Canisius College, Buffalo, N. Y., April 1963

3670. Johnson, R. E., and Lantz, C. D., "Ionization Detection Systems for Gas Chromatography," 7th Detroit Anachem Conf., Detroil, Mich., October 1959

3671. Johnston, H. L., and Grilly, E. R., "The Thermal Conductivities of Eight Common Gases [O_2, N_2, CO, NO, H_2, He, N_2O, CO_2, and CH_4] Between 80° and 380°K.," J. Chem. Phys., 14, 233-238 (1946); CA 40, 3951[6]

3672. Johnston, Jr., J. W., "Current Problems in Olefaction," Georgetown Med. Bull., 13, 112-117 (1959)

3673. Johnston, P. V., Kopaczyk, K. C., and Kummerow, F. A., "Effects of Pyridoxine Deficiency on Fatty Acid Composition of Carcass and Brain Lipids in the Rat," J. Nutrition, 74, 96-102 (1961)

3674. Johnston, P. V., and Kummerow, F. A., "Gas-Liquid Chromatography of Methyl Esters of Fatty Acids from Human and Chicken Brain Lipides," Proc. Soc. Exptl. Biol. Med., 104, 201-205 (1960); CA 54, 17493a

3675. Johnston, V. D., "Instrumental Methods of Analyses Save Time, Give More Information," Givaudanian, No. 1, 3-6 (1959)

3676. Johnston, V. D., "Instrumental Methods of Analysis," Soap, Perfumery & Cosmetics, 32, 621-623 (1959)

3677. Johnston, V. D., "Instrumental Methods of Analysis and the Perfumer," 5th Annual Symp., Am. Soc. of Perfumers, New York, N. Y., April 1959; Abstr., Perfumery Essent. Oil Record, 50, 607 (1959)

3678. Johnston, V. D., "Evaluation of a New Gas Chromatographic Instrument," Givaudanian, 2, 10 (February 1960)

3679. Johnston, V. D., "The Analyst's View. Safrole - Its Detection and Identification," Givaudanian, 2, 6-7 (September 1960)

3680. Johnston, V. D., "On Trouble Shooting in Gas Chromatographic Instruments," Givaudanian, 3, 9-10 (June 1961)

3681. Johnston, V. D., "Gas Chromatography Aids Perfumers," Instrumentation, 14, No. 3, 18-20 (1961)

3682. Johnston, V. D., "Chemical and Physical Methods of Analysis and Their Basic Significance," Am. Perfumer, 77, No. 3, 20 (1962); CA 57, 2354i

3683. Johnston, V. D., "Preparative Gas Chromatography. An Evaluation," Givaudanian, May 1963, 8-10

3684. Johnstone, R. A. W., and Douglas, A. G., "Detector for Gas-Liquid Chromatography," Chem. & Ind. (London), 1959, 154; CA 53, 20937a

3685. Johnstone, R. A. W., and Plimmer, J. R., "The Chemical Constituents of Tobacco and Tobacco Smoke," Chem. Revs., 59, 885-936 (1959); CA 54, 1809b

3686. Johnstone, R. A. W., and Quan, P. M., "Cyclic Dimers of Isoprene and Their Relation to Some Components of Tobacco Smoke," J. Chem. Soc., 1963, 2221-2224

3687. Joklik, J., "Gas-Chromatographic Separation of Ethylchlorosilanes," Collection Czechoslov. Chem. Communs., 26, 2079-2080 (1961); CA 56, 2873d

3688. Joklik, J., and Bazant, V., "Capillary Tube Crusher for Use in Gas Chromatography," Chem. listy, 53, 277-278 (1959); CA 53, 10853c

3689. Jones, A. G., "Analytical Chemistry. Some New Techniques," Academic Press, Inc., New York, 1959

3690. Jones, C. A., "A Review of the Air Pollution Research Program of the Smoke and Fumes Committee of the American Petroleum Institute," J. Air Pollution Control Assoc., 8, 268-271 (1958); CA 53, 3559c

3691. Jones, C. E. R., "The Gas Chromatography of Ethylenically Unsaturated Compounds with Particular Reference to Esters," Gas Chromatog., Proc. Symposium, 3rd, Edinburgh, 1960, 401-411; CA 56, 12291a

3692. Jones, C. E. R., "The Gas Chromatography of Ethylenically Unsaturated Compounds with Particular Reference to Esters," in "Gas Chromatography 1960," edited by R. P. W. Scott, Butterworths, London, 1960, pp. 401-411

135

3693. Jones, C. E. R., and Moyles, A. F., "Rapid Identification of High Polymers Using a Simple Pyrolysis Unit with a Gas-Liquid Chromatograph," Nature, 189, 222-223 (1961); CA 55, 27956e

3694. Jones, C. E. R., and Moyles, A. F., "Pyrolysis and Gas-Liquid Chromatography on the Microgram Scale," Nature, 191, 663-665 (1961); CA 56, 9902d

3695. Jones, E. P., Mason, L. H., Dutton, H. J., and Nystrom, R. R., "Labeling Fatty Acids by Exposure to Tritium Gas. II. Methyl Oleate and Linoleate," J. Org. Chem., 25, 1413-1417 (1960); CA 54, 24365b; 135th Natl. ACS Mtg., Boston, Mass., April 1959, Program Abstr., p. 37-0

3696. Jones, E. R. H., Lee, H. H., and Whiting, M. C., "Researches on Acetylenic Compounds, LXII. The Prepatation and Some Synthetical Applications of Penta-1, 2, 4-triene and Penta-1, 2-dien-4-yne," J. Chem. Soc., 1960, 341-346; CA 55, 15540c

3697. Jones, E. R. H., Lee, H. H., and Whiting, M. C., "Acetylenic Compounds. LXIV. The Preparation of Conjugated Octa- and Decaacetylenic Compounds," J. Chem. Soc., 1960, 3483-3489; CA 55, 19750h

3698. Jones, G. E. S., Turner, D., Sarlos, L. J., Barnes, A. C., and Cohen, R., "Determination of Urinary Pregnanediol by Gas-Liquid Chromatography," Fertility Sterility, 13, 544-549 (1962); CA 59, 1914d

3699. Jones, G. V., and Hammond, E. G., "Analysis of Glyceride Structure of Cocoa Butter by Thermal Gradient Crystallization," J. Am. Oil Chemists' Soc., 38, No. 2, 69-73 (1961); CA 55, 6887h

3700. Jones, H. G., Jones, J. K. N., and Perry, M. B., "The Gas-Liquid Partition Chromatography of Carbohydrate Derivatives. III. The Separation of Amino Glucose Derivatives and of Carbohydrate Acetal and Ketal Derivatives," Can. J. Chem., 40, 1559-1563 (1962); CA 57, 12600h

3701. Jones, H. G., and Perry, M. B., "The Gas-Liquid Partition Chromatography of Carbohydrate Derivatives. II. The Separation of Methyl Glycosides and of Acetylated Disaccharides," Can. J. Chem., 40, 1339-1343 (1962); CA 57, 12600f

3702. Jones, J. H., Fenske, M. R., Hutton, D. G., and Allendorf, H. D., "Vapor Phase Oxidation of Aromatic Hydrocarbons," 137th Natl. ACS Mtg., Cleveland, Ohio, April 1960, Program Abstr., p. 16Q; Preprints, Div. Petrol. Chem., 5, No. 2, C5

3703. Jones, J. H., and Ritchie, C. D., "A New Procedure for the Collection of Fractions in Gas Chromatography," J. Assoc. Offic. Agr. Chemists, 41, 753-756 (1958); CA 53, 1989e

3704. Jones, J. H., Ritchie, C. D., and Heine, Jr., K. S., "Gas Chromatography of Aromatic Amines and Nitro Compounds," J. Assoc. Offic. Agr. Chemists, 41, 749-752 (1958); CA 53, 1999h

3705. Jones, J. H., Ritchie, C. D., and Newburger, S. H., "Analysis of Nail Lacquers. IV. The Determination of Nail Lacquer Solvents by Gas-Liquid Chromatography," J. Assoc. Offic. Agr. Chemists, 41, 673-676 (1958); CA 53, 2542e

3706. Jones, R. A., "Molecular Sieves," in "Advances in Petroleum Chemistry Refining," edited by J. J. McKetta, Interscience Publishers, 4, 115-161 (1961); CA 57, 15405c

3707. Jones, T. E., Johns, T., and Bochinski, J. H., "Comparative Study of Blood CO_2, O_2 Determination with Gas Chromatography and the Van Slyke Procedure," 18th Southwest Regional Mtg., ACS, Dallas, Texas, December 1962

3708. Jones, W. C., "The Analysis of C_6-C_9 Aromatics by Gas-Liquid Chromatography," 132nd Natl. Mtg., New York, N. Y., September 1957, Program Abstr., p. 37B; Preprints, Div. Petrol. Chem., 2, No. 4, D117-D125 (1957)

3709. Jones, W. C., "Trace Analysis for Impurities in Hydrocarbon Concentrates," in "Gas Chromatography," edited by H. J. Noebels, R. F. Wall, and N. Brenner, Academic Press, Inc., New York, 1961, pp. 311-321

3710. Jones, W. L., "Physical Parameters in Gas Chromatography," Southwide Chemical Conference on Instrumentation, ISA, Memphis, Tenn., December 1956

3711. Jones, W. L., "Modifications to the van Deemter Equation for the Height Equivalent to a Theoretical Plate in Gas Chromatography," Anal. Chem., 33, 829-832 (1961); CA 55, 22935e

3712. Jones, W. L., Dal Nogare, S., Desty, D. H., Golay, M. J. E., Keulemans, A. I. M., Martin, A. J. P., Ober, S., Phillips, C. S. G., Thoburn, J., and Williams, E., "Standard Nomenclature Considerations and Recommendations," in "Gas Chromatography," edited by V. J. Coates, H. J. Noebels, and I. S. Fagerson, Academic Press, Inc., New York, 1958, pp. 315-317

3713. Jones, W. L., and Kieselbach, R., "Units of Measurement in Gas Chromatography," Anal. Chem., 30, 1590-1592 (1958); CA 53, 6699g

3714. Jones, W. M., "Pyrazolines. II. The Stereochemical Consequences of High Temperature Reactions of Diazomethanes with Olefins and α, β-Unsaturated Esters," J. Am. Chem. Soc., 81, 3776-3779 (1959); CA 54, 13051b

3715. Jones, W. M., "Pyrazolines. IV. Mechanism of Decomposition and Conformational Analyses of 2-Pyrazolines," J. Am. Chem. Soc., 82, 3136-3137 (1960); CA 55, 2622a

3716. Jones, W. M., "An Attempted Reductive Rearrangement of Norcamphor," J. Org. Chem., 26, 3606-3607 (1961); CA 56, 11623i

3717. Joossens, J. V., Evrard, E., Bosch, J. vanden, and Somer, P. de, "The Fatty Acid Pattern of Different Fractions of Plasma Lipids in Normolipemic and Hyperlipemic Rabbits," Verhandel. Koninkl. Vlaam. Acad. Geneesk. Belg., 23, 183-199 (1961); CA 56, 3882e

3718. Jordan, L. A., "Modern Methods of Research," Chem. & Ind. (London), 1957, 306-318

3719. Jordan, T. E., "The Vapor Pressure of Organic Compounds," Interscience Publishers, Inc., New York, 1954

3720. Judd, A. H., and Nicksic, S. W., "Analytical Hydrogenation. The Determination of Unsaturation and Carbon Skeleton Structure," 135th Natl. ACS Mtg., Boston, Mass., April 1959, Program Abstr., p. 6Q; Preprints, Div. Petrol. Chem., 4, No. 1, 37-43 (1959)

3721. Judson, C. M., McKinney, C. R., and Howard, R. F., "Use of a Conventional Mass Spectrometer for Gas Chromatography Monitoring," 14th Pittsburgh Conf. on Anal. Chem. & Appl. Spectroscopy, Pittsburgh, Pa., March 1963, Program Abstr., p. 66

3722. Juentgen, J., and Karwell, J., "Artificial Coalification of Coal," Freiberger Forschungsh., A229, 27-36 (1962); CA 58, 11136h

3723. Juneau, J., "Bibliography on Gas Chromatography and Some Related Methods," Petroleum Experiment Station, U. S. Bureau of Mines, Bartlesville, Okla., 1956; published by Consolidated Electrodynamics Corp., Pasadena, Calif.

3724. Jungermann, E., and Beck, E. C., "Analysis of ABS (Alkylbenzenesulfonate) Detergents," Soap Chem. Specialties, 38, No. 5, 72-75 (1962); CA 57, 3579e

3725. Jungermann, E., Davis, G. A., Beck, E. C., and Linfield, W. M., "Statistical Approach to Detergency Evaluation. Correlation of Performance Data with Gas Chromatographic Patterns of Alkylbenzenes," J. Am. Oil Chemists' Soc., 39, 50-53 (1962)

3726. Juranek, J., "Combined Colorimetric and Chromatographic Ultramicroanalysis of Gases. I. Analysis of Gases Forming Carbon Dioxide on Combustion," Chem. listy, 51, 2280-2286 (1957); CA 52, 4384g

3727. Juranek, J., "Colorimetric-Chromatographic Gas Ultramicroanalysis: Determination of Hydrocarbons," Chem. listy, 52, 1289-1298 (1958); CA 53, 5976i

3728. Juranek, J., "Determination of Ultramicroquantities of Hydrocarbons by Colorimetric-Chromatographic Analysis," Khromatog., ee Teoriya i Primenenie. Akad. Nauk S.S.S.R., Trudy Vsesoyuz Soveshchaniya, Moscow, 1958, 323-333 (Pub. 1960); CA 55, 14173i

3729. Juranek, J., "Combined Colorimetric and Chromatographic Ultramicroanalysis of Gases. I. Analysis of Gases Forming Carbon Dioxide on Combustion," Collection Czechoslov. Chem. Communs., 24, 135-143 (1959); CA 53, 6887i

3730. Juranek, J., "Colorimetric-Chromatographic Gas Microanalysis. Determination of Hydrocarbons," Collection Czechoslov. Chem. Communs., 24, 2306-2317 (1959); CA 53, 18757c

3731. Juranek, J., "The Significance of the Photocolorimetric Indication System for Gas Chromatography," in "Gas Chromatographie 1959," edited by R. E. Kaiser and H. G. Struppe, Akademie Verlag, Berlin, 1959, pp. 164-176

3732. Juranek, J., and Ambrova, A., "The Use of Gas Chromatography in Metallurgy. The Determination of Carbon in the Presence of Sulfur in Steel," 3rd Intern. Conf. on Anal. Chem., Prague, Czech., September 1959

3733. Juranek, J., and Ambrova, A., "Colorimetric-Chromatographic Ultramicroanalysis of Gases. III. Determination of Ultramicroamounts of Carbon in Technical Iron and Iron Alloys," Collection Czechoslov. Chem. Communs., 25, 2814-2821 (1960); CA 55, 4236e

3734. Juvet, Jr., R. S., "The Liquid Phase," 13th Annual Fisk Infrared Institute, Gas Chromatography Session, Nashville, Tenn., August 1962

3735. Juvet, Jr., R. S., "Qualitative and Quantitative Gas-Liquid Chromatography," 13th Annual Fisk Infrared Institute, Gas Chromatography Session, Nashville, Tenn., August 1962

3736. Juvet, Jr., R. S., "Special Separations and Applications, Including Non-Analytical Applications and Gas Chromatography of High Boilers," 13th Fisk Infrared Institute, Gas Chromatography Session, Nashville, Tenn., August 1962; 14th Fisk Infrared Institute, August 1963

3737. Juvet, Jr., R. S., "Theory of Gas-Liquid Chromatography: I. Retention, Resolution and Distribution Theory," 13th Annual Fisk Infrared Institute, Gas Chromatography Session, Nashville, Tenn., August 1962

3738. Juvet, Jr., R. S., "Introduction, Terminology and Practical Gas-Liquid Chromatography," 13th Annual Fisk Infrared Institute, Gas Chromatography Session, Nashville, Tenn., August 1962

3739. Juvet, Jr., R. S., "The Fourth International Gas Chromatography Symposium. Hamburg, Germany - A 10-Year Anniversary," Anal. Chem., 34, No. 10, 77A-82A (1962)

3740. Juvet, Jr., R. S., "Practical Aspects of Gas Chromatography," Gordon Research Conf., New Hampton, N. H., August 1963

3741. Juvet, Jr., R. S., "The Liquid Phase: Prediction of Retention Behavior. Liquid Phase Selection," 14th Annual Fisk Univ. Infrared Spectroscopy Institute, Gas Chromatography Session, Nashville, Tenn., August 1963

3742. Juvet, Jr., R. S., "Gas-Liquid Chromatography Introduction. Terminology and Practice," 14th

Fisk Univ. Infrared Spectroscopy Institute, Gas Chromatography Session, Nashville, Tenn., August 1963

3743. Juvet, Jr., R. S., "Theory of Gas-Liquid Chromatography: Retention, Resolution, and the Van Deemter-Golay Equations. Optimizing Column Parameters," 14th Annual Fisk Univ. Infrared Spectroscopy Institute, Gas Chromatography Session, Nashville, Tenn., August 1963

3744. Juvet, Jr., R. S., "Qualitative and Quantitative Gas-Liquid Chromatographic Analysis," 14th Annual Fisk Univ. Infrared Spectroscopy Institute, Gas Chromatography Session, Nashville, Tenn., August 1963

3745. Juvet, Jr., R. S., and Chiu, J., "Gas Chromatography. IV. Thermodynamics and Kinetics of the Alcoholysis of Acetals," J. Am. Chem. Soc., 83, 1560-1563 (1961); CA 55, 25730c; 139th Natl. ACS Mtg., St. Louis, Mo., March 1961, Program Abstr., p. 34B

3746. Juvet, Jr., R. S., and Tivin, F., "Analysis, Solution Thermodynamics, and Purification for Semi-conductor Purposes of Inorganic Compounds via Gas Chromatography," 14th Pittsburgh Conf. on Anal. Chem. & Appl. Spectroscopy, Pittsburgh, Pa., March 1963, Program Abstr., p. 58

3747. Juvet, Jr., R. S., and Tivin, F., "High Temperature Gas Chromatographic Separation of Inorganic Compounds," 14th Annual Mid-America Spectroscopy Symp., Chicago, Ill., May 1963

3748. Juvet, Jr., R. S., and Wachi, F. M., "Gas Chromatography. II. The Determination of Certain Physical Constants in the Alcoholysis of Esters," J. Am. Chem. Soc., 81, 6110-6115 (1959); CA 55, 13977f

3749. Juvet, Jr., R. S., and Wachi, F. M., "Gas Chromatographic Separation of Metal Halides by Inorganic Fused Salt Substrates," Anal. Chem., 32, 290-291 (1960); CA 55, 15211a

3750. Kabasakalian, P., and Townley, E. R., "Photolysis of Nitrile Esters in Solution. VI. Nitroso Dimers from Alicyclic Nitrites," J. Org. Chem., 27, 2918-2920 (1962); CA 58, 455d

3751. Kabasakalian, P., and Townley, E. R., "Photolysis of dl-Menthol Nitrite (Synthesis of dl-3-Hydroxy-p-menthan-10-oic Acid Lactone)," Am. Perfumer & Cosm., 78, 2, 22-23 (1963)

3752. Kabot, F. J., and Ettre, L. S., "Gas Chromatography Gives Direct Analysis of Wide Variety of Alcoholic Beverages," Perkin-Elmer Instrument News, 13, No. 4, 1, 7-9 (Summer 1962)

3753. Kachanak, S., "Equations for the Bed Height of Continuously Operating Adsorption Columns," Chem. Zvesti, 15, 575-589 (1961); CA 57, 7052h

3754. Kachanak, S., "Dynamics of Adsorption in Continuously Operating Columns from the Standpoint of the Equations of the Height of the Bed," Chem. Zvesti, 15, 590-606 (1961); CA 57, 7053a

3755. Kaderavek, G., "Gas-Chromatographic Study of Pressed Olive Oils," Olearia, 16, 5-10 (1962); CA 58, 4749d

3756. Kaesz, H. D., Phillips, J. R., and Stone, F. G. A., "Preparation and Study of Some Perfluoro-alkyl Compounds of Tin and Lead," J. Am. Chem. Soc., 82, 6228-6232 (1960); CA 55, 15336a

3757. Kaesz, H. D., Stafford, S. L., and Stone, F. G. A., "Synthesis and Cleavage of Perfluorovinyltin Compounds," J. Am. Chem. Soc., 82, 6232-6235 (1960); CA 55, 12273i

3758. Kainz, G., "Modern Methods of Gas Analysis," Osterr. Chem.-Ztg., 59, 45-51 (1958)

3759. Kainz, G., and Huber, H., "Gas-Chromatographic Studies on the Azotometer Gas Formed in the Determination of Amino Groups by van Slyke," Mikrochim. Acta, 1, 51-60 (1959); CA 55, 209c

3760. Kainz, G., and Huber, H., "Studies of Reactions Which Cause Anomalous Results in Amino-Nitrogen Determinations. Anomalies Caused by Isonitroso Compounds," Mikrochim. Acta, 1, 337-345 (1959); CA 54, 24113h

3761. Kainz, G., and Kasler, F., "Gas Chromatographic Investigation of Azotometer Gases Produced During the Dumas Determination of Nitrogen," Z. Anal. Chem., 168, 425-429 (1959); CA 54, 1164g

3762. Kaiser, R., "Current Applications of Gas Chromatography," Abhandl. Deut. Akad. Wiss. Berlin, Kl. Chem., Geol., Biol., 1959, No. 9, 87-92; CA 58, 5018b

3763. Kaiser, R., "A Thermoconductivity Cell on High Sensitivity," Abhandl. Deut. Akad. Wiss. Berlin, Kl. Chem., Geol., Biol., 1959, No. 9, 327-328; CA 58, 5019b

3764. Kaiser, R., "Gas Chromatography," Bibliographisches Institut, Mannheim, 1960; CA 58, 10735d

3765. Kaiser, R., "Chromatography in the Gas Phase. Part I. Gas Chromatography," 22/22a, Hochschultaschenbücher, Bibliographisches Inst., Mannheim, 1960; "Part II. Capillary Gas Chromatography," ibid., 23, 1961

3766. Kaiser, R., "New Developments in Gas Chromatography from the 1961 Literature," Z. Anal. Chem., 189, 1-14 (1962); CA 57, 13163e

3767. Kaiser, R., and Holzhauser, H., "Problems of Sampling into Capillary Columns in Gas Chromatography," 3rd Conference on Anal. Chem., Prague, Czech., September 1959

3768. Kaiser, R., and Kienitz, H., "Process Control Automatic Process Gas Chromatography," 4th Intern. Gas Chromatography Symp., Hamburg, Germany, June 1962; pub. in "Gas Chromatography 1962," ed. by M. van Swaay, Butterworth & Co., London, 1962

3769. Kaiser, R. E., "Detectors. A Highly Sensitive Thermal Conductivity Cell," in "Gas Chromatographie 1958," edited by H. P. Angele, Akademie Verlag, Berlin, 1959, pp. 327-338

3770. Kaiser, R. E., "Present Stand on the Application of Gas Chromatography," in "Gas Chromatog-

138

raphie 1958," edited by H. P. Angele, Akademie Verlag, Berlin, 1959, pp. 87-92

3771. Kaiser, R. E., "Gas Chromatographie," Akademische Verlagsgesellschaft Geest & Portig K-G., Leipzig, 1960

3772. Kaiser, R. E., and Struppe, H. G., (Eds.), "Gas Chromatographie 1959," Akademie Verlag, Berlin, 1959. Contains papers presented at the 2nd Symposium on Gas Chromatography, Bohlen, East Germany, October 1959

3773. Kaiser, R. E., and Struppe, H. G., "The Nomenclature of Gas Chromatography. Propositions for a Uniform Nomenclature and for the Presentation of Gas Chromatographic Data," in "Gas Chromatographie 1959," edited by R. E. Kaiser and H. G. Struppe, Akademie Verlag, Berlin, 1959, pp. 3-17

3774. Kaiser, R. E., and Struppe, H. G., "On the Theory and Experimental Conditions of Capillary Gas Chromatography," in "Gas Chromatographie 1959," edited by R. E. Kaiser and H. G. Struppe, Akademie Verlag, Berlin, 1959, pp. 177-194

3775. Kaiser, R. E., Struppe, H. G., Holzhäuzer, H., and Kuhl, M., "Gas-Chromatographic Analysis of Gasoline," Freiberger Forschungsh., A192, 205-218 (1961); CA 56, 8992a

3776. Kalab, V., and Pinkava, J., "Apparatus for the Delivery of Measured Amounts of Gases," Chem. listy, 52, 156-158 (1958); CA 52, 11481a

3777. Kalinenko, R. A., and Naimushin, N. N., "Gas-Chromatographic Analysis of Complex Mixtures of Oxygenated Compounds," Neftekhimiya, 1, 117-120 (1961); CA 57, 5304d

3778. Kallen, J., and Heilbronner, E., "The Vapor-Chromatography of a Labile Compound (the System A ⇌ B)," Helv. Chim. Acta, 43, 489-500 (1960); CA 54, 16105b

3779. Kallend, A. S., and Pitts, Jr., J. N., "Application of a High Sensitivity, Repeated Sampling, Gas Chromatographic Technique to Studies of Photochemical Systems," 144th Natl. ACS Mtg., Los Angeles, Calif., March-April 1963, Program Abstr., p. 25R

3780. Kallianos, A. G., and Mold, J. D., "Adaptation of the Iodoform Reaction to Microgram Quantities for Determination of Organic Structures," Anal. Chem., 34, 1174-1175 (1962); CA 57, 11860e

3781. Kallina, D., and Kuffner, F., "Separation of Isomeric Alcohols by Means of Gas-Liquid Chromatography. II. Separation of C_8 Alcohols," Monatsh. Chem., 91, 289-293 (1960); CA 54, 22318e

3782. Kalmanovskii, V. I., Fiks, M. M., and Yashin, Y. I., "New Detection Equipment for Automatic Chromatographic Gas Analyzers," Tr. po Khim. i Khim. Tekhnol., 3, 625-637 (1960); CA 56, 14901d

3783. Kalmanovskii, V. I., Kiselev, A. V., Lebedev, V. P., Savinov, I. M., Smirnov, N. Y., Fiks, M. M., and Shcherbakova, K. D., "Gas Chromatography in Glass Capillary Columns with Chemically Modified Surfaces," Zhur Fiz. Khim., 35, 1386-1388 (1961); CA 56, 5380g

3784. Kamibayashi, A., "Gas Chromatography and Its Applications," Hakko Kyokaishi, 17, 322-337 (1959); CA 54, 19086e

3785. Kamibayashi, A., Miki, M., and Ono, H., "Analysis of Fermentation Products by Gas Chromatography. I. Analysis of Fusel Oil," Kogyo Gijutsuin, Hakko Kenkyusho Kenkyu Hokoku, No. 18, 101-110 (1960); CA 54, 25547i

3786. Kamibayashi, A., Miki, M., and Ono, H., "Analysis of Fusel Oil by Gas Chromatography," Hakko Kyokaishi, 18, 411-416 (1960); CA 55, 20319c

3787. Kamibayashi, A., Miki, M., and Ono, H., "Acetals and Other Impurities in Methanol Fraction from the Distillation of Sulfite Waste Liquor Alcohol," Hakko Kyokaishi, 19, 233-236 (1961); CA 57, 11675b

3788. Kamibayashi, A., Miki, M., and Ono, H., "Gas Chromatographic Studies on Fermentation Products. I. Selectivity of Liquid Substrates for the Analysis of Lower Alcohols," Nippon Nogeikagaku Kaishi, 35, 968-973 (1961); Abstr., Agric. & Biol. Chem., 25, No. 10, A79 (1961); Biol. Abstr., 38, 17935

3789. Kamibayashi, A., Miki, M., and Ono, H., "Gas Chromatographic Studies on Fermentation Products. II. Analysis of Small Quantities of Lower Alcohols in Water," Nippon Nogeikagaku Kaishi, 35, 974-977 (1961); Abstr., Agric. & Biol. Chem., 25, No. 19, A79 (1961); Biol. Abstr., 38, 17936

3790. Kamer, J. H. van de, Gerritsma, K. W., and Wansink, E. J., "Gas-Liquid Partition Chromatography; the Separation and Micro-estimation of Volatile Fatty Acids from Formic Acid to Dodecanoic Acid," Biochem. J., 61, 174-176 (1956); CA 49, 15641d

3791. Kametani, F., "Gas-Chromatographic Analysis of Organic Reagents. I. Analysis of Pyridine, 2-Picoline, Piperidine, and 2-Pipecoline," Yakugaku Zasshi, 81, 489-491 (1961); CA 55, 18449i

3792. Kametani, F., and Kubota, S., "Gas-Chromatographic Analysis of Organic Reagents. II. Analysis of Pyridine Bases," Yakugaku Zasshi, 82, 659-661 (1962); CA 57, 6588h

3793. Kamio, H., and Shimo, T., "Determination of 3, 5-di-tert-Butyl-4-hydroxytoluene in Vitamin A," Bunseki Kagaku, 11, 731-734 (1962); CA 57, 9956f

3794. Kan, T., "Gas Chromatography of Heavy Hydrogen and Ortho- and Para-Hydrogen," Kagaku no Ryoiki Zokan, No. 44, 97-113 (1961); CA 56, 2i

3795. Kanazawa, J., Etchu, T., and Sato, R., "Gas Chromatography of the Nematocide, Ethylene Dibromide," Noyaku Seisan Gijutsu, 1, No. 5, 15-18 (1961); CA 55, 27751a

3796. Kanazawa, J., and Sato, R., "Gas Chromatography of Agricultural Chemicals. I. Gas Chromatography of D-D Mixture," Bunseki Kagaku, 10, 760-763 (1961); CA 58, 7307d

3797. Kanazawa, J., and Sato, R., "Gas Chromatography of the Nematocide 1, 2-Dibromo-3-chloropropane," Bunseki Kagaku, 10, 1350-1353 (1961); CA 56, 15877c

3798. Kanazawa, J., and Sato, R., "Gas-Liquid Chromatography of a Herbicide, 2, 4-Dichlorophenoxyacetic Acid Ethyl Ester," Bunseki Kagaku, 11, 523-526 (1962); CA 57, 7665i

3799. Kaneda, T., "Biosynthesis of Branched-Chain Fatty Acids. I. Isolation and Identification of Fatty Acids from Bacillus Subtilis," J. Biol. Chem., 238, 1222-1228 (1963); CA 58, 11709a

3800. Kanne, F., "Activities of Light Hydrocarbons on Different Stationary Phases," Erdol u. Kohle, 11, 339 (1958)

3801. Kannuluik, W G., and Carman, E. H., "The Thermal Conductivity of Rare Gases," Proc. Phys. Soc. (London), 65B, 701-709 (1952); CA 46, 10736i

3802. Kannuluik, W. G., and Martin, L. H., "The Thermal Conductivity of Some Gases at 0°," Proc. Roy. Soc. (London), A144, 496-513 (1934); CA 28, 4656²

3803. Karabatsos, G. J., "Infrared Spectra of Isotopically Labeled Compounds. I. Diisopropylketones," J. Org. Chem., 25, 315-318 (1960); CA 54, 17049h

3804. Karabatsos, G J., "Infrared Spectra of Isotopically Labeled Compounds. II. Compounds Possessing the 2, 4-Dimethyl-3-pentyl Skeleton," J. Org. Chem., 25, 1409-1412 (1960); CA 54, 24358c

3805. Karagounis, G., Charbonnier, E., and Floss, E., "Gas Chromatographic Resolution of Racemic Compounds," J. Chromatog., 2, 84-89 (1959); CA 53, 21361a

3806. Karagounis, G., and Lemperle, E., "Gas Chromatographic Separation of Racemic Compounds," Z. Anal. Chem., 189, 131-137 (1962); CA 58, 6689d

3807. Karagounis, G., and Lippold, G., "Vapor-Phase Chromatographic Separation of Racemic Compounds," Naturwissenschaften, 46, 145 (1959); CA 53, 16944c

3808. Karasek, F. W., "Mass Spectrometry vs. Vapor-Phase Chromatography," Chem. & Eng. News, 35, No. 18, 70 (1957)

3809. Karasek, F. W., "Stream Analysis and Data Reduction in Pilot Plants," Control Eng., 8, No. 12, 93-96 (1961)

3810. Karasek, F. W., and Ayers, B. O., "Fast Sampling Valve for Gas Chromatography," ISA Journal, 7, No. 3, 70-71 (1960); CA 54, 23460e

3811. Karasek, F. W., and Burk, M. C., "Total Analysis Digital Readout for Chromatograph," 12th Pittsburgh Conf. on Anal. Chem. & Appl. Spectroscopy, Pittsburgh, Pa., February-March 1961, Program Abstr., p. 51

3812. Karasek, F W., Burk, M. C., and Ayers, B. O., "Total Analysis Digital System for Chromatographs," Anal. Chem., 33, 1543-1546 (1961); CA 56, 9382i

3813. Karasev, K. I., and Mukhina, T. N., "Application of Tagged Atoms to the Study of the Effectiveness of Fractionation of Hydrocarbon Gases," Trudy Kom. Anal. Khim., Akad. Nauk S.S.S.R., Inst. Geokhim. i Anal. Khim., 9, 349-355 (1958); CA 53, 4011e

3814. Karashima, J., "Carrier for Gas-Chromatography," Kagaku no Ryoiki Zokan, No. 44, 39-62 (1961); CA 56, 2i

3815. Karchmer, J. H., "Gas-Liquid Partition Chromatography of Sulfur Compounds with Beta, Beta'-Iminodipropionitrile," Anal. Chem., 31, 1377-1379 (1959); CA 53, 19663d

3816. Karger, B. L, and Cooke, W. D., "The Effect of Column Length on Resolution," 145th Natl. ACS Mtg., New York, N. Y., September 1963, Program Abstr., p. 11B

3817. Karger, B. L., and Cooke, W. D., "Gas Chromatography Under Normalized Time Conditions. Effect of Particle Size on Resolution," 145th Natl. ACS Mtg., New York, N. Y., September 1963, Program Abstr., p. 11B

3818. Karlsson, B M., "Determination of Minute Quantities of Nitrogen in Argon by Gas Chromatography," Anal. Chem., 35, 1311-1312 (1963)

3819. Karmen, A., "Gas Chromatography and Lipid Metaolism," Am. Heart J., 59, 937-939 (1960)

3820. Karmen, A., "Radio Assay of Gas Chromatographic Compounds Labelled with Tritium," Univ. of Houston Intern. Symp. on Advances in Gas Chromatography, Houston, Texas, January 1963

3821. Karmen, A., and Bowman, R. L., "A Radio Frequency Glow Detector for Gas Chromatography," Ann. N. Y. Acad. Sci., 72 714-719 (1959); CA 53, 14598i

3822. Karmen, A., and Bowman, R. L., "A DC Discharge Detector for Gas Chromatography," 3rd Intern. Symposium on Gas Chromatography, ISA, East Lansing, Mich., June 1961; ISA Proc., 3, 129-131 (1961)

3823. Karmen, A., and Bowman, R. L., "A Radio Frequency Discharge Detector for Gas Chromatography," in "Gas Chromatography," edited by H. J. Noebels, R. F. Wall, and N. Brenner, Academic Press, New York, 1961, pp. 65-73

3824. Karmen, A., and Bowman, R L, "Self-Sustained Discharge Detector for Gas Chromatography," Nature, 196, 62-63 (1962); CA 57, 15778g

3825. Karmen, A., and Bowman, R. L., "A DC Discharge Detector for Gas Chromatography," in "Gas

Chromatography," edited by N. Brenner, J. E. Callen and M. D. Weiss, Academic Press, Inc., New York, 1962, pp. 189-194

3826. Karmen, A., Giuffrida, L., and Bowman, R. L., "Detection by Ionization of Atmospheric Gases During Analysis by Gas Chromatography," Nature, 191, 906-907 (1961); CA 56, 6317f

3827. Karmen, A., Giuffrida, L., and Bowman, R. L., "Comparison of Helium and Argon in Ionization Detectors," J. Chromatog., 9, 13-20 (1962)

3828. Karmen, A., Giuffrida, L., and Bowman, R. L., "Radioassay by Gas-Liquid Chromatography of Lipids Labeled with Carbon-14," J. Lipid Research, 3, 44-52 (1962); CA 56, 13031d

3829. Karmen, A., Giuffrida, L., and Tritch, H., "Radioassay of Carbon-14 Labeled Compounds by Gas Chromatography," 137th Natl. ACS Mtg., Cleveland, Ohio, April 1960, Program Abstr., p. 27B

3830. Karmen, A., McCaffrey, I., and Bowman, R. L., "A Flow-Through Method for Scintillation Counting of Carbon and Tritium in Gas-Liquid Chromatographic Effluents," J. Lipid Research, 3, 372-377 (1962); CA 57, 13390e

3831. Karmen, A., McCaffrey, I., and Bowman, R. L., "Use of Carbon Dioxide as Carrier Gas in Gas Chromatography," Nature, 193, 575-576 (1962); CA 56, 14901h

3832. Karmen, A., McCaffrey, I., Winkelman, J. W., and Bowman, R. L., "Measurement of Tritium in the Effluent of a Gas Chromatography Column," Anal. Chem., 35, 536-542 (1963); CA 59, 1092b

3833. Karmen, A., and Tritch, H. R., "Radioassay by Gas Chromatography of Compounds Labelled with Carbon-14," Nature, 186, 150-151 (1960); CA 54, 16938e

3834. Karohl, J. H., "Off-Line Machine Reduction of Gas Chromatographic Data," 7th Conf. on Anal. Chem. in Nuclear Technology, Gatlinburg, Tenn., October 1963

3835. Karp, H. R., "Chromatography Steps Up the Pace," Control Eng., 7, No. 7, 37-38 (1960)

3836. Karp, H. R., "Industrial Process Chromatographs," Control. Eng., 8, No. 6, 87-94 (1961)

3837. Karpathy, O. C., "Carbon Determination of Ferrous Alloys with a Newly Developed Gas Chromatograph," 14th Pittsburgh Conf. on Anal. Chem. & Appl. Spectroscopy, Pittsburgh, Pa., March 1963, Program Abstr., p. 50

3838. Karpathy, O. C., and Johnson, C. H., "Automatic Monitoring of Atmospheres in Blast Furnaces with a New Gas Chromatograph," 13th Pittsburgh Conf. on Anal. Chem. & Appl. Spectroscopy, Pittsburgh, Pa., March 1962, Program Abstr., p. 54

3839. Karpukhina, G. V., and Maizus, Z. K., "Gas-Liquid Chromatographic [Analysis] of the Liquid-Phase Oxidate of n-Decane," Neftekhimiya, 2, 901-905 (1962); CA 58, 8831d

3840. Karr, Jr., C., and Brown, P. M., "Retention Volumes of Phenol, Methyl-and Dimethylphenols in Gas-Liquid Partition Chromatography," 130th Natl. ACS Mtg., Atlantic City, N. J., September 1956, Program Abstr., p. 4K

3841. Karr, Jr., C., Brown, P. M., Estep, P. A., and Humphrey, G. L., "Identification and Determination of Low-Boiling Phenols in Low Temperature Coal Tar," Anal. Chem., 30, 1413-1416 (1958); CA 52, 20997e; 132nd Natl. ACS Mtg., New York, N. Y., September 1957, Program Abstr., p. 5K

3842. Karr, Jr., C., Brown, P. M., Estep, P. A., and Humphrey, G. L., "Analysis of Low-Temperature Tar Phenols Boiling Up to 234°," Fuel, 37, 227-235 (1958); CA 52, 9566b

3843. Karr, Jr., C., Childers, E. E., and Warner, W. C., "Analysis of Aromatic Hydrocarbon Samples by Liquid Chromatography with Operating Conditions Analogous to Those of Gas Chromatography," Anal. Chem., 35, 1290-1291(1963)

3844. Karr, Jr., C., Comberiati, J. R., and Estep, P. A., "Structure Determination of Resins from Pitch of Low Temperature Tar by Combined Pyrolysis and Gas-Liquid Chromatography," 142nd Natl. ACS Mtg., Atlantic City, N. J., September 1962, Program Abstr., p. 1K

3845. Karr, Jr., C., Comberiati, J. R., and Warner, W. C., "Comparison of Pitch Resins from Different Sources by Combined Pyrolysis and Gas-Liquid Chromatography," Anal. Chem., 35, 1441-1444 (1963); 144th Natl. ACS Mtg., Los Angeles, Calif., March-April 1963, Program Abstr., p. 5-O

3846. Karwat, H. H., "The Binary Mixture H_2-HD," Chem.-Ing.-Tech., 32, 605-610 (1960); CA 54, 23548e

3847. Kashima, J., and Yamazaki, T., "Gas-Chromatographic Determination of the Hydrogen Content of Molten Aluminum by a Nitrogen Circulating Procedure," Rept. Casting Research Lab., Waseda Univ. (Tokyo), No. 10, 89-94 (1959); CA 55, 7149c

3848. Kashima, J., and Yamazaki, T., "Gas Chromatographic Determination of Hydrogen in Cast Iron. Hot-Extraction Procedure by Carrier Gas Technique," Rept. Casting Research Lab., Waseda Univ. (Tokyo), No. 11, 51-57 (1960); CA 55, 17353e

3849. Katague, D. B., and Kirch, E. R., "Analysis of the Volatile Components of Ylang-Ylang Oil by Gas Chromatography," J. Pharm. Sci., 52, 252-258 (1963); CA 58, 12366d

3850. Kateman, G., "A Simple Apparatus for Keeping the Final Pressure in Gas Chromatographs Constant," J. Chromatog., 8, 278-279 (1962)

3851. Kateman, G., "An Apparatus for Estimating the Product of Peak Height and Retention Time in Gas Chromatography," J. Chromatog., 8, 280-281 (1962)

3852. Kates, M., and Baxter, R. M., "Lipids of Psychrophilic and Mesophilic Yeasts," 44th Canadian Chemical Conf. and Exhibition, CIC, Montreal, Canada, August 1961; Abstr., Chem. in Can., 13, No. 5, 32 (1961)

3853. Katnik, R. J., "Gas Chromatographic Behavior of Some C_1-C_4 Oxygenated Hydrocarbons," Univ. Microfilms (Ann Arbor, Mich), Order No. 63-2000, 109 pp.; Dissertation Abstr., 23, 3615 (1963)

3854. Katsuhara, J., and Kobayashi, M., "Myrcene in the Oil of Various Mentha Species," Nippon Kagaku Zasshi, 83, 607-608 (1962); CA 58, 7049h

3855. Katz, I., and Keeney, M., "Location of Double Bonds in the Octadecenoic Acid Fractions of Milk Fat, Beef Fat, and Lard," 58th Annual Mtg., Am. Dairy Sci. Assoc., Lafayette, Ind., June 1963; Abstr., J. Dairy Sci., 46, 605 (1963)

3856. Katz, I., and Keeney, M., "Isolation and Identification of Fatty Aldehydes in Rumen Microbial Lipid," 58th Annual Mtg., Am. Dairy Sci. Assoc., Lafayette, Ind., June 1963; Abstr., J. Dairy Sci., 46, 611 (1963)

3857. Katz, S., and Barr, J. T., "Gas Titrations," Anal. Chem., 25, 619-624 (1953); CA 47, 6193a

3858. Kauffman, F. L., Harlan, J. W., Lee, G. D., Wilding, M. D., and Beckstrom, R. D., "Gas Chromatography for Meat Flavor Research," Am. Meat Inst. Found. Mtg., Chicago, Ill., March 1960

3859. Kauffman, F. L., and Lee, G. D., "A Study of Octadecanoic Acids by Gas-Liquid Partition Chromatography and Infrared Spectrophotometry," J. Am. Oil Chemists' Soc., 37, 385-386 (1960); CA 54, 20247g

3860. Kauffman, F. L., Weiss, T. J., Lee, G. D., and Rockwood, B. N., "Gas-Liquid Partition Chromatography and Ultraviolet Absorption of Oils Before and After Hydrogenation," J. Am. Oil Chemists' Soc., 38, 495-496 (1961); CA 55, 24053c; 138th Natl. ACS Mtg., New York, N. Y., September 1960, Program Abstr., pp. 2A-3A

3861. Kaufman, H. R., and Zlatkis, A., "Butadiene-Vinylacetylene Analysis by Gas Chromatography," Chem. & Ind. (London), 1958, 1001; CA 53, 5980a

3862. Kaufman, J. J., Todd, J. E., and Koski, W. S., "Application of Gas-Phase Chromatography to the Boron Hydrides," Anal. Chem., 29, 1032-1035 (1957); CA 51, 13651a

3863. Kaufman, S., "Solubilization of Methanol by Soap Micelles as Measured by Gas Chromatography," J. Colloid Sci., 17, 231-242 (1962); CA 57, 97b; 139th Natl. ACS Mtg., St. Louis, Mo., March 1961, Program Abstr., p. 20I

3864. Kaufmann, H. P., Seher, A., and Mankel, G., "Gas Chromatography of Fats. II. Quantitative Applications," Fette, Seifen, Anstrichmittel, 64, 501-509 (1962); CA 57, 7399b

3865. Kaufmann, H. P., Mankel, G., and Lehmann, R., "Gas Chromatography of Fatty Compounds. I. General Survey," Fette, Seifen, Anstrichmittel, 63, 1109-1116 (1961); CA 56, 11729h

3866. Kaunitz, H., Johnson, R. E., and McKay, D. G., "Influence of Pregnancy and an Oxidized Lipid Diet on the Fatty Acid Composition of Blood and Tissues," Nature, 197, 600-601 (1963); CA 58, 10565h

3867. Kaunitz, H., Slanetz, C. A., Johnson, R. E., and Babayan, V. K., "The Regulation of Depot Fat by Linoleic Acid," J. Nutrition, 73, 386-390 (1961)

3868. Kaunitz, H., Slanetz, C. A., Johnson, R. E., and Herb, S. F., "Influence of Lauroyl and Myristoyl Peroxides and Oxidized Cottonseed Oil on Depot Fat and Liver Lipid Composition," J. Am. Oil Chemists' Soc., 38, 301-305 (1961); CA 55, 18914b

3869. Kauss, J. M., Peters, J., and Martin, A. J., "Design Considerations and Applications of a Fully Automatic Preparative Gas Chromatograph," 10th Detroit Anachem Conf., Detroit, Mich., October 1962, Program Abstr., p. 33

3870. Kawai, S., "Gas Chromatographic Method for Analysis of Drugs. I. Aerosol, Methanol, and Ethanol," Eisei Shikenjo Hokoku, No. 77, 2733 (1959); CA 55, 9782h

3871. Kawai, S., "Gas Chromatographic Analysis of Drugs. III. Determination of o-Dichlorobenzene," Eisei Shikenjo Hokoku, No. 79, 67-68 (1961); CA 59, 382f

3872. Kawai, S., and Takeuchi, S., "Gas Chromatographic Analysis of Drugs. II. Analysis of the Saponified Cresol Solutions and Creosotes in the Japanese Pharmacopeia [JP]," Bunseki Kagaku, 10, 473-478 (1961); CA 58, 6646h

3873. Kawamoto, H., "Determination of Pentylenetetrazole by Gas Chromatography," Kumamoto Med. J., 15, 69-72 (1962); CA 58, 10628c

3874. Kawashiro, I., and Ishii, A., "Application of Gas Chromatography to Food Analysis. III. Separation and Identification of Aliphatic Esters in Artificial Flavor," Bull. Natl. Inst. Hyg. Sci., 79, 103-105 (1961)

3875. Kawashiro, I., and Ishii, A., "Application of Gas Chromatography to Food Analysis. IV. Determination of Butyl p-Hydroxybenzoate in a Mixture of Preservatives," Shokuhin Eiseigaku Zasshi, 2, 54-56 (1961); CA 57, 7683a

3876. Kawashiro, I., Ishii, A., and Fujita, M., "Application of Gas Chromatography to Food Analysis. II. Quantitative Analysis of Sorbic Acid," Shokuhin Eiseigaku Zasshi, 2, No. 1, 50-52 (1961);

CA <u>55</u>, 27675f

3877. Kawashiro, I., Tanabe, H., and Ishii, A., "Application of Gas Chromatography to Food Analysis. I. Fatty Acids in Butter and Cheese," Shokuhin Eiseigaku Zasshi, <u>1</u>, No. 1, 78-83 (1960); CA <u>55</u>, 26291f

3878. Kawazumi, K., Kataoka, S., and Maruyama, K., "Analysis of Methylchlorosilanes by Gas Chromatography," Kogyo Kagaku Zasshi, <u>64</u>, 784-787 (1961); CA <u>57</u>, 2832h

3879. Kay, K., "Air Pollution," Anal. Chem., <u>29</u>, 589-604 (1957); CA <u>51</u>, 7218i

3880. Kaye, W. I., "The Application of Far Ultraviolet Spectroscopy to Gas Chromatography," Instr. Meas. Chem. Anal. Elec. Quant. Nucleonic Process Control, Proc. Intern. Conf. Stockholm, <u>1</u>, 455-464 (1960) (Pub. 1961); CA <u>56</u>, 6632c

3881. Kaye, W. I., "The Application of Far Ultraviolet Spectroscopy to Gas Chromatography," Analyzer, <u>2</u>, No. 3, 4-7 (July 1961)

3882. Kaye, W. I., "Far-Ultraviolet Spectroscopic Detection of Gas Chromatograph Effluent," Anal. Chem., <u>34</u>, 287-293 (1962); CA <u>56</u>, 12290b; 12th Pittsburgh Conf. on Anal. Chem. & Appl. Spectroscopy, Pittsburgh, Pa., February-March 1961, Program Abstr., p. 51

3883. Kazanskii, B. A., Zhukhovitskii, A. A., Smerligov, O. D., and Turkel'taub, N. M., "Chromatographic Analysis of Hydrocarbon Mixtures," Gazovaya Khromatografiya, Trudy 1-oi [Pervoi] Vsesoyuz. Konf., Akad. Nauk S.S.S.R., Moscow, <u>1959</u>, 244-267 (Pub. 1960); CA <u>55</u>, 26864a

3884. Kazyak, L., "An Application of Gas Chromatography to Analytical Toxicology," 12th Annual Symp. on Recent Developments in Research Methods and Instrumentation, National Institutes of Health, Bethesda, Md., October 1962; Anal. Chem., <u>35</u>, 1448-1452 (1963)

3885. Kearns, R., and Guild, L. V., "An Apparatus for Analysis by Gas Chromatography Method," 7th Pittsburgh Conf. on Anal. Chem. & Appl. Spectroscopy, Pittsburgh, Pa., February-March 1956, Program Abstr., p. 40

3886. Kebarle, P., and Bryce, W. A., "The Decomposition of 1-Butene-4-d_3 Induced by Methyl Radicals," Can. J. Chem., <u>35</u>, 576-579 (1957); CA <u>51</u>, 16273b

3887. Kecki, Z., and Wincel, H., "Determination of Radiation Yield of Hydrogen by Gas Chromatography," Nukleonika, <u>7</u>, 169-174 (1962); CA <u>57</u>, 14614i

3888. Keefer, R. M., and Andrews, L. J., "Trifluoroacetic Acid Catalyzed Chlorination of Aromatic Hydrocarbons in Carbon Tetrachloride. Inhibition by Acetic Acid," J. Am. Chem. Soc., <u>82</u>, 4547-4553 (1960); CA <u>55</u>, 3475c

3889. Keen, R. T., Baxter, R. A., Miller, L. J., Shepard, R. C., and Rotheram, M. A., "Methods for Analysis of Polyphenol Reactor Coolants," U. S. At. Energy Comm., <u>NAA-SR-4356</u>, 51 pp. (1961); CA <u>55</u>, 13160c

3890. Keenan, R. G., "New Analytical Techniques for Industrial Hygiene," A.M.A. Arch. Ind. Health, <u>21</u>, 261-267 (1960); CA <u>54</u>, 12435a

3891. Keeney, P. G., "Distearyl Triglycerides in the High Melting Fraction of Milk Fat," 55th Annual Mtg., Am. Dairy Sci. Assoc., Logan, Utah, June 1960; Abstr., J. Dairy Sci., <u>43</u>, 840 (1960)

3892. Keidel, F. A., and Lewis, C. D., "Liquid Metering Device for Process-Stream Gas Chromatographs and Small Pumps," Anal. Chem., <u>33</u>, 1456 (1961); CA <u>55</u>, 24131a

3893. Keigh, D. F., "Use of Gas Chromatography in Measuring the Ethylene Production of Stored Apples," J. Sci. Food Agr., <u>11</u>, 381-385 (1960)

3894. Kelker, H., "Retention Volumes of Some Organic Compounds (Chromatographic Data Table 47)," J. Chromatog., <u>2</u>, D28-D29 (1959); cf. Angew. Chem., <u>71</u>, 218 (1959)

3895. Kelker, H., "Beta, Beta'-Oxydipropionitrile and Hexaethylene Glycol Dimethylether as Stationary Phases in Gas Chromatography," Angew. Chem., <u>71</u>, 218-220 (1959); CA <u>53</u>, 13891f

3896. Kelker, H., "Determination of Inert Gas Components in Refrigerants by Means of Gas Chromatography," Kalte-technik, <u>11</u>, 101-103 (1959); CA <u>53</u>, 13869a

3897. Kelker, H., "Gas Chromatographic Determination of Formaldehyde," Z. Anal. Chem., <u>176</u>, 3-8 (1960); CA <u>55</u>, 3277c

3898. Kelker, H., Rohleder, H., and Weber, O., "Rational Integration Procedures for Gas Chromatography, Involving Novel Combinations of Instruments," 4th Intern. Gas Chromatography Symp., Hamburg, Germany, June 1962; pub. in "Gas Chromatography 1962," M. van Swaay, ed., Butterworth & Co., London, 1962

3899. Keller, R. A., "Chromatography," Univ. Microfilms (Ann Arbor, Mich.), Publ. No. <u>20043</u>, 110 pp.; Dissertation Abstrs., <u>17</u>, 757 (1957); CA <u>51</u>, 11810a

3900. Keller, R. A., "Gas Chromatography," Scientific American, <u>205</u>, No. 4, 58-67 (October 1961)

3901. Keller, R. A., "Gas-Liquid Chromatography of Volatile Metal Halides," J. Chromatog., <u>5</u>, 225-235 (1961); CA <u>56</u>, 5385g

3902. Keller, R. A., Bate, R., Costa, B., and Forman, P., "Changes Occurring With the Immobile Liquid Phase in Gas-Liquid Chromatography," J. Chromatog., <u>8</u>, 157-177 (1962); CA <u>58</u>, 1893f

3903. Keller, R. A., and Freiser, H., "Gas-Liquid Partition Chromatography for Separation of Metal Halides," Gas Chromatog. Proc. Symposium, 3rd, Edinburgh, <u>1960</u>, 301-307; CA <u>56</u>, 5385h

3904. Keller, R. A., and Freiser, H., "Gas-Liquid Partition Chromatography for Separation of Metal Halides," in "Gas Chromatography 1960," edited by R. P. W. Scott, Butterworths, London, 1960, pp. 301-307

3905. Keller, R. A., and Stewart, G. H., "Three Phases in Gas-Liquid Chromatography," Anal. Chem., 34, 1834-1838 (1962); CA 58, 2825a

3906. Keller, R. A., and Stewart, G. H., "Changes Occurring with the Immobile Liquid Phase in Gas Liquid Chromatography. II. The Effect on Retention Volumes," J. Chromatog., 9, 1-12 (1962); CA 58, 5019c

3907. Keller, R. A., Stewart, G. H., and Giddings, J. C., "Chromatography," Ann. Rev. Physical Chem., 11, 347-368 (1960); CA 54, 23503i

3908. Kellner, S. M. E., and Walters, W. D., "Thermal Decomposition of N-Propylcyclobutane," J. Phys. Chem., 65, 466-469 (1961); CA 56, 2339i

3909. Kemball, C., and Rooney, J. J., "The Cracking of Cyclopentene on a Silica-Alumina Catalyst," Proc. Roy. Soc. (London), A257, 132-145 (1960); CA 54, 23674b

3910. Kennedy, C. D., "Preparative Gas Chromatography. The Megachrom," 1st Annual Gas Chromatography Institute, Canisius College, Buffalo, N. Y., May 1959

3911. Kennedy, C. D., "Large Diameter Columns," 2nd Annual Gas Chromatography Institute, Canisius College, Buffalo, N. Y., April 1960

3912. Kenney, R. L., and Fisher, G. S., "Photosensitized Oxidation of Carvomenthene," 138th Natl. ACS Mtg., New York, N. Y., September 1960, Program Abstr., p. 79P

3913. Kent, J. A., and Norman, R. O. C., "The Homolytic Methylation of Naphthalene," J. Chem. Soc., 1959, 1724-1726; CA 53, 17978g

3914. Kent, T. B., "The Determination of Trace Impurities in Gases by Gas Chromatography," Chem. & Ind. (London), 1960, 1260-1261; CA 55, 13172a

3915. Keppler, J. G., "Application of Vapor-Phase Chromatography," Chem. Weekblad, 51, 911-914 (1955)

3916. Keppler, J. G., Dijkstra, G., and Schols, J. A., "Vapour Phase Chromatography," at High-Temperatures," in "Vapour Phase Chromatography," edited by D. H. Desty, Academic Press, Inc., New York, 1957, pp. 222-234

3917. Keppler, J. G., Schols, J. A., and Dijkstra, G., "Multiple Gas Chromatographic Apparatus for Use at Temperatures Up to 250°," Rev. trav. chim., 75, 965-976 (1956); CA 50, 15135g

3918. Kerenyi, E., and Keszthelyi, S., "The Analysis of the Composition of a Gasoline from Tujmaz," Magyar Kem, Folyoirat, 65, 389-394 (1959); CA 54, 13616e

3919. Kerenyi, E., and Keszthelyi, S., "Identification of the Hydrocarbons in Gasoline by Precise Rectification and Gas-Liquid Chromatography," Khim. i Tekhnol. Topliv. i Masel, 5, No. 6, 53-60 (1960); CA 54, 17857c

3920. Kergomard, A., Pigeret, F., and Renard, M., "Equilibrium in the Isomerization of Ionones," Bull. soc. chim. France, 1962, 1648-1651; CA 58, 5725e

3921. Kerr, J. A., and Calvert, J. G., "The Photolysis of Azo-n-propane; the Decomposition of the n-Propyl Radical," J. Am. Chem. Soc., 83, 3391-3396 (1961); CA 55, 27017h

3922. Kerr, J. A., and Trotman-Dickenson, A. F., "Adsorbents for Aldehydes and Olefins," Nature, 182, 466 (1958); CA 53, 2719i

3923. Kerr, J. A., and Trotman-Dickenson, A. F., "The Reactions of Alkyl Radicals. I. n-Propyl Radicals from the Photolysis of n-Butyraldehyde," Trans. Faraday Soc., 55, 572-580 (1959); CA 54, 2898c

3924. Kerr, J. A., and Trotman-Dickenson, A. F., "The Combination of Unlike Radicals in the Gas Phase," Chem. & Ind. (London), 1959, 125-126; CA 53, 12803b

3925. Kerr, J. A., and Trotman-Dickenson, A. F., "Reactions of Alkyl Radicals. III. n-Butyl Radicals from the Photolysis of n-Valeraldehyde," J. Chem. Soc., 1960, 1602-1608; CA 54, 17249i

3926. Kerr, J. A., and Trotman-Dickenson, A. F., "Reactions of Alkyl Radicals. IV. The Reaction of Methyl Radicals with Isopropyl and tert-Butyl Radicals," J. Chem. Soc., 1960, 1609-1611; CA 54, 17250b

3927. Kerr, J. A., and Trotman-Dickenson, A. F., "Reactions of Alkyl Radicals. V. Ethyl Radicals from Propionaldehyde," J. Chem. Soc., 1960, 1611-1617; CA 54, 17249i

3928. Kesterson, J. W., and Hendrickson, R., "Florida Cold-Pressed Murcott Oil," Am. Perfumer Aromat., 75, No. 11, 35-37 (1960); CA 55, 4888g

3929. Kesterson, J. W., and Hendrickson, R., "The Composition of Valencia Orange Oil as Related to Fruit Maturity," Am. Perfumer Cosmet., 77, No. 12, 21-24 (1962); CA 58, 12366b

3930. Kesterson, J. W., and Hendrickson, R., "Evaluation of Coldpressed Marsh Grapefruit Oil'" Am. Perfumer Cosmet., 78, No. 5, 32-35 (1963)

3931. Keszler, I., "Chromatographic Techniques," Chemia (Buenos Aires), 17, 36-42 (1957); CA 52, 1837b

3932. Keulemans, A. I. M., "Discussions on Katharometers as Recorders in Gas Chromatography," Analyst, 81, 57-58 (1956)

3933. Keulemans, A. I. M., "Some Fundamental Aspects of Gas Chromatography," 132nd Natl. ACS Mtg., New York, N. Y., September 1957, Program Abstr., p. 31B; Preprints, Div. Petrol. Chem., 2,

No. 4, D5-D15 (1957)

3934. Keulemans, A. I. M., "Gas Chromatography," 2nd ed., Reinhold Publishing Corp., New York, 1959

3935. Keulemans, A. I. M., "Application of Gas Chromatography in Petroleum Industry," Proc. Congr. Modern Anal. Chem. Ind., Univ. St. Andrews, Heffer, Cambridge, England, June 1959, pp. 217-227; CA 53, 113h

3936. Keulemans, A. I. M., "Advances in Gas Chromatography," Z. Anal. Chem., 170, 212-219 (1959); CA 54, 3043c

3937. Keulemans, A. I. M., "New Developments in Gas Chromatography," in "Gas Chromatographie 1959," edited by R. E. Kaiser and H. G. Struppe, Akademie Verlag, Berlin, 1959, pp. 195-197

3938. Keulemans, A. I. M., "Gas Chromatography," Symposium on Modern Methods of Analysis of Organic Compounds, German Chem. Soc., Munich, Germany, October 1960; Abstr., Angew. Chem., 73, 145 (1961)

3939. Keulemans, A. I. M., "Present Status of Gas Chromatography - A Review," Z. Anal. Chem., 181, 350-351 (1961)

3940. Keulemans, A. I. M., "Instruments Employed and Their Application in the Investigation of Fruit Juices," Intern. Fruchtsaft-Union Ber. Wis. Tech. Komm., 4, 191-204 (1962); CA 58, 7299g

3941. Keulemans, A. I. M., "Analysis of Flavors by Gas Chromatographic Techniques," Univ. of Houston Intern. Symp. on Advances in Gas Chromatography, Houston, Texas, January 1963

3942. Keulemans, A. I. M., and Cremer, E., "Gas Chromatographie," Verlag Chemie, Weinheim, Germany, 1959

3943. Keulemans, A. I. M., and Kwantes, A., "Analysis of Volatile Organic Compounds by Means of Vapor-Phase Chromatography," 4th World Petroleum Congr., Rome, June 1955; Abstr., Petroleum Times, 59, 458 (1955)

3944. Keulemans, A. I. M., and Kwantes, A., "Factors Determining Column Efficiency in Gas-Liquid Partition Chromatography," in "Vapour Phase Chromatography," edited by D. H. Desty, Academic Press, Inc., New York, 1957, pp. 15-34

3945. Keulemans, A. I. M., Kwantes, A., and Rijnders, G. W. A., "Quantitative Analysis with Thermal Conductivity Detection in Gas-Liquid Chromatography," Anal. Chim. Acta, 16, 29-39 (1957); CA 51, 7931i

3946. Keulemans, A. I. M., Kwantes, A., and Zaal, P., "The Selectivity of the Stationary Liquid in Vapour-Phase Chromatography," Anal. Chim. Acta, 13, 357-372 (1955); CA 50, 3943e

3947. Keulemans, A. I. M., and McNair, H. M., "Techniques of Gas Chromatography," in "Chromatography," edited by E. Heftmann, Reinhold Publishing Corp., New York, 1961, Chapter 8

3948. Keulemans, A. I. M., and Perry, S. G., "Identification of Hydrocarbons by Thermal Cracking," 4th Intern. Gas Chromatography Symp., Hamburg, Germany, June 1962; pub. in "Gas Chromatography 1962," M. van Swaay, ed., Butterworth & Co., London, 1962

3949. Keulemans, A. I. M., and Perry, S. G., "Qualitative Analysis of Hydrocarbons by Pyrolysis and Gas Chromatography," Nature, 193, 1073 (1962); CA 57, 2833i

3950. Keulemans, A. I. M., and Rijnders, G. W. A., "The Principles and Some Applications of Gas Chromatography," Naturwiss., 45, 301-309 (1958); CA 52, 16005a

3951. Keulemans, A. I. M., and Verver, C. G., "Gas Chromatography," Reinhold Publishing Corp., New York, 1957

3952. Keulemans, A. I. M., and Voge, H. H., "Reactivities of Naphthenes Over a Platinum Catalyst by a Gas Chromatographic Technique," J. Phys. Chem., 63, 476-480 (1959); CA 53, 15734c; 133rd Natl. ACS Mtg., San Francisco, Calif., April 1958, Program Abstr., p. 14I

3953. Keyes, F. G., "Thermal Conductivity of Gases," Trans. Am. Soc. Mech. Engrs., 76, 809-816 (1954); CA 48, 10398b

3954. Khan, M. A., "Non-Equilibrium Theory of Gas-Liquid Chromatography," Nature, 186, 800-801 (1960); CA 54, 21930g

3955. Khan, M. A., "Non-Equilibrium Theory of Capillary Columns and the Effect of Interfacial Resistance on Column Efficiency," 4th Intern. Gas Chromatography Symposium, Hamburg, Germany, June 1962; pub. in "Gas Chromatography 1962," M. van Swaay, ed., Butterworth & Co., London, 1962

3956. Khan, M. A., "Fundamental Aspects of Gas Chromatography," Lab. Practice, 11, No. 1, 26-27 (1962); CA 56, 10877i

3957. Khan, M. A., "Fundamental Aspects of Gas Chromatography. I. Basic Principles. II. General Theory. III. Applications in Analysis. IV. Applications in Thermodynamics," Lab. Practice, 10, 547-551, 562; 709-714 (1961); 11, 120-124; 195-198 (1962); CA 57, 15775c

3958. Khan, M. A., and Whitman, B. T., "Analytical Separation of the Methyl Esters of the C_8-C_{34} Straight-Chain Fatty Acids and the Detection of Odd-Carbon-Number Acids in Commercial Mixtures of Fatty Acids by Gas Chromatography," J. Appl. Chem. (London), 8, 549-552 (1958); CA 53, 5020h

3959. Kharasch, M. S., Hambling, J. K., and Rudy, T. P., "Reactions of Atoms and Free Radicals in Solution. XL. Reaction of Grignard Reagents with 1-Bromooctane in the Presence of Cobaltous Bromide," J. Org. Chem., 24, 303-305 (1959); CA 54, 9718e

145

3960. Kharin, A. N., Protasov, P. N., and Voitko, L. M., "The Chromatographic Separation of Dissolved Substances," Doklady Akad. Nauk SSSR, 80, 611-614 (1953); CA 51, 7220f

3961. Khatri, L. L., and Day, E. A., "Analysis of the Free Fatty Acids of Fresh Milk Fats and Ripened Cream Butters," 57th Annual Mtg., Am. Dairy Sci. Assoc., College Park, Md., June 1962; Abstr., J. Dairy Sci., 45, 660 (1962)

3962. Kienitz, H., "Fundamentals of Gas Chromatography," Erdol u. Kohle, 11, 338 (1958)

3963. Kienitz, H., "Modern Physical Methods of Analysis," Chem.-Ing.-Tech., 32, 641-650 (1960); CA 55, 1271e

3964. Kieselbach, R., "Theory of Plate Height in Gas Chromatography," Anal. Chem., 32, 880-881 (1960); CA 54, 21929h

3965. Kieselbach, R., "Reduction of Noise in Thermal Conductivity Detectors for Gas Chromatography," Anal. Chem., 32, 1749-1754 (1960); CA 55, 6945c

3966. Kieselbach, R., "Influence of Detector and Column Performance," 13th Annual Summer Symp. on Anal. Chem., Univ. of Houston, Houston, Texas, June 1960

3967. Kieselbach, R., "Gas Chromatography. The Effect of Gaseous Diffusion on Mass Transfer in Packed Columns," Anal. Chem., 33, 23-28 (1961); CA 55, 15006c

3968. Kieselbach, R., "Gas Chromatography. Source of the Velocity-Independent A Term in the van Deemter Equation," Anal. Chem., 33, 806-807 (1961); CA 55, 20755a

3969. Kieselbach, R., "Thermal Conductivity Detection," 3rd Annual Eastern Analytical Symp. and Instrument Exhibit, New York, N. Y., November 1961

3970. Kieselbach, R., "Gas Chromatography - 1961," in "Gas Chromatography," edited by N. Brenner, J. E. Callen, and M. D. Weiss, Academic Press, Inc., New York, 1962, pp. 139-148

3971. Kieselbach, R., "Gas Chromatography: An Experimental Study of Air Peaks," Anal. Chem., 35, 1342-1345 (1963)

3972. Kieser, M. E., and Pollard, A., "The Estimation of Some Fruit Juice Volatiles by Use of the Flame-Ionization Detector," Intern. Fruchtsaft-Union, Ber. Wiss.-Tech. Komm., 4, 249-255 (1962); CA 58, 13056f

3973. Kieser, M. E., and Sissons, D. J., "Formation of Volatile Compounds on Gas-Liquid Chromatography Columns, Nature, 185, 529 (1960); CA 55, 25177e

3974. Kikuchi, Y., "Application to Polymer Chemistry," Kagaku no Ryoiki Zokan, No. 44, 169-181 (1961); CA 56, 3a

3975. Kilheffer, J. V., and Jungermann, E., "Quantitative Gas Chromatography of Fatty Derivatives. Relative Detector Response to C_6-C_{14} Saturated Methyl Esters," J. Am. Oil Chemists' Soc., 37, 456-458 (1960); CA 54, 23369e

3976. Kim, G., "Activity of Chromia-Silica as a Dehydro-Aromatization Catalyst," Univ. Microfilms (Ann Arbor, Mich.), L. C. Card No. Mic 60-3595, 97 pp.; Dissertation Abstr., 21, 775 (1960); CA 55, 8005e

3977. Kimura, K., "Effect of Administration of High Fat Diet on Liver Damage," Nippon Naika Gakkai Zasshi, 50, 1270-1288 (1962); CA 59, 941a

3978. Kimura, M., "Recent Advances of Instrumental Analyses in the Petroleum Industry," Nenryo Kyokaishi, 38, 509-517 (1959); CA 54, 5058h

3979. Kimura, M., "Gas Chromatography of Petroleum Hydrocarbons," Yukagaku, 9, 8-13 (1960); CA 54, 13939a

3980. King, L. C., and Farber, H., "Reactions of Terpenes. IV. Reaction of α-Pinene Oxide with p-Toluenesulfonic Acid and Quinaldine," J. Org. Chem., 26, 326-329 (1961); CA 55, 12441d

3981. King, R. W., Fabrizio, F. A., and O'Donnell, A. R., "Application of Gas Chromatography to Some High-Boiling Compounds Present in Petroleum and Tar Oils," 3rd Intern. Symp. on Gas Chromatography, ISA, East Lansing, Mich., June 1961; ISA Proc., 3, 101-110 (1961)

3982. King, R. W., Fabrizio, F. A., and O'Donnell, A. R., "Application of Gas Chromatography to Some High-Boiling Compounds Present in Petroleum and Tar Oils," in "Gas Chromatography," edited by N. Brenner, J. E. Callen, and M. D. Weiss, Academic Press, Inc., New York, 1962, pp. 149-162

3983. King, W. J., Wilson, K., and Swartz, D. J., "Analysis of Automotive Exhaust Gas," J. Air Pollution Control Assoc., 12, 5-21, 47 (1962); CA 56, 7654a

3984. Kingsbury, K. J., Crossley, S. P., and Morgan, D. M., "The Fatty Acid Composition of Human Depot Fat," Biochem. J., 78, 541-550 (1961); CA 55, 8565e

3985. Kingsley, G. R., "Clinical Chemistry," Anal. Chem., 33, No. 5, 13R-32R (April, 1961); CA 55, 12137e

3986. Kingsley, G. R., "Clinical Chemistry," Anal. Chem., 35, No. 5, 11R-35R (April, 1963)

3987. Kingston, B. H., "Perfumery and Essential Oils," Mfg. Chemist, 33, No. 5, 193-196 (1962); CA 57, 6039i

3988. Kingston, C. R., and Kirk, P. L., "Separation of Components of Marijuana by Gas-Liquid Chromatography," Anal. Chem., 33, 1794-1795 (1961); CA 56, 1528i

3989. Kinsell, L. W., Michaels, G. D., Wheeler, P., Barcellini, A., and Walker, G., "Fatty Acid Composition of Plasma and Plaque Lipids in Patients Receiving Natural and Purified Poly-Unsaturated Fats," 32nd Scientific Session, Am. Health Assoc., Philadelphia, Pa., October 1959

3990. Kipling, J. J., and Peakall, D. B., "Reversible and Irreversible Adsorption of Vapors by Solid Oxides and Hydrated Oxides," J. Chem. Soc., 1957, 834-842; CA 51, 7099a

3991. Kipping, P. J., "Determination of Hydrogen in Gaseous Mixtures by Gas Chromatography," Nature, 191, 270-271 (1961); CA 56, 17a

3992. Kipping, P. J., and Jeffery, P. G., "Gas Chromatography in Fuel-Gas Analysis," Res. Develop. Ind., No. 9, 46-49 (1962); CA 58, 2304g

3993. Kircher, H. W., "Gas-Liquid Partition Chromatography of Methylated Sugars," Anal. Chem., 32, 1103-1106 (1960); CA 54, 24132f; 136th Natl. ACS Mtg., Atlantic City, N. J., September 1959, Program Abstr., p. 18D

3994. Kircher, H. W., "Gas-Liquid Partition Chromatography in Carbohydrate Chemistry," 139th Natl. ACS Mtg., St. Louis, Mo., March 1961, Program Abstr., pp. 9D-10D

3995. Kircher, H. W., "Separation of Glucose, Mannose, and Xylose by Gas-Liquid Chromatography of Their Acetylated Glycosides," Tappi, 45, 143-145 (1962); CA 57, 13168b

3996. Kircher, H. W., "Gas-Liquid Partition Chromatography of Sugar Derivatives," in "Methods of Carbohydrate Chemistry," edited by R. L. Whistler and M. L. Wolfrom, Academic Press, Inc., New York, 1, 13-20 (1962); CA 58, 12649c

3997. Kirk, A. D., "The Pyrolysis of Alkyl Hydroperoxides in the Gas Phase," Trans. Faraday Soc., 56, 1296-1303 (1960); CA 55, 5092h

3998. Kirkland, J. J., "An Apparatus for Laboratory Preparative-Scale Vapor Phase Chromatography," in "Gas Chromatography," edited by V. J. Coates, H. J. Noebels, and I. S. Fagerson, Academic Press, Inc., New York, 1958, pp. 203-222

3999. Kirkland, J. J., "Analysis of Polychlorobenzoic Acids by Gas Chromatography of the Methyl Esters," 11th Delaware Science Symp., ACS and Am. Inst. Chem. Engrs, Wilmington, Del., February 1959

4000. Kirkland, J. J., "Analysis of Sulfonic Acids and Salts by Gas Chromatography of Volatile Derivatives," Anal. Chem., 32, 1388-1393 (1960); CA 55, 4254d; 137th Natl. ACS Mtg., Cleveland, Ohio, April 1960, Program Abstr., p. 33B

4001. Kirkland, J. J., "Residue Analysis by Gas Chromatography," Agr. News Letter, 28, 13 (Fall, 1960)

4002. Kirkland, J. J., "Analysis of Polychlorinated Benzoic Acids by Gas Chromatography of Their Methyl Esters," Anal. Chem., 33, 1520-1524 (1961); CA 56, 935d

4003. Kirkland, J. J., "Trace Analysis by Programmed Temperature Chromatography. Simultaneous Determination of Monuron and Diuron Herbicide Residues," Anal. Chem., 34, 428-433 (1962); CA 56, 14664b; 140th Natl. ACS Mtg., Chicago, Ill., September 1961, Program Abstr., pp. 19A, 61B

4004. Kirkland, J. J., "A Critical Study of Fluorocarbons as Solid Supports in Gas Chromatography," 14th Pittsburgh Conf. on Anal. Chem. & Appl. Spectroscopy, Pittsburgh, Pa., March 1963, Program Abstr., pp. 54-55

4005. Kirkland, J. J., "Solid Supports in VPC," 5th Eastern Analytical Symp., New York, N. Y., November 1963

4006. Kirkland, J. J., "Fibrillar Boehmite - A New Adsorbent for Gas Solid Chromatography," Anal. Chem., 35, 1295-1297 (1963)

4007. Kirmse, W., and Doering, W. v. E., "Cyclopropanes from Alkyl Chlorides by Alpha-Elimination," Tetrahedron, 11, 266-271 (1960); CA 55, 10341i

4008. Kirsch, F. W., and Shull, S. E., "Selective Hydrogenation of Butadiene. Catalyst Poisoning and Stabilization," Prod. Res. & Develop., 2, No. 1, 48-52 (1963)

4009. Kirschman, J. C., and Coniglio, J. G., "Polyunsaturated Fatty Acids in Tissues of Growing Male and Female Rats," Arch. Biochem. Biophys., 93, 297-301 (1961); CA 55, 20124g

4010. Kirschner, M. A., and Fales, H. M., "Gas Chromatographic Analysis of 17-Hydroxycorticosteroids by Means of Their Bismethylenedioxy Derivatives," Anal. Chem., 34, 1548-1551 (1962); CA 58, 2617b

4011. Kirsten, W. J., and Andren, R. G. G., "Gas Chromatographic Separation of Pyridine Homologues," J. Chromatog., 8, 531 (1962); CA 58, 3871e

4012. Kiselev, A. V., "Surface Chemistry, Adsorption Energy and Adsorption Equilibria," Quart. Rev. (London), 15, 99-124 (1961); CA 55, 14009f

4013. Kiselev, A. V., "Modification of the Solid Phase for Gas Chromatographic Adsorbents and Capillaries," Vestnik Moskov. Univ., Ser. II, Khim., 16, No. 5, 31-51 (1961); CA 56, 6686c

4014. Kiselev, A. V., El'tekov, Y. A., and Semenova, V. N., "Adsorption of Solutions of Thiophene with n-Heptane on the Molecular Sieve 5A," Kinetika i Kataliz, 3, 421-426 (1962). CA 58, 45a

4015. Kiselev, A. V., Khrapova, E. V., and Shcherbakova, K. D., "Gas Chromatographic Determination of Heat of Adsorption of Lower Hydrocarbons on the 5A-Type Zeolites," Neftekhimiya, 2, 877-884 (1962); CA 58, 8430a

4016. Kiselev, A. V., and Pavlova, L. F., "Adsorption with the Molecular Sieve 5A from Benzene-Hexane

Solution," Kinetika i Kataliz, 2, 599-605 (1961); CA 57, 90h

4017. Kiselev, A. V., and Shcherbakova, K. D., "Chemical Modification of the Surface of Silicon Gels, Its Influence on the Adsorption Properties and Its Possible Role in Gas Adsorption Chromatography," in "Gas Chromatographie 1959," edited by R. E. Kaiser and H. G. Struppe, Akademic Verlag, Berlin, 1959, pp. 198-232

4018. Kishimoto, K., "Flame-Ionization Detector for Gas Chromatography," Kagaku no Ryoiki Zokan, No. 44, 247-256 (1961); CA 56, 3b

4019. Kishimoto, K., and Yasumori, Y., "Preparative Gas Chromatograph," Bunseki Kagaku, 12, 125-130 (1963); CA 58, 12183c

4020. Kishimoto, Y., and Radin, N. S., "Isolation and Determination Methods for Brain Cerebrosides, Hydroxy Fatty Acids, and Unsaturated and Saturated Fatty Acids," J. Lipid Research, 1, 72-78 (1959); CA 54, 11204a

4021. Kisic, A., and Prostenik, M., "Sphingolipids Series. XIX. Note on the Distribution of C_{18}- and C_{20}- Phytosphingosine in Yeast Cerebrin," Croatica Chem. Acta, 32, 229-230 (1960); CA 55, 23359f

4022. Kistiakowsky, G. B., and Sauer, K., "Reactions of Methylene. II. Ketene and Carbon Dioxide," J. Chem. Soc., 80, 1066-1071 (1958); CA 52, 11591d

4023. Kitagawa, I., Sugai, M., and Kummerow, F. A., "Infrared Spectra and Gas Chromatography of Some Oxygenated Fatty Derivatives," J. Am. Oil Chemists' Soc., 39, 217-222 (1962); CA 56, 14416b

4024. Kitahara, K., "Study of Photochemical Changes in Unsaturated Fatty Acids by Gas Chromatography," Yakugaku Zasshi, 80, 1661 (1960); CA 55, 7007d

4025. Kitahara, K., "Unsaturated Fatty Acids. I. Quantitative Analysis of Methyl Oleate and Its Ozonide by Gas Chromatography," Yakugaku Zasshi, 80, 1624-1627 (1960); CA 55, 20459c

4026. Kitahara, K., "Unsaturated Fatty Acids. II. Decomposition Products of Unsaturated Fatty Acids Ozonides by Gas Chromatography," Yakugaku Zasshi, 80, 1628-1631 (1960); CA 55, 20459e

4027. Kitahara, K., "Unsaturated Fatty Acids. III. Investigation by Gas Chromatography of the Photochemical Conversion of Unsaturated Fatty Acids," Yakugaku Zasshi, 81, 126-129 (1961); CA 55, 23323b

4028. Kitahara, K., and Tanaka, Y., "Behavior of Unsaturated Fatty Acids and Liquid Organic Compounds in a Gamma-Ray Field," Yakugaku Zasshi, 83, 203-206 (1963); CA 58, 12102h

4029. Kitahara, M., and Konishi, T., "Estimation of Carbonyl Compounds by Gas Chromatography," Rika Gaku Kenkyusho Hokoku, 38, 90-94 (1962); CA 58, 5021e

4030. Kjaer, A., and Jart, A., "Isothiocyanates. XXIX. Separation of Volatile Isothiocyanates by Gas Chromatography," Acta Chem. Scand., 11, 1423 (1957); CA 52, 9959e

4031. Klaas, P. J., "Gas Chromatographic Determination of Sulfur Compounds in Naphthas Employing a Selective Detector," Anal. Chem., 33, 1851-1854 (1961); CA 56, 11883a

4032. Klasse, J. M., and Hampton, W., "A New and Versatile Gas Chromatograph for the Research Laboratory," ISA Winter Instrument-Automation Conf., Houston, Texas, February 1960

4033. Klaver, R. F., and LeTourneau, R. L., "A Fast Mass Spectrum Digitizer for a Gas Chromatograph-Mass Spectrometer Combination," 28th Midyear Mtg., API Div. of Refining, Philadelphia, Pa., May 1963

4034. Kleber, W., and Schmid, P., "Gas-Chromatographic Determination of Trichlorethylene in Paste-Forming Substances," Brauwissenschaft, 16, No. 2, 36-39 (1963); CA 59, 1100a

4035. Klein, A. K., and Gajan, R. J., "Determination of Pentachloronitrobenzene in Vegetables," J. Assoc. Offic. Agr. Chemists, 44, 712-719 (1961); CA 56, 745h

4036. Klein, E., and Barter, Jr., C. J., "Separations of Methylated Sugars by Gas-Liquid Partition Chromatography," Textile Res. J., 31, 486-487 (1961); CA 56, 6199b

4037. Klemm, L. H., and Reed, D., "Correlation Between Chromatographic Adsorbability on Alumina and Polarographic Reducibility of Conjugated Hydrocarbons," 133rd Natl. ACS Mtg., San Francisco Calif., April 1958, Program Abstr., p. 6P

4038. Klemm, L. H., Solomon, W. C., and Kohlik, A. J., "Coplanarity Effects on the Spectral, Gas Chromatographic, Polarographic, and Diels-Alder Characteristics of 1-Alkyl-1-(2-naphthyl) ethenes," J. Org. Chem., 27, 2777-2786 (1962); CA 57, 9753d

4039. Klenk, E., and Brucker-Voigt, L., "The Eicosapolyenoic Acids of Herring Oil," Z. Physiol. Chem. 324, 1-11 (1961); CA 56, 3579h

4040. Klenk, E., and Eberhagen, D., "The C_{20} and C_{22} Polyene Acids of Glycerol Phosphatides from Bovine Adrenals," Z. Physiol. Chem., 322, 258-266 (1960); CA 55, 10629c

4041. Klenk, E., and Eberhagen, D., "Composition of Fatty Acid Mixtures of Various Fish Oils," Z. Physiol. Chem., 328, 180-188 (1962); CA 57, 14303d

4042. Klenk, E., and Mohrhauer, H., "Metabolism of Polyene Fatty Acids in the Rat," Z. Physiol. Chem., 320, 218-222 (1960); CA 55, 2839d

4043. Klenk, E., and Oette, K., "Nature of the C_{20}- and C_{22}-Polyenoic Acids in Liver Phosphatides After Feeding Linoleic and Linolenic Acid to Rats Raised on a Fat-Free Diet," Z. Physiol. Chem., 318,

86-99 (1960); CA 54, 17628e

4044. Klenk, E., Oette, K., Köhler, J., and Schöll, H., "The Metabolism of Polyenoic Fatty Acids," Z. Physiol. Chem., 323, 270-277 (1961); CA 56, 15932a

4045. Klesper, E., Corwin, A. H., and Turner, D. A., "Porphyrin Studies. XX. High Pressure Gas Chromatography Above Critical Temperatures," J. Org. Chem., 27, 700-701 (1962); CA 57, 1511b

4046. Klima, J., "Methods of Gas Analysis," Paliva, 32, 275-282 (1952); CA 50, 8391b

4047. Kliman, B., and Foster, D. W., "Analysis of Aldosterone by Gas-Liquid Chromatography," Anal. Biochem., 3, 403-407 (1962); CA 57, 10129a

4048. Klingman, C. L., "Semi-Continuous Analysis of Helium in Natural Gas," 6th Instrumental Methods of Analysis Symp., ISA, Montreal, Canada, June 1960

4049. Klingman, C. L., "Modified Chromatograph to Record the Helium Content of Natural-Gas Streams," Rev. Sci. Instr., 32, 822-824 (1961); CA 56, 14531c

4050. Klinkenberg, A., "Chromatography of Substances Undergoing Slow Reversible Chemical Reactions," Chem. Eng. Sci., 15, 255-259 (1961); CA 56, 8042g

4051. Klinkenberg, A., and Sjenitzer, F., "Holding-Time Distributions of the Gaussian Type," Chem. Eng. Sci., 5, 258-270 (1956); CA 51, 778e

4052. Klinkenberg, A., and Sjenitzer, F., "'Eddy' Diffusion in Chromatography," Nature, 187, 1023 (1960); CA 55, 4102c

4053. Klouwen, M. H., and Heide, R. ter, "Identification of Monoterpene Hydrocarbons by Means of Gas Chromatography," 3rd Intern. Symposium on Gas Chromatography, ISA, East Lansing, Mich., June 1961 (see Item 4055)

4054. Klouwen, M. H., and Heide, R. ter, "Identification of Monoterpenes by Means of Gas Chromatography," Gas Chromatog., Intern. Symp., 1961, 3, 485-505 (Pub. 1962); CA 58, 3873g

4055. Klouwen, M. H., and Heide, R. ter, "Identification of Monoterpene Hydrocarbons by Means of Gas Chromatography," in "Gas Chromatography," edited by N. Brenner, J. E. Callen, and M. D. Weiss, Academic Press, Inc., New York, 1962, pp. 485-506 (see Item 4053)

4056. Klouwen, M. H., and Heide, R, ter, "Studies on Terpenes. I. A Systematic Analysis of Monoterpene Hydrocarbons by Gas-Liquid Chromatography," J. Chromatog., 7, 297-310 (1962); CA 57, 11837i

4057. Klouwen, M. H., and Heide, R. ter, "Separation of Stereoisomeric Menthols by Means of Gas-Liquid Chromatography," Soap, Perfumery Cosmetics, 35, 1082-1083 (1962); CA 58, 8846f

4058. Knapman, C. E. H., "Gas Chromatography in Medical Research," Nature, 192, 717-718 (1961) Review of Gas Chromatography Discussion Group, Institute of Petroleum, Mtg., September 1960

4059. Knapman, C. E. H., and Scott, C. G., (Eds.), "Gas Chromatography Abstracts 1959," Butterworths, Inc., Washington, D. C., 1959

4060. Knapman, C. E. H., and Scott, C. G., (Eds.), "Gas Chromatography Abstracts 1959," Butterworths, Inc., Washington, D. C., 1960

4061. Knapman, C. E. H., and Scott, C. G., (Eds.), "Gas Chromatography Abstracts 1960," Butterworths, Inc., Washington, D. C., 1961

4062. Knapman, C. E. H., and Scott, C. G., (Eds.), "Gas Chromatography Abstracts 1961," Butterworths, Inc., Washington, D. C., 1962

4063. Kniebes, D. V., "Utility Gas Analysis by Gas Chromatography," Inst. Gas Technol., Tech. Rept. No. 4, 32 pp., 1962; CA 57, 16965c

4064. Knight, A. R., and Gunning, H. E., "Primary Methoxy Radical Formation in the Reaction of Methanol Vapor with Hg $6(^3P_1)$ Atoms," Can. J. Chem., 39, 1231-1238 (1961); CA 55, 25727h

4065. Knight, H. S., "Gas Chromatography of Olefins. Determination of Pentenes and Hexenes in Gasoline," Anal. Chem., 30, 9-14 (1958); CA 52, 8521i

4066. Knight, H. S., "Gas-Liquid Chromatography of Hydroxyl and Amino Compounds. Production of Symmetrical Peaks," Anal. Chem., 30, 2030-2032 (1958); CA 53, 3834b

4067. Knight, H. S., "Stationary Phase," in "Principles and Practice of Gas Chromatography," edited by R. L. Pecsok, John Wiley & Sons, Inc., New York, 1959, pp. 48-55

4068. Knight, H. S., "Column Conditions," in "Principles and Practice of Gas Chromatography," edited by R. L. Pecsok, John Wiley & Sons, Inc., New York, 1959, pp. 56-62

4069. Knight, H. S., "Peak Distortions," in "Principles and Practice of Gas Chromatography," edited by R. L. Pecsok, John Wiley & Sons, Inc., New York, 1959, pp. 63-68

4070. Knight, H. S., "Column Selection," in "Principles and Practice of Gas Chromatography," edited by R. L. Pecsok, John Wiley & Sons, Inc., New York, 1959, pp. 69-82

4071. Knight, H. S., "Column Construction," in "Principles and Practice of Gas Chromatography" edited by R. L. Pecsok, John Wiley & Sons, Inc., New York, 1959, pp. 83-85

4072. Knight, H. S., and Groennings, S., "Indicator Chromatographic Analysis of Organic Mixtures," Anal. Chem., 26, 1549-1553 (1954); CA 49, 2254i

4073. Knight, H. S., and Weiss, F. T., "Determination of Traces of Water in Hydrocarbons. A Calcium Carbide-Gas Liquid Chromatography Method," Anal. Chem., 34, 749-751 (1962); CA 57, 9212h

4074. Knight, J. A., McDaniel, R. L., and Sicilio, F., "Radiolysis Products of C_8 and Greater Carbon Content from 2, 2, 4-Trimethylpentane," J. Phys. Chem., 67, 921-923 (1963); CA 59, 1212e

4075. Knight, J. A., Sicilio, F., and Houze, N., "Gas Chromatographic Analysis in Fractional Distillation of Multi-Component Systems," J. Chromatog., 5, 179-181 (1961); CA 56, 5386b

4076. Knight, J. D., and House, R., "Analysis of Surfactant Mixtures. I.," J. Am. Oil Chemists' Soc., 36, 195-200 (1959); CA 53, 12712d

4077. Knights, B. A., and Thomas, G. H., "R_M Values in the Gas-Liquid Chromatography of Steroids," Nature, 194, 833-835 (1962); CA 57, 6589b

4078. Knights, B. A., and Thomas, G. H., "Effect of Substituents on Relative Retention Times in Gas Chromatography of Steroids," Anal. Chem., 34, 1046-1048 (1962); CA 57, 13168a

4079. Knights, B. A., and Thomas, G. H., "ΔR_{Mg} Values in the Gas Chromatography of Steroids," Chem. & Ind. (London), 1963, 43-44

4080. Knowles, J. R., Norman, R. O. C., and Radda, G. K., "A Quantitative Treatment of Electrophilic Aromatic Substitution," J. Chem. Soc., 1960, 4885-4896; CA 55, 22218a

4081. Knox, J. H., "Application of Gas Phase Partition Chromatography to Competitive Chlorination Reactions," Chem. & Ind. (London), 1955, 1631-1632; CA 50, 13713d

4082. Knox, J. H., "Gas Chromatography," Sci. Progr., 45, 227-244 (1957); CA 51, 9398g

4083. Knox, J. H., "The Application of Gas Chromatography to Gas Reaction Kinetics," Mtg, Scottish Section, Society for Analytical Chemistry, Glasgow, Scotland, November 1958; Abstr., Analyst, 84, 75 (1959)

4084. Knox, J. H., "Constant Flow Device for Temperature-Programmed Gas Chromatography," Chem. & Ind. (London), 1959, 1085-1086; CA 54, 2832c

4085. Knox, J. H., "Gaseous Products from the Oxidation of Propane at 318°C.," Trans. Faraday Soc., 56, 1225-1234 (1960); CA 55, 4338g

4086. Knox, J. H., "The Speed of Analysis by Gas Chromatography," J. Chem. Soc., 1961, 433-441; CA 55, 9145d

4087. Knox, J. H., "Gas Chromatography," John Wiley & Sons, Inc., New York, 1962

4088. Knox, J. H., "The Spreading of 'Air Peaks' in Capillary and Packed Gas Chromatographic Columns," Univ. of Houston Intern. Symp. on Advances in Gas Chromatography, Houston, Texas, January 1963

4089. Knox, J. H., and McLaren, L., "The Spreading of Air Peaks in Capillary and Packed Gas Chromatographic Columns," Anal. Chem., 35, 449-454 (1963)

4090. Knox, J. H., and Nelson, R. L., "Competitive Chlorination Reaction in the Gas Phase: Hydrogen and C_1-C_5 Saturated Hydrocarbons," Trans. Faraday Soc., 55, 937-946 (1959); CA 54, 7521d

4091. Knox, J. H., Smith, R. F., and Trotman-Dickenson, A. F., "Competitive Oxidations. I. Ethane + Propane Mixtures," Trans. Faraday Soc., 54, 1509-1514 (1958); CA 53, 12803d

4092. Knox, J. H., and Trotman-Dickenson, A. F., "Synthesis of Activated Molecules: Methyl Cyclopropane," Chem. & Ind. (London). 1957, 1039; CA 52, 1076i

4093. Knox, J. H., Trotman-Dickenson, A. F., and Wells, C. H. J., "Reactions of Methylene with Isobutene," J. Chem. Soc., 1958, 2897-2898; CA 53, 880e

4094. Knox, W. R., and Libers, R., "Gas Chromatographic Analysis of C_8-C_{10} Aromatic Hydrocarbons on Benzoquinoline," 16th Southwest Regional Mtg., ACS, Oklahoma, Oklahoma City, Okla., December 1960

4095. Kobashi, Y., "Separation and Identification of the Bases of Pyridine and Nicotine by Gas Chromatography," Nippon Kagaku Zasshi, 82, 1262-1265 (1961); CA 58, 11411e

4096. Kobashi, Y., and Watanabe, M., "Determination of Pyridine and Nicotine Homologs by Gas Chromatography," Nippon Kagaku Zasshi, 82, 1262-1265 (1961); CA 58, 11676h

4097. Kobayashi, M., Yoshigi, H., and Katsuhara, J., "The Composition of Japanese Mint Oil," Koryo, No. 60, 27-35 (1960); CA 57, 6042b

4098. Koberstein, E., "The Catalytic Activity of Gamma-Aluminum Oxides Toward Hydrocarbon Reactions," Z. Elektrochem., 64, 906-919 (1960); CA 55, 76a

4099. Koch, R. C., and Grandy, G. L., "Xenon-Krypton Separation by Gas Chromatography," Nucleonics, 18, No. 7, 76-80 (1960); CA 55, 59i

4100. Koch, R. C., and Grandy, G. L., "Rapid Method for Separation and Analysis of Radioactive Fission Gases," Anal. Chem., 33, 43-48 (1961); CA 55, 8163f; 11th Pittsburgh Conf. on Anal. Chem. & Appl. Spectroscopy, Pittsburgh, Pa., February-March 1960, Program Abstr., pp. 44-45

4101. Koch, S. D., Kliss, R. M., Lopiekes, D., and Wineman, R. J., "Synthesis of Polycyclic Hydrocarbons Containing Cyclopropyl Groups," J. Org. Chem., 26, 3122-3125 (1961); CA 56, 5848h

4102. Kochloefl, K., Schneider, P., Rericha, R., Horak, M., and Bazant, V., "Isoprenoid-Skeleton Hydrocarbons in Low-Temperature Brown Coal Tar," Chem. & Ind. (London), 1963, 692; CA 59, 348f

4103. Kocirik, K., and Grubner, O., "Regarding Some Questions on Capillary Gas Chromatography," Gas Chromatog. Symp., Brno, Czech., June 1962; Abstr., J. Gas Chromatog., 1, No. 4, 6 (1963)

4104. Koegl, F., Gier, J. de, Mulder, I., and Deenan, L. L. M. van, "Metabolism and Functions of Phosphatides. Specific Fatty Acids Composition of the Red Blood Cell Membranes," Biochim. Biophys. Acta, 43, 95-103 (1960); CA 55, 3772a

4105. Koegler, H., "Analysis of Mineral-Oil Products," Abhandl. Deut. Akad. Wiss. Berlin, Kl. Chem., Geol., Biol., 1959, No. 9, 261-270; CA 58, 5019a

4106. Kofler, W., "Separation and Purification of Organic Compounds by Means of Sublimation Through Adsorptive Substances," Monatsh. Chem., 80, 694-701 (1949); CA 44, 3407b

4107. Kofron, W. G., Kirby, F.B., and Hauser, C. R., "Reactions of Tetrahalomethanes with Potassium tert-Butoxide and Potassium Amide," J. Org. Chem., 28, 873-875 (1963); CA 58, 11200c

4108. Kogler, H., "Thermo-Gas-Chromatography, a New Method for the Separation and Determination of Mixtures," Chem. Tech. (Berlin), 9, 400-403 (1957); CA 52, 154d

4109. Kogler, H., "Investigations on Petroleum Products with Gas Chromatography," in "Gas Chromatographie 1958," edited by H. P. Angele, Akademie Verlag, Berlin, 1959, pp. 261-270

4110. Kogler, H., Hultschig, M., Fischer, J., and Weidenbach, G., "Automatic Gas Analysis in Industrial Control," Chem. Tech. (Berlin), 9, 220 (1957); CA 51, 13359d

4111. Kokes, R. J., Tobin, Jr., H., and Emmett, P. H., "New Microcatalytic-Chromatographic Technique for Studying Catalytic Reactions," J. Am. Chem. Soc., 77, 5860-5862 (1955); CA 50, 5382c

4112. Kolbel, H., Kuschel, J., and Hammer, H., "Formation of Oxygen Containing Compounds in the Synthesis of Hydrocarbons from Carbon Monoxide and Steam with Cobalt Catalysts," Ann. Chem., 632, 8-15 (1960); CA 54, 22337f

4113. Kolesnikova, L. P., Kamzolkin, V. V., and Khotimskaya, M. I., "Determination of the Composition of Isomeric Alcohols by Gas Chromatography," Neftekhimiya, 2, 355-358 (1962); CA 58, 3871b

4114. Kolloff, R. H., "Septum Bleeding in Flame Ionization Programmed Temperature, Gas Liquid Chromatography," Anal. Chem., 34, 1840-1841 (1962); CA 58, 2825b

4115. Kolobikhin, V. A., "Chromatographic Analysis of the Contact Gas (Formed in the Catalytic Dehydrogenation of Butane and Butenes)," Zavodskaya Lab., 25, 154-157 (1959); Ind. Lab. 25, 164-166 (1959); CA 54, 19289h

4116. Kolobikhin, V. A., "Determination of 1,3-Butadiene by a Chromathermographic Method," Zavodskaya Lab., 26, 814-815 (1960); CA 54, 20670c

4117. Komers, R., and Bazant, V., "The Analysis of Mixtures of Dimethyl Esters of Benzenedicarboxylic Acids by Means of Gas-Liquid Chromatography," Doklady Akad. Nauk S.S.S.R., 126, 1268-1269 (1959); CA 53, 21793a

4118. Komers, R., and Bazant, V., "Separation of the Dimethyl Esters of Benzenedicarboxylic Acids by Gas-Liquid Chromatography," Gazovaya Khromatografiya, Trudy 1-oi [Pervoi] Vsesoyuz. Konf., Akad. S.S.S.R., Moscow, 1959, 313-314 (Pub. 1960); CA 56, 5386c

4119. Komers, R., and Bazant, V., "Separation of Methyl Esters of Benzoic Acid and Benzene Dicarbonic Acids," Gas Chromatog. Symp., Brno, Czech., June 1962; Abstr., J. Gas Chromatog., 1, No. 4, 9 (1963)

4120. Komers, R., and Kochloefl, K., "Study on Gas Chromatographic Separation of the Stereoisomers of Alkylcyclohexanol," Collection Czechoslov. Chem. Communs., 28, 46-54 (1963)

4121. Komers, R., Kochloefl, K., and Bazant, V., "Gas-Liquid Partition Chromatography of Stereo-Isomeric Methyl Cyclohexanols," Chem. & Ind. (London), 1958, 1405-1406; CA 54, 13023f

4122. Komers, R., Kochloefl, K., and Bazant, V., "A Study of the Separation of Alkyl Cyclohexanols by Means of Gas-Liquid Chromatography," Gas Chromatog. Symp. Brno, Czech., June 1962; Abstr., J. Gas Chromatog., 1, No. 4, 9 (1963)

4123. Kondrat'ev, D. A., Markov, M. A., and Minachev, K. M., "Analysis of Mixtures of C_5-C_7 Hydrocarbons by the Method of Gas-Liquid Chromatography," Zavodskaya Lab., 25, 1301-1304 (1959); CA 54, 8468a

4124. Kontorovich, L. M., Iogansen, A. V., Levchenko, G. T., Semina, G. N., Bobrova, V. P., and Stepanova, V. A., "Chromatographic Analysis of Acetylic Hydrocarbons," Zavodskaya Lab., 28, 146-148 (1962); CA 57, 1517h

4125. Kooiman, P.. and Adams, G. A., "Constitution of Glucomannan from Tamarack (Larix Laricina)," Can. J. Chem., 39, 889-896 (1961); CA 55, 20974i

4126. Korn, E. D., "The Fatty Acid and Positional Specificities of Lipoprotein Lipase," J. Biol. Chem., 236, 1638-1642 (1961); CA 55, 24872f

4127. Korol, A. N., "The Use of Gas Adsorption Chromatography for the Analysis of Gases in the Production of Phenol, Acetone, and Synthetic Alcohol," Khim. i Tekhnol. Topliv. i Masel, 5, No. 6, 60-63 (1960); CA 54, 18176f

4128. Korol, A. N., "Analysis of Products of Phenol and Synthetic Ethanol by the Method of Gas-Liquid Chromatography," Zavodskaya Lab., 26, 51-54 (1960); CA 54, 10629f

4129. Korotov, S. Y., and Vyrodov, V. A., "Effect of Gas Velocity on the Plate Efficiency in Absorption Processes," Izvest. Vyssikh Ucheb. Zavedenii, Lesnoi Zhur., 4, No. 1, 124-127 (1961); CA 55, 24139c

4130. Korte, F , and Stiasni, M , "Insecticides in Metabolism III. Microsynthesis of C^{14}-Labeled Telodrin," Ann. 656, 140-144 (1962); CA 58, 6777d

4131. Korvzee, A E , "The Second Virial Coefficient of Benzene " Rec. trav. chim , 72, 483-489 (1953); CA 47, 9702f

4132. Koster, R , and Bruno, G., "Organometallic Compounds XXXIII. Exchange of Hydrocarbon Groups Between Organic Aluminum and Boron Compounds," Ann , 629, 89-103 (1960); CA 54 22331h

4133. Kotel'nikov, B. P , and Datskevich, A. A , "Method for the Determination of the Composition of Fatty Acids and Fatty Alcohol Mixtures, " Masloboino-Zhirovaya Prom., 26, No. 4, 20-26 (1960); CA 54, 18990c

4134. Kovacs, A. S., and Wolf, H. O , "Chromatography of Fruits and Preserves," Ind. Obst- Gemuss- verwert. 47, 159-161 (1962); CA 57, 6378g

4135. Kováts, E., "Gas-Chromatographic Characterization of Organic Compounds. I. Retention Indexes of Aliphatic Halides, Alcohols, Aldehydes, and Ketones," Helv. Chim. Acta, 41, 1915-1932 (1958); CA 53, 8765g

4136. Kováts, E., "Qualitative Analysis by Gas Chromatography," Symp. on Modern Methods for Analysis of Organic Compounds, Munich, Germany, October 1960; Abstr., Angew. Chem., 73, 145 (1961)

4137. Kováts, E., "Relation Between Structure and Gas-Chromatographic Data for Organic Compounds," Z Anal. Chem., 181, 351-366 (1961); CA 56, 7c

4138. Kováts, E., and Heilbronner, E., "Standard-Substance Technique for the Gas Chromatographic Characterization of Organic Compounds," Chimia (Switz.), 10, 288-289 (1956);   CA 51, 5638e

4139. Kováts, E., Simon, W., and Heilbronner, E., "Program-Controlled Gas Chromatography for Pre- parative Separation of Organic Compounds. II.," Helv. Chim. Acta, 41, 275-288 (1958); CA 52, 8826a

4140. Krajkeman, A. J., "Perfumery and Essential Oils, " Mfg. Chemist, 30, 329-332 (1959); CA 54, 826h; ibid., 32, 69-71 (1961); CA 55, 11766d

4141. Krajkeman, A. J., "Perfumery and Essential Oils, Progress Report," Mfg. Chemist, 31, 22-26 (1960); CA 54, 9213a

4142. Kramer, G. M., and Schriesheim, A., "Heptane Isomerization," J. Phys. Chem., 64, 849-850 (1960); CA 55, 5321i; 137th Natl. ACS Mtg., Cleveland, Ohio, April 1960, Program Abstr., p. 9Q; Preprints, Div. Petrol. Chem., 5, No. 1, 129-132 (1960)

4143. Kramer, K., and Wright, A. N., "Ethylidene Insertion in Phenylsilane," Tetrahedron Letters, 1962, 1095-1096; CA 58, 12590g

4144. Kramlich, W. E., and Pearson, A. M., "Separation and Identification of Cooked Beef Flavor Com- ponents," Food Research, 25, 712-719 (1961); CA 55, 23864i

4145. Kranz, Z. H., Lamberton, J. A., Murray, K. E., and Redcliffe, A. H., "Sugar Cane Wax. II. Ex- amination of the Constituents of Sugar Cane Cuticle Wax by Gas Chromatography," Australian J. Chem., 13, 498 (1960); CA 55, 9911g

4146. Krapcho, A. P., and Bothner-By, A. A., "Kinetics of the Metal-Ammonia-Alcohol Reductions of Benzene and Substituted Benzenes," J. Am. Chem. Soc., 81, 3658-3666 (1959); CA 54, 7308h

4147. Kratz, P., Jacobs, M., and Mitzner, B. M., "A Smoke-Eliminating Device for Vapor-Phase Chro- matographic Fraction Collector," Analyst, 84, 671-672 (1959); CA 54, 17975h

4148. Kratzl, K., Billek, G., Puschmann, G., Simon, H., Gruber, K., and Bland, D. E., "Radiochemi- cal and Gas Chromatographic Methods for the Study of Reversible Alkoxylations and Alkylations (in Biogenesis)," Z. Anal. Chem., 181, 550 (1961); CA 56, 5048c

4149. Kratzl, K., and Gruber, K., "Analysis of Lacquer Solvents by Gas-Distribution Chromatography. I. Quantitative Analyses," Holzforsch. u. Holzverwert., 9, 87-90 (1957); CA 52, 7734f

4150. Kratzl, K., and Gruber, K., "The Quantitative Separation and Identification of Alkoxy Groups by Gas-Liquid Chromatography," Monatsh., 89, 618-624 (1958); CA 53, 12954e

4151. Kraus, H., "Measurement of Gas Density," Gas-u. Wasserfach, 97, 840 (1956); CA 51, 1663f

4152. Kraus, J. W., and Calvert, J. G., "Some Gas-Phase Reactions of Butyl Free Radicals," 131st Natl. ACS Mtg., Miami, Fla., April 1957, Program Abstr., p. 43R

4153. Krejci, M., "The Application of Zeolites in Gas Chromatography. IV. Separation and Estimation of Krypton in the Presence of Nitrogen and Methane," Collection Czechoslov. Chem. Communs., 25, 2457-2458 (1960); CA 54, 24112i

4154. Krejci, M., and Janak, J., "Methods of Quantitative Interpretation of Thermal Conductivity and Other Differential Registration in Gas Chromatography," Chemie (Prague), 10, 264-272 (1958); CA 53, 21040i

4155. Krejci, M., and Janak, J., "Application of Zeolites in Gas Chromatography. II. The Determina- tion of Argon in the Presence of Oxygen and of Other Permanent Gases," Collection Czechoslov. Chem. Communs., 24, 3887-3892 (1959); CA 54, 10653h

4156. Krejci, M., and Tesarik, K., "Gas Chromatographic Separation of Mixture of Some Rare Gases," 3rd Intern. Conf. on Anal. Chem., Prague, Czech., September 1959

4157. Krejci, M., and Tesarik, K., "Use of Zeolites in Gas Chromatography. III. Separation and Determination of Helium, Neon, and Hydrogen at Ambient Temperatures," Collection Czechoslov. Chem. Communs., 25, 691-694 (1960)

4158. Krejci, M., Tesarik, K., and Janak, J., "Rapid Chromatographic Separation and Determination of Helium-Neon-Hydrogen and Nitrogen-Krypton-Methane Mixtures and Determination of Argon in the Presence of Oxygen at Room Temperature," Gas Chromatog. Intern. Symp., 2nd, East Lansing, Mich., 1959, 255-261 (Pub. 1960); CA 55, 19599f

4159. Krejci, M., Tesarik, K., and Janak, J., "Rapid Chromatographic Separation and Determination of Helium-Neon-Hydrogen and Nitrogen-Krypton-Methane Mixtures and Determination of Argon in the Presence of Oxygen at Room Temperature," in "Gas Chromatography," edited by H. J. Noebels. R. F. Wall, and N. Brenner, Academic Press, Inc., New York, 1961, pp. 255-261

4160. Krell, E., "Automatized Physical Analytical Apparatus for Laboratory and Chemical Plant," Chem. Tech. (Berlin), 12, 575-580 (1960)

4161. Kremer, E., Kraus, T., and Bechtold, E., "Use of a Highly Sensitive, Selective, Halogen Detector in Gas Chromatography," Chem.-Ing.-Tech., 33, 632-633 (1961); CA 56, 286b

4162. Krepinsky, J., and Herout, V., "Plant Substances. XVIII. Isolation of Terpenic Compounds from Solidago Canadensis," Collection Czechoslov. Chem. Communs., 27, 2459-2462 (1962); CA 58, 749f

4163. Kretschmer. F., "Gas Density and Means of Measurement," Gas-u. Wasserfach., 97, 461-465 (1956); CA 50, 11643g

4164. Kreuchunas, A., "Transfer Device for Gas Chromatography," Chemist-Analyst, 49, 82-83 (1960); CA 57, 2827e

4165. Kreyenbuhl, A., "Construction of Tungsten Wire Katharometers for Gas Chromatography," J. Chromatog., 4, 130-137 (1960); CA 55, 4060c

4166. Kreyenbuhl, A., "Construction of Heat-Resistant Glass Capillary Columns for Gas Chromatography," Bull. soc. chim. France, 1960, 2125-2127; CA 55, 17113f

4167. Kreyenbuhl, A., and Weiss, H., "Analysis of Phenols by Gas Chromatography," Bull. soc. chim. France, 1959, 1880-1885; CA 55, 5243g

4168. Krhut, A., Brhacek, L., and Kalandra, O., "Chromatographic Determination of Gases in Iron and Steel," 3rd Intern. Conf. on Anal. Chem., Prague, Czech., September 1959

4169. Krhut, A., Kalandra, O., and Brhacek, L., "Chromatographic Determination of Gases in Iron and Steel," Hutnicke listy, 15, 133-136 (1960); CA 55, 20766a

4170. Krichmar, S. I., "The Application of Chromatographic Analysis in the Control of Ethylbenzene Manufacture," Khim. Prom., 1961, 465-467; CA 56, 935b

4171. Krichmar, S. I., and Beilina, L. I., "Experience With a Chromatographic Analyzer Using Thermal Conductivity With Combustion of the Components to Carbon Dioxide," Zavodskaya Lab., 26, 1171-1172 (1960); Ind. Lab., 26, 1355-1356 (1960)

4172. Krijkenian, A. J., "Perfumery and Essential Oils," Mfg. Chemist, 32, 69-71 (1961); CA 55, 11766d

4173. Kring, E. V., Jenkins, G. I., and Bacchetta, V. L., "The Application of Gas-Liquid Partition Chromatography to Problems in Chemical Kinetics; Acid-Catalyzed Methanolysis of Enol Acetates," J. Phys. Chem., 64, 947-949 (1960); CA 55, 15339a; 136th Natl. ACS Mtg., Atlantic City, N. J., September 1959, Program Abstr., p. 35S

4174. Kritchevsky, D., and Holmes, W. L., "Occurrence of Demosterol in Developing Rat Brain," Biochem. Biophys. Research Communs., 7, 128-131 (1962); CA 57, 10383f

4175. Kritchevsky, D., Langan, J., Markowitz, J., Berry, J. F., and Turner, D. A., "Cholesterol Vehicle in Experimental Atherosclerosis. III. Effects of Absence or Presence of Fatty Vehicle," J. Am. Oil Chemists' Soc., 38, 74-76 (1961); CA 55, 7573b

4176. Kritz, W. R., "Automatic Gas Chromatograph for Monitoring of Reactor-Fuel Failures. I. Design," U. S. At. Energy Comm., DP-356, 10 pp. (1959); CA 54, 17086h

4177. Kritz, W. R., "Chromatographic Analysis of Radioactive Gases," U. S. At. Energy Comm., TID-7606, 268-275 (1960); CA 55, 23173h

4178. Kritz, W. R., "Gas Chromatograph Monitors Reactor for Fuel Failures," Nucleonics, 19, No. 4, 106, 108 (1961)

4179. Krivoruchko, F. D., and Turkel'taub, N. M., "Chromatographic Method of Separate Determination of Butadiene, Ethylbenzene, and Styrene in Air," Zavodskaya Lab., 22, 1408 (1956); CA 51, 4879e

4180. Kroman, H. S., and Bender, S. R., "Gas Chromatography of Estrogens," Gas-Chrom Newsletter (Applied Science Laboratories, State College, Pa.), 3, No. 4, 4 (December 1962)

4181. Kroman, H. S., and Bender, S. R., "An Improved Method for the Gas Chromatographic Separation of Estrogens," J. Chromatog., 10, 111-112 (1963)

4182. Kronmueller, G., "Automatic Fraction Collector for Gas Chromatography," Gas Chromatog., Intern. Symposium, 2nd, East Lansing, Mich., 1959, 199-206 (Pub. 1961); CA 55, 18203g

4183. Kronmueller, G., "Automatic Fraction Collector for Gas Chromatography," in "Gas Chromatography," edited by H. J. Noebels, R. F. Wall, and N. Brenner, Academic Press, Inc., New York, 1961, pp. 199-206

4184. Krumphanzl, V., and Dyr, J., "Application of Gas Chromatography to Wine Industry and in Testing Spirits," Prumsyl Potravin, 13, 436-439 (1962); CA 57, 17207a

4186. Krupp, H., "Magnetic Oxygen Determination with a Heated Wire Arrangement," Chem.-Ing.-Tech., 27, 79-83 (1955); CA 49, 5893c

4187. Krylov, B. K., and Kalmanovskii, V. I., "Mass Spectrometer Detector for Chromatographic Analysis," Tr. po Khim. i Khim. Tekhnol., 4, 747-752 (1961); CA 58, 7346h

4188. Krzeminski, Z. S., and Angyal, S. J., "Separation of Inositols and Their Monomethyl Ethers by Gas Chromatography," J. Chem. Soc., 1962, 3251-3252; CA 57, 12607h

4189. Kubiczkova, H., Rezl, V., and Kucharczyk, N., "Separation of Nicotine and Iso-nicotine Acid by Means of Gas Chromatography," Gas Chromatog. Symp., Brno, Czech., June 1962; Abstr., J. Gas Chromatog., 1, No. 4, 10 (1963)

4190. Kubinova, M., "Gas Chromatographic Determination of Alcohols," 3rd Intern. Conf. on Anal. Chem., Prague, Czech., September 1959

4191. Kubinova, M., "Isoprene Estimation in Reaction Mixtures," Chem. prumsyl, 9, 160-164 (1959); CA 53, 20874d

4192. Kubinova, M., "Application of the Mathematical Statistics for the Evaluation of Chromatographic Calibration Curves," in "Gas Chromatographie 1959," edited by R. E. Kaiser and H. G. Struppe, Akademie Verlag, Berlin, 1959, pp. 233-235

4193. Kubinova, M., "Carriers for Gas-Liquid Distribution Chromatography," Chem. listy, 53, 850-856 (1959); CA 53, 21362f

4194. Kubinova, M., "Diatomaceous Earths and Firebrick Materials as Solid Stationary Phases for Gas-Liquid Chromatography," Gazovaya Khromatografiya, Trudy 1-oi [Pervoi] Vsesoyuz. Konf., Akad. Nauk S.S.S.R., Moscow, 1959, 231-239 (Pub. 1960); CA 55, 25576h

4195. Kubinova, M., "Chromatographic Estimation of Diethyl Ether in Condensates from the Lebedev Synthesis of Butadiene," Chem. prumsyl, 10, 359-360 (1960); CA 55, 248e

4196. Kubinova, M., and Mikl, O., "Application of Gas Chromatography in Synthetic Rubber Research," Przemysl Chem., 39, 552-555 (1960); CA 55, 11897a

4197. Kuck, J. A., "Perspectives in Quantitative Organic Microanalysis," Anal. Chem., 30, 1552-1556 (1958); CA 52, 19711i

4198. Kudo, S., Shimomura, K., and Negishi, K., "Noncatalytic Oxidation of Paraffinic Hydrocarbons," Kogyo Kagaku Zasshi, 65, 1372-1378 (1962); CA 58, 6685h

4199. Kudo, S., Shimomura, K., Negishi, K., and Fukui, K., "Position of Attack on the Carbon Chain of n-Paraffinic Hydrocarbons by Molecular Oxygen in Liquid-Phase Autoxidation," Kogyo Kagaku Zasshi, 65, 1379-1384 (1962); CA 58, 6686c

4200. Kuffner, F., and Kallina, D., "Separation of Isomeric Alcohols by Means of Gas Liquid Chromatography. Separation of Saturated C_5-Alcohols," Monatsh., 90, 463-466 (1959); CA 54, 3173d

4201. Kuhl, M., "Electronic Integration of Gas Chromatographic Curves," in "Gas Chromatographie 1959," edited by R. E. Kaiser and H. G. Struppe, Akademie Verlag, Berlin, 1959, pp. 236-245

4202. Kuhl, M., "Principles and Adaptation of an Electronically Stabilized Equipment Used as Power Supply for Thermal Conductivity Cells in Gas Chromatography," in "Gas Chromatographie 1958," edited by H. P. Angele, Akademie Verlag, Berlin, 1959, pp. 254-260

4203. Kuhl, M., "Construction of an Electronically Stabilized Network-Fed Current Supply for the Heating of Thermoconductivity Cells," Abhandl. Deut. Akad. Wiss. Berlin, Kl. Chem., Geol., Biol., 1959, No. 9, 254-260; CA 58, 5018h

4204. Kuhn, W., Narten, A., and Thuerkauf, M., "Continuous Gas Chromatography. I. Procedure for Continuous Separation of a Mixture of Several Components in Two-Phase Countercurrent Operation with Temperature Gradient," Helv. Chim. Acta, 41, 2135-2148 (1958); CA 53, 8765e

4205. Kuhn, W., Narten, A., and Thuerkauf, M., "Continuous Separation of Multicomponent Mixtures by Gas Chromatography," World Petrol. Congr., Proc. 5th, N. Y., 1959, 45-53 (Pub. 1960); CA 59, 433c

4206. Kuhnhanss, G., Rosner, H., Huttig, E., Wagner, M., and Tischendorf, G., "Analysis of Hydrocarbon Mixtures," Erdol u. Kohle, 10, 372-375 (1957); CA 51, 16205i

4207. Kuhns, L. J., Braman, R. S., and Graham, J. E., "Evaluation of Analytical Methods for Decaborane," Anal. Chem., 34, 1700-1702 (1962); CA 58, 6189c; 138th Natl. ACS Mtg., New York, N. Y., September 1960, Program Abstr., p. 24B

4208. Kuksis, A., McCarthy, M. J., and Beveridge, J. M. R., "Quantitative Gas-Liquid Chromatographic Analysis of Butterfat Triglycerides," 36th Fall Mtg., American Oil Chemists' Soc., Toronto, Ontario, October 1962

4209. Kulcsar, G. J., and Kulcsar-Novakova, M., "Automatic Gas Flow Rate Regulator," Acad. rep.

populare Romine, Filiala cluj, Studii ceretari chim., 7, No. 1-4, 119-132 (1956); CA 52, 8632f

4210. Kuley, C. J., "Gas Chromatographic Analysis of C$_1$ to C$_4$ Hydrocarbons in the Parts Per Million Range in Air and in Vaporized Liquid Oxygen," Anal. Chem., 35, 1472-1475 (1963)

4211. Kumar, D., Goodno, J. A., and Barnes, A. C., "Isolation of Progesterone from Human Pregnant Myometrium," Nature, 195, 1204 (1962); CA 58, 1687b

4212. Kung, J. T., Bambara, P., and Perkins, Jr., F., "Gas Chromatographic Analysis of Lemon Oil," 136th Natl. ACS Mtg., Atlantic City, N. J., September 1959, Program Abstr., p. 24B

4213. Kung, J. T., Whitney, J. E., and Cavagnol, J. C., "Analysis of Aqueous Solutions by Gas Chromatography," Anal. Chem., 33, 1505-1507 (1961); CA 56, 4076a; 12th Pittsburgh Conf. on Anal. Chem. & Appl. Spectroscopy, Pittsburgh, Pa., February-March 1961, Program Abstr., p. 59

4214. Kunst, E. D., "New Routes in Petroleum Research. Modern Analytical Methods," Chem. Weekblad, 55, 605-611 (1959)

4215. Kunugi, T., Ikeda, M., and Miyazaki, H., "Special Gas Chromatography," Yukagaku, 9, 2-7 (1960); CA 54, 1398i

4216. Kuo, H-F., T'ang, H-Y., Kuan, T-S., Yang, H-J., and Tai, L., "Determination of Composition of Tar from Bituminous Shale by Composite Gas-Liquid Chromatography and Infrared Spectroscopy," Jan Liao Hsueh Pao, 5, No. 2, 95-107 (1960); CA 57, 12799d

4217. Kurkchi, G. A., and Iogansen, A. V., "Gas-Chromatographic Determination of the Solubility of Gases and Vapors in Liquids," Dokl. Akad. Nauk SSSR, 145, 1085-1088 (1962); CA 57, 15879c

4218. Kurn, R. W., Hughes, K. J., and Chase, J. O., "Application of Gas Chromatography to Analysis of Exhaust Gases," S.A.E. (Soc. Automotive Engrs.) Preprint 11C, 31 pp. (1958); CA 54, 4239e

4219. Kuroda, T., "Chemical and Botanical Properties of Artemisia Kurramensis Cultivated in Japan. III. Components of the Essential Oil," Yakugaku Zasshi, 82, 179-181 (1962); CA 57, 2329h

4220. Kurono, G., Kimura, H., Sakai, T., Kitade, K., and Ozaki, Y., "Fatty Acids from the Seed Oil of Cornaceae. I. Fatty Acids from the Seed Oil of Cornus Controversa," Kanazawa Daigaku Yakugakubu Kenkya Nempo, 12, 24-27 (1962); CA 58, 11584h

4221. Kusy, V., "Chromatographic Separation of Hydrogen, Paraffins, and Olefins by Using Flame-Detector Indication," Chem. listy, 54, 1168-1172 (1960); CA 55, 9974c

4222. Kusy, V., "The Influence of Carrier Gas on the Quantitative Evaluation of Chromatograms," in "Gas Chromatographie 1959," edited by R. E. Kaiser and H. G. Struppe, Akademie Verlag, Berlin, 1959, pp. 246-257

4223. Kuwada, D. M., "Determination of Water in Hydrazine by Gas Chromatography," J. Gas. Chromatog., 1, No. 3, 11-13 (1963)

4224. Kuzdzal-Savoic, S., and Paquot, C., "Gas Chromatography and Lipid Chemistry. III. Identification of Abnormal Ingredients in Butter," Ann. Fals. Expert. chim., 55, 9-11 (1962); CA 57, 12985f

4225. Kvitkovskii, L. N., and Grushetskaya, E. V., "Determination of Normal Paraffin Hydrocarbons in Gasoline by Use of Molecular Sieves," Khim. i Tekhnol. Topliva i Masel, 7, 61-64 (1962); CA 57, 1146d

4226. Kwan, T., "Determination of Argon Present in the Air by Gas Adsorption Chromatography," J. Research Inst. Catalysis, Hokkaido Univ., 8, 14-17 (1960); CA 55, 226a

4227. Kwan, T., "Separation of Parahydrogen, Orthohydrogen, Hydrogen Deuteride, Orthodeuterium, and Paradeuterium by Gas Absorption Chromatography," J. Research Inst. Catalysis, Hokkaido Univ., 8, 18-28 (1960); CA 55, 60c

4228. Kwantes, A., and Rijnders, G. W. A., "The Determination of Activity Coefficients at Indefinite Dilution by Gas-Liquid Chromatography," in "Gas Chromatography 1958," edited by D. H. Desty, Academic Press, Inc., New York, 1958, pp. 125-136

4229. Kwie, W. W., and Gardiner, Jr., W. C., "Thermal Isomerization of Dimethyl Maleate," Tetrahedron Letters, 1963, 405-408

4230. Kyryacos, G., "Cool Flame Combustion Studies of Hydrocarbons by Gas Chromatography," Univ. Microfilms (Ann Arbor, Mich.), Publ. No. 21442, 133 pp.; Dissertation Abstr., 17, 1216 (1957); CA 51, 14235f

4231. Kyryacos, G., and Boord, C. E., "Separation of Hydrogen, Oxygen, Nitrogen, Methane, and Carbon Monoxide by Gas Adsorption Chromatography," Anal. Chem., 29, 787-788 (1957); CA 51, 10177h

4232. Kyryacos, G., and Boord, C. E., "Gas Adsorption Chromatography in the Analysis of Cool-Flame Combustion Products," 131st Natl. ACS Mtg., Miami, Fla., April 1957, Program Abstr., p. 35B

4233. Kyryacos, G., Menapace, H. R., and Boord, C. E., "Gas-Liquid Chromatographic Analysis of Some Oxygenated Products of Cool-Flame Combustion," Anal. Chem., 31, 222-225 (1959); CA 53, 9903c

4234. Lacey, R. A. S., "Perfumery Chemicals and Odour," Chem. & Ind. (London), 1959, 132

4235. Lacey, R. A. S., "Perfumery Chemicals," J. Soc. Cosmetic Chemists, 11, 2-13 (1960); CA 54, 11387h

4236. Lacey, R. N., "6-Acetylcyclohex-2-enones - Condensation of β-Diketones with α, β-Unsaturated

Ketones," J. Chem. Soc., <u>1960</u>, 1625-1633; CA <u>54</u>, 19535e

4237. Lacey, R. N., "Acid-Catalyzed Heterolysis of Amides with Alkyl-Nitrogen Fission (A_{AL})," J. Chem. Soc., <u>1960</u>, 1633-1639; CA <u>54</u>, 19578d

4238. Lacey, R. N., "Hydrolysis of 6-Acylcyclohex-2-enones and the Base-Catalyzed Rearrangement of Cyclohex-2-enones," J. Chem. Soc., <u>1960</u>, 1639-1648; CA <u>54</u>, 19536g

4239. Lacy, J., and Hill, R. V., "Determination of Argon and Methane in Gas Mixtures. Accuracy and Precision of Analysis Using a Simple Gas Chromatographic Technique," Chem. & Ind. (London), <u>1959</u>, 1148-1149

4240. Lacy, J., and Woolmington, K. G., "The Determination of Impurities in Chlorine Gas by Gas Chromatography," Analyst, <u>86</u>, 350-355 (1961); CA <u>55</u>, 20771h

4241. Lacy, J., and Woolmington, K. G., "Apparatus for the Preparation of Standard Gas Mixtures," Analyst, <u>86</u>, 547-548 (1961); CA <u>56</u>, 7971h

4242. Lacy, J., Woolmington, K. G., and Hill, R. V., "The Application of Gas Chromatography to the Analysis of Ammonia Synthesis Gas," S. African Ind. Chemist, <u>14</u>, 47-51 (1960); CA <u>54</u>, 21670c

4243. Lada, Z., Waclawik, J., and Waszak, S., "Gas Analysis," Chem. Anal., <u>3</u>, 329-348 (1958); CA <u>53</u>, 5961f

4244. LaFace, F., "Essential Oils from Citrus Fruits," France et ses parfums, <u>3</u>, No. 15, 44-51 (1960); CA <u>54</u>, 20092e

4245. LaFace, F., "Recent Work Concerning the Composition and Analysis of Citrus Fruit Juices," Essenze Deriv. Agrumari, <u>32</u>, 81-91 (1962); CA <u>57</u>, 14250a

4246. Lafon, M., and Baraud, J., "Oxidation of Alcohols," Bull. soc. chim. France, <u>1960</u>, 943-948; CA <u>55</u>, 1419g

4247. Laitinen, H. A., "Chemical Analysis. An Advanced Text and Reference," McGraw-Hill Book Co., Inc., New York, 1960, pp. 500-511

4248. Lake, R. D., and Corson, B. B., "Chloromethylation of 1,2,4-Trimethylbenzene," J. Org. Chem., <u>24</u>, 1823-1825 (1959); CA <u>55</u>, 13343a

4249. Lakshminarayana, G., Kruger, F. A., Cornwell, D. G., and Brown, J. B., "Chromatographic Studies on the Composition of Commercial Samples of Triolein-I^{131} and Oleic Acid-I^{131}, and the Distribution of the Label in Human Serum Lipides Following Oral Administration of These Lipides," Arch. Biochem. Biophys., <u>88</u>, 318-327 (1960); CA <u>54</u>, 19908g

4250. Lalau-Keraly, F. X., "Application of Gas-Phase Chromatography to the Study of High Polymers," Peintures, Pigments, Vernis, <u>39</u>, 4-10 (1963); CA <u>58</u>, 11473c

4251. Lambert, J. D., Roberts, G. A. H., Rowlinson, J. S., and Wilkinson, V. J., "Second Virial Coefficients of Organic Vapors," Proc. Roy. Soc. (London), <u>A196</u>, 113-125 (1949); CA <u>44</u>, 10411g

4252. Lamberton, J. A., "The Dibasic Acids of Japan Wax," Australian J. Chem., <u>14</u>, 323-324 (1961); CA <u>55</u>, 20948d

4253. Lambertsen, G., and Holman, R. T., "Partial Characterization of the Hydrocarbons of Herring Oil," Acta Chem. Scand., <u>17</u>, No. 1, 281-282 (1963); CA <u>58</u>, 12778a

4254. Lamparsky, D., "The Use of Gas Chromatography in the Analysis of Essential Oils," Reichstoffe u. Aromen, <u>9</u>, No. 7, 201-206, 241-245 (1959)

4255. Lamprey, P. S., and Amell, A. R., "The Radiation-Induced Exchange of Carbon-14 Between Ethane and Methylamine," 139th Natl. ACS Mtg., St. Louis, Mo., March 1961, Program Abstr., p. 37R

4256. Landault, C., and Guiochon, G., "Study of the Use of Teflon as a Support in Gas-Liquid Chromatography. Application to the Separation of Strongly Polar Compounds," J. Chromatog., <u>9</u>, 133-146 (1962); CA <u>58</u>, 6165h

4257. Landegren, G. F., "Heat Bridge for Measuring Thermal Conductivity [of Solids]," Am. J. Phys., <u>25</u>, 532-534 (1957); CA <u>52</u>, 7788g

4258. Landis, P. S., and Haag, W. O., "Formation of Hexamethylbenzene from Phenol and Methanol," J. Org. Chem., <u>28</u>, 585 (1963); CA <u>58</u>, 10104g

4259. Landowne, R. A., and Bergmann, W., "Marine Products. L. Phospholipides of Sponges," J. Org. Chem., <u>26</u>, 1257-1261 (1961); CA <u>55</u>, 20975i

4260. Landowne, R. A., and Lipsky, S. R., "Detection of Certain Brominated Long-Chain Fatty Acid Esters by Gas Liquid Chromatography," Nature, <u>182</u>, 1731-1732 (1958); CA <u>53</u>, 9905e

4261. Landowne, R. A., and Lipsky, S.R., "Gas-Chromatographic Analysis of Permanent Gases Using Standard Ionization Detector Equipment," Nature, <u>189</u>, 571-572 (1961); CA <u>55</u>, 11182h

4262. Landowne, R. A., and Lipsky, S. R., "Use of Capillary Columns for the Separation of Some Closely Related Positional Isomers of Methyl Linoleate by Gas Chromatography," Biochim. et Biophys. Acta, <u>46</u>, 1-6 (1961); CA <u>55</u>, 17113c

4263. Landowne, R. A., and Lipsky, S. R., "A Simple Method for Distinguishing Between Unsaturated and Branched Fatty Acid Isomers by Gas Chromatography," Biochim. et Biophys. Acta, <u>47</u>, 589-592 (1961); CA <u>55</u>, 22466g

4264. Landowne, R. A., and Lipsky, S. R., "Electron Capture Spectrometry, an Adjunct to Gas Chromatography. Quantitative Study of Operating Parameters and the Qualitative and Quantitative Distinction Between Compounds Containing the Same Heteroatom," Anal. Chem., 34, 726-730 (1962); CA 57, 2821h

4265. Landowne, R. A., and Lipsky, S. R., "The Electron Capture Spectrometry of Haloacetates: A Means of Detecting Ultramicro Quantities of Sterols by Gas Chromatography," Anal. Chem., 35, 532-535 (1963)

4266. Landsman, D. A., and Lane, E. S., "Inert Diluents for Use in Nuclear Energy Extraction Plants. I. Vapor Phase Chromatography of Selected Solvents," J. Appl. Chem. (London), 12, 24-33 (1962); CA 57, 2829d

4267. Lang, K. F., and Zander, M., "Pyrolysis of Acenaphthene," Chem. Berichte, 94, 1871-1876 (1961)

4268. Langenau, E. E., and Rogers, J. A., "Instrumental Analysis of Essential Oils and Related Products," Am. Perfumer Aromat., 75, No. 3, 38, 45-48 (1960); CA 54, 11386d

4269. Langer, A., "A Gas-Blending System," Rev. Sci. Instr., 18, 101-103 (1947); CA 41, 2613e

4270. Langer, S. H., and Pantages, P., "Microsyringe for Small Liquid Volumes," Anal. Chem., 30, 1889-1890 (1958); CA 53, 1866g

4271. Langer, S. H., and Pantages, P., "Peak-Shift Technique in Gas-Liquid Chromatography: Trimethylsilyl Ether Derivatives of Alcohols," Nature, 191, 141-142 (1961); CA 56, 4105a

4272. Langer, S. H., Pantages, P., and Wender, E., "Gas-Liquid Chromatographic Separation of Phenols as Trimethylsilyl Esters," Chem. & Ind. (London), 1958, 1664-1665; CA 53, 10097f

4273. Langer, S. H., and Purnell, J. H., "A Gas-Liquid Chromatographic Study of the Thermodynamics of Solution of Some Aromatic Compounds," J. Phys. Chem., 67, 263-270 (1963); CA 58, 5058a

4274. Langer, S. H., and Zahn, C., "Bibliography on Gas Chromatography," in "Gas Chromatography," edited by V. J. Coates, J. J. Noebles, and I. S. Fagerson, Academic Press, Inc., New York, 1958, pp. 287-313

4275. Langer, S. H., Zahn, C., and Pantazoplos, G., "Gas-Liquid Chromatographic Resolution of m- and p-Xylenes: Tetrahalophthalate Liquid Phase," Chem. & Ind. (London), 1958, 1145-1147; CA 53, 11274c

4276. Langer, S. H., Zahn, C., and Pantazoplos, G., "Selective Gas-Liquid Chromatographic Separation of Aromatic Compounds with Tetrahalophthalate Esters," J. Chromatog., 3, 154-167 (1960); CA 54, 22481b

4277. Langer, S. H., Zahn, C., and Vial, M. H., "Preparation and Molecular Complexes of Tetrahalophthalate Esters," J. Org. Chem., 24, 423-425 (1959); CA 54, 9836a

4278. Langermeersch, A. van, Cornu, A., and Joly, D., "French Contribution to the Application of Spectrographic and Chromatographic Methods in the Petroleum Industry," 5th World Petrol. Congr., New York, June 1959

4279. Langlois, B. E., Stemp, A. R., and Liska, B. J., "Rapid Method for the Extraction and Detection of Certain Insecticide Residues in Milk," 58th Annual Mtg., Am. Dairy Sci. Assoc., Lafayette, Ind., June 1963; Abstr., J. Dairy Sci., 46, 606 (1963)

4280. Langvad, T., "Theory of Chromatography," Acta Chem. Scand., 10, 1649-1662 (1956); CA 52, 13363e

4281. Lansbury, P. T., and Mesehke, R. W., "Thermal Carbonylation of Cyclohexene," J. Org. Chem., 24, 104-106 (1959); CA 54, 6658b

4282. Lantz, C. D., and Rushneck, D. R., "Gas Chromatography Characteristics of the Argon Ionization Detector," 8th Detroit Anachem Conf., Detroit, Mich., October 1960

4283. Lapidus, L., and Amundson, N. R., "Mathetmatics of Adsorption in Beds. VI. The Effect of Longitudinal Diffusion in Ion Exchange and Chromatographic Columns," J. Phys. Chem., 56, 984-988 (1952); CA 47, 9065c

4284. Lapiere, C., "Vapor Chromatography," Farmaco (Pavia) Ed. prat., 15, 12-33 (1960); CA 54, 12490h

4285. Laramy, R. E., and Lively, L. D., "Quantitative Analysis of Alkanol and Alkane Systems by Programmed Temperature Gas Chromatography," 12th Pittsburgh Conf. on Anal. Chem. & Appl. Spectroscopy, Pittsburgh, Pa., February-March 1961, Program Abstr., p.59

4286. Laramy, R. E., Lively, L. D., and Perkins, Jr., G., "The Application of the Electron Capture Detector to the Analysis of Lead Alkyls in Gasoline," 13th Pittsburgh Conf. on Anal. Chem. & Appl. Spectroscopy, Pittsburgh, Pa., March 1962, Program Abstr., p. 47

4287. Laramy, R. E., Lively, L. D., and Perkins, Jr., G., "Response of the Katharometer to Polar Compounds as Compared to Saturated Hydrocarbons," 13th Pittsburgh Conf. on Anal. Chem. & Appl. Spectroscopy, Pittsburgh, Pa., March 1962, Program Abstr., p. 58

4288. Lard, E. W., and Horn, R. C., "Separation and Determination of Argon, Oxygen, and Nitrogen by Gas Chromatography," Anal. Chem., 32, 878-879 (1960); CA 54, 18176g

4289. Latham, D. R., Ferrin, C. R., and Ball, J. S., "Identification of Fluorenones in Wilmington Petroleum by Gas-Liquid Chromatography and Spectrometry," Anal. Chem., 34, 311-313 (1962); CA 56, 11882g; 140th Natl. ACS Mtg., Chicago, Ill., September 1916, Program Abstr., p. 3S

4290. Laur, M. H., "Application of Gas Chromatography to the Study of the Fatty Acids of the Rhodophyceae," Comp. rend., 253, 966-968 (1961); CA 56, 7707c

4291. Lawrey, D. M. G., and Cerato, C. C., "Determination of Trace Amounts of Methane in Air," Anal. Chem., 31, 1011-1012 (1959); CA 53, 14828g

4292. Lawrie, T. D. V., McAlpine, S. G., Pirrie, R., Rifkind, B. M., and Blades, J., "The Fatty Acids of the Serum in Hypothyroidism," J. Endocrinol., 25, 29-34 (1962); cf. Clin. Sci., 20, 255-261 (1961); CA 58, 12971f

4293. Lawrie, T. D. V., McAlpine, S. G., Rifkind, B. M., Robinson, J. F., and Alice, H. W. Mc., "Serum Fatty Acid Patterns in Coronary-Artery Disease," Lancet, 1961-I, 421-424; CA 55, 11625d

4294. Lawson, A., "Thermal Conductivity Detectors," 14th Annual Fisk Univ. Infrared Spectroscopy Inst., Gas Chromatography Session, Nashville, Tenn., August 1963

4295. Lawson, D. D., and Havlik, A. J., "FORTRAN Source Program for Reduction of Gas-Liquid Retention Data," J. Gas Chromatog., 1, No. 5, 17-20 (1963)

4296. Lawson, Jr., W. H., and Johnson, Jr., R. L., "Gas Chromatography in Measuring Pulmonary Blood Flow and Diffusing Capacity," J. Appl. Physiol, 17, 143-147 (1962); CA 56, 14556i

4297. Lawton, D., and Powell, H. M., "The Structure of Molecular Compounds. XII. Molecular Compounds of Tri-o-thymotide," J. Chem. Soc., 1958, 2339-2357; CA 52, 17082d

4298. Lebbe, J., "Recent Applications of Gas Chromatography to the Studies of Air Pollution," Bull. soc. chim. France, 1963, 462-464; CA 59, 2094c

4299. LeBel, N. A., "The Stereochemistry of Addition to Olefins. I. The Free Radical Addition of Hydrogen Bromide to 2-Bromo-2-norborene," J. Am. Chem. Soc., 82, 623-627 (1960); CA 54, 14295g

4300. Lederer, E., (Ed.), "Chromatography in Organic and Biological Chemistry. Vol. I. General Principles, Application in Organic Chemistry," Masson et Cie, Paris, 1959/1960

4301. Lederer, E., and Lederer, M., "Chromatography," Elsevier Publishing Co., New York, 1955

4302. Lederer, E., and Lederer, M., "Chromatography, A Review of Principles and Applications," D. Van Nostrand Co., Inc., Princeton, N. J., 1957

4303. Lee, E. H., and Oliver, G. D., "Use of Two or More Internal Standards in Gas Chromatography," Anal. Chem., 31, 1925 (1959); CA 54, 3044e

4304. Lee, F. A., and Mattick, L. R., "Fatty Acids of the Lipides of Vegetables. I. Peas (Pisum Sativum)," J. Food Sci., 26, 273-275 (1961); CA 55, 25083g

4304a. Lee, G. D., Kauffman, F. L., Harlan, J. W., and Niezabitowski, W., "Application of Gas Chromatography to a Study of Nutmeg Oil Flavor," 3rd Intern. Symp. on Gas Chromatography, ISA, East Lansing, Mich., June 1961; ISA Proc., 1961, 301-304

4305. Lee, G. D., Kauffman, F. L., Harlan, J. W., and Niezabitowski, W., "Application of Gas Chromatography to a Study of Nutmeg Oil Flavor," Intern. Symp., 1961, 3, 475-484 (Pub. 1962); CA 58, 4974b

4306. Lee, J. B., and Price, M. J., "Oxidation of Cyclic Olefins and Unsaturated Terpenes with Thallium (III) Salts," Tetrahedron Letters, 1962, 1155-1159; CA 58, 11405c

4307. Lee, J. K., Lee, E. K. C., Musgrave, B., Tang, Y-N., Root, J. W., and Rowland, F. S., "Proportional Counter Assay of Tritium in Gas Chromatographic Streams," Anal. Chem., 34, 741-747 (1962); CA 57, 4261a

4308. Lee, J. K., Musgrave, B., and Rowland, F. S., "Hot Atom Reactions and Radiation-Induced Effects in the Reactions of Recoil Tritium with Cyclopropane," Can. J. Chem., 38, 1756-1768 (1960); CA 55, 20711h

4309. Lee, J. K., Musgrave, B., and Rowland, F. S., "Inter-Molecular Isotope Effect in Recoil Tritium Reactions with Hydrogen," J. Chem. Phys., 32, 1266-1267 (1960); CA 54, 20430i

4310. Lee, J. K., Musgrave, B., and Rowland, F. S., "Isotope Effect in Recoil Tritium Abstraction Reactions with Methane," J. Phys. Chem., 64, 1950-1951 (1960); CA 55, 20711f

4311. Lee, T. G., "Electron Attachment Coefficients of Some Hydrocarbon Flame Inhibitors," J. Phys. Chem., 67, 360-366 (1963)

4312. Leffler, A. J., "Fluorination of Hexachlorobenzene with Antimony Pentafluoride," J. Org. Chem., 24, 1132-1133 (1959); CA 54, 5517i

4313. LeFort, D., Paquot, C., and Pourchez, A., "Gas Chromatography and Lipochemistry. II. Study of Some Compounds With a Functional Ester Group by Gas Chromatography," Oleagineux, 16, 253-259 (1961); cf. Rev. franc, corps gras, 7, 391-395 (1960); CA 55, 17046i

4314. LeFort, D., Paquot, C., and Pourchez, A., "Gas Chromatography and Lipochemistry. VI. Comparison of the Methyl, Propyl, and Isopropyl Esters of Fatty Acids by Gas Chromatography," Oleagineux, 17, 629-630 (1962); CA 57, 15781a

4315. Lefort, M., "Gas Microanalysis," Bull. soc. chim. France, 1960, 239-242; CA 54, 16271f

4316. Lefort, M., "An Expression for the Specific Retention Volume, Vg, as a Function of the Temperature of the Column and the Thermodynamic Properties of the Eluted Compound in Gas-Liquid Chromatography," Publ. Groupe. Advance. Methodes Spectrog., 1960, 289-295; CA 56, 5417i

4317. Lefort, M., and Tarrago, X., "Separation and Microanalysis of Nitrogen and Nitrogen Oxide in Their Mixture on an Adsorption Column," J. Chromatog., 2, 218-220 (1959); CA 53, 16804a

4318. Legate, C. E., and Burnham, H. D., "Micropyrolytic-Gas Chromatographic Technique for the Analysis of Organic Phosphates and Thiophosphates," Anal. Chem., 32, 1042-1045 (1960); CA 54, 18198h; 11th Pittsburgh Conf. on Anal. Chem. & Appl. Spectroscopy, Pittsburgh, Pa., February-March 1960, Program Abstr., p. 45

4319. Leggoe, Jr., J. H., Brewer, J. E., and Hoffmann, N. L., "Analysis of Mixed Utility Gases by Gas Chromatography," 135th Natl. ACS Mtg., Boston, Mass., April 1959, Program Abstr., p. 20B

4320. Leggon, H. W., "A KBr Powder Trap for Gas Chromatographs for Obtaining Infrared Spectra," Anal. Chem., 33, 1295-1296 (1961); CA 55, 26557e

4321. Leghissa, S., and Carazzola, G. A., "Application of Vapor Phase Chromatography to Analysis of Plastic Materials," Annali di Chimica, 49, 1621-1631 (1959); CA 54, 14749e

4322. Lehmann, F. A., and Brauer, G. M., "Analysis of Pyrolyzates of Polystyrene and Poly(methyl Methacrylate) by Gas Chromatography," Anal. Chem., 33, 673-676 (1961); CA 56, 13078b

4323. Lehmann, H., "Thermal Conductivity of Gas Mixtures," Chem. Tech. (Berlin), 9, 530-537 (1957); CA 52, 2484h

4324. Lehrle, R. S., and Robb, J. C., "Direct Examination of Degradation of High Polymers by Gas Chromatography," Nature, 183, 1671 (1959); CA 54, 953c

4325. Lehtinen, O., Karkkainen, V. J., and Antila, M., "5, 9, 12-Octadecatrienoic Acid in Finnish Pinewood and Tall Oil," Suomen Kemistilehti, 35B, 179-180 (1962); CA 58, 2564f

4326. Leibetseder, F., and Ahrens, Jr., E. H., "The Fatty Acid Composition of Red Cells in Paroxysmal Nocturnal Hemoglobinuria," Brit. J. Haematol., 5, 356-364 (1959); CA 55, 21319c

4327. Leibnitz, E., Hager, W., and Kraus, U., "Products of Paraffin Oxidation. V. Constituents of Fatty Acids from Paraffin Oxidation," J. prakt. chem., 9, 267-274 (1959); CA 55, 366b

4328. Leibnitz, E., Heinze, G., and Konnecke, H. G., "Automatic Preparative Scale Gas Chromatography," Gas Chromatog. Symp., Brno, Czech., June 1962; Abstr., J. Gas Chromatog., 1, No. 4, 8 (1963)

4329. Leibnitz, E., Hrapia, H., and Konnecke, H. G., "An Integral-Detector for Chromatographic Gas Analysis," Brennstoff-Chem., 38, 14-16 (1957); CA 51, 4877d

4330. Leighton, W. B., "Instrumentation for Aerosols," Soap Chem. Specialities, 34, No. 8, 79-81, 83, 89, 91 (1958); CA 52, 17822i

4331. Leipnitz, W., and Mohnke, M., "Glass Capillary Tubes for Gas Chromatography," Chem. Tech. (Berlin), 14, 753-754 (1962); CA 58, 10703f

4332. Leithe, W., "Practical Applications of Gas Chromatography," Osterr. Chem.-Ztg., 58, 141-148 (1957); CA 51, 16191d

4333. Leithe, W., "Developments in Gas Chromatography Since 1957," Osterr. Chem.-Ztg., 61, No. 2, 33-37 (1960); CA 54, 10453i

4334. Lemmich, E., and Gjaldbaek, J. C., "Analysis of Anesthetics by Gas Chromatography. I. Chloroform," Dansk Tidskr. Farm., 37, 1-8 (1963); CA 58, 10039e

4335. Lemmon, R. M., Mazzetti, F., Reynolds, F. L., and Calvin, M., "Labeling of Benzene with a Carbon-14 Ion Beam," J. Am. Chem. Soc., 78, 6414-6415 (1956); CA 51, 4973a

4336. Lenoir, J. M., and Comings, E. W., "Thermal Conductivity of Gases. Measurement at High Pressures," Chem. Engr. Progr., 47, 223-231 (1951); CA 45, 5463i

4337. Lenoir, J. M., Junk, W. A., and Comings, E. W., "Measurement and Correlation of Thermal Conductivities of Gases at High Pressure," Chem. Engr. Progr., 49, 539-542 (1953); CA 48, 417h

4338. Leonard, N. J., and Musker, W. K., "Unsaturated Amines. XVI. An Oxidative Cyclization Route to Oxazolidines and Tetrahydro-1, 3-oxazines," J. Am. Chem. Soc., 82, 5148-5155 (1960); CA 55, 5497g

4339. Lepley, A. R., "Gas Chromatographic Determination of C_2-C_{10} Products from the Gamma Irradiation of Liquid Cyclopentane," Anal. Chem., 34, 322-325 (1962); CA 56, 13547d

4340. Lerner, M., "Marihuana; Tetrahydrocannabinol and Related Compounds," Science, 140, 175-176 (1963); CA 59, 904h

4341. Lerner, M., Mills, A. L., and Mount, S. F., "Narcotics Analysis - A Simple Approach," J. Forensic Sci., 8, 126-131 (1963); CA 58, 6648f

4342. LeRosen, H. D., "Gas Chromatographic Analysis of Olefin-Free Naphthas for n-Pentane and Lighter Components Using Ethyl Chloride as an Internal Standard," 136th Natl. ACS Mtg., Atlantic City, N. J., September 1959, Program Abstr., p. 23B

4343. LeRosen, H. D., "Gas Chromatographic Determination of n-Pentane and Lighter Components in Olefin-Free Naphthas Using Ethyl Chloride as an Internal Standard," Anal. Chem., 32, 444-445 (1960); CA 54, 14647b

4344. LeRosen, H. D., "Recovery and Identification of Mercaptans from Aqueous Alkaline Solutions by Gas Chromatography," Anal. Chem., 33, 973-974 (1961); CA 55, 18453f

4345. LeSech, M., "Gas Chromatography of Phosphorus Halogen Derivatives," Journees Intern. Etude Methodes Separation Immediate Chromatog., Paris, 1961, 290-298 (Pub. 1962); CA 59, 1080e

4346. Leslie, S. W., and Jensen, R., "Identification of Volatile Flavor Components in Food," 13th Annual Mid-America Spectroscopy Symp., Chicago, Ill., April-May 1962

4347. Lesser, J. M.,"Device for Isolation of Components Separated by Gas Chromatography," Anal. Chem., 31, 484 (1959); CA 53, 8723e

4348. Lesser, R., "Detection by Means of an Ionization Detector of Very Small Amounts of Inorganic Gases Separated by Gas Chromatography," German Bunsen Society for Physical Chemistry, Bonn, Germany, May 1960; Abstr., Angew. Chem., 72, 631 (1960)

4349. Lesser, R., "Determination of Very Small Amounts of Inorganic Gases with an Ionization Detector," Angew. Chem., 72, 775-777 (1960); CA 55, 6949d

4350. Lesser, R., and Gruber, H., "Determination of Gases in Metals with the Hot Extraction Method and a Gas Chromatograph," Z. Metallk., 51, 495-501 (1960); CA 55, 220d

4351. Lester, D., "Concentration of Apparent Endogenous Ethanol," Quart. J. Studies Alc., 23, 17-25 (1962); CA 56, 16013f

4352. LeTourneau, R. L., "Petroleum," Anal. Chem., 29, 684-697 (1957); CA 51, 7219b

4353. LeTourneau, R. L., "Review of Industrial Applications of Analysis, Control, and Instrumentation. Petroleum," Anal. Chem., 31, 730-749 (1959); CA 53, 8912h

4354. LeTourneau, R. L., "Petroleum," Anal. Chem., 33, No. 5, 92R-112R (1961); CA 55, 12137g

4355. LeTourneau, R. L., "Analysis of Petroleum," 140th Natl. ACS Mtg., Chicago, Ill., September 1961, Program Abstr., p. 14S; Preprints, Div. Petrol. Chem., 6, No. 3-A, 15-25 (1961); CA 58, 12341g

4356. Levadie, B., "The Determination of Organic Vapors in Air by Gas Chromatography Using a Direct Injection Method," Am. Ind. Hygiene Assoc. J., 21, No. 4, 322-324 (1960); CA 54, 23142i

4357. Levadie, B., and Harwood, J. F., "An Application of Gas Chromatography to Analysis of Solvent Vapors in Industrial Air," Am. Ind. Hyg. Assoc. J., 21, 20-24 (1960); CA 54, 14517d

4358. Levi, L., Nigam, I. C., and Davies, L., "Essential Oils Analysis by Coupled Gas-Liquid-Thin Layer Chromatography," 144th Natl. ACS Mtg., Los Angeles, Calif., March-April 1963, Program Abstr., p. 18A

4359. Levit, A. M., "Effect of Carbon Dioxide and Hydrogen Sulfide on the Results from Total and Component Analysis of Hydrogen Gases," Razved. i Promysl. Geofiz., Sb., Moscow, 1961, No. 40, 72-75; CA 57, 1151e

4360. Levy, E. J., Doyle, R. R., Brown, R. A., and Melpolder, F. W., "Identification of Components in Paraffin Wax by High-Temperature Gas Chromatography and Mass Spectrometry," 138th Natl. ACS Mtg., New York, N. Y., September 1960, Program Abstr., p. 6R; Preprints, Div. Petrol. Chem., 5, No. 3, 171-183 (1960); CA 55, 20402c

4361. Levy, E. J., Doyle, R. R., Brown, R. A., and Melpolder, F. W., "Identification of Components in Paraffin Wax by High Temperature Gas Chromatography and Mass Spectrometry," Anal. Chem., 33, 698-704 (1961); CA 56, 11890b

4362. Levy, E. J., Galbraith, F. J., and Melpolder, F. W., "Interpretive Techniques for the Determination of Paraffin Wax Composition by Mass Spectrometry and Gas Chromatography," Advan. Mass Spectrometry, Proc. Conf., 2nd, Oxford, 1961, 2, 395-407 (Pub. 1963); CA 59, 1099a

4363. Levy, E. J., Lawrey, D. M. G., Herk, Jr., L. P., and Stahl, W. H., "The Application of Isolative Vapor Phase Chromatography and Mass Spectrometry to Problems in Odor Research," 4th Annual Mtg., ASTM Committee E-14 on Mass Spectrometry, Cincinnati, Ohio, May-June 1956

4364. Levy, E. J., Melpolder, F. W., and Galbraith, F. J., "Determination of Molecular Structure in Paraffin Wax by Mass Spectrometry," 141st Natl. ACS Mtg., Washington, D. C., March 1962, Program Abstr., p. 2B

4365. Levy, E. J., Miller, E. D., and Beggs, W. S., "Application of Time of Flight Mass Spectrometry and Gas Chromatography to Reaction Studies," Anal. Chem., 35, 946-949 (1963)

4366. Levy, E. J., and Paul, D. G., "Application of Dual Flame Ionization Gas Chromatography to the Analysis of Paraffin Waxes," Facts & Methods (F & M Scientific Corp.), 4, No. 1, 10 (1963)

4367. Lewin, S. Z., "Proper Utilization of Analytical Instrumentation," Anal. Chem., 33, No. 3, 23A-32A, 37A-43A (1961)

4368. Lewin, S. Z., "Chemical Instrumentation. 15. Gas Chromatographs," J. Chem. Educ., 38, No. 17, A869-A884 (Dec. 1961); ibid., 39, A5-A26 (Jan. 1962); ibid., 39, A83-A115 (Feb. 1962); ibid., 39, A161-A202 (Mar. 1962)

4369. Lewin, S. Z., "Chromatographic Glossary. A. Terminology of Thin Layer Chromatography; B. Terminology of Gas Chromatography," J. Chem. Educ., 40, No. 3, A167-A168, A170, A172, A174, A178, A184 (March 1963); Reproduced in J. Gas Chromatog., 1, No. 5, 21-24 (1963) Terminology is given in English, German, French, and Spanish.

4370. Lewin, S. Z., Connor, J., and Konigsbacher, K., "The Evaluation of Instrumental Techniques for the Analysis of Trace Constituents," 138th Natl. ACS Mtg., New York, N. Y., September 1960, Program Abstr., pp. 17A-18A

4371. Lewis, E. S., and Herndon, W. C., "The Determination of Gaseous Chloroformates. I. The Rates of Simple Alkyl Compounds, " J. Am. Chem. Soc., 83, 1955-1958 (1961); CA 55, 27009f

4372. Lewis, E. S., Herndon, W. C., and Duffey, D. C., "The Decomposition of Gaseous Chloroformates. II. Substitution and Elimination Stereochemistry, " J. Am. Chem. Soc., 83, 1959-1961 (1961); CA 55, 27009g

4373. Lewis, H. R., "Paraffins in Low Temperature Tar from a Fluidized Carbonization Process, " Chem. & Ind. (London), 1959, 1049-1050; CA 54, 3921d

4374. Lewis, J. S., McCloud, G. T., and Schirmer, Jr., W., "Storage of Gas Chromatographic Data Using Key-Punched IBM Cards, " J. Chromatog., 5, 541-542 (1961)

4375. Lewis, J. S., and Patton, H. W., "Analysis of Ester-Type Plasticizers by Gas-Liquid Chromatography, " in "Gas Chromatography, " edited by V. J. Coates, H. J. Noebels, and I. S. Fagerson, Academic Press, Inc., New York, 1958, pp. 145-153

4376. Lewis, J. S., Patton, H. W., and Kaye, W. I., "Qualitative Gas Chromatographic Analysis Using Two Columns of Different Characteristics, " Anal. Chem., 28, 1370-1373 (1956); CA 50, 16533a

4377. Lewis, K. G., and Stimson, V. R., "Catalysis by Hydrogen Halides in the Gas Phase. II. tert-Butyl Alcohol and Hydrogen Chloride, " J. Chem. Soc., 1960, 3087-3089; CA 54, 24337c

4378. Lewis, L. L. and Melnick, L. M., "Vacuum Fusion-Gas Chromatographic Determination of Oxygen and Nitrogen in Metals, " Anal. Chem., 34, 868-869 (1962); CA 57, 2846a

4379. Lewis, W. K., Gilliland, E. R., Chertow, B., and Cadogan, W. P., "Adsorption Equilibria. Hydrocarbon Gas Mixtures, " Ind. Eng. Chem., 42, 1319-1326 (1950)

4380. Libbey, L. M., Bills, D. D., Day, E. A., and Young, J. O., "Preliminary Observations on the Volatile Fraction of Cheddar Cheese, " 57th Annual Mtg., Am. Dairy Sci. Assoc., College Park, Md., June 1962; Abstr., J. Dairy Sci., 45, 660 (1962)

4381. Liberti, A., "Coulometry Applied to Gas-Phase Chromatography, " Anal. chim. Acta, 17, 247-253 (1957); CA 52, 7005g

4382. Liberti, A., "Vapor Phase Chromatography of Methyl Esters of Fatty Acids and Their Quantitative Determination by Automatic Coulometry, " Ann. chim. (Rome). 48, 40-49 (1958)

4383. Liberti, A., and Cartoni, G., "Coulometry Combined with Gas-Phase Chromatography in the Analysis of Volatile Mixtures, " Atti accad. nazl. Lincei, Rend., Classe sci., fis., mat. e nat., 20, 787-794 (1956); CA 51, 12733f

4384. Liberti, A., and Cartoni, G., "Coulometric Determination of Mercaptans Separated by Gas Chromatography, " Chim. e Ind. (Milan), 39, 821-824 (1957); CA 52, 2664g

4385. Liberti, A., and Cartoni, G. P., "Analysis of Essential Oils by Gas Chromatography, " Gas Chromatog., 1958, 321-328; CA 53, 16476e

4386. Liberti, A., and Cartoni, G. P., "Analysis of Essential Oils by Gas Chromatography, " in "Gas Chromatography 1958, " edited by D. H. Desty, Academic Press, Inc., New York, 1958, pp. 321-329

4387. Liberti, A., and Cartoni, G. P., "Gas Phase Chromatography of Terpenic Hydrocarbons, " Ricerca Sci., 28, 1192-1198 (1958); CA 54, 21660d

4388. Liberti, A., Cartoni, G., and Pallotta, U., "Coulometric Microdetermination of Volatile Fatty Acids in Dairy Products as Separated by Vapor-phase Chromatography, " Latte, 30, 581-584 (1956); CA 51, 13253c

4389. Liberti, A., Cartoni, G. P., and Pallotta, U., "Coulometric Determination of Volatile Fatty Acids in Cheese Products, as Separated by Vapor Phase Chromatography, " Pubbl. univ. cattolica S. Cuore, Ann. fac. agrar. Ser. 5 Atti convegno appl. tec. chromatogr. prod. agr., 53, 138-149 (1957); CA 52, 20737d

4390. Liberti, A., Cartoni, G. P., and Pallotta, U., "Vapor-Phase Chromatography of the Methyl Esters of Fatty Acids and Their Quantitative Determination in Fats by Automatic Electrometric Titration, " Ann. chim. (Rome), 48, 40-49 (1958); CA 52, 9868e

4391. Liberti, A., and Conti, L., "Application of Gas-Phase Chromatography to the Study of Essential Oils, " Riv. ital. essenze, profumi, piante offic., oil vegetali, saponi, 39, 128-134 (1957); CA 51, 12440a

4392. Liberti, A., Conti, L., and Crescenzi, V., "Molecular-Weight Determination of Components by Gas-Phase Chromatography, " Nature, 178, 1067-1069 (1956); CA 51, 4064b

4393. Liberti, A., Conti, L., and Crescenzi, V., "Estimate of Molecular Weights of Constituents Identified with Gas-Phase Chromatography, " Atti accad. nazl. Lincei, Rend., Classe sci.,fis., mat. e nat., 20, 623-629 (1956); CA 51, 8508e

4394. Liberti, A., Costa, G., and Pauluzzi, E., "Applicability of the Infrared Spectrophotometer as Analyser in Gas Chromatography, " Chim. e ind. (Milan), 38, 674-677 (1956); CA 50, 16534g

4395. Lichtenfels, D. H., Fleck, S. A., and Burow, F. H., "Gas-Liquid Partition Chromatography, " Anal. Chem., 27, 1510-1513 (1955); CA 50, 4700g; 6th Pittsburgh Conf. on Anal. Chem. & Appl. Spectroscopy, Pittsburgh, Pa., February-March 1955, Program Abstr., p. 49

4396. Lichtenfels, D. H., Fleck, S.A., Burow, F. H., and Coggeshall, N. D., "Gas Partition Analysis of Light Ends in Gasoline, " Anal. Chem., 28, 1376-1379 (1956); CA 50, 17401a; 7th Pittsburgh

Conf. on Anal. Chem. & Appl. Spectroscopy, Pittsburgh, Pa., February-March 1956, Program Abstr., p. 41

4397. Lieberman, M., and Craft, C. C., "Ethylene Production by Cytoplasmic Particles from Apple and Tomato Fruits in the Presence of Thiomalic and Thioglycolic Acid," Nature, 189, 243 (1961); CA 55, 10598d

4398. Liggett, R. W., Feazel, C. E., and Ellenberg, J. Y., "Browning Reaction Initiated by Gamma Radiation," J. Agr. Food Chem., 7, 277-280 (1959); CA 53, 17350f

4399. Lightfoot, E. N., "Equilibrium Operation of Chromatographic Columns with Longitudinal Diffusion: Final Form Fronts," J. Phys. Chem., 61, 1686 (1957); CA 52, 5088c

4400. Lijinsky, W., and Domsky, I. I., "Chromatographic Determination of Trace Amounts of Polynuclear Hydrocarbons in Mixtures," 141st Natl. ACS Mtg., Washington, D. C., March 1962, Program Abstr., p. 23B

4401. Lijinsky, W., Domsky, I. I., and Mason, G., "Analysis of Tars and Pyrolyzates by Gas Chromatography Programmed at Moderate Temperatures," 144th Natl. ACS Mtg., Los Angeles, Calif., March-April, 1963, Program Abstr., p. 21B

4402. Lijinsky, W., Domsky, I. I., Mason, G., Ramahi, H. Y., and Safavi, T., "The Chromatographic Determination of Trace Amounts of Polynuclear Hydrocarbons in Petrolatum, Mineral Oil, and Coal Tar," Anal. Chem., 35, 952-956 (1963); Correction, ibid., p. 1397

4403. Lillard, D. A., Montgomery, M. E., and Day, E. A., "Flavor Threshold Values of Certain Carbonyl Compounds in Milk," 57th Annual Mtg., Am. Dairy Sci. Assoc., College Park, Md., June 1962; Abstr., J. Dairy Sci., 45, 660 (1962)

4404. Lille, U., "The Use of Gas Chromatography for the Analysis of High-Boiling Oil-Shale Gases," Trudy Tallin Politekh. Inst. Ser. A, 1958, No. 153, 115-122; CA 55, 21553h

4405. Lin, T. H., Rubinstein, R., and Holmes, W. L., "Gas Liquid Chromatography Study of the Effect of D- and L-3, 5, 3'-Triiodothyronine on Bile Acid Excretion in Rats," 46th Annual Mtg., Federation of Am. Soc. for Exper. Biol., Atlantic City, N. J., April 1962; Abstr., Federation Proc., 21, 298 (1962)

4406. Lincoln, R. M., Rogers, R. L., Burwasser, H., and Keenan, V. J., "Irradiation of Petroleum Hydrocarbons," Ind. Eng. Chem., 51, 547-548 (1959); CA 53, 12646c

4407. Linde, H. W., and Rogers, L. B., "The Analysis of Volatile Liquid Mixtures by Thermal Conductivity Measurements on Their Vapors," Anal. Chim. Acta, 19, 347-353 (1958); CA 54, 1161e

4408. Lindeman, L. P., "Combination of Mass Spectrometer, Gas Chromatograph, and Catalytic Hydrogenation for the Analysis of Olefins," 142nd Natl. ACS Mtg., Atlantic City, N. J., September 1962, Program Abstr., p. 1S

4409. Lindeman, L. P., and Annis, J. L., "Use of a Conventional Mass Spectrometer as a Detector for Gas Chromatography," Anal. Chem., 32, 1742-1749 (1960); CA 55, 6239h; 137th Natl. ACS Mtg., Cleveland, Ohio, April 1960, Program Abstr., p. 25B

4410. Lindgren, F. T., Nicholas, A. V., Freeman, N. K., and Wills, R. D., "Potential Contamination in the Analysis of Methyl Esters of Fatty Acids by Gas-Liquid Chromatography," J. Lipid Research, 3, 390-391 (1962); CA 57, 13905f

4411. Lindgren, F. T., Nichols, A. V., and Wills, R. D., "Fatty Acid Distributions in Serum Lipids and Serum Lipoproteins," Am. J. Clin. Nutrition, 9, 13-23 (1961); CA 55, 11573g

4412. Lindquist, K., and Brunner, J. R., "Composition of the Free Fat of Spray-Dried Whole Milk," 57th Annual Mtg., Am. Dairy Sci. Assoc., College Park, Md., June 1962; Abstr., J. Dairy Sci., 45, 661 (1962)

4413. Lineweaver, H., Pippen, E. L., and Nonaka, M., "[Gas] Chromatography of Chicken Flavor Volatiles," World's Poultry Congr. Proc., 12th, Sydney, 1962, 405-408; CA 58, 10664b

4414. Link, W. E., "General Methods of Analysis of Drying Oils," J. Am. Oil Chemists' Soc., 36, 477-483 (1959); CA 53, 22989e

4415. Link, W. E., Hickman, H. M., and Morrissette, R. A., "Gas-Liquid Chromatography of Fatty Derivatives. I. Separation of Homologous Series of n-Olefins, n-Hydrocarbons, n-Nitriles, and n-Alcohols," J. Am. Oil Chemists' Soc., 36, 20-23 (1959); CA 53, 4773i

4416. Link, W. E., Hickman, H. M., and Morrissette, R. A., "Gas-Liquid Chromatography of Fatty Derivatives. II. Analysis of Fatty Alcohol Mixtures by Gas-Liquid Chromatography," J. Am. Oil Chemists' Soc., 36, 300-303 (1959); CA 53, 16561e

4417. Link, W. E., and Morrissette, R. A., "Gas-Liquid Chromatography of Fatty Derivatives. IV. Quantitative Analysis of n-Alcohols," J. Am. Oil Chemists' Soc., 37, 668-671 (1960); CA 55, 4011h; 51st Annual Mtg., Am. Oil Chemists' Soc., April 1960

4418. Link, W. E., Morrissette, R. A., Cooper, A. D., and Smullin, C. F., "Gas-Liquid Chromatography of Fatty Derivatives. III. Analysis of Fatty Amines," J. Am. Oil Chemists' Soc., 37, 364-366 (1960); CA 54, 17917d

4419. Linko, Y. Y., Miller, B. S., and Johnson, J. A., "Quantitative Determination of Certain Carbonyl Compounds in Preferments," Cereal Chem., 39, 263-272 (1962); Biol. Abstr., 40, 13936

4420. Lins, G., and Raudonat, H. W., "Measurement by Gas Chromatography of Residual Alcohol on Mucous Membranes of the Mouth After Alcohol Imbibition," Deut. Z. Ges. Gerichtl. Med., 52, 242-245 (1962); CA 57, 2730f

4421. Lipsky, S. R., "Argon Detectors," Chem. & Eng. News, 38, No. 16, 5 (April 18, 1960)

4422. Lipsky, S. R., "Theory and Practice of Ionization Techniques in Gas Chromatography," 2nd Eastern Analytical Symp., New York, N. Y., November 1960

4423. Lipsky, S. R., "The Application of Gas Chromatography to the Analysis of Substances of Biochemical Interest," 13th Annual Mtg., Am. Assoc. Clin. Chemists, August 1961; Abstr., Clin. Chem., 7, 562 (1961)

4424. Lipsky, S. R., "Recent Advances in Analysis of Lipids by Gas Chromatography," 11th Annual Instrument Symp. and Research Equipment Exhibit, National Institutes of Health, Bethseda, Md., October 1961

4425. Lipsky, S. R., "Electron Capture Spectrometry of Biological Compounds," Univ. of Houston Intern. Symp. on Advances in Gas Chromatography, Houston, Texas, January 1963

4426. Lipsky, S. R., and Landowne, R. A., "The Separation of Lipides by Gas-Liquid Chromatography," 134th Natl. ACS Mtg., Chicago, Ill., September 1958, Program Abstr., p. 16B

4427. Lipsky, S. R., and Landowne, R. A., "New Partition Agent for Use in the Rapid Separation of Fatty Acid Esters by Gas-Liquid Chromatography," Biochim. et Biophys. Acta, 27, 666-667 (1958); CA 52, 10796d

4428. Lipsky, S. R., and Landowne, R. A., "Evaluation of a Stationary Phase for Fatty Acid Analysis by Gas-Liquid Chromatography," Ann. N. Y. Acad. Sci., 72, 666-674 (1959); CA 53, 14818i

4429. Lipsky, S. R., and Landowne, R. A., "Gas Chromatography - Biochemical Applications," Ann. Rev. Biochem., 29, 649-668 (1960); CA 54, 19815i

4430. Lipsky, S. R., and Landowne, R. A., "Effects of Varying the Chemical Composition of the Stationary Phase on the Separation of Certain C_{19}, C_{21}, and C_{27} Steroids by Gas Chromatography," Anal. Chem., 33, 818-828 (1961); CA 55, 18454g

4431. Lipsky, S. R., Landowne, R. A., and Godet, M. R., "Effects of Varying the Chemical Composition of the Stationary Liquid on the Resolution of the Long-Chain Saturated and Unsaturated Fatty Acid Esters by Gas-Liquid Chromatography," Biochim. et Biophys. Acta, 31, 336-347 (1959); CA 53, 10899g

4432. Lipsky, S. R., Landowne, R. A., and Lovelock, J. E., "Separation of Lipides by Gas-Liquid Chromatography," Anal. Chem., 31, 852-856 (1959); CA 53, 21361g; 134th Natl. ACS Mtg., Chicago Ill., September 1958, Program Abstr., p. 16B

4433. Lipsky, S. R., Lovelock, J. E., and Landowne, R. A., "Use of High-Efficiency Capillary Columns for the Separation of Certain Cis-Trans Isomers of Long-Chain Fatty Acid Esters by Gas Chromatography," J. Am. Chem. Soc., 81, 1010 (1959); CA 53, 13992f

4434. Lipsky, S. R., Lovelock, J. E., and Landowne, R. A., "Fatty Acid Analysis Using High Temperature Capillary Column Gas Chromatography," 50th Annual Mtg., Am. Soc. Biological Chemists, Atlantic City, N. J., April 1959; Abstr., Federation Proc., 18, 275 (1959)

4435. Lipsky, S. R., and Shahin, M. M., "Sensitive Ionization System for the Detection of Permanent Gases and Organic Vapors by Gas Chromatography," Nature, 197, 625-626 (1963); CA 58, 11934e

4436. Lis, E. W., Tinoco, J., and Okey, R. A., "A Micromethod for Fractionation of Lipides by Silicic Acid Chromatography," Anal. Biochem., 2, 100-106 (1961); CA 55, 22454i

4437. Litchfield, C., "The Analysis of Cis-Trans Fatty Acid Isomers Using Gas-Liquid Chromatography," 53rd Spring Mtg., Am. Oil Chemists' Soc., New Orleans, La., May 1962; Biochemical Gas Chromatography Seminar, Wilkens Instrument & Research, Inc., New Orleans, La., July 1963

4438. Litchfield, C., Isbell, A. F., and Reiser, R., "Analysis of the Geometric Isomers by Methyl Linoleate by Gas Chromatography," J. Am. Oil Chemists' Soc., 39, 330-334 (1962); CA 57, 7399d

4439. Litchfield, C., Reiser, R., and Isbell, A. F., "The Analysis of Cis-Trans Fatty Acid Isomers Using Gas-Liquid Chromatography," J. Am. Oil Chemists' Soc., 40, 302-309 (1963)

4440. Little, L. H., Klauser, H. E., and Amberg, C. H., "Infrared Study of the Adsorption of Butenes on Surfaces of Porous Vycor Glass," Can. J. Chem., 39, 42-60 (1961); CA 55, 11081f

4441. Littlewood, A. B., "Techniques Used in a Study of the Boron and Silicon Hydrides," Physical and Microchemistry Groups with Scottish Section of the Soc. for Analytical Chemistry Mtg., Symp. on Gas Chromatography, Stevenson, Scotland, May 1955; Abstr., Anal. Chem., 27, 1667 (1955)

4442. Littlewood, A. B., "An Examination of Column Efficiency in Gas-Liquid Chromatography Using Columns of Wetted Glass Beads," Gas Chromatog., Proc. Symposium, Amsterdam, 1958, 23-331 CA 53, 14636c

4443. Littlewood, A. B., "An Examination of Column Efficiency in Gas-Liquid Chromatography Using Columns of Wetted Glass Beads," in "Gas Chromatography 1958," edited by D. H. Desty, Academic Press, Inc., New York, 1958, pp. 23-33

4444. Littlewood, A. B., "Sensitivity of Katharometers in Gas Chromatography and the Thermal Conduct-
      ivity of Binary Gas Mixtures," Nature, 184, 1631-1632 (1959); CA 54, 16989f

4445. Littlewood, A. B., "Compensated Wheatstone's Bridge Circuit for Gas Chromatographic Katharo-
      metry," J. Sci. Instr., 37, 185-188 (1960)

4446. Littlewood, A. B., "Gas Chromatography. Principles. Techniques, and Applications," Academic
      Press, Inc., New York, 1962

4447. Littlewood, A. B., "Gas-Chromatographic Specific Retention Volumes of Many Common Organic
      Compounds on Several Common Stationary Phases at 80°C.," J. Gas Chromatog., 1, No. 5, 6-8
      (1963)

4448. Littlewood, A. B., "Gas Chromatography Symposium, London, England, April 1963," J. Gas
      Chromatog., 1, No. 5, 28-31 (1963)

4449. Littlewood, A. B., Phillips, C. S. G., and Price, D. T., "The Chromatography of Gases and Vapors.
      V. Partition Analysis with Columns of Silicone 702 and of Tritolyl Phosphate," J. Chem. Soc.,
      1955, 1480-1489; CA 49, 10800h

4450. Lityaeva, Z. A., Markosov, P. I., and Zaichenko, V. N., "Gas-Chromatographic Determination of
      Carbon Monoxide, Methane, and Acetylene in High-Purity Ethylene," Tr. Vses. Neftegaz.
      Nauchn.-Issled. Inst. Krasnodarsk. Filial, 1962, No. 8, 110-125; CA 58, 6189d

4451. Lloyd, D. I., "Continuous Gas-Liquid Chromatography," Birmingham Univ. Chem. Engr., 13, 103-
      104 (1962); CA 58, 7345e

4452. Lloyd, H. A., Fales, H. M., Highet, P. F., and VandenHeuvel, W. J. A., "Separation of Alkaloids
      by Gas Chromatography," J. Am. Chem. Soc., 82, 3791 (1960)

4453. Lloyd, H. A., Kielar, E. A., Highet, R. J., Uyeo, S., Fales, H. M., and Wildman, W. C., "Posi-
      tion of Aromatic Methoxyl in Alkaloids Related to Powelline," J. Org. Chem., 27, 373-377 (1962);
      CA 57, 879h

4454. Lloyd, M. I., "Temperature Programmed Instruments," 2nd Gas Chromatography Institute,
      Canisius College, Buffalo, N. Y., April 1960

4455. Lloyd, M. I., and Guild, L. V., "Sampling Methods in Gas Chromatography," 12th Pittsburgh Conf.
      on Anal. Chem. & Appl. Spectroscopy, Pittsburgh, Pa., February-March 1961, Program Abstr.,
      p. 56

4456. Lochte, H. L., and Pittman, A. G., "The Nitrogen Compounds of Petroleum Distillates. XXVIII.
      Isolation of 2-Methyl-7, 7-dihydro-1, 5-pyridine. Preparation of Some Methyl-dihydro-pyridines,"
      J. Am. Chem. Soc., 82, 469-472 (1960); CA 54, 8816c

4457. Lochte, H. L., and Pittman, A. G., "Nitrogen Compounds of Petroleum Distillates. XXIX. Identi-
      fication of 5-Methyl-6, 7-dihydro-1, 5-pyridine," J. Org. Chem., 25, 1462-1464 (1960); CA 54,
      24707e

4458. Locke, L. N., "Use of Chromatography as a Laboratory Control Instrument for Plant Operation,"
      37th Annual Mtg., Natural Gasoline Assoc. Am., Dallas, Texas, April 1958, Tech. Papers, 37,
      14-15 (1958); CA 54, 1834i

4459. Locke, L. N., "Here's How Gas Chromatography Works in the Laboratory," Oil Gas J., 56, No. 16,
      120-121 (1958); CA 52, 15883e

4460. Lockhart, E. E., "Chemistry of Coffee," Chemistry of Natural Flavors, Report of Symposium,
      May 1957, pp. 174-191

4461. Lockhart, E. E., "Problems 'Flavor Prints' Can Solve," Food Eng., 31, No. 7, 82 (1959)

4462. Lodge, Jr., J. P., "Air Pollution," Anal. Chem., 33, No. 5, 3R-13R (1961); CA 55, 12137e

4463. Loffler, W., and Nausch, E., "Micro-Gas-Analysis for Series Studies," Mikrochim. Acta, 1955,
      950-953; CA 50, 3953c

4464. Loft, J. T., and York, W. B., "The Application of Gas Chromatography to the Determination of
      Naphthalene and Alkylnaphthalenes in Petroleum Fractions," Oklahoma Tetrasectional ACS
      Mtg., Tulsa, Okla., March 1963

4465. Lohman, F., "Aromatics and Natural Raw Materials," 5th Annual Symp., Am. Soc. Perfumers,
      New York, N. Y., April 1959; Abstr., Perfumery Essent. Oil Record, 50, 606 (1959)

4466. Lohr, L. J., "Modification of F & M Model 500 Gas Chromatograph for Collecting Eluted Compo-
      nents," Facts & Methods (F & M Scientific Corp.), 2, No. 2, 1 (Fall 1961)

4467. Lohr, L. J., and Warren, R. W., "Gas Chromatography of Certain Oximes," J. Chromatog., 8,
      127-129 (1962); CA 58, 5558b

4468. Lombard, R., and Kress, A., "Action of Boron Fluoride on Terpene Hydrocarbons," Bull. soc.
      chim. France, 1959, 1415-1419; CA 54, 11071h

4469. Lombardo, J., Molinari, M. A., and Lires, O. A., "Partition Chromatography of Saturated Hydro-
      carbons," Anales asoc. quim. arg., 48, 140-152 (1960); CA 55, 21988d

4470. Long, C. N. H., "Gas Chromatography. Retention Volumes of Compounds of Interest," in "Bio-
      chemists' Handbook," edited by C. N. H. Long, E. J. King, and W. M. Sperry, D. Van Nostrand
      Co., Inc., Princeton, N. J., 1961, pp. 118-120

4471. Long, D. R., McBride, H. D., Tuemmler, F. D., Heinrich, B. J., Alford, D. O., Edwards, R. T.,

Johnson, J. W., Borup, R. E., Wronka, J. A., Walker, J., Boulet, G. A., Farrell, R. E., King, R. W., Haines, W. E., Patterson, G. H., Marantette, J. C., and Meador, G. R., "Petroleum," Anal. Chem., <u>35</u>, No. 5, 111R-142R (1963)

4472. Longenecker, W. H., "Simplified Partition Chromatographic Procedures. Resolution of Sulfon-amides, Sulfones, and Their Metabolic Products from Biological Materials," Anal. Chem., <u>21</u>, 1402-1405 (1949); CA <u>44</u>, 3065e

4473. Longhetti, A., and Cadman, W. J., "Identification of Gasolines by Gas Chromatography," Spring Seminar, California Assoc. of Criminalistics, Los Angeles, Calif., April 1959

4474. Longmire, C. L., "Methods for Determining Thermal Conductivity at High Temperatures," Rev. Sci. Instr., <u>28</u>, 904-906 (1957); CA <u>52</u>, 12538a

4475. Looney, R. W., "The Radiation Chemistry of Methane," Univ. Microfilms (Ann Arbor, Mich.), L. C. Card No. <u>Mic 60-5762</u>, 144 pp.; Dissertation Abstr., <u>21</u>, 1391 (1960); CA <u>55</u>, 6113e

4476. Lorant, M., "New and Rapid Gas-Chromatographic Analysis of Copolymer Systems," Chem. Rundschau (Solothurn), <u>13</u>, 429-430 (1960); CA <u>54</u>, 24089a

4477. Lorenz, D. H., and Becker, E. I., "Hydrogenolyses of Chloromethanes with Triphenyltin Hydride," J. Org. Chem., <u>27</u>, 3370-3371 (1962); CA <u>58</u>, 1336g

4478. Lorz, W., Mills, G. A., Shalit, H., and Michael, T. C., "Olefins from Neohexane and Diisopropyl: Neohexene and 2,3-Dimethylbutenes," Ind. Eng. Chem., <u>53</u>, 873-876 (1961); CA <u>56</u>, 3333b; 139th Natl. ACS Mtg., St. Louis, Mo., March 1961, Program Abstr., p. 10Q

4479. Losse, A., "The Structural Influence of the Polar Support in Gas-Liquid Partition Chromatography," 3rd Intern. Conf. on Anal. Chem., Prague, Czech., September 1959

4480. Losse, G., Losse, A., and Stoeck, J., "Separation of N-Formylamino Acid Methyl Esters by Gas Chromatography," Z. Naturforsch., <u>17b</u>, 785-786 (1962); CA <u>58</u>, 9221b

4481. Lotz, J. R., and Willingham, C. B., "Gas-Phase Chromatography," J. Chem. Educ., <u>33</u>, 485-489 (1956)

4482. Lotz, J. R., and Willingham, C. B., "Gas Chromatography: A Means for Separation and Analysis of Volatile Materials," Ind. Hyg. Foundation Am. Trans. Bull., No. 30, 21st Ann. Mtg., 195-200 (1956); CA <u>51</u>, 11157b

4483. Louedec, A., "Renewal of the Fatty Acids in Cholesterol Esters and Glycerides in the Liver and Plasma of the Normal Rat," Compt. rend., <u>246</u>, 1619-1622 (1958); CA <u>52</u>, 16524b

4484. Louedec, A., and Pascaud, M., "Nature of the Fatty Acids in the Different Hepatic Lipids in the Rat," Compt. rend., <u>247</u>, 1408-1411 (1958); CA <u>53</u>, 7357i

4485. Louloudes, S. J., Thompson, M. J., Monroe, R. E., and Dobbins, W. E., "Conversion of Choles-tanol to Δ^7-Cholestenol by the German Cockroach," Biochem. Biophys. Res. Commun., <u>8</u>, 104-106 (1962); CA <u>57</u>, 14152e

4486. Loury, M., "Mechanism of Formation of Erythro-9,10-dihydroxystearic Acid from the Autoxida-tion Products of Oleic Acid," Compt. rend., <u>255</u>, 2456-2458 (1962); CA <u>58</u>, 7798g

4487. Lovelock, J. E., "The 'Argon' Detector," in "Gas Chromatography 1958," edited by D. H. Desty, Academic Press, Inc., New York, 1958, pp. 320-332

4488. Lovelock, J. E., "Measurement of Low Vapour Concentration by Collision with Excited Rare Gas Atoms," Nature, <u>181</u>, 1460-1462 (1958); CA <u>52</u>, 19312d

4489. Lovelock, J. E., "Detector for Use with Capillary Tube Columns in Gas Chromatography," Nature, <u>182</u>, 1663-1664 (1958); CA <u>53</u>, 9740c

4490. Lovelock, J. E., "A Sensitive Detector for Gas Chromatography," J. Chromatog., <u>1</u>, 35-46 (1958); CA <u>53</u>, 16609f

4491. Lovelock, J. E., "Ionization Methods for the Measurement of Low Vapor Concentrations in Gas Chromatography," 10th Pittsburgh Conf. on Anal. Chem. & Appl. Spectroscopy, Pittsburgh, Pa., March 1959, Program Abstr., p. 54

4492. Lovelock, J. E., "Argon Detectors," Gas Chromatog. Proc. Symposium, 3rd, Edinburgh, <u>1960</u>, 16-29; CA <u>56</u>, 4073c

4493. Lovelock, J. E., "Argon Detectors," in "Gas Chromatography 1960," edited by R. P. W. Scott, Butterworths, London, 1960, pp. 16-29

4494. Lovelock, J. E., "An Ionization Detector for Permanent Gases," Nature, <u>187</u>, 49-50 (1960)

4495. Lovelock, J. E., "A Photoionization Detector for Gases and Vapours," Nature, <u>188</u>, 401 (1960); CA <u>55</u>, 6059d

4496. Lovelock, J. E., "Ionization Methods for the Analysis of Gases and Vapors," Anal. Chem., <u>33</u>, 162-178 (1961); CA <u>55</u>, 9157b

4497. Lovelock, J. E., "Affinity of Organic Compounds for Free Electrons with Thermal Energy: Its Possible Significance in Biology," Nature, <u>189</u>, 729-732 (1961); CA <u>55</u>, 22079e

4498. Lovelock, J. E., "A Sensitive Ionization Cross-Section Detector for Gas Chromatography," Univ. of Houston Intern. Symp. on Advances in Gas Chromatography, Houston, Texas, January 1963

4499. Lovelock, J. E., "Electron Absorption Detectors and Technique for Use in Quantitative and Quali-tative Analysis by Gas Chromatography," Anal. Chem., <u>35</u>, 474-481 (1963)

4500. Lovelock, J. E., and Gregory, N. L., "Electron Capture Ionization Detectors," in "Gas Chromatography," edited by N. Brenner, J. E. Callen, and M. D. Weiss, Academic Press, Inc., New York, 1962, pp. 219-229

4501. Lovelock, J. E., and Gregory, N. L., "Electron Capture Ionization Detectors," Gas Chromatog., Intern. Symp., 1961, 3, 219-229 (Pub. 1962); CA 58, 3865h

4502. Lovelock, J. E., James, A. T., and Piper, E. A., "New Type of Ionization Detector for Gas Chromatography," Ann. N. Y. Acad. Sci., 72, 720-730 (1959); CA 53, 16609h

4503. Lovelock, J. E., and Lipsky, S. R., "Electron Affinity Spectroscopy - A New Method for the Identification of Functional Groups in Chemical Compounds Separated by Gas Chromatography," J. Am. Chem. Soc., 82, 431-433 (1960); CA 54, 18159f

4504. Lovelock, J. E., Shoemake, G. R., and Zlatkis, A., "Sensitive Ionization Cross-Section Detector for Gas Chromatography," Anal. Chem., 35, 460-465 (1963)

4505. Lovelock, J. E., Simmonds, P. G., and VandenHeuvel, W. J. A., "Affinity of Steroids for Electrons with Thermal Energies," Nature, 197, 249-251 (1963); CA 59, 1706e

4506. Lovelock, J. E., and Zlatkis, A., "A New Approach to Lead Alkyl Analysis: Gas Phase Electron Absorption for Selective Detection," Anal. Chem., 33, 1958-1959 (1961); CA 56, 7575b

4507. Low, M. J. D., "Kinetics of Chemisorption of Gases on Solids," Chem. Rev., 60, 267-312 (1960); CA 54, 14865c

4508. Lowe, A. E., and Moore, D., "Scintillation Counter for Measuring Radioactivity of Vapours," Nature, 182, 133-134 (1958); CA 53, 80d

4509. Lowe, H. J., "Determination of Blood Anesthetic Concentrations," Facts & Methods (F & M Scientific Corp.), 3, No. 3, 5-6 (Winter 1962-1963)

4510. Lowe, H. J., "Studies on Anesthetics," 5th Annual Gas Chromatography Institute, Canisius College, Buffalo, N. Y., April 1963

4511. Lowe, H. J., "Determination of Blood Anesthetic Concentrations," Double Bond (publication of Western New York Section of ACS), 35, No. 5, 84-85 (1963)

4512. Lowry, R. R., "A Solid Sample Injector System for Gas Chromatography," 18th Annual Northwest Regional Mtg., ACS, Bellingham, Wash., June 1963

4513. Loyd, R. J., Ayers, B. O., and Karasek, F. W., "The Effect of Carrier Gas on High Speed Chromatographic Separations," 27th Mtg., Gulf Coast Spectroscopic Group, Houston, Texas, October 1958

4514. Loyd, R. J., Ayers, B. O., and Karasek, F. W., "Optimization of Resolution-Time Ratio with Packed Chromatographic Columns," Anal. Chem., 32, 698-701 (1960); CA 54, 14814i

4515. Lu, P-C., "Parameters of Gas-Chromatographic Separations in Gas Phase," Gazovaya Khromatografiya, Trudy 1-oi [Pervoi] Vsesoyuz Konf., Akad. Nauk S.S.S.R., Moscow, 1959, 172-182 (Pub. 1960); CA 55, 25577a

4516. Lu, P-C., and K'ang, T., "Analysis of Gas Mixtures by Chromatography. I. Active Carbon as Adsorbent for the Volumetric Chromatographic Method of Analysis of the Gaseous Saturated Hydrocarbons," Jan Liao Hsueh Pao, 1, 47-53 (1960); CA 54, 25704h

4517. Lu, P-C., Kuang, Y-D., and Khan, T., "Chromatographic Method for the Analysis of a Mixture of Saturated and Unsaturated Hydrocarbons," Jan Liao Hsueh Pao, 1, 146-148 (1956); CA 54, 25702d

4518. Lu, P-C., and Kwan, Y., "The Selectivity Index and Efficiency of the Chromatographic Column," Jan Liao Hsueh Pao, 3, 77-84 (1958); CA 52, 16115g

4519. Lu, P-C., Lu, T-F., and Wu, H-F., "Factors Determining the Efficiency of a Gas-Liquid Chromatographic Column," Jan Liao Hsueh Pao, 3, 255-276 (1958); CA 55, 1141g

4520. Lu, P-T., Lu, T-F., and Li, H. C., "Factors Determining the Efficiency of a Gas-Liquid Chromatographic Column," K'o Hsueh T'ung Pao, 1957, 699; CA 53, 18592f

4521. Lucas, D. M., "The Identification of Petroleum Products in Forensic Science by Chromatography," J. Forensic Sci., 5, 236-247 (1960); CA 54, 14647f

4522. Lucchesi, P. J., and Bartok, W., "The Controlled Chain Oxidation of n-Hexane Induced by Co-60 Radiation," J. Am. Chem. Soc., 82, 4528-4530 (1960); CA 55, 5323a

4523. Lucchesi, P. J., and Heath, C. E., "Radiation-Induced Chain Alkylation of Propylene with Isobutane," J. Am. Chem. Soc., 81, 4770-4773 (1959); CA 54, 2903b

4524. Lucchesi, P. J., Heath, C. E., and Baeder, D. L., "Radiation-Induced Carbonium Ion Reactions and the Chain Nature of Acid Catalyzed Isomerization," J. Am. Chem. Soc., 82, 4530-4533 (1960); CA 55, 5323c

4525. Luce, C. C., Humphrey, E. F., Norrish, H. H., and Guild, L. V., "Analysis of Polyester Resins by Gas Chromatography," 14th Pittsburgh Conf. on Anal. Chem. & Appl. Spectroscopy, Pittsburgh, Pa., March 1963, Program Abstr., p. 68

4526. Luce, R. L., Felter, R. E., and Currie, L. A., "Tritium Recoil Chemistry and Radiation Chemistry of n-Pentane," 139th Natl. ACS Mtg., St. Louis, Mo., March 1961, Program Abstr., p. 4R

4527. Luchsinger, W., "Separation of C$_4$ Hydrocarbons," Abhandl. Deut. Akad. Wiss. Berlin, Kl. Chem.,

Geol., Biol., 1959, No. 9, 118-135; CA 58, 5018d

4528. Lück, H., and Kohn, R., "Formation of Fatty Acids of Intermediate Chain-Length in Electron-Ir-radiated Fats," Experientia, 17, 109-110 (1961); CA 55, 24052d

4529. Lück, H., and Kohn, R., "The Structure of the Carbonyl Compounds in Fats After the Action of an Electron Beam," Experientia, 18, No. 2, 62-63 (1962)

4530. Lück, H., Purr, A., and Kohn, R., "Detection of Extraneous Fats in Cocoa Butter by Means of Di-electric Infrared Spectroscopic, Gas, and Column Chromatographic Methods," Rev. intern. chocolat., 16, 106-116 (1961); CA 55, 15961d

4531. Luddy, F. E., Barford, R. A., and Riemenschneider, R. W., "Direct Conversion of Lipid Compo-nents to Their Fatty Acid Methyl Esters," J. Am. Oil Chemists' Soc., 37, 447-451 (1960); CA 54, 23371d; 50th Mtg., Am. Oil Chemists' Soc., New Orleans, La., April 1959, Program Abstr., p. 30

4532. Luebbe, Jr., R. H., and Willard, J. E., "Temperature and Phase Effects on the Photolysis of Ethyl Iodide," J. Am. Chem. Soc., 81, 761-769 (1959); CA 54, 4120b

4533. Luh, B. S., and Chaudhry, M. S., "Gas Chromatography of CO_2, H_2, O_2, and N_2 in Processed Foods," Food Technol., 15, 52-54 (1961); CA 56, 6420b

4534. Lukas, D. S., and Ayres, S. M., "Determination of Blood Oxygen by Gas Chromatography," J. Appl. Physiol., 16, 371-374 (1961); CA 55, 22470d

4535. Lukes, V., and Herout, V., "Apparatus for Preparative Gas-Liquid Chromatography," Collection Czechoslov. Chem. Communs., 25, 2770-2776 (1960); CA 55, 4060d

4536. Lukes, V., and Herout, V., "Use of Gas Chromatography for the Analysis of Essential Oils," Gas Chromatog., Symp., Brno, Czech., June 1962; Abstr., J. Gas Chromatog., 1, No. 4, 9 (1963)

4537. Lukes, V., Komers, R., and Herout, V., "Ground Unglazed Tile. A New Support for Gas-Liquid Chromatography," J. Chromatog., 3, 303-307 (1960); CA 54, 21930d

4538. Lulova, N. I., Piguzova, L. I., Tarasov, A. I., and Fedosova, A. K., "Examination of Synthetic Zeolites by Gas Chromatography," Sintetich, Tseolity, Poluchenie, Issled. i Preimenenie, Akad. Nauk SSSR, Otd. Khim. Nauk, 1962, 59-64; CA 58, 10759g

4539. Lulova, N. I., Tarasov, A. I., Kuz'mina, A. V., and Koroleva, N. M., "Gas-Chromatographic Analysis of a Gas Stream in the Production of Ethylene," Neftekhimiya, 2, 885-891 (1962); CA 58, 8395a

4540. Lunaas, T., "Free Butyric Acid as a Possible Source of Off-Flavor of the Cow's Milk After Ad-ministration of Oestrogens," Acta Chem. Scand., 14, 773-775 (1960)

4541. Lund, N. A., "Application of Gas Chromatography in the Confectionery Industry," Perfumery Essent. Oil Record, 50, 147-152 (1959)

4542. Lund, P. G., and Shorb, M. S., "Steroid Requirements of Trichomonads," J. Protozool., 9, 151-154 (1962); CA 58, 767f

4543. Lundberg, W. O., and Peifer, J. J., "Essential Fatty Acids and Fat Biosynthesis," 50th Mtg., Am. Oil Chemists' Soc., New Orleans, La., April 1959, Program Abstr., p. 25

4544. Lundeen, A., "The Isomerization of Trialkylacetic Acids in Sulfuric Acid," J. Am. Chem. Soc., 82, 3228 (1960); CA 54, 24363c

4545. Luskina, B. M., Syavtsillo, S. V., Terent'ev, A. P., and Turkel'taub, N. M., "Microdetermina-tion of Carbon and Hydrogen in Organic Compounds with Gas Chromatography," Dokl. Akad. Nauk SSSR, 141, 869-871 (1961); CA 57, 50a

4546. Luther, H., and Klose, A., "Oxidation Studies on Mineral Oils. III. Results of Oxidation Experi-ments and Oxygen Balance Studies," Erdol u. Kohle, 12, 898-903 (1959); CA 54, 7122b

4547. Lutwick, G. D., and Harris, W. E., "Column Efficiency in Gas-Liquid Chromatography at Constant Average Pressure," 141st Natl. ACS Mtg., Washington, D. C., March 1962, Program Abstr., p. 19B

4548. Lutz, C. A., and Ritter, D. M., "Observations on Alkylboranes," Can. J. Chem., 41, 1344-1358 (1963)

4549. Lutz, E. F., "The Cyclic Trimerization of Acetylene Over a Ziegler Catalyst," J. Am. Chem. Soc., 83, 2551-2554 (1961); CA 55, 20999a

4550. Luukkainen, T., VandenHeuvel, W. J. A., Haahti, E. O. A., and Horning, E. C., "Gas-Chromato-graphic Behavior of Trimethylsilyl Ethers of Steroids," Biochim. et Biophys. Acta, 52, 599-601 (1961); CA 56, 10478b

4551. Luukkainen, T., VandenHeuvel, W. J. A., and Horning, E. C., "Estrogen Determination Method Using Gas Chromatography," Biochim. et Biophys. Acta, 62, 153-159 (1962); CA 57, 12811h

4552. Lynch, C. T., "Cadmium N-Methylglucamine System - Specific Adsorbents in Gas Chromatography," Univ. Microfilms (Ann Arbor, Mich.), L. C. Card No. Mic 61-162, 144 pp.; Dissertation Abstr., 21, 3609-3610 (1961); CA 55, 23150g

4553. Lynch, E. R., and McCall, E. B., "Syntheses of Some Monoalkylbenzenes," J. Chem. Soc., 1960, 1254-1262; CA 54, 17296i

4554. Lynch, J., and Burkhalter, T. S., "The Gas Chromatography of Organic Impurities in Silicon

Halides," 11th Pittsburgh Conf. on Anal. Chem. & Appl. Spectroscopy, Pittsburgh, Pa., February-March 1960, Program Abstr., p. 45

4555. Lynn, T. R., Sweeny, R. F., Rose, A., and Supina, W. R., "Vapor Phase Chromatography of High Molecular Weight, Low Volatility Substances," 12th Pittsburgh Conf. on Anal. Chem. & Appl. Spectroscopy, Pittsburgh, Pa., February-March 1961, Program Abstr., p. 56

4556. Lyons, J. M., McGlasson, W. B., and Pratt, H. K., "Ethylene Production, Respiration, and Internal Gas Concentrations in Cantaloupe Fruits at Various Stages of Maturity," Plant Physiol., 37, 31-36 (1962); CA 56, 10596e

4557. Lyons, M. J., "Vehicular Exhausts: Identification of Further Carcinogens of the Polycyclic Aromatic Hydrocarbon Class," Brit. J. Cancer, 13, 126-131 (1959); CA 53, 17393c

4558. Lysyj, I., "Gas Chromatographic Analysis of Nitriles," Anal. Chem., 32, 771 (1960); CA 54, 18162a

4559. Lysyj, I., "A Parallel Dual Column System for Gas Chromatographic Separations," 3rd Intern. Symp. on Gas Chromatography, ISA, East Lansing, Mich., June 1961; ISA Proc., 3, 279-281 (1961)

4560. Lysyj, I., "A Parallel Dual Column System for Gas Chromatographic Separations," in "Gas Chromatography," edited by N. Brenner, J. E. Callen, and M. D. Weiss, Academic Press, Inc., New York, 1962, pp. 443-448

4561. Lysyj, I., "Parallel Dual Column System for Gas-Chromatographic Separations," Gas Chromatog., Intern. Symp., 1961, 3, 443-448 (Pub. 1962); CA 58, 3866e

4562. Lysyj, I., and Newton, P. R., "Evaluation of Gas Chromatographic Columns for the Separation of Fluorinated Materials," Anal. Chem., 35, 90-92 (1963); CA 58, 6166a

4563. Ma, A. P., Hotchkiss, D. K., and Allen, R. S., "Early Effects of Dietary Glycerides and Fatty Acids on Serum Fatty Acids (Free and Combined) Levels in Young Dairy Calves," 58th Annual Mtg., Am. Dairy Sci. Assoc., Lafayette, Ind., June 1963; Abstr., J. Dairy Sci., 46, 644 (1963)

4564. Maccoll, A., and Stone, R. H., "Gas-Phase Eliminations. II. The Pyrolysis of sec-Butyl Chloride. The Direction of Elimination from sec-Butyl Compounds," J. Chem. Soc., 1961, 2756-2761; CA 56, 15360g

4565. Maccoll, A., and Stone, R. H., "Gas-Phase Eliminations. III. Pyrolysis of Some sec- and tert-Alkyl Acetates," J. Chem. Soc., 1962, 335-340; CA 56, 15360h

4566. MacDonell, H. L., Noonan, J. M., and Williams, J. P., "Porous Glass as an Adsorption Medium for Gas Chromatography," Anal. Chem., 35, 1253-1255 (1963)

4567. Macfarlane, M. G., "Cardiolipin and Other Phospholipides in Ox Liver," Biochem. J., 78, 44-51 (1961); CA 55, 6638a

4568. Macfarlane, M. G., Gray, G. M., and Wheeldon, L. W., "Fatty Acids of Phospholipids from Mitrochondria and Microsomes of Rat Liver," Biochem. J., 74, 43P-44P (1960); CA 56, 9255b

4569. Macfarlane, M. G., Gray, G. M., and Wheeldon, L. W., "Fatty Acid Composition of Phospholipides from Subcellular Particles of Rat Liver," Biochem. J., 77, 626-631 (1960); CA 55, 7576a

4570. Machacek, Z., and Laita, Z., "The Laboratory Purification of Ethylene," Chem. prumysl., 10, 251-252 (1960); CA 54, 24331e

4571. Machata, G., "Application of Gas Chromatography to Toxicological Analysis (Intoxication with Paraldehyde and Oil of Turpentine)," Arch. Toxicol., 18, 338-346 (1960); CA 56, 7646f

4572. Machata, G., "Routine Determination of Blood Alcohol Concentration with the Gas Chromatograph," Mikrochim. Acta, 1962, 691-700; CA 57, 6247f

4573. Machiroux, R., "Detectors in Gas Chromatography," Ind. chim. belge, 25, 1061-1072 (1960)

4574. Maciw, B. O., and Chmielinski, J. G., "Analysis of Process Hydrocarbon Streams by Gas Chromatography," 6th Symp., Instrumental Methods of Analysis, ISA, Montreal, Canada, June 1960

4575. Mackay, D. A. M., "Coming - An Analytical Revolution," Food Eng., 31, No. 7, 82-83 (1959)

4576. Mackay, D. A. M., "Trace Analysis by Gas Chromatography," 3rd Annual Eastern Analytical Symp. and Instrument Exhibit, New York, N. Y., November 1961

4577. Mackay, D. A. M., "Electron Capture - Technique and Applications," 2nd Symp. on Gas Chromatography, Toronto Section, CIC, Toronto, Ontario, February 1962

4578. Mackay, D. A. M., and Berdick, M., "Gas Chromatography in Evaluation of Foods," 139th Natl. ACS Mtg., St. Louis, Mo., March 1961, Program Abstr., p. 37B

4579. Mackay, D. A. M., and Connor, J., "The Use of Gas Chromatography in Food Additive Analysis," Instrumental Methods for Analysis of Food Additives, Symp. and Workshop, East Lansing, Mich., March 1960, Proc., pp. 235-239

4580. Mackay, D. A. M., and Hewitt, E. J., "Application of Flavor Enzymes to Processed Foods. II. Comparison of the Effect of Flavor Enzymes from Mustard and Cabbage Upon Dehydrated Cabbage," Food Research, 24, 253-261 (1959); CA 53, 22555b

4581. Mackay, D. A. M., Land, D. A., and Berdick, M., "The Objective Measurement of Odor. Gas Chromatography of Volatiles from Onion and Other Foods Using Ionization Detector," 135th Natl. ACS Mtg., Boston, Mass., April 1959, Program Abstr., p. 12A

4582. Mackay, D. A. M., Lang, D. A., and Berdick, M., "Measuring and Deodorizing Breath Odor,"

Drug & Cosmetic Ind., 86, 46-48, 105-107, 166-167, 250, 252, 264 (1960)

4583. Mackay, D. A. M., Lang, D. A., and Berdick, M., "Objective Measurement of Odor. Ionization Detection of Food Volatiles," Anal. Chem., 33, 1369-1374 (1961); CA 56, 739f

4584. Mackay, G., "Odor and Flavor Analysis," 13th Annual Summer Symp. on Anal. Chem., Univ. of Houston, Houston, Texas, June 1960

4585. Mackie, A., and Mieras, D. G., "Seeds of Selected Tropical Plants. I. Component Acids of the Fats or Oils," J. Sci. Food Agr., 12, 202-205 (1961); CA 55, 19279b

4586. Mackle, H., Mayrick, R. G., and Rooney, J. J., "Measurement of Heats of Vaporization by the Method of Gas-Liquid Chromatography," Trans. Faraday Soc., 56, 115-117 (1960); CA 54, 12722e

4587. MacLachlan, A., "Indirect Chemical Effects of High-Energy Radiation in Organic Solutions," J. Am. Chem. Soc., 82, 3309-3314 (1960); CA 55, 14032a

4588. Maczek, A. O. S., and Phillips, C. S. G., "Retention Times and Molecular Shape; the Use of Trio-thymotide in Column Liquids," in "Gas Chromatography 1960," edited by R. P. W. Scott, Butterworths, London, 1960, pp. 284-288

4589. Madan, M. P., "Simple Bridge Method for the Measurement of Thermal Conductivity of Gases and Gas Mixtures," J. Franklin Inst., 263, 207-212 (1957); CA 51, 7826h

4590. Madden, W. F., Quigg, R. K., and Kemball, C., "Method of Improving the Null-Point of Thermal Conductivity Cells for Gas-Liquid Chromatography," Chem. & Ind. (London), 1957, 892; CA 51, 14329b

4591. Madison, J. J., "Analysis of Fixed and Condensable Gases by Two-Stage Gas Chromatography," Anal. Chem., 30, 1859-1862 (1958); CA 53, 2923b; 132nd Natl. ACS Mtg., New York, N. Y., September 1957, Program Abstr., p. 35B

4592. Maeda, T., and Fujii, M., "Gas Chromatographic Determination of Aromatic Components in Platformate," Kogyo Kagaku Zasshi, 62, 649-652 (1959); CA 58, 391f, 5427f

4593. Magee, E. M., "The Course of a Reaction in a Chromatographic Column," Fundamentals, 2, No. 1, 32-36 (1963)

4594. Magidman, P., "Gas-Liquid Chromatography for the Analysis of Fats and Oils," 3rd Delaware Valley Regional Mtg., ACS, Symp. on Vapor Phase Chromatography, Philadelphia, Pa., February 1960

4595. Magidman, P., Herb, S. F., Barford, R. A., and Riemenschneider, R. W., "Fatty Acids of Cow Milk. I. Techniques Employed in Supplementing Gas-Liquid Chromatography for Identification of Fatty Acids," J. Am. Oil Chemists' Soc., 39, 137-142 (1962); CA 56, 15889g

4596. Magidman, P., Herb, S. F., Luddy, F. E., and Riemanschneider, R. W., "Fatty Acids of Lard. B. Quantitative Estimation by Silicic Acid and Gas-Liquid Chromatography," J. Am. Oil Chemists' Soc., 40, 86-88 (1963); CA 58, 13056b

4597. Magnolet, J. C. P., "Simple and Multiple Gas Micro Introducers for Adsorption Studies," J. Sci. Instr., 30, 15-17 (1953); CA 47, 3624c

4598. Magritte, H., "Chromatography in Gaseous Phase," Ind. chim. belge, 24, 887-900 (1959); CA 54, 1023e

4599. Maher, T. P., "Gas Chromatography Applied to the Study of the Liquid Products of the Low-Temperature Fluidized-Bed Carbonization of Coal," J. Chromatog., 10, 324-337 (1963)

4600. Mahrwald, R., Huttig, E., Schwerdt, R., and Wagner, M., "Benzene from Brown-Coal Light Oil. Analytical Investigations of Benzene Contained in Brown-Coal Light Oils and Fractions of HTM (High-Temperature Medium-Pressure) Process," Chem. Tech. (Berlin), 12, 266-271 (1960); CA 55, 23992g

4601. Maier, H. J., "Vapor [Phase] Fractometry as Applied to Continuous Analysis," Inst. Soc. Am., Conf. Preprint 127-59, 11 pp. (1959); CA 57, 3231d

4602. Maier, H. J., "High Speed Process Chromatography," 6th Instrumental Methods of Analysis Symp., ISA, Montreal, Canada, June 1960

4603. Maier, H. J., "18 Process Streams Successfully Analyzed by Chromatography," Control Eng., 8, No. 8, 89-90 (1961)

4604. Maier, H. J., Bossart, C. J., and Heller, H., "Performance and Application of the Automatic Gas Chromatography Distillation Analyzer," 18th Annual Instrument-Automation Conference & Exhibit, ISA, Chicago, Ill., September 1963

4605. Maier, H. J., and Claudy, H. N., "Continuous and Automatic Analysis of Process Streams by Gas Chromatography," Ann. N. Y. Acad. Sci., 87, 864-867 (1960); CA 55, 19354b

4606. Maier, H. J., and Karpathy, O. C., "Prediction of Separation and Specifications of [Gas] Chromatographic Columns," J. Chromatog., 8, 308-318 (1962); CA 58, 2824f

4607. Maier, H. J., Karpathy, O. C., and Brenner, N., "Prediction of Retention Times," 13th Pittsburgh Conf. on Anal. Chem. & Appl. Spectroscopy, Pittsburgh, Pa., March 1962, Program Abstr., p. 59

4608. Maier, H. J., Stebens, C. R., and Briscoe, F. J., "New Components for Extending the Usefulness of Process Gas Chromatographic Sensing Units, " 8th Natl. Analysis Instrumentation Symp., ISA, Charleston, W. Va., April-May 1962

4609. Mair, B. J., Kronskop, N. C., and Mayer, T. J., "Composition of the Branched Paraffin-Cyclo-paraffin Portion of the Light Gas Oil Fraction, " J. Chem. Eng. Data, 7, 420-426 (1962); CA 57, 10105d; 141st Natl. ACS Mtg., Washington, D. C., March 1962, Program Abstr., p. 4Q

4610. Mair, B. J., and Shamaiengar, M., "Fractionation of Certain Aromatic Hydrocarbons with Molecular Sieve Adsorbents, " Anal. Chem., 30, 276-279 (1958); CA 52, 7671c; 132nd Natl. ACS Mtg., New York, N. Y., September 1957, Program Abstr., p. 7R

4611. Mairanovskii, V. G., and Yanotovskii, M. T., "Polarographic Detection in Gas-Liquid Chromatography. I. Description of the Method and of the Apparatus, " Zh. Fiz. Khim., 37, 705-707 (1963); CA 59, 1060c

4612. Majer, J. R., Mile, B., and Robb, J. C., "Mercury-Photosensitized Decomposition of Ethylene, " Trans. Faraday Soc., 57, 1342-1355 (1961); CA 56, 6821e

4613. Majhofer-Orescanin, B., and Prostenik, M., "Studies in the Sphingolipid Series. XXI. C_{20}-Sphingo-sine, A New Long-Chain Base of Animal Origin, " Croat. Chem. Acta, 33, 219-228 (1961); CA 58, 3302f

4614. Makita, M., and Wells, W.W., "Quantitative Analysis of Fecal Bile Acids by Gas Liquid Chromatography, " Anal. Biochem., 5, 523-530 (1963)

4615. Malafeev, N. A., Ydina, I. P., Nevskaya, E. M., and Zhavoronkov, N. M., "Separation of High-Boiling Compounds by Low-Temperature Gas Chromatography, " Khim. Prom., 1962, 320-322; CA 57, 15780g

4616. Malafeev, N. A., Ydina, I. P., and Zhavoronkov, N. M., "High-Temperature Gas-Liquid Chromatography, " Uspekhi Khim., 31, 710-723 (1962); CA 57, 11880c

4617. Malgiolio, J., Limoncelli, E.A., and Cleary, R. E., "Analysis of Trace Impurities in Helium Using Gas-Chromatographic Techniques, " U. S. At. Energy Comm., TID-7606, 140-151 (1960); CA 55, 19599d

4618. Malgiolio, J., Limoncelli, E. A., and Cleary, R. E., "Purification and Gas-Chromatographic Analysis of Helium, " U. S. At. Energy Comm., PWAC-352, 43 pp. (1961); CA 56, 1141d

4619. Malin, L., "The Determination of Fatty Acid Composition by Gas Chromatography and U.V. Spectro-photometric Methods, " Perfumery Essential Oil Record, 50, 505-507 (1959)

4620. Malin, L., "The Determination of Fatty Acid Composition by Gas Chromatography and U.V. Spectro-photometric Methods, " Soap, Perfumery & Cosmetics, 32, 597-599 (1959); CA 53, 18512i

4621. Malin, L., "The Application of Gas Chromatography to the Analytical Problems of the Fat and Fatty Acid Industry, " ISI (Indian Standards Inst.) Bull., 12, 74 (1960); CA 55, 3092g

4622. Malinowska, K., "Gas Chromatography, " Chem. Anal. (Warsaw), 7, 1017-1041 (1962); CA 58, 9605g

4623. Malins, D. C., "Fatty Acids and Glyceryl Ethers in Alkoxydiglycerides of Dogfish Liver Oil, " Chem. & Ind. (London), 1960, 1359-1360; CA 55, 17515c

4624. Malins, D. C., and Mangold, H. K., "Analysis of Complex Lipid Mixtures by Thin Layer Chromatography and Complementary Methods, " J. Am. Oil Chemists' Soc., 37, 576-578 (1960); CA 55, 2148c

4625. Manfredi, G., Corsini, F., Paolucci, G., Salvioli, Jr., G. P., and Babini, B., "Gas Chromatographic Determination of Erythrocyte Fatty Acids in Children, " Boll. Soc. Ital. Biol. Sper., 38, 720-728 (1962); CA 58, 4881a

4626. Mangan, Jr., G. F., Merritt, Jr., C., and Walsh, J. T., "Chemical Components of the Odor of Fish, " 135th Natl. ACS Mtg., Boston, Mass., April 1959, Program Abstr., p. 13A

4627. Mangold, G. B., and Murray, G. S., "Chromatographic Analysis and Oil Exploration, " Petroleum (London), 22, 304-307 (1959); CA 54, 17854h

4628. Mangold, H. K., and Kammereck, R., "Separation, Identification and Quantitative Analysis of Fatty Acids by Thin-Layer Chromatography and Gas-Liquid Chromatography, " Chem. & Ind. (London), 1961, 1032-1034; CA 55, 24394d

4629. Mangold, H. K., and Kammereck, R., "Analyzing Industrial Aliphatic Lipids, " J. Am. Oil Chemists' Soc., 39, 201-206 (1962); CA 56, 14413f; 52nd Annual Mtg., Am. Oil Chemists' Soc., St. Louis, Mo., April-May 1961

4630. Manka, D. P., "Complete Gas Chromatographic Analysis of Fixed Gases with One Detector and Argon as Gas Carrier, " 14th Pittsburgh Conf. on Anal. Chem. & Appl. Spectroscopy, Pittsburgh, Pa., March 1963, Program Abstr., p. 50

4631. Manka, D. P., "Treatment of Columns in Gas Chromatographic Analysis of Aqueous Solutions, " 14th Pittsburgh Conf. on Anal. Chem. & Appl. Spectroscopy, Pittsburgh, Pa., March 1963, Program Abstr., p. 62

4632. Manning, R. J., "Preparative Scale Gas-Liquid Chromatography; Trapping for Instrumental (e.g., Infrared) Characterization, " 13th Annual Fisk Infrared Inst., Gas Chromatography Session, Nashville, Tenn., August 1962

4633. Mansur, R. H., Pero, B. F., and Krause, L. A., "Vapor Phase Chromatography in Quantitative Determination of Air Samples Collected in the Field," Am. Ind. Hyg. Assoc. J., 19, 175-182 (1959); CA 54, 4978h

4634. Mantell, C. L., "Adsorption," 2nd ed., McGraw-Hill Book Co., Inc., New York, 1951

4635. Mantell, C. L., "Adsorption as an Engineering Tool. I.," Chem. in Can., 7, No. 5, 90 (1955); "II.," ibid., 7, No. 6, 60 (1955)

4636. Manuel, T. A., "Reactions of Monoolefins with Iron Carbonyls," J. Org. Chem., 27, 3941-3945 (1962); CA 58, 10105e

4637. Mapelson, W. W., "The Rate of Uptake of Halothane Vapor in Man," Brit. J. Anes., 34, 11-18 (1962)

4638. Marcali, K., "Trace Analysis by Temperature-Gradient Gas Chromatography," 14th Pittsburgh Conf. on Anal. Chem. & Appl. Spectroscopy, Pittsburgh, Pa., March 1963, Program Abstr., pp. 62-63

4639. Marco, G. J., Machlin, L. J., Emery, E., and Gordon, R. S., "Dietary Effects of Fats Upon Fatty Acid Composition of the Mitochondria," Arch. Biochem. Biophys., 94, 115-120 (1961); CA 55, 24954e

4640. Marcolongo, F., Cugudda, E., Gragnoli, G., Giovannelli, G., and Pasquinucci, D., "Behavior of the Fatty Acids of the Various Lipid Fractions in Diabetes Mellitus," Recenti Progr. Med., 32, 228-261 (1962); CA 57, 13098e

4641. Marcus, A. J., "The Study of the Fatty Acids and Aldehydes of Human Platelet Phosphatides by Gas Chromatography," 4th Annual Eastern Analytical Symp., New York, N. Y., November 1962, Program Abstr., p. 25

4642. Marcus, A. J., Ullman, H. L., and Ballard, H. S., "Fatty Acids of Human Platelet Phospholipids," Proc. Soc. Biol. & Med., 107, 483-486 (1961); CA 55, 23730i

4643. Marcus, A. J., Ullman, H. L., Safier, L. B., and Ballard, H. S., "Platelet Phosphatides; Their Fatty Acid and Aldehyde Composition and Activity in Different Clotting Systems," J. Clin. Invest., 41, 2198-2212 (1962); CA 58, 7070d

4644. Mare, H. E. de la, and Rust, F. F., "Intramolecular Radical Reactions. Decomposition of Pure Bis(2-Methyl-2-Hexyl) Peroxide in the Liquid Phase," J. Am. Chem. Soc., 81, 2691-2694 (1959); CA 54, 3172g

4645. Marechal, J., "Introduction to Vapor-Phase Chromatography," Ind. chim. belge, 22, 675-682 (1957)

4646. Marechal, J., Convent, L., and Rysselberge, J. van, "Study of Catalytic Reactions by Vapor-Phase Chromatography," Rev. inst. franc. petrole et Ann. combustibles Liquides, 12, 1067-1074 (1957); CA 52, 4299g

4647. Maresh, C., Sundberg, O. E., Hofstader, R. A., and Gerhardt, G. E., "Microdetermination of Carbon, Hydrogen, and Nitrogen by Gas Chromatography," Microchem. J. Symp. Ser., 2, 387-396 (1962); CA 58, 6191f; Intern. Symp. on Microchemical Techniques, University Park, Pa., August 1961

4648. Maricq, L., and Molle, L., "Investigations on the Determination of Alcoholism by Gas Chromatography," Bull. Acad. Roy. Med. Belg., 6, 199-232 (1959); CA 56, 7632f

4649. Maricq, L., and Molle, L., "Gas Chromatographic Analysis of So-Called Ether Bonbons," J. pharm. Belg., 14, 156-158 (1959); CA 54, 2658i

4650. Maricq, L., and Molle, L., "Simultaneous Determination of Diethyl Ether and Ethanol in Blood and Viscera by Gas Chromatography. Application to a Medico-Legal Case," Annales pharm. franc., 18, 811-816 (1960)

4651. Marinetti, G. V., Ford, T., and Stotz, E., "The Structure of Cerebrosides in Gaucher's Disease," J. Lipid Research, 1, 203-207 (1960); CA 54, 16617c

4652. Marino, G., and Brown, H. C., "Relative Rates and Isomer Distributions in the Acetylation of the Methylbenzenes by Acetyl Chloride-Aluminum Chloride in Ethylene Dichloride Solution," J. Am. Chem. Soc., 81, 5929-5933 (1959); CA 54, 9819f

4653. Marks, M. L., Dickholtz, R., and Hall, L. G., "A Comparison of Various Types of Detectors for Gas-Liquid Partition Chromatography," 9th Pittsburgh Conf. on Anal. Chem. & Appl. Spectroscopy, Pittsburgh, Pa., March 1958, Program Abstr., p. 50

4654. Martens, R. J., and Hertog, H. J. den, "Indications for the Occurrence of 2,3-Pyridine as an Intermediate," Tetrahedron Letters, 1962, 643-645; cf. Rec. Trav. Chim., 80, 1376 (1961); CA 58, 7902a

4655. Martin, A. E., and Smart, J., "Gas Phase Chromatography," Nature, 175, 422-423 (1955); CA 49, 9334h

4656. Martin, A. J., "Programmed-Temperature Gas Chromatography," 13th Annual Fisk Infrared Inst., Gas Chromatography Session, Nashville, Tenn., August 1962

4657. Martin, A. J., Bennett, C. E., and Martinez, Jr., F. W., "Linear Programmed Temperature Gas Chromatography," 1st Annual Gas Chromatography Institute, Canisius College, Buffalo, N. Y., May 1959

4658. Martin, A. J., Bennett, C. E., and Martinez, Jr., F. W., Linear Programmed Temperature

Gas Chromatograph Operable to 500°C.," 12th Delaware Sci. Symp., Wilmington, Del., January 1960

4659. Martin, A. J., Bennett, C. E., and Martinez, Jr., F. W., "Linear Programmed Temperature Gas Chromatography Using Flame Ionization Detection," 8th Detroit Anachem Conf., Detroit, Mich., October 1960

4660. Martin, A. J., Bennett, C. E., and Martinez, Jr., F. W., "Linear Programmed Temperature Gas Chromatography," Gas Chromatog., Intern. Symposium, 2nd, East Lansing, Mich., 1959, 363-374 (Pub. 1961); CA 55, 18205d

4661. Martin, A. J., Bennett, C. E., and Martinez, Jr., F. W., "Linear Programmed Gas Chromatography," in "Gas Chromatography," edited by H. J. Noebels, R. F. Wall, and N. Brenner, Academic Press, Inc., New York, 1961, pp. 363-374

4662. Martin, A. J. P., "The Principles of Chromatography," Endeavour, 6, 21-28 (1947); CA 41, 3677f

4663. Martin, A. J. P., "Development of Partition Chromatography," Prix Nobel, 1952, 110-121; CA 48, 8112c

4664. Martin, A. J. P., "Chromatography. Introduction," Brit. Med. Bull., 10, 161-162 (1954); CA 49, 9428b

4665. Martin, A. J. P., "Gas-Liquid Partition Chromatography. A Technique for Analysis of Volatile Materials," Kolloid-Z., 136, No. 1, 5 (1954)

4666. Martin, A. J. P., "Gas-Liquid Chromatography," Symp. of Gas Chromatography, Physical Methods and Microchemistry Groups with Scottish Section of Society for Anal. Chem., Stevenson, Scotland, May 1955; Abstr., Anal. Chem., 27, 1667 (1955)

4667. Martin, A. J. P., "Gas-Liquid Chromatography," Experientia, Suppl. No. 5, 21-32 (1956); CA 51, 7221f

4668. Martin, A. J. P., "Trends in Gas Chromatography," in "Vapour Phase Chromatography," edited by D. H. Desty, Academic Press, Inc., New York, 1957, pp. 1-14

4669. Martin, A. J. P., "Past, Present, and Future of Gas Chromatography," ISA Journal, 4, 563-566 (1957); CA 52, 9713i

4670. Martin, A. J. P., "Past, Present, and Future of Gas Chromatography," in "Gas Chromatography," edited by V. J. Coates, H. J. Noebels, and I. S. Fagerson, Academic Press, Inc., New York, 1958, pp. 237-247

4671. Martin, A. J. P., "Recent Trends and New Developments," in "Gas Chromatography 1958," edited by D. H. Desty, Academic Press, Inc., New York, 1958, pp. 139-141

4672. Martin, A. J. P., "Progress in Gas-Liquid Chromatography," 11th Annual Symp. on Modern Methods of Analytical Chemistry, Louisiana State Univ., Baton Rouge, La., January 1959

4673. Martin, A. J. P., "Future Possibilities in Microanalysis," Discussion Group, Inst. of Petroleum, Mtg., Salford, Eng., September 1962; cf. Nature, 196, 817 (1962)

4674. Martin, A. J. P., and James, A. T., "Gas-Liquid Chromatography. Gas Density Meter for Detection of Vapors in Flowing Gas Streams," Biochem. J., 63, 138-142 (1956); CA 50, 13660g

4675. Martin, A. J. P., and Synge, R. L. M., "A New Form of Chromatogram Employing Two Liquid Phases, 1. A Theory of Chromatography, 2. Application to the Microdetermination of the Higher Monoamine Acids in Proteins," Biochem. J., 35, 1358-1368 (1941); CA 36, 5197[4]

4676. Martin, D. B., Horning, M. G., and Vagelos, P. R., "Fatty Acid Synthesis in Adipose Tissue. I. Purification and Properties af a Long Chain Fatty Acid-Synthesizing System," J. Biol. Chem. 236, 663-668 (1961); CA 55, 13513g; 138th Natl. ACS Mtg., New York, N. Y., September 1960, Program Abstr., p. 45C

4677. Martin, F., Courteix, J., and Vertalier, S., "Determination of Benzene Toluene," Bull. soc. chim. France, 1958, 494-496; CA 52, 16126c

4678. Martin, F., and Vertalier, S., "Selective Microdetermination of the Alkoxy Group by Gas-Liquid Chromatography," 15th Intern. Congr., Pure and Appl. Chem., Lisbon, Portugal, September 1956

4679. Martin, F., Vertalier, S., and Camier, J., "Gas-Chromatographic Determination of 2-Methyl-phenoxyacetic, 2-Methylphenoxyisopropionic, 2-Methylphenoxybutyric Acids and Their 4- or 6-Chloro and 4,6-Dichloro Derivatives," Bull. soc. chim. France, 1960, 2067-2071; CA 55, 19103i

4680. Martin, G. E., Caggiano, G., and Beck, J. E., "Determination of Methanol by Gas Liquid Chromatography," J. Assoc. Offic. Agr. Chemists, 46, 297-298 (1963)

4681. Martin, H. F., "Separation of Estrogens by Gas Chromatography," Univ. Microfilms (Ann Arbor, Mich.), Order No. 61-3371, 190 pp.; Dissertation Abstr., 22, 997 (1961); CA 56, 2672f

4682. Martin, H. F., Driscoll, J. L., and Gudzinowicz, B. J., "A Method for Calculating Gas Chromatographic Relative Retention Times for Phenothiazine Derivatives on Non-Polar Liquid Stationary Phases," 145th Natl. ACS Mtg., New York, N. Y., September 1963, Program Abstr., pp. 10B-11B

4683. Martin, J. C., and Drew, E. H., "Molecule-Induced Homolytic Decompositions. I. Oxygen-18

Labeling Studies on the Reaction Yielding Cyclohexyl Acetate from Cyclohexene and Acetyl Peroxide," J. Am. Chem. Soc., 83, 1232-1237 (1961); CA 55, 22171h

4684. Martin, J. L., "Elimination of the Water Effect on Argon Ionization Detectors Fitted to Pye Chromatographs," Analyst, 88, 326-327 (1963)

4685. Martin, L., Smith, D. M., and Farmilo, C. G., "Essential Oils from Fresh Cannabis Sativa and Its Use in Identification," Nature, 191, 774-776 (1961); CA 56, 1539b

4686. Martin, R. L., "Gas Chromatographic Analysis of Olefinic Naphthas by a Subtraction Technique," 10th Pittsburgh Conf. on Anal. Chem. & Appl. Spectroscopy, Pittsburgh, Pa., March 1959, Program Abstr., p. 57

4687. Martin, R. L., "Changes in Selectivity of Gas-Chromatographic Columns Due to Adsorption on the Liquid-Phase Surface," 137th Natl. ACS Mtg., Cleveland, Ohio, April 1960, Program Abstr., p. 23B

4688. Martin, R. L., "Gas Chromatographic Analysis of Olefinic Naphthas in the Three- to Six-Carbon Range with the Aid of a Subtraction Technique," Anal. Chem., 32, 336-338 (1960); CA 54, 23286i

4689. Martin, R. L., "Adsorption on the Liquid Phase in Gas Chromatography," Anal. Chem., 33, 347-352 (1961); CA 55, 11175b

4690. Martin, R. L., "Determination of Hydrocarbon Types in Gasoline by Gas Chromatography," Anal. Chem., 34, 896-899 (1962); CA 57, 6205g; 141st Natl. ACS Mtg., Washington, D. C., March 1962, Program Abstr., p. 19B ·

4691. Martin, R. L., "Adsorption of Solutes at the Liquid-Gas Interface as Measured by Gas Chromatography and Gibbs Equation," Anal. Chem., 35, 116-117 (1963); CA 58, 10706b

4692. Martin, R. L., and Winters, J. C., "Composition of Crude Oil Through Seven Carbons as Determined by Gas Chromatography," Anal. Chem., 31, 1954-1960 (1959); CA 54, 7116c; 136th Natl. ACS Mtg., Atlantic City, N. J., September 1959, Program Abstr., p. 4R; Preprints, Div. Petrol. Chem., 4, No. 3, 109-118 (1959)

4693. Martin, R. L., and Winters, J. C., "Determination of Hydrocarbons in Crude Oil by Capillary-Column Gas Chromatography," 145th Natl. ACS Mtg., New York, N. Y., September 1963

4694. Martin, S. B., "Gas Chromatography. Application to the Study of Rapid Degradative Reactions in Solids," J. Chromatog., 2, 272-283 (1959); CA 54, 8418h

4695. Martin, S. B., "Flash Pyrolysis of Cellulose," 138th Natl. ACS Mtg., New York, N. Y., September 1960, Program Abstr., p. 18E

4696. Martin, S. B., and Ramstad, R. W., "Compact Two-Stage Gas Chromatograph for Flash Pyrolysis Studies," Anal. Chem., 33, 982-985 (1961); CA 56, 13524b; 137th Natl. ACS Mtg., Cleveland, Ohio, April 1960, Program Abstr., p. 24B

4697. Martin, S. B., and Ramstad, R. W., "A Compact Two-Stage Gas Chromatograph for Flash Pyrolysis Studies," U. S. At. Energy Comm., USNRDL-TR-467, 18 pp. (1960); Nuclear Sci. Abstracts, 15, 1432 (1961)

4698. Martinelli, M., "Liver Lipids and Their Composition in Rats Submitted to Various Diets," Epatologia, 7, 476-488 (1961); CA 57, 1352f

4699. Martinelli, M., Schlemmer, W., Turchetto, E., Re, M., and Moruzzi, G., "Fatty Acids of Myocardium Lipids. His Bundle, and Valvular Tissue of Calf Heart," Giorn. Biochem., 10, 506-512 (1961); CA 57, 2701b

4700. Martinelli, M., Turchetto, E., Schlemmer, W., and Moruzzi, G., "Fatty Acids of Lipides of Muscle Tissue Studied by Means of Gas-Phase Chromatography," Boll. soc. ital. biol. sper., 36, 1566-1570 (1960); CA 55, 23748e

4701. Martinelli, M., Turchetto, E., and Sechi, A. M., "Preliminary Observations on Lipide Fatty Acids and the Lipide Fraction of Normal Human Blood by Means of Gas-Phase Chromatography," Boll. soc. ital. biol. sper., 36, 1700-1703 (1960); CA 55, 23748g

4702. Martinez, C. J. L., and Zazurca, L. G., "Separation of Organic Volatile Compounds by Vapor Phase Chromatography," Combustibles, 16, 65-75 (1956); Fuel Abstr., 20, No. 3425 (1956)

4703. Martinez, C. J. L., and Zazurca, L. G., "Gas Chromatography of Organic Substances," Quim. e Ind., 5, 3 (1958)

4704. Martire, D. E., "Choice of a Substrate in a GLC Separation Through Application of the Theory of Solutions," 139th Natl. ACS Mtg., St. Louis, Mo., March 1961, Program Abstr., pp. 5Q-6Q

4705. Martire, D. E., "Applications of the Theory of Solutions to the Choice of Solvent for Gas-Liquid Chromatography," Anal. Chem., 33, 1143-1147 (1961); CA 57, 7d

4706. Martire, D. E., Pollara, L. Z., and Funke, P. T., "Prediction of Activity Coefficients and G.L.C. Solvent Selectivities," 145th Natl. ACS Mtg., New York, N. Y., September 1963, Program Abstr., p. 10B

4707. Maruyama, M., and Seno, S., "Gas Chromatography. I. Isooctylphenyl Polyoxyethylene as a Stationary Liquid Phase," Takamine Kenkyusho Nempo, 12, 177-180 (1960); CA 55, 6240i

4708. Maruyama, M., and Seno, S., "Gas Chromatography. II. Determination of Methanol in Ethanol and Some Abnormal Phenomena in Evaluating Chromatograms." Takamine Kenkyusho Nempo, 12, 181-185 (1960); CA 55, 6241a

4709. Maruyama, M., and Seno, S., "Gas Chromatographic Analysis of Methyl Esters of Long Chain Fatty Acids by Using Polyvinyl Acetates as Partition Media," Kogyo Kagaku Zasshi, 64, 777-780 (1961); CA 57, 4783e

4710. Marvel, C. S., and Gall, E. J., "Pyrolytic Cleavage of 2, 6-Diphenyl-1, 7-diacetoxyheptane and 2, 6-Diphenylheptadiene," J. Org. Chem., 24, 1494-1497 (1959); CA 54, 5567a

4711. Marvel, C. S., and Woolford, R. G., "Formation of a Cyclic Recurring Unit in the Polymerization of Diallyldimethylsilane," J. Org. Chem., 25, 1641-1643 (1960); CA 55, 8279h

4712. Marvel, E. N., Richardson, B., Anderson, R., and Stephenson, J. L., "The Claisen Rearrangement of Allyl 2-Alkyphenyl Ethers," 138th Natl. ACS Mtg., New York, N. Y., September 1960, Program Abstr., p. 85P

4713. Marvillet, L., and Tranchant, J., "Qualitative and Quantitative Analysis by Gas-Solid Chromatography of Mixtures Containing Nitrogen Oxides," Gas Chromatog. Proc. Symposium, 3rd, Edinburgh, 1960, 321-332; CA 56, 6662h

4714. Marvillet, L., and Tranchant, J., "Qualitative and Quantitative Analysis, by Gas-Solid Chromatography, of Mixtures Containing Nitrogen Oxides," in "Gas Chromatography 1960," edited by R. P. W. Scott, Butterworths, London, 1960, pp. 321-332

4715. Marvillet, L., and Tranchant, J., "Analysis by Gas-Liquid Chromatography of Products Solid at Ordinary Temperatures: Dinitrotoluenes," Journees Intern. Etude Methodes Separation Immediate Chromatog., Paris, 1961, 265-276, Discussion 276 (Pub. 1962); CA 59, 1098e

4716. Mashiko, Y., Araki, S., Maruyama, M., and Takanishi, T., "GCDC [Gas Chromatographic Data Committee of Japan] Cards Coding System," 14th Pittsburgh Conf. on Anal. Chem. & Appl. Spectroscopy, Pittsburgh, Pa., March 1963, Program Abstr., p. 64

4617. Mashiko, Y., and Echizen, A., "Gas Chromatographic Rapid Analysis of Polymers," 14th Pittsburgh Conf. on Anal. Chem. & Appl. Spectroscopy, Pittsburgh, Pa., March 1963, Program Abstr., p. 68

4618. Mashiko, Y., Kanbayashi, U., Nukada, K., Suzuki, T., Takeda, I., and Tomita, H., "Analysis of a Mixture of Chloropropanes and Chloropropenes Using a Preparative Scale Gas Chromatograph and a High Resolution NMR Spectrometer," 14th Pittsburgh Conf. on Anal. Chem. & Appl. Spectroscopy, Pittsburgh, Pa., March 1963, Program Abstr., p. 63

4719. Mason, E. A., and Saxena, C. S., "Approximate Formula for the Thermal Conductivity of Gas Mixtures," Phys. Fluids, 1, 361-369 (1958); CA 53, 6717h

4720. Mason, K. G., Sperry, J. A., and Stern, E. S., "The Formation of an Alkyl Fluoride from Silicon Tetrafluoride," J. Chem. Soc., 1963, 2558-2559

4721. Mason, L. H., Dutton, H. J., and Bair, L. R., "Ionization Chamber for High-Temperature Gas Chromatography," J. Chromatog., 2, 322-323 (1959); CA 54, 8163f

4722. Mason, M., and Waller, G. R., "Dimethoxy Propane Induced Interesterification of Lipides in the Preparation and GLC Analysis of Methyl Esters," 18th Southwest Regional Mtg., ACS, Dallas, Texas, December 1962

4723. Massart, R., and Missa, L., "Determination of Dissolved Gases in Water and Steam by Gas Chromatography," Centre belge etude et document eaux, Bull. trimestr. CEBEDEAU, 47, 43-49 (1960); CA 55, 9734f

4724. Masuko, Y., and Takenishi, T., "Documentation for the Gas Chromatographic Data," Yuki Gosei Kagaku Kyokai Shi, 19, 142-147 (1960); CA 55, 10189a

4725. Mathews, W. S., Pickering, G. B., and Umoh, A. T., "Acetylenic Ethers in the Essential Oil from Litsea Odorifera," Chem. & Ind. (London), 1963, 122-123; CA 58, 8847e

4726. Matousek, S., "Ionization Detector for Gas Chromatography," Chem. prumysl., 10, 16-21 (1960); CA 54, 7246b

4727. Matousek, S., "Comparison of Integral and Differential Ionizatoin Detectors for Gas Chromatography," Gas Chromatog. Proc. Symposium, 3rd, Edinburgh, 1960, 75-80; CA 56, 3306b

4728. Matousek, S., "Comparison of Integral and Differential Ionization Detectors for Gas Chromatography," in "Gas Chromatography 1960." edited by R. P. W. Scott, Butterworths, London, 1960, pp. 65-80

4729. Matousek, S., "Argon Detector with Several Electrodes," Gas Chromatog. Symp., Brno, Czech., June 1962; Abstr., J. Gas Chromatog., 1, No. 4, 8 (1963)

4730. Matsubara, I., and Kinoshita, S., "Rapid Analysis of Solvents in Acetone-Butanol Fermentation Broth by Gas Chromatography," Bunseki Kagaku, 10, 29-33 (1961); CA 58, 7335g

4731. Matsuda, H., and Matsuda, S., "Gas Chromatography of Organotin Compounds," Kogyo Kagaku Zasshi, 63, 1960-1964 (1960); CA 57, 8598h

4732. Matsuda, T., and Yatsugi, H., "Gas Chromatography. II. Some Behaviors of Two-Components Stationary Phases in Gas Chromatography," Bunseki Kagaku, 11, 1116-1120 (1962); CA 58, 6165f

4733. Matsumura, Y., and Soda, R., "Gas Chromatographic Analysis of Atmospheric Pollutants in Industries," Bull. Natl. Inst. Ind. Health (Japan), 4, No. 4, 44-53 (1960); CA 56, 6315e

4734. Matsuura, T., Aratani, T., Komae, H., and Hayashi, S., "Catalytic Effect of Support for Gas-Liquid Chromatography on Samples," Kogyo Kagaku Zasshi, 64, 795-799 (1961); CA 57, 2825g

4735. Matsuura, T., Aratani, T., Komae, H., and Hayashi, S., "Crushed Crystallized Quartz Support for Gas-Liquid Chromatography," Kogyo Kagaku Zasshi, 64, 799-802 (1961); CA 57, 2875h

4736. Matsuura, T., Komae, H., Aratani, T., and Hayashi, S., "Analysis of Terpenes and Essential Oils by Gas Chromatography. I. Analysis of Terpene Hydrocarbons by Gas Chromatography," Kogyo Kagaku Zasshi, 63, 1761-1765 (1960); CA 57, 9215g

4737. Matsuura, T., Komae, H., Aratani, T., and Hayashi, S., "Analysis of Terpene Hydrocarbons by Gas Chromatography," Kogyo Kagaku Zasshi, 64, 791-795 (1961); CA 57, 6588i

4738. Matteson, D. S., Drysdale, J. J., and Sharkey, W. H., "Thermal Rearrangement of 5-Methylenebicyclo (2, 2, 1)hept-2-ene," J. Am. Chem. Soc., 82, 2853-2857 (1960)

4739. Matthews, J. S., Burow, F. H., and Snyder, R. E., "Separation and Identification of C_8 Aldehydes. Use of Gas-Liquid Chromatography, Nuclear Magnetic Resonance, and Infrared Spectroscopy," Anal. Chem., 32, 691-693 (1960); CA 54, 16290f

4740. Matthews, R. F., "Gas and Paper Chromatography of Volatile Flavor Constituents of Several Vegetables," Univ. Microfilms (Ann Arbor, Mich.), L. C. Card No. Mic 60-6507, 113 pp.; Dissertation Abstr., 21, 1693-1694 (1961); CA 55, 9715d

4741. Mattick, L. R., Barry, D. L., Antenucci, F. M., and Avens, A. W., "The Disappearance of Endrin Residues on Cabbage," J. Agr. Food Chem., 11, 54-55 (1963); CA 58, 4967a

4742. Mattick, L. R., and Lee, F. A., "Fatty Acids of the Lipides of Vegetables. II. Spinach," J. Food Sci., 26, 356-358 (1961); CA 56, 3864f; 20th Annual Mtg., Inst. of Food Tech., San Francisco, Calif., May 1960, Program Abstr., p. 30

4743. Mattick, L. R., Moyer, J. C., and Schollenberger, R. S., "A Volatile Acidic Flavor Component of Apple Sauce," Food Tech., 12, 613-615 (1958)

4744. Mattson, F. H., and Volpenhein, R. A., "The Specific Distribution of Fatty Acids in the Glycerides of Vegetable Fats," J. Biol. Chem., 236, 1891-1894 (1961); CA 55, 26148h

4745. Mattsson, S., "Fatty Acids in Milk Phospholipids," Intern. Dairy Congr., Proc. 16th, Copenhagen, 1962, 2, Pt. 1, 537-544; CA 58, 11894e

4746. Maurel, A., "Recent Applications of Gas Chromatography to the Analysis of Alcohols," Bull. soc. chim. France, 1963, 316-319; CA 58, 13088h

4747. Maurel, R., "Experimental Accuracy in Quantitative Analysis by Gas Chromatography," Compt. rend., 244, 3157-3159 (1957); CA 51, 14465e

4748. Maurel, R., Bassery, L., and Germain, J.-E., "Kinetics and Equilibria in the Isomerization of Cyclohexene and the Methylcyclopentenes," Bull. soc. chim. France, 1962, 1688-1694; CA 58, 7801c

4749. Mawson, J., "Process Control by Gas Chromatography," Ind. Eng. Chem., 52, No. 2, 85A (1960)

4750. Mawson, J., "Development and Application of Plant Stream Analyzers," 6th Instrumental Methods of Analysis Symp., ISA, Montreal, Canada, June 1960

4751. Mayer, S. W., and Tompkins, E. R., "Ion Exchange as a Separation Process. IV. A Theoretical Analysis of the Column Separation Process," J. Am. Chem. Soc., 69, 2866-2874 (1947); CA 42, 3237f

4752. Mazitova, F. N., Ermakova, S. K., and Virobyants, R. A., "The Analysis of Gaseous Hydrocarbons by Chromatographic Adsorption on Alumina," Khim. i Tekhnol. Topliv i Masel, 7, No. 4, 66-69 (1962); CA 57, 5285f

4753. Mazitova, F. N., Virobyants, R. A., and Ermakova, S. K., "Analysis of Light Petroleum Hydrocarbon Fractions by Gas-Liquid Chromatography," Izv. Akad. Nauk SSSR, Otd. Khim. Nauk, 1962, 1546-1550; CA 58, 2303h

4754. Mazliak, P., "Vapor Phase Chromatographic Study of the Fatty Acids in 'Oil' from the Skin of Calville White Apples," Compt. rend., 250, 182-184 (1960); CA 54, 14370e

4755. Mazliak, P., "Gas Phase Chromatography of Fatty Acids from the Wax of Apples," Compt. rend., 250, 2255-2257 (1960); CA 55, 4992d

4756. Mazliak, P., "Gas-Chromatographic Study of the Long-Chain Paraffins and Alcohols of the Cuticular Wax of Apples," Compt. rend., 251, 2393-2395 (1960); CA 55, 9714c

4757. Mazliak, P., "The Existence of Paraffins with Even Numbers of Carbon Atoms in the Liquid Wax of Apple Cuticle," Compt. rend., 252, 1507-1509 (1961); CA 55, 13553g

4758. Mazliak, P., "Diols of Apple and Carnuba Waxes," Phytochemistry, 1, 79-85 (1962); CA 57, 8678a

4759. Mazliak, P., and Combes, R., "A Gas Chromatographic Study of the Long Chain Paraffins and Alcohols in the Cuticle Wax of Potatoes," Compt. rend., 251, 2393-2395 (1960)

4760. McAdams, W. H., "A Method of Applying Gas Chromatography to Process Monitoring," Fall Instrument-Automation Conf. & Exhibit, ISA, New York, N. Y., September 1960

4761. McBride, W. R., and Bens, E. M., "Alkylhydrazines. III. Dimerization of Certain Substituted

1, 1-Dialkyldiazenes to Tetraalkyltetrazenes," J. Am. Chem. Soc., 81, 5546-5550 (1959); CA 54, 7532f

4762. McCallum, J. D., "Preparative Gas Chromatography," in "Progress in Industrial Gas Chromatography," Vol. 1, edited by H. A. Szymanski, Plenum Press, Inc., New York, 1961, pp. 125-145

4763. McCallum, J. D., "Systematic Calculation of Gas Chromatography Temperature Programs," in "Lectures on Gas Chromatography 1962," edited by H. A. Szymanski, Plenum Press, Inc., New York, 1963, pp. 153-176

4764. McCallum, J. D., "Integration of Gas Chromatography with Auxiliary Analytical Techniques," in "Lectures on Gas Chromatography 1962," edited by H. A. Szymanski, Plenum Press, Inc., New York, 1963, pp. 199-226

4765. McCallum, J. D., "Calibration Methods in Quantitative Analyses," 5th Annual Gas Chromatography Institute, Canisius College, Buffalo, N. Y., April 1963

4766. McCarthy, M. J., Kuksis, A., and Beveridge, J. M. R., "Gas-Liquid Chromatographic Analysis of the Triglyceride Composition of Molecular Distillates of Butter Oil," Can. J. Biochem. Physiol., 40, 1693-1703 (1962); CA 58, 5908e

4767. McCarthy, R. D., Patton, S., and Evans, L., "Structure and Synthesis of Milk Fat. II. The Fatty Acid Distribution in the Triglycerides of Milk and Other Animal Fats," J. Dairy Sci., 43, 1196-1201 (1960); CA 54, 25354c

4768. McCarthy, W. C., and Sullivan, J. B., "Aminothiophenes," 144th Natl. ACS Mtg., Los Angeles, Calif., March-April 1963, Program Abstr., p. 13L

4769. McCasland, G. E., Furuta, S., Johnson, L. F., and Shoolery, J. N., "Synthesis of the Five Diastereomeric 1, 2, 4, 5-Cyclohexanetetrols. Nuclear Magnetic Resonance Configurational Proofs," J. Org. Chem., 28, 894-900 (1963)

4770. McComas, D. B., and Goldflen, A., "A Device for the Introduction of Submicrogram Quantities of Solids into a Gas Chromatograph," Anal. Chem., 35, 263-264 (1963); CA 58, 10496g

4771. McCoy, L. L., "Three Membered Rings. II. The Stereoselective Formation of Difunctional Cyclopropanes," J. Am. Chem. Soc., 82, 6416-6417 (1960); CA 55, 9299i

4772. McCoy, L. L., "Three Membered Rings. II. The Stereochemistry of Formation of Some 1, 1, 2, 2-Tetrasubstituted Cyclopropanes," J. Org. Chem., 25, 2078-2082 (1960); CA 55, 10340h

4773. McCoy, L. L., and Zagalo, A., "Lead Tetraacetate Oxidation of Some Glutaric Acids. Formation of Gamma-Lactones," J. Org. Chem., 25, 824-826 (1960); CA 55, 2566f

4774. McCreadie, S. W. S., and Williams, A. F., "Quantitative Measurement and Transfer of Samples in Gas Chromatography," J. Appl. Chem. (London), 7, 47-48 (1957); CA 51, 13471i

4775. McDermott, P. S., and Cooper, C. V., "A Simple Modification of Thermoconductivity Cells to Produce the Derivative of Elution Chromatograms," 10th Pittsburgh Conf. on Anal. Chem. & Appl. Spectroscopy, Pittsburgh, Pa., March 1959, Program Abstr., p. 58

4776. McDonald, H. J., "Developments in Chromatography," Research/Development," 11, No. 6, 4-6, 8-9 (1960)

4777. McDonnell, H., "Cost-Conscious Chromatographs," Chem. Eng. Progr., 55, No. 6, 108, 110 (1959); CA 53, 22857a

4778. McDonough, E. G., Mackay, D. A. M., and Berdick, M., "Cosmetic Knowledge Through Instrumental Techniques," J. Soc. Cosmetic Chemists," 8, 126-138 (1957); CA 51, 11660a

4779. McDowell, C. A., and Sifniades, S., "Isomerization as a Primary Process in the Photolysis of Crotonaldehyde," J. Am. Chem. Soc., 84, 4606-4607 (1962); CA 58, 10054d

4780. McEwen, D. J., "Separation of C_2-C_4 Hydrocarbons by Gas Liquid Partition Chromatography," Symp. on Instrumental Methods of Analysis, Anal. Chem. Div., CIC, Sarnia, Ontario, April 1959; Abstr., Chem. in Can., 11, No. 3, 45 (1959)

4781. McEwen, D. J., "The Separation of the Light Hydrocarbons by Gas Chromatography," Chem. in Can., 11, No. 10, 35-37 (1959); CA 54, 5047e

4782. McEwen, D. J., "Improved Sampling Valve for Gas Chromatography," J. Chromatog., 9, 266-269 (1962)

4783. McEwen, D. J., "Gas Chromatographic Analysis of Dilute Hydrocarbon Mixtures Using Temperature Programmed Capillary Columns," 10th Detroit Anachem Conf., Detroit, Mich., October 1962, Program Abstr., p. 32

4784. McEwen, D. J., "Temperature-Programmed Columns in Gas Chromatography. Comparison of Packed and Capillary Columns for the Analysis of Dilute Hydrocarbon Gas Mixtures," 144th Natl. ACS Mtg., Los Angeles, Calif., March-April 1963, Program Abstr., p. 19B

4785. McEwen, D. J., "Backflush and Two-Stage Operation of Capillary Columns in Gas Chromatography," 11th Detroit Anachem Conf., Detroit, Mich., October 1963, Program Abstr., pp. 27-28

4786. McEwen, J. I., "Continuous Analysis of Radioactive Gas and Liquid Streams," U. S. At. Energy Comm., TID-7606, 291-310 (1960); CA 55, 23163b

4787. McFadden, J. L., "Gas Chromatography and the Thermal Conductivity Detector," Z. Anal. Chem., 170, 232 (1959)

4788. McFadden, J. L., "Gas Analysis by Thermal Conductivity," Instruments & Control Systems, 34, 2055-2057 (1961)

4789. McFadden, J. L., "Thermal Conductivity Detectors," 13th Annual Fisk Infrared Institute, Gas Chromatography Session, Nashville, Tenn., August 1962

4790. McFadden, W. H., "Use of Mixed Stationary Liquids in Gas-Liquid Chromatography," Anal. Chem., 30, 479-481 (1958); CA 52, 12466b

4791. McFadden, W. H., McIntosh, R. G., and Harris, W. E., "Chemical Effects of the (n, γ) Activation of Bromine in the Alkyl Bromides; Isomerization in the Bromobutanes," J. Phys. Chem., 64, 1076-1078 (1960); CA 55, 3166i

4792. McFarland, C. H., "Coupling of Columns to Gas Chromatographs," Chemist-Analyst, 50, No. 4, 122 (December 1961)

4793. McGinness, J. D., Whittenbaugh, J. A., and Lucchesi, C. A., "Composition of Guaiacol Fraction of Tar Acids," Tappi, 43, 1027-1029 (1960); CA 55, 10886e

4794. McGovern, L. J., and Carlisle, Jr., L. J., "The Application of the Gas Chromatograph to Petroleum Refinery," 13th Annual Instrument Automation Conf., ISA, Philadelphia, Pa., September 1958

4795. McGovern, L. J., and Carlisle, Jr., L. J., "Gas Chromatograph Improves Distillation Operation by 18%," ISA Journal, 6, 60, 63 (1959)

4796. McGovern, L. J., and Carlisle, Jr., L. J., "Chromatographic Control of Magnolia's Alkylation Unit," Refinery Eng., 31, C6-C9 (1959); CA 53, 6585h

4797. McGreer, D. E., "Pyrazolines," J. Org. Chem., 25, 852-853 (1960); CA 55, 1584c

4798. McGregor, R. F., Newland, J., and Cornatzer, W. E., "Fatty Acid Composition of Milk Phospholipids of Mice With and Withoug the Mammary Tumor Virus," Nature, 198, 482-483 (1963)

4799. McGregor, R. F., Ward, D. N., Cooper, J. A., and Creech, B. G., "Estimation of Diethylstilbesterol and Identification of Related Synthetic Estrogens by Gas Chromatography," Anal. Biochem., 2, 441-446 (1961); CA 56, 6284g

4800. McGugan, W. A., and Howsam, S. G., "A Method for the Analysis of Cheese Volatiles," 56th Annual Mtg., Am. Dairy Sci. Assoc., Madison, Wisc., June 1961; Abstr., J. Dairy Sci., 44, 1169 (1961)

4801. McGugan, W. A., and Howsam, S. G., "Analysis of Neutral Volatiles in Cheddar Cheese," J. Dairy Sci., 45, 495-500 (1962); CA 57, 7687e

4802. McInnes, A. G., "Practical Notes on Gas-Liquid Chromatography as Applied to the Estimation of Volatile Fatty Acids," in "Vapour-Phase Chromatography," edited by D. H. Desty, Academic Press, Inc., New York, 1957, pp. 304-315

4803. McInnes, A. G., Ball, D. H., Cooper, F. P., and Bishop, C. T., "Separation of Carbohydrate Derivatives by Gas-Liquid Partition Chromatography," J. Chromatog., 1, 556-557 (1958)

4804. McInnes, A. G., Hansen, R. P., and Jessop, A. S., "The Volatile Acids of Mutton Fat," Biochem. J., 63, 702-704 (1956); CA 50, 15795i

4805. McInnes, A. G., and Lemieux, R. U., "The Preparation of Sucrose Monoesters," 44th Canadian Chemical Conf. and Exhibition, CIC, Montreal, Canada, August 1961; Abstr., Chem. in Can., 13, No. 5, 43-44 (1961)

4806. McInnes, A. G., Tattrie, W. H., and Kates, M., "Application of Gas-Liquid Partition Chromatography to the Quantitative Estimation of Monoglycerides," J. Am. Oil Chemists' Soc., 37, 7-11 (1960); CA 54, 5129f

4807. McIntyre, E. A., Rooney, T. B., Curren, W. J., and Aznavourian, W., "Determining Conditions for Low Temperature Analysis of High Boiling Compounds," 14th Pittsburgh Conf. on Anal. Chem. & Appl. Spectroscopy, Pittsburgh, Pa., March 1963, Program Abstr., p. 53

4808. McIntyre, H. C., "Canadian Plastics Plant Uses Process Chromatography," Instr. Plastics Plant Uses Process Chromatography," Instr. Practice, 14, 1192-1195 (1960)

4809. McKay, D. K., Seligson, D., and Taylor, B. W., "The Measurement of Carbon Dioxide in Serum by Gas Chromatography," 11th Annual Mtg., Am. Assoc. Clinical Chemists, Cleveland, Ohio, August 1959; Abstr., Clin. Chem., 5, No. 3, 260 (1959)

4810. McKee, H. C., Rhoades, J. W., and McMahon, W.A., "Use of Drying Agents in Atmosphere Sampling," 136th Natl. ACS Mtg., Atlantic City, N. J., September 1959, Program Abstr., p. 24U

4811. McKee, H. C., Rhoades, J. W., Wheeler, R. J., and Burchfield, H. P., "Gas Chromatographic Measurement of Trace Contaminants in a Simulated Space Cabin," NASA (Natl. Aeron. Space Admin.), Tech. Note, TN D-1825, 39 pp. (1963); CA 58, 11888h

4812. McKelvey, J. M., and Hoelscher, H. E., "Apparatus for Preparation of Very Dilute Gas Mixtures," Anal. Chem., 29, 123 (1957); CA 51, 4767c

4813. McKenna, Jr., T. A., and Idleman, J. A., "Gas Chromatographic Analysis of Hydrocarbon Streams from Butane Dehydrogenation," Anal. Chem., 31, 1021-1023 (1959); CA 53, 15537e

4814. McKenna, Jr., T. A., and Idleman, J. A., "Separation of C_4 and Lighter Hydrocarbons by Gas-Liquid Chromatography," Anal. Chem., 31, 2000-2003 (1959); CA 54, 6098d

4815. McKenna, Jr., T. A., and Idleman, J. A., "Gas-Solid Chromatographic Separation of Some Light

Hydrocarbons," Anal. Chem., 32, 1299-1301 (1960); CA 55, 956b

4816. McKennis, Jr., H., Turner, R. A., Turnball, L. B., Bowman, E. R., Muelder, W. W., Neidhardt, M. P., Hake, C. L., Henderson, R., Nadeau, H. G., and Spencer, S., "The Excretion and Metabolism of Triethylene Glycol," Toxicol. Appl. Pharmacol., 4, 411-431 (1962)

4817. McKinley, Jr., J. J., "Sub-Ambient Gas Chromatography," 8th Natl. Analysis Instrumentation Symp., ISA, Charleston, W. Va., April-May 1962

4818. McKinney, C. B., Durr, R. A., and Carle, D. W., "A Study with Several Gas Chromatograph Detectors," 6th Symp. on Instrumental Methods of Analysis, ISA, Montreal, Canada, June 1960

4819. McKinney, C. B., Durr, R. A., and Carle, D. W., "The Design and Application of Two High Sensitivity Gas Chromatography Detectors," 6th Symp. on Instrumental Methods of Analysis, ISA, Montreal, Canada, June 1960

4820. McKinney, R. W., and Baird, J. H., "Storage and Retrieval of Data from Preston Gas Chromatography Abstract Cards," J. Gas Chromatog., 1, No. 5, 16 (1963)

4821. McMichael, H. E., "Flue Gas Analysis - Sampling, Measurement, Utility," Natl. Power Instr. Symp., 5, 91-97 (1962); CA 58, 3883c

4822. McMillan, G., and Wijnen, M. H. J., "Reactions of Alkoxy Radicals. V. Photolysis of Di-tert-butyl Peroxide," Can. J. Chem., 36, 1227-1232 (1958); CA 53, 5120c

4823. McMillan, G. R., "Photolysis of Diisopropyl Peroxide," J. Am. Chem. Soc., 83, 3018-3023 (1961); CA 56, 3053h

4824. McMillan, G. R., and Noyes, W. A., "Photochemical Studies. LIII. Isopropyl Iodide," J. Am. Chem. Soc., 80, 2108-2111 (1958); CA 52, 12570b

4825. McMillan, Jr., W. G., and Teller, E., "The Role of Surface Tension in Multilayer Gas Adsorption," J. Chem. Phys., 19, 25-32 (1951); CA 45, 6454b

4826. McNair, H., "Ionization Detectors," 3rd Annual Eastern Analytical Symp. & Instrument Exhibit, New York, N. Y., November 1961

4827. McNair, H. M., "Efficiency of Solvents in Gas Chromatography," Univ. Microfilms (Ann Arbor, Mich.), L. C. Card No. Mic 59-6498, 80 pp.; Dissertation Abstr., 20, 2523 (1960); CA 54, 9435a

4828. McNair, H. M., Cramers, K. A. M. G., and Keulemans, A. I. M., "Evaluation of the Flame-Ionization Detector and the Micro Argon β-Ray Detector," 139th Natl. ACS Mtg., St. Louis, Mo., March 1961, Program Abstr., p. 7Q; Div. Petrol. Chem. Symp., 6, No. 2B, 55-61 (1961); CA 57, 14411g

4829. McNair, H. M., and DeVries, T., "Efficiency of Solvents in Gas Chromatography Correlated with Interaction Forces," 139th Natl. ACS Mtg., St. Louis, Mo., March 1961, Program Abstr., p. 5Q; Div. Petrol. Chem., Symp., 6, No. 2B, 5-10 (1961); CA 57, 14410f

4830. McNair, H. M., and DeVries, T., "1, 2, 3-Tris(2-cyanoethoxy)-propane, a Stationary Liquid for Gas Chromatography Columns," Anal. Chem., 33, 806 (1961); CA 55, 19613b

4831. McNamara, L. S., "Dehydration of Alcohols. Cis- and Trans-2-Benzylcyclopentanols," Univ. Microfilms (Ann Arbor, Mich.), L. C. Card No. Mic 60-3599, 128 pp.; Dissertation Abstr., 21, 1063 (1960)

4832. McNeil, D., "Composition of Wood Preserving Cresote," Record Ann. Conv. Brit. Wood Preserving Assoc., 1959, 136-150; CA 54, 10214a

4833. McNesby, J. R., Drew, C. M., and Gordon, A. S., "Synthesis of Butane-2, 2, 3, 3-d_4 by the Photolysis of Diethylketone-2, 2, 4, 4-d_4," J. Phys. Chem., 59, 988-989 (1955); CA 50, 7717c

4834. McNesby, J. R., Drew, C. M., and Gordon, A. S., "Mechanism of the Decomposition of Primary and Secondary n-Butyl-Free Radicals," J. Chem. Phys., 24, 1260 (1956); CA 50, 12612f

4835. McNesby, J. R., and Gordon, A. S., "Mechanism of the Isomerization of Cyclopropane," J. Chem. Phys., 25, 582-583 (1956); CA 51, 52g

4836. McNesby, J. R., and Gordon, A. S., "The Reaction of Methyl-d_3 Radicals with Cyclopropane and Cyclopentane," J. Am. Chem. Soc., 79, 825-826 (1957); CA 51, 10387a

4837. McNesby, J. R., and Gordon, A. S., "Reactions of CD_3 Radicals with the Butenes," J. Am. Chem. Soc., 79, 5902-5906 (1957); CA 52, 3533h

4838. McNesby, J. R., and Gordon, A. S., "Photolysis and Pyrolysis of 2-Pentanone-1, 1, 1-3, 3-d_5," J. Am. Chem. Soc., 80, 261-264 (1958); CA 52, 6949i

4839. McReynolds, W. O., "Development of Gas Chromatographic Methods for the Determination of Formaldehyde," 12th Pittsburgh Conf. on Anal. Chem. & Appl. Spectroscopy, Pittsburgh, Pa., February-March 1961, Program Abstr., p. 56

4840. McTigue, P. H., and Buchanan, A. S., "Photolysis of Methyl Halides in the Presence of Metals. II. Methyl Bromide," Trans. Faraday Soc., 55, 1160-1164 (1959); CA 54, 9454a

4841. McWilliam, I. G., "Comparison of Detectors for Gas Chromatography," J. Appl. Chem., (London), 9, 379-388 (1959); CA 53, 19475d

4842. McWilliam, I. G., "Linearity and Response Characteristics of the Flame Ionization Detector," J. Chromatog., 6, 110-117 (1961); CA 56, 9382h

4843. McWilliam, I. G., "Retention Temperature in Programmed Temperature Gas Chromatography," J. Chromatog., 6, 359-360 (1961); CA 56, 9450f

4844. McWilliam, I. G., "Applications of Gas Chromatography," Revs. Pure and Appl. Chem. (Australia), 11, 33-62 (1961); CA 55, 15054b

4845. McWilliam, I. G., and Dewar, R. A., "Flame Ionization Detector for Gas Chromatography," Nature, 181, 760 (1958); CA 52, 11647b

4846. McWilliam, I. G., and Dewar, R. A., "Flame Ionization Detector for Gas Chromatography," Gas Chromatog., Proc. Symposium, Amsterdam, 1958, 142-147; CA 53, 20940a

4847. McWilliam, I. G., and Dewar, R. A., "Flame Ionization Detector for Gas Chromatography," in "Gas Chromatography 1958," edited by D. H. Desty, Academic Press, Inc., New York, 1958, pp. 142-152

4848. Mead, J. F., and Gouze, M. L., "Alterations in Aorta Lipids with Advancing Atherosclerosis," Proc. Soc. Exptl. Biol. & Med., 106, 4-7 (1961); CA 55, 8607d

4849. Mead, J. F., Kayama, M., and Reiser, R., "Biogenesis of Polyunsaturated Acid in Fish," J. Am. Oil Chemists' Soc., 37, 438-440 (1960); CA 54, 23092c

4850. Mead, J. F., and Levis, G. M., "α-Oxidation of Brain Fatty Acids," Biochem. Biophys. Res. Commun., 9, 231-234 (1962); CA 58, 7204d

4851. Mead, J. F., and Nevenzel, J. C., "The Question of Biohydrogenation of Fatty Acids," J. Lipid Research, 1, 305-310 (1960); CA 55, 4685a

4852. Meadows, G. E., Hubbard, G. L., and Busey, H. M., "Determination of Impurities in Helium," U. S. At. Energy Comm., LA-2540, 23 pp. (1961); CA 55, 20771e

4853. Meakins, G. D., and Swindells, R., "Structures of Two Acids from Olive Leaves," J. Chem. Soc., 1959, 1044-1047; CA 53, 17897a

4854. Mechelynck-David, C., "Preparative Gas-Phase Chromatography," Ind. chim. belge, 26, 196-202 (1961)

4855. Mecke, R., and DeVries, M., "Gas Chromatographic Studies of Alcoholic Beverages," Z. Anal. Chem., 170, 326-332 (1959); CA 54, 3846b

4856. Mecke, R., Schindler, R., and DeVries, M., "Gas Chromatography. An Effective Analytical Method for Beverage Research," Wein-Wiss., Beil. Fachz. Deut. Weinbau, 8, 151-157 (1960); CA 55, 2007e

4857. Mecke, R., Schindler, R., and DeVries, M., "Investigations of Wines by Gas Chromatography," Wein-Wiss., Beil. Fachz. Deut. Weinbau, 15, 183-191 (1960); CA 55, 7754b

4858. Mecke, R., and Zirker, K., "Calculation of the Surface Factor for Gas Chromatography with the Aid of Kinetic Gas Theory," J. Chromatog., 7, 1-12 (1962); CA 57, 6582c

4859. Medlock, R. S., and Wilson, H., "Industrial Gas Sampling," Automatic Measurement Quality Process Plants, Proc. Conf., Swansea, 1957, (Academic Press, Inc., London), 72-103; CA 53, 5019e

4860. Medvedeva, N. I., and Torsueve, E. S., "Chromatographic Method of Separating Products from Cracking of Hydrocarbons," Trudy Komissii Anal. Khim., Akad. Nauk S.S.S.R., 6, 88-96 (1955); CA 50, 8185a

4861. Meek, J. M., and Craggs, J. D., "Electrical Breakdown of Gases," Clarendon Press, Oxford, 1953; CA 48, 9814b

4862. Mehlenbacher, V. C., "Newer Analytical Methods for the Fat and Oil Industry," J. Am. Oil Chemists' Soc., 37, 613-617 (1960); CA 55, 4011c

4863. Mehlitz, A., and Gierschner, K., "Further Investigations on Aroma Concentrates from Fruit Juices," Intern. Fruchtsaft Union Ber. Wiss. Tech. Komm., No. 4, 301-318 (1962); CA 58, 9562f

4864. Meigh, D. F., "Nature of the Olefins Produced by Apples," Nature, 184, 1072 (1959); CA 54, 10073b

4865. Meigh, D. F., "Ethylene Production by Tomato and Apple Fruits," Nature 186, 902-903 (1960); CA 55, 6721a

4866. Meigh, D. F., "Use of Gas Chromatography in Measuring the Ethylene Production of Stored Apples," J. Sci. Food Agr., 11, 381-385 (1960); CA 54, 21531i

4867. Meilgaard, M., "Hop Analysis, Cohumulone Factor, and the Bitterness of Beer - Review and Critical Evaluation," J. Inst. Brewing, 66, No. 1, 35-50 (1960); CA 54, 11371f

4868. Meinertz, H., and Dole, V. P., "Radioassay of Low-Activity Fractions Encountered in Gas-Liquid Chromatography of Long-Chain Fatty Acids," J. Lipid Research, 3, 140-144 (1962); CA 56, 13182b

4869. Melamed, N., and Renard, M., "Analysis of Mixtures of Amino Acids by Gas Chromatography," J. Chromatog., 4, 339-346 (1960); CA 55, 4254e

4870. Melikadze, L. D., and Eliava, T. A., "Investigation of the High-Molecular Fraction of Petroleum by Chromatography," Trudy Inst. Khim. im. P. G. Melikishvili Akad. Nauk Gruzin. S.S.R., 13, 145-164 (1957); CA 52, 21013i

4871. Melkonian, G. A., and Reps, B., "Isotope Displacement by Adsorption and Desorption on Silica Gels at Low Temperatures and Pressures," Z. Elektrochem., 58, 616-619 (1954); CA 49, 7922a

4872. Mellado, G. L., and Kobayashi, R., "Get K-Values by Chromatography," Petrol. Refiner, 39, 125-128 (1960); CA 54, 7117a

4873. Mellon, E. F., Herb, S. F., Barford, R. A., Viola, S. J., and Luddy, F. E., "The Fatty Acid Composition of Lipids from Different Layers of Fresh Steer Hides," J. Am. Leather Chemists' Assoc., 57, 26-35 (1962); CA 56, 13059f

4874. Mellon, M. G., "Automation in Analytical Chemistry," Anal. Chem., 30, No. 12, 25A, 27A-30A, 34A (1958)

4876. Mellor, N., "Factors Affecting Katharometer Sensitivity and Column Efficiency in Vapour-Phase Partition Chromatography," in "Vapour-Phase Chromatography," edited by D. H. Desty, Academic Press, Inc., New York, 1957, pp. 63-73

4877. Meloan, C. E., and Kiser, R. W., "Problems and Experiments in Instrumental Analysis," Charles E. Merrill Books, Inc., Columbus, Ohio, 1963; Gas Liquid Partition Chromatography, pp. 219-233

4878. Melton, G. E., Hurst, G. S., and Bortner, T. E., "Ionization Produced by 5-m. e. v. Alpha-Particles in Argon Mixtures," Phys. Rev., 96, 643-645 (1954); CA 49, 2884c

4879. Menapace, H. R., "Gas Chromatographic Study of the Cool Flame and Motored Engine Combustion of Hydrocarbons," Univ. Microfilms (Ann Arbor, Mich.), L. C. Card No. Mic 59-401, 124 pp.; Dissertation Abstr., 19, 2234-2235 (1959); CA 53, 10716h

4880. Menapace, H. R., Kyryacos, G., and Boord, C. E., "Gas Absorption Chromatographic Determination of Some Oxygenated Products in Cool-Flame Combustion," 132nd Natl. ACS Mtg., New York N. Y., September 1957, Program Abstr., p. 40B; Preprints, Div. Petrol. Chem., 2, No. 4, D-153 (1957)

4881. Merits, I., "Gas-Liquid Chromatography of Adrenal Cortical Steroid Hormones," J. Lipid Research, 3, 126-127 (1962); CA 56, 14853f

4882. Meriwether, L. S., Colthup, E. C., and Kennerly, G. W., "Polymerization of Acetylenes by Nickel Carbonyl-Phosphine Complexes. II. Proof of Structure of the Linear Low Polymers of Monosubstituted Acetylenes," J. Org. Chem., 26, 5163-5169 (1961); CA 57, 1048i

4883. Merritt, Jr., C., "Qualitative Gas Chromatographic Analysis," 144th Natl. ACS Mtg., Los Angeles, Calif., March-April 1963, Program Abstr., pp. 8B-9B

4884. Merritt, Jr., C., Bresnick, S. R., Bazinet, M. L., Walsh, J. T., and Angelini, P., "Techniques for the Determination of Volatile Components of Food Stuffs. Use of Gas Chromatography and Mass Spectrometry," 134th Natl. ACS Mtg., Chicago, Ill., September 1958, Program Abstr., p. 32A

4885. Merritt, Jr., C., Bresnick, S. R., Bazinet, M. L., Walsh, J. T., and Angelini, P., "Foodstuff Volatiles. Determination of Volatile Components of Foodstuffs. Techniques and Their Application to Studies of Irradiated Beef," J. Agr. Food Chem., 7, 784-787 (1959); CA 54, 2625f

4886. Merritt, Jr., C., and Mendelsohn, J. M., "Flavor in Fish," 142nd Natl. ACS Mtg., Atlantic City, N. J., September 1962, Program Abstr., p. 14A

4887. Merritt, Jr., C., and Walsh, J. T., "Qualitative Functional Group Analysis of Gas Chromatographic Effluents," 11th Pittsburgh Conf. on Anal. Chem. & Appl. Spectroscopy, Pittsburgh, Pa., February-March 1960, Program Abstr., p. 39; 8th Detroit Anachem Conf., Detroit, Mich., October 1960

4888. Merritt, Jr., C., and Walsh, J. T., "A Two-Column Method of Qualitative Gas Chromatographic Analysis," 12th Pittsburgh Conf. on Anal. Chem. & Appl. Spectroscopy, Pittsburgh, Pa., February-March 1961, Program Abstr., p. 56

4889. Merritt, Jr., C., and Walsh, J. T., "Qualitative Gas Chromatographic Analysis by Means of Retention Volume Constants," Anal. Chem., 34, 903-907 (1962); CA 57, 10511a

4890. Merritt, Jr., C., and Walsh, J. T., "Simultaneous Dual Column Gas Chromatography," Anal. Chem., 34, 908-911 (1962); CA 57, 10511g

4891. Merritt, Jr., C., and Walsh, J. T., "Programmed Cryogenic Temperature Gas Chromatography Applied to the Separation of Complex Mixtures," Anal. Chem., 35, 110-113 (1963); CA 58, 7294g

4892. Merritt, J., Comendant, F., Abrams, S. T., and Smith, V. N., "Process Gas Chromatograph with Ultraviolet Detector," Anal. Chem., 35, 1461-1464 (1963); 14th Pittsburgh Conf. on Anal. Chem. & Appl. Spectroscopy, Pittsburgh, Pa., March 1963, Program Abstr., pp. 65-66

4893. Merten, F., "Measuring Equipment for the Laboratory," Chem.-Ing.-Tech., 33, 271-280 (1961); CA 55, 13931f

4894. Messner, A. E., Rosie, D. M., and Argabright, P. A., "Correlation of Thermal Conductivity Cell Response with Molecular Weight and Structure. Quantitative Gas Chromatographic Analysis," Anal. Chem., 31, 230-233 (1959); CA 53, 8913b

4895. Metcalf, R. L., "Advances in Pest Control Research," John Wiley & Sons, Inc., New York, 1962

4896. Metcalfe, L. D., "Gas Chromatography of Unesterified Fatty Acids Using Polyester Columns Treated with Phosphoric Acid," Nature, 188, 142-143 (1960); CA 55, 6365c

4897. Metcalfe, L. D., "Direct Analysis of Fatty Acids," Facts & Methods (F&M Scientific Corp.), 2, No. 1, 1-3 (Spring 1961)

4898. Metcalfe, L. D., "The Role of Separations in Organic Analysis," Anal. Chem., 33, 1559-1562 (1961); CA 56, 1977g

4899. Metcalfe, L. D., "The Direct Gas Chromatographic Analysis of Long Chain Quaternary Ammonium Compounds," J. Am. Oil Chemists' Soc., 40, 25-27 (1963); 53rd Spring Mtg., Am. Oil Chemists' Soc., New Orleans, La., May 1962

4900. Metcalfe, L. D., "The Gas Chromatography of Fatty Acids and Related Long Chain Compounds on Phosphoric Acid Treated Columns," J. Gas Chromatog., 1, No. 1, 7-11 (1963)

4901. Metcalfe, L. D., Germanos, G. A., and Schmitz, A. A., "The Gas Chromatography of Long Chain Acid Amides," J. Gas Chromatog., 1, No. 5, 32-33 (1963)

4902. Metcalfe, L. D., and Schmitz, A. A., "The Rapid Preparation of Fatty Acid Esters for Gas Chromatographic Analysis," Anal. Chem., 33, 363-364 (1961); CA 56, 9378h

4903. Meyer, Jr., A. S., White, J. C., and Rubin, I. B., "Gas-Chromatographic Determination of Helium in Neutron-Irradiated Beryllium Oxide," U. S. At. Energy Comm., TID-7606, 158-173 (1960); CA 55, 20771i

4904. Meyer, H., and Garcia, F. G., "Vapor-Phase Oxidation of Toluene with Betonite as a Catalyst," Anales real. soc. espaff. fís. y quím. (Madrid), Ser. B, 53, 785-800 (1957); CA 52, 11764c

4905. Meyer, R. A., "Routine Gas Chromatographic Analysis for Trace Materials," 8th Pittsburgh Conf. on Anal. Chem. & Appl. Spectroscopy, Pittsburgh, Pa., March 1957, Program Abstr., p. 34

4906. Meyer, R. A., "The Analysis of High Vapor Pressure Natural Gasolines," in "Gas Chromatography," edited by V. J. Coates, H. J. Noebels, and I. S. Fagerson, Academic Press, Inc., New York, 1958, pp. 93-98

4907. Meyer zu Reckendorf, W., "Polyesters as the Stationary Phase in Gas Chromatography," Z. Anal. Chem., 175, 350-355 (1960); CA 55, 3929e

4908. Meyerson, S., McCollum, J. D., and Rylander, P. N., "Organic Ions in the Gas Phase. VIII. Bicycloheptadiene," J. Am. Chem. Soc., 83, 1401-1403 (1961); CA 55, 25798g

4909. Michaels, G. D., Wheeler, P., Fukayama, G., and Kinsell, L. W., "Plasma Cholesterol Fatty Acids in Human Subject as Determined by Alkaline Isomerization and by Gas Chromatography," Ann. N. Y. Acad. Sci., 72, 633-640 (1959); CA 53, 15179c

4910. Michalek, W., "Analysis of the Gases Leaving the Reaction Vessel in the Direct Synthesis of Methylchlorosilanes by the Method of Adsorption Chromatography," Tworzywa-Guma-Lakiery 4, No. 1, 6-12 (1959); CA 53, 16812b

4911. Michelina, M. V., and Pilleri, R., "Separation and Determination of Benzene, Thiophene, Carbon Disulfide, and Toluene by the Gas-Liquid Chromatographic Method," Rass. chim., 10, No. 5, 19-20 (1958); CA 53, 17773c

4912. Mickel, J. P., "Versatile Microcatalytic-Chromatographic Instrument for Use in the Study of Catalysts," 6th Detroit Anachem Conf., Detroit, Mich., October 1958

4913. Mickel, J. P., "The Use of Gas Chromatography to Study Gases Evolved During Vacuum-Arc Casting of Refractory Metals," 3rd Intern. Symp. on Gas Chromatography, ISA, East Lansing, Mich., June 1961; ISA Proc., 3, 271-273 (1961)

4914. Mickel, J. P., "The Use of Gas Chromatography to Study the Gases Evolved During Vacuum-Arc Casting of Refractory Metals," in "Gas Chromatography," edited by N. Brenner, J. E. Callen, and M. D. Weiss, Academic Press, Inc., New York, 1962, pp. 431-436

4915. Mickel, J. P., "Gas Chromatography to Study the Gases Evolved During Vacuum-Arc Casting of Refractory Metals," Gas Chromatog., Intern. Symp., 1961, 3, 431-435 (Pub. 1962); CA 58, 3879h

4916 Middlehurst, J., and Kennett, B., "An A. C. Modulated Flame Ionization Triode," Nature, 190, 142-143 (1961); CA 55, 17110h

4917. Middlehurst, J., and Kennett, B., "Flame Ionization Detectors," J. Chromatog, 10, 294-302 (1963)

4918. Miedlich, H., "Application of the U. R. A. S., as a Detector for Gas Chromatography," in "Gas Chromatographie 1959," edited by R. E. Kaiser and H. G. Struppe, Akademie Verlag, Berlin, 1959, pp. 258-274

4919. Miettinen, J. K., "Gas Chromatography – A Review," Suomen Kemistilehti, 31A, 149-174 (1958); CA 53, 3973a

4920. Mignolet, J. C. P., "Simple and Multiple Gas Microintroducers for Adsorption Studies," J. Sci. Instr., 30, 15-17 (1953); CA 47, 3624c

4921. Mikhailova, E. A., and D'yachenko, A. I., "Separation of Stereoisomeric Hexenes by the Method of Preparative Gas-Liquid Chromatography," Dokl. Akad. Nauk SSSR, 144, 1056-1058 (1962); CA 57, 13594h

4922. Mikkelsen, L., "The Analysis of Trace Impurities in Vinyl Chloride by Gas Chromatography," 11th Pittsburgh Conf. on Anal. Chem. & Appl. Spectroscopy, Pittsburgh, Pa., February-March 1960,

Program Abstr., p. 39

4923. Mikkelsen, L., "Quantitative Interpretation of Gas Chromatographic Data," 4th Delaware Valley Regional Mtg., ACS, Philadelphia, Pa., January 1962

4924. Mikkelsen, L., "Special Techniques in Gas Chromatography, Including Trace Analysis, Solid Samples, and Polymers," 4th Annual Gas Chromatography Institute, Canisius College, Buffalo, N. Y., April 1962

4925. Mikkelsen, L., "Quantitative Interpretation of Gas Chromatographic Data," 8th Natl. Analysis Instrumentation Symp., ISA, Charleston, W. Va., April-May 1962

4926. Mikkelsen, L., "Applications and Techniques," 5th Annual Gas Chromatography Institute, Canisius College, Buffalo, N. Y., April 1963; see entry No. 4929

4927. Mikkelsen, L., and Beck, M. G., "The Characterization of High Molecular Weight Substances by Gas Chromatography," 13th Pittsburgh Conf. on Anal. Chem. & Appl. Spectroscopy, Pittsburgh, Pa., March 1962, Program Abstr., p. 59

4928. Mikkelsen, L., and Richmond, R. S., "Gas Chromatographic Determination of Oxidation Products of Isobutylene. Hydroxy Acids, Nitrato Acids, and Related Compounds," Anal. Chem., 34, 74-76 (1962); CA 56, 9424e; 12th Pittsburgh Conf. on Anal. Chem. & Appl. Spectroscopy, Pittsburgh, Pa., February-March 1961, Program Abstr., p. 59

4929. Mikkelsen, L., and Spencer, S. F., "Special Techniques in Gas Chromatography," in "Lectures on Gas Chromatography 1962," edited by H. A. Szymanski, Plenum Press, Inc., New York, 1963, pp. 177-198; see entry No. 4926

4930. Mikkelsen, L., Wisniewski, J. V., and Beck, M. G., "Analytical Application of Dual Column Gas Chromatography," Southwest Regional Mtg., ACS, Dallas, Texas, December 1962

4931. Miles, H. T., and Fales, H. M., "Application of Gas Chromatography to Analysis of Nucleosides," Anal. Chem., 34, 860-861 (1962)

4932. Miller, A. J., "Gas Chromatography: For Natural Gas and Gasoline Analysis – 1," Oil Gas J., 54, No. 85, 126-127 (1956)

4933. Miller, A. J., "How Gasoline Plants Are Using Gas Chromatography," Oil Gas J., 56, No. 9, 88-91 (1958); CA 52, 13237i

4934. Miller, A. J., "What You Ought to Know About Gas Chromatography," Oil Gas J., 57, No. 17, 105-110 (1959); CA 53, 17484e

4935. Miller, A. J., "Analysis of Natural Gas by Gas Chromatography," Rev. inst. franç. pétrole et ann. combustibles liquides, 17, 1317-1327 (1962); CA 58, 11141d

4936. Miller, B., "Phosphorothioates. I. The Reaction of Phosphorochloridothioates with Carboxylate and t-Butoxide Anions," J. Am. Chem. Soc., 82, 3924-3928 (1960); CA 55, 11287b

4937. Miller, B. S., Johnson, J. A., and Robinson, R. J., "Identification of Carbonyl Compounds Produced in Preferments," Cereal Chem., 38, 507-515 (1961)

4938. Miller, C. T., "Gas Chromatography," Res. & Eng., 2, No. 2, 26-27 (1956)

4939. Miller, D. B., "Higher Alpha, Omega-Dienes in Paraffin Pyrolysates," 144th Natl. ACS Mtg., Los Angeles, Calif., March-April 1963, Program Abstr., pp. 9-10

4940. Miller, D. L., Samsel, E. P., and Cobler, J. G., "Determination of Acrylate and Maleate Esters in Polymers by Combined Zeisel and Gas Chromatographic Analysis," Anal. Chem., 33, 667-680 (1961); CA 56, 8918c

4941. Miller, J., Gregoriou, G., and Mosher, H. S., "Relative Rates of Grignard Addition and Reduction Reactions," J. Am. Chem. Soc., 83, 3966-3971 (1961); CA 56, 3339h

4942. Miller, J. J., "1-Ethynyl-2-vinylbenzene," J. Org. Chem., 26, 3583-3585 (1961)

4943. Miller, L. D., "The Lipids of Bovine Spermatozoa," Univ. Microfilms (Ann Arbor, Mich.), L. C. Card No. Mic $60-4046$, 78 pp.; Dissertation Abstr., 21, 1359-1360 (1960); CA 55, 5693h

4944. Miller, R., and Hyer, H., "Calibration of Gas Chromatography Using Impure Materials," 1st Annual Gas Chromatography Institute, Canisius College, Buffalo, N. Y., May 1959

4945. Milligan, I. B., Bradow, R. L., Rose, J. E., Hubbert, H. E., and Roe, A., "Photochemical Interchange of Halogens in Aromatic Compounds," J. Am. Chem. Soc., 84, 158-162 (1962); CA 58, 1329c

4946. Million, J. G., Weber, C. W., and Kuehn, P. R., "Gas Chromatography of Corrosive Halogen-Containing Gases," Southeastern Regional Mtg., ACS, Gatlinburg, Tenn., November 1962

4947. Mil'vitskaya, E. M., and Plate, A. F., "Structural Isomerization of Cycloheptatriene Under the Conditions of the Diels-Alder Reaction," Zh. Obshch. Khim., 32, 2566-2576 (1962); CA 58, 8927e

4948. Minkoff, G. J., "Micro Gas Analysis," Analyst, 85, 164 (1960)

4949. Minter, C. C., and Burdy, L. M. J., "Thermal Conductivity Bridge for Gas Analysis," Anal. Chem., 23, 143-147 (1951); CA 45, 3202e

4950. Minter, C. C., and Schuldiner, S., "Thermal Conductivity of Equilibrated Mixtures of H_2, D_2, and HD," J. Chem. Eng. Data, 4, No. 3, 223 (1959); CA 54, 9479c

4951. Miquel, R., and Benard, C., "Chromatographic Determination of Hydrogen and Light Hydrocarbons," Compt. rend., 256, 940-941 (1963); CA 58, 8402b

4952. Miras, C. J., and Contaxis, C. C., "Gas-Liquid Chromatographic Detection of Aldosterone and Cortisone," Chim. Chronika (Athens, Greece), 27, No. 2, 43-44 (1962); CA 57, 1213c

4953. Mirev, D., Elenkov, D., and Balarev, K., "Rate of Absorption of Pure Gases," Compt. rend. acad. bulgare sci., 14, 263-266 (1961); CA 55, 24135a

4954. Mirocha, C. J., and DeVay, J. E., "Rapid Gas Chromatographic Method for Determining Fumaric Acid in Fungus Cultures and Diseased Plant Tissues," Phytopathology, 51, 274-276 (1961)

4955. Mironov, V. A., Mavrov, M. V., and Elizarova, A. N., "Substituted Cyclopentadienes and Related Compounds. I. 1,3-Dimethylcyclopentadiene," Zh. Obshch. Khim., 32, 2723-2731 (1962); CA 58, 8917f

4956. Miropol'skaya, M. A., Pedotova, N. I., Veinberg, A. Y., Yanotovskii, M. T., and Samokhvalov, G. I., "Synthetic Studies of Polyene Compounds. XVIII. Selective Hydrogenation of 6-Methyl-3,5-heptadien-2-one and 6-Methyl-3,5-heptadien-2-ol Over Pd-CaCO₃," Zh. Obshch. Khim., 32, 2214-2217 (1962); CA 58, 8892d

4957. Misic, D., and Thodos, G., "Thermal Conductivity of Hydrocarbon Gases at Normal Pressures," A. I. Ch. E. Journal, 7, 264-271 (1961); CA 56, 8031f

4958. Miskalis, A. J., "Gas-Chromatographic Analysis of Coke Oven Benzene-Toluene-Xylene (BTX), and Benzene for Minor Components," Am. Chem. Soc., Preprints, Div. Gas Fuel Chem., April 1960, 52-62; CA 57, 6224h; 137th Natl. ACS Mtg., Cleveland, Ohio, April 1960, Program Abstr., p. 2J

4959. Mistretta, A. G., "Progress in Fractionation Procedures. Differential Migration Methods," Microchem. J., 3, 305-314 (1959); CA 54, 8435i

4960. Mistretta, A. G., "Progress in Fractionation Procedures. Differential Migration Methods: 1959," Microchem. J., 4, 289-305 (1960); CA 55, 7967c

4961. Mistrik, I. E. J., and Polievka, M., "Hydrogenation of Oxo Synthesis Products," Chem. průmysl, 12, 123-128 (1962); CA 58, 3306a

4962. Mitchell, Jr., J., "Gas Chromatography. Detector Systems," in "Submicrogram Experimentation," edited by N. D. Cheronis, Interscience Publishers, Inc., New York, 1961, pp. 201-218; CA 56, 911g

4963. Mitchell, Jr., J., "Gas Chromatography. Detector Systems," Microchem. J., Symp. Ser., 1, 201-218 (1961); CA 58, 5015b

4964. Mitchell, R., "Problem Applications in Gas Chromatography of Aromatic Compounds," 1st Annual Gas Chromatography Institute, Canisius College, Buffalo, N. Y., May 1959

4965. Mitchell, T., "How Chromatography Can Improve Gasoline-Plant Profits," Oil Gas J., 58, No. 2, 73-75 (1960)

4966. Mitzner, B. M., "Adapting Vapor Fractometers for 3/16-Inch Columns," Perkin-Elmer Instr. News, 12, No. 3, 12 (1961)

4967. Mitzner, B. M., "An Improved Trapping Device for Gas Chromatography," Chemist-Analyst, 51, No. 1, 20 (March 1962)

4968. Mitzner, B. M., and Freeman, S. K., "Sectional Spiral Glass Columns for Vapor Phase Chromatography," Chemist-Analyst, 51, No. 2, 54 (July 1962)

4969. Mitzner, B. M., and Friedlander, M., "The Application of Logarithmic Recording by Gas Chromatography," 10th Pittsburgh Conf. on Anal. Chem. & Appl. Spectroscopy, Pittsburgh, Pa., March 1959, Program Abstr., p. 53

4970. Mitzner, B. M., and Gitoneas, P., "Multiple-Column Gas Chromatograph Utilizing a Single Detector and Recorder," Anal. Chem., 34, 589-590 (1962); CA 57, 1514g

4971. Mitzner, B. M., and Jacobs, M., "Simple Effect Trap for Gas Chromatography," Chemist-Analyst, 48, 104 (1959)

4972. Mitzner, B. M., and Jacobs, M., "The Application of Capillary Columns to the Essential Oil Industry. I. Terpenes and Hydrocarbons," 137th Natl. ACS Mtg., Cleveland, Ohio, April 1960, Program Abstr., p. 34B

4973. Mitzner, B. M., and Jacobs, M., "The Application of Capillary Columns to the Essential Oil Industry. II. Terpene Alcohols and Oils," 137th Natl. ACS Mtg., Cleveland, Ohio, April 1960. Program Abstr., p. 35B

4974. Mitzner, B. M., and Theimer, E. T., "β-Terinene and β-Phellandrene from the Pyrolysis of Sabinene," J. Org. Chem., 27, 3359 (1962); CA 58, 2473f

4975. Miwa, T. K., "Identification of Peaks in Gas-Liquid Chromatography," J. Am. Oil Chemists' Soc., 40, 309-313 (1963); 53rd Spring Mtg., Am. Oil Chemists' Soc., New Orleans, La., May 1962

4976. Miwa, T. K., Earle, F. R., Miwa, G. C., and Wolff, I. A., "Biosynthesis of Epoxyoleic Acid in Maturing Vernonia Anthelmintica Seeds," 139th Natl. ACS Mtg., St. Louis, Mo., March 1961, Program Abstr., p. 29C

4977. Miwa, T. K., Earle, F. R., Miwa, G. C., and Wolff, I. A., "Fatty Acid Composition of Maturing Vernonia Anthelmintica (L.) Willd. Seeds. Dihydroxyoleic Acid – A Possible Precursor of Epoxyoleic Acid," J. Am. Oil Chemists' Soc., 40, 225-229 (1963)

4978. Miwa, T. K., Mikolajczak, K. L., Earle, F. R., and Wolff, I. A., "Gas Chromatographic Charac-

terization of Fatty Acids. Identification Constants for Mono- and Dicarboxylic Methyl Esters,"
Anal. Chem., 32, 1739-1742 (1960); CA 55, 18453d; 137th Natl. ACS Mtg., Cleveland, Ohio,
April 1960, Program Abstr., p. 36B

4979. Miyahara, S., "Gas-Liquid Chromatographic Separation and Determination of Volatile Fatty Acids
in Fish Meat During Spoilage," Bull. Japanese Soc. Sci. Fish, 27, 42-47 (1961)

4980. Miyahara, S., "Gas-Chromatographic Separation and Determination of Methylamines," Nippon Kagaku Zasshi, 82, 1108-1110 (1961); CA 57, 2856h

4981. Miyake, H., "Gas Chromatography," Kagaku no Ryoiki, 11, 171-180 (1957); CA 52, 16835a

4982. Miyake, H., "Analysis of Hydrocarbon Gases by Gas Chromatography," Kagaku no Ryoiki, 12, 627-
630 (1958); CA 52, 19339b

4983. Mlejnek, O., "Identification of Polymeric Substances by Gas-Liquid Chromatography of the Products
of their Thermal Decomposition," Chem. průmysl. 11, 604-607 (1961); CA 57, 1037i

4984. Mlejnek, O., "Identification of Polyhydric Alcohols in Polyesters with the Aid of Gas-Liquid Chromatography," Chem. průmysl, 13, 105-107 (1963); CA 59, 1069b

4985. Mlejnek, O., and Pauliny, M., "Glass Head for Column for Gas-Liquid Chromatography," Chem.
Zvesti, 15, 462-464 (1961); CA 55, 24135e

4986. Mlejnek, O., and Seckarova, H., "A Simple Method for Testing the Thermal Stability of Stationary
Phases in Gas-Liquid Chromatography," Chem. Zvesti, 15, 607-611 (1961); CA 56, 5383g

4987. Moffat, A. J., and Solomon, P. W., "Gas-Solid Chromatographic Separation of Polyphenyls on LiCl-
Firebrick," U. S. At. Energy Comm., IDO-16732, 10 pp. (1961); CA 56, 5386d

4988. Moger, G. G., and Sergeev, G. B., "Analysis of Products of Low-Temperature Bromination and
Hydrobromination of Olefins by Gas Chromatography," Vestn. Mosk. Univ., Ser. II, Khim., 18,
No. 2, 14-16 (1963); CA 59, 1071f

4989. Moghadame, P. E., "Vapor-Phase Partition Chromatography. II.," Rev. inst. franç. pétrole et ann.
combustibles liquides, 12, 58-66 (1957); CA 51, 17302a

4990. Moghadame, P. E., "Partition Chromatography Between Gaseous and Liquid Phases," Chim. Anal.,
40, 149-155 (1958); CA 52, 13516a

4991. Mohnke, M., "Adsorption-Capillary Chromatography by Stable Isotopes," 4th Intern. Gas Chromatography Symp., Hamburg, Germany, June 1962; pub. in "Gas Chromatography 1962," M. van
Swaay, ed., Butterworth & Co., London, 1962

4992. Mohnke, M., and Renker, K., "A New Electromechanical Integrator, Specifically for Gas Chromatographic Analysis," Chem. Tech. (Berlin), 12, 493-494 (1960)

4993. Mohnke, M., and Saffert, W., "Katharometer for Detecting Small Quantities of Gas," Chem. Tech.
(Berlin), 13, 685-686 (1961); CA 57, 2826e

4994. Mohnke, M., and Saffert, W., "Gas-Solid Capillary Chromatography of Hydrogen Isotopes," Gas
Chromatog. Symp., Brno, Czech., June 1962; Abstr., J. Gas Chromatog., 1, No. 4, 7 (1963)

4995. Mohr, W., "Studies on Cacao Aroma with Special Reference to Processing Chocolate Paste," Fette,
Seifen, Anstrichmittel, 60, 661-669 (1958)

4996. Mokrushin, S. G., "Definition of Chromatography," Nature, 178, 1244-1245 (1956); CA 51, 5501e

4997. Moll, F., "Application of Gas Chromatography to Pharmaceutical and Food Chemistry Problems,"
Deut. Apotheker Ztg., 101, 1357-1360 (1960); CA 56, 8837h

4998. Molle, L., and Vogelenzang, E. H., "Gas Chromatography and Possible Uses in the Analysis of
Medicinals," Pharm. Weekblad, 93, 348-358 (1958); CA 52, 14085e

4999. Molnar, W. S., and Yarborough, V. A., "Polyethylene Capillary Tubes as Micro Liquid Cells for
Infrared Spectroscopy," Appl. Spectroscopy, 12, 143-145 (1958); CA 53, 7765d

5000. Momigny, J., "Ionization Potentials of cis- and trans-Dichloro- and Dibromoethylene," Nature,
191, 1089-1090 (1961); CA 56, 4210a

5001. Momyer, Jr., F. F., "The Radiochemistry of the Rare Gases," U. S. At. Energy Comm., NAS-NS-
3025, 55 pp. (1960); CA 55, 14083c

5002. Monet, G. P., "Sorption Processes – Adsorption, Dialysis and Ion Exchange," Chem. Eng. Progr.,
53, 514-517 (1957); CA 52, 1698h

5003. Monkman, J. L., "Some Applications of Gas Chromatography to Commonly Encountered Analytical
Problems," in "Gas Chromatography," edited by V. J. Coates, H. J. Noebels, and I. S. Fagerson, Academic Press, Inc., New York, 1958, pp. 111-115

5004. Monkman, J. L., and Dubois, L., "The Determination of Halogenated Hydrocarbons by Gas Chromatography and Flame Photometry," Gas Chromatog. Intern. Symposium, 2nd, East Lansing, Mich.,
1959, 333-337 (Pub. 1961); CA 55, 19616h

5005. Monkman, J. L., and Dubois, L., "The Determination of Halogenated Hydrocarbons by Gas Chromatography and Flame Photometry," in "Gas Chromatography," edited by H. J. Noebels, R. F.
Wall, and N. Brenner, Academic Press, Inc., New York, 1961, pp. 333-337

5006. Monkman, J. L., Dubois, L., and Techman, T., "Confirmatory Tests in Gas Chromatography. III.
Aldehydes," Chem. in Can., 13, No. 5, 92-96 (1961); CA 55, 24392c

5007. Mooney, J. B., and Garbini, L. J., "A Sensitive Carbon Determination by Combustion and Gas
Chromatography," 13th Pittsburgh Conf. on Anal. Chem. & Appl. Spectroscopy, Pittsburgh, Pa.,

March 1962, Program Abstr., p. 53

5008. Mooney, T. F., Barrow, R. E., and Smith, R. G., "The Identity of Some Organic Vapors in Urban Air," 140th Natl. ACS Mtg., Chicago, Ill., September 1961, Program Abstr., p. 23W

5009. Moore, A. D., "Electron Capture with an Argon Ionization Detector in Gas Chromatographic Analysis of Insecticides," J. Econ. Entomol., 55, 271-272 (1962); Biol. Abstr., 39, 12207

5010. Moore, C. E., and Meinstein, S., "Determination of Benzoic Acid in Phthalic Anhydride by Gas Liquid Chromatography," Anal. Chem., 34, 1503-1504 (1962); CA 58, 22c

5011. Moore, D. R., "Reaction of Benzyl Methyl Ether with Sodium Metal," J. Org. Chem., 26, 3596-3597 (1961); CA 56, 10009i

5012. Moore, D. R., and Kossoy, A. D., "Quantitative Analysis of Menthol Stereoisomers by Gas-Liquid Chromatography," Anal. Chem., 33, 1437 (1961); CA 55, 24394h

5013. Moore, W. R., "Vapor-Phase Chromatography," 128th Natl. ACS Mtg., Minneapolis, Minn., September 1955, Program Abstr., p. 30O

5014. Moore, W. R., "Gas Chromatography," 137th Natl. ACS Mtg., Cleveland, Ohio, April 1960, Program Abstr., p. 2F

5015. Moore, W. R., and Ward, H. R., "Separation of Ortho-Hydrogen and Para-Hydrogen," J. Am. Chem. Soc., 80, 2909-2910 (1958); CA 52, 17893g

5016. Moore, W. R., and Ward, H. R., "Reaction of Dibromocyclopropanes with Alkyllithium Reagents. Formation of Allenes, Spiropentanes, and a Derivative of Bicyclopropylidene," J. Org. Chem., 25, 2073 (1960); CA 55, 13332e

5017. Moore, W. R., and Ward, H. R., "Gas-Solid Chromatography of H_2, HD, and D_2. Isotropic Separation and Heats of Adsorption on Alumina," J. Phys. Chem., 64, 832 (1960); CA 54, 23600a

5018. Moore, W. R., and Ward, H. R., "Formation of Allenes from gem-Dihalocyclopropanes by Reaction with Alkyllithium Reagents," J. Org. Chem., 27, 4179-4181 (1962); CA 58, 10098e

5019. Morgan, D. J., "Construction and Operation of a Simple Flame-Ionization Detector for Gas Chromatography," J. Sci. Instr., 38, 501-502 (1961); CA 56, 12290g

5020. Morgan, M. E., "Exploratory Analyses of Silage Volatiles Which May Affect Milk Flavor," 55th Annual Mtg., Am. Dairy Sci. Assoc., Logan, Utah, June 1960; Abstr., J. Dairy Sci., 43, 848-849 (1960)

5021. Morgan, M. E., and Pereira, R. L., "Identity of the Grassy Aroma Constituents of Some Green Forages," 58th Annual Mtg., Am. Dairy Sci. Assoc., Lafayette, Ind., June 1963; Abstr., J. Dairy Sci., 46, 613-614 (1963)

5022. Morgan, T. O., "Chromatographic Analysis of Combustion Gases: A Simple, Accurate Sampling System That Permits Calibration with Pure Gases:" 11th Pittsburgh Conf. on Anal. Chem. & Appl. Spectroscopy, Pittsburgh, Pa., February-March 1960, Program Abstr., p. 46

5023. Morgantini, M., "The Use of Gas Chromatography for the Analysis of Vinegars. I. Concentration of Ethanol," Boll. Lab. Chim. Prov. (Bologna), 13, 117-121 (1962); CA 57, 10356a

5024. Morgantini, M., "Programmed Flow Gas Chromatography Applied to the Analysis of Volatile Fatty Acids," Boll. Lab. Chim. Provinciali (Bologna), 13, 545-554 (1962); CA 59, 846d

5025. Morgantini, M., "Gas Chromatographic Determination of Trace Amounts of Nitrogen and Oxygen in Nitrous Oxide," J. Chromatog., 9, 536-538 (1962)

5026. Morgareidge, K., "Review of Applied Analysis. Food," Anal. Chem., 31, 691-696 (1959); CA 53, 8912g

5027. Mori, M., Horrii, T., and Iwakiri, Y., "Analysis of Fats and Oils by Gas-Liquid Chromatography. I. Saturated Fatty Alcohols," Nippon Suisan Chuo Kenkyusho Hokoku, 9, 52-59 (1961); CA 58, 3619f

5028. Morice, I. M., "Seed Fats of the New Zealand Agavaceae," J. Sci. Food Agr., 13, 666-669 (1962); CA 58, 7050g

5029. Morice, I. M., and Shorland, F. B., "Occurrence of n-Odd-Numbered Monoethylenic Fatty Acids in the Liver Oil of the New Zealand School Shark (Galerorhinus Australis Macleay)," Nature, 190, 443 (1961); CA 55, 17930c

5030. Morin, R. J., Bernick, S., Mead, J. F., and Alfin-Slater, R. B., "The Influence of Exogenous Cholesterol on Hepatic Lipid Composition of the Rat," J. Lipid Research, 3, 432-438 (1962)

5031. Morley, H. V., Cooper, F. P., and Holt, A. S., "Separation and Identification of Degradation Products of Porphyrins by Gas-Liquid Partition Chromatography," Chem. & Ind. (London), 1959, 1018; CA 54, 7406d

5032. Morley, H. V., and Holt, A. S., "Studies on Chlorobium Chlorophylls. II. The Resolution of Oxidation Products of Chlorobium Pheophorbide (660) by Gas-Liquid Partition Chromatography," Can. J. Chem., 39, 755-760 (1961); CA 56, 2452e

5033. Morozova, O. E., Zemskova, Z. K., Osityanskaya, L. Z., Kislinskii, A. N., and Petrov, A. A., "Catalytic Dehydroisomerization of Alkylcyclopentanes," Neftekhimiya, 2, 676-680 (1962); CA 59, 1498f

5034. Morris, L. J., "Chromatography of Lipids," in "Chromatography," edited by E. Heftmann, Reinhold Publishing Corp., New York, 1961, Chapter 16

5035. Morris, L. J., Hayes, H., and Holman, R. T., "Naturally-Occurring Epoxy Acids. III. Methods for Their Isolation," J. Am. Oil Chemists' Soc., 38, 316-321 (1961); CA 55, 19277f

5036. Morris, L. J., Holman, R. T., and Fontell, K., "Vicinally Unsaturated Hydroxy Acids in Seed Oils," J. Am. Oil Chemists' Soc., 37, 323-327 (1960); CA 54, 17920a

5037. Morris, L. J., Holman, R. T., and Fontell, K., "Alteration of Some Long-Chain Esters During Gas-Liquid Chromatography," J. Lipid Research, 1, 412-420 (1960); CA 55, 10307c

5038. Morris, L. J., Holman, R. T., and Fontell, K., "Naturally Occurring Epoxy Acids. I. Detection and Evaluation of Epoxy Fatty Acids by Paper, Thin-Layer, and Gas-Liquid Chromatography," J. Lipid Research, 2, 68-76 (1961); CA 55, 14942f

5039. Morris, R. A., and Chapman, R. L., "Flame Ionization Hydrocarbon Analyzer," J. Air Pollution Control Assoc., 11, 467-469, 489 (1961); CA 56, 1307d

5040. Morrison, R. L., "Determination of Ethanol in Wine by Gas-Liquid Partition Chromatography," Am. J. Enol. & Viticult., 12, No. 3, 101-106 (1961)

5041. Morrison, R. L., "The Determination of Acetaldehyde in High-Proof Fortifying Spirits, Beverage Brandy, and Wine," Am. J. Enol. Viticult., 13, 158-168 (1962); CA 58 9594d

5042. Morrison, W. R., Lawrie, T. D. V., and Blades, J., "By-Products Formed During the Methylation of Long Chain Fatty Acids with Diazomethane," Chem. & Ind. (London), 1961, 1534-1535; CA 56, 3342c

5043. Morrow, H. N., and Buckley, K. B., "Need Help with Gas Chromatography," Petrol. Refiner, 36, No. 8, 157-161 (1957); CA 51, 15338c

5044. Mortensen, E. M., and Eyring, H., "Potential Energy Barrier for the Rotation and the Condensation Coefficients of H_2 and D_2 on Alumina by Gas Chromatography," J. Phys. Chem., 64, 433-434 (1960); CA 54, 16986b

5045. Mortimer, J. V., and Gent, P. L., "Use of Modified 'Bentone-34' for the Gas Chromatographic Separation of Aromatic Hydrocarbons," Nature, 197, 789-790 (1963); CA 58, 10706e

5046. Moruzzi, G., Martinelli, M., and Caldarera, C. M., "Cholesterol Lowering Activity of Lipids in Wheat Gluten," Arch. Biochem. Biophys., 91, 328-329 (1960); CA 55, 8558d

5047. Moryashchev, A. K., and Voronin, V. G., "Determination of the Composition of Some Etheral Oils by Gas-Liquid Chromatography," Zh. Analit. Khim., 18, 401-405 (1963); CA 58, 12365d

5048. Mosen, A. W., and Buzzelli, G., "Determination of Impurities in Helium by Gas Chromatography," Anal. Chem., 32, 141-142 (1960); CA 54, 24112h

5049. Mosher, W. A., and DeSimone, G. J., "The Copolymerization of tert-Butyl Alcohol and 3-Pentanol," 138th Natl. ACS Mtg., New York, N. Y., September 1960, Program Abstr., p. 86P

5050. Moshier, R. W., Schwarberg, J. E., Morris, M., and Sievers, R. E., "The Quantitative Aspects of the Gas Chromatography of Certain Metal Coordinate Compounds," 14th Pittsburgh Conf. on Anal. Chem. & Appl. Spectroscopy, Pittsburgh, Pa., March 1963, Program Abstr., p. 58

5051. Mossini, F., and Vitali, T., "Gas Chromatography of Indole Compounds," Ricerca Sci. Rend. Sez. A, 1, 244-246 (1961); CA 57, 6617g

5052. Motl, O., Herout, V., and Sorm, F., "On Terpenes. CXII. The Composition of the Oil from Juniperus Oxycedrus L. Berries," Collection Czechoslov. Chem. Commun., 25, 1656-1661 (1960); CA 54, 21175b

5053. Mottlau, A. Y., "Some Analytical Aspects of Hydrogen Exchange Between Hydrocarbons and Sulfuric Acid," Anal. Chem., 33, 293-297 (1961); CA 55, 9147d

5054. Mourgues, L. de, "Applications of Gas Phase Chromatography to the Study of Solid Catalysts," Chim. Anal. (Paris), 45, 103-110 (1963); CA 59, 1419d

5055. Mourgues, L. de, and Capony, J., "Study of Different Properties of Aluminosilicate Catalysts by the Microcatalytic Chromatographic Technique," Journées Intern. Etude Méthodes Separation Immediate Chromatog., Paris, 1961, 163-174 (Pub. 1962); CA 59, 1133d

5056. Mourgues, L. de, and Rochina, V., "A Thermistor Katharometer for Gas-Phase Chromatography," Bull. soc. chim. France, 1962, 729-732; CA 57, 5284b

5057. Moussebois, C., and Duyckaerts, G., "Vapor-Phase Radiochromatography," J. Chromatog., 1, 200-201 (1958); CA 52, 18067g

5058. Mousseron, M., Mousseron-Canet, M., Philippe, G., and Wylde, J., "9,10-Octahydronaphthalene and Some of Its Derivatives," Compt. rend., 256, 51-53 (1963); CA 59, 468h

5059. Muccini, G. A., and Schuler, R. H., "Radiation Chemistry of Cyclopentane-Cyclohexane Mixtures," J. Phys. Chem., 64, 1436-1438 (1960); CA 55, 13020g

5060. Mueller, E., Fiedler, G., Huber, H., Narr, B., Suhr, H., and Witts, K., "Comparative Investigations of the Clemmensen and Wolff-Kishner Reduction of Medium Ring Detones," Z. Naturforsch., 18b, 5-7 (1963); CA 58, 11234d

5061. Mueller, E. P., and Freund, W., "Use of Gas Chromatography in Natural Gas and Petroleum Prospecting and Production," Z. Angew. Geol., 8, 304-307 (1962); CA 57, 12781c

5062. Muenster, H. K., "Chromatographic Analysis of Gases and Vapors," Lab. sci. (Milan), 6, 12-23

(1958); CA <u>52</u>, 16005a

5063. Muhlstadt, M., "Double Mannich Condensation with Cyclic Detones on α,α'-Dimethylenecycloalkanones," Chem. Ber., <u>93</u>, 2638-2648 (1960)

5064. Muhs, M. A., and Wiess, F. T., "Determination of Equilibrium Constants of Silver-Olefin Complexes Using Gas Chromatography," J. Am. Chem. Soc., <u>84</u>, 4697-4705 (1962); CA <u>58</u>, 5470g

5065. Mukhina, T. N., and Itsek, S. E., "Relation of the Distillation Curve of Gasolines to the Products of Their Pyrolysis," Neftekhimiya, <u>2</u>, 723-729 (1962); CA <u>58</u>, 7766d

5066. Mullen, R. T., "The Chemical Interaction of Accelerated Carbon-14 Ions with Benzene," U. S. At. Energy Comm., <u>UCRL-9603</u>, 263 pp. (1961); CA <u>56</u>, 2098i

5067. Muller, G., "Sampling Apparatus for Small Gas Quantities," Chem. Tech. (Berlin), <u>13</u>, 237 (1961); CA <u>55</u>, 18207g

5068. Müller, R. H., "Portable Apparatus Analyzes Multicomponent Mixtures by Fractional Separation of Vapors in Partition Column," Anal. Chem., <u>27</u>, No. 6, 33A-36A (1955)

5069. Müller, R. H., "Instrumentation: Vapor-Phase Chromatography Used to Perform Automatic Analyses," Anal. Chem., <u>29</u>, No. 3, 55A-57A (1957)

5070. Müller, R. H., "Vapor Phase Chromatography Advancing Rapidly," Anal. Chem., <u>29</u>, No. 10, 67A-68A (1957)

5071. Müller, R. H., "Relatively Simple Analog Computer Facilitates Rapid Calculation of Complete Analysis of Vapor Phase Chromatogram," Anal. Chem., <u>31</u>, No. 7, 67A-68A (1959)

5072. Müller, R. H., "Developments in Ultrasonic Machining, Vapor Phase Chromatography, and Analytical Techniques Described," Anal. Chem., <u>31</u>, No. 9, 91A-94A (1959)

5073. Müller, R. H., "'Break-Through' in Recorders Results from Use of Electromechanical Strain Gage and Design Improvements," Anal. Chem., <u>33</u>, No. 6, 101A-102A (1961)

5074. Mumpower, R. C., Lewis, J. S., and Touey, G. P., "Determination of Carbon Monoxide in Cigarette Smoke by Gas Chromatography," Tobacco Sci., <u>6</u>, 140-143 (Pub. in Tobacco, <u>155</u>, No. 8, 30-33 (1962)); CA <u>57</u>, 17091d

5075. Munch, R. H., "Vapor-Phase Chromatography," Chem. & Eng. News, <u>33</u>, No. 15, 1510-1511 (April 11, 1955); 127th Natl. ACS Mtg., Cincinnati, Ohio, March-April 1955, Program Abstr., p. 8B

5076. Munch, R. H., "Detection of Sample Fractions in Vapor-Phase Chromatography," 129th Natl. ACS Mtg., Dallas, Texas, April 1956, Program Abstr., p. 14B

5077. Munch, R. H., "Vapor-Phase Chromatography," Record Chem. Prog., <u>18</u>, 69-101 (1957); CA <u>51</u>, 12598d

5078. Munch-Petersen, J., Bretting, C., Jorgensen, R. M., Refn, S., Andersen, V. K., and Jart, A., "Conjugate Additions of Grignard Reagents to α,β-Unsaturated Esters. IX. 1, 4- and 1, 6-Additions of n-Butyl Magnesium Bromide to the sec-Butyl Ester of Sorbic Acid," Acta Chem. Scand., <u>15</u>, 277-292 (1961); CA <u>55</u>, 27040e

5079. Mund, W., and Meerssche, M. van, "Thermoconductometric Analysis of Gases," Bull. classe sci., Acad. roy. belg., <u>39</u>, 676-683 (1953); CA <u>48</u>, 2516a

5080. Munday, C. W., and Primavesi, G. R., "Properties of the Martin Gas Density Balance and Possible Modifications Thereof," in "Vapour-Phase Chromatography," edited by D. H. Desty, Academic Press, Inc., New York, 1957, pp. 146-153

5081. Mungall, T. G., Mitchen, J. H., and Johnson, D. E., "Determination of Microgram Amounts of Carbon in Sodium Metal," 145th Natl. ACS Mtg., New York, N. Y., September 1963, Program Abstr., p. 18B

5082. Munnecke, D. E., Domsch, K. H., and Eckert, J. W., "Fungicidal Activity of Air Passed Through Columns of Soil Treated with Insecticides," Phytopathology, <u>52</u>, 1298-1306 (1962); CA <u>58</u>, 6139f

5083. Murad, E., "The Photolysis of Ethyl Vinyl Ether," J. Am. Chem. Soc., <u>83</u>, 1327-1330 (1961); CA <u>55</u>, 15080b

5084. Murad, E., and Noyes, Jr., W. A., "Photochemical Studies. LV. The Nitrous Oxide-Ethane System. Liquid Products and Competitive Rates," J. Am. Chem. Soc., <u>81</u>, 6405-6408 (1959); CA <u>54</u>, 10467c

5085. Murakami, Y., "Rapid Gas Chromatographic Analysis of a Mixture of Oxygen, Nitrogen, Methane, Carbon Monoxide, and Carbon Dioxide," Bull. Chem. Soc. Japan, <u>32</u>, 316-317 (1959); CA <u>54</u>, 3045e

5086. Muraosa, K., and Iwaya, K., "Carbon Deposition in the Manufacture of Synthetic Gas. XII. Analysis of Methane-Steam Reformed Gas and Ethane and Propane by Gas Chromatography," Kogyo Kagaku Zasshi, <u>65</u>, 1550-1554 (1962); CA <u>58</u>, 12333f

5087. Murata, Y., and Takenishi, T., "Direct Gas-Chromatographic Analysis of Organic Components in Aqueous Solution," Kogyo Kagaku Zasshi, <u>64</u>, 787-791 (1961); CA <u>57</u>, 2824h

5088. Murdoch, I.A., "Determination of Nitrogen in Argon by Gas Chromatography," Analyst, <u>86</u>, 856-857 (1961); CA <u>57</u>, 7898a

5089. Murfee, Jr., J. A., "Swamping Catalyst Effect in the Bromination of Acetophenone and the Development of Gas Chromatographic Analysis to Establish Reaction Products," Univ. Microfilms (Ann

Arbor, Mich.), Order No. 61-6270, 154 pp.; Dissertation Abstr., 22, 1830-1831 (1961); CA 56, 8599b

5090. Murray, K. E., "New Design of the Martin and James Gas Density Meter," Australian J. Appl. Sci., 10, 156-168 (1959); CA 53, 14601i

5091. Murray, K. E., "A Method for the Determination of the Structure of Saturated Branched-Chain Fatty Acids," Australian J. Chem., 12, 657-670 (1959); CA 54, 8612a

5092. Murray, K. E., "Waxes. XXI. The Branched-Chain Acids of the Preen Gland of the Goose," Australian J. Chem., 15, 510-520 (1962); CA 57, 16775a

5093. Murray, R. W., and Trozzolo, A. M., "Reaction of Diphenylmethylene with the Carbon-Halogen Bond," J. Org. Chem., 27, 3341-3344 (1962); CA 58, 461c

5094. Murray, W. J., and Williams, A. F., "The Determination of Small Amounts of γ-Picoline in Aqueous Solutions of β-Picoline by Vapor-Phase Chromatography," Chem. & Ind. (London), 1956, 1020-1021; CA 51, 2473i

5095. Murto, J., "Kinetics of the Alkaline Hydrolysis and Alcoholysis of Methyl Iodide in Dimethyl Sulfoxide-Water and Dimethyl Sulfoxide-Alcohol Mixtures," Suomen Kemistilehti, B34, 92-98 (1961); CA 58, 7803a

5096. Murty, N. L., Williams, M. C., and Reiser, R., "Nonsynthesis of Linoleic Acid from Acetate-1-C^{14} by the Laying Hen," J. Nutrition, 72, 451-454 (1960)

5097. Musaev, I. A., Ku, C-W., Topchiev, A. V., and Sanin, P. I., "Separation of C_{8-14} Aromatic Hydrocarbons by Gas Chromatography," Neftekhimiya, 1, 459-472 (1961); CA 56, 14904h

5098. Musgrave, W. K. R., "Thermistor Detectors for Gas Chromatography," Chem. & Ind. (London), 1959, 46; CA 53, 9734h

5099. Musso, H., "Modern Separation Methods in Chemistry," Naturwissenschaften, 45, 97-104 (1958); CA 52, 11518f

5100. Muysers, K., Siehoff, F., and Worth, G., "Blood and Respiratory Gas Analyses with the Aid of Gas Chromatography," Klin. Wochschr., 39, 83-87 (1961); CA 55, 11520c

5101. Muzyczuk, J., "Disturbance Waves in Adsorption Gas Chromatography with the Thermal Conductivity Detector," Chem. Anal. (Warsaw), 7, 1083-1094 (1962); CA 59, 1063g

5102. Myddleton, W. W., "Cosmetic and Toilet Preparations," Mfg. Chemist, 28, 330-332 (1957); CA 51, 17106c

5103. Myers, H. W., and Putnam, R. F., "Determination of Chloroboranes, Diborane (6), and Hydrogen Chloride by Gas Chromatography," Anal. Chem., 34, 664-668 (1962); CA 57, 1547d

5104. Myers, H. W., and Putnam, R. F., "Application of Gas Chromatography to the Study of Chloroborane Intermediates in the Synthesis of Diborane," 14th Pittsburgh Conf. on Anal. Chem. & Appl. Spectroscopy, Pittsburgh, Pa., March 1963, Program Abstr., p. 58

5105. Myers, L. J., "Gas Chromatographic Analysis of Liquids Containing Nonvolatile Viscous Materials," Anal. Chem., 35, 119-120 (1963); CA 58, 6165d

5106. Nabivach, V. M., "Utilization of Aromatic Nitro Compounds for the Separation of Hydrocarbons of the Benzene Series," Zh. Prikl. Khim., 35, 2114-2115 (1962); CA 58, 5548b

5107. Nabivach, V. M., "Immobile Phases for Gas-Chromatographic Separation of Isomeric Xylenes," Neftekhimiya, 2, 906-910 (1962); CA 58, 8831c

5108. Nachbauer, E., and Engelbrecht, A., "Gas Chromatographic Separation of Nitrogen Trifluorides from Carbon Tetrafluoride," J. Chromatog., 2, 562-564 (1959); CA 54, 13960a

5109. Nadeau, H. G., and Oaks, Jr., D. M., "Determination of Ethylene and Propylene Glycols in Mixtures by Gas Chromatography," Anal. Chem., 32, 1760-1762 (1960); CA 55, 7168c

5110. Nadeau, H. G., and Oaks, Jr., D. M., "Separation and Analysis of Substituted Aromatic Compounds in Mixtures by Gas Chromatography," 12th Pittsburgh Conf. on Anal. Chem. & Appl. Spectroscopy, Pittsburgh, Pa., February-March 1961, Program Abstr., p. 56

5111. Nadeau, H. G., and Oaks, Jr., D. M., "Separation and Analysis of Chlorobenzenes in Mixtures by Gas Chromatography," Anal. Chem., 33, 1157-1159 (1961); CA 56, 1982h

5112. Nair, G. V., and Rudloff, E. von, "Chemical Composition of the Heartwood Extractives of Tamarack (Larix Laricina)," Can. J. Chem., 37, 1608-1613 (1959); CA 54, 13650d

5113. Nair, G. V., and Rudloff, E. von, "The Chemical Composition of the Heartwood Extractives of Larix Lyalli Parl," Can. J. Chem., 38, 177-181 (1960); CA 54, 16549a

5114. Nair, III, J. H., Dugan, P. R., Sprague, M. L., and O'Neill, R. D., "Pesticide Analysis in a New York Watershed, An Application of Electron Capture Gas-Liquid Chromatography," 145th Natl. ACS Mtg., New York, N. Y., September 1963

5115. Nair, P. P., and Turner, D. A., "The Application of Gas Chromatography to the Determination of Vitamins E and K," 36th Fall Mtg., Am. Oil Chemists' Soc., Toronto, Ontario, October 1962

5116. Nakagawa, T., Inoue, H., and Kuriyama, K., "Determination of Impurities in Nonionic Detergents by Gas Chromatography," Anal. Chem., 33, 1524-1526 (1961); CA 56, 8867c

5117. Nakagawa, T., and Nakata, I., "Partition Chromatography of Polyethylene Glycols," Kogyo Kagaku Zasshi, 59, 710-712 (1956); CA 52, 4418c

5118. Nakagawa, T., and Tori, K., "Solubilization of Long-Chain Alkyl Compounds by Non-Ionic Surfactants and Cloud Formation in Such Compounds," Kolloid-Z., 168, 132-139 (1960); CA 54, 16991b

5119. Nakahara, Y., and Matsumoto, M., "Gas Chromatography," Yuki Kagaku no Shimpo, 13, 14-56 (1959); CA 55, 6357c

5120. Nakanishi, T., and Nakae, T., "Studies on the Fatty Acid Constituents of Butter Fat Produced in Japan by Gas Chromatography. I. Preparation of Methyl Esters of Fatty Acid from Butter Fat," Original Abstr., Nippon Nogeikagaku Kaishi, 36, 361-364 (1962); Abstr., Agr. & Biol. Chem., 26, No. 5, A33 (1962); Biol. Abstr., 39, 22390

5121. Nakanishi, T., and Nakae, T., "Studies on the Fatty Acid Constituents of Butter Fat Produced in Japan by Gas Chromatography. II. Separation of Methyl Esters of the Lower Fatty Acids by Chromatography," Orig. Abstr., Nippon Nogeikagaku Kaishi, 36, 364-369 (1962); Abstr., Agr. & Biol. Chem., 26, No. 5, A33-A34 (1962); Biol. Abstr., 39, 22391

5122. Nakano, T., and Djerassi, C., "Terpenoids. XLVI. Copalic Acid," J. Org. Chem., 26, 167-173 (1961); CA 55, 18797e

5123. Napier, D. H., and Simonson, J. R., "Metering Valve for Capillary Chromatography Samples," Chem. & Ind. (London), 1962, 1831-1832; CA 58, 251c

5124. Napier, Jr., E. A., "2-Methylalkanoic Acids as Internal Standards in Gas Liquid Chromatographic Assay of Fatty Acids," Anal. Chem., 35, 1294-1295 (1963)

5125. Napier, I. M., and Rodda, H. J., "Multiple Fraction Collector for Gas Chromatography," Chem. & Ind. (London), 1958, 1319; CA 53, 5771f

5126. Narasimhachari, N., and Rudloff, E. v., "Gas-Liquid Chromatography of Some Flavonoid Compounds and Hydroxybiphenyls," Can. J. Chem., 40, 1123-1129 (1962); CA 57, 6588f

5127. Naro, P. A., "The Synthesis and Properties of Some Spiro Compounds," Univ. Microfilms (Ann Arbor, Mich.), L. C. Card No. Mic 60-2166, 203 pp.; Dissertation Abstr., 21, 58 (1960); CA 55, 2516b

5128. Naro, P. A., and Dixon, J. A., "The Synthesis and Dehydration of 7-Hydroxyspiro(5, 6)dodecane, a Neopentyl System," J. Org. Chem., 26, 1021-1024 (1961); CA 55, 19824f

5129. Nasini, A. G., Saini, G., Trossarelli, L., and Campi, E., "Polyethylidine from Diazomethane with Metal Catalysts: Chromatographic Analysis of Evolved Gases and Mechanisms of the Reaction," Mezhdunarod. Simpozium po Makromol. Khim., Doklady, Moscow, 1960, Sektsiya, 1, 38-46; CA 55, 8278i

5130. Natelson, S., and Stellate, R. L., "Apparatus for Extraction of Gases for Injection into the Gas Chromatograph. Application to Oxygen and Nitrogen in Jet Fuel and Blood," Anal. Chem., 35, 847-851 (1963)

5131. Naudet, M., and Derbesy, M., "On the Fusion Point of Some Fatty Esters of Glycols," Bull. soc. chim. France, 1963, 173

5132. Naudet, M., and Vezinet, P., "Composition of the Residues of Pyrolysis of Industrial Methyl Ricinoleate," Rev. franç. corps gras, 7, 385-391 (1960); CA 54, 21793h

5133. Naughton, J. J., Heald, E. F., and Barnes, Jr., I. L., "The Chemistry of Volcanic Gases. I. Collection and Analysis of Equilibrium Mixtures by Gas Chromatography," J. Geophys. Res., 68, 539-544 (1963); CA 58, 6591g

5134. Naves, Y. R., "Citronellol (Rhodinol) and Geraniol in Geranium and Rose Oils," Perfumery Essent. Oil Record, 48, 118-120 (1957); CA 51, 12439d

5135. Naves, Y. R., "Volatile Plant Substances. CXLV. Iso-α-irone and Iso-α-irols and Neo-iso-α-irols," Helv. Chim. Acta, 40, 1123-1129 (1957); CA 52, 5345d

5136. Naves, Y. R., "Determination of cis-Anethole in Essential Oils and in Preparations of Anethole," Compt. rend., 246, 1734-1736 (1958); CA 52, 14091h

5137. Naves, Y. R., "Analytical Distinctions Between Lavender and Lavandin Oils," Compt. rend., 246, 2163-2165 (1958); CA 52, 14979f

5138. Naves, Y. R., "Volatile Plant Materials. CLI. The Presence of Iso-alpha-Irone in the Essential Oil of Iris," Helv. Chim. Acta, 41, 653-657 (1958); CA 53, 1641c

5139. Naves, Y. R., "Nature of the Fixed Phase or of the Carrier in Gas-Liquid Partition Chromatography of Essential Oils and Aromatics," J. Soc. Cosmet. Chem., 9, No. 2, 101-103 (1958); Am. Perfumer Aromat., 71, No. 5, 38 (1958); CA 52, 10507g, 12330e

5140. Naves, Y. R., "Composition of Anethole," Parfums, cosmet., savons, 1, 219 (1958); CA 52, 15831g

5141. Naves, Y. R., "The Analysis of Rose Oils by Vapor-Phase Partition Chromatography and by Infrared Spectroscopy," Perfumery Essent. Oil Record, 49, 290-296 (1958); CA 52, 19020i

5142. Naves, Y. R., "Studies on Volatile Materials of Vegetable Origin. CLIII. Characteristics of Δ_3-Carene, Δ_4-Carene and the Caranes," Bull. soc. chim. France, 1959, 554-557; CA 53, 22057b

5143. Naves, Y. R., "Studies on Volatile Materials of Vegetable Origin. CLIX.(I). The Essential Oil of Carqueja from Santa Catarina, Brazil," Bull. soc. chim. France, 1959, 1871-1879; CA 54, 14297d

5144. Naves, Y. R., "Studies on Volatile Materials of Vegetable Origin. CLXIV.(I). Syntheses of Iso-tetrahydro-Carquejone (Ortho-menthene-1-one-3)," Bull. soc. chim. France, 1959, 1940-1942; CA 54, 14297i

5145. Naves, Y. R., "Volatile Plant Materials. CLV. Composition of Jasmin Absolute," Helv. Chim. Acta, 42, 1237-1238 (1959); CA 54, 3486a

5146. Naves, Y. R., "Studies on Volatile Vegetable Materials. CLVIII. Composition of the Absolute Essence of Ylang-Ylang," Helv. Chim. Acta, 42, 1692-1695 (1959); CA 54, 6037e

5147. Naves, Y. R., "Studies on Volatile Materials of Vegetable Origin. CLXI. Presence of Ledol in the Essential Oil of Carqueja," Helv. Chim. Acta, 42, 1996-1998 (1959); CA 54, 14298b

5148. Naves, Y. R., "Studies on Volatile Materials of Vegetable Origin. CLXIII. Presence of Borneol and Camphor, But Not of Bornyl Acetate, in Essential Oils of Lavender and Lavandin," Helv. Chim. Acta, 42, 2744-2746 (1959); CA 54, 25594d

5149. Naves, Y. R., "Vapor-Phase Partition Chromatography of Thyme Oils," France et ses parfums, 2, No. 8, 23-27 (1959); CA 53, 14424f

5150. Naves, Y. R., "Volatile Vegetable Materials. CLXXI. (-)-Lavandulal," Bull. soc. chim. France, 1960, 1741-1742; CA 55, 11766g

5151. Naves, Y. R., "Volatile Plant Materials. CLXV. Synthesis of cis-Anethole Starting from trans-Anethole," Helv. Chim. Acta, 48, 230-232 (1960); CA 54, 17450a

5152. Naves, Y. R., "The Essential Oils of Gingergrass (Cymbopogon Martini Variety Sofia) and Cymbopogon Densiflorus," Perfumery Essent. Oil Record, 51, 242-245 (1960); CA 54, 18891c

5153. Naves, Y. R., "Errors in Application of Instrumentation to the Analysis of Perfumery Raw Materials," J. Soc. Cosmet. Chemists, 14, 29-44 (1963); CA 58, 8847a

5154. Naves, Y. R., Ardizio, P., and Favre, C., "Studies on Volatile Plant Materials. CL. Isolation and Characterization of cis-Anethole," Bull. soc. chim. France, 1956, 566-569; CA 52, 17154c

5155. Naves, Y. R., Gottlieb, O. R., and Magalhaes, M. T., "Essential Oils. CLXXVII. Essential Oil of Ocotea Teleiandra," Helv. Chim. Acta, 44, 1121-1123 (1961); CA 55, 27783i

5156. Naves, Y. R., and Grampoloff, V. A., "Studies on Volatile Plant Materials. CLVI.(I). Cis- and trans-Isoeugenols and Their Methyl Esters," Bull. soc. chim. France, 1959, 1233-1237; CA 54, 16416g

5157. Naves, Y. R., and Grampoloff, V. A., "The Composition of the Essential Oil of the Fruit of Litsea Citrata," Compt. rend., 248, 2029-2031 (1959); CA 54, 13558b

5158. Naves, Y. R., and Grampoloff, V. A., "p-Menthadienols from the Essential Oils of Cymbopogon Martini Stapf Var. Sofia (Essence of Gingergrass) and C. Densiflorus Stapf," Compt. rend., 249, 307-308 (1959); CA 55, 20336a

5159. Naves, Y. R., and Grampoloff, V. A., "Studies on Volatile Materials of Vegetable Origin. CLX.(I). The p-Menthadienols in the Essential Oils of Cymbopogon Martini Stapf Var. Sofia (Essence of Gingergrass) and of C. Densiflorus Stapf," Bull. soc. chim. France, 1960, 37-46; CA 54, 17450d

5160. Naves, Y. R., and Grampoloff, V. A., "Essential Oils. CLXXV. The Products Obtained by Ozonolysis of (+)-3-Carene," Helv. Chim. Acta, 44, 637 (1961); CA 55, 27783f

5161. Naves, Y. R., Lamparsky, D., and Ochsner, P., "Essential Oils. CLXXIV. The Presence of Tetrahydropyrans in Oil of Geranium," Bull. soc. chim. France, 1961, 645; CA 55, 27783d

5162. Naves, Y. R., and Ochsner, P., "Volatile Plant Materials. CLXVI. Presence of β-Decahydroelsholtzione in the Essential Oil of Elsholtzia Oldhami Hemsl," Helv. Chim. Acta, 43, 406-410 (1960); CA 54, 17450c

5163. Naves, Y. R., and Ochsner, P., "Volatile Plant Substances. CLXVII. β-Dehydroelsholtzione and Some Derivatives of Elsholtzione," Helv. Chim. Acta, 43, 568-573 (1960); CA 54, 19632f

5164. Naves, Y. R., Ochsner, P., and Tullen, P., "Volatile Plant Materials. CLXVIII. Epoxides in the Essential Oil of Lavender," Helv. Chim. Acta, 43, 1616-1619 (1960); CA 55, 11766d

5165. Naves, Y. R., and Odermatt, A., "Volatile Plant Substances. CXLIX. Application of Gas-Liquid Partition Chromatography to the Analysis of Mixtures of Citrals and of Certain of Their Isomeric Derivatives," Bull. soc. chim. France, 1958, 377-384; CA 52, 12968a

5166. Naves, Y. R., and Odermatt, A., "Analysis of Essential Oils by Vapor-Liquid Partition Chromatography in General, and of Essential Oils of Citronella and Lemon Grass in Particular," Compt. rend., 247, 687-689 (1958); CA 53, 3608f

5167. Naves, Y. R., and Odermatt, A., "Application of Gas-Liquid Chromatography to the Determination of Eugenol in Various Essential Oils," France et ses parfums, 1, 10-16 (1958); CA 53, 13512d

5168. Naves, Y. R., and Odermatt, A., "Essential Oils. CLXXVI. Free Fatty Acids with Uneven Carbon Atom Numbers in the Essential Oil of Iris Root," Helv. Chim. Acta, 44, 999-1001 (1961); CA 55, 27783h

5169. Naves, Y. R., and Tucakov, J., "Presence of Anetholes in Jugoslavian Oil of Fennel," Compt. rend., 248, 843-845 (1959); CA 53, 14424i

5170. Naves, Y. R., and Tullen, P., "Volatile Plant Materials. CLXX. Presence of β-Myrcene, 3-Carene and (+)-1-Terpinen-4-ol in the Essential Oil of Lavender," Bull. soc. chim. France, 1960, 2123-2124; CA 55, 13777h

5171. Naves, Y. R., and Tullen, P., "Volatile Plant Materials. CLXIX. The Terpenes of the Essential Oil of Lavender," Helv. Chim. Acta, 43, 1619-1623 (1960); CA 55, 11766f

5172. Naves, Y. R., and Tullen, P., "Essential Oils. CLXXII. The Presence of (+)-Nopinene and Sabinene in the Essential Oil of Lavender," Helv. Chim. Acta, 43, 2150-2152 (1960); CA 58, 11165e

5173. Naves, Y. R., and Tullen, P., "Essential Oils. CLXXIII. Terpenes from the Essential Oil of Lavender: Ocimene, α-Pinene, Camphene," Helv. Chim. Acta, 44, 316-319 (1961); CA 55, 27783b

5174. Nawar, W. W., Cancel, L. E., and Fagerson, I. S., "Heat-Induced Changes in Milk Fat," J. Dairy Sci., 45, 1172-1177 (1962); CA 58, 2779b; 56th Annual Mtg., Am. Dairy Sci. Assoc., Madison, Wisc., June 1961; Abstr., J. Dairy Sci., 44, 1171 (1961)

5175. Nawar, W. W., and Fagerson, I. S., "Technique for Collection of Food Volatiles for Gas Chromatographic Analysis," Anal. Chem., 32, 1534-1535 (1960); CA 55, 5050b

5176. Nawar, W. W., and Fagerson, I. S., "Direct Gas Chromatographic Analysis as an Objective of Flavor Measurement," Food Technol., 16, No. 11, 107-109 (1962); CA 58, 6123f

5177. Nawar, W. W., Sawyer, F. M., Beltran, E. G., and Fagerson, I. S., "An Injection System for Gas Chromatography," Anal. Chem., 32, 1534 (1960); CA 55, 5050f

5178. Nawar, W. W., Sawyer, F. M., Fletcher, W., and Fagerson, I. S., "Application of Gas Chromatography to the Study of Volatiles Formed in Milk Fat Upon Heating," 55th Annual Mtg., Am. Dairy Sci. Assoc., Logan, Utah, June 1960; Abstr., J. Dairy Sci., 43, 839 (1960)

5179. Nayak, U. R., and Dev, S., "Longicyclene, the First Tetracyclic Sesquiterpene," Tetrahedron Letters, 1963, 243-246; CA 59, 670d

5180. Nazarov, I. N., Ivanova, L. N., and Rudenko, B. A., "Dehydrogenation of Unsymmetrical Methylisopropylethylene and Its Mixtures with Tetramethylethylene," Doklady Akad. Nauk S.S.S.R., 122, 242-245 (1958); CA 53, 2068f

5181. Nazarova, N. M., Freidlin, L. K., Shafran, R. N., and Litvin, E. F., "The Thermal Alkylation of Methylcyclohexane by Olefins Under Pressure," Neftekhimiya, 1, 613-618 (1961); CA 57, 3696f

5182. Nebbia, L., and Pagani, B., "Chromatographic Determination of Acetylene and Diacetylene in the Presence of Monosubstituted Acetylenes," Chim. et ind., 37, No. 3, 200-201 (1955); CA 49, 11503h

5183. Nedorost, M., "Polarographic Vessel for Continuous Gas Analysis," Chem. listy, 50, 317-318 (1956); CA 50, 6209d

5184. Neely, E. E., "Analysis of Chlorine Cell Gas by Gas Chromatography," Anal. Chem., 32, 1382-1383 (1960); CA 55, 4231e

5185. Neely, W. Brock, Nott, J., and Roberts, C. B., "Examination of Distribution of Substituents in Partially Methylated Cellulose by Gas Liquid Partition Chromatography," Anal. Chem., 34, 1423-1425 (1962); CA 57, 15392f

5186. Neeman, M., Caserio, M. C., Roberts, J. D., and Johnson, W. S., "Methylation of Alcohol with Diazomethane," Tetrahedron, 6, 36-47 (1959); CA 53, 17877a

5187. Negri, R. G., "Trends in Gas Chromatography," Afinidad, 18, 383-394 (1961); CA 57, 608a

5188. Nel, W., Mortimer, J., and Pretorius, V., "A Combustion-Thermal Conductivity Meter for Gas Chromatography," S. African Ind. Chemist, 13, 68-70 (1959); CA 53, 19476d

5189. Nel. W., Wet, W. J. de, and Pretorius, V., "Sample Inlet Systems for Gas Chromatography," S. African Ind. Chemist, 13, 44-45 (1959); CA 53, 18564f

5190. Neligan, R. E., Mader, P. P., and Chambers, L. A., "Exhaust Composition in Relation to Fuel Composition," J. Air Pollution Control Assoc., 11, 178-186 (1961); CA 55, 12719d

5191. Nelsen, F. M., and Eggertsen, F. T., "Determination of Surface Area. Adsorption Measurements by a Continuous Flow Method," Anal. Chem., 30, 1387-1390 (1958); CA 52, 16833c

5192. Nelsen, F. M., Eggertsen, F. T., and Holst, J. J., "Determination of Volatile Hydrocarbons in Aqueous Emulsions and Latexes by Gas Chromatography," Anal. Chem., 33, 1150-1152 (1961); CA 55, 27952d; 139th Natl. ACS Mtg., St. Louis, Mo., March 1961, Program Abstr., p. 33B

5193. Nelsen, F. M., and Groennings, S., "Determination of Carbon in Hydrogen Peroxide by Combustion-Gas Chromatography," Anal. Chem., 35, 660-663 (1963); CA 59, 1075e

5194. Nelson, D. C., and Paull, D. L., "Absolute Calibration of Gas Chromatographic Apparatus from Ion Chamber Measurements," 14th Pittsburgh Conf. on Anal. Chem. & Appl. Spectroscopy, Pittsburgh, Pa., March 1963, Program Abstr., p. 66

5195. Nelson, D. C., Ressler, Jr., P. C., and Hawes, R. C., "Performance of an Instrument for Simultaneous Gas Chromatographic and Radioactivity Analysis," 1st Annual Pacific Regional Mtg. for Appl. Spectroscopy and Anal. Chem., Pasadena, Calif., October 1962

5196. Nelson, D. F., and Kirk, P. L., "Identification of Substituted Barbituric Acids by Gas Chromatography of Their Pyrolysis Products," Anal. Chem., 34, 899-903 (1962); CA 57, 6025e

5197. Nelson, G. J., "Studies on Human Serum Lipoprotein Phospholipids and Phospholipid Fatty Acid Composition by Silicic Acid Chromatography," J. Lipid Research, 3, 71-79 (1962); CA 56, 14770b

5198. Nelson, G. J., "The Lipid Composition of Normal Mouse Liver," J. Lipid Research, 3, 256-262 (1962); CA 57, 6464d

5199. Nelson, G. J., and Freeman, N. K., "The Phospholipide and Phospholipide-Fatty Acid Composition of Human Serum Lipoprotein Fractions, " J. Biol. Chem., 235, 578-583 (1960); CA 54, 15604e

5200. Nelson, J. P., "Gas Chromatography of Selected Pregnenes and Pregnanes," J. Gas Chromatog., 1, No. 3, 27-29 (1963)

5201. Nelson, J. P., and Milun, A., "Gas Chromatography of High-Molecular-Weight Fatty Primary Amines," Chem. & Ind. (London), 1960, 663-664; CA 54, 23599e

5202. Nelson, J. P., and Milun, A. J., "Anomalous Response of a Beta-Ray Ionization Detector to Steroids," Chem. & Ind. (London), 1962, 1722-1723; CA 58, 9379h

5203. Nelson, K. H., Hines, W. J., Grimes, M. D., and Smith, D. E., "Gas Chromatographic Determination of C_6, C_7, and C_8 Olefins According to Their Carbon Structures," Anal. Chem., 32, 1110-1114 (1960); CA 54, 24131i

5204. Nelson, N. A., Fassnacht, J. H., and Piper, J. U., "Cycloheptatrienes from the Solvolysis of 1,4-Dihydrobenzyl-p-toluenesulfonates," J. Am. Chem. Soc., 83, 206-213 (1961); CA 55, 8371a

5205. Nelson, W. R., Werthessen, N. T., Holman, R. L., Hadaway, H., and James, A. T., "Changes in Fatty Acid Composition of Human Aorta Associated with Fatty Streaking," Lancet, 1961-I, 86-88; CA 55, 7640b

5206. Nemeckova, A., Janak, J., Pelikan, V., and Santavy, F., "Analysis of Intestinal Gases by Gas Chromatography," Ceskoslov. Fysiol., 10, 461-463 (1961); CA 56, 11943g

5207. Nenitzescu, C. D., Necsoiu, I., Glatz, A., and Zalman, M., "Aluminum Chloride Catalyses. A New Rearrangement of the Phenyl Alkanes," Chem. Ber., 92, 10-17 (1959); CA 53, 10079e

5208. Nerheim, A. G., "Gas-Liquid Chromathermography," Anal. Chem., 32, 436-437 (1960); CA 54, 12877i

5209. Nerheim, A. G., "A New Gas-Density Detector for Gas Chromatography," 137th Natl. ACS Mtg., Cleveland, Ohio, April 1960, Program Abstr., p. 24B

5210. Nesmeyanov, A. N., and Avodonina, E. N., "Determination of Amines by Gas Chromatography," Vestn. Mosk. Univ., Ser. II, Khim., 17, No. 5, 38-40 (1962); CA 58, 2849f

5211. Neubauer, N. R., Skreckoski, G., White, R. G., and Kane, A. J., "Gas Chromatographic Analysis of Free Toluene Diisocyanate in Adducts with Trimethylolpropane," 14th Pittsburgh Conf. on Anal. Chem. & Appl. Spectroscopy, Pittsburgh, Pa., March 1963, Program Abstr., p. 62

5212. Neudert, W., and Huber, J., "Double-Freezing-Point and Gas Chromatographic Methods for the Assay of DL-Menthol," Arch. Pharm., 295, 67-73 (1962); CA 56, 11710b

5213. Neumann, E. W., and Nadeau, H. G., "Analysis of Polyether and Polyolefin Polymers by Gas Chromatographic Determination of the Volatile Products Resulting from Controlled Pyrolysis," Anal. Chem., 35, 1454-1457 (1963); 14th Pittsburgh Conf. on Anal. Chem. & Appl. Spectroscopy, Pittsburgh, Pa., March 1963, Program Abstr., p. 68

5214. Neureiter, N. P., "Pyrolysis of 1,1-Dichloro-2-vinylcyclopropane. Synthesis of 2-Chlorocyclopentadiene," J. Org. Chem., 24, 2044-2046 (1959); CA 54, 10887c

5215. Neville, G., "Electron Affinity Chromatography," 10th Detroit Anachem Conf., Detroit, Mich., October 1962, Program Abstr., p. 34

5216. Newman, M. S., and Arkell, A., "New Reactions on Decomposition of a Hindered Alpha-Diazoketone," J. Org. Chem., 24, 385-387 (1959); CA 54, 9753a

5217. Neylan, D. N., and Phillips, T. R., "A Statistical Investigation of Factors Affecting Column Performance in the Chromatography of Inorganic Gases," 4th Intern. Gas Chromatography Symp., Hamburg, Germany, June 1962; Phillips, T. R., and Owens, D. R., "The Gas-Chromatographic Analysis of Inorganic Halogen Compounds on Capillary Columns,"Gas Chromatog. Proc. Symposium, 3rd, Edinburgh, 1960, 308-315; CA 56, 5385e

5218. Nichols, A. V., Rehnborg, C. S., and Lindgren, F. T., "Gas Chromatographic Analysis of Fatty Acids from Dialyzed Lipoproteins," J. Lipid Research, 2, 203-207 (1961)

5219. Nicolaides, N., "The Use of Silicone Rubber Gums or Grease in Low Concentration as Stationary Phase for the High Temperature Gas Chromatographic Separation of Lipids," J. Chromatog., 4, 496-499 (1960); CA 55, 15363f

5220. Nicolaides, N., "Gas Chromatographic Analyses of the Waxes of Human Skin Surface Fat," J. Invest. Dermatol., 37, 507-511 (1961); CA 56, 14764g

5221. Niegisch, W. D., and Stahl, W. H., "The Onion: Gaseous Emanation Products," Food Research, 21, 657-665 (1956); CA 53, 19213d

5222. Nield, E., Stephens, R., and Tatlow, J. C., "Fluorocyclohexanes. V. cis-1H,4H-trans-2H-cis--5H-, cis-1H,5H-trans-2H-cis-4H-, cis-1H,2H-trans-4H-cis-5H-, and cis-1H,2H,4H-trans-5H-Octafluorocyclohexane and Derived Compounds," J. Chem. Soc., 1959, 159-166; CA 53, 17004d

5223. Nield, E., Stephens, R., and Tatlow, J. C., "Aromatic Polyfluoro-Compounds. I. The Synthesis of Aromatic Polyfluoro-Compounds from Pentafluorobenzene," J. Chem. Soc., 1959, 166-171; CA 53, 9127c

5224. Nield, E., Stephens, R., and Tatlow, J. C., "Fluorocyclohexanes. VI. Some Hexa- and Pentafluorocyclohexenes and Their Dehydrofluorination," J. Chem. Soc., 1960, 3800-3806; CA 56, 7152b

5225. Nigam, I. C., and Levi, L., "Gas-Liquid Partition Chromatography of Sesquiterpene Compounds," Can. J. Chem., 40, 2083-2087 (1962); CA 58, 407h

5226. Nigam, I. C., and Levi, L., "Preparation and Isolation of Isomeric Ketones by the Girard Reaction," Anal. Chem., 35, 1087-1088 (1963)

5227. Nigam, I. C., and Levi, L., "Sabinene Hydrate and Sabinene Acetate: Two New Constituents of American Spearmint Oil," J. Agr. Food Chem., 11, 276 (1963)

5228. Nigam, I. C., Sahasrabudhe, M., Davis, T. W. Mc., Bartlet, J. C., and Levi, L., "Collection of Gas-Chromatographic Fractions [of Essential Oils] for Infrared Analysis," Perfumery Essent. Oil Record, 53, 614-615 (1962); CA 58, 3265g

5229. Nigam, I. C., Sahasrabudhe, M., and Levi, L., "Coupled Gas-Liquid-Thin-Layer Chromatography. Simultaneous Determination of Piperitone and Piperitone Oxide in Essential Oils," Can. J. Chem., 41, 1535-1539 (1963); CA 59, 379e

5230. Nigam, I. C., Skakum, W., and Levi, L., "Determination of Trace Constituents of Oil of Ajowan," Perfumery Essent. Oil Record, 54, 25-28 (1963); CA 58, 12365e

5231. Nightingale, C. F., and Walker, J. M., "Simultaneous Carbon-Hydrogen-Nitrogen Determination by Gas Chromatography," Anal. Chem., 34, 1435-1437 (1962); CA 57, 15805g

5232. Nikelly, J. G., "Programmed Temperature Gas Chromatography with Glass Microbeads. Quantitative Analysis of a Homologous Series of Alcohols and Hydrocarbons," Anal. Chem., 34, 472-475 (1962); CA 56, 13527d; 13th Pittsburgh Conf. on Anal. Chem. & Appl. Spectroscopy, Pittsburgh, Pa., March 1962, Program Abstr., p. 58

5233. Nikelly, J. G., "Gas Chromatography of Unesterfied Fatty Acids," 142nd Natl. ACS Mtg., Atlantic City, N. J., September 1962, Program Abstr., p. 11B

5234. Niklasch, F., "Gas Chromatography," Deut. Lebensm-Rundschau, 55, No. 3, 71 (1959)

5235. Nikolina, V. Y., Neimark, I. E., and Piontkovskaya, M. A., "Molecular Sieves," Uspekhi Khim., 29, 1088-1111 (1960); CA 55, 2952g

5236. Nikolinski, P., Mladenov, I., and Dimov, N., "Gas Chromatography in the Analysis of the Pyrolysis Products of Vulcanizates," Khim. i Ind. (Sofia), 33, No. 3, 69-72 (1961); CA 56, 3611i

5237. Nishi, S., "Gas-Chromatographic Determinations of Carbonyl Compounds by Pyrolysis of Their 2, 4-Dinitrophenylhydrazones with Dicarboxylic Acids," Bunseki Kagaku, 11, 415-420 (1962); CA 57, 1554i

5238. Nishimura, S., and Ichizuka, I., "Gas Chromatographic Analysis of Mixtures of Chlorinated Propanes and Prophylenes," Kogyo Kagaku Zasshi, 64, 780-783 (1961); CA 57, 4004a

5239. Nixon, A. C., and Thorpe, R. E., "Radiation Chemistry of Cyclohexane," J. Chem. Phys., 28, 1004-1005 (1958); CA 52, 15268g

5240. Noble, F. W., "An Ultrasonic Detector for Gas Chromatography," ISA Journal, 8, No. 6, 54-57 (1961); 13th Annual Conf. on Electrical Techniques in Medicine and Biology, Washington, D. C., October 1960, Program Abstr., pp. 12-13

5241. Nodop, G., "Gas Chromatographic Analysis of Very Pure Ethylene," Z. Anal. Chem., 164, 120-127 (1958); CA 53, 6887c

5242. Noebels, H. J., "Application of Gas Chromatography to Control Methods in Industry," Dechema Monograph., 35, No. 528-555, 105-113 (1959); CA 57, 3999e

5243. Noebels, H. J., "Application of Gas Chromatography to Process Control," Erdol u. Kohle, 13, 774-776 (1960)

5244. Noebels, H. J., Wall, R. F., and Brenner, N., (Eds.), "Gas Chromatography," Academic Press, Inc., New York, 1961. Proceedings of the Second International Symposium on Gas Chromatography, ISA, East Lansing, Mich., June 1959

5245. Noehren, T. H., and Cudmore, J. W., "Ethyl Ether Content in Blood as Determined by Gas Chromatography," Anesthesiology, 22, 519-524 (1961); CA 57, 15427g

5246. Noel, C. J., "The Thermal Degradation of Poly(Isopropyl Methacrylate)," 144th Natl. ACS Mtg., Los Angeles, Calif., March-April 1963, Program Abstr., p. 20Q

5247. Noguchi, S., "Gas Chromatography," Kagaku To Seibutsu, 1, No. 2, 48-55 (1963); CA 59, 1063d

5248. Noland, J. S., "Isolation and Identification of the Odorous Constituents of Corn – Alkyl N(2-Pseudo-thiouroniumethyl)carbamate Halides," Univ. Microfilms (Ann Arbor, Mich.), L. C. Card No. Mic 60-1567, 115 pp.; Dissertation Abstr., 20, 4527 (1960); CA 54, 17563h

5249. Nomura, T., and Nukada, M., "Life of Adsorption Columns in Process Gas Chromatography. An Attempt to Maintain the Separation Characteristics of Molecular Sieve 13X Columns," Kogyo Kagaku Zasshi, 64, 810-814 (1961); CA 57, 4944c

5250. Nonaka, M., and Pippen, E. L., "Application of Gas-Liquid Chromatography to the Study of Poultry Flavor," 137th Natl. ACS Mtg., Cleveland, Ohio, April 1960, Program Abstr., p. 13A

5251. Nonaka, M., and Pippen, E. L., "Gas Chromatography of Fried Chicken Volatiles: The Effect of Storage on Composition of the Volatile Fraction," 144th Natl. ACS Mtg., Los Angeles, Calif., March-April 1963, Program Abstr., p. 17A

5252. Norem, S. D., "A Combustion Device for Use in Conjunction with Chromatographic Columns," in

193

"Gas Chromatography," edited by V. J. Coates, H. J. Noebels, and L. S. Fagerson, Academic Press, Inc., New York, 1958, pp. 191-194

5253. Norem, S. D., "Behavior of Inert Gas Packets in Chromatographic Columns," Anal. Chem., 34, 40-42 (1962); CA 56, 9377i; 12th Pittsburgh Conf. on Anal. Chem. & Appl. Spectroscopy, Pittsburgh, Pa., February-March 1961, Program Abstr., p. 58

5254. Norikov, Y. D., "Gas-Liquid Chromatographic Analysis of Oxidation Products of Low-Molecular-Weight Hydrocarbons," Zavodskaya Lab., 27, No. 1, 28-30 (1961); CA 55, 26863i

5255. Norman, R. O. C., "2,4,7-Trinitrofluorenone as a Stationary Phase in Gas Chromatography," Proc. Chem. Soc. (London), 1958, 151; CA 52, 17892f

5256. Norman, R. O. C., and Radda, G. K., "The Ortho:Para-Ratio in Aromatic Substitution. I. The Nitration of Methyl Phenethyl Ether," J. Chem. Soc., 1961, 3030-3037; CA 55, 25826a

5257. Norman, V., Newsome, J. R., and Keith, C. H., "Vapor Phase Analysis of Tobacco Smoke," 145th Natl. ACS Mtg., New York, N. Y., September 1963

5258. Norris, T. G., and Crosser, O. K., "Non-Ideal Effects Upon Elution Curve Shape in Vapor Partition Chromatography," 132nd Natl. ACS Mtg., New York, N. Y., September 1957, Program Abstr., p. 31B; Preprints, Div. Petrol. Chem., 2, No. 4, D15-D26 (1957)

5259. Norris, W. P., "Preparation of Propene-1-d," J. Org. Chem., 24, 1579-1580 (1959); CA 54, 6511h

5260. Norrish, R. G. W., and Porter, K., "The Gas Phase Oxidation of n-Butenes," Proc. Roy. Soc. (London), Ser. A, 272, 164-191 (1963); CA 58, 11181e

5261. Norrish, R. G. W., and Purnell, J. H., "The Decomposition of n-Hexane. I. By Mercury Photosensitization," Proc. Roy. Soc. (London), 243, 435-448 (1958); CA 52, 10863c

5262. Northington, M., and Owens, G., "New Method for Blood-Ether Analysis," 5th Annual Gas Chromatography Institute, Canisius College, Buffalo, N. Y., April 1963

5263. Norton, C. J., "Catalytic Activities of Synthetic Molecular Sieves," Chem. & Ind. (London), 1962, 258-259; CA 56, 12339i

5264. Norton, J., and Moss, T. E., "Oxidative Dealkylation of Alkylaromatic Hydrocarbons. Naphthalene as Product in Vapor-Phase Air Oxidations of Monomethylnaphthalenes and Dimethylnaphthalenes Over Vanadia," Ind. Eng. Chem., Process Design Develop., 2, 140-147 (1963); CA 58, 8871f

5265. Notari, B., and Pines, H., "Relative Side-Chain Alkylation of 2- and 4-Alkylpyridines," J. Am. Chem. Soc., 82, 2945-2948 (1960); CA 55, 7415g

5266. Novak, J., "Simple Preparation of Definite Gaseous Mixtures for Calibration in Gas-Chromatographic Analysis of Trace Amounts of Compounds," Chem. listy, 54, 1189-1193 (1960); CA 55, 8147e

5267. Novak, J., "Elimination of the Pressure Oscillations of Carrier Gas in Gas Chromatography," Collection Czechoslov. Chem. Commun., 27, 411-423 (1962); CA 56, 15310d

5268. Novak, J., and Janak, J., "The Nonlinearity of the Signal Response and the Inversion Effect in Flame Ionization Detection," J. Chromatog., 4, 249-251 (1960); CA 55, 3274i

5269. Novak, J., and Janak, J., "Effectivity of Ionization in the Flame Ionization Detector," Gas Chromatog. Symp., Brno, Czech., June 1962; Abstr., J. Gas Chromatog., 1, No. 4, 6 (1963)

5270. Novak, J., and Janak, J., "Operating Parameters of the High-Temperature Gas Chromatograph Chrom-2 with Flame Ionization Detector," Gas Chromatog. Symp., Brno, Czech., June 1962; Abstr., J. Gas Chromatog., 1, No. 4, 8 (1963)

5271. Novak, J., and Rusek, M., "Working Properties of the High Temperature Chromatographic Apparatus with the Flame Ionization Detector Made by 'Laboratorni Pristroje n.p.'," 3rd Intern. Conf. on Anal. Chem., Prague, Czech., September 1959

5272. Novak, J., Rusek, M., and Janak, J., "Working Properties of a High-Temperature Gas Chromatograph with Flame-Ionization Detector 'Chrom I'," Chem. listy, 54, 1173a-1183 (1960); CA 55, 9974h

5273. Novotny, L., and Herout, V., "Plant Substances. XIX. The Constituents of Petasites Spurius Rhizomes," Collection Czechoslov. Chem. Communs., 27, 2462-2464 (1962); CA 58, 749h

5274. Nowak, P., and Klemm, H., "Determination of Volatile Hydrocarbons in Polystyrene Containing Expanding Agents," Kunstoffe, 52, 604-605 (1962); CA 58, 2543e

5275. Nowakowska, J., Melvin, E. H., and Wiebe, R., "Separation of Esters of Mono- and Dicarboxylic Acids by Gas Liquid Chromatography," 30th Fall Mtg., Am. Oil Chemists' Soc., Chicago, Ill., September 1956; Abstr., Anal. Chem., 28, 2027 (1956)

5276. Nowakowska, J., Melvin, E. H., and Wiebe, R., "Separation of the Oxidation Products of Fatty Acids by Means of Gas-Liquid Partition Chromatography," J. Am. Oil Chemists' Soc., 34, 411-414 (1957); CA 51, 15149g

5277. Nunez, L. J., Armstrong, W. H., and Cogswell, H. W., "Analysis of Hydrocarbon Blends by Gas-Liquid Partition Chromatography," Anal. Chem., 29, 1164-1165 (1957); CA 51, 15930e

5278. Nunziata, L., Rodrigo, A. R., and Parada, J. F., "Oxygen Analyzer," Semana Med. (Buenos Aires), 1954, II, 847-848; CA 49, 2785h

5279. Nuss, G., "Process Gas Chromatography Applications of Current Interest," Analyzer, 3, No. 3, 16 (July 1962)

5280. Nuttall, R. L., and Ginnings, D. C., "Thermal Conductivity of Nitrogen from 50° to 500°C and 1 to 100 Atmospheres," J. Research NBS, 58, 271-278 (1957); CA 51, 11834d

5281. Nystrom, R. F., and Berger, C. R. A., "Separation of Allylic Bromides Without Isomerization by Gas Chromatographic Techniques," Chem. & Ind. (London), 1958, 559-560; CA 53, 2069d

5282. Nystrom, R. F., and Brown, W. G., "Reduction of Organic Compounds by Lithium Aluminum Hydride. I. Aldehydes, Ketones, Esters, Acid Chlorides, and Acid Anhydrides," J. Am. Chem. Soc., 69, 1197-1199 (1947); CA 41, 4772g

5283. Nystrom, R. F., and Brown, W. G., "Reduction of Organic Compounds by Lithium Aluminum Hydride. II. Carboxylic Acids," J. Am. Chem. Soc., 69, 2548-2549 (1947); CA 42, 111a

5284. Nystrom, R. F., Mason, L. H., Jones, E. P., and Dutton, H. J., "Labeling Fatty Acids by Exposure to Tritium Gas," J. Am. Oil Chemists' Soc., 36, 212-214 (1959); CA 53, 12710c

5285. Oae, S., Kitao, T., and Kitaoka, Y., "Rearrangements of Tertiary Amine Oxides. IV. Mechanism of the Reaction of 4-Picoline N-Oxide with Acetic Anhydride," J. Am. Chem. Soc., 84, 3362-3365 (1962); CA 58, 1328h

5286. Ober, S. S., "The Interrelationship of Column Efficiency and Resolving Power in Gas Chromatography," in "Gas Chromatography," edited by V. J. Coates, H. J. Noebels, and I. S. Fagerson, Academic Press, Inc., New York, 1958, pp. 41-50

5287. Obolentsev, R. D., and Aivazov, B. V., "The Separation of Hydrocarbon Mixtures and Organic Sulfur Compounds by Vapor-Phase Chromatography," Khim. Sera-Org. Soedinenii, Soderzhashch. v Neft i Nefteprodukt, Akad. Nauk S.S.S.R. Bashkir. Filial Doklady 3-ei (Tret'ei) Nauch. Sessii, Ufa, 1957, 110-124 (Pub. 1959); CA 55, 3041g

5288. O'Brien, L., and Scholly, P. R., "Gas Chromatographic Separation of Meta- and Para-Xylenes in Aromatic Mixtures," Nature, 181, 1794 (1958)

5289. Ocker, H. D., "Volatile Flavor Constituents of Bread Investigated by Gas Chromatography," Ber. Getreide-Chemiker Tagung Detmold, 1961, 172-182; CA 56, 9170e

5290. O'Connor, J. G., Burow, F. H., and Norris, M. S., "Determination of Normal Paraffins in C_{20} to C_{32} Paraffin Waxes by Molecular Sieve Adsorption. Molecular Weight Distribution by Gas-Liquid Chromatography," Anal. Chem., 34, 82-85 (1962); CA 56, 9001h

5291. O'Connor, J. G., and Norris, M. S., "Molecular Sieve Adsorption. Application to Hydrocarbon Type Analysis," Anal. Chem., 32, 701-706 (1960); CA 54, 13615c

5292. Ode, W. H., "Solid and Gaseous Fuels. Industrial Applications of Analysis, Control, and Instrumentation," Anal. Chem., 29, 657-669 (1957); CA 51, 7219a

5293. Ode, W. H., and Christos, T., "Solid and Gaseous Fuels," Anal. Chem., 33, No. 5, 61R-66R (1961); CA 55, 12137f

5294. Oehlmann, G., Schroder, E., and Leibnitz, E., "The Formation of Ketones in the Gas Phase Oxidation of n-Heptane at Low Temperature Area," Chem. Ber., 93, 2567-2572 (1960); CA 55, 4347e

5295. Oette, K., and Ahrens, Jr., E. H., "Quantitative Gas-Liquid Chromatography of Short-Chain Fatty Acids as 2-Chloroethanol Esters," Anal. Chem., 33, 1847-1850 (1961); CA 56, 8007h

5296. Ofner, A., Kimel, W., Holmgren, A., and Forrester, F., "The Synthesis of Nerolidol and Related C_{15} Alcohols," Helv. Chim. Acta, 42, 2577-2584 (1959); CA 54, 13166a

5297. Oganesyan, G. A., "Acoustic Method of Gas Analysis," Primenie Ul'traakustiki k Issledovan Veshchestva, 1956, No. 3, 139-145; CA 51, 5635h

5298. Ogawa, M., and Hirose, Y., "Chemical Structure of Some Furan Derivatives Isolated from Sweet Potato Fusel Oil," Nippon Kagaku Zasshi, 83, 747-748 (1962); CA 59, 1565b

5299. Ogilvie, J. L., Simmons, M. C., and Hinds, Jr., G. P., "Exploratory Studies of High Temperature Gas-Liquid Chromatography," Anal. Chem., 30, 25-27 (1958); CA 52, 7925g; 131st Natl. ACS Mtg., Miami, Fla., April 1957, Program Abstr., p. 19Q

5300. Ognyanov, I., and Vlakhov, R., "Terpene Hydrocarbons in Bulgarian Peppermint Oil," Compt. Rend. Acad. Bulgare Sci., 14, 459-462 (1961); CA 56, 7450e

5301. Ohkoshi, S., Fujita, Y., and Kwan, T., "Gas Chromatographic Separation of Hydrogen Isotopes D_2 and HD," Bull. Chem. Soc. Japan, 31, 770-771 (1958); CA 53, 13869g

5302. Ohkoshi, S., Tenma, S., Fujita, Y., and Kwan, T., "Gas Chromatography as a New Tool for Analyses of Hydrogen Isotopes," Bull. Chem. Soc. Japan, 31, 772 (1958); CA 53, 9883i

5303. Ohkoshi, S., Tenma, S., Fujita, Y., and Kwan, T., "Enrichment of Deuterium at Low Temperature Gas Adsorption Chromatography," Bull. Chem. Soc. Japan, 31, 773-774 (1958); CA 53, 9842f

5304. Ohline, R. W., "Chromathermography – The Application of Moving Thermal Gradients to Gas-Liquid Partition Chromatography," Univ. Microfilms (Ann Arbor, Mich.), L. C. Card Mic 60-4783, 103 pp.; Dissertation Abstr., 21, 1352-1353 (1960); CA 55, 6990a

5305. Ohline, R. W., and DeFord, D. D., "Chromathermography, the Application of Moving Thermal Gradients to Gas Liquid Partition Chromatography," Anal. Chem., 35, 227-234 (1963); CA 58, 8392a

5306. Ohloff, G., "Thermal Reactions. I. Pyrolysis of (+)-3-Hydroxymethyl-Δ^4-carene," Chem. Ber., 93, 2673-2681 (1960); CA 55, 4567e

195

5307. Okey, R., Lee, M., Hampton, M. C., and Miljanich, P., "Effect of Safflower and Coconut Oil upon Plasma Cholesterol and Lipid Fractions," Metabolism. 9, 791-799 (1960)

5308. Okuno, L, Morris, J. C., and Haines, W. E., "Microdetermination of Sulfur by Hydrogenation and Gas Chromatography," Anal. Chem., 34, 1427-1431 (1962); CA 58, 21h; 140th Natl. ACS Mtg., Chicago, Ill., September 1961, Program Abstr., p. 26B

5309. Olah, G. A., and Kuhn, S. J., "Organic Fluorine Compounds. XXVII. Preparation of Acyl Fluorides with Anhydrous Hydrogen Fluoride," J. Org. Chem., 26, 237-238 (1961); CA 55, 19779c

5310. Olah, G. A., and Kuhn, S. J., "Aromatic Substitution. XII. Steric Effects in Nitronium Salt Nitrations of Alkylbenzenes and Halobenzenes," J. Am. Chem. Soc., 84, 3684-3687 (1962); CA 58, 7799c

5311. Olah, G. A., and Tolgyesi, W. S., "Thermal Decomposition and Electrophilic Arylations with Aryldiazonium Tetrachloroborates and Tetrabromoborates. Mechanism of the Schiemann Reaction," J. Org. Chem., 26, 2053-2055 (1961); CA 55, 24615f

5312. Olah, K., and Schay, G., "Progress and Shape of Gas Chromatographic Front," Acta Chem. Acad. Sci. Hung., 14, 453-470 (1958); CA 53, 17630e

5313. Olander, C. J., and Sunner, S., "Equilibrium Studies on Disulfides Using Vapor-Phase Chromatography," Pure Appl. Chem., 2, 117-120 (1961); CA 56, 2034c

5314. Oldenkamp, R. D., "Some Theoretical and Experimental Aspects of Gas-Solid Chromatography," Univ. Microfilms (Ann Arbor, Mich.), Order No. 62-5117, 253 pp.; Dissertation Abstr., 23, 2046 (1962); CA 58, 7345f

5315. Oldenkamp, R. D., and Houghton, G., "An Experimental Study of Adsorption Chromatography with a Nonlinear Isotherm Using the System Isobutylene-Activated Alumina," J. Phys. Chem., 67, 597-600 (1963); CA 58, 8429a

5316. Oldfield, J. F. T., "Formation of Volatile Organic Acids in Factory Juices and the Estimation of These Acids by Gas-Liquid Chromatography," Compt. rend. assemblée comm. intern. tech. sucrerie, 10, London, 1957, 56-66; CA 55, 9917h

5317. Olive, T. R., and Danatos, S., "Guide to Process Instrument Elements," Chem. Eng., 64, No. 6, 287-320 (1957); CA 51, 11775b

5318. Oliver, T., Smith, J. C., and Fenning, C. M., "Isopropenyl and Isopropylidene Groups. The Terpene Problem," Chem. & Ind. (London), 1959, 1575; CA 54, 8589e

5319. Olivier, K. L., and Young, W. G., "Allylic Rearrangements. XLVI. The Thermal Decomposition of the Butenyl Chloroformates," J. Am. Chem. Soc., 81, 5811-5817 (1959); CA 54, 7543h

5320. Olson, A. C., "Isomerization in the Alkylation of Aromatics with n-1-Olefins," 136th Natl. ACS Mtg., Atlantic City, N. J., September 1959, Program Abstr., p. 3R; Preprints, Div. Petrol. Chem., 4, No. 3, 89-95 (1959)

5321. Olson, A. C., "Alkylation of Aromatics with 1-Alkenes," Ind. Eng. Chem., 52, 833-836 (1960); CA 55, 7335e

5322. O'Neill, H. J., Cutscher, R. E., Dynako, A., and Boquist, C., "Pyrolysis Studies of Furfuryl Alcohol Resins by Gas Chromatography," J. Gas Chromatog., 1, No. 2, 28-35 (1963)

5323. O'Neill, H. J., and Gershbein, L. L., "Determination of Cholesterol and Squalene by Gas Chromatography," Anal. Chem., 33, 182-185 (1961); CA 55, 13532a

5324. O'Neill, L. A., and Rybicka, S. M., "Examination of Unsaturated Glyceride Oils by Gas Chromatography," Chem. & Ind. (London), 1963, 390-392; CA 58, 10413e

5325. Onely, J. H., and Mills, P. A., "Detection and Estimation of Chlorinated Pesticides in Eggs," J. Assoc. Offic. Agr. Chemists, 45, 983-987 (1962); CA 58, 6125c

5326. Ongkiehong, L., "Investigation of the Hydrogen Flame Ionization Detector," Gas Chromatog. Proc. Symposium, 3rd, Edinburgh, 1960, 7-15; CA 56, 3305i

5327. Ongkiehong, L., "Investigation of the Hydrogen Flame Ionization Detector," in "Gas Chromatography 1960," edited by R. P. W. Scott, Butterworths, London, 1960, pp. 7-15

5328. Ookuni, L, and Fry, A., "Disproportionation in Acid-Catalyzed Ketone Rearrangements," Tetrahedron Letters, 1962, 989-992; CA 58, 8867e

5329. Orlando, Jr., C. M., and Weiss, K., "Chemistry of Cycloheptatriene. III. Carbon-Carbon Bond Formation in Catalytic Hydrogenolysis: Differences in Behavior of Adsorbed Isomeric C_7H_7 Radicals," J. Org. Chem., 27, 4714-4715 (1962); CA 58, 11195a

5330. Ormerod, E. C., and Scott, R. P. W., "Gas Chromatography of Polar Solutions with a Nonpolar Liquid Phase," J. Chromatog., 2, 65-68 (1959); CA 53, 21362a

5331. Oro, J. F., Guidry, C. L., and Zlatkis, A., "The Odor of Methional Purified by Gas Chromatography," Food Research, 24, 240-241 (1959)

5332. Orr, C. H., "Recorder-Integrator Errors in Gas Chromatography Area Measurements," Anal. Chem., 33, 158-159 (1961); CA 55, 6241b

5333. Orr, C. H., and Callen, J. E., "Separation of Polyunsaturated Fatty Acid Methyl Esters by Gas Chromatography," J. Am. Chem. Soc., 80, 249 (1958); CA 52, 7032f

5334. Orr, C. H., and Callen, J. E., "Gas Chromatographic Separation of Methyl Esters of Fatty Acids,"

Ann. N. Y. Acad. Sci., 72, 649-665 (1959); CA 53, 14599i

5335. Osborne, A. D., and Skirrow, G., "The Gas Phase Oxidation of Crotonaldehyde," J. Chem. Soc., 1960, 2750-2758; CA 54, 20853h

5336. Oscik, J., and Waksmundzki, A., "Adsorption Selectivity in Adsorption Chromatography," Ann. Univ. Mariae Curie-Sklodowska, Lublin-Polonia, Sect. AA, 9, 9-34 (1954)(Pub. 1956); CA 51, 5501f

5337. Oshima, S., Katsumata, A., and Dan, T., "Mass Spectrometric Analysis of Hydrocarbons Separated by Gas Chromatography," Kogyo Kagaku Zasshi, 62, 646-649 (1959); CA 57, 7882d

5338. Oster, H., "Conditions for Attaining Highly Accurate Measurements in Gas Chromatographic Analysis," German Chemical Society Mtg., Div. Anal. Chem., Freiburg, West Germany, April 1959

5339. Oster, H., "Attainment of High Accuracy in Gas Chromatographic Analysis," Z. Anal. Chem., 170, 264-271 (1959); CA 54, 3043i

5340. Oster, H., "Precision Gas Chromatograph for the Works Laboratory," Siemens-Z., 33, No. 3, 137-144 (1959); CA 54, 13759f

5341. Ostman, B., "Orientation in the Nitration of 2-Nitrothiophene," Arkiv Kemi, 19, No. 37, 527-529 (1962); CA 58, 5611a

5342. Oswald, A. A., Griesbaum K., Thaler, W. A., and Hudson, Jr., B. E., "Organic Sulfur Compounds. VIII. Addition of Thiols to Conjugated Diolefins," J. Am. Chem. Soc., 84, 3897-3904 (1962); CA 58, 6684a

5343. Ott, D. E., and Gunther, F. A., "Forced Volatilization Cleanup of Butterfat for Gas Chromatographic Evaluation of Organochlorine Insecticide Residues," 144th Natl. ACS Mtg., Los Angeles, Calif., March-April 1963, Program Abstr., p. 7A

5344. Ottenstein, D. M., "The New Fisher Gas Analyzer Model 25," 1st Annual Gas Chromatography Institute, Canisius College, Buffalo, N. Y., May 1959

5345. Ottenstein, D. M., "The Influence of the Chromatographic Support on the Separation of Polar Compounds," 7th Detroit Anachem Conf., Detroit, Mich., October 1959

5346. Ottenstein, D. M., "The Fisher Gas Partitioner," Fisher Scientific Co., Pittsburgh, Pa., 1960

5347. Ottenstein, D. M., "The Packed Chromatographic Column," in "Progress in Industrial Gas Chromatography," Vol. 1, edited by H. A. Szymanski, Plenum Press, Inc., New York, 1961, pp. 51-72

5348. Ottenstein, D. M., "Analysis of Blast Furnace Gases," 9th Detroit Anachem Conf., Detroit, Mich., October 1961, Program Abstr., p. 27

5349. Ottenstein, D. M., "Analysis of Fixed Gases, Hydrocarbons and Related Compounds by Gas Chromatography," 13th Pittsburgh Conf. on Anal. Chem. & Appl. Spectroscopy, Pittsburgh, Pa., March 1962, Program Abstr., p. 53

5350. Ottenstein, D. M., "Column Supports and Substrates," 5th Annual Gas Chromatography Institute, Canisius College, Buffalo, N. Y., April 1963

5351. Ottenstein, D. M., "Column Support Materials for Use in Gas Chromatography," J. Gas Chromatog., 1, No. 4, 11-23 (1963)

5352. Ottenstein, D. M., "Solid Supports and the Gas Phase," 14th Annual Fisk Infrared Spectroscopy Inst., Gas Chromatog. Session, Nashville, Tenn., August 1963

5353. Otto, K., and Doubek, M., "Gas-Chromatography Apparatus," Chem. prumysl, 10, 476-478 (1960); CA 55, 5050h

5354. Otto, P. P. H. L., and Zanten, B. van, "Synthesis of 1-Pentene-4-C^{14} and 1-Pentene-5-C^{14}," Rec. Trav. Chim., 81, 380-384 (1962); CA 58, 2354g

5355. Otvos, J. W., and Stevenson, D. P., "Cross-Sections of Molecules for Ionization by Electrons," J. Am. Chem. Soc., 78, 546-551 (1956); CA 50, 7577c

5356. Overberger, C. G., and Borchert, A. E., "Ionic Polymerization. XVI. Reactions of 1-Cyclopropylethanol-Vinylcyclopropane," J. Am. Chem. Soc., 82, 4896-4899 (1960); CA 55, 10342h

5357. Overberger, C. G., and Halek, G. W., "Monomers and Polymers. A Synthesis of Vinyl Cyclopropane and Dicyclopropyl," J. Org. Chem., 28, 867-868 (1963); CA 58, 11225d

5358. Owens, E. J., and Thodos, G., "Thermal Conductivity-Reduced-State. Correlation for the Inert Gases," A.I.Ch.E. Journal, 3, 454-461 (1957); CA 52, 3438i

5359. Owens, E. J., and Thodos, G., "Thermal Conductivity: Reduced-State Correlation for Ethylene and Its Application to Gaseous Aliphatic Hydrocarbons and Their Derivatives at Moderate Pressures," A.I.Ch.E. Journal, 6, 676-681 (1960); CA 56, 8031d

5360. Owens, P. J., Long, R., and Garner, F. H., "Benzene Formation During Oil Gasification in Hydrogen," Ind. Eng. Chem., 53, 10-14 (1961); CA 56, 3726i

5361. Ozeris, S., and Bassette, R., "Quantitative Study of Gas Chromatographic Analysis of Head Space Gas of Dilute Aqueous Solutions," Anal. Chem., 35, 1091 (1963)

5362. Oziraner, S. N., Gaziev, G. A., Yanovskii, M. I., and Kornyakov, V. S., "Ionization Detector with Promethium-147 for Gas Chromatography," Zavodskaya Lab., 25, 760-761 (1959); cf. Ind. Lab., 25, 791-792 (1959)

5363. Oziraner, S. N., Gaziev, G. A., Yanovskii, M. I., and Kornyakov, V. S., "An Ionization Detector for

Gas Chromatography Based on Promethium-147," Gazovaya Khromatografiya, Trudy 1-oi (Pervoi) Vsesoyuz. Konf., Akad. Nauk S.S.S.R., Moscow, 1959, 199-203 (Pub. 1960); CA 56, 913g

5364. Oziraner, S. N., Gaziev, G. A., Yanovskii, M. I., Kornyakov, V. S., and Kapshaninov, Y. I., "Use of Pm[147] in a Highly Sensitive Ionization Gas Analyzer, "Radioaktiv. Izotopy i Yadernye Izlucheniya v Narod, Khoz. S.S S.R., Tr. Vses. Soveshch. Riga, 1960, No. 1, 278-282 (1961); CA 57, 7e

5365. Paasivirta, J., "1-Methylnorticyclene Compounds," Ann. Acad. Sci. Fennicae, Ser. A II, 116, 59 pp. (1962); CA 59, 464g

5366. Paglis, J. P., "Determination of Dissolved Oxygen in Petroleum Liquids," in "Gas Chromatography," edited by H. J. Noebels, R. F. Wall, and N. Brenner, Academic Press, Inc., New York, 1961, pp. 351-355

5367. Pain, J., Hugel, M. F., and Barbier, M., "On the Constituents of the Attractive Mixture of the Mandibular Glands of Queen Bees (Apis Mellifica) at Different Stages of Their Lives," Compt. rend., 250, 1046-1048 (1960); CA 55, 4808h

5368. Pallotta, U., "Characterization of the Methyl Esters of the Higher Saturated Fatty Acids by Infrared Spectroscopy and Their Separation by Vapor-Phase Chromatography," Riv. ital. sostanze grasse, 38, 191-197 (1961); CA 55, 23322d

5369. Palmer, P. E., and Weaver, E. R., "Thermal Conductivity Method for Gas Analysis," U.S. Bur. Standards Tech Papers 18, No. 249, 100 pp. (1924); CA 18, 1099

5370. Palmer, R. C., Davis, D. K., and Van Willis, W., "Solenoid-Operated Gas Sampler for Use in Gas Chromatography," Anal. Chem., 32, 894-895 (1960); CA 54, 19031d

5371. Panetti, M., and Musso, G., "Apparatus for the Determination of the Partial Vapor Pressures of the Components of a Mixture by Vapor Phase Chromatography," Ann. chim. (Rome), 52, 472-481 (1962); CA 57, 13569g

5372. Paoletti, R., Paoletti, P., and Garattini, S., "Incorporation of Different Precursors in Conditions of Increased Cholesterol and Fatty Acid Biosynthesis," Biochem. J., 71, No. 1,4P (1959); CA 53, 4468h

5373. Paolini, F., and Pascucci, E., "Anomalous Properties Characteristic of Olive Oils. Olive Oils from Tunis," Riv. ital. sostanze grasse, 38, 265-274 (1961); CA 56, 1543i

5374. Paolino, V. J., "Gas Chromatography as Applied to the Industrial Separation of Nitrogen and Helium from Neon," Proc. Cryogenic Eng. Conf., 2nd, Boulder, Colo., Sept. 1956; pub. in "Advances in Cryogenic Engineering," Vol. 2, edited by K. D. Timmerhaus, pp. 197-202, Plenum Press, N. Y. (1957); CA 52, 19034a

5375. Paolucci, G., Salvioli, Jr., G. P., Corsini, F., Babini, B., and Manfredi, G., "Gas-Chromatographic Determination of Plasma and Erythrocyte Fatty Acids in Nephrosis," Boll. Soc. Ital. Biol. Sper., 38, 732-734 (1962); CA 58, 6075c

5376. Papariello, G. J., Slack, S. C., and Mader, W. J., "Pharmaceuticals and Related Drugs," Anal. Chem., 33, 113R-126R (1961); CA 55, 12137g

5377. Paquot, C., "Some New Techiques Used in Fat Analysis," Bull. assoc. franç. chim. ind. cuir et doc. sci. et tech. ind. cuir, 24, 221-230 (1962); CA 57, 13903b

5378. Paquot, C., LeFort, D., and Pourchez, A., "Gas Chromatography and Lipochemistry. I. Analysis of Fatty Alcohols," Rev. franç. corps gras, 7, 391-395 (1960)

5379. Parish, R. V., and Parsons, W. H., "The 'Helium Detector' for Gas Chromatography," Chem. & Ind. (London), 1961, 1951; CA 56, 4552c

5380. Park, J. D., Dick, J. R., and Lacher, J. R., "Reaction of 1-Ethoxy-2-chloro-3,3,4,4-tetrafluorocyclo-butene with Alkoxide Ion; Evidence for the SN2' Mechanism," J. Org. Chem., 28, 1154-1155 (1963)

5381. Park, J. D., Groves, J. D., and Lacher, J. R., "Synthesis of Fluorine Containing Organosilanes," J. Org. Chem., 25, 1628-1632 (1960); CA 55, 8310h

5382. Park, J. D., Rogers, F. E., and Lacher, J. R., "Free Radical Catalyzed Addition of Unsaturated Alcohols to Perhaloalkanes," J. Org. Chem., 26, 2089-2095 (1961); CA 55, 24552b

5383. Parker, K. D., Fontan, C. R., and Kirk, P. L., "Separation and Identification of Tranquilizers by Gas Chromatography," Anal. Chem., 34, 757-760 (1962); CA 57, 3566c

5384. Parker, K. D., Fontan, C. R., and Kirk, P. L., "Separation and Identification of Some Sympathomi-metic Amines by Gas Chromatography," Anal. Chem., 34, 1345-1346 (1962); CA 57, 12638i

5385. Parker, K. D., Fontan, C. R., and Kirk, P. L., "Rapid Gas Chromatographic Method for Screening of Toxicological Extracts for Alkaloids, Barbiturates, Sympathomimetic Amines, and Tranquilizers," Anal. Chem., 35, 356-359 (1963)

5386. Parker, K. D., Fontan, C. R., and Kirk, P. L., "Improved Gas Chromatographic Column for Barbi-turates," Anal. Chem., 35, 418-419 (1963); CA 58, 10039h

5387. Parker, K. D., Fontan, C. R., Yee, J. L., and Kirk, P. L., "Gas Chromatographic Determination of Ethyl Alcohol in Blood for Medicolegal Purposes. Separation of Other Volatiles from Blood or Aqueous Solution," Anal. Chem., 34, 1234-1236 (1962); CA 57, 12811d

5388. Parker, K. D., and Kirk, P. L., "Separation and Identification of Barbiturates by Gas Chromatogra-phy," Anal. Chem., 33, 1378-1381 (1961); CA 56, 1699h

5389. Parker, L. F., and Irwin, G. A., "Canadian Petrofina Moves Toward Optimum Production. Process Gas Chromatograph Improves Operation of Refinery's Alkylation Process," Analyzer, 4, No. 3, 8-10 (July 1963)

5390. Parker, W. W., and Hudson, R. L., "A Simplified Chromatographic Method for Separation and Identification of Mixed Lead Alkyls in Gasoline," Anal. Chem., 35, 1334-1335 (1963)

5391. Parker, W. W., Smith, G. Z., and Hudson, R. L., "Determination of Mixed Alkyls in Gasoline by Combined Gas Chromatographic and Spectrophotometric Techniques," Anal. Chem., 33, 1170-1171 (1961); CA 56, 6237c

5392. Parks, E. J., and Linnig, F. J., "Natural and Synthetic Rubbers," Anal. Chem., 35, No. 5, 160R-178R (1963)

5393. Parks, J. C., and Hinkle, E. A., "A New Calibration Method for Process Gas Chromatographs," 8th Natl. Analysis Instrumentation Symp., ISA, Charleston, W. Va., April-May 1962

5394. Parks, M. L., "Gas Chromatography: An Automatic Instrument for Hazardous Areas," Oil Gas J., 54, No. 85, 136, 138, 140 (1956)

5395. Parks, O. W., Keeney, M., and Schwartz, D. P., "Bound Aldehydes in Butteroil," J. Dairy Sci., 44, 1940-1943 (1961)

5396. Parriss, W. H., and Holland, P. D., "New Uses for Gas-Liquid Chromatography in Plastics," Brit. Plastics, 33, 372-375 (1960); CA 54, 23406i

5397. Parry, R. M., Sampugna, J., and Jensen, R. M., "Some Effects of Feeding Safflower Oil on the Fatty Acid Composition of Milk Fat," 58th Annual Mtg., Am. Dairy Sci. Assoc., Lafayette, Ind., June 1963; Abstr., J. Dairy Sci., 46, 605 (1963)

5398. Parsons, J. S., Prescott, W. B., and Lawrence, H. C., "Effect of Air on Hot Wire Thermal Conductivity Detectors," Anal. Chem., 34, 1337 (1962)

5399. Parsons, J. S., Tsang, S. M., and DiGiaimo, M. P., with Feinland, R., and Paylor, R. A. L., "Separation and Determination of Mono- and Dinitrotoluene Isomers by Gas-Liquid Chromatography," Anal. Chem., 33, 1858-1859 (1961); CA 56, 8010f

5400. Parsons, M. L., Pennington, S. N., and Walker, J. M., "A Rapid Combustion Method for Nitrogen Determination Utilizing Gas Chromatography," Anal. Chem., 35, 842-844 (1963)

5401. Parsons, M. L., Pennington, S. N., and Walker, J. M., "A Rapid Method for Nitrogen Analysis in Organic Compounds and Associated Materials by Gas Chromatography," 144th Natl. ACS Mtg., Los Angeles, Calif., March-April 1963, Program Abstr., p. 20B

5402. Parsons, T. D., Silverman, M. B., and Ritter, D. M., "Alkenylboranes. I. Preparation and Properties of Some Vinyl and Propenylboranes," J. Am. Chem. Soc., 79, 5091-5101 (1957); CA 52, 4474b

5403. Pascaud, M., "Contaminations, in Gas Chromatography, of Radioactive Esters of Fatty Acids on Polyester Stationary Phase," J. Chromatog., 10, 125-130 (1963)

5404. Pasch, E., "Instrumentation Increases Efficiency in Gas Chromatography," Chem. Processing, 21, No. 9, 132-133, 138 (1958)

5405. Paterson, A. R., "Gas Chromatographic Separations of Close Boiling Isomers," in "Gas Chromatography," edited by H. J. Noebels, R. F. Wall, and N. Brenner, Academic Press, Inc., New York, 1961, pp. 323-326

5406. Patil, V. S., and Hansen, A. E., "Effect of Diets With and Without Fat at Low and High Caloric Levels on Fatty Acids in Blood Cells and Plasma of Dogs," J. Nutrition, 78, 167-172 (1962)

5407. Patrick, C. R., "Large Columns," Gas Chromatography Discussion Group Informal Symp., Bristol, England, September 1959

5408. Patterson, G. D., "Review of Industrial Applications of Analysis, Control, and Instrumentation. Automatic Operations in Analytical Chemistry," Anal. Chem., 31, 646-655 (1959); CA 53, 8911i

5409. Patton, H. W., "Use of Adsorbents in Vapor-Phase Chromatography," 129th Natl. ACS Mtg., Dallas, Texas, April 1956, Program Abstr., p. 14B

5410. Patton, H. W., "Gas Chromatography," 10th Annual Symp. on Modern Methods of Anal. Chem., Baton Rouge, La., January 1957; Abstr., Anal. Chem., 29, 313 (1957)

5411. Patton, H. W., "Introduction," in "Principles and Practice of Gas Chromatography," edited by R. L. Pecsok, John Wiley & Sons, Inc., New York, 1959, pp. 1-7

5412. Patton, H. W., "Fundamental Principles," in "Principles and Practice of Gas Chromatography," edited by R. L. Pecsok, John Wiley & Sons, Inc., New York, 1959, pp. 8-20

5413. Patton, H. W., "Theory of Gas-Liquid Chromatography: II. Van Deemter-Golay-Jones Equations," 13th Annual Fisk Infrared Inst., Gas Chromatog. Session, Nashville, Tenn., August 1962

5414. Patton, H. W., and Lewis, J. S., "Fractionation and Analysis on a Micro Scale by Gas Chromatography," 3rd Natl. Air Pollution Symp., Pasadena, Calif., April 1955; Abstr., Anal. Chem., 27, 1034 (1955)

5415. Patton, H. W., Lewis, J. S., and Kaye, W. I., "Separation and Analysis of Gases and Volatile Liquids by Gas Chromatography," Anal. Chem., 27, 170-174 (1955); CA 49, 9910b

5416. Patton, H. W., and Touey, G. P., "Gas Chromatography Determination of Some Hydrocarbons in Cigarette Smoke," Anal. Chem., <u>28</u>, 1685-1688 (1956); CA <u>51</u>, 3094d

5417. Patton, S., "Chemical Aspects of Flavor Research on Milk and Its Products," Perfumery Essent. Oil Record, <u>49</u>, 396-390 (1956); CA <u>52</u>, 18940f

5418. Patton, S., "Chemical Aspects on Flavor Research on Milk and Its Products," in "Flavor Research and Food Acceptance," Arthur D. Little, Inc., Cambridge, Mass., 1958, pp. 315-323

5419. Patton, S., "Gas Chromatographic Analysis of Milk Fat," J. Dairy Sci., <u>43</u>, 1350-1354 (1960)

5420. Patton, S., "Gas Chromatographic Analysis of Flavor in Processed Milk," J. Dairy Sci., <u>44</u>, 207-214 (1961)

5421. Patton, S., Evans, L., and McCarthy, R. D., "The Action of Pancreatic Lipase on Milk Fat," J. Dairy Sci., <u>43</u>, 95-96 (1960); CA <u>54</u>, 11317f

5422. Patton, S., Forss, D. A., and Day, E. A., "Methyl Sulfide and the Flavor of Milk," J. Dairy Sci., <u>39</u>, 1469-1470 (1960)

5423. Patton, S., and Keeney, P. G., "The High-Melting Glyceride Fraction from Milk Fat," J. Dairy Sci., <u>41</u>, 1288-1289 (1958)

5424. Patton, S., McCarthy, R. D., Evans, L. E., and Lynn, T. R., "The Structure and Synthesis of Milk Fat. I. Gas Chromatographic Analysis," J. Dairy Sci., <u>43</u>, 1187-1195 (1960); CA <u>54</u>, 25353i

5425. Paulsen, S. R., and Huck, G., "Contribution to the Chemistry of Diazacyclopropanes," Chem. Ber., <u>94</u>, 968-975 (1961); CA <u>55</u>, 22274h

5426. Pavlova, S. N., Driatskaya, Z. V., and Mkhchiyan, M. A., "Determination of Normal Paraffinic Hydrocarbons [in Crude Oil] by Use of Molecular Sieves," Khim. i Tekhnol. Topliva i Masel, <u>7</u>, No. 3, 58-59 (1962); CA <u>57</u>, 1146c

5427. Paylor, R. A. L., and Feinland, R., "Determination of Trace Quantities of Acetylene in Ethylene by Gas Chromatography," Anal. Chem., <u>33</u>, 808-809 (1961); CA <u>55</u>, 15225g

5428. Payn, D. S., "Routine Analysis of Phenols by Gas-Liquid Chromatography," Chem. & Ind. (London), <u>1960</u>, 1090; CA <u>55</u>, 8148h

5429. Pearl, I. A., and Harrocks, J. A., "Studies on the Chemistry of Aspenwood. X. Neutral Materials from the Benzene Extractives of Populus Tremuloides," J. Org. Chem., <u>26</u>, 1578-1583 (1961); CA <u>55</u>, 25734g

5430. Pecsok, R. L., (Ed.), "Principles and Practice of Gas Chromatography," John Wiley & Sons, Inc., New York, 1959

5431. Pecsok, R. L., "Analytical Methods," in "Principles and Practice of Gas Chromatography," edited by R. L. Pecsok, John Wiley & Sons, Inc., New York, 1959, pp. 135-150

5432. Pecsok, R. L., "Present Status of Gas Chromatography in Russia," 137th Natl. ACS Mtg., Cleveland, Ohio, April 1960, Program Abstr., p. 26B

5433. Pecsok, R. L., "Gas Chromatography. Basic Principles and New Developments," J. Chem. Educ., <u>38</u>, 212-216 (1961)

5434. Pelick, N., Supina, W. R., and Rose, A., "Triglyceride Elution by Gas Chromatography," J. Am. Oil Chemists' Soc., <u>38</u>, 506 (1961); CA <u>55</u>, 24056d

5435. Peng, C. T., "Fatty Acid Composition of Tissue Lipides in Normal and Tumor-Bearing Rats and Mice," 138th Natl. ACS Mtg., New York, N. Y., September 1960, Program Abstr., p. 63C

5436a. Peng, S-Y., Huang, H-F., and Yu, H-F., "Separation and Analysis of Hydrocarbon Mixtures by Chromatography," Jan Liao Hsueh Pao, <u>1</u>, No. 2, 109-129 (1957); CA <u>52</u>, 6770f

5436b. Penney, W. H., and Windey, J. P., "A Remote Injection and Fraction Collection Apparatus for Preparative Chromatography," 13th Ann. Mid-America Spectr. Symp., Chicago, Ill., April-May 1962; pub. in "Developments in Applied Spectroscopy," Vol. 2, edited by J. R. Ferraro and J. S. Ziomek, Plenum Press, New York, 1963, pp. 409-414

5437. Penther, G. J., and Hickling, J. W., "New Liquid-Sampling Valve Extends Usefulness of Process Chromatography," Oil Gas J., <u>59</u>, No. 20, 130-133 (1961); CA <u>55</u>, 18090i

5438. Peoletti, P., and Grossi, E., "Fatty Acid Composition of Lipides from Various Organs of the Rat. Analysis by Gas Liquid Chromatography," Ricerca sci., <u>30</u>, 726-730 (1960); CA <u>55</u>, 14040b

5439. Pepper, J. M., "The Interaction of Vinyl Esters and Benzenoid Hydrocarbons," 44th Canadian Chem. Conf. and Exhibition, CIC, Montreal, Canada, August 1961; Abstr., Chem. in Can., <u>13</u>, No. 5, 44 (1961)

5440. Pepper, J. M., Manolopoulo, M., and Burton, R., "Gas-Liquid Chromatographic Analysis of Lignin Oxidation Products," Can. J. Chem., <u>40</u>, 1976-1980 (1962); CA <u>57</u>, 15393i

5441. Percival, D. F., "Analysis of Polyester Resins by Gas Chromatography," Anal. Chem., <u>35</u>, 236-238 (1963); CA <u>58</u>, 8089d

5442. Percival, W. C., "Quantitative Determination of Fluorinated Hydrocarbons by Gas Chromatography," Anal. Chem., <u>29</u>, 20-24 (1957); CA <u>51</u>, 11939e; 7th Pittsburgh Conf. on Anal. Chem. & Appl. Spectroscopy, Pittsburgh, Pa., February-March 1956, Program Abstr., p. 41

5443. Percy, L. E., "Sample-System Design," Instruments & Control Systems," <u>33</u>, 1755-1757 (1960)

5444. Perila, O., and Bishop, C. T., "Enzymic Hydrolysis of a Glucomannan from Jack Pine (Pinus Bankziana Lamb)," Can. J. Chem., <u>39</u>, 815-826 (1961); CA <u>55</u>, 16991g

5445. Perkin-Elmer Corp., "Gas Chromatography: Successful Use by Essential Oil and Flavor Manufacturers," Essential Oils and Aromatics, Monthly Reporter, 9, 5-7 (November 1959)

5446. Perkins, E. G., Endres, J. G., and Kummerow, F. A., "The Metabolism of Fats. I. Effect of Dietary Hydroxy Acids and Their Triglycerides on Growth, Carcass, and Fecal Fat Composition in the Rat," J. Nutrition, 73, 291-298 (1961)

5447. Perkins, E. G., Endres, J. G., and Kummerow, F. A., "Effect of Ingested Thermally Oxidized Corn Oil on Fat Composition in the Rat," Proc. Soc. Exptl. Biol. Med., 106, 370-372 (1961); CA 55, 11568c

5448. Perkins, Jr., G., Laramy, R. E., and Lively, L. D., "Flame Response in the Quantitative Determination of High Molecular Weight Paraffins and Alcohols by Gas Chromatography," Anal. Chem., 35, 360-362 (1963); CA 58, 9633f; 13th Pittsburgh Conf. on Anal. Chem. & Appl. Spectroscopy, Pittsburgh, Pa., March 1962, Program Abstr., p. 58

5449. Perkins, Jr., G., Rouayheb, G. M., Lively, L. D., and Hamilton, W. C., "Response of the Gas-Chromatographic Flame Ionization Detector to Different Functional Groups," Gas Chromatog., Intern, Symp., 1961, 3, 269-285 (Pub. 1962); CA 58, 3865a

5450. Perkins, Jr., G., Rouayheb, G. M., Lively, L. D., and Hamilton, W. C., "Response of the Gas Chromatographic Flame Ionization Detector to Different Functional Groups," in "Gas Chromatography," edited by N. Brenner, J. E. Callen, and M. D. Weiss, Academic Press, Inc., New York, 1962, pp. 269-286

5451. Perkins, W. C., and Koski, W. S., "Chemical Reactions of Nitrogen-13 Recoils in Some Alcohols," 138th Natl. ACS Mtg., New York, N. Y., September 1960, Program Abstr., p. 9N

5452. Perrett, R. H., and Purnell, J. H., "Evaluation of Liquid Mass Transfer Coefficients in Gas Chromatography," Anal. Chem., 34, 1336-1337 (1962); CA 57, 11834h

5453. Perrett, R. H., and Purnell, J. H., "A Study of the Reaction of Hexamethyldisilazane with Some Common Gas-Liquid Chromatographic Solid Supports and Its Effect on Their Adsorptive Properties," J. Chromatog., 7, 455-466 (1962); CA 58, 948h

5454. Perrett, R. H., and Purnell, J. H., "Contribution of Diffusion and Mass Transfer Processes to Efficiency of Gas Liquid Chromatography Columns," Anal. Chem., 35, 430-439 (1963)

5455. Perrine, W. L., "A Precision Integrator for Gas Chromatography," Gas Chromatog. Intern. Symposium, 2nd, East Lansing, Mich., 1959, 119-125 (Pub. 1960); CA 55, 20530c

5456. Perrine, W. L., "A Precision Integrator for Gas Chromatography," in "Gas Chromatography," edited by H. J. Noebels, R. F. Wall, and N. Brenner, Academic Press, Inc., New York, 1961, pp. 119-125

5457. Perry, E., and Ory, H. A., "Reaction Between Triethylaluminum and 1-Octene," J. Org. Chem., 25, 1685-1686 (1960); CA 55, 10304a

5458. Perry, J., "Seminar on Preparative Chromatography," 13th Annual Mid-America Spectroscopy Symp., Chicago, Ill., April-May 1962; publ. as "A Review of Preparative Gas Chromatography" in "Developments in Applied Spectroscopy," Vol. 2, edited by J. R. Ferraro and J. S. Ziomek, Plenum Press, New York, 1963, pp. 368-382

5459. Perry, J. A., "Versatile Gas Chromatographic Instrument," Gas Chromatog., Intern. Symposium, 2nd, East Lansing, Mich., 1959, 183-197 (Pub. 1961); CA 55, 18203h

5460. Perry, J. A., "A Versatile Gas Chromatographic Instrument," in "Gas Chromatography," edited by H. J. Noebels, R. F. Wall, and N. Brenner, Academic Press, Inc., New York, 1961, pp. 183-197

5461. Perry, M. V., "Chromatography for Gasoline Plant Control," 38th Annual Mtg., Natural Gasoline Assoc. of America, Dallas, Texas, April 1959

5462. Perry, W. H., and Tabor, E. C., "Industrial Health Foundation. VI. National Air Sampling Network Measurements of SO_2 and NO_2," Arch. Environmental Health, 4, 254-264 (1962); CA 57, 17012e

5463. Peterkin, M. E., and Loveland, J. W., "Carbonyl Test Detects Wax Odor," Oil Gas J., 59, No. 27, 121-124 (1961); CA 55, 23989h

5464. Peters, K., and Weil, K., "The Separation of Gases by Adsorption on Charcoal," Z. angew. Chem., 43, 608-612 (1930; CA 24, 5565

5465. Peters, R. A., Hall, R. J., Ward, P. F. V., and Sheppard, N., "Chemical Nature of the Toxic Compounds Containing Fluorine in the Seeds of Dichapetalum Toxicarum," Biochem. J., 77, 17-23 (1960); CA 54, 22859c

5466. Peterson, D. L., and Lundberg, G. W., "Rapid Analysis and Sample Introduction in Gas Chromatography," Anal. Chem., 33, 652-653 (1961); CA 55, 13157h

5467. Peterson, D. L., and Redlich, O., "Sorption of Normal Paraffins by Molecular Sieves Type 5A," J. Chem. Eng. Data, 7, Pt. 2, 570-574 (1962); CA 58, 2868b

5468. Peterson, J. I., Kindley, L. M., and Podall, H. E., "Synthesis and Gas Chromatographic Purification of Organo-metallic Compounds for Semiconductor Applications," Ultrapurif. Semicond. Mater., Proc. Conf., Boston, Mass., 1961, 253-263 (Pub. 1962); CA 57, 5758d

5469. Peterson, J. I., Kindley, L. M., and Podall, H. E., "The Synthesis and Purification of Organogermanium Compounds for Semiconductor Application," 14th Pittsburgh Conf. on Anal. Chem. &

Appl. Spectroscopy, Pittsburgh, Pa., March 1963, Program Abstr., p. 58

5470. Peterson, M. L., and Hirsch, J., "A Calculation for Locating the Carrier Gas Front of a Gas-Liquid Chromatogram," J. Lipid Research, 1, 132-134 (1959); CA 54, 10454a

5471. Peterson, P. E., "Solvents of Low Nucleophilicity. I. Reactions of Hexyl Tosylates and Hexenes in Trifluoroacetic Acid and Other Acids," J. Am. Chem. Soc., 82, 5834-5837 (1960); CA 55, 20915e

5472. Peterson, P. E., and Allen, G., "Addition of Trifluoroacetic Acid to Hexanes and Cyclic Alkenes," 139th Natl. ACS Mtg., St. Louis, Mo., March 1961, Program Abstr., p. 34-O

5473. Petho, A., Fejes, P., and Engelhardt, J., "Calculation of Gas-Chromatographic Elution Waves with Regard to Variability of Flow Rate," Acta Chim. Acad. Sci. Hung., 30, No. 1, 63-70 (1962); CA 57, 6582b

5474. Petitjean, D. L., "New Direct Method of Calibration for Gas Chromatography," 9th Pittsburgh Conf. on Anal. Chem. & Appl. Spectroscopy, Pittsburgh, Pa., March 1958, Program Abstr., p. 50

5475. Petitjean, D. L., "Peak Area vs. Peak Height in Quantitative Gas Chromatography," 9th Pittsburgh Conf. on Anal. Chem. & Appl. Spectroscopy, Pittsburgh, Pa., March 1958, Program Abstr., p. 50

5476. Petitjean, D. L., and Lantz, C. D., "Electron Attachment Determination of Thiophosphate Pesticides in the Picogram Range," J. Gas Chromatog., 1, No. 2, 23 (1963)

5477. Petitjean, D. L., and Leftault, Jr., C. J., "Oxide-Coated Aluminum Tubing for Capillary Gas Chromatography," J. Gas Chromatog., 1, No. 3, 18-21 (1963); 13th Pittsburgh Conf. on Anal. Chem. & Appl. Spectroscopy, Pittsburgh, Pa., March 1962, Program Abstr., p. 50

5478. Petranek, J., "Gas-Chromatographic Separation of Organic Sulfides," J. Chromatog., 5, 254-258 (1961); CA 56, 5385i

5479. Petranek, J., and Slosar, J., "3,5-Dinitrobenzene as Stationary Phases for Separation of Aromatic Hydrocarbons by Gas Chromatography," Collection Czech. Chem. Commun., 26, 2667-2669 (1961); CA 58, 2826g

5480. Petrocelli, J. A., and Lichtenfels, D. H., "Determination of Dissolved Gases in Petroleum Fractions by Gas Chromatography," Anal. Chem., 31, 2017-2019 (1959); CA 54, 7118d; 136th Natl. ACS Mtg., Atlantic City, N. J., September 1959, Program Abstr., p. 4R; Preprints, Div. Petrol. Chem., 4, No. 3, 103-118 (1959)

5480a. Petrova, R. S., Khrapova, E. V., and Shcherbakova, K. D., "Study of Physico-Chemical Adsorption Characteristics by Gas Chromatographic Methods," Gas Chromatography, Intern. Symp., 4th, Hamburg, Germany, June 1962; pub. in "Gas Chromatography 1962," ed. by M. van Swaay, Butterworth & Co., London, 1962

5481. Petrowitz, H. J., "Gas Chromatography of Terpene Hydrocarbons," Riechstoffe u. Aromen, 12, 397-402 (1962); CA 59, 1683f

5482. Petrowitz, H. J., Nerdel, F., and Ohloff, G., "Gas Partition Chromatography of the Stereoisomeric Menthols," J. Chromatog., 3, 351-358 (1960); CA 54, 22702e

5483. Petrowitz, H. J., Nerdel, F., and Ohloff, G., "Gas Chromatography of Terpene Hydrocarbons from Geranium Oil," Riechstoffe u. Aromen, 11, 389-395 (1961); CA 56, 10299h

5484. Petrowitz, H. J., Nerdel, F., and Ohloff, G., "Gas Chromatography of the Terpenes of Canada Balsam Oil," Riechstoffe u. Aromen, 12, 1-6 (1962); CA 56, 11726h

5485. Petzelt, B., and Nemec, L., "High-Temperature Chromatograph, Chrom-2, for Packed and Capillary Columns," Gas Chromatog. Symp., Brno, Czech., June 1962; Abstr., J. Gas Chromatog., 1, No. 4, 7 (1963)

8486. Peurifoy, P. V., Ogilvie, J. L., and Dvoretzky, I., "An Apparatus for Preparative-Scale Gas-Liquid Chromatography," J. Chromatog., 5, 418-429 (1961); CA 56, 11385f

5487. Peyron, L., "Study of Essences from Liquors Obtained by the Steam Distillation of Morocco Rose Geranium," Compt. rend., 255, 2981-2982 (1962); CA 58, 4369h

5488. Peyrot, P., "Vapor-Phase Partition Chromatography," Rev. franç. corps gras., 3, 552 (1956)

5489. Peyrot, P., "Use of Vapor-Phase Chromatography for Rapid Analysis of Mixtures of Solvent," Chim. & ind. (Paris), 78, 3-8 (1957); CA 52, 153h

5490. Pfenninger, H., "The Separation of Isomeric, Saturated, Aliphatic Alcohols Up to Five Carbon Atoms by Gas Chromatography," Helv. Chim. Acta, 45, 460-462 (1962); CA 57, 646d

5491. Pfenninger, H., "Comprehensive Reviews. Gas-Chromatographic Studies of Fusel Oil from Various Fermentation Products. I. The Problem and Literature Review," Z. Lebensm.-Untersuch. u. -Forsch., 119, 401-405 (1963); CA 59, 1047h

5492. Phelps, F. T., "An Application of Gas Chromatography to Solid State Physics: Determination of the Gases Evolved When Colored Potassium Chloride Dissolves," 14th Annual Mid-America Spectroscopy Symp., Chicago, Ill., May 1963

5493. Phillips, C. S. G., "The Chromatography of Gases and Vapours - Part 1," Disc. Faraday Soc., No. 7, 241-248 (1949); CA 45, 1460e

5494. Phillips, C. S. G., "Adsorption and Partition Methods," Mtg. of Physical Methods and Microchemistry Groups with Scottish Section, Soc. for Anal. Chem., Stevenson, Scotland, May 1955; Abstr., Anal. Chem., 27, 1667 (1955)

5495. Phillips, C. S. G., "Gas Chromatography," in "Physical Methods in Chemical Analysis," Vol. III, edited by W. G. Berl, Academic Press, Inc., New York, 1956, pp. 1-28; CA 51, 3352i

5496. Phillips, C. S. G., "Gas Chromatography," Academic Press, Inc., New York, 1956

5497. Phillips, C. S. G., "Gas Chromatography," Discovery, 18, 472-476 (1957)

5498. Phillips, C. S. G., "Gas Chromatography," Nature, 180, 840-841 (1957)

5499. Phillips, C. S. G., "Gas-Liquid Chromatography," Svensk. Kem. Tidskr., 69, 199-212 (1957); CA 51, 11809h

5500. Phillips, C. S. G., "Gas Chromatography Instrumentation for the Laboratory," in "Gas Chromatography," edited by V. J. Coates, H. J. Noebels, and I. S. Fagerson, Academic Press, Inc., New York, 1958, pp. 51-65

5501. Phillips, C. S. G., "Gas-Liquid Chromatography," Petroleum, 22, 91-93, 110 (1959)

5502. Phillips, C. S. G., "Summing Up of Symposium," in "Gas Chromatography 1960," edited by R. P. W. Scott, Butterworths, London, 1960, pp. 433-435

5503. Phillips, C. S. G., "Gas Chromatography and Inorganic Chemistry," Univ. of Houston Intern. Symp. on Advances in Gas Chromatography, Houston, Texas, January 1963

5504. Phillips, C. S. G., Powell, P., and Semlyen, J. A., "The Chromatography of Gases and Vapors. VII. Substituted Borazoles," J. Chem. Soc., 1963, 1202-1207; CA 58, 8885a

5505. Phillips, C. S. G., and Timms, P. L., "Molecular Weight Determination with the Martin Density Balance," J. Chromatog., 5, 131-136 (1961); CA 56, 6634f

5506. Phillips, C. S. G., and Timms, P. L., "Some Applications of Gas Chromatography in Inorganic Chemistry," Anal. Chem., 35, 505-510 (1963); CA 59, 1064g

5507. Phillips, D. D., Pollard, G. E., and Soloway, S. B., "Thermal Isomerization of Endrin and Its Behavior in Gas Chromatography," 18th Intern. Congress on Pure and Applied Chemistry, Montreal, Canada, August 1961; Abstr., Chem. in Can., 13, No. 5, 61 (1961); J. Agr. Food Chem., 10, 217-221 (1962)

5508. Phillips, G., "An Electronic Method of Detecting Impurities in the Air," J. Sci. Instr., 28, 342-347 (1951); CA 46, 3330e

5509. Phillips, J. R., and Stone, F. G. A., "Purification of Fluorocarbon Organometallic and Other Sensitive Compounds by Vapor-Phase Chromatography," 14th Pittsburgh Conf. on Anal. Chem. & Appl. Spectroscopy, Pittsburgh, Pa., March 1963, Program Abstr., pp. 58-59

5510. Phillips, R. E., "Natural Gas Analysis by Column Chromatography," Gas (Los Angeles), 36, No. 9, 108-109, 112, 116 (1960); CA 54, 21710i

5511. Phillips, T. R., and Neylan, D., "A Statistical Investigation of Factors Affecting Column Performance in the Chromatography of Inorganic Gases," in "Gas Chromatography 1962," ed. by M. van Swaay, Butterworth & Co., London 1962

5512. Phillips, T. R., and Owens, D. R., "The Gas-Chromatographic Analysis of Inorganic Halogen Compounds on Capillary Columns," Gas Chromatog. Proc. Symposium, 3rd, Edinburgh, 1960, 308-315; pub. in "Gas Chromatography 1960," edited by R. P. W. Scott, Butterworths, London, 1960, pp. 308-320; CA 56, 5385e

5513. Phillips, T. R., and Owens, D. R., "Purification of Hydrogen Isotopes by Displacement Gas Chromatography," U. K. At. Energy Authority, Prod. Group, PG Rept. 386(CA), 7 pp. (1962); CA 58, 6418a

5514. Pichat, L., Baret, C., Guermont, J. P., and Audinot, M., "Examples of the Use of Preparative Gas-Phase Chromatography for Producing Tagged Molecules," Comm. energie at. (France), Rappt. No. 1787, 16 pp. (1960); CA 55, 18552 g

5515. Pichler, H., Herlan, A., and Schulz, H., "Combined Application of Gas Chromatography and Mass Spectrometry for the Investigation of Coking Gas," Brennstoff-Chem., 43, 269-276 (1962); CA 58, 386a

5516. Pichler, H., and Schulz, H., "Continuous Separation of Gases by a New Process of Counter-Current Distribution," Brennstoff-Chem., 39, 148-153 (1958); CA 52, 12510i

5517. Pichler, H., and Sotiropoulos, B. A., "Cracking of n-Paraffins in the Presence of a SiO_2-Al_2O_3 Catalyst," Ann., 618, 241-250 (1958); CA 53, 8602i

5518. Pierotti, G. J., Deal, C. H., and Derr, E. L., "Activity Coefficients and Molecular Structure," Ind. Eng. Chem., 51, 95-102 (1959); CA 53, 5830d

5519. Pierotti, G. J., Deal, C. H., Derr, E. L., and Porter, P. E., "Solvent Effects of Gas-Liquid Partition Chromatography," J. Am. Chem. Soc., 78, 2989-2998 (1956); CA 50, 14311h

5520. Pieterse, M. J., and Hertog, H. J. D., "Rearrangements During Aminations of Halopyridines, Presumably Involving a Pyridine Intermediate," Rec. trav. chim., 80, 1376-1380 (1961); CA 58, 5625e

5521. Pietsch, H., "Determination of Very Small Amounts of Oxygen, Carbon Monoxide, Methane, and Nitrogen in Purest Ethylene by Adsorption Chromatography," Erdol u. Kohle, 11, 157-159 (1958); CA 52, 10800b

5522. Pietsch, H., "The Determination of Gases by Means of Chromatography," Erdol u. Kohle, 11, 702-705 (1958); CA 53, 4707g

5523. Piez, K. A., and Saroff, H. A., "Chromatography of Amino Acids and Peptides," in "Chromatography," edited by E. Heftmann, Reinhold Publishing Corp., New York, 1961, Chapter 14

5524. Pillai, C. N., and Pines, H., "High Pressure Thermal Alkylation of Xylenes and Related Compounds by Propylene," J. Am. Chem. Soc., 83, 983-985 (1961); CA 55, 24605a

5525. Pilleri, R., and Vietti-Michelina, M., "Gas-Chromatographic Analysis of Pyridine-Nicotine Mixtures," Z. Anal. Chem., 174, 172-174 (1960); CA 55, 2347c

5526. Pilo, C., and Runeberg, J., "The Chemistry of the Natural-Order Cupressales. XXV. Heartwood Constituents of Juniperus Chinensis," Acta Chem. Scand., 14, 353-358 (1960); CA 56, 13250h

5527. Pinchin, F. J., and Prichard, E., "Gas-Liquid Chromatography of Anthracene Oil," Chem. & Ind. (London), 1962, 1753-1754; CA 58, 8816g

5528. Pinder, A. R., "Physical Methods in Organic Chemistry. I. Practical Techniques," Chem. & Ind. (London), 1961, 758-770

5529. Pinder, A. R., "Physical Methods in Organic Chemistry. II. Physicochemical Measurements," Chem. & Ind. (London), 1961, 1180-1192; CA 56, 2895h

5530. Pines, H., and Benoy, G., "Alumina: Catalyst and Support. II. Hydroisomerization and Aromatization of Hydrocarbons in the Presence of Molybdena-Alumina Catalysts," J. Am. Chem. Soc., 82, 2483-2487 (1960); CA 54, 22404f

5531. Pines, H., and Chen, C-T., "Alumina: Catalyst and Support. IV. Aromatization of 1,1-Dimethyl-cyclohexane, 4,4-Dimethylcyclohexene and of Methylcycloheptane Over Chromia-Alumina Catalysts," J. Am. Chem. Soc., 83, 3562-3566 (1960); CA 55, 1481h

5532. Pines, H., and Chen, C-T., "Alumina: Catalyst and Support. VII. Aromatization of Heptane 1-C^{14} Over Chromia-Alumina Catalysts," J. Org. Chem., 26, 1057-1061 (1961); CA 55, 19830e

5533. Pines, H., and Greenlee, T. W., "Alumina: Catalyst and Support. VI. Aromatization of 1,1-Dimethylchclohexane, Methylcycloheptane and Related Hydrocarbons Over Platinum-Alumina Catalysts," J. Org. Chem., 26, 1052-1057 (1961); CA 55, 19829c

5534. Pines, H., and Haag, W. O., "Alumina: Catalyst and Support. I. Alumina, Its Acidity and Catalytic Activity," J. Am. Chem. Soc., 82, 2471-2483 (1960); CA 54, 22403h

5535. Pines, H., and Pillai, C. N., "High Pressure Thermal Alkylation of Monoalkylbenzenes by Butenes," J. Am. Chem. Soc., 81, 3629-3633 (1959); CA 54, 2220d

5536. Pines, H., and Pillai, C. N., "Phenyl Migration During Decomposition of Peroxides in Alkylbenzenes," J. Am. Chem. Soc., 82, 2921-2925 (1960); CA 55, 13342g

5537. Pines, H., and Schaap, L., "Base-Catalyzed Reactions. XIII. The Relative Rates of Side-Chain Ethylation of Aromatic Hydrocarbons," J. Am. Chem. Soc., 80, 3076-3079 (1958); CA 52, 19984e

5538. Pines, H., and Shabtai, J., "Synthesis of Cyclobutanes by Dimerization of β-Alkylstyrenes," J. Am. Chem. Soc., 83, 2781-2782 (1961); CA 56, 15387d

5539. Piozzi, F., and Merlini, L., "Pyrolysis of Tryptophan," Gazz. chim. ital., 92, 1105-1117 (1962); CA 58, 12665c

5540. Piozzi, F., and Vita-Finizi, P., "Gas Chromatography of Phenanthrenic Hydrocarbons," Atti accad. nazl. Lincei. Rend., Classe sci. fis., mat. e nat., 29, 549-554 (1960); CA 56, 8010g

5541. Piringer, O., and Pascalau, M., "A New Detector for Gas Chromatography," J. Chromatog., 8, 410-412 (1962)

5542. Pisano, J. J., "Biochemical Applications of Gas Chromatography," Metropolitan Regional Mtg., ACS, New York, N. Y., January 1963

5543. Pisano, J. J., VandenHeuvel, W. J. A., and Horning, E. C., "Gas Chromatography of Phenylthiohydant (PTH) and Dinitrophenyl Derivatives of Amino Acids," Biochem. Biophys. Res. Communs., 7, 82-86 (1962); Biol. Abstr., 39, 9465

5544. Pitkethly, R. C., "Low Pressure Electric Discharge Detectors," Anal. Chem., 30, 1309-1314 (1958); CA 52, 16801i; 132nd Natl. ACS Mtg., New York, N. Y., September 1957, Program Abstr., p. 33B; Preprints, Div. Petrol. Chem., 2, No. 4, D-67 (1957)

5545. Pitkethly, R. C., and Goble, A. G., "Techniques for Catalyst Studies at Low Surface Coverage," 135th Natl. ACS Mtg., Boston, Mass., April 1959, Program Abstr., p. 4Q; Preprints, Div. Petrol. Chem., 4, No. 2C, 103 (1959)

5546. Pitts, Jr., J. N., DeFord, D. D., and Becktenwald, G. W., "Manometric Gas Analysis Apparatus," Anal. Chem., 24, 1566-1568 (1952); CA 47, 355d

5547. Pitts, Jr., J. N., and Osborne, A. D., "Structure and Reactivity in the Radiolysis of Ketones," J. Am. chem. Soc., 83, 3011-3014 (1961); CA 56, 1323d

5548. Platek, J., "Device for Development of Chromatograms with [Gaseous] Toxic Substances," Roczniki Chem., 31, 685-686 (1957); CA 52, 2463e

5549. Plessis, L. A. du, and Spong, A. H., "Effect of Carrier Humidity in the Gas Chromatography of Ammonia," Chem. & Ind. (London), 1959, 1246; CA 54, 2879b

5550. Plessis, L. A. du, and Spong, A. H., "Behavior of Ammonia in Gas Chromatography. I. Ammine Formation in Columns Containing Solutions of Silver Nitrate in Benzyl Cyanide," J. Chem. Soc., 1959, 2027-2031; CA 53, 15701h

5551. Podbielniak, W. J., and Preston, S. T., "Hydrocarbon Analysis Made More Accurate, Speedy," Petroleum Engr., 27, No. 5, C17-C25 (1955); CA 49, 9909i

5552. Podbielniak, W. J., and Preston, S. T., "New Tool–Vapor-Phase Chromatography," Petroleum Refiner, 34, No. 11, 165-169 (1955); CA 50, 559i

5553. Podbielniak, W. J., and Preston, S. T., "Two New Instruments Developed to Aid in Analyzing Hydrocarbon Mixtures. I. New Thermal Conductivity Cell for Distillation Equipment. II. Vapor-Phase Chromatographic Equipment," Oil Gas J., 54, No. 50, 211-212, 215-219 (1956); CA 51, 4773e

5554. Podbielniak, W. J., and Preston, S. T., "Vapor-Phase Chromatography," Petroleum Refiner, 35, No. 4, 215-220 (1956); CA 50, 8368f

5555. Podbielniak, W. J., and Preston, S. T., "Gas Chromatography: What of the Future?" 31st Annual Mtg., California Nat. Gas Assoc., Los Angeles, Calif., November 1956; Abstr., Oil Gas J., 54, No. 85, 140 (1956)

5556. Podbielniak, W. J., and Preston, S. T., "The Future Possibilities of Gas Chromatography," Gas, 34, No. 2, 119-120, 123, 126 (1958); CA 52, 4383g

5557. Podbielniak, W. J., Preston, S. T., and Turkal, P. J., "Analysis of Fatty Acids and Their Esters by Vapor-Phase Chromatography," 30th Fall Mtg., Am. Oil Chemists' Soc., Chicago, Ill., September 1956; Abstr., Anal. Chem., 28, 2027 (1956)

5558. Podkletnov, N. E., "The Use of Gas-Liquid Chromatography for a Closer Study of the Chemical Composition of Petroleum," Izvest. Sibir. Otdel., Akad. Nauk S.S.S.R., 1961, No. 5, 70-79; CA 55, 26418c

5559. Podkletnov, N. E., and Bryanskaya, E. K., "Gas-Liquid Chromatography of Liquid Petroleum Hydrocarbons," Uspekhi Khim., 24, 1354-1360 (1958); CA 53, 4707f

5560. Poe, R. W., and Kaelble, E. F., "Quantitative Analysis of 1-Olefins by Programmed Temperature Gas Chromatography," 36th Fall Mtg., Am. Oil Chemists' Soc., Toronto, Ontario, October 1962

5561. Polak, L. S., Topchiev, A. V., and Cherniak, N. Y., "Radiolysis of Heptane and Some Other Alkanes," Doklady Akad. Nauk S.S.S.R., 119, 307-310 (1958); CA 53, 194i

5562. Polgár, A., and Jungnickel, J. L., "Determination of Olefinic Unsaturation," in "Organic Analysis," Vol. 3, edited by J. Mitchell, Jr., I. M. Kolthoff, E. S. Proskauer, and A. Weissberger, Interscience Publishers, Inc., New York, 1956, pp. 341-359

5563. Polgár, A. G., Holst, J. J., and Groennings, S., "Determination of Alkanes and Cycloalkanes Through C_8 and Alkenes Through C_7 by Capillary Gas Chromatography," Anal. Chem., 34, 1226-1234 (1962)

5564. Polgár, A. G., Holst, J. J., and Groennings, S., "Determination of Saturates Through C_8's by Gas Chromatography," 141st Natl. ACS Mtg., Washington, D. C., March 1962, Program Abstr., p. 23B

5565. Polgár, N., and Smith, W., "Constituents of the Lipids of Tubercle Bacilli. IX.," J. Chem. Soc., 1962, 4262-4263; CA 58, 5506g

5566. Pollard, F. H., "Gas Chromatography," British Assoc. Mtg., Section B (Chemistry), Bristol, England, September 1955; cf. Nature, 176, 1188 (1955)

5567. Pollard, F. H., "Separation of Volatile Substances by Differing Rates of Migration," Chem. & Ind. (London), 1955, 1290-1291

5568. Pollard, F. H., and Hardy, C. J., "Effect of Temperature of Injection Upon the Separation of Liquid Mixtures of Gas-Phase Chromatography," Chem. & Ind. (London), 1955, 1145-1146; CA 51, 3370i

5569. Pollard, F. H., and Hardy, C. J., "The Application of Vapor-Phase Chromatography to the Preparation of Pure Materials," Chem. & Ind. (London), 1956, 527-528; CA 50, 16530i

5570. Pollard, F. H., and Hardy, C. J., "The Analyses of Halogenated Hydrocarbons by Vapor-Phase Chromatography," Anal. chim. acta, 16, 135-143 (1957); CA 51, 7239d

5571. Pollard, F. H., and Hardy, C. J., "A Preliminary Study of Some Factors Influencing the Order of Elution of Halogenated Methanes, the Degree of Separation, and the Reproducibility of Retention Volumes in Gas-Liquid Partition Chromatography," in "Vapour-Phase Chromatography," edited by D. H. Desty, Academic Press, Inc., New York, 1957, pp. 115-126

5572. Pollard, F. H., and Hardy, C. J., "Corrected Retention Volumes of Chloro and Fluoro Compounds (Chromatographic Data Tables XLVII, XLVIII)," J. Chromatog., 1, No. 5, xxxii (1958); cf. "Gas Chromatography," edited by D. H. Desty, Academic Press, Inc., New York, 1957, p. 117

5573. Pollard, F. H., Pedler, A. E., and Hardy, C. J., "Effect of Nitrogen Dioxide on the Thermal Decomposition of Ethyl Nitrite," Nature, 174, 979 (1954); CA 49, 5085f

5574. Polyakova, T. A., Sokolova, T. A., and Tsarfin, Y. A., "Chromatographic Determination of Furan and Carbon Dioxide in Products of Decarbonylation of Furfural," Zavodskaya Lab., 29, No. 1, 18-19 (1963); CA 59, 1096h

5575. Ponomarev, A. S., "High-Sensitivity Catharometer for the Chromatographic Analysis of Gases," Zavodskaya Lab., 26, 634-636 (1960); CA 56, 4074a

5576. Pop, A., Barbul, M., and Beschea, C., "Analysis of the Dehydrogenation Products of Ethylbenzene by Gas Chromatography," Rev. chim. (Bucharest), 12, 497-498 (1961); CA 56, 2892c

5577. Pop, A., Barbul, M., and Beschea, C., "Gas-Liquid Partition Chromatographic Determination of Relative Volatilities of Isopentane Dehydrogenation Products in the Presence of Solvents," Rev. chim. (Bucharest), 13, No. 6, 363-367 (1962); CA 57, 16965i

5578. Pop, A., Barbul, M., and Popescu, R., "Analysis of Some C_5-C_7 Hydrocarbons," Rev. chim. (Bucharest), 12, 173-176 (1961)

5579. Pop, A., and Munteanu, I., "Analysis of Cracked Gases by Means of Gas-Liquid Chromatography," Rev. chim. (Bucharest), 12, 347-349 (1961); CA 56, 14526e

5580. Pope, C. G., "Gas Solid Elution Chromatography with Graphitized Carbon Black," Anal. Chem., 35, 654-658 (1963); CA 59, 1064a

5581. Popják, G., "Biosynthesis of Derivatives of Allylic Alcohols from Mevalonate-2-C^{14} in Liver Enzyme Preparations and Their Relation to the Synthesis of Squalene," Tetrahedron Leters, 1959, No. 19, 19-28; CA 54, 8911f

5582. Popják, G., Cornforth, J. W., Cornforth, R. H., Ryhage, R., and Goodman, D. S., "Biosynthesis of Cholesterol. XVI. Chemical Synthesos of trans,-trans-Farnesyl-1-D_2-2-C^{14} and -1-H_2^3-2-C^{14} Pyrophosphates and Their Utilization in Squalene Biosynthesis," J. Biol. Chem., 237, 56-61 (1962); CA 56, 11991a

5583. Popják, G., and Cornforth, R. H., "Gas-Liquid Chromatography of Allylic Alcohols and Related Branched-Chain Acids," J. Chromatog., 4, 214-221 (1960); CA 55, 5566h

5584. Popják, G., Goodman, D. S., Cornforth, J. W., Cornforth, R. H., and Ryhage, R., "Biosynthesis of Cholesterol. XV. Mechanism of Squalene Biosynthesis from Farnesyl Pyrophosphate and from Mevalonate," J. Biol. Chem., 236, 1934-1947 (1961); CA 56, 691g

5585. Popják, G., Lowe, A. E., and Moore, D., "Scintillation Counter for Simultaneous Assay of H^3 and C^{14} in Gas-Liquid Chromatographic Vapors," J. Lipid Research, 3, 364-371 (1962); CA 57, 14671f

5586. Popják, G., Lowe, A. E., Moore, D., Brown, L., and Smith, F. A., "Scintillation Counter for the Measurement of Radioactivity of Vapors in Conjunction with Gas-Liquid Chromatography," J. Lipid Research, 1, 29-39 (1959); CA 54, 10409f

5587. Popják, G., Lowe, A. E., Moore, D., Brown, L., and Smith, F. A., "Gas-Liquid Radiochromatography," Biochem. J., 73, No. 2, 33P (1959)

5588. Porath, J., "Charcoal Chromatography with a Step-Graded Adsorption Column," Arkiv Kemi, 7, 535-537 (1955); CA 49, 7437c

5589. Porcaro, P. J., "Observations on the Use of 'Empty' Copper Tubular Capillary Columns," J. Gas Chromatog., 1, No. 6, 17-19 (1963)

5590. Porcaro, P. J., and Johnston, V. D., "Primary Amyl Alcohols Determined by Gas Chromatography," Anal. Chem., 33, 361-362 (1961); CA 56, 9387i

5591. Porcaro, P. J., and Johnston, V. D., "Determination of Menthols by Gas Chromatography," Anal. Chem., 33, 1748-1751 (1961); CA 56, 4107e

5592. Porcaro, P. J., and Johnston, V. D., "Determination of Thymol Isomers by Gas-Liquid Chromatography Using Lanolin," Anal. Chem., 34, 1071-1073 (1962); CA 57, 11837f

5593. Porcaro, P. J., and Johnston, V. D., "Extension of the Use of Lanolin as a Substrate for Gas-Liquid Chromatography," Anal. Chem., 34, 1845-1846 (1962); CA 58, 3871d

5594. Porsch, F., and Farnow, H., "Geranial and Neral," Dragoco Report, 7, 83-96 (1960); CA 57, 2351i

5595. Porsch, F., and Farnow, H., "Isocitral and Isogeraniol," Dragoco Report, 7, No. 10, 215-230 (1960); ibid., No. 7, 3, 83 (1960); CA 55, 26375c

5596. Porter, J. W., and Anderson, D. G., "Chromatography of Terpenes, Carotenoids, and Fat-Soluble Vitamins," in "Chromatography," edited by E. Heftmann, Reinhold Publishing Corp., New York 1961, Chapter 17

5597. Porter, K., and Volman, D. H., "Flame Ionization Detection of Carbon Monoxide for Gas Chromatographic Analysis," Anal. Chem., 34, 748-749 (1962)

5598. Porter, M., Saville, B., and Watson, A. A., "Reduction by Lithium Aluminum Hydride in the Analysis of Mixtures of Alk(en)yl Mono-, Di- and Polysulfides," J. Chem. Soc., 1963, 346-352; CA 58, 11960h

5599. Porter, P. E., Deal, C. H., and Stross, F. H., "The Determination of Partition Coefficients from Gas-Liquid Partition Chromatography," J. Am. Chem. Soc., 78, 2999-3006 (1956); CA 50, 14312b

5600. Porter, R. P., and Noyes, W. A., "Photochemical Studies. LIV. Methanol Vapor," J. Am. Chem. Soc., 81, 2307-2311 (1959); CA 53, 18604e

5601. Porter, R. S., Hinkins, R. L., Tornheim, L., and Johnson, J. F., "Computer Optimization of Mixed Liquid Phases by Gas Chromatography," 144th Natl. ACS Mtg., Los Angeles, Calif., March-April 1963, Program Abstr., p. 19B

5602. Porter, R. S., Hoffman, A. S., and Johnson, J. F., "Gravimetric Introduction Device for Gas Chromatography, Application to Pyrolysis Studies," Anal. Chem., 34, 1179-1180 (1962); CA 57, 10512c

5603. Porter, R. S., and Johnson, J. F., "Gas Chromatographic Determination of Fuel Dilution in Lubricating Oils," Anal. Chem., 31, 866-869 (1959); CA 53, 15540e

5604. Porter, R. S., and Johnson, J. F., "Circular Gas Chromatograph," Nature, 183, 391-392 (1959); CA 53, 12761c

5605. Porter, R. S., and Johnson, J. F., "Volatile Liquid Partition Chromatography," Nature, 184, 978-979 (1959); CA 54, 11634c

5606. Porter, R. S., and Johnson, J. F., "Extractive Distillation Studies by Circular Gas Chromatography," Ind. Eng. Chem., 52, 691-694 (1960); CA 55, 23985h; 136th Natl. ACS Mtg., Atlantic City, N. J., September 1959, Program Abstr., p. 5R; Preprints, Div. Petrol. Chem., 4, No. 3, 51-58 (1959)

5607. Porter, R. S., and Johnson, J. F., "Here's Better Test for Flash Point Contaminants," Petroleum Refiner, 39, No. 6, 193-195 (1960); CA 54, 20169c

5608. Porter, R. S., and Johnson, J. F., "Improved Gas Chromatographic Analysis for Fuel Dilution and Volatile Contaminants," Anal. Chem., 33, 1129-1130 (1961); CA 56, 8992d

5609. Porter, R. S., and Johnson, J. F., "Low Temperature Gas Chromatography," Anal. Chem., 33, 1152-1155 (1961); CA 56, 3i; 139th Natl. ACS Mtg., St. Louis, Mo., March 1961, Program Abstr., p. 33B

5610. Potter, R. A., "A Unitized Gas Chromatograph for Repetitive Analyses," 13th Pittsburgh Conf. on Anal. Chem. & Appl.Spectroscopy, Pittsburgh, Pa., March 1962, Program Abstr., p. 46

5611. Potter, R. A., and Stokes, E. M., "Detector Bypass for Trace Analysis with Flame Ionization Detector," Perkin-Elmer Instrument News, 13, No. 3, 12 (1962)

5612. Potterat, M., "Gas Chromatographic Analysis of Fruit Aroma Concentrates," Intern. Fruchsaft-Union Ber. Wiss. Tech. Komm., 4, 205-224 (1962); CA 58, 9562b

5613. Poukka, R., "Gas Chromatography," Finsk. Veterinarstidskr., 68, 340-347 (1962); CA 58, 2823b

5614. Poukka, R., Vasenius, L., and Turpeinen, O., "Catalytic Hydrogenation of Fatty Acid Methyl Esters for Gas-Liquid Chromatography," J. Lipid Research, 3, 128-129 (1962); CA 56, 13034c

5615. Poulos, T. J., "Fisher Prep/Partitioner," 4th Annual Gas Chromatography Institute, Canisius College, Buffalo, N. Y., April 1962; see Item 1178 for publication data

5616. Poulos, T. J., "Factors Affecting Performance of Preparative Gas Chromatography Columns," 13th Annual Mid-America Spectroscopy Symp., Chicago,Ill., April-May 1962

5617. Pourchez, A., Paquot, C., and Lefort, D., "Gas Chromatography and Lipid Chemistry. VIII. Variation of Retention Indexes and Volumes as a Function of Temperature for Methyl Esters of Saturated Linear Aliphatic Acids," Rev. franç. corps gras, 9, 681-683 (1962); CA 58, 9341b

5618. Poutsma, M., and Wolthuis, E., "Clemmensen Reduction of Acetophenone," J. Org. Chem., 24, 875-877 (1959); CA 54, 376g

5619. Povinelli, R. J., and Szymanski, H. A., "A New Collecting Cell for Vapor Phase Chromatography," 1st Annual Gas Chromatography Institute, Canisius College, Buffalo, N. Y., May 1959

5620. Powe, W. C., and Marple, W. L., "The Fatty Acid Composition of Clothes Soil," J. Am. Oil Chemists' Soc., 37, 136-138 (1960); CA 54, 9325f

5621. Powell, A. L., Swain, C. G., and Morgan, C. R., "Mechanism of the Canizzaro Reaction," Tritium Phys. Biol. Sci., Proc. Symp. Detection, Use;Vienna, Austria, 1961, 1, 153-160 (Pub. 1962); CA 59, 1453g

5622. Powell, H., and Thomas, W. H., "Modern Trends in Petroleum Analysis," J. Inst. Petrol., 44, 19-28 (1958); CA 52, 7671c

5623. Powell, J. W., and Whiting, M. C., "The Decomposition of Sulfonylhydrazone Salts. I. Mechanism and Sterochemistry," Tetrahedron, 7, 305-310 (1959); CA 54, 7619g

5624. Powell, J. W., and Whiting, M. C., "Analysis of Mixtures of Octahydronaphthalenes," Tetrahedron, 12, 163-167 (1961); CA 55, 16494d

5625. Powell, J. W., and Whiting, M. C., "Decomposition of Sulfonylhydrazone Salts. Mechanism and Stereochemistry. II. Pyrolysis of the Diazodecahydronaphthalenes," Tetrahedron, 12, 168-172 (1961); CA 55, 16498a

5626. Powers, G. W., and Piehl, F. J., "Rapid Chromatographic Analysis of Soap-Thickened Lubricating Greases," Anal. Chem., 30, 28-31 (1958); CA 52, 8524a

5627. Poy, F., "Vapor Chromatography and Its Application to Pharmacy," Boll. chim. farm., 96, 553-555 (1957); CA 52, 5745f

5628. Poy, F., "Behavior of the Stationary Phase in Analysis of Methyl Esters of Fatty Acids with Particular Regard to the Linolenic Acid-Arachidic Acid Separation and on the Use of a New Gas Chromatographic Column," Boll. lab. chim. provinciali (Bolonga), 12, 445-451, discussion, 451-457 (1961); CA 57, 3577e

5629. Poy, F., "On the Behavior of the Stationary Phase in Vapor Phase Chromatography of the Methyl Esters of Fatty Acids," Riv. ital. sost. grasse, 3, 137-139 (1962)

5630. Poy, F., and Gagliardi, P., "Gas Chromatography, a New Technique for Analytical and Research Laboratories," Ist. "Carlo Erba" ricerche terap., raccolta pubbl. chim., biol. e med., 3, 331-354 (1960); CA 56, 652d

5631. Pozdeev, V. V., Nesmeyanov, A. N., and Dzantiev, B. G., "Synthesis of Tritium-Labeled Organic Compounds by the Method of Recoil Atoms," Metody Polucheniya Radioaktiva, Preparov, Sb. Statei, 1962, 89-93; CA 59, 1505h

5632. Prabucki, A. L., "Application of Gas Chromatography to Determination of Fatty Acids in Naturally Occurring Fats," Mitt. Gebiete Lebensm. u. Hyg., 51, 509-514 (1960); CA 57, 7402f

5633. Prabucki, A. L., and Lenz, F., "Gas-Chromatographic Separation of Aromatic Aldehydes," Helv. Chim. Acta, 45, 2012-2014 (1962); CA 58, 8395b

5634. Prabucki, A. L., and Pfenninger, H., "Gas Chromatographic Separation of the Isomeric Saturated Aliphatic Alcohols Up to Five-Carbon Atoms," Helv. Chim. Acta, 44, 1284-1286 (1961); CA 58, 3915b

5635. Praill, P. F. G., "Some Observations on the Acid Catalyzed Reaction Between Acid Anhydrides and Tertiary Alcohols," Chem. & Ind. (London), 1959, 1123-1124; CA 54, 3173i

5636. Pratt, G. L., and Purnell, J. H., "Sampling Valve for Use in Gas Chromatographic Analysis of the Products of Gaseous Reactions," Anal. Chem., 32, 1213 (1960); CA 54, 23460b

5637. Pratt, G. L., and Purnell, J. H., "Pyrolysis of Acetaldoxime," Proc. Roy. Soc. (London), A260, 317-332 (1961); CA 55, 16504f

5638. Pratt, T. H., and Wolfgang, R., "The Self-Induced Exchange of Tritium Gas with Methane," J. Am. Chem. Soc., 83, 10-17 (1961); CA 55, 16392a

5639. Prelog, V., and Smith, H. E., "Reactions with Microorganisms. XI. The Reduction and Oxidation of (±)-trans- and (±)-cis-2-Decalone with Curvularia Falcata," Helv. Chim. Acta, 42, 2624-2636 (1959); CA 54, 14200f

5640. Preston, Jr., S. T., "Vapor-Liquid Equilibrium Constants from Gas-Chromatography Data," Proc. Ann. Conv. Nat. Gasoline Assoc. Am., Tech. Papers, 38, 33-37 (1959); CA 54, 2879b

5641. Preston, Jr., S. T., "Bibliography on Gas Chromatography, III," in "Gas Chromatography," edited by N. Brenner, J. E. Callen, and M. D. Weiss, Academic Press, Inc., New York, 1962, pp. 571-707

5642. Preston, Jr., S. T., "Temperature Limitations of Stationary Phases Used in Gas Chromatography," J. Gas Chromatog., 1, No. 3, 8-10 (1963)

5643. Preston, Jr., S. T., and Hyder, G., "Gas Chromatography Patents 1952-1963. Part I - United States Patents," J. Gas Chromatog., 1, No. 3, 22-26 (1963); ibid., 1, No. 4, 24-30 (1963)

5644. Pretorius, V., "Gas Chromatography," S. African Ind. Chemist, 10, 303-308 (1956); CA 51, 5501f

5645. Prévot, A., "Difficulties Encountered in Analysis by Gas-Phase Chromatography," Bull. soc. chim. France, 1962, 667-671; CA 57, 1505h

5646. Prévot, A., "Recent Applications of Gas Chromatography to the Analysis of Lipids," Bull soc. chim. France, 1963, 314-316

5647. Prévot, A., and Cabeza, F., "Determination of Traces of Hydrocarbons in Extracted Oils," Rev. franç. corps gras, 7, 34-38 (1960); CA 54, 7184h

5648. Prévot, A., and Cabeza, F., "Effervescences of Concentrated Translucid Soaps. Investigation of the Phenomenon by Gas-Phase Chromatography," Rev. franç. corps gras, 7, 262-266 (1960); CA 54, 15961h

5649. Prévot, A., and Cabeza, F., "The Direct Determination of Water and Short-Chain Fatty Acids by Gas Chromatography," Rev. franç. corps gras, 8, 632-636 (1961); CA 56, 6110g

5650. Prévot, A., and Cabeza, F., "Gas Chromatography of a Number of Less Common Fats and Oils," Rev. franç. corps gras, 9, 149-152 (1962); Abstr., J. Am. Oil Chemists' Soc., 39, No. 7, 25 (1962)

5651. Price, A. R., "Analysis of Light Hydrocarbon Mixtures by Gas Chromatography," 3rd Annual Tech. Mtg., Saline Area Sect., A.I.Ch.E., and Texas-Louisiana Gulf Sect., ACS, Beaumont, Texas, March 1956

5652. Price, S. J. W., and Kutschke, K. O., "The Reactions of Perfluoroethyl Radicals with Hydrogen and Methane," Can. J. Chem., 38, 2128-2136 (1960); CA 55, 6358h

5653. Price, S. J. W., and Trotman-Dickenson, A. F., "Metal-Carbon Bonds. I. Pyrolysis of Dimethyl-mercury and Dimethylcadmium," Trans. Faraday Soc., 53, 939-944 (1957); CA 52, 9950h

5654. Price, S. J. W., and Trotman-Dickenson, A. F., "Metal-Carbon Bonds. II. Pyrolysis of Dimethyl-zinc," Trans. Faraday Soc., 53, 1208-1213 (1957); CA 52, 11736a

5655. Prigogine, I., and Woelbrock, F., "Heat Conductivity and Chemical Reactions in Gases," Brit. Chem. Eng., 2, 596 (1957); CA 52, 4302h

5656. Primavesi, G. R., "Analysis of Light Hydrocarbons," Nature, 184, 2010-2011 (1959); CA 54, 14648d

5657. Primavesi, G. R., Oldham, G. F., and Thompson, R. J., "Hydrogen Flame Detector Using Nitrogen as Carrier Gas," Gas Chromatog., Proc. Symposium, Amsterdam, 1958, 165-174; CA 53, 16610d

5658. Primavesi, G. R., Oldham, G. F., and Thompson, R. J., "Study of the Hydrogen Flame Detector, Using Nitrogen as Carrier Gas," in "Gas Chromatography 1958," edited by D. H. Desty, Academic Press, Inc., New York, 1958, pp. 165-177

5659. Prince, R. G. H., and Ince, J. H., "The Measurement of Intensity of Odour," J. Appl. Chem., 8, 314 (1958); CA 52, 17825g

5660. Prinzler, H., and Hanel, R., "Constituents of Petroleum and Related Products. I. Sulfur Compounds in Tuimasy (USSR) Gasoline," Wiss. Z. Tech. Hochsch. Chem. Leuna-Merseburg, 3, 107-110 (1960/61); CA 55, 27862h

5661. Priori, O., and Panetti, M., "Gas-Chromatographic Analysis of Phenoplasts," Poliplasti, 8, No. 37, 19-22 (1960); CA 57, 1044f

5662. Pritchard, F. W., "Modified 'Thermal Conductivity' Gas Analyzer for Measuring Methane in Air or

Carbon Dioxide," J. Sci. Instr., <u>29</u>, 116-117 (1952); CA <u>46</u>, 5898a

5663. Pritchard, F. W., and Walton, W. H., "A New Technique of Gas Sampling," Chem. & Ind. (London), <u>1952</u>, 166-167; CA <u>46</u>, 8565c

5664. Pritkin, L., "Recording a Variable and Its Integral With One Pen," Control Eng., <u>7</u>, No. 10, 149, 151 (1960)

5665. Privett, O. S., and Blank, M. L., "A Method for the Structural Analysis of Triglycerides and Lecithins," J. Am. Oil Chemists' Soc., <u>40</u>, 70-75 (1963); CA <u>58</u>, 8151b

5666. Privett, O. S., Nadenicek, J. D., Weber, R. P., and Pusch, F. J., "Petrosalinic Acid and Nonsaponifiable Constituents of Parsley Seed Oil," J. Am. Oil Chemists' Soc., <u>40</u>, 28-30 (1963)

5667. Privett, O. S., and Nickell, E. C., "Determination of Structure of Unsaturated Fatty Acids Via Reductive Ozonolysis," J. Am. Oil Chemists' Soc., <u>39</u>, 414-419 (1962); CA <u>58</u>, 1651h

5668. Privett, O. S., Weber, R. P., and Nickell, E. C., "Preparation and Properties of Methyl Arachidonate from Pork Liver," J. Am. Oil Chemists' Soc., <u>36</u>, 443-449 (1959); CA <u>53</u>, 23003d

5669. Proctor, C. M., and Blumer, M., "Discussion of Thermometric Monitor for Chromatographic Streams," Anal. Chem., <u>32</u>, 1897-1898 (1960); CA <u>55</u>, 6944h

5670. Proehl, S., and Bredel, H., "Proof and Interpretation of Hydrocarbon Anomalies," Ber. Geol. Ges. Deut. Demokrat. Rep. Gesamtgebiet Geol. Wiss., <u>7</u>, 354-366 (1962); CA <u>58</u>, 4349b

5671. Profft, E., and Solf, G., "Gas Chromatography of the Chlorothiophenes," Ann., <u>649</u>, 100-103 (1961); CA <u>57</u>, 57b

5672. Prosser, T. J., "The Rearrangement of Allyl Ethers to Propenyl Ethers," J. Am. Chem. Soc., <u>83</u>, 1701-1704 (1961); CA <u>55</u>, 20944a

5673. Provvedi, F., and Ciallella, G., "Gas Chromatography of Some Edible Fats," Riv. ital. sost. grasse, <u>38</u>, 361-366 (1961); CA <u>56</u>, 10647i

5674. Pudov, V. S., and Neiman, M. B., "Decomposition Kinetics of Isotactic Polypropylene Peroxides Studied by Chromatography," Neftekhimiya, <u>2</u>, 918-923 (1962); CA <u>58</u>, 9243d

5675. Pulsford, E. W., "Analysis of Binary Gas Mixtures by a Sonic Method," J. Brit. Inst. Radio Engrs., <u>15</u>, 117 (1955); CA <u>49</u>, 14574c

5676. Pungor, E., "New Developments in Instrumental Analysis," Magyar Kem. Lapja, <u>13</u>, 101-107 (1958); CA <u>52</u>, 13518d

5677. Purcell, J. R., Draper, J. W., and Weitzel, D. H., "Unique Thermal Conductivity Gas Analyzer," Proc. Cryogenic Eng. Conf., 3rd, Boulder, Colo., August, 1957, pub. in "Advances in Cryogenic Engineering," Vol. 3, edited by K. D. Timmerhaus, Plenum Press, N. Y., pp. 191-195; CA <u>52</u>, 13325h

5678. Purcell, J. R., and Keeler, R. N., "Sensitive Thermal Conductivity Gas Analyzer," Rev. Sci. Instr., <u>31</u>, 304-306 (1960); CA <u>55</u>, 8955i

5679. Purchase, I. F. H., "Estimation of Halothane Tensions in Blood by Gas Chromatography," Nature, <u>198</u>, 895-896 (1963)

5680. Purnell, J. H., "A Basis for the Comparison and Choice of Solvents in Vapour Phase Partition Chromatography," in "Vapour-Phase Chromatography," edited by D. H. Desty, Academic Press, Inc., New York, 1957, pp. 52-62

5681. Purnell, J. H., "Physics in the Service of Chemistry. Gas Chromatography," J. Roy. Inst. Chem., <u>82</u>, 586-608 (1958); CA <u>52</u>, 19269a

5682. Purnell, J. H., "Development of Highly Efficient Gas-Liquid Chromatographic Columns," Ann. N. Y. Acad. Sci., <u>72</u>, 592-605 (1959); CA <u>53</u>, 14598i

5683. Purnell, J. H., "Comparison of Efficiency and Separating Power of Packed and Capillary Gas Chromatographic Columns," Nature, <u>184</u>, 2009 (1959); CA <u>54</u>, 14814g

5684. Purnell, J. H., "Correlation of Separating Power and Efficiency of Gas-Chromatographic Columns," J. Chem. Soc., <u>1960</u>, 1268-1274; CA <u>54</u>, 12674g

5685. Purnell, J. H., "Gas Chromatography," John Wiley & Sons, Inc., New York, 1962

5686. Purnell, J. H., "Diffusion and Mass Transfer Coefficient in Packed Gas Chromatographic Columns," Univ. of Houston Intern. Symp. on Adv. in Gas Chromatography, Houston, Tex., January 1963

5687. Purnell, J. H., and Quinn, C. P., "An Approach to Higher Speeds in Gas-Liquid Chromatography," in "Gas Chromatography 1960," edited by R. P. W. Scott, Butterworths, London, 1960, pp. 184-198

5688. Purnell, J. H., and Quinn, C. P., "Higher Speeds in Gas-Liquid Chromatography," Gas Chromatog. Proc. Symposium, 3rd, Edinburgh, <u>1960</u>, 184-198; CA <u>56</u>, 5380e

5689. Purnell, J. H., and Quinn, C. P., "Nature of the Reactions Involved in the Pyrolysis of n-Butane Inhibited by Propylene," Nature, <u>189</u>, 656-658 (1961); CA <u>55</u>, 19749b

5690. Purnell, J. H., and Spencer, M. S., "Use of Solubilizing Agents in Gas-Phase Partition Chromatography," Nature, <u>175</u>, 988-989 (1955); CA <u>49</u>, 12175h

5691. Pursglove, L. A., "Decyanoethylation of N,N-bis(2-Cyanoethyl)amines," J. Org. Chem., <u>24</u>, 576-577 (1959); CA <u>54</u>, 385a

5692. Puschmann, H., "Gas Chromatographic Study of Polyethylene Glycols," Fette, Seifen, Anstrichmittel, <u>65</u>, 1-5 (1963); CA <u>58</u>, 10706a

5693. Puschmann, H., and Miller, J. E., "Determination of Humectants in Tobacco," Z. Lebensm.-Untersuch. u. -Forsch., <u>114</u>, 297-301 (1961); CA <u>55</u>, 14831d

5694. Pust, H. W., Moberg, M. L., and Nishibayashi, M., "Separation of Water, Methyl Hydrazines and
Related Compounds by Gas-Liquid Chromatography," 135th Natl. ACS Mtg., Boston, Mass., April
1959, Program Abstr., p. 10Q; Preprints, Div. Petrol. Chem., <u>4</u>, No. 1, 91-93 (1959)

5695. Putnam, R. E., and Castle, J. E., "Fluorodienes. III. 1,1,2-Trifluoro-3-trifluoromethylbutadiene,"
J. Am. Chem. Soc., <u>83</u>, 389-391 (1961); CA <u>55</u>, 10347d

5696. Putnam, R. F., and Myers, H. W., "Determination of Boron-Hydrogen Bonding by Chemically Active
Gas Chromatography," Anal. Chem., <u>34</u>, 486-488 (1962); CA <u>57</u>, 46f

5697. Pyke, B. H., and Swinbourne, E. S., "Heats of Solution from Vapor-Liquid Partition Chromatography,"
Australian J. Sci., <u>12</u>, 104-106 (1959); CA <u>53</u>, 9807a

5698. Pyke, M., and Hill, D., "Gas Chromatography," Science Digest, <u>51</u>, No. 4, 13-17 (June 1962)

5699. Pypker, J., "Packing Methods for Preparative Columns," in "Gas Chromatography 1960," edited by
R. P. W. Scott, Butterworths, London, 1960, pp. 240-241

5700. Quin, L. D., "Separation of Some Tobacco Alkaloids by Gas Chromatography," Nature, <u>182</u>, 865
(1958); CA <u>53</u>, 5596i

5701. Quin, L. D., "Gas Chromatography in the Analysis of Nonvolatile Acids of Cigarette Smoke," Perkin-
Elmer Instrument News, <u>9</u>, No. 4, 1, 4-5 (Spring 1958)

5702. Quin, L. D., "Alkaloids of Tobacco Smoke. I. Fractionation of Some Tobacco Alkaloids and of the
Alkaloid Extract of Burley Cigarette Smoke by Gas Chromatography," J. Org. Chem., <u>24</u>, 911-914
(1959); CA <u>54</u>, 21660h

5703. Quin, L. D., "Alkaloids of Tobacco Smoke. II. Identification of Some of the Alkaloids in Burley
Cigarette Smoke," J. Org. Chem., <u>24</u>, 914-916 (1959); CA <u>54</u>, 21661b

5704. Quin, L. D., George, W., and Menetee, B. S., "Semiquantitative Gas Chromatographic Studies on the
Organic Acids of Tobacco and Its Smoke," J. Assoc. Offic. Agr. Chemists, <u>44</u>, 367-373 (1961);
CA <u>55</u>, 18021e

5704. Quin, L. D., George, W., and Menetee, B. S., "Semiquantitative Gas Chromatographic Studies on the
Organic Acids of Tobacco and Its Smoke," J. Assoc. Offic. Agr. Chemists, <u>44</u>, 367-373 (1961);
CA <u>55</u>, 18021e

5705. Quin, L. D., and Hobbs, M. E., "Analysis of the Nonvolatile Acids in Cigarette Smoke by Gas Chromatography of Their Methyl Esters," Anal. Chem., <u>30</u>, 1400-1405 (1958); CA <u>52</u>, 17623e

5706. Quin, L. D., Menefee, B. S., and Pappas, N. A., "Alkaloids of Tobacco Smoke. III. Methyl- and
Ethyl-3-pyridyl Ketone as Constituents of Burley Tobacco Cigarette Smoke," J. Org. Chem., <u>26</u>,
267-268 (1961); CA <u>55</u>, 18792c

5707. Quin, L. D., and Pappas, N. A., "Quantitative Determination of Individual Alkaloids in Tobacco by
Gas Chromatography," J. Agr. Food Chem., <u>10</u>, 79-82 (1962); CA <u>57</u>, 12918a

5708. Quin, L. D., Wilder, Jr., P., and Hobbs, M. E., "Chromatographic Determination of Steam-Volatile
Acids in Cigarette Smoke," Anal. Chem., <u>30</u>, 546-547 (1958); CA <u>52</u>, 13195g

5709. Quiram, E. R., "Applications of Wide-Diameter Open Tubular Columns in Gas Chromatography,"
Anal. Chem., <u>35</u>, 593-595 (1963); Metropolitan Regional Mtg., ACS, New York, N. Y., January 1963

5710. Quiram, E. R., and Biller, W. F., "Determination of Trace Quantities of Hydrocarbons in the Atmosphere," Anal. Chem., <u>30</u>, 1166-1171 (1958); CA <u>52</u>, 16662d

5711. Rabinovitch, B. S., and Michel, K. W., "The Thermal Unimolecular Cis-Trans Isomerization of cis-
Butene-2," J. Am. Chem. Soc., <u>81</u>, 5065-5071 (1959); CA <u>54</u>, 3171b

5712. Rabjohn, N., and Stapp, P. R., "Alkylation of β,β,β-Trialkylpropionitriles," J. Org. Chem., <u>26</u>, 45-50
(1961); CA <u>55</u>, 19772b

5713. Rabovskaya, N. S., and Vyakhirev, D. A., "Gas-Chromatographic Analysis of Gaseous Hydrocarbon
Mixtures," Gazovaya Khromatografiya, Trudy 1-oi [Pervoi] Vsesoyuz. Konf., Akad. Nauk S.S.S.R.,
Moscow, <u>1959</u>, 225-230 (Pub. 1960); CA <u>55</u>, 25603c

5714. Rachinskii, V. V., "Basic Rule of Chromatography, Classification of Forms of Chromatography, and
Methods of Chromatographic Separation," Trudy Komissii Anal. Khim., Akad. Nauk S.S.S.R., Inst.
Geokhim.; Anal. Khim., <u>6</u>, 21-29 (1955); CA <u>50</u>, 7648f

5715. Radell, E. A., and Strutz, H. C., "Identification of Acrylate and Methacrylate Polymers by Gas
Chromatography," Anal. Chem., <u>31</u>, 1890-1891 (1959); CA <u>54</u>, 4237f

5716. Rader, C. P., Wicks, Jr., G. E., Aaron, H. S., "Stereochemistry of Hydroxyquinolizidines,"
144th Natl. ACS Mtg., Los Angeles, Calif., March-April 1963, Program Abstr., p. 40M

5717. Radin, N. S., and Akahori, Y., "Fatty Acids of Human Brain Cerebrosides," J. Lipid Research, <u>2</u>,
335-341 (1961); CA <u>56</u>, 5245a

5718. Radin, N. S., Hajra, A. K., and Akahori, Y., "Preparation of Methyl Esters," J. Lipid Research, <u>1</u>,
250-251 (1960); CA <u>54</u>, 18343h

5719. Radzitzky, P. de, and Brandli, J., "Gas Formation During Autoxidation of Organic Compounds in   e
the Liquid Phase," Ind. chim. belge, <u>24</u>, 1051-1059 (1959); CA <u>54</u>, 2145c

5720. Ragelis, E. P., and Gajan, R. J., "Determination of Styrene Monomers in Polystyrene Resins by Gas
Chromatography and Polarography," J. Assoc. Offic. Agr. Chemists, <u>45</u>, 918-921 (1962); CA <u>58</u>,
3551f

5721. Rajbenbach, A., and Szwarc, M., "Addition of Methyl Radicals to Isolated, Conjugated and Cumulated Dienes," Proc. Roy Soc. (London), A251, 394-406 (1959); CA 54, 18321g

5722. Ralls, J. W., "Flash Hydrazone-Ketoacid Exchange Gas Chromatography," 135th Natl. ACS Mtg., Boston, Mass., April 1959, Program Abstr., p. 12A

5723. Ralls, J. W., "Flash Exchange Gas Chromatography for the Analysis of Potential Flavor Components of Peas," J. Agr. Food Chem., 8, 141-153 (1960); CA 55, 4816d

5724. Ralls, J. W., "Rapid Method for Semiquantitative Determination of Volatile Aldehydes, Ketones, and Acids. Flash Exchange Gas Chromatography," Anal. Chem., 32, 332-336 (1960); CA 54, 22174d

5725. Ralls, J. W., "Reactions of Lactams with Diazoalkanes," J. Org. Chem., 26, 66-68 (1961); CA 55, 18705c

5726. Ralls, J. W., Lundin, R. E., and Bailey, G. F., "Preparation and Thermal Rearrangement of Alkenyl 3-Alkenyloxy-2-butenoates. Catalysis of the Aliphatic Claisen Rearrangement by Ammonium Chloride," 144th Natl. ACS Mtg., Los Angeles, Calif., March-April 1963, Program Abstr., p. 11M

5727. Ramaradhya, J. M., and Freeman, G. R., "Radiolysis of Cyclohexane. III. Vapor Phase," J. Chem. Phys., 34, 1726-1729 (1961); CA 55, 24593e

5728. Ramsey, L. H., "Analysis of Gas in Biological Fluids by Gas Chromatography," Science, 129, 900-901 (1959); CA 53, 14200e

5729. Ramsey, L. H., and Sell, G., "The Use of Gas Chromatography in Detection and Localization of Left-to-Right Shunts in Man," 32nd Scientific Session, Am. Heart Assoc., Philadelphia, Pa., October 1959

5730. Ramshaw, E. H., "Gas Chromatography of Vinyl Ketones," J. Chromatog., 10, 303-308 (1963)

5731. Ramshaw, E. H., and Stark, W., "Modern Instruments for Dairy Manufacturing Research. II. Gas Chromatography," J. Dairy Technol., 16, 243-248 (1961); CA 56, 15888c

5732. Rangel Trevine, E., "Partition Chromatography in the Gaseous Phase. I.," Rev. soc. quim. Mex., 2, 69-80 (1958);   "II.," ibid., 145-160; CA 53, 8913g

5733. Rao, C. N. R., and Balasubrhmanyam, S. N., "Mechanism of the Isomerization of Methyl Thiocyanate to the Isothiocyanate," Chem. & Ind. (London), 1960, 625-626; CA 54, 21959a

5734. Raposo, M. R., Estevens., C., and Ralha, A. J. C., "Gas-Liquid Chromatography of Spirits," Rev. Port. Farm., 11, No. 2, 79-88 (1961); CA 56, 5217c

5735. Rapport, M. M., Skipski, V. P., and Sweeley, C. C., "The Lipid Residues in Cytolipin. H.," J. Lipid Research, 2, 148-151 (1961); CA 55, 15563f

5736. Rath, G. A., "Preparative Gas Chromatography, Essential Oils, Mineral Oil," Instr. Meas., Chem. Anal., Elec. Quant., Nucleonics Process Control, Proc. Intern. Conf., Stockholm, 1960, 1, 498-506 (Pub. 1961); CA 57, 6588b

5737. Rath, G. A., "Determining the Usefulness of a New Preparative Chromatograph," Reichstoffe u. Aromen, 10, 293-297 (1960)

5738. Ratney, R. S., and English, Jr., J., "Formation of Cyclopropane Derivatives from 4-Bromocrotonic Esters," J. Org. Chem., 25, 2213-2215 (1960); CA 55, 9335b

5739. Ratusky, J., and Bastar, L., "Separation of Esters of Benzenecarboxylic Acids by Gas-Liquid Chromatography with β-Cyanoethyl Ethers of Polyhydric Alcohols," Chem. & Ind. (London), 1962, 650; CA 57, 2829e

5740. Raulet, C., and Levas, M., "The Action of Acid Halides on Polyhaloacroleins in the Presence of Aluminum Halides. A Route to Polyhalopropenes," Compt. rend., 225, 1406-1408 (1962); CA 58, 1322h

5741. Raulin, J., and Lefort, D., "Importance and Characteristics of Intestinal and Fecal Phospholipides Studied in the Rat. Influence of the Alimentary Regime and of Bacteriostats," Arch. sci. physiol., 14, 239-255 (1960); CA 55, 10613h

5742. Raulins, N. R., and Sibert, L. A., "Formation of Phenols in the Reductive Ozonolysis of Aryl Allyl Ethers. Product Analysis with Gas-Liquid Chromatography," J. Org. Chem., 26, 1382-1386 (1961); CA 55, 23406c; 136th Natl. ACS Mtg., Atlantic City, N. J., September 1959, Program Abstr., p. 79P

5743. Raupp, G., "Choice of the Stationary Phase for Qualitative Gas Chromatographic Analysis," Z. Anal. Chem., 164, 135-146 (1958); CA 53, 6867b

5744. Raupp, G., "Gas Chromatographic Separation of the Lower Fatty Acids," Angew. Chem., 71, 284-285 (1959); CA 53, 15601i

5745. Raupp, G., "Retention Volumes of Organic Compounds Relative to n-Pentane (= 1) on Different Stationary Phases (Chromatographic Data Table 20)," J. Chromatog., 2, D10-D15 (1959); cf. Z. Anal. Chem., 164, 135 (1958)

5746. Ray, N. H., "Gas Chromatography. I. The Separation and Estimation of Volatile Organic Compounds by Gas-Liquid Partition Chromatography," J. Appl. Chem., 4, 21-25 (1954); CA 49, 2257h

5747. Ray, N. H., "Gas Chromatography. II. The Separation and Analysis of Gas Mixtures by Chromatographic Methods," J. Appl. Chem., 4, 82-85 (1954); CA 48, 11244d

5748. Ray, N. H., "A Rapid Chromatographic Method for the Determination of Impurities in Ethylene,"

Analyst, 80, 853-860 (1955); CA 50, 4714c

5749. Ray, N. H., "Rapid Chromatographic Method for the Determination of Bromine-Inert Impurities in Ethylene," Mtg. of Physical Methods and Microchemistry Groups with the Scottish Section, Soc. for Anal. Chem., Stevenson, Scotland, May 1955; Abstr., Anal. Chem., 27, 1667 (1955)

5750. Ray, N. H., "Gas Chromatography," Mtg. of Birmingham and Midland Sect. of S.C.I. with R.I.C., December 1955; cf., Chem. & Ind. (London), 1956, 75

5751. Ray, N. H., "Gas Chromatography," Nature, 180, 403-405 (1957)

5752. Ray, N. H., "Effect of Carrier Gas on the Sensitivity of a Thermal Conductivity Detector in Gas Chromatography," Nature, 182, 1663 (1958); CA 53, 9738d

5753. Ray, N. H., "Gas Chromatography," J. Soc. Glass Technol., 43, 100T-112T (1959); CA 53, 15509a

5754. Ray, N. H., "Effect of the Carrier Gas on the Sensitivity of a Thermal-Conductivity Detector in Gas Chromatography," Nature, 183, 674 (1959); CA 53, 20926b

5755. Ray, N. H., "Carrier Gas and Sensitivity in Gas Chromatography," Nature, 184, 54 (1959)

5756. Ray, N. H., "The Effect of Temperature, Pressure, and Type of Carrier Gas on the Sensitivity of a Thermal Conductivity Detector," Gas Chromatog. Intern. Symposium, 2nd, East Lansing, Mich., 1959, 127-132 (Pub. 1960); CA 55, 20526c

5757. Ray, N. H., "The Effects of Temperature, Pressure, and Type of Carrier Gas on the Sensitivity of a Thermal Conductivity Detector," in "Gas Chromatography," edited by H. J. Noebels, R. F. Wall, and N. Brenner, Academic Press, Inc., New York, 1961, pp. 127-132

5758. Redlich, O., Gable, C. M., Beason, L. R., and Millar, R. W., "Addition Compounds of Thiourea," J. Am. Chem. Soc., 72, 4161-4162 (1950); CA 45, 1850e

5759. Redlich, O., Gable, C. M., Dunlop, A. K., and Millar, R. W., "Addition Compounds of Urea and Organic Substances," J. Am. Chem. Soc., 72, 4153-4160 (1950); CA 45, 3807c

5760. Reece, M. P., "A Simple and Inexpensive Emission Regulator for Ionization Gauges," J. Sci. Instr., 34, 513-514 (1957)

5761. Reed, S. F., "3-Pentadienone," J. Org. Chem., 27, 4116-4117 (1962); CA 58, 7820a

5762. Reed, T. B., and Breck, D. W., "Crystalline Zeolites. II. Crystal Structure of Synthetic Zeolite, Type A," J. Am. Chem. Soc., 78, 5972-5977 (1956); CA 51, 5498d

5763. Reed, III, T. M., "Gas-Liquid Partition Chromatography of Fluorocarbons," Anal. Chem., 30, 221-228 (1958); CA 52, 6892h

5764. Reed, III, T. M., "The Separation of C_6F_{14} Isomers by Gas Chromatography and the Effect of Stationary Phase Concentration," J. Chromatog., 9, 419-430 (1962)

5765. Reed, III, T. M., Walter, J. F., Cecil, R. R., and Dresdner, R. D., "Quantity Purification of Fluorocarbons by Gas-Liquid Chromatography," Ind. Eng. Chem., 51, 271-274 (1959); CA 53, 11942a

5766. Rehm, C. R., and Luders, R. C., "The Gas Chromatographic Behavior of Antihistamines," Metropolitan Regional Mtg., ACS, New York, N. Y., January 1963

5767. Rehnborg, C. S., Ashikawa, J. K., and Nichols, A. V., "Comparison of the Effects of Whole-Body X-Irradiation and Fasting on the Plasma Lipids of Mice," Radiation Res., 16, 860-866 (1962); CA 57, 11495d

5768. Rehnborg, C. S., Nichols, A. V., and Ashikawa, J. D., "Fatty Acid Composition of Mouse Lipides and Lipoproteins," Proc. Soc. Exptl. Biol. Med., 106, 547-549 (1961); CA 55, 14620d

5769. Reid, L. J., "Effect of Molecular Properties on Gas-Phase Mass-Transfer Coefficients," Univ. Microfilms (Ann Arbor, Mich.), Publ. No. 23278, 217 pp.; Dissertation Abstr., 17, 2541 (1957); CA 52, 2467c

5770. Reilley, C. N., "Some Effects of Column Packing and Sample Input in Gas Chromatography," Southeastern Regional Mtg., ACS, Gatlinburg, Tenn., November 1962

5771. Reilley, C. N., Hildebrand, G. P., and Ashley, Jr., J. W., "Responses to Impulse Functions in Gas Chromatography," 141st Natl. ACS Mtg., Washington, D. C., March 1962, Program Abstr., p. 18B

5772. Reilley, C. N., Hildebrand, G. P., and Ashley, Jr., J. W., "Gas Chromatographic Response as a Function of Sample Input Profile," Anal. Chem., 34, 1198-1212 (1962); CA 57, 13166g

5773. Reilley, C. N., Hildebrand, G. P., Papa, L. J., and Norteman, Jr., W. E., "Gas Chromatography with Mixed Bed Packing," 144th Natl. ACS Mtg., Los Angeles, Calif., March-April 1963, Program Abstr., p. 18B

5774. Reilley, C. N., and Sawyer, D. T., "Experiments for Instrumental Methods," McGraw-Hill Book Co., Inc., New York, 1961, pp. 246-259

5775. Rein, H. T., Miville, M. E., and Fainberg, A. H., "Separation of Oxygen and Nitrogen by Packed Column Chromatography at Room Temperature," Anal. Chem., 35, 1536 (1963)

5776. Reinertson, R. P., and Wheatley, V. R., "The Chemical Composition of Human Epidermal Lipids," J. Invest. Dermatol., 32, 49-60 (1959); CA 53, 20383d

5777. Reisch, J. C., Robison, C. H., and Wheelock, T. D., "Permeability of Gas-Liquid Chromatography Columns," in "Gas Chromatography," edited by N. Brenner, J. E. Callen, and M. D. Weiss, Academic Press, Inc., New York, 1962, pp. 91-104

5778. Reiser, R., Choudbury, R. B. R., and Leighton, R. E., "On the Origin of Stearic Acid in Ruminant

Depot Fat," J. Am. Oil Chemists' Soc., 36, 129-130 (1959); CA 53, 9400f

5779. Reiser, R., Murty, N. L., and Rakoff, H., "Biosynthesis of Linoleic Acid from cis-20-Octenoic Acid 1-C^{14} by the Laying Hen," J. Lipid Research, 3, 56-59 (1962); CA 56, 13365d

5780. Reist, E. J., Junga, I. G., and Baker, B. R., "Potential Anticancer Agents. XXXVII. Monofunctional Aziridines Related to Tetramin," J. Org. Chem., 25, 1673-1674 (1960); CA 55, 9410a

5781. Reist, E. J., Junga, I. G., Wain, M. E., Crews, O. P., Goodman, L., and Baker, B. R., "Potential Anticancer Agents. LII. meso-1,4-bis(1-aziridinyl)-2,3-butanediol," J. Org. Chem., 26, 2139-2142 (1961); CA 55, 25958d

5782. Reitsema, R. H., and Allphin, N. L., "Determination of Nitrogen with Gas Chromatography," Anal. Chem., 33, 355-359 (1961); CA 55, 9157c

5783. Renfrow, W. B., and Hawkins, P. J., "Organic Chemistry Laboratory Operations," The Macmillan Co., New York, 1962

5784. Renshaw, A., and Biran, L. A., "A Method for the Introduction of Samples of Long Chain Fatty Acid Methyl Esters on to Gas Chromatography Columns," J. Chromatog., 8, 343-348 (1962); CA 58, 1896e

5785. Ressner, J., "Studies on the Detection of Narcotics in Human Urine by Gas Liquid Chromatography," 14th Annual Mid-America Spectroscopy Symp., Chicago, Ill., May 1963

5786. Reynolds, B. L., Calfee, R. K., and Stricklen, H. L., "Applications of Gas Chromatography to Surgery," Facts & Methods (F&M Scientific Corp.), 3, No. 3, 1-2, 12 (Winter 1962-1963)

5787. Reynolds, J. G., "A Rapid Method for the Identification of Traces of Chlorinated Pesticides," Chem. & Ind. (London), 1962, 729

5788. Reynolds, J. G., and Elgar, K., "Analysis, by Gas Liquid Chromatography, of Extracts from Corps, Soil, and Biological Tissue for Traces of Some Halogenated Pesticides," 18th Intern. Congress of Pure and Applied Chemistry, Montreal, Canada, August 1961

5789. Reynolds, L. M., and Smyth, R. B., "The Analysis of Carboxylic Acid Congeners in Distilled Alcoholic Beverages by Gas Chromatography," 14th Pittsburgh Conf. on Anal. Chem. & Appl. Spectroscopy, Pittsburgh, Pa., March 1963, Program Abstr., p. 62

5790. Reynolds, R. G., and Monkman, J. L., "The Calibration of Flame Ionization Detectors," 14th Pittsburgh Conf. on Anal. Chem. & Appl. Spectroscopy, Pittsburgh, Pa., March 1963, Program Abstr., p. 65

5791. Rezl, V., "Determination of Naphthalene in Products of Coal Tar Rectification by Gas-Liquid Chromatography," Ropa Uhlie, 3, 296-298 (1961); CA 56, 10470g

5792. Rezl, V., Kubiczkova, H., and Kucharczyk, N., "Gas Chromatography of Products from the Amooxidation of Lutidines," Gas Chromatog. Symp., Brno, Czech., June 1962; Abstr., J. Gas Chromatog., 1, No. 4, 10 (1963)

5793. Rezl, V., and Liska, D., "Crude Benzene Analysis," Ropa a Uhlie, 3, 40-42 (1961); CA 56, 14525e

5794. Rhoades, J. W., "Sampling Methods for Analysis of Coffee Volatiles by Gas Chromatography," Food Research, 23, 254-261 (1958); CA 52, 20761i, 53, 19216d

5795. Rhoades, J. W., "Sampling Method for Analysis of Coffee Volatiles by Gas Chromatography," Coffee Brewing Inst. Pub. No. 34, New York, N. Y., June 1958

5796. Rhoades, J. W., "Coffee Volatiles. Analysis of the Volatile Constituents of Coffee," J. Agr. Food Chem., 8, 136-141 (1960); CA 55, 4817c

5797. Ribereau-Gayon, P., "Determination of Ethyl Acetate in Wines by Gas Chromatography," Chim. anal., 43, 161-164 (1961); CA 55, 15825b

5798. Ricciardi, A. I. A., Burgos, J. L., and Cassano, A. E., "Gas-Chromatographic Separation of Light Fractions of Essential Oils," Rev. fac. ing. quim., Univ. nacl. litoral, Santa Fe, Arg., 30, 37-45 (1961); CA 59, 379b

5799. Ricciardi, A. I. S., Cassano, A. E., and Burgos, J. L., "Volatile Essential Oils of the Argentine Seacoast," Rev. fac. ing. quim., Univ. nacl. litoral, Santa Fe, Arg., 30, 27-36 (1961); CA 59, 378h

5800. Richardson, C. W., Johnson, H. D., Gehrke, C. W., and Goerlitz, D. F., "Effects of Environmental Temperature and Humidity on the Fatty Acid Composition of Milk Fat," J. Dairy Sci., 44, 1937-1940 (1961)

5801. Richardson, D. B., Simmons, M. C., and Dvoretzky, I., "The Reactivity of Methylene from Photolysis of Diazomethane," J. Am. Chem. Soc., 82, 5001-5002 (1960); CA 55, 2459i

5802. Richardson, D. B., Simmons, M. C., and Dvoretzky, I., "The Reactivity of Methylene from Photolysis of Diazomethane," J. Am. Chem. Soc., 83, 1934-1937 (1961); CA 55, 23308f

5803. Richardson, R. D., "Process Gas Chromatography for the Measurement of Impurities in Hydrogen as Low as 0.5 p.p.m.," 14th Pittsburgh Conf. on Anal. Chem. & Appl. Spectroscopy, Pittsburgh, Pa., March 1963, Program Abstr., p. 50

5804. Richardson, R. D., and Tarrant, P., "Fluoroolefins. VIII. Preparation of 2-Perfluoroalkyl-1,3-butadienes," J. Org. Chem., 25, 2254-2256 (1960); CA 55, 11282d

5805. Richardson, T., Tappel, A. L., and Gruger, Jr., E. H., "Essential Fatty Acids in Mitochondria," Arch. Biochem. Piophys., 94, 1-6 (1961); CA 55, 27464f

5806. Richardson, T., Tappel, A. L., Smith, L. M., and Houle, C. R., "Polyunsaturated Fatty Acids in Mito-chondria," J. Lipid Research, 3, 344-350 (1962); CA 57, 14154c

5807. Richmond, A. B., "Separation of Nitrogen Trifluoride from Carbon Tetrafluoride by Gas Chromatography," Anal. Chem., 33, 1806 (1961); CA 56, 7974a

5808. Riddle, V. M., "Determination of Traces of Organic Acids in Small Samples. Gas Liquid Chromatography of the Total Concentrate," Anal. Chem., 35, 853-859 (1963)

5809. Ridgway, Jr., J. A., and Schoen, W., "Hexane Isomer Equilibrium," 135th Natl. ACS Mtg., Boston, Mass., April 1959, Program Abstr., p. 15Q; Preprints, Div. Petrol. Chem., 4, No. 2 A5-A11 (1959)

5810. Ridgway, Jr., J. A., and Schoen, W., "New Data for Estimating Hexane Isomer Equilibrium," Ind. Eng. Chem., 51, 1023-1026 (1959); CA 54, 4351c

5811. Riedel, O., and Uhlmann, E., "Gas Chromatographic ('Fraktometric') Analysis of Deuterium in a Stream of Hydrogen," Z. Anal. Chem., 166, 433-439 (1959); CA 53, 16810g

5812. Rieker, A., Mueller, E., and Beckert, W., "Aroxyls as Electron Acceptors - Their Behavior Towards Grignard Compounds," Z. Naturforsch., 17b, 718-722 (1962); CA 59, 487d

5813. Riemenschneider, R. W., "Fractionation and Analysis of Small Amounts of Tissue Lipids," 37th Natl. ACS Mtg., Cleveland, Ohio, April 1960, Program Abstr., pp. 10A-11A

5814. Riemenschneider, R. W., "Methods and Techniques for Analysis of Complex Lipids," Chem. & Ind. (London), 1960, 1593-1594

5815. Riesz, P., and Wilzbach, K. E., "Labeling of Some C_6 Hydrocarbons by Exposure to Tritium," J. Phys. Phys. Chem., 62, 6-9 (1958); CA 52, 7895c; 132nd Natl. ACS Mtg., New York, N. Y. September 1957, Program Abstr., p. 14R

5816. Riganesis, M. D., "The Chromatographic Method for the Study of Essential Oils and Their Components," Essenze deriv. agrumari, 26, 29-61 (1956); CA 50, 15027b

5817. Rigby, C. A., "Chromatograph for Polymer Research," Brit. Plastics, 35, 517 (1962); CA 58, 578c

5818. Rigby, F. L., and Bethune, J. L., "Rapid Methods for the Determination of Total Hop Bitter Substances (Iso-Compounds) in Beer," J. Inst. Brewing, 61, 325-332 (1955); CA 54, 10234g

5819. Rigby, F. L., and Bethune, J. L., "Analysis of Hop Oil by Gas-Liquid Partition Chromatography," J. Inst. Brewing, 63, 154-161 (1957); CA 54, 10230a

5820. Rigby, F. L., Sihto, E., and Bars, A., "Rapid Method for Detailed Analysis of the α-Acid Fraction of Hops by Gas Chromatography," J. Inst. Brewing, 66, 242-249 (1960); CA 54, 18874f

5821. Rigby, F. L., Sihto, E., and Bars, A., "Additional Constituents of Hops," J. Inst. Brewing, 68, 60-65 (1962); CA 56, 12087h

5822. Righi, H., "Process Monitoring by Continuous Gas Chromatography," Dechema Monograph., 35, No. 528-555, 114-124 (1959); CA 57, 6205d

5823. Rijnders, G. W. A., "Separation Possibilities in Gas-Liquid Chromatography," Chem. Weekblad, 54, 669-674 (1958); CA 54, 1160f

5824. Rijnders, G. W. A., "Measurement of Retention Volumes," Gas Chromatography Discussion Group Informal Symp., Bristol, England, September 1959

5825. Riley, B., "Gas Analysis by the Use of the Discharge Tube," Gas Chromatog., Proc. Symposium, 3rd, Edinburgh, 1960, 81-87; CA 55, 25387a

5826. Riley, B., "Gas Analysis by the Use of the Discharge Tube," in "Gas Chromatography 1960," edited by R. P. W. Scott, Butterworths, 1960, pp. 81-87

5827. Rillaers, G., and Verzele, M., "Prehumulone, a New α-Acid," Bull. soc. chim. Belges, 71, 438-445 (1962); CA 58, 5525d

5828. Rinehart, Jr., K. L., Goldberg, S. I., Tarimu, C. L., and Culbertson, T. P., "Cyclopropene Rear-rangement in the Polymerization of Sterculic Acid," J. Am. Chem. Soc., 83, 225-231 (1961); CA 55, 10308i

5829. Rinetti, M., and Giovetti, G. L., "Influence of Heat Treatment on Chemical and Biological Character-istics of Alimentary Fats. Spectrophotometric and Gas-Chromatographic Variations," Minerva Dietol., 2, No. 3, 131-136 (1962); CA 58, 4972d

5830. Ring, R. D., "The Effect of the Ratio of Partition Liquid to Inert Support on the Separation of (1) Mono-, Di-, and Tri-ethylene Glycols, and (2) Cis- and Trans-2,5-dimethylpiperazine," in "Gas Chromatography," edited by V. J. Coates, H. J. Noebels, and I. S. Fagerson, Academic Press, Inc., New York, 1958, pp. 195-201

5831. Ring, R. D., and Riley, F. W., "Gas-Liquid Chromatographic Analysis of Amine Mixtures," 6th Detroit Anachem Conf., Detroit, Mich., October 1958

5832. Rio, A., Ripa, D., and Tribastone, S., "Application of Gas Chromatography to the Continuous Control of Industrial Plants. I. Analysis of a Mixture of C_1-C_4 Hydrocarbons," Chim. e ind. (Milan), 41, 1185-1188 (1959); CA 54, 15763d

5833. Rippere, R. E., "Isolation and Identification of a Low Concentration Component by Gas Chromatography and Mass Spectrometry," in "Gas Chromatography," edited by V. J. Coates, H. J. Noebels, and I. S. Fagerson, Academic Press, Inc., New York, 1958, pp. 223-224

5834. Risinger, G. E., and Mach, E. E., "Thermal Decomposition of Methyl Phenylacetate," Nature, 196, 1091-1092 (1962)

5835. Ritter, H., and Schnier, H., "Gas Chromatographic Analysis of Coal Tar By-Products, with Special Reference to Those Having High-Boiling Constituents," Z. Anal. Chem., 170, 310-317 (1959); CA 54, 3922b

5836. Ritter, W., and Hanni, H., "The Synthesis of Volatile Fatty Acids from Amino Acids by Micrococci," Pathol. et microbiol., 23, 669-680 (1960); CA 55, 6597i

5837. Roaldi, A., "Quantitative Analysis in Gas Chromatography," 4th Annual Gas Chromatography Institute, Canisius College, Buffalo, N. Y., April 1962

5838. Robb, J. C., and Vofsi, D., "The Quantitative Separation of Mixtures Containing Vinyl Acetate and Bromotrichloromethane by Vapour-Phase Chromatography," in "Vapour-Phase Chromatography," edited by D. H. Desty, Academic Press, Inc., New York, 1957, pp. 428-431

5839. Roberts, A. L., and Ward, C. P., "Application of Modern Analytical Technique in the Gas Industry. I. Analysis of Primary Flash Distillate (p.f.d.) Feedstock, Using Gas Chromatography," Gas Council, (Gt. Brit.) Research Commun. No. GC78, 6 pp. (1961); CA 56, 8993c

5840. Roberts, C. B., "The Separation of Mixtures of High-Boiling Compounds Using Very Short Columns," Aerograph Research Notes (Wilkens Instrument & Research, Inc., Walnut Creek, Calif.), p. 5, Summer 1963

5841. Roberts, H. L., and Ray, N. H., "Sulfur Chloride Pentafluoride: Preparation and Some Properties," J. Chem. Soc., 1960, 665-667; CA 54, 12864a

5842. Roberts, J. B., "Determination of the Cohumulone Content of Hop Resins," J. Inst. Brewing, 67, 337-339 (1961); CA 55, 26358d

5843. Roberts, J. B., "Hop Oil. I. Preliminary Investigations of the Oxygenated Fraction," J. Inst. Brewing, 68, 197-200 (1962); CA 57, 1278g

5844. Roberts, J. B., "Elution Sequence as a Function of Temperature in Gas Chromatography," Nature, 193, 1071-1072 (1962); CA 57, 2352g

5845. Roberts, L. R., and McKetta, J. J., "Analysis of Multicomponent Hydrocarbon Systems Containing Large Amounts of Nitrogen," J. Gas Chromatog., 1, No. 3, 14-17 (1963)

5846. Roberts, R. M., and Han, Y. W., "Alkylbenzenes. XI. Rearrangements of Pentylbenzenes Induced by Aluminum Chloride," J. Am. Chem. Soc., 85, 1168-1171 (1963); CA 58, 12387b

5847. Roberts, R. M., and Shiengthong, D., "Alkylbenzenes. VIII. Rearrangements Accompanying Friedel-Crafts Alkylations with Propyl and Butyl Chlorides," J. Am. Chem. Soc., 82, 732-735 (1960); CA 54, 24467h

5848. Robertson, G. R., and Jacobs, T. L., "Laboratory Practice of Organic Chemistry," 4th ed., The Macmillan Co., New York, 1962

5849. Robertson, J. A., and Harper, W. J., "Factors Affecting Distribution of Fatty Acids Hydrolyzed from Milk Fat by Milk Lipase," 57th Annual Mtg., Am. Dairy Sci. Assoc., College Park, Md., June 1962; Abstr., J. Dairy Sci., 45, 646 (1962)

5850. Robertson, R. H. S., "Sepiolite: A Versatile Raw Material," Chem. & Ind. (London), 1957, 1492-1495; CA 52, 8479d

5851. Robin, J., and Blandenet, G., "Porosity of Supports for Gas-Liquid Chromatography," Compt. rend., 255, No. 6, 1113-1115 (1962); CA 57, 15777i

5852. Robinson, B., "The Autoxidation of 2-Hydroxy-1,3,3-trimethyl-2-t-butylindoline and 1,3,3-Trimethyl-2-methyleneindoline," J. Chem. Soc., 1963, 586-590

5853. Robinson, J. W., "Determination of Monohydric Alcohols by Gas Chromatography," Anal. Chim. Acta, 27, 377-380 (1962); CA 58, 22g

5854. Robinson, M. A., and McKinney, C. B., "The Influence of Temperature on High Speed Analyses in Gas-Liquid Partition Chromatography," 12th Pittsburgh Conf. on Anal. Chem. & Appl. Spectroscopy, Pittsburgh, Pa., February-March 1961, Program Abstr., p. 58

5855. Robinson, M. A., and Said, A. S., "Theory of Gas Chromatography. Plate Theory and Its Application to the Nonlinear Adsorption Isotherm," 138th Natl. ACS Mtg., New York, N. Y., September 1960, Program Abstr., p. 5B

5856. Robinson, R. H., "Automatic Chemical Analysis. I. An Automated Dissolved Hydrogen Instrument," 11th Pittsburgh Conf. on Anal. Chem. & Appl. Spectroscopy, Pittsburgh, Pa., February-March 1960, Program Abstr., p. 44

5857. Robinson, R. H., and Conklin, D. B., "Pre-Integrated Gas Thermal Conductometry," U. S. At. Energy Comm., WAPD-BT-16, 147-150 (1959); CA 54, 14818b

5858. Robinson, R. J., "Gas Chromatography as a Recovery Process," Univ. Microfilms (Ann Arbor, Mich.), L. C. Card No. Mic 59-4979; Dissertation Abstr., 20, 1706 (1959); CA 54, 2832b

5859. Robson, J. W., and Askew, W. B., "Gas Chromatographic Separation of Sulfur Tetrafluoride and Thionyl Fluoride," J. Chromatog., 7, 409-411 (1962); CA 57, 14691d

5860. Robson, P., Stacey, M., Stephens, R., and Tatlow, J. C., "Aromatic Polyfluoro-compounds. VI. Penta- and 2,3,5,6-Tetrafluorothiophenol," J. Chem. Soc., 1960, 4754-4760; CA 55, 9325b

5861. Rock, H., "Gas Chromatography and Extractive Distillation," Chem.-Ing.-Tech., 28, 489-495 (1956); CA 50, 14311e

5862. Rock, H., "Symposium on Chromatography of Gases and Vapors, London, 1956," Brennstoff-Chem., 37, 347-349 (1956)

5863. Rock, H., "Gas Chromatography," in "Selected Modern Separation Techniques for the Purification of Organic Compounds," Steinkopff, Darmstadt, 1957, pp. 72-137

5864. Rockoff, C. V., "Quantitative Analysis of the Inorganic Gases and Lower Hydrocarbons Using Gas Chromatography," 9th Detroit Anachem Conf., Detroit, Mich., October 1961, Program Abstr., p. 27

5865. Rockoff, C. V., "Calibration of Gas Chromatographs for Quantitative Gas Analysis," 13th Pittsburgh Conf. on Anal. Chem. & Appl. Spectroscopy, Pittsburgh, Pa., March 1962, Program Abstr., p. 54

5866. Rodig, O. R., and Ellis, L. C., "Conformational Studies. I. The Relative Stabilities of the cis-2-Decalols," J. Org. Chem., 26, 2197-2202 (1961); CA 55, 27223d

5867. Roe, E. T., and Swern, D., "Branched Carboxylic Acids from Long-Chain Unsaturated Compounds and Carbon Monoxide at Atmospheric Pressure," J. Am. Oil Chemists' Soc., 37, 661-668 (1960); CA 55, 5331b

5868. Rogers, J. A., "Instrumental Evaluation of Citronella Oils and Palmarosa Oil. Geraniol, Citronellol and Other Constituents," Proc. Sci. Sect., Tiolet Goods Assoc., No. 32, 9-13 (December 1959); CA 54, 11387g

5869. Rogers, J. A., "Instrumentation Techniques as Applied to Quality Standards of Perfume Materials," Am. Perfumer, 77, No. 3, 21-28 (1962); CA 57, 2351e

5870. Rogers, J. A., and Toth, Z. E., "Instrumental Evaluation of Commercial Citronellols and Geraniols and Their Esters," Proc. Sci. Sect., Toilet Goods Assoc., 35, 29-36 (1961); CA 55, 19143e

5871. Rogers, L. B., "Use of Inorganic Support Materials in VPC," 5th Eastern Analytical Symposium, New York, N. Y., November 1963

5872. Rogers, L. B., and Altenau, A. G., "New Adsorbents for Gas Chromatography," Anal. Chem., 35, 915-916 (1963)

5873. Rogers, L. B., and Reilley, C. N., "Minimizing Time and Optimizing Resolution in Gas Chromatography," 138th Natl. ACS Mtg., New York, N. Y., September 1960, Program Abstr., p. 6B

5874. Rogers, L. B., and Spitzer, J. C., "Crushed Unfused Vycor as a Support for Gas Chromatography," Anal. Chem., 33, 1959-1960 (1961); CA 56, 7968e

5875. Rogers, L. H., "Relative Merits of Gas Chromatography, Colorimetry, and Spectrometry for Air Pollution Studies," Symp. on Instrumentation in Atmospheric Analysis, ASTM Spec. Tech. Pub. No. 250, 42-48 (1958); CA 54, 3811g

5876. Roginskii, S. Z., Yanovskii, M. I., Zhabrova, G. M., Kadenatsii, B. M, and Vinogradova, O. M., "Application of Gas Chromatography to Catalyst Studies," Gazovaya Khromatografiya, Trudy 1-oi [Pervoi] Vsesoyuz Konf., Akad. Nauk S.S.S.R., Moscow, 1959, 135-143 (Pub. 1960); CA 56, 4132c

5877. Roginskii, S. Z., Yanovskii, M. I., Zhabrova, G. M., Vinogradova, O. M., Kadenatsii, B. M., and Markova, Z. A., "Catalytic Synthesis of Unsaturated Hydrocarbons of the C_4-Series Containing C^{14} by Vapor Phase Distribution Radiochromatography," Doklady Akad. Nauk SSSR, 121, 674-677 (1958); CA 54, 24330b

5878. Rogozinski, M., Shorr, L. M., and Warshawsky, A., "Observations on the Gas Chromatographic Analysis of Aqueous Alcohols," J. Chromatog., 8, 429-432 (1962); CA 58, 3870h

5879. Rogozinski, M., Shorr, L. M., and Warshawsky, A., "Gas Chromatographic Analysis of Aqueous Alcohols. II. Quantitative Analysis of Aqueous Butanol Solutions Containing Non-Volatile Salts," J. Chromatog., 10, 114-116 (1963)

5880. Rogozinski, M., and Warshawsky, A., "Comments on Cyanoethylated Polyols as Gas Chromatographic Stationary Phases," J. Gas Chromatog., 1, No. 3, 30 (1963)

5881. Rohleder, H., "Gas-Chromatographic Determination of the o-Cresol Component in Tritolyl Phosphate," Z. Anal. Chem., 180, 32-36 (1961); CA 55, 18450h

5882. Rohrschneider, L., "'Polarity' of the Stationary Phase in Gas Chromatography," Z. Anal. Chem., 170, 256-263 (1959); CA 54, 3045i

5883. Rombaut, J., and Fodderie, C., "Chromatography of Gases Containing Light Hydrocarbons and Inert Gases (Carbon Dioxide, Carbon Monoxide, Nitrogen, Oxygen, and Hydrogen)," Ind. Chim. (Belge), 23, 845-854 (1958); CA 52, 21013h

5884. Romeike, A., "Chromatography of Alkaloids," in "Chromatography," edited by E. Heftmann, Reinhold Publishing Corp., New York, 1961, Chapter 20

5885. Romovacek, J., and Novotny, Z., "Analysis of Crude Benzene by Gas Chromatography," Brennstoff-Chem., 42, 161-165 (1961); CA 55, 23186b

5886. Root, M. J., "Quality Control of Chlorofluoromethanes and Chlorofluoroethanes by Gas Chromatography," in "Gas Chromatography," edited by V. J. Coates, H. J. Noebels, and I. S. Fagerson, Academic Press, Inc., New York, 1958, pp. 99-103

5887. Root, M. J., and Maury, M. J., "Gas-Chromatographic Analysis of Aerosol Products," J. Soc. Cosmetic Chemists, 8, 92-107 (1957); CA 51, 9095b

5888. Root, M. J., and Maury, M. J., "Analysis of Volatile Aerosol Constituents," Soap Chem. Specialties, <u>33</u>, No. 3, 75-78, 109; No. 4, 101, 103, 105, 107 (1957); CA <u>51</u>, 8457c

5889. Roper, Jr., J. N., "A Heater for Gas Chromatographic Columns," Anal. Chem., <u>32</u>, 447-448 (1960); CA <u>54</u>, 12674f

5890. Roper, W. A., Doscher, T., and Kobayashi, R., "Evidence of Chromatographic Effect During Flow of Gases Through Oil Field Cores," J. Petrol. Tech., <u>10</u>, No. 3, 61-63 (1958); CA <u>52</u>, 8515e

5891. Rose, A., "Separation Processes and the Petroleum Industry Before and After 1961," 140th Natl. ACS Mtg., Chicago, Ill., September 1961, Program Abstr., pp. 14S-15S

5892. Rose, A., and Schrodt, V. N., "Correlation and Prediction of Vapor Pressures of Homologs. Structure Parameters and Gas-Chromatography Data," J. Chem. Eng. Data, <u>8</u>, 9-13 (1963); CA <u>58</u>, 6207h

5893. Rose, A., and Sumantri, R. B., "Rigorous Calculations for Prediction of Scaled-Up Gas Chromatography," 144th Natl. ACS Mtg., Los Angeles, Calif., March-April 1963, Program Abstr., p. 18J

5894. Rose, B. A., "Gas Chromatography and Its Analytical Applications - A Review," Analyst, <u>84</u>, 574-595 (1959); CA <u>54</u>, 10630d

5895. Rosen, A. A., Musgrave, L. R., and Lichtenberg, J. J., "Characterization of Coastal Oil Pollution by Submarine Seeps," Calif. State Water Pollution Control Board Publ. No. <u>21</u>, No. 2, 47-61 (1959); CA <u>54</u>, 18838b

5896. Rosenfeld, R. S., "Analyses of Urinary Steroids by Gas Chromatography," 4th Annual Eastern Analytical Symp., New York, N. Y., November 1962, Program Abstr., p. 24

5897. Rosenfeld, R. S., and Lebeau, M. C., "Analysis of Steroid Hormone Metabolites by Vapor Phase Chromatography," 46th Annual Mtg., Federation of Am. Soc. for Exper. Biol., Atlantic City, N. J., April 1962; Abstr., Federation Proc., <u>21</u>, 281 (1962)

5898. Rosenfeld, R. S., Lebeau, M. C., Jandorek, R. D., and Salumaa, T., "Analysis of Urinary Extracts by Gas Chromatography. $3\alpha,17$-Dihydroxypregnan-20-one, Pregnane-$3\alpha,17,20\alpha$-triol, and Δ^5-Pregnene-$3\beta,17,20\alpha$-triol," J. Chromatog., <u>8</u>, 355-358 (1962); CA <u>58</u>, 5953a

5899. Rosenfeld, R. S., Lebeau, M. C., Shulman, S., and Seltzer, J., "Analysis of Fecal Sterols by Gas Chromatography," J. Chromatog., <u>7</u>, 293-296 (1962); CA <u>57</u>, 14097b

5900. Rosenfeld, W. D., and Silverman, S. R., "Carbon Isotope Fractionation in Bacterial Production of Methane," Science, <u>130</u>, 1658-1659 (1959); CA <u>54</u>, 7827b

5901. Rosengren, K., "A Systematic Study of the Photolysis of Some Dialkyl Disulfides in a Rigid Glass at 77°K," Acta Chem. Scand., <u>16</u>, 1401-1417 (1962); CA <u>57</u>, 14618f

5902. Rosie, D. M., and Grob, R. L., "Thermal Conductivity Behavior. Importance in Quantitative Gas Chromatography," Anal. Chem., <u>29</u>, 1263-1264 (1957); CA <u>52</u>, 4382a; 131st Natl. ACS Mtg., Miami, Fla., April 1957, Program Abstr., p. 10B

5903. Ross, G. N., "The Mass Spectrometer in the Chemical Industry," Brit. Chem. Eng., <u>2</u>, 614-619 (1957); CA <u>52</u>, 3423e

5904. Ross, J. M., "Gas Chromatography Using Molecular Sieve 5A," Chem. & Ind. (London), <u>1961</u>, 1523

5905. Ross, R. A., and Stimson, V. R., "Catalysis by Hydrogen Halides in the Gas Phase. III. Isopropyl Alcohol and Hydrogen Bromide," J. Chem. Soc., <u>1960</u>, 3090-3094; CA <u>54</u>, 24337d

5906. Rossi, B. B., and Staub, H. H., "Ionization Chambers and Counters," McGraw-Hill Book Co., Inc., New York, 1949

5907. Rossi, C., Munari, S., Cengarle, L., and Tealdo, G. F., "Gas Chromatography of C_1-C_4 Hydrocarbons on Organophilic Silica," Chim. e ind. (Milan), <u>42</u>, 724-727 (1960); CA <u>55</u>, 21986c

5908. Roth, J. F., and Ellwood, R. J., "Determination of Surface Area Using a Gas Chromatograph," Anal. Chem., <u>31</u>, 1738-1739 (1959); CA <u>55</u>, 9972h

5909. Rothman, L. A., and Becker, E. I., "Hydrogenolyses of Aromatic Halides with Triphenyltin Hydride," J. Org. Chem., <u>25</u>, 2203-2206 (1960); CA <u>55</u>, 9337f

5910. Rotini, O. T., "Analysis of Tunisian Olive Oils," Oleari, <u>15</u>, 65-70 (1961); CA <u>55</u>, 25294f

5911. Rotini, O. T., Baragli, S., and Gentili, M., "Variation of Lipids Composition in Tunisian Olives During Ripening," Chim. e ind. (Milan), <u>44</u>, 1126-1129 (1962); CA <u>58</u>, 749d

5912. Rotzsche, H., "Gas-Liquid Chromatography of Fluorocarbon Compounds," in "Gas Chromatographie 1959," edited by R. E. Kaiser and H. G. Struppe, Akademie Verlag, Berlin, 1959, pp. 275-303

5913. Rotzsche, H., "Fluorine Chemistry. IV. Gas Chromatography of Fluorocarbons," Z. Anal. Chem., <u>175</u>, 338-350 (1960); CA <u>55</u>, 4233i

5914. Rotzsche, H., "A New Type Polar Stationary Phase with Adjustable Selectivity Coefficient," 4th Intern. Gas Chromatography Symp., Hamburg, Germany, June 1962; pub. in "Gas Chromatography, 1962." M. van Swaav. ed.. Butterworth & Co., London, 1962

5915. Rotzsche, H., and Roesler, H., "Silicones. LI. Gas-Chromatographic Study of Methylhydrocyclo-siloxanes," Z. Anal. Chem., <u>181</u>, 407-417 (1961); CA <u>57</u>, 10006i

5916. Rouayheb, G. M., Folmer, O. F., and Hamilton, W. C., "Quantitative Analysis of Aliphatic, Aromatic and Alicyclic Hydrocarbon Systems by Gas Chromatography Using Capillary Columns," 11th Pittsburgh Conf. on Anal. Chem. & Appl. Spectroscopy, Pittsburgh, Pa., February-March 1960, Program Abstr., p. 38

5917. Rouayheb, G. M., Folmer, O. F., and Hamilton, W. C., "Quantitative Gas Chromatographic Analysis of Hydrocarbon Systems Using the Lovelock Diode Detector and Capillary Columns," Anal. Chim. Acta, 26, 378-390 (1962); CA 57, 1517d

5918. Rouayheb, G. M., and Hamilton, W. C., "Deterioration of Solid-Coated Capillary Columns on Standing," Nature, 191, 801-802 (1961); CA 56, 4558g

5919. Rouayheb, G. M., Perkins, Jr., G., Lively, L. D., and Hamilton, W. C., "Response of the Gas Chromatographic Flame Ionization Detector to Different Functional Groups," 3rd Intern. Symp. on Gas Chromatography, ISA, East Lansing, Mich., June 1961; ISA Proc., 3, 185-194 (1961)

5920. Rouayheb, G. M., Smith, E. E., Carel, A. B., and Folmer, O. F., "Determination of Water in Organic Compounds by Gas Chromatography," 16th Southwest Regional Mtg., ACS, Oklahoma City, Okla., December 1960

5921. Roubal, W. T., "Tuna Fatty Acids. I. Initial Studies on the Composition of the Light and Dark Meats of Bluefin Tuna (Thunnus Thynnus) - Structural Isomers of the Monoenoic Fatty Acids," J. Am. Oil Chemists' Soc., 40, 213-215 (1963)

5922. Roubal, W. T., "Tuna Fatty Acids. II. Investigations of the Composition of Raw and Processed Domestic Tuna," J. Am. Oil Chemists' Soc., 40, 215-218 (1963)

5923. Rouit, C., "The Chromatographic Method for Analysis of Gases and Vapors," Rev. inst. franç. petrole et Ann. combustibles liquides, 11, 213-230 (1956); CA 50, 13660f

5924. Rouit, C., "The Analysis and Control of Refinery-Gas Streams Using the Chromatographic Technique (Janak Method)," in "Vapour-Phase Chromatography," edited by D. H. Desty, Academic Press, Inc., New York, 1957, pp. 291-303

5925. Roussel, J., "Chromatographic Analysis of Residual Gas from the Manufacture of Formaldehyde and Acetaldehyde," Mem. poudres, 42, 457-472 (1960); CA 55, 12149f

5926. Rovesti, G., and Rovesti, P., "Chromatographic Deterpenation of Citrus Oils and Their Uses in Perfumes," Riv. Ital. Essenze-Profumi, Piante Offic.-Oli Vegetali-Saponi, 44, 236-238 (1962); CA 57, 11327c

5927. Rovesti, P., and Variati, G. L., "Lemon Grass Extract in Italian Somaliland," France et ses Parfums, 3, No. 19, 39-44 (1960); CA 55, 7765a

5928. Rowan, Jr., R., "Prediction of Retention Temperatures in Programmed Temperature Gas Chromatography. A Descriptive Equation and Computational Method," Anal. Chem., 33, 510-515 (1961); CA 55, 16259g; 138th Natl. ACS Mtg., New York, N. Y., September 1960, Program Abstr., p. 5B

5929. Rowan, Jr., R., "Identification of Hydrocarbon Peaks in Gas Chromatography by Sequential Application of Class Reactions," Anal. Chem., 33, 658-665 (1961); CA 55, 15211b; 137th Natl. ACS Mtg., Cleveland, Ohio, April 1960, Program Abstr., p. 28B

5930. Rowan, Jr., R., "Programmed Temperature Gas Chromatography at Constant Pressure," Anal. Chem., 34, 1042-1046 (1962); CA 57, 10512b

5931. Rowe, C. E., "The Biosynthesis of Phospholipids by Human Blood Cells," Biochem. J., 73, 438-442 (1959); CA 54, 10083d

5932. Rowe, C. E., "The Phospholipids of Human Blood Plasma and Their Exchange with the Cells," Biochem. J., 76, 471-475 (1960); CA 54, 22899f

5933. Rowland, F. S., Lee, J. K., and White, R. M., "Gas Counting of Tritium Labeled Compounds," Oklahoma Conf. on Radioisotopes in Agriculture, Stillwater, Okla., April 1959, Rept. TID-7578, pp. 39-48 (1960); CA 54, 18106b

5934. Rowlinson, J. S., and Thacker, R., "The Physical Properties of Some Fluorine Compounds and Their Solutions. III. Perfluorocyclohexane and Perfluoromethylcyclohexane," Trans. Faraday Soc., 53, 1-8 (1957); CA 51, 10156a

5935. Roxburgh, J. M., "Determination of Oxygen Utilization in Fermentations by Gas Chromatography," Can. J. Microbiol., 8, 221-227 (1962); Biol. Abstr., 39, 15946

5936. Royals, E. E., "Pyrolysis of Esters. I. Nonselectivity in the Direction of Elimination by Pyrolysis," J. Org. Chem., 23, 1822 (1958); CA 53, 21609g

5937. Ruby, E. D., "High Temperature Gas Chromatography," 6th Detroit Anachem Conf., Detroit, Mich., October 1958

5938. Rudenko, B. A., and Kucherov, V. F., "NaCl as a Solid Support in Gas-Liquid Chromatography," Doklady Akad. Nauk S.S.S.R., 145, 577-579 (1962); CA 57, 14410e

5939. Rudenko, B. A., Kucherov, V. F., Smit, V. A., and Semenovskii, A. V., "Gas Chromatography of Isoprenoid Compounds," Izv. Akad. Nauk S.S.S.R., Otd. Khim. Nauk, 1962, 236-243; CA 57, 4038a

5940. Rudenko, B. A., Yufit, S. S., Ivanova, L. N., and Kucherov, V. F., "Gas-Liquid Chromatography for the Analysis of Some Hydrocarbon Mixtures," Izvest. Akad. Nauk S.S.S.R., Otdel. Khim. Nauk, 1960, No. 7, 1147-1152; CA 55, 4251f

5941. Rudloff, E. von, "The Wax of the Leaves of Picea Pungens (Colorado Spruce)," Can. J. Chem., 37, 1038-1042 (1959); CA 53, 19862e

5942. Rudloff, E. von, "Separation of Some Terpenoid Compounds by Gas-Liquid Chromatography," Can. J. Chem., 38, 631-640 (1960); CA 54, 18020h

5943. Rudloff, E. von, "Gas-Liquid Chromatography of Terpenes. II. Dehydration Products of α-Terpineol," Can. J. Chem., 39, 1-12 (1961); CA 55, 13470a

5944. Rudloff, E. von, "Gas-Liquid Chromatography of Terpenes. III. The Use of Oleic Esters as Liquid Phases," Can. J. Chem., 39, 1190-1199 (1961); CA 55, 23938e

5945. Rudloff, E. von, "Gas-Liquid Chromatography of Terpenes. IV. The Analysis of the Essential Oil of the Leaves of Eastern White Cedar," Can. J. Chem., 39, 1200-1206 (1961); CA 55, 23938f

5946. Rudloff, E. von, "A Simple Reagent for the Specific Dehydration of Terpene Alcohols," Can. J. Chem., 39, 1860-1864 (1961); CA 56, 1483f

5947. Rudloff, E. von, "The Isolation of Guaiol and α-Cadinol from the Wood Oil of Neocallitropsis Araucarioides," Chem. & Ind. (London), 1962, 743-744; CA 57, 4699g

5948. Rudloff, E. von, "Gas-Liquid Chromatography of Terpenes. VI. The Volatile Oil of Thuja Plicata," Phytochemistry, 1, 195-202 (1962); CA 57, 15257f

5949. Rudloff, E. von, "Gas-Liquid Chromatography of Terpenes. VII. The Dehydration of p-Menthan-8-ol," Can. J. Chem., 41, 1-8 (1963); CA 58, 3461e

5950. Rudloff, E. von, and Erdtman, H., "Stereochemistry of Occidentalol and Its Hydrogenation Products," Tetrahedron, 18, 1315-1329 (1962); CA 58, 6870g

5951. Ruggieri, S., and Cioni, P., "Gas-Chromatographic Studies on the Composition of the Fatty Acids of the Liver of the Rat Intoxicated with Carbon Tetrachloride," Boll. Soc. Ital. Biol. Sper., 38, 28-29 (1962); CA 59, 1023d

5952. Ruhlmann, K., and Giesecke, W., "Gas Chromatography of Silylated Amino Acids," Angew. Chem., 73, 113 (1961); CA 55, 11191i

5953. Runeberg, J., "The Chemistry of the Natural Order Cupressales. XXVII. Heartwood Constituents of Juniperus Utahensis Lemm," Acta Chem. Scand., 14, 797-804 (1960); CA 56, 13250i

5954. Runeberg, J., "The Chemistry of the Natural Order Cupressales. XXVIII. Constituents of Juniperus Virginiano L.," Acta Chem. Scand., 14, 1288-1294 (1960); CA 56, 13251a

5955. Runeberg, J., "The Chemistry of the Natural Order Cupressales. XXIX. Constituents of Juniperus Thurifers," Acta Chem. Scand., 14, 1985-1990 (1960); CA 56, 13251b

5956. Runeberg, J., "The Chemistry of the Natural Order Cupressales. XXX. Heartwood Constituents of Juniperus Cedrus," Acta Chem. Scand., 14, 1991-1994 (1960); CA 56, 13251c

5957. Runeberg, J., "The Chemistry of the Natural Order Cupressales. XXXI. Heartwood Constituents of Juniperus Phoenicea," Acta Chem. Scand., 14, 1995-1998 (1960); CA 56, 13251c

5958. Runge, H., "Gas-Chromatographic Analysis of Inorganic Gases," Z. Anal. Chem., 189, 111-124 (1962); CA 57, 13167i

5959. Runti, C., and Bruni, G., "Application of Gas Chromatography in the Analysis of Thymus Oil," Boll. chim. farm., 99, 435-447 (1960); CA 54, 21651f

5960. Rusek, M., and Krejci, M., "Chromatographic Determination of Carbon Monoxide in Concentration Down to 10^{-4} Volume Per Cent," Gas Chromatog. Symp., Brno, Czech., June 1962; Abstr., J. Gas Chromatog., 1, No. 4, 8 (1963)

5961. Rushing, D. E., "Gas Chromatography in Industrial Hygiene and Air Pollution Problems," Am. Ind. Hyg. Assoc. J., 19, 238-245 (1958)

5962. Russell, A. S., and Cochrane, C. N., "Surface Areas of Heated Alumina Hydrates," Ind. Eng. Chem., 42, 1336-1340 (1950)

5963. Russell, C. D., and Anson, F. C., "Analysis of Products from the Electrolytic Oxidation of Acetate Ion in Acetonitrile," Anal. Chem., 33, 1282-1284 (1961); CA 55, 26784e

5964. Russell, D. S., "An Infrared Cell for Small Gas Samples (Obtained in Gas Chromatography)," Can. J. Chem., 36, 1745-1746 (1958); CA 53, 6701h

5965. Russell, D. S., and Bednas, M. E., "Transfer Cell for Gas Chromatography," Anal. Chem., 29, 1562 (1957); CA 52, 1692i

5966. Russell, G. A., "Directive Effects in Aliphatic Substitution. XI. Solvent Effects in the Reactions of Free Radicals and Atoms. 2. Effects of Solvents on the Position of Attack of Chlorine Atoms Upon 2,3-Dimethylbutane, Isobutane, and 2-Deuterio-2-methylpropane," J. Am. Chem. Soc., 80, 4987-4996 (1958); CA 54, 7525c

5967. Russell, G. A., "Directive Effects in Aliphatic Substitution. XII. Solvent Effects in the Reactions of Free Radicals and Atoms. 3. Effect of Solvents in the Competitive Photochlorination of Hydrocarbons and Their Derivatives," J. Am. Chem. Soc., 80, 4997-5001 (1958); CA 54, 7525h

5968. Russell, G. A., "Directive Effects in Aliphatic Substitution. XIII. Solvent Effects in the Reactions of Free Radicals and Atoms. 4. Effects of Aromatic Solvents in Sulfuryl Chloride Chlorinations," J. Am. Chem. Soc., 80, 5002-5003 (1958); CA 54, 7526c

5969. Russell, G. A., "Catalysis by Metal Halides. I. Mechanism of the Disproportionation of Ethyltrimethylsilane," J. Am. Chem. Soc., 81, 4815-4825 (1959); CA 54, 7313i

5970. Russell, G. A., "Catalysis by Metal Halides. II. The Disproportionation of Trimethylsilane, Phenyltrimethylsilane and Bromotrimethylsilane," J. Am. Chem. Soc., 81, 4825-4831 (1959); CA 54, 7314d

5971. Russell, G. A., "Catalysis by Metal Halides. IV. Relative Efficiencies of Freidel-Crafts Catalysts in Cyclohexane-Methylcyclopentane Isomerism, Alkylation of Benzene and Polymerization of Styrene," J. Am. Chem. Soc., 81, 4834-4838 (1959); CA 54, 7314i

5972. Russell, G. A., "Solvent Effects in the Reaction of Free Radicals and Atoms. V. Effects of Solvents on the Reactivity of t-Butoxy Radicals," J. Org. Chem., 24, 300-302 (1959); CA 54, 10899a

5973. Ruys, A. H., and Heide, R. ter, "Use of Gas Chromatography in the Cosmetic and Soap Industries," Seifen-Ole-Fette-Wachse, 86, 35-38 (1960)

5974. Ruzicka, D. J., and Bryce, W. A., "The Pyrolysis of Diallyl(1,5-hexadiene)," Can. J. Chem., 38, 827-834 (1960); CA 54, 20439b

5975. Rybicka, S. M., "The Quantitative Analysis of Drying Oil Fatty Acids by Gas-Liquid Chromatography," Chem. & Ind. (London), 1960, 1594-1595

5976. Ryce, S. A., and Bryce, W. A., "Analysis of Volatile Organic Sulfur Compounds by Gas Partition Chromatography," Anal. Chem., 29, 925-928 (1957); CA 51, 11935i

5977. Ryce, S. A., and Bryce, W. A., "Ionization Gage Detector for Gas Chromatography," Can J. Chem., 35, 1293-1297 (1957); CA 52, 3414e

5978. Ryce, S. A., and Bryce, W. A., "Ionization Gage Detector for Gas Chromatography," Nature 179, 541 (1957); CA 51, 10135e

5979. Ryce, S. A., and Bryce, W. A., "A New Design for Gas Chromatography Columns," Anal. Chem., 33, 654 (1961); CA 55, 11951a

5980. Ryce, S. A., and Bryce, W. A., "Some Reactions of Methyl Radicals with 1-Butene," Trans. Faraday Soc., 57, 943-950 (1961)

5981. Ryce, S. A., Kebarle, P., and Bryce, W. A., "Thermal Conductivity Cell for Gas Chromatography," Anal. Chem., 29, 1386-1387 (1957); CA 52, 5891c

5982. Ryder, Jr., W. S., "Gas Chromatography Targets Gourmet Qualities," Food Eng., 31, No. 7, 81-82 (1959)

5983. Ryhage, R. R., Stallberg-Stenhagen, S., and Stenhagen, E., "Studies on Phthiocerol. II. The Nature of the Acidic Products Obtained by Oxidation of Phthiocerol by Chromic Acid," Arkiv Kemi, 14, 247-257 (1959); CA 54, 7545g

5984. Ryhage, R. R., Stallberg-Stenhagen, S., and Stenhagen, E., "Studies on Phthiocerol. III. Identification of Phthiocerane as a Mixture of 4-Methyldotriacontane and 4-Methyltetratriacontane," Arkiv Kemi, 14, 259-265 (1959); CA 54, 7545g

5985. Rysselberge, J. van, "Determination of the Light Hydrocarbons in Crude Oils by Gas Chromatography," Ind. chim. belge, 24, 1023-1036 (1959); CA 54, 1833h; 5th World Petrol. Congr., New York, N. Y., June 1959

5986. Rysselberge, J. van, and Van Der Stricht, M., "Complete Separation of Xylenes and Ethylbenzene by Gas Chromatography," Nature, 193, 1281-1282 (1962); CA 58, 2827d

5987. Rysselberge, J. van, Vaerman, J., and Van Der Stricht, M., "Gas Chromatography of Exhaust Gases," Ind. chim. belge, 26, 780-787 (1961); CA 56, 6315i

5988. Sachs, V., and Droegemeir, G., "Determination of Carbon Monoxide in the Blood Post Mortem by Gas Chromatography," Deut. Z. Ges. Gerichtl. Med., 51, 627-629 (1961); CA 56, 5046e

5989. Sadowski, F., and Wagner, H., "Gas Chromatography and Its Uses in the Analysis of Lacquer Solvents," Plaste u. Kautschuk., 7, 103 (1960); CA 54, 15955h

5990. Safranski, L. W., and Dal Nogare, S., "Gas Chromatography. The Fine Touch in Separations," Chem. & Eng. News, 39, No. 26, 102-110 (June 26, 1961); CA 55, 20572c

5991. Sagert, N. H., and Laidler, K. J., "Kinetics and Mechanism of the Pyrolysis of n-Butane. I. The Uninhibited Decomposition," Can. J. Chem., 41, 838-847 (1963); CA 58, 9657h

5992. Said, A. S., "Theoretical Plate Concept in Chromatography," A.I.Ch.E. Journal, 2, 477-481 (1956); CA 51, 2450c

5993. Said, A. S., "Nonhomogeneous Chromatographic Column," Chem. Eng. Sci., 9, 153-163 (1958); CA 54, 979e

5994. Said, A. S., "Theoretical Plate Concept in Chromatography. II.," A.I.Ch.E. Journal, 5, 69-72 (1959)

5995. Said, A. S., "Chromatographic Columns Containing a Large Number of Theoretical Plates," A.I.Ch.E. Journal, 5, 223-224 (1959); CA 56, 2297h

5996. Said, A. S., "The Theory of Programmed Temperature Gas Chromatography," 3rd Intern. Symp. on Gas Chromatography, East Lansing, Mich., June 1961; ISA Proc., 3, 65-71 (1961)

5997. Said, A. S., "The Theory of Programmed Temperature Gas Chromatography," in "Gas Chromatography," edited by N. Brenner, J. E., Callen, and M. D. Weiss, Academic Press, Inc., New York, 1962, pp. 79-90

5998. Said, A. S., "Efficiency of Chromatographic Separations. Correction of the Gluekauf Equation," J. Gas Chromatog., 1, No. 6, 20-24 (1963)

5999. Said, A. S., and Johns, T., "Systematic Temperature Programming Gas Chromatography," 13th Pittsburgh Conf. on Anal. Chem. & Appl. Spectroscopy, Pittsburgh, Pa., March 1962, Program Abstr., p. 50

6000. Saifer, A., and Goldman, L., "The Free Fatty Acids Bound to Human Serum Albumin," J. Lipid Research, 2, 268-270 (1961); CA 55, 22520a

6001. Saifer, A., and Vecsler, F., "The Gas Chromatographic Analysis of the Free Fatty Acids Bound to Human Serum Albumin," Joint Mtg. of Am. Assoc. of Clin. Chemists and Am. Assoc. Adv. of Science, December 1960; Abstr., 7, 308 (1961)

6002. Saint-James, D., "Automatic Gas Analysis," Chim. anal., 37, 31-38 (1955); CA 49, 6657b

6003. St. Pierre, L. E., and Dewhurst, H. A., "The Effect of Oxygen on the Radiolysis of Silicones," J. Phys. Chem., 64, 1060-1062 (1960); CA 55, 3167e

6004. Saint-Rat, L. de, and Bertrand, D., with Carr, A., "First International Symposium on the Analysis of Foods by Gas Chromatography," Bordeaux, France, October 1962, J. Gas Chromatog., 1, No. 4, 31-33 (1963)

6005. Sakai, T., Nishimura, K., and Hirose, Y., "Analysis of Essential Oil of Narcissus Tazetta Var. Chinensis," Nippon Kagaku Zasshi, 82, 1716-1718 (1961); CA 58, 10037b

6006. Sakaida, R. R., Rinker, R. G., Cuffel, R. F., and Corcoran, W. H., "Determination of Nitric Oxide in Nitric Oxide-Nitrogen System by Gas Chromatography," Anal. Chem., 33, 32-34 (1961); CA 55, 10192a

6007. Salinis, C. A., "Notes Describing Gas Chromatographic Qualitative-Quantitative Calibration, Column Packing, and Data Transfer," in "Progress in Industrial Gas Chromatography," Vol. 1, edited by H. A. Szymanski, Plenum Press, Inc., New York, 1961, pp. 37-50

6008. Salinis, C. A., "The Effect of Particle Size and Multiple Column Reversal on Component Resolution," in "Progress in Industrial Gas Chromatography," Vol. 1, edited by H. A. Szymanski, Plenum Press, Inc., New York, 1961, pp. 209-219

6009. Salinis, C. A., "The Selection of Columns in Gas Chromatography," 4th Annual Gas Chromatography Institute, Canisius College, Buffalo, N. Y., April 1962

6010. Salinis, C. A., and Davis, A., "The Use of Various Types of Solid Supports and Liquid Substrates," 1st Annual Gas Chromatography Institute, Canisius College, Buffalo, N. Y., May 1959

6011. Salinis, C. A., and Davis, A., "The Effect of Particle Size on Resolution," 1st Annual Gas Chromatography Institute, Canisius College, Buffalo, N. Y., May 1959

6012. Salo, T., "Gas Chromatography of the Isopropanol-Ethanol-Methanol-Water System," Z. Lebensm.-Untersuch. u. -Forsch., 115, 54-56 (1961); CA 55, 20755e

6013. Salomaa, P., and Laiho, S., "Cis-trans Isomerism and the Kinetics of Hydrolysis of 2,5-Dimethyl-1,3-dioxolan-4-one," Suomen Kemistilehti, 35B, 92-96 (1962); CA 58, 4394d

6014. Saltzman, B. E., "Preparation and Analysis of Calibrated Low Concentrations of Sixteen Toxic Gases: Ammonia, Arsine, Bromine, Carbon Dioxide, Carbon Monoxide, Chlorine, Chlorine Dioxide, Ethylene Oxide, Hydrogen Chloride, Hydrogen Cyanide, Hydrogen Fluoride, Monoethanolamine, Nitric Oxide, Nitrogen Dioxide, Phosgene, and Stibine," Anal. Chem., 33, 1100-1112 (1961); CA 55, 27724c

6015. Salvioli, Jr., G. P., Manfredi, G., Corsini, F., Babini, B., and Paolucci, G., "Gas-Chromatographic Determination of Plasma and Erythrocyte Fatty Acids in Obesity of Children," Boll. Soc. Ital. Biol. Sper., 38, 730-732 (1962); CA 58, 6072h

6016. Sambasivarao, K., "The Nature of the Octadecadienoic and Octadecatrienoic Acids of Summer Butter Fat," Univ. Microfilms (Ann Arbor, Mich.), L. C. Card No. Mic 60-4128, 79 pp.; Dissertation Abstr., 21, 1047 (1960)

6017. Samm, J., and Speier, J., "The Addition of Silicon Hydrides to Olefinic Double Bonds. VI. Addition to Branched Olefins," J. Am. Chem. Soc., 83, 1351-1355 (1961); CA 55, 16399g

6018. Sammons, H. G., and Wiggs, S. M., "Quantitative Separation of Fatty Acids from Unsaponifiable Matter," Analyst, 85, 417-418 (1960); CA 55, 4012e

6019. Samsel, E. P., Aldrich, J. C., "Sample Injection Valve for Gas Chromatography," Anal. Chem., 31, 1288 (1959); CA 53, 19476c

6020. Sand, D. M., and Schlenk, H., "Acylated Cyclodextrins as Polar Stationary Phases for Gas-Liquid Chromatography," Anal. Chem., 33, 1624-1625 (1961); CA 56, 1983a

6021. Sand, P. T., "Physical Analysis of Gases," Can. Chem. Processing, 41, No. 5, 114-120 (1957); CA 51, 17599g

6022. Sandermann, W., and Bruns, K., "The Biogenesis of Limonenes in Pinus Pinea," Naturwissenschaften, 49, 258 (1962); CA 57, 7594d

6023. Sandermann, W., and Weissmann, G., "Gas Chromatography of Tall Oil Fatty Acids," Fette, Seifen, Anstrichmittel, 64, 807-813 (1962); CA 58, 1652b

6024. Sandermann, W., and Weissmann, G., "Gas-Chromatographic Study of Pyrocatechol Derivatives. Phenols from Tall Oil First Fractions," Z. Anal. Chem., 189, 137-148 (1962); CA 57, 15396b

6025. Sandler, S., and Beech, J. A., "Quantitative Analysis of Combustion Products by Gas Chromatography. The Oxidation of a Rich n-Pentane-Air Mixture in a Flow System," Can. J. Chem., 38, 1455-1466 (1960); CA 55, 2347c

6026. Sandler, S., and Strom, R., "Determination of Formaldehyde by Gas Chromatography," Anal. Chem.,

<u>32</u>, 1890-1891 (1960); CA <u>55</u>, 4234a

6027. Sandulescu, D., "Experimental Gas Chromatography. III. Apparatus for Gas Chromatography," Rev. Chim. (Bucharest), <u>12</u>, 224-228 (1961); CA <u>57</u>, 6577d

6028. Sandulescu, D., "Experimental Gas Chromatography. IV. The Introduction of the Sample in the Gas-Chromatographic Apparatus," Rev. Chim. (Bucharest), <u>12</u>, 290-295 (1961); CA <u>57</u>, 6577d

6029. Sandulescu, D., "Experimental Gas Chromatography. V. 1. The Stationary Phase," Rev. Chim. (Bucharest), <u>12</u>, 341-345 (1961); CA <u>57</u>, 6577d

6030. Sandulescu, D., "Experimental Gas Chromatography. V. 2. Stationary Phase," Rev. Chim. (Bucharest), <u>12</u>, 406-412 (1961); CA <u>57</u>, 6577e

6031. Sandulescu, D., "Experimental Gas Chromatography. VII. Detectors," Rev. Chim. (Bucharest), <u>12</u>, 549-556 (1961); CA <u>57</u>, 6577e

6032. Sandulescu, D., "Experimental Gas Chromatography," Rev. Chim. (Bucharest), <u>12</u>, 655-659 (1961); CA <u>57</u>, 6577f

6033. Sandulescu, D., Pete, O., Stanescu, L., Hanes, A., and Enache, M., "Determination of Hydrocarbons by Gas Chromatography. I. Determination of C_1-C_4 Hydrocarbons by Gas Chromatography," Rev. Chim. (Bucharest), <u>12</u>, 297-302 (1961); CA <u>55</u>, 26864d

6034. Sandulescu, D., Stanescu, L., and Ionescu, A. G., "Study of Catalysts by Means of Gas Chromatography. I. The Chromatographic Characteristics of Oxide Catalysts for Olefin Polymerization," Rev. Chim. (Bucharest), <u>11</u>, 151-155 (1960); CA <u>56</u>, 11779e

6035. Sarancha, E. T., and Dubovikova, A. P., "Analysis of Products of Isobutyl and Butyl Alcohol Synthesis by Gas Chromatography," Zavodskaya Lab., <u>27</u>, 398-399 (1961); CA <u>57</u>, 6588e

6036. Saroff, H. A., and Karmen, A., "Gas Chromatography of the N-Trifluoroacetylmethyl Esters of the Amino Acids," Anal. Biochem., <u>1</u>, 344-350 (1960); CA <u>55</u>, 20063f

6037. Saroff, H. A., Karmen, A., and Healy, J. W., "Gas Chromatography of the Amino Acid Esters in Ammonia," J. Chromatog., <u>9</u>, 122-123 (1962); CA <u>58</u>, 4801f

6038. Sasaki, N., Tominaga, K., and Aoyagi, M., "Micro Gas Chromatograph," Nature, <u>186</u>, 309-310 (1960); CA <u>54</u>, 20356b

6039. Sass, S., "Analytical Aspects of Chemical and Biological Assay in the U. S. Army CBR Agency," Anal. Chem., <u>35</u>, No. 6, 25A-28A, 30A-37A (1963)

6040. Sasse, W. H. F., "Synthetical Applications of Activated Metal Catalysts. VIII. The Action of Degassed Raney Nickel on Quinoline and Some of Its Derivatives," J. Chem. Soc., <u>1960</u>, 526-533; CA <u>54</u>, 12137g

6041. Sassenberg, W., and Wrabetz, K., "Bis(3,3,5-trimethylcyclohexyl)phthalate as the Stationary Phase for Gas Chromatography of Phenol Derivatives," Z. Anal. Chem., <u>184</u>, 423-427 (1961); CA <u>56</u>, 9387b

6042. Sassiver, M. L., and English, Jr., J., "Epoxidation of Some Allylic Alcohols," J. Am. Chem. Soc., <u>82</u>, 4891-4895 (1960); CA <u>55</u>, 14299a

6043. Sato, S., "Identification of Substances by Gas Chromatography," Kagaku no Ryoiki Zokan, No. <u>44</u>, 79-96 (1961); CA <u>56</u>, 2i

6044. Sato, S., and Cvetanovic, R. J., "The Effect of Molecular Oxygen on the Reaction of Oxygen Atoms with cis-2-Pentene," Can. J. Chem., <u>37</u>, 953-965 (1959); CA <u>54</u>, 14216h

6045. Sato, T., Ikegami, A., and Fujino, Z., "Rapid Analysis of Monochloroacetic, Dichloroacetic, and Acetic Acid Mixtures by Gas Chromatography," Bunseki Kagaku, <u>10</u>, 854-858 (1961); CA <u>56</u>, 6666d

6046. Sato, Y., and Momotani, M., "Gas-Chromatographic Determination of Fatty Acids," Bunseki Kagaku, <u>10</u>, 196-198 (1961); CA <u>57</u>, 6049d

6047. Saucy, G., Marbet, R., Lindlar, H., and Isler, O., "A New Synthesis of Citral and Related Compounds," Chimia (Switz.), <u>12</u>, 326-327 (1958); CA <u>53</u>, 9268c

6048. Sauer, Jr., M. C., and Willard, J. E., "Effects of Ethylene, Hydrogen and Radiation Dosage on the Tritiated Products Resulting from the $He^3(n,p)H^3$ Reaction in Gaseous Hydrocarbons," J. Phys. Chem., <u>64</u>, 359-362 (1960); CA <u>54</u>, 17005g

6049. Sauerland, H. D., "The Importance of Temperature Programming for the Gas Chromatographic Analysis of Coal Tar," Facts & Methods (F&M Scientific Corp.), <u>4</u>, No. 1, 5-7 (1963)

6050. Sauerland, H. D., "Recent Developments in Gas Chromatography of Bituminous Coal Tar Products," Brennstoff-Chem., <u>44</u>, No. 2, 37-43 (1963); CA <u>58</u>, 10010e

6051. Sauers, C. K., and Cotter, R. J., "New Synthesis of β-Cyanoesters," J. Org. Chem., <u>26</u>, 6-10 (1961); CA <u>55</u>, 18642d

6052. Sauers, R. R., "The Effect of Structure on the Course of Phosphoryl Chloride-Pyridine Dehydration of Tertiary Alcohols," J. Am. Chem. Soc., <u>81</u>, 4873-4876 (1959); CA <u>54</u>, 18320h

6053. Sauers, R. R., and Kwiatkowski, G. T., "Stereochemistry of Reactions of the Norbornyl Grignard Reagent," J. Org. Chem., <u>27</u>, 4049-4050 (1962); CA <u>58</u>, 7844b

6054. Sauers, R. R., and Landesberg, J. M., "Rearrangements During Phosphoryl Chloride-Pyridine Dehydrations," J. Org. Chem., <u>26</u>, 964-966 (1961); CA <u>55</u>, 23577e

6055. Sauers, R. R., and Tucker, R. J., "Ring Enlargement Reactions of Bicyclo[2,2,1]heptane Derivatives," J. Org. Chem., <u>28</u>, 876-878 (1963); CA <u>58</u>, 11231e

6056. Saunders, M., and Murray, R. W., "Reaction of Dichlorocarbene with Amines," Tetrahedron, 11, 1-10 (1960); CA 55, 4349f

6057. Saunders, W. H., and Carges, G. L.,"'1,3 Shifts' in Rearrangements of 3,4,4-Trimethyl-2-pentyl Derivatives," J. Am. Chem. Soc., 82, 3582-3585 (1960); CA 55, 13294b

6058. Sauvage, J. F., "The Stereochemistry of the Catalytic Hydrogenation of Cycloalkanes," Univ. Microfilms (Ann Arbor, Mich.), L. C. Card No. Mic 60-4793, 74 pp.; Dissertation Abstr., 21, 1066-1067 (1960)

6059. Sauvage, J. F., Baker, R. H., and Hussey, A. S., "Hydrogenation of Cyclohexenes Over Platinum Oxide," J. Am. Chem. Soc., 82, 6090-6095 (1960); CA 55, 11330i

6060. Savary, P., Constantin, M. J., and Desnuelle, P., "Structure of the Triglycerides of Rat Lymph Chylomicrons," Biochim. et Biophys. Acta, 48, 562-571 (1961); CA 55, 23745a

6061. Savidan, L., "La Chromatographie," Dunod, 92 Rue Bonaparte, Paris 6, France, 1958

6062. Savidan, L., "Chromatography of High-Molecular Weight Aliphatic Compounds," Oleagineux, 13, 207-208 (1958); CA 52, 11518h

6063. Savitsky, A., "Continuous Composition Analyzers," 13th Annual Instrument Automation Conf., ISA, Philadelphia, Pa., September 1958

6064. Savitsky, A., "Laboratory Instrumentation Moves Into the Plant," Anal. Chem., 30, No. 3, 17A-22A (1958)

6065. Sawicki, E., "Analysis for Airborne Particulate Hydrocarbons; Their Related Proportions as Affected by Different Types of Pollution," Natl. Cancer Inst., Monograph No. 9, 201-220 (1962); CA 58, 4963a

6066. Sawyer, D. T., "Considerations for Idealizing Preparative Gas Chromatography," 145th Natl. ACS Mtg., New York, N. Y., September 1963, Program Abstr., p. 12B

6067. Sawyer, D. T., and Barr, J. K., "Theory and Practice of Low-Loaded Columns in Gas Chromatography," Anal. Chem., 34, 1052-1057 (1962); CA 57, 10510e

6068. Sawyer, D. T., and Barr, J. K., "Evaluation of Several Integrators for Use in Gas Chromatography," Anal. Chem., 34, 1213-1216 (1962); CA 57, 13167a

6069. Sawyer, D. T., and Barr, J. K., "Evaluation of Support Materials for Use in Gas Chromatography," Anal. Chem., 34, 1518-1520 (1962); CA 58, 4d

6070. Sax, K. J., and Stross, F. H., "Squalane, A Standard," Anal. Chem., 29, 1700-1702 (1957); CA 52, 5916b

6071. Saxena, S. C., "Thermal Conduction and Gas Analysis," Nature, 178, 1462 (1956); CA 51, 7237g

6072. Saxena, S. C., "Thermal Conductivity of Binary and Ternary Mixtures of Helium, Argon, and Xenon," Indian J. Phys., 31, 597-606 (1957); CA 52, 13399i

6073. Scarpellino, R., "Application of Gas-Chromatographic Techniques to the Study of Volatile Components in Cheddar Cheese," Univ. Microfilms (Ann Arbor, Mich.), L. C. Card No. Mic 61-2440, 63, pp.; Dissertation Abstr., 22, 421 (1961); CA 55, 23863i

6074. Scarpellino, R., and Kosikowski, F. V., "A Method for Concentrating Cheese Volatiles for Gas Chromatography," J. Dairy Sci., 44, 10-15 (1961); CA 55, 8689e

6075. Scarpellino, R., and Kosikowski, F. V., "Evolution of Volatile Compounds in Ripening Raw and Pasteurized Milk Cheddar Cheese Observed by Gas Chromatography," J. Dairy Sci., 45, 343-348 (1962); CA 57, 3835e

6076. Schaefer, C. A., and Thodos, G., "Thermal Conductivity of Diatomic Gases: Liquid and Gaseous States," A. I.Ch.E. Journal, 5, 367-371 (1959); CA 56, 8031a

6077. Schaefer, F. C., "Synthesis of the s-Triazine System. V. Cotrimerization of Imidates," J. Org. Chem., 27, 3362 (1962); CA 58, 1457e

6078. Schaeppi, W. H., "Purification of Fatty Acids by Zone Melting," Chimia (Switz.), 16, 291-295 (1962); CA 58, 7051g

6079. Schamp. N., "Gas Chromatography,"Mededel. Vlaam. Chem. Ver., 19, 53 (1957)

6080. Schamp. N., "New Technique in Gas Chromatography," Mededel. Vlaam. Chem. Ver., 24, 35-53 (1962); CA 57, 6577c

6081. Schams, F., "Problems Encountered in Gas Chromatography," 1960 Conf., German Chemical Society, Stuttgart, Germany, April 1960, Program Abstr., p. 46; Abstr., Angew. Chem., 72, 593 (1960)

6082. Scharfe, G., "Cracking Gas Analysis with Gas Chromatography," Proc. Gas-Kolloquium, Hamburg, West Germany, November 1956, pp. 26-27

6083. Scharfe, G., "Methods for Anaylsis of Hydrocarbon Gas Mixtures by Gas Chromatography," Erdol u. Kohle, 12, 723-728 (1959); CA 54, 3920a

6084. Scharpenseel, H. W., "Combined Gas Chromatography and Radioactivity Measurement for the Determination of C^{14}- and H^3-Labeled Substances," Angew. Chem., 73, 615-619 (1961); CA 56, 3295g

6085. Scharpenseel, H. W., and Menke, K. H., "Radiochromatography with Weak Beta-Ray Emitters," Z. Anal. Chem., 180, 81-96 (1961); CA 55, 20755d

223

6086. Schay, G., "Several Problems in Elution Gas Chromatography," Magy. Tud. Akad. Kem. Tud. Oszt. Kozlemen., 11, 135-145 (1959); CA 57, 14456g

6087. Schay, G., "Theoretical Principles of Gas Chromatography," VEB Deutscher Verlag der Wissenschaften, Berlin, 1960

6088. Schay, G., Fejes. P., Halasz, I., and Kiraly, J., "Determination of Adsorption Isotherms by Frontal Gas Chromatography," Acta Chim. Sci. Hung., 11, 381-393 (1957); CA 51, 15215i

6089. Schay, G., Fejes, P., Halasz, I., and Kiraly, J., "Determination of Adsorption Isotherms by Vapor Phase Chromatography," Magyar Kem. Folyoirat, 63, 143-149 (1957); CA 52, 17892h

6090. Schay, G., Fejes, P., and Szathmary, J., "Adsorption of Gas Mixtures. I. Statistical Theory of Adsorption of the Langmuir Type in Multicomponent Systems," Acta Chim. Acad. Sci. Hung., 12, 299-306 (1957); CA 52, 8674e

6091. Schay, G., Petho, A., and Fejes, P., "Further Contributions to the Solution of the System of Differential Equations of a Gas Chromatographic Model," Acta Chim. Acad. Sci. Hung., 22, 285-299 (1960); CA 55, 2240a

6092. Schay, G., and Székely, G., "Gas-Adsorption Measurements in Flow Systems," Acta Chim. Acad. Sci. Hung., 5, 167-182 (1954); CA 50, 7540f

6093. Schay, G., Szekely, G., and Fejes, P., "Determination of Adsorption Isotherms of Gases by Frontal Chromatography," Hua Hsueh Pao, 23, 421-427 (1957); CA 52, 19338h

6094. Schay, G., Székely, G., and Szigetvary, G., "Adsorption of Gas Mixtures. New Chromatographic Method for Determination of Mixed Adsorption. Adsorption of CO_2-C_2H_2 Mixtures on Charcoal," Acta Chim. Acad. Sci. Hung., 12, 309-323 (1957); CA 52, 8674g

6095. Schay, G., Székely, G., and Traply, G., "Experiences with a Catalytic Combustion Cell for the Determination of Trace Substances," 4th Intern. Gas Chromatography Symp., Hamburg, Germany, June 1962; pub. in "Gas Chromatography 1962," M. van Swaay, ed., Butterworth & Co., London, 1962

6096. Schechter, M. S., and Westlake, W. E., "Chemical Residues and the Analytical Chemist," Anal. Chem., 34, No. 1, 25A-28A, 30A-32A (1962)

6097. Scheidegger, A. E., "The Physics of Flow Through Porous Media," Macmillan Co., Inc., New York, 1957

6098. Schenck, G. O., Foldiak, G., and Meder, W. A., "Boundary Temperature in the Irradiated Chemical-Thermal Cracking of Paraffins," Naturwissenschaften, 48, 571-572 (1961); CA 57, 8802a

6099. Schenck, G. O., and Steinmetz, R., "Photofragmentation of Dehydronorcamphor into Cyclopentadiene and Ketene and the Photosensitized Synthesis of Pentacyclo[5.2.1.0^{2,6}.0^3.9.0^{4,8}]decane," Ber., 96, 520-525 (1963); CA 58, 12435h

6100. Schepartz, A. I., and McDowell, P. E., "Gas Chromatographic Retention Time of Formaldehyde," Anal. Chem., 32, 723 (1960); CA 54, 22174e

6101. Schiller, R., "Gas Chromatograph for the Investigation of the Radiolysis of Organic Compounds," Magy. Tud. Akad. Kozp.Fiz. Kut. Int. Kozlemen., 9, 263-272 (1961); CA 57, 5287c

6102. Schimelpfenig, C. W., "Gas Chromatography Analysis for the Elementary Organic Laboratory," J. Chem. Educ., 39, 310 (1962); CA 57, 6576e

6103. Schlater, J. M., Mikkelsen, L., and Beck, M. G., "Programmed Temperature Gas Chromatography," in "Lectures on Gas Chromatography 1962," edited by H. A. Szymanski, Plenum Press, Inc., New York, 1963, pp. 105-131

6104. Schlegel, W., and Noller, C. R., "Relation of Fabacein to Curcurbitacin B," Tetrahedron Letters, 1959, No. 13, 16-19; CA 54, 4666a

6105. Schlenk, H., "Analysis and Isolation of Fatty Acids by Gas Chromatography," 137th Natl. ACS Mtg., Cleveland, Ohio, April 1960, Program Abstr., p. 10A

6106. Schlenk, H., and Gellerman, J. L., "Esterification of Fatty Acids with Diazomethane on a Small Scale," Anal. Chem., 32, 1412-1414 (1960); CA 55, 2469f

6107. Schlenk, H., Gellerman, J. L., and Sand, D. M., "Cyclodextrin Esters as Stationary Phases in GLC," 142nd Natl. ACS Mtg., Atlantic City, N. J., September 1962, Program Abstr., p. 8B

6108. Schlenk, H., Gellerman, J. L., and Sand, D. M. M., "Acylated Cyclodextrins as Stationary Phases for Comparative Gas Liquid Chromatography," Anal. Chem., 34, 1529-1532 (1962); CA 58, 3863h

6109. Schlenk, H., Mangold, H. K., Gellerman, J. L., Link, W. E., Morrissette, R. A., Holman, R. T., and Hayes, H., "Comparative Analytical Studies of Fatty Acids of the Alga Chlorella Pyrenoidosa," J. Am. Oil Chemists' Soc., 37, 547-552 (1960); CA 55, 2145b

6110. Schlenk, H., and Sand, D. M., "Acylate Cyclodextrins as Polar Stationary Phases for Gas-Liquid Chromatography," 140th Natl. ACS Mtg., Chicago, Ill., September 1961, Program Abstr., p. 26B

6111. Schlenk, H., and Sand, D. M., "Collection of Gas-Liquid Chromatography Fractions by Gradient Cooling," Anal. Chem. 34, 1676 (1962); CA 58, 3864g

6112. Schleyer, P. v. R., and Donaldson, M. M., "The Relative Stability of Bridged Hydrocarbons. II. endo- and exo-Trimethylenenorbornane. The Formation of Adamantane," J. Am. Chem. Soc., 82, 4645-4651 (1960); CA 55, 17527e

224

6113. Schmauch, L. J., "Response Time and Flow Sensitivity of Detectors for Gas Chromatography," Anal. Chem., 31, 225-230 (1959); CA 53, 9884i

6114. Schmauch, L. J., and Dinerstein, R. A., "Effect of the Carrier Gas on the Sensitivity of a Thermal-Conductivity Dectector in Gas Chromatography," Nature, 183, 673-674 (1959); CA 53, 20925i

6115. Schmauch, L. J., and Dinerstein, R. A., "Response of Thermal-Conductivity Cells in Gas Chromatography," Anal. Chem., 32, 343-352 (1960); CA 54, 21930a; 136th Natl. ACS Mtg., Atlantic City, N. J., September 1959, Program Abstr., p. 5R; Preprints, Div. Petrol. Chem., 4, No. 3, 119-138 (1959)

6116. Schmeltz, I., Miller, R. L., and Stedman, R. L., "Gas Chromatographic Study of the Steam-Volatile Fatty Acids of Various Tobaccos," J. Gas Chromatog., 1, No. 8, 27-28 (1963)

6117. Schmeltz, I, and Schlotzhauer, W. S., "Nonvolatile Acids of Cigar Smoke," Tobacco Sci., 6, 88-89 (Pub. in Tobacco, 154, No. 20, 32-33) (1962); CA 57, 11566e

6118. Schmid, C. E., and Maier, M. J., "A Versatile Programmer for Process Gas Chromatography," 8th Natl. Analysis Instrumentation Symp. ISA, Charleston, W. Va., April-May 1962

6119. Schmidbaur, H., and Schmidt, M., "Germanosiloxanes. I. Preparation and Properties of Alkyl-germanosiloxanes," Chem. Ber., 94, 1138-1142 (1961); CA 55, 20925a

6120. Schmidt, A., "Apparatus for the Exact Analysis of Gaseous Mixtures," Explosivestoffe, 5, 1-7 (1957); CA 51, 7769a

6121. Schmidt, J. J. E., Krimmel, J. A., and Farrell, T. J., "Alkylpolyphenols. I. 4'-Alkyl-m-terphenyls," J. Org. Chem., 25, 252-256 (1960); CA 54; 15299e

6122. Schmidt-Bleek, F., Stocklin, G., and Herr, W., "Heterogeneous Exchange on Gas Chromatographic Columns as a Method for Radioactive Labeling of Organic Halogen Compounds," Angew. Chem., 72, 778 (1960)

6123. Schmidt-Collerus, J. J., and Frank, A. J., "Analysis of Gases in Metals," U. S. Dept. Com., Office Tech. Serv., AD 260, 980, 49 pp. (1961); CA 58, 7360a

6124. Schmidt-Kuster, W. J., "Gas Chromatographic Resolving Powers of Silica Gels," Symp. on the Use of Physical-Chemical Methods for Qualitative and Quantitative Analysis, German Chemical Society, Freiburg Breisgau, Germany, April 1959, Program Abstr., pp. 16-17

6125. Schmidt-Kuster, W. J., "Special Problems in Development and Application of Ionization Chamber Detectors in Gas Chromatography," Erdol u. Kohle, 13, 829 (1960); 13th Annual Mtg., German Society for Petroleum and Coal Chemistry, Frankfurt, Germany, October 1960; Abstr., Brennstoff-Chem., 41, 342 (1960)

6126. Schmied, H., and Koski, W. S., "Chemistry of N^{13} Recoils in Some Carbon Compounds," J. Am. Chem. Soc., 82, 4766-4770 (1960); CA 55, 3234e

6127. Schneider, C. R., and Freund, H., "Determination of Low Level Hydrocyanic Acid in Solution Using Gas Chromatography," Anal. Chem., 34, 69-74 (1962); CA 56, 9419c; 140th Natl. ACS Mtg., Chicago, Ill., September 1961, Program Abstrs., pp. 40B, 16W

6128. Schneider, I. A., "Chromatographic Analysis of Some Gas Mixtures," Acad. rep. populare Romîne, Studii cercetări chim., 8, 451-463 (1960); CA 55, 18435c

6129. Schneider, W., Teubel, J., and Mahrwald, R., "Attempts to Dehydroisomerize Five-Ring Naphthenes with Catalyst 8376. I. Method and Works Connected with the Temperature Dependence of the Dehydroisomerization," Chem. Tech. (Berlin), 13, 139-143 (1961); CA 58, 3325a

6130. Schneyder, J., "Gas Chromatographic Investigations: Wine Yeast-Oil," Mitt. Klosternberg, Ser. A, Rebe u. Wein, No. 8, 186-192 (1958); CA 53, 8531i

6131. Schoedler, C., "Detection Methods in Gas Chromatography," Bull. soc. chim. France, 1962, 2323-2324; CA 58, 6163c

6132. Schoedler, C., "Automatic Analysis of Blast Furnace Gases by Gas Chromatography," Chim. Ind. (Paris), 87, 231-239 (1962); CA 58, 921g

6133. Schoenfelder, C. W., "The Purification of Dinitrogen Tetrafluoride by Gas Chromatography, Its Mass Spectrum, Infrared Spectrum and Vapor Pressure Curve," J. Chromatog., 7, 281-287 (1962); CA 57, 12286i

6134. Schogt, J. C. M., Haverkamp-Begemann, P., and Recourt, J. H., "Composition of Aldehydes Derived from Some Bovine Lipids," J. Lipids Research, 2, 142-147 (1961); CA 55, 15563e

6135. Scholfield, C. R., and Cowan, J. C., "Cyclization of Linolenic Acids by Alkali Isomerization," J. Am. Oil Chemists' Soc., 36, 631-635 (1959); CA 54, 3186d

6136. Scholfield, C. R., Jones, E. P., Butterfield, R. O., and Dutton, H. J., "Argentation in Counter-current Distribution to Separate Isologues and Geometric Isomers of Fatty Acid Esters," Anal. Chem., 35, 1588-1591 (1963)

6137. Scholfield, C. R., Jones, E. P., and Dutton, H. J., "Hydrogenation of Linolenate. Comparison of Products from Trilinolenin and Methyl Linolenate by Use of Countercurrent Distribution and Gas-Liquid Chromatography," Anal. Chem., 33, 1745-1745 (1961); CA 57, 7098g; 139th Natl. ACS Mtg., St. Louis, Mo., March 1961, Program Abstr., p. 13B

6138. Scholfield, C. R., Jones, E. P., Nowakowska, J., and Dutton, H. J., "Hydrogenation of Linolenate.

II. Hydrazine Reduction," J. Am. Oil Chemists' Soc., 38, 208-211 (1961); CA 55, 11883a

6139. Scholfield, C. R., Jones, E. P., Nowakowska, J., Selke, E., Sreenivasan, B., and Dutton, H. J., "Hydrogenation of Linolenate. I. Fractionation and Characterization Studies," J. Am. Oil Chemists' Soc., 37, 579-582 (1960); CA 55, 2144d

6140. Scholfield, C. R., Nowakowska, J., and Dutton, H. J., "Hydrogenation of Linolenate. IV. Kinetics of Catalytic and Homogeneous Chemical Reduction," J. Am. Oil Chemists' Soc., 39, 90-95 (1962); CA 56, 8862h; 52nd Annual Mtg., Am. Oil Chemists' Soc., St. Louis, Mo., April-May 1961

6141. Scholl, F., "Gas Chromatography and Its Application to the Coating Field," Deut. Farben-Z., 16, 93-96, 146-155 (1962); CA 57, 2364f

6142. Scholl, F., "Determination of Surface Area and Surface Quality of Solids by Gas Chromatographic Methods," Trans. Intern. Ceram. Congr., 8th, Copenhagen, 1962, 75-83; CA 58, 948g

6143. Scholly, P. R., and Brenner, N., "Comparative Retention Values of Representative Sample Types on Standard Gas Chromatography Columns," Gas Chromatog., Intern. Symposium, 2nd, East Lansing, Mich., 1959, 263-309 (Pub. 1961); CA 55, 18204c

6144. Scholly, P. R., and Brenner, N., "Comparative Retention Values of Representative Sample Types on Standard Gas Chromatography Columns," In "Gas Chromatography," edited by H. J. Noebels, R. F. Wall, and N. Brenner, Academic Press, Inc., New York, 1961, pp. 263-309

6145. Schols, J. A., "Complementary Use of Gas Chromatography and Spectroscopic Methods," 2nd Alberta Gas Chromatography Discussion, Edmonton, Alberta, February 1959

6146. Schols, J. A., "Quantitative Analysis for Carbonyl Sulfide in Natural Gas by Gas-Liquid Chromatography," Anal. Chem., 33, 359-360 (1961); CA 55, 18086c

6147. Schols, J. A., Stock, C. E., and Young, D. M., "Analyzing Gas Mixtures on the Spot," ISA Journal, 9, 44-46 (1962); CA 57, 1514g

6148. Scholz, R. G., "Solid Support Effects in Gas Chromatography," Univ. Microfilms (Ann Arbor, Mich.) Order No. 61-5761, 165 pp.; Dissertation Abstr., 22, 2166 (1962); CA 56, 10879g

6149. Scholz, R. G., and Brandt, W. W., "The Effect of Solid Supports on Retention Volumes," 3rd Intern. Symp. on Gas Chromatography, ISA, East Lansing, Mich., June 1961; ISA Proc., 3, 9-19 (1961)

6150. Scholz, R. G., and Brandt, W. W., "The Effect of Solid Supports on Retention Volumes," in "Gas Chromatography," edited by N. Brenner, J. E. Callen, and M. D. Weiss, Academic Press, Inc., New York, 1962, pp. 7-26

6151. Schomburg, G., "Quantitative Use of the Fraktogram in Partition Chromatography (Thermal Conductivity Method)," Z. Anal. Chem., 164, 147-158 (1958); CA 53, 6868a

6152. Schomburg, G., "The Importance of Gas Chromatographic Methods for the Chemistry of Boron Alkyls and Hydrides," 4th Intern. Gas Chromatography Symp., Hamburg, Germany, June 1962; pub. in " Gas Chromatography 1962," ed. by M. van Swaay, Butterworth & Co., London, 1962

6153. Schomburg, G., "New Developments in Gas-Chromatographic Detectors," Z. Anal. Chem., 189, 14-50 (1962); CA 57, 13163f

6154. Schomburg, G., Koster, R., and Henneberg, D., "Gas Chromatographic Investigations on Boron Carbon Compounds in Combination with Mass Spectrometric Measurements," Mtg. of German Chemical Soc., Div of Anal. Chem., Freiburg, West Germany, April 1959

6155. Schomburg, G., Koster, R., and Henneberg, D., "Boron Compounds. II. Gas Chromatographic Analysis of Trialkyl Boron Compounds in Combination with Mass Spectrometric Measurements," Z. Anal. Chem., 170, 285-301 (1959); CA 54, 5210c

6156. Schormuller, J., and Langner, H., "Quantitative Determination of Organic Acids in Foods," Z. Lebensm.-Untersuch. u. -Forsch, 113, 104-112 (1960); CA 55, 1953e

6157. Schrade, W., Biegler, R., and Böhle, E., "Fatty Acid Distribution in the Lipid Fractions of Healthy Persons of Different Age, Patients with Atherosclerosis and Patients with Idiopathic Hyperlipemia," J. Atherosclerosis Research, 1 47-61 (1961); CA 56, 5296e

6158. Schrade, W., Böhle, E., and Biegler, R., "The Component Fatty Acids of Blood," Fette, Seifen, Anstrichtemittel, 62, 673-679 (1960); CA 55, 2836a

6159. Schrade, W., Böhle, E., and Biegler, R., "Humoral Changes in Arteriosclerosis. Investigation of Lipides, Fatty Acids, Ketone Bodies, Pyruvic Acid, Lactic Acid and Glucose in the Blood," Lancet, 1960-II, 1409-1416; CA 55, 7640c

6160. Schrade, W., Böhle, E., and Biegler, R., "Recent Advances in Fat-Metabolism Investigations and Their Clinical Significance," Deut. med. Wochschr., 86, 781-791 (1961); CA 55, 17876h

6161. Schrade, W., Böhle, E., Biegler, R., Meder, V., and Teicke, R., "Fatty Acids in Human Serum Examined by Gas Chromatography. I. Composition of the Fatty Acids in Healthy, Arteriosclerosis, and Diabetic Subjects," Klin. Wochschr., 38, 126-134 (1960); CA 55, 5739e

6162. Schrade, W., Böhle, E., Biegler, R., and Sabel, C., "Gas Chromatographic Studies of Fatty Acids in Human Serum. II. Contributions to the Regulation of Unesterified Fatty Acids," Klin. Wochschr., 38, 707-716 (1960); CA 54, 25127i

6163. Schrade, W., Böhle, E., Biegler, R., Teicke, R., and Ullrich, B., "Gas Chromatographic Studies of Serum Fatty Acids in Man. III. The Fatty Acids of Cholesterol Esters, Phospholipides, and Triglycerides, and the Unesterified Fatty Acids in Healthy and Arteriosclerotic Subjects," Klin. Wochschr., 38, 739-753 (1960); CA 54, 25192g

6164. Schrader, R., Ackermann, G., and Grund, H., "New Determination of the Gas Content of Salts. II. Isolation of Gases for Gas-Chromatographic Determination," Acta Chim. Acad. Sci. Hung., 33, 31-38 (1962); CA 58, 3883a

6165. Schramm, R. M., and Langlois, G. E., "The Alkali, Metal Catalyzed Alkylation of Toluene with Propylene," J. Am. Chem. Soc., 82, 4912-4918 (1960); CA 55, 9310e

6166. Schreiber, H. P., and Waldman, M. H., "Rapid Determination of Some Surface Properties of Solids," Can. J. Chem., 37, 1782-1785 (1959); CA 54, 6281h

6167. Schroter, M., and Metzner, K., (Eds.), "Gas-Chromatographie 1961," Akademie-Verlag, Berlin, 1962. Lectures and Discussions Presented at the 3rd Symp. on Gas Chromatography, Schkopan, Germany, May 1961

6168. Schroter, M., and Prenkschas, W., "Examples for the Application of Gas Chromatography for the Analysis of C₄-C₅ Hydrocarbons," in "Gas Chromatographie 1959," edited by R. E. Kaiser and H. G. Struppe, Akademie Verlag, Berlin, 1959, pp. 304-317

6169. Schuberth, H., "New Apparatus for the Determination of Vapor-Liquid Phase Equilibria," Z. Chem., 1, 312-314 (1961); CA 56, 8495b

6170. Schuck, E. A., Ford, H. W., and Stephens, E. R., "Air Pollution Effects of Irradiated Automobile Exhaust as Related to Fuel Composition," Air Pollution Foundation, San Francisco, Calif., Rept. No. 26, 1958

6171. Schuck, E. A., and Renzetti, N. A., "Eye Irritants Formed During Photooxidation of Hydrocarbons in the Presence of Oxides of Nitrogen" J. Air Pollution Control Assoc., 10, 389-392 (1960); CA 55, 863b

6172. Schuetz, R. D., Taft, D. D., O'Brien, J. P., Shea, J. L., and Mork, H. M., "Fluorinated Heterocyclic Compounds," J. Org. Chem., 28, 1420-1422 (1963); CA 59, 1567d

6173. Schuhknecht, W., "Gas Chromatographic Analysis. A Modern Rapid Method for Gas Analysis," Arch. Eisenhuttenw., 29, 101-106 (1958); CA 52, 7936h

6174. Schuhknecht, W., "Chromatographic Gas Analysis," Techn. Mitt. 51, No. 3, 1-8 (1958)

6175. Schulek, E., Pungor, E., and Trompler, J., "New Ideas in Gas Analysis. I. Tension Measurement on a Microscopic Scale," Mikrochim. Acta, 1956, 1005-1022; CA 50, 9231d

6176. Schulek, E., Trompler, J., and Pungor, E., "Vapor Analysis of Multicomponent Systems. V. Determination of the Partial Pressure of Ethanol Over Ethanol-Water-Perchloric Acid Solutions," Mikrochim. Acta, 1, 18-21 (1959); CA 55, 69c

6177. Schultz, D. R., and Lykken, L., "New Trends in Determination of Pesticide Chemical Residue in Foods and Commodities," 10th Detroit Anachem Conf., Detroit, Mich., October 1962, Program Abstr., p. 23

6178. Schultze, G. R., and Schmidt-Kuster, W. J., "Separation Abilities of Silica Gels in Gas Chromatography," Z. Anal. Chem., 170, 232-238 (1959); CA 54, 2879g

6179. Schulz, G., "Gas Chromatography in Industrial Practice," Z. Anal. Chem., 181, 390-395 (1961); CA 56, 24i

6180. Schulz, H., "Continuous Material Separation in Counter Current Under the Restrictions of Gas-Elution-Chromatography," 4th Intern. Gas Chromatography Symp., Hamburg, Germany, June 1962; pub. in "Gas Chromatography 1962," M. van Swaay, ed., Butterworth & Co., London, 1962

6181. Schwartz, C. E., and Smith, J. M., "Flow Distribution in Packed Beds," Ind. Eng. Chem., 45, 1209-1218 (1953); CA 47, 8425e

6182. Schwartz, R. D., "Capillary Adsorption Columns," Univ. of Houston Intern. Symp. on Advances in Gas Chromatography, Houston, Texas, Januray, 1963

6183. Schwartz, R. D., and Brasseaux, D. J., "Resolution of Complex Hydrocarbon Mixtures by Capillary Column Gas-Liquid Chromatography: Composition of the 28-114°C. Portion of Petroleum," Anal. Chem., 35, 1374-1382 (1963); 145th Natl. ACS Mtg., New York, N. Y., September 1963

6184. Schwartz, R. D., Brasseaux, D. J., and Shoemake, G. R., "Sol-Coated Capillary Adsorption Columns for Gas Chromatography," Anal. Chem., 35, 496-499 (1963); CA 58, 11933a

6185. Schwartz, R. D., Brasseaux, D. J., and Shoemake, G. R., "Capillary Column Gas-Liquid Chromatography with Thermal Conductivity Detectors," J. Gas Chromatog., 1, No. 1, 32-33 (1963); Correction, ibid., 1, No. 2, 17 (1963)

6186. Schwarz, H. P., Dreisbach, L., Barrionuevo, M., Kleschick, A., and Kostyk, I., "The Effect of Ionizing Irradiation on the Lipide Composition of the Liver Mitochondria of Rats," Arch. Biochem. Biophys., 92, 133-139 (1961); CA 55, 11491i

6187. Schwarz, H. P., Dreisbach, L., Barrionuevo, M., Kleschick, A., and Kostyk, I., "Chromatography of Sphingolipids of Human Brain," J. Lipid Research, 2, 208-214 (1961); CA 55, 22456h

6188. Schwenk, U., and Hachenberg, H., "Gas Chromatography as Aid for Identifying the Reversibility of Metal Salt-Gas Addition Compounds," Brennstoff-Chem., 41, 183-184 (1960); CA 54, 19086a

6189. Schwenk, U., and Hachenberg, H., "Safety Precautions in Connection with the Use of Hydrogen as a Gas Chromatographic Carrier Gas," Brennstoff-Chem., 42, No. 3, 72-74 (1961); CA 55, 17113e

6190. Schwenk, U., Hachenberg, H., and Foerderreuther, M., "Trace Analysis by Using the Flame Ionization Detector and Packed Columns in Gas Chromatography. I. Determination of Traces of Acetylene in Ethylene and Detection of Trace Components in Water," Brennstoff-Chem., 42, 192-199 (1961); CA 55, 22935d

6191. Schwenk, U., Hachenberg, H., and Foerderreuther, M., "Trace Analysis with Flame-Ionization Detectors and Packed Columns in Gas Chromatography. II. Determination of 1 p.p.m. of Carbon Monoxide Carbon Dioxide, and Propane in Ethylene," Brennstoff-Chem., 42, 295-296 (1961); CA 56, 7973f

6192. Schwenk, U., Hachenberg, H., and Schneck, E., "Use of Integrators in Gas Chromatography," Brennstoff-Chem., 42, 33-38 (1960); CA 54, 9380f

6193. Schwenk, U., and Weber, E., "Replacement of Podbielniak Distillation by a Combination of Gas Chromatography and Weighing," Z. Anal. Chem., 164, 159-164 (1958); CA 53, 6887a

6194. Scott, B. A., and Williamson, A. G., "Effect of the Carrier Gas on the Sensitivity of Thermal Conductivity Detectors in Gas Chromatography," Nature, 183, 1322-1323 (1959); CA 53, 20926e

6195. Scott, C. G., "Alumina as a Column Packing in Gas Chromatography," J. Inst. Petrol., 45, 118-122 (1959); CA 53, 10900b

6196. Scott, C. G., "Progress in Gas Chromatography," Nature, 184, 1199 (1959)

6197. Scott, C. G., "A New Approach to the Study and Assessment of Medicinal White Oil Stability," in "Gas Chromatography 1960," edited by R. P. W. Scott, Butterworths, London, 1960, pp. 372-386

6198. Scott, C. G., "Gas Chromatography: Definitions and Retention Parameters," Nature, 189, 280-281 (1961)

6199. Scott, C. G., "Linear Gas-Solid Chromatography," 4th Intern. Gas Chromatography Symp., Hamburg, Germany, June 1962; pub. in "Gas Chromatography 1962," M. van Swaay, ed., Butterworth & Co., London, 1962

6200. Scott, C. G., "Adsorption Isotherms and Linear Gas-Solid Chromatography," Nature, 193, 159-160 (1962); CA 56, 10941i

6201. Scott, C. G., "Gas Chromatography," Nature, 196, 817-819 (1962). Report on the meeting of the Gas Chromatography Discussion Group, Inst. of Petrol., September 1962

6202. Scott, C. G., and Rowell, D. A., "Gas-Solid Chromatographic Separation of Hydrocarbons of High Molecular Weight," Nature, 187, 143-144 (1960); CA 54, 21930c

6203. Scott, D. S., and Han, A., "A Modified Thermal Conductivity Cell Independent of Flow Rate," Anal. Chem., 33, 160 (1961); CA 55, 24133i

6204. Scott, C. P., Soong, C. C., and Reynolds, J. L., "Telomers and Cotelomers of Acrylate Esters with Mercaptans," 144th Natl. ACS Mtg., Los Angeles, Calif., March-April 1963, Program Abstr., p. 4Q

6205. Scott, P. G. W., "Gas-Liquid Chromatography," Chem. Age (London), 82, 167-169 (1959)

6206. Scott, R. P. W., "A New Detector for Vapor-Phase Partition Chromatography," Nature, 176, 793 (1955); CA 50, 13519h

6207. Scott, R. P. W., "A New Detector for Vapour-Phase Partition Chromatography," in "Vapour-Phase Chromatography," edited by D. H. Desty, Academic Press, Inc., New York, 1957, pp. 131-145

6208. Scott, R. P. W., "The Construction of High-Efficiency Columns for the Separation of Hydrocarbons," Gas Chromatog., Proc. Symposium, Amsterdam, 1958, 186-196; CA 53, 16608d

6209. Scott, R. P. W., "Recent Developments in Gas Chromatography. I.," Mfg. Chemist, 29, 411-416 (1958); CA 53, 1979d

6210. Scott, R. P. W., "Recent Developments in Gas Chromatography. II.," Mfg. Chemist, 29, 517-522 (1958); CA 53, 3973d

6211. Scott, R. P. W., "The Construction of High-Efficiency Columns for the Separation of Hydro-carbons," in "Gas Chromatography 1958," edited by D. H. Desty, Academic Press, Inc., New York, 1958, pp. 189-199

6212. Scott, R. P. W., "Nylon Capillary Columns for Use in Gas-Liquid Chromatography," Nature, 183, 1753-1754 (1959); CA 53, 20936g

6213. Scott, R. P. W., "Cathode Ray Presentation of Chromatograms," Nature, 185, 312-313 (1960); CA 55, 8954g

6214. Scott, R. P. W., (Ed.), "Gas Chromatography 1960," Butterworths, London, 1960. Proceedings of Symposium of Society for Analytical Chemistry and Institute of Petroleum, Edinburgh, June 1960" Butterworths, London, 1960. Proceedings of Sumposium

6215. Scott, R. P. W., "An Introductory Lecture on Apparatus and Technique," in "Gas Chromatography 1960," edited by R. P. W. Scott, Butterworths, London, 1960, pp. 3-6

6216. Scott, R. P. W., "Gas-Liquid Chromatography; the Micro-Argon Detector," Benzole Producers Ltd. (London), Research Paper 1960, No. 2, 1-8; CA 55, 16027i

6217. Scott, R. P. W., "Process Monitoring by Gas Chromatography," Research, 14, 113-117 (1961)

6218. Scott, R. P. W., "The Effect of Temperature on the Efficiency, Resolution, and Analysis Time of Capillary Columns," J. Inst. Petrol., 47, 284-290 (1961); CA 55, 25223f

6219. Scott, R. P. W., "New Column System for Gas Chromatography," Univ. of Houston Intern. Symp. on Advances in Gas Chromatography, Houston, Texas, January 1963

6220. Scott, R. P. W., "Gas Liquid Chromatography. Temperature Changes During Passage of a Solute Through a Theoretical Plate," Anal. Chem., 35, 481-488 (1963)

6221. Scott, R. P. W., "Gas Chromatography - New Equipment and New Uses," Mfg. Chemist, 34, No. 2, 55-59, 75 (1963); CA 58, 11931a

6222. Scott, R. P. W., "Preparative Scale Chromatography with Analytical Columns," Nature, 198, 782-783 (1963)

6223. Scott, R. P. W., and Cheshire, J. D., "High-Efficiency Columns for the Analysis of Hydrocarbons by Gas-Liquid Chromatography," Nature, 180, 702-703 (1957); CA 52, 3414d

6224. Scott, R. P. W., and Cumming, C. A., "Gas-Liquid Chromatography; Cathode Ray Presentation of Chromatograms," Benzole Producers Ltd. (London), Research Paper 1960, No. 7, 3-15; CA 55, 16027h

6225. Scott, R. P. W., and Cumming, C. A., "Cathode Ray Presentation of Chromatograms," in "Gas Chromatography 1960," edited by R. P. W. Scott, Butterworths, London, 1960, pp. 117-135

6226. Scott, R. P. W., and Girling, G. W., "Evolution of Volatile Hydrocarbons from Coal During Storage," Chem. & Ind. (London), 1961, 1570-1571; CA 56, 13168i

6227. Scott, R. P. W., and Hazeldean, G. S. F., "Some Factors Affecting Column Efficiency and Resolution of Nylon Capillary Columns," in "Gas Chromatography 1960," edited by R. P. W. Scott, Butterworths, London, 1960, pp. 144-161

6228. Scott, R. P. W., and Hazeldean, G. S. F., "Gas-Liquid Chromatography; Some Factors Affecting Column Efficiency and Resolution Obtained from Nylon Capillary Columns," Benzole Producers Ltd. (London), Research Paper 1960, No. 6, 23 pp.; CA 55, 24191a

6229. Scott, R. P. W., and Maggs, R. J, "Separation of Pure Aromatic Hydrocarbons from Coal Gas by a Continuous Gas-Chromatographic Process," Benzole Producers Ltd. (London), Research Paper 1960, No. 5, 3-23; CA 55, 16966g

6230. Scott, R. R., and Stannard, B. W., "Calibration of the Argon Detector for Volatile Solvents," Chem. & Ind. (London), 1960, 1259-1260; CA 55, 15002c

6231. Scott, R. R., and Stannard, B. W., "Determination of Solvent Recovery Adsorber Efficiencies by Gas-Liquid Chromatography (G.L.C.)," Chem. & Ind. (London), 1962, 604-605; CA 57, 7070d

6232. Scott, T. W., White, L. G., and Annison, E. F., "Fatty Acids in Semen," Biochem. J., 78, 740-742 (1961); CA 55, 8565g

6233. Scott, W. E., Herb, S. F., Magidman, P., and Riemenschneider, R. W., "Composition of Edible Fats. Unsaturated Fatty Acids of Butter Fat," J. Agr. Food Chem., 7, 125-128 (1959); CA 53, 18324h

6234. Searles, Jr., S., Liepins, R., and Kash, H. M., "Reactions of Sodium Methoxide with 2-Alkyl-2,3-dichloroaldehydes. II. Methacrolein Dichloride," J. Org. Chem., 26, 36-40 (1961); CA 55, 18581a

6235. Sechenov, G. P., Bunina, N. N., and Al'tshuler, V. S., "A Method for Analyzing Low-Molecular-Weight Gases by a Combination of Chromatography and Volumetric Analysis," Tr. Inst. Goryuch. Iskop., Akad. Nauk S.S.S.R., 18, 253-258 (1962); CA 58, 6624d

6236. Seely, G. R., Oliver, J. P., and Ritter, D. M., "Gas-Liquid Chromatographic Analysis of Mixtures Containing Methyldiboranes," Anal. Chem., 31, 1993-1995 (1959); CA 54, 10630d

6237. Segal, H. S., and Sutherland, M. L., "Comparison of Flame Ionization and Electron Capture Detectors for the Gas Chromatographic Evaluation of Herbicide Residues," 144th Natl. ACS Mtg., Los Angeles, Calif., March-April 1963, Program Abstr., p. 4A

6238. Seher, A., "Chromatographic Analysis Procedures in Research and Production Control," J. Soc. Cosmetic Chemists, 13, 385-403 (1962); CA 58, 5449g

6239. Seidel, C. F., Felix, D., Eschenmoser, A., Biemann, K., Palluy, E., and Stoll, M., "Rose Oil. II. The Constitution of the Oxide $C_{10}H_{18}O$ from Bulgarian Rose Oil," Helv. Chim. Acta, 44, 598-606 (1961); CA 55, 18019i

6240. Seifert, W. K., and Condit, P. C., "Selective Catalytic Hydrogenation of Nitroolefins," J. Org. Chem., 28, 265-267 (1963); CA 58, 10097h

6241. Seiyama, T., Kato, A., Fujiishi, K., and Nagatani, M., "New Detector for Gaseous Components Using Semiconductive Thin Films," Anal. Chem., 34, 1502-1503 (1962); CA 57, 14411i

6242. Sederka, B., Spevak, A., and Friedrich, K., "Infrared Indication in Gas Chromatography," Chem. prumysl, 7, 602-604 (1957); CA 52, 13515i

6243. Self, R., "Enrichment Trap for Use with Capillary [Chromatographic] Columns," Nature, 189, 223 (1961); cf. Mackay, et al., Proc. Sci. Sect. Toilet Goods Assoc., No. 31, 7 (1959); CA 55, 25385d

6244. Seligman, P. C., and Woolmington, K. G., "Construction of Gas Chromatographic Equipment," S.

African J. Sci., 57, 39-44 (1961); CA 55, 17113a

6245. Seligman, R. B., "Vapor-Phase Chromatography in Tobacco Research," Symposium on Recent Developments in Research Methods and Instrumentation, Bethesda, Md., May 1956; Abstr., Anal. Chem., 28, 1058 (1956)

6246. Seligman, R. B., and Gager, Jr., F. L., "Recent Advances in Gas Chromatography Detectors," in "Advances in Analytical Chemistry and Instrumentation," Vol. 1, edited by C. N. Reilley, Interscience Publishers, Inc., New York, 1960, pp. 119-150

6247. Seligman, R. B., Resnik, F. E., O'Keefe, A. E., Holmes, J. C., Morrell, F. A., Murrill, D. P., and Gager, Jr., F. L., "Gas Chromatography in Tobacco Research," Tobacco Sci., 1, 124-129, published in Tobacco, 145, No. 9, 24-29 (1957); CA 51, 17107i; 129th Natl. ACS Mtg., Dallas, Texas, April 1956, Program Abstr., p. 19B

6248. Selin, T. G., "Stereochemistry of Trichlorosilane Additions to Olefins," Univ. Microfilms (Ann Arbor, Mich.), L. C. Card No. Mic 61-669, 181 pp.; Dissertation Abstr., 21, 2898-2899 (1961); CA 55, 17533h

6249. Selwitz, C. M., and Stanmyer, Jr., J. L., "Dehydrogenation of Pentanes with Oxygen," 140th Natl. ACS Mtg., Chicago, Ill., September 1961, Program Abstr., p. 5S

6250. Sen, B., "Gas-Liquid Partition Chromatography. Column Behavior and Its Reproduction. Criteria for Favorable Column Behavior," Anal. Chim. Acta, 22, 130-142 (1960); CA 54, 8215i; 136th Natl. ACS Mtg., Atlantic City, N. J., September 1959, Program Abstr., p. 24B

6251. Senn, Jr., W. L., and Drushel, H. V., "Organic Composition Analysis by Combination of Gas Chromatography with Infrared Spectrophotometry," Anal. Chim. Acta, 25, 328-333 (1961); CA 56, 2867i

6252. Serfass, E. J., and Baker, W. J., "An Ultra High Purity Hydrogen Source for Use in Gas Chromatography," 12th Pittsburgh Conf. on Anal. Chem. & Appl. Spectroscopy, Pittsburgh, Pa., February-March 1961, Program Abstr., p. 51

6253. Seris, G., Vernotte, P., Clave, A. M., and Kohlmann, P., "Apparatus for the Analysis of Gaseous Impurities in Gases," Chim. anal., 42, 28-32 (1960); CA 54, 8165i

6254. Serpinet, J., "Analysis of the Heavy Products from Pyrolysis of Chlorofluoromethane by Preparative Gas Chromatography and Mass Spectrometry," Chim. anal., 41, 146-151 (1959); CA 53, 14817b

6255. Serpinet, J., "Trace Analysis by a Physical Method After Concentration by Gas Chromatography," Chim. anal., 42, 433-436 (1960); CA 55, 4232b

6256. Serpinet, J., "The Very Sensitive Detection of Permanent Gas in Gas Chromatography with a Metastable Helium Detector," Anal. Chim. Acta, 25, 505-506 (1961); CA 56, 5401h

6257. Serpinet, J., "Analysis of Mixtures of Diamines and Aminoalcohols by Gas Chromatography," Journées Intern. Etude Méthodes, Separation Immediate Chromatog., Paris, 1961, 307-318 (Pub. 1962); CA 59, 2165c

6258. Setinek, K., and Cernysev, J. A., "Thermal Decomposition of Trichlorosilane," Chem. prumysl, 12, 419-422 (1962); CA 58, 4190g

6259. Setzkorn, E. A., and Carel, A. B., "The Analysis of Alkylarenesulfonates by Micro Desulfonation and Gas Chromatography," J. Am. Oil Chemists' Soc., 40, 57-59 (1963); CA 58, 9337a

6260. Setzkorn, E. A., Carel, A. B., Smith, H. F., and Hamilton, W. C., "The Microanalysis of Alkyl Aryl Sulfonates by a Combined Chemical-Gas Chromatographic-Infrared Method," 11th Pittsburgh Conf. on Anal. Chem. & Appl. Spectroscopy, Pittsburgh, Pa., February-March 1960, Program Abstr., p. 45

6261. Sevenster, P. G., "Recording Nitrometer Detector for Gas Chromatography," S. African Ind. Chemist, 12, 75-76 (1958); CA 52, 17822d

6262. Sevenster, P. G., "Gas Analysis Apparatus Based on Gas-Chromatographic and Chemical Principles," S. African Ind. Chemist, 12, 151-158 (1958); CA 53, 4828e

6263. Shabtai, J., and Gil-Av, E., "Mechanism of Thermal Aromatization of Butadiene-Ethylene and Butadiene-Propylene Mixtures," Tetrahedron, 18, 87-94 (1962); CA 57, 634e

6264. Shabtai, J., Herling, J., and Gil-Av, E., "Gas-Liquid Partition Chromatography of Isomeric Alkylcyclopentenes and Alkylidenecyclopentanes," J. Chromatog., 2, 406-410 (1959); CA 54, 4269f

6265. Shabtai, J., Pinchas, S., Herling, J., Gruener, Ch., and Gil-Av, E., "The Preparation and Infrared Spectra of 3- and 4-Ethylcyclohexene and 3- and 4-Isopropylcyclohexene," J. Inst. Petrol., 48, 13-17 (1962); CA 57, 16420c

6266. Shahin, M. M., and Lipsky, S. R., "The Mechanisms of Operation of a New and Highly Sensitive Ionization System for the Detection of Permanent Gases and Organic Vapors by Gas Chromatography," Anal. Chem., 35, 467-474 (1963)

6267. Shahin, M. M., and Lipsky, S. R., "The Role of Argon Metastable Atoms in the Ionization of Organic Molecules," Anal. Chem., 35, 1562-1566 (1963)

6268. Shank, R. S., and Shechter, H., "Simplified Zinc-Copper Couple for Use in Preparing Cyclopropanes from Methylene Iodide and Olefins," J. Org. Chem., 24, 1825-1826 (1959); CA 55, 12313i

6269. Shapiro, H., Wexler, A., and Brako, F., "Determination of Residual Monomers in Polymer Lattices by Gas Chromatography," 14th Pittsburgh Conf. on Anal. Chem. & Appl. Spectroscopy, Pittsburgh, Pa., March 1963, Program Abstr., p. 68

6270. Shaposhnikov, Y. K., Vedeneev, K. P., Vodzinskii, Y. V., and Lazareva, N. K., "Determination of Butanol in Butyl Acetate by Gas Chromatography," Gidroliz. i Lesokhim. Prom., 15, No. 6, 22-24 (1962); CA 58, 23a

6271. Shapras, P., and Claver, G. C., "Determination of Unreacted Monomers in Aqueous Emulsions of Interpolymers," Anal. Chem., 34, 433 (1962)

6272. Shaw, P. D., "Gas Chromatography of Trimethylsilyl Derivatives of Compounds Related to Chloramphenicol," Anal. Chem., 35, 1580-1582 (1963)

6273. Shaw, R., and Trotman-Dickenson, A. F., "The Reactions of Methoxyl Radicals," Proc. Chem. Soc., 1959, 61-62; CA 53, 13986b

6274. Shaw, R., and Trotman-Dickenson, A. F., "The Reaction of Methoxy Radicals with Alkanes," J. Chem. Soc., 1960, 3210-3215; CA 55, 5321g

6275. Shcherbakov, P. M., Kotel'nikov, B. P., and Ganin, Y. V., "Analysis of Synthetic C_{17}-C_{20} [Fatty Acids] by Gas Chromatography," Maslov.-Zhir. Prom., 27, No. 12, 25-27 (1961); CA 57, 1556g

6276. Shearer, D. A., and Stone, B. C., "An Alternative Packed Column Mode for Vapor Fractometers," Perkin-Elmer Instrument News, 13, No. 3, 12 (Spring 1962)

6277. Shearer, D. A., Stone, B. C., and McGugan, W. A., "Collection of Gas-Chromatographic Fractions for Identification by Infrared Spectrophotometry," Analyst, 88, 147-149 (1963)

6278. Shelton, J. R., and Champ, A., "Reaction of tert-Butoxy Radical with 4-Vinylcyclohexene," J. Org. Chem., 28, 1393-1394 (1963); CA 59, 1451h

6279. Shelton, J. R., and Henderson, J. N., "Reactions of Free Radicals with Olefins. Reactions of tert-Butoxy and tert-Butylperoxy Radicals with 4-Vinylcyclohexene," J. Org. Chem., 26, 2185-2190 (1961); CA 55, 25792d

6280. Shemyakin, F. M., "Contemporary Development of Chromatographic Analysis," Khim. Nauka i Prom., 4, No. 4, 216-223 (1959); CA 53, 15851f

6281. Shepherd, M., Rock, S. M., Howard, R., and Stormes, J., "Isolation, Identification and Estimation of Gaseous Pollutants of Air," Anal. Chem., 23, 1431-1440 (1951); CA 46, 671b

6282. Sheppard, A. J., and Douglass, C. D., "Suitability of Lipid Extraction Procedures for Gas-Liquid Chromatography," 53rd Spring Mtg., Am. Oil Chemists' Soc., New Orleans, La., May 1962

6283. Sheppard, W. A., and Harris, Jr., J. F., "Dichlorofluoromethanesulfenyl Chloride," J. Am. Chem. Soc., 82, 5106-5107 (1960); CA 55, 6361b

6284. Sherman, W. V., and Williams, G. H., "Homolytic Reactions of Aromatic Side-Chains. III. The Effect of Solvent on the Relative Rates of α-Hydrogen Abstraction by t-Butoxy Radicals," J. Chem. Soc., 1963, 1442-1444; CA 58, 10050b

6285. Sherwood, A. E., "A Sample Injection Device," Lab. Pract., 11, 546 (1962); CA 58, 6c

6286. Sherwood, P. W., "Molecular Sieves as New Refining Aid," Brennstoff-Chem., 40, 354-358 (1960); CA 54, 3929f

6287. Shibazaki, T., "Determination of BHC in a Mixed Dust of DDT and BHC. I. Separation and Determination of a Mixed Drug by Gas Chromatography," Bunseki Kagaku, 9, 544-547 (1960); CA 56, 10629h

6288. Shibazaki, T., "Reactions in a Current of Steam. I. Determination of Acids with Diethylamine," Bunseki Kagaku, 10, 842-846 (1961); CA 56, 10887i

6289. Shibazaki, T., "Separation and Determination of Mixed Drugs by Gas Chromatography. II. Determination of BHC in Mixed Insecticides," Eisei Shikensho Hokoku, No. 79, 69-71 (1961); CA 59, 382d

6290. Shimada, A., "Application of Gas Chromatography to Petroleum Chemistry," Kagaku no Ryoiki Zokan, No. 44, 115-126 (1961); CA 56, 3a

6291. Shimada, A., and Kawakami, H., "Gas-Liquid Chromatography. Stationary Phase and Number of Theoretical Plates in a Two-Stage Column," Kogyo Kagaku Zasshi, 64, 806-809 (1961); CA 57, 2875f

6292. Shimauchi, T., "Preparation of Samples for Infrared Spectral Analysis by Gas Chromatography," Kagaku no Ryoiki Zokan, No. 44, 63-78 (1961); CA 56, 2i

6293. Shimizu, H., and Kirihara, T., "Analysis of Chlorine, Carbon Monoxide, and Carbon Dioxide by Vapor-Phase Chromatography," Nagoya Kogyo Gijutsu Shikensho Hokoku, 9, 453-457 (1960); CA 54, 22136g

6294. Shimizu, S., Ueda, H., and Ikeda, N., "The Essential Oils of the Interspecific Hybrids in the Genus Mentha. IV. 3-Octanone as a Principal Component of a European Mentha Arvensis," Agr. Biol. Chem., 25, 263-264 (1961); CA 55, 13778c

6295. Shinohara, T., Ohkusa, T., and Okada, Y., "Determination of Impurities in Liquid Oxygen by Infrared Spectrophotometry and Gas Chromatography," Bunseki Kagaku, 10, 241-245 (1961); CA 56, 14909f

6296. Shipman, G. F., "Gas-Solid Chromatography of Mixtures of Hydrogen Isotopes," Anal. Chem., 34, 877-878 (1962); CA 57, 2829c

6297. Shipotofsky, S. H., and Moser, H. C., "Gas-Chromatographic Analysis of Some Phosphorus Compounds," Anal. Chem., 33, 521-523 (1961); CA 55, 18421i

6298. Shiveley, J. H., Norris, T. A., and Roberts, F. M., "Identify the Components of Gasoline," Petrol. Refiner, 39, No. 3, 171-175 (1960); CA 54, 13616c

6299. Shoda, Y., Obata, S., and Nishida, Y., "Gas Chromatography of Perfume. I. Gas Chromatography of Essential Oils on a New Solid Phase," Koryo, No. 55, 34-44 (1960); CA 55, 13778b

6300. Shorland, F. B., "Acetone-Soluble Lipides of Grasses and Other Forage Plants. II. Properties of the Lipides with Special Reference to the Yield of Fatty Acids," J. Sci. Food Agr., 12, 39-43 (1961); CA 55, 18896h

6301. Shorland, F. B., and Gass, J. P., "Fatty Acid Composition of the Depot Fats of the Kiwi (Apteryx Australis Mantelli)," J. Sci. Food Agr., 12, 174-177 (1961); CA 55, 20134e

6302. Shuikin, N. I., An, V. V., and Lebedev, B. L., "Analysis of Mixtures of Homologs of Furan and Tetrahydrofuran by Means of Gas-Liquid Chromatography," Zavodskaya Lab., 27, 976-977 (1961); CA 56, 6669g

6303. Shuikin, N. I., Lebedev, B. L., and An, V. V., "Analysis of Mixtures of Cyclic Ethers by Gas Chromatography," Izv. Akad. Nauk S.S.S.R., Otd. Khim. Nauk, 1962, 1868-1869; CA 58, 3889c

6304. Shuikin, N. I., and Naryshkina, T. I., "Catalytic Dehydrogenation of Petroleum Methylcyclopentane," Neftekhimiya, 2, 473-479 (1962); CA 58, 7769e

6305. Shuikin, N. I., Tulupova, E. D., and Ostapenko, E. G., "Catalytic Synthesis of Methylcyclopentenes from Petroleum Methylcyclopentane," Izv. Akad. Nauk S.S.S.R., Otd. Khim. Nauk, 1962, 2204-2209; CA 58, 12432e

6306. Shulgin, A. T., "Composition of the Myristicin Fraction from Oil of Nutmeg," Nature, 197, 379 (1963); CA 58, 8846h

6307. Shulman, G. P., Trusty, M., and Vickers, J. H., "Thermal Decomposition of Aluminum Alkoxides," J. Org. Chem., 28, 907-910 (1963)

6308. Shumaker, Jr., J. H., "Arc Source for High-Temperature Gas Studies," Rev. Sci. Instr., 32, 65-67 (1961); CA 55, 20626h

6309. Sicilio, F., Bull, III, H., Palmer, R. C., and Knight, J. A., "An Inexpensive Versatile Multi-Column Gas Chromatograph for Students," J. Chem. Educ., 38, 506-508 (1961); CA 55, 25398h

6210. Sicilio, F., and Knight, J. A., "Increased Efficiency for Large-Volume Gas Chromatographic Samples," J. Chromatog., 6, 243-247 (1961); CA 56, 5382g

6311. Sicilio, F., Knight, J. A., and Alexander, E. L., "Device for Increasing Efficiency for Large-Volume Gas Samples in Chromatography," Anal. Chem., 33, 1136 (1961); CA 56, 9382e

6312. Siddiqui, I. R., and Adams, G. A., "3,5-Di-O-methyl-D-galactose, a New Methylated Sugar from a Fully Methylated Polysaccharide of Gibberella Fujikuroi (Fusarium Moniliforme)," Can. J. Chem., 38, 2029-2032 (1960); CA 56, 4841d

6313. Sideman, S., "Separation with Infinite Circular Columns in Gas Chromatography," Chem. Eng. Sci., 18, 95-98 (1963); CA 58, 11932h

6314. Sideman, S., and Giladi, J., "Gas Sample Introduction Valve," Instr. Meas., Chem. Anal., Elec. Quant., Nucleonics Process Control, Proc. Intern. Conf. Stockholm, 1, 226-230 (1960) (Pub. 1961); CA 56, 9902e

6315. Sideman, S., and Giladi, J., "Automatic Sample Introduction and Fraction Collection in Gas Chromatography, ISA, East Lansing, Mich., June 1961; ISA Proc., 3, 217-224 (1961); pub. in "Gas Chromatography," ed. by N. Brenner, J. E. Callen, and M. D. Weiss, Academic Press, Inc., New York, 1962, pp. 339-348.

6316. Sideman, S., and Giladi, J., "Determination of Separation Factors from Unresolved Two-Component Chromatographic Peaks," Gas Chromatog. Intern. Symp., 4, 260-272 (1962) (Pub. 1963); CA 57, 5753f

6317. Sidorov, R. I., Baboshin, B. K., and Rudakov, G. A., "Determination of Terpene Mixtures by Gas Chromatography. I. Conditions of Separation," Gidrolizn. i Lesokhim. Prom., 16, No. 2, 12-14 (1963); CA 59, 379c

6318. Siegel, S., and McCaleb, S., "The Stereochemistry of Hydrogenation of Isomers of Methyl Tetrahydrophthalate and Methyl Phthalate," J. Am. Chem. Soc., 81, 3655-3658 (1959); CA 54, 22479d

6319. Siegel, S., and Smith, G. V., "Stereochemistry and the Mechanism of Hydrogenation of Cycloolefins on a Platinum Catalyst," J. Am. Chem. Soc., 82, 6082-6087 (1960); CA 55, 11327g

6320. Siegel, S., and Smith, G. V., "The Stereochemistry of the Hydrogenation of Cycloolefins on Supported Palladium Catalysts," J. Am. Chem. Soc., 82, 6087-6090 (1960); CA 55, 11327i

6321. Sievers, R. E., "Gas-Phase Chromatographic Separation of Metal Chelates," 16th Annual Summer Symp., ACS, Div. Anal. Chem., Tucson, Arizona, June 1963

6322. Siggia, S., "Continuous Analysis of Chemical Process Systems," John Wiley & Sons, Inc., New York, 1959; Gas Chromatography, pp. 213-234; Thermal Conductivity, pp. 245-276

232

6323. Sihto, E., and Arkima, V., "Proportions of Some Fusel Oil Components in Beer and Their Effect on Aroma," J. Inst. Brewing, 69, 20-25 (1963); CA 58, 9598c

6324. Sihto, E., Nykanen, L., and Suomalaiuen, H., "Gas Chromatography of the Aroma Compounds of Alcoholic Beverages," Teknillisen Kemian Aikakauslehti, 19, 753-762 (1962); CA 58, 13089a

6325. Silbert, M. D., and Tomlinson, R. H., "Gas Chromatographic Apparatus for the Study of the Hot Atom Chemistry of Organic Halides," Can. J. Chem., 39, 706-719 (1961); CA 55, 16027e

6326. Silva, P., Bernal, P. G., and Leyton, G. R., "Application of the Chromatographic Technique and Electrophoretic Analysis in Taxonomic Studies," Revista Medica de Cordoba, 48, 144-149 (1960); CA 55, 26089i

6327. Silver, A. H., Adam, W. H., Gardner, H. M., and Keily, H. J., "Correlation of the Mean Molecular Weights of Commercial Alkylbenzenes with Gas-Liquid Chromatographic Data," J. Am. Oil Chemists' Soc., 38, 674-677 (1961); CA 56, 4883b; 52nd Annual Mtg., Am. Oil Chemists' Soc., St. Louis, Mo., April-May 1961

6328. Silver, M. S., "The Behavior of the t-Pentyl Cation as Produced in Deaminations and Halide Solvolyses," J. Am. Chem. Soc., 83, 3482-3489 (1961); CA 57, 2042h

6329. Silver, M. S., Caserio, M. C., Rice, H. E., and Roberts, J. D., "Small-Ring Compounds. XXXV. Studies of Rearrangements in the Nitrous Acid Deaminations of Methyl-Substituted Cyclobutyl-, Cyclopropylcarbinyl- and Allylcarbinylamines," J. Am. Chem. Soc., 83, 3671-3678 (1961); CA 56, 4628h

6330. Simek, M., and Tesarik, K., "Gas-Chromatographic Determination of Combustion Products in the Microdetermination of Nitrogen," Collection Czechoslov. Chem. Commun., 26, 1337-1342 (1961); CA 56, 927f

6331. Simmonds, P. G., and Lovelock, J. E., "Ionization Cross-Section Detector as a Reference Standard in Quantitative Analysis by Gas Chromatography," Anal. Chem., 35, 1345-1348 (1963)

6332. Simmons, H. E., and Smith, R. D., "A New Synthesis of Cyclopropanes," J. Am. Chem. Soc., 81, 4256-4264 (1959); CA 54, 5495d

6333. Simmons, M. C., "Aromatics Analysis by Capillary Gas Chromatography," Symp. on Capillary Gas Chromatography, R. D. IV. ASTM D-2, 64th Annual Mtg., ASTM, Atlantic City, N. J., June 1961

6334. Simmons, M. C., "Affect of Sample Injection on Efficiency," 8th Natl. Analysis Instrumentation Symp., ISA, Charleston, W. Va., April-May 1962

6335. Simmons, M. C., "Gas Chromatographic Data Processing," 5th Annual Gas Chromatography Institute, Canisius College, Buffalo, N. Y., April 1963

6336. Simmons, M. C., and Kelley, T. R., "Quantitative Analysis of Complex Samples by Gas-Liquid Chromatography and Mass Spectrometry," Gas Chromatog., Intern. Symposium, 2nd, East Lansing, Mich., 1959, 225-236 (Pub. 1961); CA 55, 22786d

6337. Simmons, M. C., and Kelley, T. R., "Quantitative Analysis of Complex Samples by Gas-Liquid Chromatography and Mass Spectrometry," in "Gas Chromatography," edited by H. J. Noebels, R. F. Wall, and N. Brenner, Academic Press, Inc., New York, 1961, pp. 225-236

6338. Simmons, M. C., and Koons, P. D., "Quantitative Analyses with Capillary Columns Using a Hydrogen Flame Ionization Detector," 16th Southwest Regional Mtg., ACS, Oklahoma City, Okla., December 1960

6339. Simmons, M. C., and Koons, P. D., "Quantitative Analyses Using a Hydrogen Flame Ionization Detector," 12th Pittsburgh Conf. on Anal. Chem. & Appl. Spectroscopy, Pittsburgh, Pa., February-March 1961, Program Abstr., p. 57

6340. Simmons, M. C., Richardson, D. B., and Dvoretzky, I., "Structural Analysis of Hydrocarbons by Capillary Chromatography in Conjunction with the Methylene Insertion Reaction," Gas Chromatog., Proc. Symposium, 3rd, Edinburgh, 1960, 211-223; CA 56, 4598d

6341. Simmons, M. C., Richardson, D. B., and Dvoretzky, I., "Structural Analysis of Hydrocarbons by Capillary Gas Chromatography in Conjunction with the Methylene Insertion Reaction," in "Gas Chromatography 1960," edited by R. P. W. Scott, Butterworths, London, 1960, pp. 211-223

6342. Simmons, M. C., and Snyder, L. R., "Two-Stage Gas-Liquid Chromatography," Anal. Chem., 30, 32-35 (1958); CA 52, 8519h; 131st Natl. ACS Mtg., Miami, Fla., April 1957, Program Abstr., p. 18Q

6343. Simmons, M. C., Taylor, L. M., and Nager, M., "Portable Carbon Dioxide-Conversion Apparatus for Gas-Liquid Chromatography," Anal. Chem., 32, 731-732 (1960); CA 54, 13759c

6344. Simon, J., Szepesy, L., and Simon, P., "Gas-Chromatographic Analysis of Various Liquid Hydrocarbons and Oxygen-Containing Compounds," Acta Chim. Acad. Sci. Hung., 27, 321-333 (1961); CA 55, 21986h

6345. Simon, J., Szepesy, L., and Simon, P., "Investigation of Various Liquid Hydrocarbons and Oxygen Compounds by Gas Chromatography," Magy. Asvanyolaf Foldgaz Kiserl. Int. Kozlemen, 3, 171-183 (1962); CA 57, 12784h

6346. Simon, P., Szepesy, L., and Simon, J., "Analysis of Liquids by Gas-Chromatography," Acta Chim. Acad. Sci. Hung., 27, 311-319 (1961); CA 55, 21957h

6347. Simon, R. H. M., "Gas Chromatography," Chem. & Eng. News, 34, 2350 (1956)

6348. Simon, W., "The Application of Gas Chromatography in the Organic Chemical Laboratory," Chimia (Switz.), 14, No. 6, 189-201 (1960); CA 55, 5321f

6349. Simon, W., "Use of Gas Phase Chromatography in an Organic Chemical Laboratory," Chim. anal., 42, 85 (1960); CA 55, 5321f

6350. Simon, W., and Lyssy, G. H., "Flowmeter for Small Gas Flow Rates," Mikrochim. Acta, 1960, 113-117

6351. Sims, R. P. A., "A Note on the Use of Partially-Overloaded β-Ray Ionization Detectors in Gas Chromatography," J. Chromatog., 8, 538-540 (1962)

6352. Sinclair, H. M., "Cholesteryl Ester Fatty Acids in Atheroma and Plasma," Lancet, 1959-II, 789-790; CA 54, 3692b

6353. Sinfelt, J. H., Hurwitz, H., and Shulman, R. A., "Kinetics of Methyl-cyclohexane Dehydrogenation Over Platinum-Alumina," J. Phys. Chem., 64, 1559-1562 (1960); CA 55, 11043c

6354. Sinfelt, J. H., and Rohrer, J. C., "Kinetics of the Catalytic Isomerization-Dehydrogenation of Methylcyclopentane," J. Phys. Chem., 65, 978-981 (1961); CA 55, 20988a

6355. Singer, P. D., "Gas Chromatography: Sample Introduction, Detectors, and Recorders – 4," Oil Gas J., 54, No. 85, 132-134 (1956)

6356. Singliar, M., Bobak, A., and Brida, J., "Gas-Chromatographic Separation of Propane-Propylene Chloro Derivatives," Chem. zvesti, 14, 209-214 (1960); CA 54, 18197i

6357. Singliar, M., Bobak, A., and Brida, J., "The Chromatographic Separation of Chlorinated Propanes and Propylenes," NML (Natl. Met. Lab.) Tech. J. (Jameshedpur, India), 3, No. 3, 20-24 (1961); CA 56, 915g

6358. Singliar, M., Bobak, A., Brida, J., and Lukacovic, L., "Mixed Stationary Phases for Gas-Liquid Chromatography," Z. Anal. Chem., 177, 161-166 (1960); CA 55, 19614b

6359. Singliar, M., and Brida, J., "Device for Quantitative Liquid Sample Introduction into Chromatographic Columns," Chem. prumysl, 8, 588-589 (1958); CA 54, 12674e

6360. Singliar, M., and Brida, J., "Simple Injection System for a Preparative Vapor-Phase Chromatography Column," Chem. & Ind. (London), 1960, 225-226; CA 54, 11590a

6361. Singliar, M., Brida, J., and Bobak, A., "Estimation of Butyraldehydes and Butyl Alcohols in Their Mixtures," Chem. prumysl, 10, 530-532 (1960); CA 55, 5242c

6362. Singliar, M., and Lukacovic, L., "Separation of Hexenes by Means of Gas–Liquid Chromatography," Gas Chromatog. Symp., Brno, Czech., June 1962; Abstr., J. Gas Chromatog., 1, No. 4, 9 (1963)

6363. Singliar, M., and Usakov, A., "Separation of Oxo Synthesis Products by Gas-Liquid Chromatography," Chem. prumysl, 11, 524-525 (1961); CA 56, 2873e

6364. Sirat, N., "Purified Carbon Dioxide for the Janak Chromatograph," Bull. soc. chim. France, 1962, 1388-1389; CA 57, 14412f

6365. Sjovall, J., "Bile Acids and Steroids. CXXIV. Qualitative Analysis of Bile Acids by Gas Chromatography," Acta Chem. Scand., 16, 1761-1764 (1962); CA 58, 801h

6366. Sjovall, J., Meloni, C. R., and Turner, D. A., "Separation of Substituted Cholanic Acids by Gas-Liquid Chromatography," J. Lipid Research, 2, 317-320 (1961); CA 56, 5386g

6367. Skarstrom, C. W., "Use of Adsorption Phenomena in Automatic Plant-Type Gas Analyzers," Ann. N. Y. Acad. Sci., 72, 751-763 (1959); CA 53, 15538d

6368. Skarstrom, C. W., "Continuous Plant Analyzers in Oil Refineries," Chem. in Can., 11, 45 (1959)

6369. Skell, P. S., and Allen, R. G., "Stereospecific Trans Radical Addition of DBr to the 2-Butenes. Syntheses of Erythro- and Threo-3-Deuterio-2-bromobutanes," J. Am. Chem. Soc., 81, 5383-5385 (1959); CA 54, 8591b

6370. Skellon, J. H., Thorburn, S., Spence, J., and Chatterjee, S. N., "Fatty Acids of Neem Oil and Their Reduction Products," J. Sci. Food Agr., 13, 639-643 (1962); CA 58, 7050h

6371. Skellon, J. H., and Windsor, D. A., "Fatty Acid Composition of Egg Yolk Lipids in Relation to Dietary Fats," J. Sci. Food Agr., 13, 300-303 (1962); CA 57, 10308b

6372. Skrinde, R. T., Caskey, J. W., and Gillespie, C. K., "Detection and Quantitative Estimation of Synthetic Organic Pesticides by Chromatography," J. Am. Water Works Assoc., 54, 1407-1423 (1962); CA 58, 7721b

6373. Slater, C. A., "Composition of Natural Lime Oil," Chem. & Ind. (London), 1961, 833-835; CA 56, 2521f

6374. Slater, C. A., "Citrus Essential Oils. I. Evaluation of Natural and Terpeneless Lemon Oils," J. Sci. Food Agr., 12, 257-264 (1961); CA 55, 18021a

6375. Slater, C. A., "Citrus Essential Oils. II. Composition of Distilled Oil of Limes," J. Sci. Food Agr., 12, 732-734 (1961); CA 56, 6104d

6376. Slavicek, J., "An Electromechanical Integrator," Gas Chromatog. Symp., Brno, Czech., June 1962; Abstr., J. Gas Chromatog., 1, No. 4, 8 (1963)

6377. Sloane, H., Johns, T., Ulrich, W., and Cadman, J., "Infrared Examination of Micro Samples from Gas Chromatographic Effluent," 14th Pittsburgh Conf. on Anal. Chem. & Appl. Spectroscopy, Pittsburgh, Pa., March 1963, Program Abstr., p. 63

6378. Sloman, K. G., and Borker, E., "Vapor Phase Chromatography – A Preliminary Investigation of Applications in a Food Laboratory," 129th Natl. ACS Mtg., Dallas, Texas, April 1956, Program Abstr., p. 18B

6379. Smit, W. M., and Hoek, A. van den, "Theory of Chromatography. I.," Rec. trav. chim., 76, 561-576 (1957); CA 51, 15216a

6380. Smith, B., "Recent Developments in the Field of Gas Chromatography," Svensk Kem. Tidskr., 71, 489-504 (1959); CA 54, 1023d

6381. Smith, B., "Analysis of Aqueous Mixtures by Gas Chromatography," Acta Chem. Scand., 13, 480-488 (1959); CA 55, 4231e

6382. Smith, B., "Analysis of Aromatic Solvents by Gas Chromatography," Acta Chem. Scand., 13, 877-883 (1959); CA 55, 16282g

6383. Smith, B., and Carlsson, O., "Gas Chromatographic Analysis of Polyhydric Organic Compounds (as Trimethylsilyl Ethers)," Acta Chem. Scand., 17, 455-460 (1963)

6384. Smith, B., and Erwik, M., "Determination of Hydrocarbon Constituents in the Benzene Pre-Run," Acta Chem. Scand., 17, 283-295 (1963); CA 59, 1417f

6385. Smith, B., Larson, G., and Ryden, J., "Quantitative Hydrogenation of Unsaturated Hydrocarbons Isolated by Gas Chromatography," Anal. Chim. Acta, 17, 313-318 (1963)

6386. Smith, B., and Ohlson, R., "Gas Chromatographic Separation of 3- and 4-Methyl-1-pentene," Acta Chem. Scand., 13, 1253 (1959); CA 55, 25725e

6387. Smith, B., and Ohlson, R., "Hydrogenation as an Aid in the Identification of Unsaturated Hydrocarbons by Gas Chromatography," Acta Chem. Scand., 14, 1317-1324 (1960); CA 56, 6632d

6388. Smith, B., and Ohlson, R., "Gas-Chromatographic Separation of Unsaturated Hydrocarbons Using Silver Nitrate in Ethylene Glycol as the Stationary Phase," Acta Chem. Scand., 16, 351-358 (1962); CA 57, 5285i

6389. Smith, B., Ohlson, R., and Olson, A. M., "Oxidation as an Aid in the Identification of Unsaturated Hydrocarbons by Gas Chromatography," Acta Chem. Scand., 16, 1463-1467 (1962); CA 57, 15806h

6390. Smith, B. J. K., and Patrick, C. R., "Cis- and Trans-Perfluorodecalin," Proc. Chem. Soc., 1961, 138; CA 55, 19872g

6391. Smith, C. F., Corman, B. G., and Lampe, F. W., "Hydrogen Inhibition of the Rare Gas Sensitized Radiolysis of Cyclopropane," J. Am. Chem. Soc., 83, 3559-3562 (1961); CA 56, 3055b

6392. Smith, Jr., C. R., Bagby, M. O., Miwa, T. K., Lohmar, R. L., and Wolff, I. A., "Unique Fatty Acids from Limnanthes Douglasii Seed Oil. The C_{20} and C_{22} Monenes," J. Org. Chem., 25, 1770-1774 (1960); CA 55, 2467g; 137th Natl. ACS Mtg., Cleveland, Ohio, April 1960, Program Abstr., p. 46O

6393. Smith, Jr., C. R., Wilson, T. L., and Mikolajczak, K. L., "Occurrence of Malvalic, Sterculic and Dihydrosterculic Acids Together in Seed Oils," Chem. & Ind. (London), 1961, 256-258; CA 55, 25785h

6394. Smith, Jr., C. R., Wilson, T. L., Miwa, T. K., Zobel, H., Lohmar, R. L., and Wolff, I. A., "Lesquerolic Acid. A New Hydroxy Acid from Lesquerella Seed Oil," J. Org. Chem., 26, 2903-2905 (1961); CA 56, 309h; 139th Natl. ACS Mtg., St. Louis, Mo., March 1961, Program Abstr., p. 29C

6395. Smith, D. E., and Coffman, J. R., "Analysis of Food Flavors by Gas-Liquid Chromatography. II. Separation and Identification of the Neutral Components from Bread Pre-ferment Liquid," Anal. Chem., 32, 1733-1737 (1960); CA 55, 6716b; 137th Natl. ACS Mtg., Cleveland, Ohio, April 1960, Program Abstr., p. 33B

6396. Smith, D. E., Favre, J. A., and Hines, W. J., "The Analysis of Light Hydrocarbon Mixtures by Gas Chromatography," 13th Southwest Regional Mtg., ACS, Tulsa, Okla., December 1957

6397. Smith, D. H., and Clark, F. E., "Some Useful Techniques and Accessories for Adaptation of the Gas Chromatograph to Soil Nitrogen Studies," Soil Sci. Soc. Am. Proc., 24, 111-115 (1960); CA 54, 16711i

6398. Smith, D. H., Nakayama, F. S., and Clark, F. E., "Gas-Solid Chromatographic Determination of NO_2 in the Presence of Oxygen," Soil Sci. Soc. Am. Proc., 24, 145-146 (1960); CA 54, 16272g

6399. Smith, D. M., Bartlet, J. C., and Levi, L., "Sucrose Acetate Isobutyrate as a New Ester Liquid Phase for Gas-Liquid Partition Chromatography," Anal. Chem., 32, 568-569 (1960); CA 54, 11805d

6400. Smith, D. M., and Campbell, R. G., "Soap-Bubble Flowmeter for Gas Chromatography," Chemist-Analyst, 50, 80-81 (1961); CA 56, 1981c

6401. Smith, D. M., and Levi, L., "Analysis of Spearmint Oils by Gas-Liquid Partition Chromatography," 138th Natl. ACS Mtg., New York, N. Y., September 1960, Program Abstr., p. 6B

6402. Smith, D. M., and Levi, L., "Treatment of Compositional Data for the Characterization of Essential Oils. Determination of Geographical Origins of Peppermint Oils by Gas Chromatographic Analysis," J. Agr. Food Chem., 9, 230-244 (1961); CA 55, 19144a

6403. Smith, D. M., Skakum, W., and Levi, L., "Determination of Botanical and Geographical Origin of Spearmint Oils by Gas Chromatographic and Ultraviolet Analysis," J. Agr. Food Chem., 11, 268-276 (1963); CA 59, 379f

235

6404. Smith, E. D., "Preparation of Gas Chromatographic Packings by Frontal Analysis, " Anal. Chem., 32, 1049 (1960); CA 54, 19086e

6405. Smith, E. D., "Simplified Chromatographic Separation and Analysis of C_4 Through C_{12} Dibasic Acids, " Anal. Chem., 32, 1301-1304 (1960); CA 55, 246h

6406. Smith, E. D., "Effect of Silica Flour in Dow Corning High Vacuum Grease on Gas-Chromatographic Retention Volumes, " Anal. Chem., 33, 1625-1626 (1961); CA 56, 2003i

6407. Smith, E. D., and Gosnell, A. B., "Gas Chromatographic Analysis of Fatty and Chlorinated Fatty Acids," Anal. Chem., 34, 438-439 (1962); CA 56, 14416g

6408. Smith, E. D., and Gosnell, A. B., "Effect of Preheater Contamination on Gas Chromatographic Analysis of Strongly Adsorbed Substances, " Anal. Chem., 34, 646-648 (1962); CA 57, 1511d

6409. Smith, E. D., and Johnson, J. L., "Measurement and Use of Substrate and Partition Liquid Selectivities in Gas Chromatography," Anal. Chem., 35, 1204-1207 (1963)

6410. Smith, E. D., and Radford, R. D., "Modification of Gas Chromatographic Substrates for the Separation of Aliphatic Diamines, " Anal. Chem., 33, 1160-1162 (1961); CA 56, 1982i; 138th Natl. ACS Mtg., New York, N. Y., September 1960, Program Abstr., p. 6B

6411. Smith, E. D., Riddick, E. B., Shook, T. E., and Slaton, B. L., "Gas Chromatographic Analysis of Amino-Acid Derivatives," 18th Southwest Regional Mtg., ACS, Dallas, Texas, December 1962

6412. Smith, G. G., and Kosters, B., "Effect of the Structure on the Pyrolysis of Esters. III. Pyrolysis Aryl and Benzyl Carbonates, " Chem. Ber., 93, 2400-2404 (1960); CA 55, 3502b

6413. Smith, G. G., Wetzel, W. H., and Kosters, B., "Qualitative and Quantitative Analysis of Products from Pyrolysis," Analyst, 86, 480-483 (1961); CA 56, 6632b

6414. Smith, G. W., and Williams, H. D., "Reactions of Adamantane and Adamantane Derivatives, " J. Org. Chem., 26, 2207-2212 (1961); CA 55, 25790i

6415. Smith, H. A., and Carter, E. H., "The Separation of Hydrogen, Tritium, and Tritium Hydride by Gas Chromatography, " Tritium Phys. Biol. Sci., Proc. Symp., Vienna, Austria, 1961, 1, 121-133 (Pub. 1962); CA 57, 13369c

6416. Smith, H. A., and Hunt, P. P., "Separation of Hydrogen, Hydrogen Deuteride, and Deuterium by Gas Chromatography," J. Phys. Chem., 64, 383-384 (1960); CA 54, 16989e

6417. Smith, I., "Chromatography as an Analytical Tool, " Chem. & Ind. (London), 1959, 650-651

6418. Smith, J. C. B., "Analysis of Organosilicon Compounds with Special Reference to Silanes and Siloxanes. A Review," Analyst, 85, 465-474 (1960); CA 55, 2362e

6419. Smith, J. E., "Gas Chromatographic Investigations of Chlorinated Hydrocarbons and Attendant Air-Sampling Techniques, " Univ. Microfilms (Ann Arbor, Mich.), L. C. Card No. Mic 61-809, 115 pp.; Dissertation Abstr., 21, 2871-2872 (1961); CA 55, 16075a

6420. Smith, J. F., "Relative G. L. C. Retention Data Using a Single Standard," Chem & Ind. (London), 1960, 1024-1025; CA 55, 1142a

6421. Smith, J. F., "Non-Coincidence of 'Gas Hold-Up' and Retention of Air in Gas-Liquid Chromatography," Nature, 193, 679 (1962); CA 56, 14901a

6422. Smith, J. R., and Hamilton, L. H., "D_{LCO} Measurements with Gas Chromatography," J. Appl. Physiol., 17, 856-860 (1962); cf. Physiologist, 3, 145 (1960); CA 59, 983a

6423. Smith, L. M., "Quantitative Fatty Acid Analysis of Milk Fat by Gas-Liquid Chromatography, " J. Dairy Sci., 44, 607-622 (1961); CA 55, 15766a

6424. Smith, L. M., and Lowry, R. R., "Fatty Acid Composition of the Phospholipids and Other Lipids in Milk," J. Dairy Sci., 45, 581-588 (1962); CA 58, 3735h

6425. Smith, L. M., and Ronning, M., "Comparison of Fatty Acid Composition of Milk Fats Produced by Cows Fed Alfalfa, Oat, or Ground Pelleted Alfalfa Hay," 56th Annual Mtg., Am. Dairy Sci. Assoc., Madison, Wisc., June 1961; Abstr., J. Dairy Sci., 44, 1170 (1961)

6426. Smith, R. N., Swinehart, J., and Lesnini, D. G., "Chromatographic Analysis of Gas Mixtures Containing Nitrogen, Nitrous Oxide, Nitric Oxide, Carbon Monoxide, and Carbon Dioxide, " Anal. Chem., 30, 1217-1218 (1958); CA 52, 14147f

6427. Smith, R. P., and Tatlow, J. C., "Fluorocyclohexanes. I. Cis- and trans-1H, 2H-Decafluorocyclohexanes," J. Chem. Soc., 1957, 2505-2511; CA 51, 14568a

6428. Smith, S. R., and Gordon, A. S., "Studies of Diffusion Flames: Some Ketone Flames, " 131st Natl. ACS Mtg., Miami, Fla., April 1957, Program Abstr., p. 51R

6429. Smith, S. R., Gordon, A. S., and Hunt, M. H., "Diffusion Flames. III. The Diffusion Flames of the Butanols, " J. Phys. Chem., 61, 553-558 (1957); CA 51, 13530f

6430. Smith, T. E., and Calvert, J. G., "The Thermal Decomposition of 2-Nitropropane, " J. Phys. Chem., 63, 1305-1309 (1959); CA 54, 2149d

6431. Smith, V. N., and Merritt, J., "A Negative Ion Gas Analysis Technique," 13th Pittsburgh Conf. on Anal. Chem. & Appl. Spectroscopy, Pittsburgh, Pa., March 1962, Program Abstr., p. 46

6432. Smith, W. B., and Anderson, J. D., "Some Evidence Regarding Free Radical Rearrangement Reactions," J. Am. Chem. Soc., 82, 656-658 (1960); CA 55, 24607e

6433. Smith, W. B., and Gilde, H-G., "The Kolbe Electrolysis as a Source of Free Radicals in Solution, "

J. Am. Chem. Soc., 81, 5325-5329 (1959); CA 54, 8590d

6434. Smith, W. B., and Gilde, H-G., "The Kolbe Electrolysis as a Source of Free Radicals in Solution. III. Some Aspects of the Stereochemistry of the Electrode Process," J. Am. Chem. Soc., 83, 1355-1358 (1961); CA 55, 27003a

6435. Smith, W. R., and Gudzinowicz, B. J., "High Temperature Gas Chromatography of Polyphenol Ethers," Gas Chromatog. Intern. Symp., 3, 163-168 (1961) (Pub. 1962); CA 59, 488d; 3rd Intern. Symp. on Gas Chromatography, ISA, East Lansing, Mich., June 1961, ISA Proc., 3, 111-115 (1961)

6436. Smith, W. R., and Gudzinowicz, B. J., "High Temperature Gas Chromatography of Polyphenol Ethers," in "Gas Chromatography," edited by N. Brenner, J. E. Callen, and M. D. Weiss, Academic Press, Inc., New York, 1962, pp. 163-168

6437. Smith, Jr., W. T., and King, G. G., "N-Sulfinyl Amines. Reaction with Carboxylic Acids," J. Org. Chem., 24, 976-978 (1959); CA 54, 22425i

6438. Smullin, C. F., and Olsanski, V. L., "Chromatographic Determination of Mono-, Di- and Triglycerol Esters of Fatty Acids and Free Glycerol in Monoglyceride Mixtures," 50th Annual Mtg., Am. Oil Chemists' Soc., New Orleans, La., April 1959

6439. Sneen, R. A., and Rosenberg, A. M., "An Example of Ester Cleavage by Mechanism $B_{A1}2$," J. Org. Chem., 26, 2099-2101 (1961); CA 55, 24644i

6440. Snowden, F. C., and Eanes, R. D., "Thermal Conductivity Detectors for Process Control," Ann. N. Y. Acad. Sci., 72, 764-778 (1959); CA 53, 14606g

6441. Soemantri, R. M., "Gas Chromatography and Selective Hydro Treatment," Proefschr. tech. Hogeschool Delft, 1958, 80 pp.; CA 55, 18672c

6442. Soemantri, R. M., "Gas Chromatography. I. Introduction, Principles and Apparatus. II. Applications in the Pharmaceutical Field," Suara Pharm. Madjalah, 4, No. 5, 139-146, No. 6, 157-161 (1959); CA 58, 12369e

6443. Soemantri, R. M., and Waterman, H. I., "The Separation of Naphthalene and Its Hydrogenated Products by Vapor-Phase Chromatography," J. Inst. Petroleum, 43, 94-99 (1957); CA 51, 7239f

6444. Sojak, L., Hrusovsky, M., and Matas, M., "Gas-Chromatographic Analysis of the Products from the Cracking of Higher Straight-Chain Paraffins," Ropa Uhlie, 4, 234-237 (1962); CA 58, 2304g

6445. Sokol, L., "Gas-Adsorption Chromatography with Thermal Conductivity Indication," Chem. průmysl, 7, 189-190 (1957); CA 52, 2639d

6446. Sokol, L., "Analysis of Phenols and Hydrocarbons by Means of Gas-Liquid Chromatography," Chem. listy, 52, 1726-1734 (1958); CA 53, 4012g

6447. Sokol, L., "Analysis of Phenols and Hydrocarbons by Gas-Liquid Chromatography," Collection Czechoslov. Chem. Communs., 24, 437-447 (1959); CA 53, 12098b

6448. Sokol, L., "Combination of Gas Chromatography and Infrared Spectroscopy for the Analysis of Complex Organic Mixtures," in "Gas Chromatographie 1959," edited by R. E. Kaiser and H. G. Struppe, Akademie Verlag, Berlin, 1959, pp. 318-336

6449. Sokol, L., "Gas Chromatography in the I. V. Stalin Plant in Litvinov," Gazovaya Khromatografiya, Trudy 1-oi [Pervoi] Vsesoyuz Konf., Akad. Nauk S.S.S.R., Moscow, 1959, 240-243 (Pub. 1960); CA 55, 25577d

6450. Sokol, L., "Combination of Gas Chromatography and Absorption Spectral Methods for Analysis of Organic Materials. I. Description and Properties of Preparative-Scale Chromatographic Columns," Collection Czechoslov. Chem. Communs., 25, 906-911 (1960)

6451. Sokol, L., "Determination of Nonaromatic Hydrocarbons in Technically Pure Aromatics by Means of Gas Chromatography," Sb. Praci Vyzkumu Chem. Vyuziti Uhli, Dehtu, Ropy, 1960, No. 1, 124-127; CA 58, 5430e, 8830h

6452. Sokol, L., Zavazal, J., and Karas, V., "Analysis of 100-30° Gasoline Fractions from High-Pressure Hydrocracked Distillate Cuts," Sb. Praci Vyzkumu Chem. Vyuziti Uhli, Dehtu, Ropy, 1960, No. 1, 128-133; CA 58, 7771f

6453. Sokolov, V. A., "Methods and Apparatus for Gas Analysis and Ways of Their Improvement," Izdatel Akad. Nauk S.S.S.R., 1958, 211-221 (Pub. 1959); CA 54, 11863g

6454. Sokolov, V. A., Alekseeva, F. A., Bars, E. A., Geodekyan, A. A., Mogilevskii, G. A., Yirovskii, Y. M., and Yasenev, B. P., "Investigation of Oil Detection Methods," World Petrol. Congress, 5th, New York, 1959, Proc. Sect. I, 667-687 (Pub. 1960); CA 54, 17855e

6455. Sokolov, V. A., Andronikashvili, T. G., Kuz'mina, L. P., and Shishkova, V. P., "Application of Some Minerals with Different Adsorption Capacities for Chromatographic Analysis of Gases," Khim. i Tekhnol. Topliva i Masel, 1957, No. 10, 61-65; CA 52, 4397b

6456. Sokolov, V. A., and Kolesnikova, L. P., "Separation of Alcohols by Gas-Liquid Chromatography," Neftekhimiya, 1, 564-566 (1961); CA 56, 14904f

6457. Sokolov, V. A., and Kuz'mina, L. P., "Chromatographic Analysis of C_1-C_2 Hydrocarbons and Some Inorganic Gases," Zavodskaya Lab., 23, 1034-1037 (1957); CA 53, 2939g

6458. Solo, A. J., and Pelletier, S. W., "The Gas-Liquid Chromatography of Phenanthrenes: Its Application to the Separation and Identification of Dehydrogenation Mixtures from Natural Products," Chem. & Ind. (London), 1961, 1755-1756; CA 56, 5406g; 140th Natl. ACS Mtg., Chicago, Ill., September 1961, Program Abstr., p. 68Q

6459. Solo, A. J., and Pelletier, S. W., "Gas-Liquid Chromatography of Phenanthrenes. Use in Identification of the Components of Mixtures of Alkyl Phenanthrenes," Anal. Chem., 35, 1584-1587 (1963)

6460. Soma, S., and Takeuchi, Y., "New Type of Thermal Conductivity Gas Detector," J. Phys. Soc. Japan, 15, 233-236 (1960); CA 54, 17978f

6461. Sonntag, F., "Use of Gas Chromatography in Analytical Problems Encountered in the Production of Benzene," Brennstoff-Chem., 42, 123-129 (1961); CA 55, 22778g

6462. Sorenson, I., and Soltoft, P., "Low Resistance Vapor-Phase Chromatograph Column," Acta Chem. Scand., 10, 1673-1674 (1956); CA 52, 12464d

6463. Sorenson, I., and Soltoft, P., "Gas-Liquid Chromatographic Determination of the Fatty Acid Composition in Oil from Seeds of Ulmus," Acta Chem. Scand., 12, 814-822 (1958)

6464. Sosnovsky, G., and O'Neill, H. J., "Reactions of tert-Butyl Peresters. VI. Pyrolysis of α-Acryloxy Derivatives of Aliphatic Sulfides," J. Org. Chem., 27, 3469-3471 (1962); CA 57, 16375i

6465. Southwick, P. L., Munsell, M. W., and Bartkus, E. A., "Direct Aromatic Carboxymethylation with Chloroacetylpolyglycolic Acids. Orientation of Dibenzofuran as Evidence of Free Radical Mechanism," J. Am. Chem. Soc., 83, 1358-1368 (1961); CA 56, 7250b

6466. Spacklen, S. B., "The Development of Gas Chromatographs," ISA Journal, 4, 514-517 (1957)

6467. Sparagana, M., Keutmann, E. H., and Mason, W. B., "Quantitative Determination of Individual $C_{19}O_2$ and $C_{19}O_2$ Urinary 17-Ketosteroids by Gas Chromatography," Anal. Chem., 35, 1231-1238 (1963)

6468. Sparagana, M., Mason, W. B., and Keutmann, E. H., "Preliminary Isolation of 17-Ketosteroids from Urine for Analysis by Gas Chromatography," Anal. Chem., 34, 1157-1159 (1962); CA 57, 10129d

6469. Spauschus, H. O., and Olsen, R. S., "Gas Analysis, a New Tool for Determining the Chemical Stability of Hermetic Systems," Refrigerating Eng., 67, No. 2, 25-30 (1959); CA 53, 19478h

6470. Spears, A. W., "Quantitative Determination of Phenol in Cigarette Smoke," Anal. Chem., 35, 320-322 (1963); CA 58, 11692e

6471. Spears, A. W., Lassiter, C. W., and Bell, J. H., "Quantitative Determination of Alkanes in Cigarette Smoke," J. Gas Chromatog., 1, No. 4, 34-37 (1963)

6472. Spence, D. W., and Pope, Jr., H. L., "The Servomechanical Integrator in Laboratory and Process Gas Chromatography," 13th Pittsburgh Conf. on Anal. Chem. & Appl. Spectroscopy, Pittsburgh, Pa., March 1962, Program Abstr., p. 47

6473. Spencer, C. F., Baumann, F., and Johnson, J. F., "Gas Odorants Analysis by Gas Chromatography," Anal. Chem., 30, 1473-1474 (1958); CA 52, 20993h

6474. Spencer, C. F., and Johnson, J. F., "Punched Card Storage of Gas Chromatographic Data," Anal. Chem., 30, 893-894 (1958); CA 52, 11493f

6475. Spencer, C. F., and Johnson, J. F., "Analysis of Phenylalkanes by Gas Chromatography," J. Chromatog., 4, 244-248 (1960); CA 55, 5381a

6476. Spencer, S., and Mikkelsen, L., "Alcohols and Glycols by Gas Chromatography," Facts & Methods (F&M Scientific Corp.), 3, No. 1, 5 (Spring 1962)

6477. Spencer, S., and Nadeau, H. G., "Determination of Trace Amounts of Diethylene Glycol in Triethylene Glycol by Gas Chromatography," Anal. Chem., 33, 1626-1627 (1961); CA 56, 933d

6478. Spencer, S. F., "Applications of Gas Chromatography in the Biomedical Field," Facts & Methods (F&M Scientific Corp.), 3, No. 3, 7-8 (Winter 1962-1963)

6479. Spencer, S. F., "Rapid Separation of Xylenes and Ethylbenzene by Gas Chromatography Using Packed Columns," Anal. Chem., 35, 592 (1963)

6480. Spencer, S. F., and Kauss, J. M., "Applications of Automatic Preparative Gas Chromatography," 14th Pittsburgh Conf. on Anal. Chem. & Appl. Spectroscopy, Pittsburgh, Pa., March 1963, Program Abstr., pp. 63-64

6481. Spencer, S. F., and Mikkelsen, L., "The Analysis of Glycols, Alcohols, and Related Compounds by Gas Chromatography," 10th Detroit Anachem Conf., Detroit, Mich., October 1962, Program Abstr., p. 32

6482. Spengler, G., and Wollner, E., "Adsorption Chromatographic Separation of Waxes and Their Components," Fette, Seifen, Anstrichmittel, 56, 775-784 (1954); CA 50, 7482g

6483. Spingler, H., and Markert, F., "Gas Chromatographic Micro Determination of Formyl- and Acetyl-Groups, with Special Reference to Digitalis Glycosides," Mikrochim. Acta, 1959, 122-128; CA 55, 3009i

6484. Spitsyn, V. I., Vereshchinskii, I. V., Glazunov, P. Y., Ryabchikova, G. G., and Sibirskaya, G. K., "Radiolysis of Propane at High Temperature," Tr. 2-go (Vtorogo) Vses. Soveshch. po Radiats. Khim., Akad. Nauk S.S.S.R., Otd. Khim. Nauk, Moscow, 1960, 308-311 (Pub. 1962); CA 58, 6359d

6485. Spolnicki, J., and Crooks, W. M., "A Chromatography Unit with Automatic Sampling for Kinetic Studies," J. Appl. Chem. (London), 13, 12-16 (1963); CA 58, 7405g

6486. Sporek, K. F., and Danyi, M. D., "Detection and Identification of Alcohols, Alkoxy Groups, Lignin, and Wood by Gas Liquid Chromatography," Anal. Chem., 34, 1527-1529 (1962); CA 58, 661g

6487. Sporek, K. F., and Danyi, M. D., "Detection and Identification of Mercaptans by Gas Liquid Chromatography," Anal. Chem., 35, 956-958 (1963); 14th Pittsburgh Conf. on Anal. Chem. & Appl. Spectroscopy, Pittsburgh, Pa., March 1963, Program Abstr., p. 63

6488. Spracklen, S. B., "Measurement of Chlorine Purity by a Plant-Type Vapor Fraction (Chromatographic) Analyzer," Southwide Chemical Conference, Memphis, Tenn., December 1956

6489. Spracklen, S. B., "Development of Gas Chromatographs," ISA Journal, 4, 514-517 (1957); CA 52, 8632f

6490. Spracklen, S. B., "Development of a Process Gas Chromatographic Analyzer," ISA Journal, 5, 81 (1958)

6491. Spracklen, S. B., "Review of Trends in Data Presentation for Gas Chromatographs," 13th Annual Instrument Automation Conf., ISA, Philadelphia, Pa., September 1958

6492. Sreenivasan, B., Brown, J. B., Jones, E. P., Davison, V. L., and Nowakowska, J., "Preparation and Purity of Linoleic Acid from Commercial Corn, Cottonseed and Safflower Oils," J. Am. Oil Chemists' Soc., 39, 255-259 (1962); CA 57, 4781g; 52nd Annual Mtg., Am. Oil Chemists' Soc., St. Louis, Mo., April-May 1961

6493. Srinivasan, R., "Photoisomerization Processes in Cyclic Ketones. IV. Cycloheptanone," J. Am. Chem. Soc., 81, 5541-5542 (1959); CA 54, 8671f

6494. Srinivasan, R., "Photoisomerization of 5-Hexen-2-one," J. Am. Chem. Soc., 82, 775-778 (1960); CA 54, 17250e

6495. Srinivasan, R., "The Photochemistry of 1,3-Butadiene and 1,3-Cyclohexadiene," J. Am. Chem. Soc., 82, 5063-5066 (1960); CA 55, 15370c

6496. Srinivasan, R., "Photochemistry of Cyclopentanone. I. Details of the Primary Process," J. Am. Chem. Soc., 83, 4344-4347 (1961); CA 56, 3059b

6497. Srivastava, B. N., and Saxena, S. C., "Thermal Conductivity of Binary and Ternary Rare Gas Mixtures," Proc. Phys. Soc. (London), 70B, 369-378 (1957); CA 52, 13400b

6498. Stacey, F. W., and Harris, Jr., J. F., "Radical Addition of Hydrogen Bromide to Hexafluoropropene," J. Org. Chem., 27, 4089-4090 (1962); CA 58, 7813e

6499. Stadler, P. A., "Gas Chromatographic Study of the Ketones of Lavandin Oil," Helv. Chim. Acta, 43, 1601-1612 (1960); CA 55, 7765b

6500. Stadler, P. A., Eschenmoser, A., Sundt, E., Winter, M., and Stoll, M., "Isoprenoid C_5 Alcohols in Essential Oils," Experientia, 16, 283-284 (1960); CA 54, 21659d

6501. Stadler, P. A., and Oberhansli, P., "Iodine Catalyzed Cyclization of trans-Geranic Acid Methyl Ester," Helv. Chim. Acta, 42, 2597-2603 (1959); CA 54, 13165c

6502. Staff of Arthur D. Little, Inc., (Eds.), "Flavor Research and Food Acceptance," Reinhold Publishing Corp., New York, 1958

6503. Stahl, E., "Chemical Varieties of Plants Containing Terpenoids," Essenze deriv. agrumari, 27, 188-207 (1957); CA 52, 10507d

6504. Stahl, E., and Trennheuser, L., "Gas-Phase Chromatography of Terpene and Hydroxyphenylpropane Substances," Arch. Pharm., 293, 826-837 (1960); CA 55, 909d

6505. Stahl, W. H., "Techniques and Methods for Research in Flavors – Gas Chromatography and Mass Spectrometry in the Study of Flavor," Quartermaster Food and Container Inst. Surveys Progr. Military Subsistence Problems, Ser. I, No. 9, 58-76 (1957); CA 52, 2290i

6506. Stahl, W. H., "Instrumental Methods Used in the Analysis of Odorants," Proc. Ann. Symposium on Problems of Air Pollution, Franklin Institute, Philadelphia, Pa., October 1958

6507. Stahl, W. H., and Levy, E. J., "Application of Isolative Vapor Phase Chromatography to Problems in Food Research," Symp. on Recent Developments in Research Methods and Instrumentation, Bethesda, Md., May 1956; Abstr., Anal. Chem. 28, 1058 (1956)

6508. Stahl, W. H., Voelker, W. A., and Sullivan, J. H., "A Gas Chromatographic Method for Determining Gases in the Headspace of Cans and Flexible Packages," Food Technol., 14, 14-16 (1960)

6509. Stainer, C., and Gloesener, E., "Analysis of Cocaine, Arecoline, and Meperidine Hydrochloride by Gas Chromatography," Farmaco (Pavia) Ed. pract., 15, 721-731 (1960); CA 55, 18011b

6510. Stairs, R. A., Diaper, D. G. M., and Gatzke, A. L., "Reaction of Chromyl Chloride with Some Olefins. I. The Products from Cyclohexene, Cyclopentene, 1-Hexene, and 2-Methyl-1-pentene," Can. J. Chem., 41, 1059-1064 (1963)

6511. Stalkup, F. I., and Deans, H. A., "Perturbation Velocities in Gas-Liquid Partition Chromatographic Columns," A.I.Ch.E. (Am. Inst. Chem. Engrs.) Journal, 9, 106-108 (1963); CA 58, 7626e

6512. Stalkup, F. I., and Kobayashi, R., "High-Pressure Phase-Equilibrium Studies by Gas-Liquid Partition Chromatography," A.I.Ch.E. (Am. Inst. Chem. Engrs.) Journal, 9, 121-128 (1963); CA 58, 7419f

6513. Stanford, F. G., "Sample-Injection Method for Gas-Liquid Chromatography," Analyst, 84, 321-322 (1959); CA 53, 16609d

6514. Stanford, F. G., "Analysis of Phosphorus Trichloride and Oxychloride Mixtures," J. Chromatog., 4, 419 (1960); CA 55, 13168g

6515. Stanley, C. W., and Peterson, W. R., "Polymer Analysis by Using Gas-Chromatographic Separation of Pyrolyzates on a Capillary Column," SPE (Soc. Plastics Engrs.) Trans., 2, 298-301 (1962); CA 58, 2507e

6516. Stanley, R. G., "Terpene Formation in Pine from Mevalonic Acid," Nature, 182, 738-739 (1958); CA 53, 5414f

6517. Stanley, R. G., "Terpene Formation in Pine," Proc. Intern. Congr. Biochem., 4th Congr., Vienna, 1958, 2, 48-55 (Pub. 1959); CA 54, 15536b

6518. Stanley, R. G., and Mirov, N. T., "Terpene Analysis by Gas Chromatography," 133rd Natl. ACS Mtg., San Francisco, Calif., April 1958, Program Abstr., p. 7A

6519. Stanley, W. L., "Citrus Flavors," Candy Ind. & Confectioner's J., March 18, 1958

6520. Stanley, W. L., "Citrus Flavors," in "Flavor Research and Food Acceptance," Arthur D. Little, Inc., Cambridge, Mass., Reinhold Publishing Corp., New York, 1958, pp. 344-368

6521. Stanley, W. L., "Citrus Oils: Analytical Methods and Compositional Characteristics," Intern. Fruchtsaft-Union, Ber. Wiss.-Tech. Komm., 4, 91-102 (1962); CA 59, 379d

6522. Stanley, W. L., Ikeda, R. M., and Cook, S., "Hydrocarbon Composition of Lemon Oils and Its Relation to Optical Rotation," Food Technol., 15, 381-385 (1961); CA 56, 553b

6523. Stanley, W. L., Ikeda, R. M., Vannier, S. H., and Rolle, L. A., "Determination of the Relative Concentrations of the Major Aldehydes in Lemon, Orange, and Grapefruit Oils by Gas Chromatography," J. Food Sci., 26, 43-48 (1961); CA 57, 6433c

6524. Starbuck, W. C., and Busch, H., "Programming System for the Automatic Amino Acid Analyzer," Anal. Chem., 34, 875-876 (1962); CA 57, 2826c

6525. Stasch, A. R., "Selection of Liquid Stationary Phases for the Separation of Volatile Mixtures by Chromatography," Bull. Georgia Acad. Sci., 16, No. 3, 38-45 (1958); CA 53, 15701e

6526. Staszewski, R., and Janak, J., "Solid Supports for Gas-Liquid Chromatography, Especially Porous Teflon," Collection Czechoslov. Chem. Communs., 27, 532-545 (1962); CA 56, 14903d

6527. Stedman, R. L., and Dymicky, M., "Composition Studies on Tobacco. VI. Phthalates from Flue-Cured Leaves," Tobacco (New York), 148, 196-198 (1959); Biol. Abstr., 33, 38845

6528. Stedman, R. L., Swain, A. P., Dymicky, M., and D'Iorio, B. I., "Acids and Esters of Flue-Cured Tobacco," U. S. Dept. Agr. ARS 73-31, 16 pp. (1960); CA 55, 16916e

6529. Steele, E. L., "Analytical Instrumentation in Space Exploration," Anal. Chem., 35, No. 9, 23A-30A, 32A, 35A-38A (1963)

6530. Stehling, F. C., and Anderson, R. C., "Thermal Reactions of Acetylene and Acetylene-Aromatic Hydrocarbon Mixtures," 135th Natl. ACS Mtg., Boston, Mass., April 1959, Program Abstr., p. 31R

6531. Stehling, F. C., Coats, F. H., and Anderson, R. C., "Mass Spectrographic and Gas Chromatographic Study of the Thermal Reactions of Acetylene," 132nd Natl. ACS Mtg., New York, N. Y., September 1957, Program Abstr., p. 23S

6532. Stein, A. A., Opalka, E., and Peck, F., "Fatty Acid Analysis of Brain Tumors by Gas Phase Chromatography," Arch. Neurol., 8, No. 1, 50-55 (1963); CA 58, 12956b

6533. Stein, K. C., Feenan, J. J., Thompson, G. P., Schultz, J. F., Hofer, L. J. E., and Anderson, R. B., "An Approach to Air Pollution Control. Catalytic Oxidation of Hydrocarbons," Ind. Eng. Chem., 52, 671-674 (1960)

6534. Stein, K. C., Feenan, J. J., Thompson, G. P., Schultz, J. P., Hofer, L. J. E., and Anderson, R. B., "The Oxidation of Hydrocarbons on Simple Oxide Catalysts," J. Air Pollution Control Assoc., 10, 275-281 (1960); CA 54, 21725h; 135th Natl. ACS Mtg., Boston, Mass., April 1959, Program Abstr., p. 4Q; Preprints, Div. Petrol. Chem., 4, No. 2, C115-C122

6535. Stein, R. A., "Gas Chromatography of Some Brominated Methyl Octadecanoates," J. Chromatog., 6, 118-121 (1961); CA 56, 9388d

6536. Stein, R. A., and Nicolaides, N., "Structure Determination of Methyl Esters of Unsaturated Fatty Acids by Gas-Liquid Chromatography of the Aldehydes Formed by Triphenylphosphine Reduction of the Ozonides," J. Lipid Research, 3, 476-478 (1962); CA 58, 6687g

6537. Stein, S. S., "The Design, Construction, and Evaluation of a Continuous Pilot-Plant Hypersorption Unit for the Vapor-Phase Separation of Carbon Disulfide from Hydrogen Sulfide," Univ. Microfilms (Ann Arbor, Mich.), Publ. No. 4625, 543 pp.; Dissertation Abstr., 13, 208-209 (1953)

6538. Steingiser, S., David, D. J., and Baumann, G. P., "Considerations in the Design and Analytical Applications of a Triple Stage Gas Chromatograph," 8th Natl. Analysis Instrumentation Symp., ISA, Charleston, W. Va., April-May 1962

6539. Stemp, A. R., Langlois, B. E., and Liska, B. J., "Factors Involved in Determination of Chlorinated Insecticide Residues in Milk by Electron Capture Gas Chromatography," 58th Annual Mtg., Am. Dairy Sci. Assoc., Lafayette, Ind., June 1963; Abstr., J. Dairy Sci., 46, 606 (1963)

6540. Stenhagen, E., "Mass Spectrometry in Determination of Structure of Organic Compounds, Especially Lipids and Peptides," Z. Anal. Chem., 181, 462-480 (1961); CA 56, 2309h

6541. Stephens, E. R., Darley, E. F., Taylor, O. C., and Scott, W. E., "Photochemical Reaction Products in Air Pollution," Intern. J. Air Pollution, 4, 79-100 (1961); CA 55, 19082h

6542. Stephens, E. R., and Schuck, E. A., "Air Pollution Effects of Irradiated Auto Exhaust as Related to Fuel Composition," Chem. Eng. Progr., 54, No. 11, 71-77 (1958); CA 53, 6496c

6543. Stephens, R., and Tatlow, J. C., "Pentafluorobenzene," Chem. & Ind. (London), 1957, 821-822; CA 51, 16322b

6544. Stephens, R., and Wiseman, E. H., "Fluorocyclohexanes. VII. Prototropic Migration in 3H,4H/-Octafluorocyclohexene," J. Chem. Soc., 1963, 2083-2085

6545. Stephens, R. L., and Teszler, A. P., "Quantitative Estimation of Low Boiling Carbonyls by a Modified α-Ketoglutaric Acid-2,4-Dinitrophenylhydrazone Exchange Procedure," Anal. Chem., 32, 1047 (1960); CA 54, 18199i

6546. Sterescu, M., "Chromatography, General Principles and Applications," Rev. chim. (Bucharest), 8, 535-540 (1957); CA 52, 1725f

6547. Sterligov, O. D., Turkel'taub, N. M., Zhukhovitskii, A. A., and Kazanskii, B. A., "Quantitative Gas-Chromatographic Analysis of C5 Hydrocarbon Mixtures," Khromatog., ee Teoriya i Primenenie, Akad. Nauk S.S.S.R., Trudy Vsesoyuz. Soveshchaniya, Moscow, 1958, 287-290 (Pub. 1960); CA 55, 15227d

6548. Stern, R., Atkinson, E. R., and Jennings, Jr., F. C., "Preparative Scale Separation of Diastereoisomers by Gas-Liquid Chromatography," Chem. & Ind. (London), 1962, 1758-1759; CA 58, 11200f

6549. Sternberg, J. C., "Ionization Detectors," 14th Annual Fisk University Infrared Spectroscopy Institute, Gas Chromatography Session, Nashville, Tenn., August 1963

6550. Sternberg, J. C., and Bochinski, J., "Gaseous Electronic Detectors," 14th Pittsburgh Conf. on Anal. Chem. & Appl. Spectroscopy, Pittsburgh, Pa., March 1963, Program Abstr., p. 65

6551. Sternberg, J. C., and Carson, L. M., "Continuous Recording Chromatography with Liquid Eluents," 11th Pittsburgh Conf. on Anal. Chem. & Appl. Spectroscopy, Pittsburgh, Pa., February-March 1960, Program Abstr., p. 45

6552. Sternberg, J. C., Gallaway, W. S., and Jones, D. T. L., "Mechanism of Response of Flame Ionization Detectors," Gas Chromatog., Intern. Symp., 1961, 3, 231-267 (Pub. 1962); CA 58, 3865d; 3rd Intern. Symp. on Gas Chromatography, ISA, East Lansing, Mich., June 1961, ISA Proc., 3, 159-184

6553. Sternberg, J. C., Gallaway, W. S., and Jones, D. T. L., "The Mechanism of Response of Flame Ionization Detectors," in "Gas Chromatography," edited by N. Brenner, J. E. Callen, and M. D. Weiss, Academic Press, Inc., New York, 1962, pp. 231-268

6554. Sternberg, J. C., and Poulson, R. E., "A Telsa Discharge Detector for Gas Chromatography," J. Chromatog., 3, 406-410 (1960); CA 55, 5050e

6555. Sternberg, J. C., and Poulson, R. E., "Principles of Low Pressure Drop Packed Columns," 14th Pittsburgh Conf. on Anal. Chem. & Appl. Spectroscopy, Pittsburgh, Pa., March 1963, Program Abstr., p. 53

6556. Sterrett, F. S., "The Nature of Essential Oils. II. Chemical Constituents, Analysis," J. Chem. Educ., 39, 246-251 (1962)

6557. Stevens, R., "Beer Flavor. I. Volatile Products of Fermentation: A Review," J. Inst. Brewing, 66, 453-471 (1960); CA 55, 9777e

6558. Stevens, R., "Gas-Chromatographic Retention Data Using Glycerol as Stationary Phase with Particular Reference to Formaldehyde," Anal. Chem., 33, 1126-1127 (1961); CA 56, 9424g

6559. Stevens, R., "Beer Flavor. III. 2-Phenylethanol," J. Inst. Brewing, 67, 329-331 (1961); CA 55, 27763d

6560. Stevens, R., "Formation of Phenethyl Alcohol and Tyrosol During Fermentation of a Synthetic Medium Lacking Amino Acids," Nature, 191, 913-914 (1961); CA 56, 1840b

6561. Stevens, R. K., and Mold, J. D., "A Gas Chromatographic Trap Designed to Collect Compounds Which Tend to Form Aerosols," J. Chromatog., 10, 398-399 (1963)

6562. Stevens, T. E., "Fluorination of Some Nitriles and Ketones with Bromine Trifluoride," J. Org. Chem., 26, 1627-1630 (1961); CA 55, 23310d

6563. Stevens, T. E., "Amine Oxidations with Iodine Pentafluoride: Preparation of Azoisobutane," J. Org. Chem., 26, 2531-2533 (1961); CA 56, 2313e

6564. Stevenson, G. W., and Luck, J. M., "Bromodecarboxylation of Amino Acids: Formation of Nitriles," J. Biol. Chem., 236, 715-717 (1961); CA 55, 17486d

6565. Stewart, A. T., and Squires, G. L., "Analysis of Ortho- and Para-Hydrogen Mixtures by the Thermal-Conductivity Method," J. Sci. Instr., 32, 26-29 (1955); CA 49, 7309h

6566. Stewart, B. A., Porter, L. K., and Beard, W. E., "Determination of Total Nitrogen by a Combined Dumas-Gas Chromatographic Technique," Anal. Chem., 35, 1331-1332 (1963)

6567. Stewart, G. H., "Plate Height in Programmed Temperature Gas Chromatography," Anal. Chem., 32, 1205 (1960); CA 54, 23460c

6568. Stewart, G. H., Seager, S. L., and Giddings, J. C., "Influence of Pressure Gradients on Resolution in Gas Chromatography," Anal. Chem., 1738 (1959); CA 55, 9974g

6569. Stewart, J. E., Brace, R. O., Johns, T., and Ulrich, W. F., "Microquantitative Infrared Analysis: Application to Gas-Chromatographic Fractions," Nature, 186, 628-629 (1960); CA 54, 18155c

6570. Stewart, J. E., and Gallaway, W. S., "Micro Infrared Analyses as Applied to Gas Chromatography

241

Fractions," Symp. on Spectroscopy, Am. Assoc. of Spectrographers, Chicago, Ill., June 1958; Abstr., Spectrochim. Acta, 13, (1/2), 154 (1958)

6571. Stewart, R. D., Swank, J. D., Roberts, C. B., and Dodd, H. C., "Detection of Halogenated Hydrocarbons in the Expired Air of Human Beings Using the Electron Capture Detector," Nature, 198, 696-697 (1963)

6572. Stiles, A. R., Reilly, C. A., Pollard, G. R., Tieman, C. H., Ward, Jr., L. F., Phillips, D. D., Soloway, S. B., and Whetstone, R. R., "Preparation, Physical Properties, and Configuration of the Isomers in Phosdrin Insecticides," J. Org. Chem., 26, 3960-3968 (1961); CA 56, 8544c

6573. Stimson, V. R., and Watson, E. J., "Catalysis by Hydrogen Halides in the Gas Phase. VIII. Methyl t-Butyl Ether and Hydrogen Bromide," J. Chem. Soc., 1963, 524-527

6574. Stirling, P. H., "Advances in Process Chromatography," 5th Natl. Symp. on Instrumental Methods of Analysis, ISA, Houston, Texas, May 1959

6575. Stirling, P. H., "Recent Knowledge of the Dynamics of Process Chromatographs and Automatic Chemical Analysis Devices," 8th Natl. Analysis Instrumentation Symp., ISA, Charleston, W. Va., April-May 1962

6576. Stirling, P. H., and Ho, H., "Instrumentation Moves Ahead – The 1960 ISA-IMA Symposium," Ind. Eng. Chem., 52, No. 9, 63A-64A, 66A, 68A (1960)

6577. Stirling, P. H., and Ho, H., "Versatility Unlimited – Gas Chromatography 1960. From Odor Analysis to Pilot Plant Process," Ind. Eng. Chem., 52, No. 10, 63A-64A (1960)

6578. Stirling, P. H., and Ho, H., "Ionization Detectors for Gas Chromatography," Ind. Eng. Chem., 52, No. 11, 61A-62A, 64A (1960)

6579. Stirling, P. H., and Ho, H., "Successful Sampling. Systems Approach Simplifies Analyzer-Sample Handling," Ind. Eng. Chem., 53, No. 3, 57A-59A, 62A (1961)

6580. Stirling, P. H., and Ho, H., "Analysis – 1961," Ind. Eng. Chem., 53, No. 7, 51A-53A (1961)

6581. Stirling, P. H., and McSloan, D., "Designing a Katharometer for Gas Chromatography," 5th Natl. Symp. on Instrumental Methods of Analysis, ISA, Houston, Texas, May 1959

6582. Stock, L. M., and Brown, H. C., "Relative Rates and Isomer Distributions in the Halogenation of t-Butylbenzene and Some of Its Derivatives. Partial Rate Factors for Non-catalytic Bromination and Chlorination in Acetic Acid," J. Am. Chem. Soc., 81, 5615-5620 (1959); CA 56, 3379b

6583. Stock, L. M., and Brown, H. C., "Rates of Bromination of Anisole and Certain Derivatives. Partial Rate Factors for the Bromination Reaction. The Application of the Selectivity Relation to the Substitution Reactions of Anisole," J. Am. Chem. Soc., 82, 1942-1947 (1960); CA 55, 19845h

6584. Stock, L. M., and Himoe, A., "Rates and Isomer Distributions in the Chlorination of Benzene, Toluene, and Butylbenzene in Aqueous Acetic Acid Solvents. The Influence of Solvent on the Reaction and the Baker-Nathan Effect," J. Am. Chem. Soc., 83, 1937-1944 (1961); CA 55, 20999e

6585. Stock, R., "Determination of Surface Area by Gas Chromatography," Anal. Chem., 33, 966-967 (1961); CA 55, 17156e

6586. Stocklin, G., Schmidt-Bleek, F., and Herr, W., "Specific Marking with Tritium in a Heterogeneous Exchange on a Gas Chromatographic Column," Angew. Chem., 73, 220 (1961); CA 55, 16453g

6587. Stoffel, W., and Ahrens, Jr., E. H., "Isolation and Structure of the C_{16} Unsaturated Fatty Acids in Menhaden Body Oil," J. Am. Chem. Soc., 80, 6604-6608 (1958); CA 53, 5708e

6588. Stoffel, W., and Ahrens, Jr., E. H., "Separation and Structural Determination of Highly Unsaturated Fatty Acids," 50th Annual Mtg., Am. Oil Chemists' Soc., New Orleans, La., April 1959

6589. Stoffel, W., and Ahrens, Jr., E. H., "The Unsaturated Fatty Acids in Menhaden Body Oil. The C_{18}, C_{20}, and C_{22} Series," J. Lipid Research, 1, 139-146 (1960); CA 54, 11516e

6590. Stoffel, W., Chu, F., and Ahrens, Jr., E. H., "Analysis of Long-Chain Fatty Acids by Gas-Liquid Chromatography. Micromethod for Preparation of Methyl Esters," Anal. Chem., 31, 307-308 (1959); CA 53, 13992d

6591. Stoffel, W., Insull, Jr., W., and Ahrens, Jr., E. H., "Gas-Liquid Chromatography of Highly Unsaturated Fatty Acid Methyl Esters," Proc. Soc. Exptl. Biol. Med., 99, 238-241 (1958); CA 53, 3973c

6592. Stolow, R. D., and Boyce, C. B., "Non-Chair Conformations. Equilibrium of cis- and trans-2,5-Di-t-butyl-1,4-cyclohexanedione," J. Am. Chem. Soc., 83, 3722-3723 (1961); CA 56, 8583f

6593. Storrs, E. E., and Burchfield, H. P., "Internal Standards in the Analysis of Chlorine-Containing Herbicides by Gas Chromatography," Contrib. Boyce Thompson Inst., 21, 423-437 (1962); CA 58, 3834g

6594. Storto, T., and Di Prima, A., "Gas Chromatography of Solvents and Diluents in Perfumes and in Cosmetic, Pharmaceutical and Similar Preparations," Riv. ital. essenze, profumi, piante offic., oli vegetali, saponi, 42, 537-547 (1960); CA 55, 8771i

6595. Storto, T., and Di Prima, A., "The Chromatographic Relation Linalool-Linalyl Acetate in Lavender Essences," Riv. ital. essenze, profumi, piante offic., oli vegetali, saponi, 43, 157-160 (1960); CA 55, 23940g

6596. Story, P. R., "7-Substituted Norbornadienes," J. Org. Chem., 26, 287-290 (1961); CA 55, 13337h

6597. Strackenbrock, K. H., "Gas Chromatographic Measurement of Aroma Quality of Fruit Juices as Influenced by Various Technological Methods," Intern. Fruchsaft-Union Ber. Wiss. Tech. Komm., No. 4, 287-299 (1962); CA 58, 9562e

6598. Strain, H. H., "Chromatography. Analysis by Differential Migration," Anal. Chem., 30, 620-629 (1958); CA 52, 9847a

6599. Strain, H. H., "Chromatography," Anal. Chem., 32, 3R-18R (1960); CA 54, 11804f

6600. Strain, H. H., and Sato, T. R., "Chromatography and Electrochromatography," U. S. At. Energy Comm., TID-7512, 175-182 (1956); CA 50, 7001b

6601. Strange, J. P., "The Selective Measurement of One Gas in a Multicomponent Mixture," Can. Chem. Proc., 40, No. 9, 95-96, 100, 102 (1956)

6602. Strange, J. P., "A Thermal Conductivity Instrument for Multicomponent Gas Mixtures," 7th Pittsburgh Conf. on Anal. Chem. & Appl. Spectroscopy, Pittsburgh, Pa., February-March 1956, Program Abstr., p. 45

6603. Strange, J. P., "Sensors for Automatic Analyses," Instruments and Control Systems, 35, No. 11, 91-95 (1962)

6604. Stranks, D. R., "A Scintillation Counter for the Assay of Radioactive Gases," J. Sci. Instr., 33, 1-4 (1956); CA 50, 6102c

6605. Strassburger, J., Brauer, G. M., Tryon, M., and Forziati, A. F., "Analysis of Methyl Methacrylate Copolymers by Gas Chromatography," Anal. Chem., 32, 454-457 (1960); CA 56, 6160a; 136th Natl. ACS Mtg., Atlantic City, N. J., September 1959, Program Abstr., p. 18T

6606. Straten, H. A. C. van, "Chromatographic Gas Analysis," Zakenwereld (Amsterdam), 34, 46 (1956)

6607. Straten, H. A. C. van, "Chromatographic Gas Analysis," Chem. en Pharm. Tech. (Dordrecht), 12, 6 (1957)

6608. Strating, J., and Venema, A., "Gas-Chromatographic Study of an Aroma Concentrate from Beer. I. Qualitative Features," J. Inst. Brewing, 67, 525-528 (1961); CA 56, 9232f

6609. Streim, H. G., Boyce, E. A., and Smith, J., "Determination of Water in 1,1-Dimethylhydrazine, Diethylenetriamine, and Mixtures," Anal. Chem., 33, 85-89 (1961); CA 55, 8151c

6610. Strickler, A., and Gallaway, W. S., "A New Method of Integration Recording," 9th Pittsburgh Conf. on Anal. Chem. & Appl. Spectroscopy, Pittsburgh, Pa., March 1958, Program Abstr., p. 45

6611. Strickler, A., and Gallaway, W. S., "Continuous Recording Analog Integrator for Gas Chromatography," 11th Pittsburgh Conf. on Anal. Chem. & Appl. Spectroscopy, Pittsburgh, Pa., February-March 1960, Program Abstr., p. 41

6612. Strickler, A., and Gallaway, W. S., "Analog Integration Techniques in Chromatographic Analysis," J. Chromatog., 5, 185-193 (1961); CA 56, 6631f

6613. Strickler, H., and Kováts, E., "Influence of Experimental Conditions on Peak Resolution in Gas Chromatography," J. Chromatog., 8, 289-302 (1962)

6614. Strouts, C. R. N., Wilson, H. N., and Parry-Jones, R. T., with Gilfillan, J. H., (Eds.), "Chemical Analysis, The Working Tools," Vol. I, Oxford University Press, London, 1962

6615. Struppe, H. G., "Evaluation and Production of Selective Separation Columns," in "Gas Chromatographie 1958," edited by H. P. Angele, Akademie Verlag, Berlin, 1959, pp. 28-86

6616. Struppe, H. G., "Preparation and Evaluation of Efficient Columns," Abhandl. Deut. Akad. Wiss. Berlin, Kl. Chem. Geol., Biol., 1959, No. 9, 28-86; CA 58, 5018b

6617. Struppe, H. G., "Application of Capillary Gas Chromatography for the Separation of C_5-C_7 Olefins," Gas Chromatog. Symp., Brno, Czech., June 1962; Abstr., J. Gas Chromatog., 1, No. 4, 7 (1963)

6618. Stull, D. R., "Vapor Pressure of Pure Substances. Organic Compounds," Ind. Eng. Chem., 39, 517-540 (1947); CA 41, 2945g

6619. Stull, D. R., "Vapor Pressure of Pure Substances. Inorganic Compounds," Ind. Eng. Chem., 39, 540-550 (1947); CA 41, 2945g

6620. Stull, J. W., and Brown, W. H., "Seasonal and Breed Variation in the Fatty Acid Content of Milk Fat," 57th Annual Mtg., Am. Dairy Sci. Assoc., College Park, Md., June 1962; Abstr., J. Dairy Sci., 45, 661 (1962)

6621. Stuve, W., "A Simple Katharometer for Use with the Combustion Method," in "Gas Chromatography 1958," edited by D. H. Desty, Academic Press, Inc., New York, 1958, pp. 178-188

6622. Stuve, W., "Gas Chromatography of Free Fatty Acids. The Analysis of Saturated and Unsaturated C_6 to C_{20} Fatty Acids," Fette, Seifen, Anstrichmittel, 63, 325-329 (1961); CA 55, 17047g

6623. Styring, Jr., R. E., "Application of Gas Chromatography in a Crude Oil Production Research Laboratory," Proc. Ann. Conv. Natl. Gasoline Assoc. Am., Tech. Papers, 37, 24-27 (1958); CA 54, 1833g

6624. Sudario, E., "Chromatographic Determination of Sorbic Acid Used as a Preservative in Beverages," Chim. e ind. (Milan), 39, 811-814 (1957); CA 52, 3251e

6625. Suffis, R., and Dean, D. E., "Identification of Alcohol Peaks in Gas Chromatography by a Nonaqueous Extraction Technique," Anal. Chem., 34, 480-483 (1962); 140th Natl. ACS Mtg., Chicago, Ill., September 1961, Program Abstr., p. 25B

6626. Sugisawa, H., MacGregor, D. R., and Matthews, J. S., "Apple Juice Volatiles," Intern. Fruchsaft-

Union, Ber. Wiss.-Tech. Komm., 4, 351-364 (1962); CA 58, 13056h

6627. Suhr, H., and Zollinger, H., "Secondary Hydrogen Isotope Effects in the Nitration of Toluene," Helv. Chim. Acta, 44, 1011-1016 (1961); CA 55, 25813i

6628. Sullivan, J. H., Walsh, J. T., and Merritt, Jr., C., "Improved Separation in Gas Chromatography by Temperature Programming. Application to Mercaptans and Sulfides," Anal. Chem., 31, 1826-1828 (1959); CA 54, 24087i; 134th Natl. ACS Mtg., Chicago, Ill., September 1958, Program Abstr., p. 24B

6629. Sullivan, L. J., Lotz, J. R., and Willingham, C. B., "Retention Volumes of Isomeric Hexenes and Hexanes in Gas-Liquid Partition Chromatography Using Phthalate Esters as Liquid Phase," Anal. Chem., 28, 495-498 (1956); CA 50, 9822b

6630. Sullivan, L. J., Lotz, J. R., and Willingham, C. B., "Relative Retention Volumes of Isomeric Hexenes and Hexanes (Chromatographic Data Table XLI)," J. Chromatog., 1, xxvi (1958); cf. Anal. Chem., 28, 495 (1956) (n-Pentane = 1, Temperature = 20°)

6631. Sullivan, R. F., Egan, C. J., Langlois, G. E., and Sieg, R. P., "A New Reaction That Occurs in the Hydrocracking of Certain Aromatic Hydrocarbons," J. Am. Chem. Soc., 83, 1156-1160 (1961); CA 55, 18083f; 138th Natl. ACS Mtg., New York, N. Y., September 1960, Program Abstr., p. 11R

6632. Sulovsky, J., and Zavodsky, R., "Application of Gas Chromatography to the Determination of Products of Synthesis of Isophytol and Thiopentabarbitol," Gas Chromatog. Symp., Brno, Czech., June 1962; Abstr., J. Gas Chromatog., 1, No. 4, 10 (1963)

6633. Sumansky, L. W., "Analysis of Benzene, Toluene, and Xylene in Process Absorption Oil by Dual-Column Gas Chromatography," Gas Chromatog. Intern. Symp., 1961, 3, 449-455 (Pub. 1962); CA 58, 5417b; 3rd Intern. Symp. on Gas Chromatography, ISA, East Lansing, Mich., June 1961

6634. Sumansky, L. W., "The Analysis of Benzene, Toluene and Xylene in Process Absorption Oil by Dual Column Gas Chromatography," in "Gas Chromatography," edited by N. Brenner, J. E. Callen, and M. D. Weiss, Academic Press, Inc., New York, 1962, pp. 449-456

6635. Sumimoto, S., and Ishizuka, I., "Synthesis of Isoxazole Derivatives. I. Gas-Chromatographic Determination of the Binary System of 5-Methyl-3-carbethoxyisoxazole – 3-Methyl-5-carbethoxyisoxazole," Kogyo Kagaku Zasshi, 64, 1820-1822 (1961); CA 57, 2855e

6636. Sumimoto, S., and Ishizuka, I., "Synthesis of Isoxazole Derivatives. II. Gas-Chromatographic Determination of Ethyl 4-Hydroxyiminoacetopyruvate in 3-Methyl-5-carbethoxyisoxazole," Kogyo Kagaku Zasshi, 64, 2076-2077 (1961); CA 57, 2855g

6637. Sumimoto, S., and Ishizuka, I., "Synthesis of Isoxazole Derivatives. III. Gas-Chromatographic Determination of Ethyl Oxalate in Ethyl Acetopyruvate," Kogyo Kagaku Zasshi, 64, 2077-2078 (1961); CA 57, 2855h

6638. Summa, A. F., "The Analysis of Manila Elemi Oil," Univ. Microfilms (Ann Arbor, Mich.), L. C. Card No. Mic 60-5257, 62 pp.; Dissertation Abstr., 21, 1772 (1961); CA 55, 8759a

6639. Summers, F. W., and Adriani, J., "Gas Chromatography: An Analytical Method for Anesthesiology Research," Anesthesiology, 22, 100-107 (1961); CA 57, 3727c

6640. Sun, S. C., and Argyle, P., "Gas Chromatography Solves an Age-Old Problem in Analyzing Flotation Collectors," Mineral Ind. Penn. State Univ., 30, No. 1, 1-6 (1960); CA 56, 6631a

6641. Sundberg, O. E., and Maresh, C., "Application of Gas Chromatography to Microdetermination of Carbon and Hydrogen," Anal. Chem., 32, 274-277 (1960); CA 54, 8417i

6642. Sungren, A., Rauhala, V. T., and Juusela, S., "Composition of Certain Natural Gases Found in Finland," Valtion Tek. Tutkimuslaitos Tiedotus, Sarja IV, No. 40, 35 pp. (1962); CA 58, 3246f

6643. Sunner, S., Karrman, K. J., and Sunden, V., "Separation of Mercaptans by Gas-Liquid Partition Chromatography," Mikrochim. Acta, 1956, 1144-1151; CA 50, 8396e

6644. Supina, W. R., "The Use of Organosilicon Polyesters as Stationary Phases for Gas Chromatography," 14th Pittsburgh Conf. on Anal. Chem. & Appl. Spectroscopy, Pittsburgh, Pa., March 1963, Program Abstr., p. 54

6645. Supina, W. R., "Columns and Column Support Materials," in "Lectures on Gas Chromatography 1962," edited by H. A. Szymanski, Plenum Press, Inc., New York, 1963, pp. 33-51

6646. Supina, W. R., "New Column Materials," 5th Annual Gas Chromatography Institute, Canisius College, Buffalo, N. Y., April 1963

6647. Supina, W. R., "Cyclohexanedimethanol Polyesters as Stationary Phases for Gas Chromatographic Analysis of Ketosteroids," Anal. Chem., 35, 1304 (1963)

6648. Surovy, J., "Recording of Gaseous Chromatographic Fractions by Means of Mercury Drop Potentiometer," Chem. listy, 54, 263-268 (1960); CA 54, 14815b

6649. Susz, B. P., Naves, Y. R., and Collaud, C., "Studies of Volatile Materials of Vegetable Origin. CLVII. Ultraviolet Absorption Spectrum of S-Gaiazulene," Helv. Chim. Acta, 42, 1375-1376 (1959); CA 54, 3486b

6650. Sutherland, M. D., and Waters, O. J., "Terpenoid Chemistry. VI. The Structure of Humulene," Australian J. Chem., 14, 596 (1961); CA 57, 16664a

6651. Suzuki, S., "Gas-Chromatographic Determination of Gases in Metals," Bunseki Kagaku, 11, 618-621

(1962); CA 57, 5295d

6652. Suzuki, S., and Amaha, M., "Gas-Chromatographic Study on the Oxygen Content of Neck-Space Air in Packaged Beer," J. Inst. Brewing, 68, 508-514 (1962); CA 58, 9598g

6653. Svensen, A., "Estimation of Volatile Fatty Acids in Cheese by Gas Chromatography," Meierposten, 50, 263-269, 285-289, 306-313 (1961); CA 57, 15565e

6654. Sventsitskii, E. I., Lulova, N. I., Tarasov, A. I., and Zemskova, E. I., "Analysis of Hydrocarbon Gases by the Chromathermographic Method," Zavodskaya Lab., 22, 1399-1403 (1956); CA 51, 18561a

6655. Sverak, J., and Reiser, P. L., "Gas Chromatographic Detection of Small Amounts of Ethers in Ethylene," Mikrochim. Acta, 1958, 159-168; CA 53, 982e

6656. Sverak, J., and Reiser, P. L., "Gas Chromatographic Determination of Carbon Disulfide in Carbon Dioxide and Inert Gases," Mikrochim. Acta, 1958, 426-433

6657. Swain, A. P., and Stedman, R. L., "Analytical Studies on the Higher Fatty Acids of Tobacco," J. Assoc. Offic. Agr. Chemists, 45, 536-540 (1962)

6658. Swain, C. G., Wiles, R. A., and Bader, R. F. W., "Use of Substituent Effects on Isotope Effects to Distinguish Between Proton and Hydride Transfers. I. Mechanism of Oxidation of Alcohols by Bromine in Water," J. Am. Chem. Soc., 83, 1945-1950 (1961); CA 55, 20899f

6659. Swann, M. H., Adams, M. L., and Esposito, G. G., "Coatings," Anal. Chem., 33, 33R-37R (1961); CA 55, 12137e

6660. Swann, M. H., Adams, M. L., and Esposito, G. G., "Coatings," Anal. Chem., 35, No. 5, 35R-39R (1963)

6661. Swann, W. B., and Dux, J. P., "New Technique for Pyrolyzing Samples for Gas Chromatographic Analysis," Anal. Chem., 33, 654-655 (1961); CA 55, 13158e

6662. Sweeley, C. C., "A Gas Chromatographic Method for Sphingosine Assay," Biochim. et Biophys. Acta, 36, 268-270 (1959); CA 54, 4785a

6663. Sweeley, C. C., "Analysis of Urinary Aromatic Acids by Gas Chromatography," Joint Mtg., Am. Assoc. of Clin. Chemists and Am. Assoc. Adv. of Science, December 1960; Abstr., Clin. Chem., 7, 306 (1961)

6664. Sweeley, C. C., and Chang, T-C. L., "Gas Chromatography of Steroids. Relation of Structure to Molar Response in an Argon Ionization Detector," Anal. Chem., 33, 1860-1863 (1961); CA 56, 8012d

6665. Sweeley, C. C., and Horning, E. C., "Microanalytical Separation of Steroids by Gas Chromatography," Nature, 187, 144-145 (1960); CA 54, 23198g

6666. Sweeley, C. C., and Moscatelli, E. A., "Qualitative Microanalysis and Estimation of Sphingolipid Bases," J. Lipid Research, 1, 40-47 (1959); CA 54, 11127f

6667. Sweeley, C. C., and Williams, C. M., "Microanalytical Determination of Urinary Aromatic Acids by Gas Chromatography," Anal. Biochem., 2, 83-86 (1961); CA 55, 13523e

6668. Sweeting, J. W., "Modification of Agla Micrometer Hypodermic Syringe for Use in Vapor-Phase Chromatography," Chem. & Ind. (London), 1959, 1150; CA 54, 7d

6669. Sweeting, J. W., and Wilshire, J. F. K., "The Selective Separation of High-Boiling Aromatic Compounds," J. Chromatog., 6, 385-395 (1961); CA 56, 12293d

6670. Sweeting, J. W., and Wilshire, J. F. K., "The Pyrolysis of ω,ω'-Diphenylalkanes," Australian J. Chem., 15, 89-105 (1962); CA 58, 3294e

6671. Swell, L., Field, Jr., H., Schools, Jr., P. E., and Treadwell, C. R., "Fatty Acid Composition of Tissue Cholesterol Esters in Elderly Humans with Atherosclerosis," Proc. Soc. Exptl. Biol. Med., 103, 651-655 (1960); CA 54, 14422i

6672. Swell, L., Field, Jr., H., Schools, Jr., P. E., and Treadwell, C. R., "Lipide Fatty Acid Composition of Several Areas of the Aorta in Subjects with Atherosclerosis," Proc. Soc. Exptl. Biol. Med., 105, 662-665 (1960); CA 55, 8606i

6673. Swell, L., Field, Jr., H., and Treadwell, C. R., "Correlation of Arachidonic Acid of Serum Cholesterol Esters in Different Species with Susceptibility to Atherosclerosis," Proc. Soc. Exptl. Biol. Med., 104, 325-328 (1960); CA 54, 18716i

6674. Swell, L., Field, Jr., H., and Treadwell, C. R., "Relation of Age and Race to Serum Cholesterol Ester Fatty Acid Composition," Proc. Soc. Exptl. Biol. Med., 105, 129-131 (1960); CA 55, 4692h

6675. Swell, L., Law, M. D., Field, Jr., H., and Treadwell, C. R., "Cholesterol Ester Fatty Acids in Serum and Liver of Normal and Lymph-Fistula Rats," J. Biol. Chem., 235, 1960-1962 (1960); CA 54, 22954g

6676. Swell, L., Law, M. D., Field, Jr., H., and Treadwell, C. R., "Composition of Lymph Cholesterol Ester Fatty Acids after Feeding of Cholesterol and Oleic Acid," Proc. Soc. Exptl. Biol. Med., 104, 7-8 (1960); CA 54, 16573g

6677. Swell, L., Law, M. D., Schools, Jr., P. E., and Treadwell, C. R., "Tissue Lipid Fatty Acid Composition in Pyridoxine-Deficient Rats," J. Nutrition, 74, 148-156 (1961)

6678. Swift, L. J., "Determination of Linalool and α-Terpineol in Florida Orange Products," J. Agr. Food

245

Chem., 9, 298-301 (1961); CA 56, 5174c

6679. Swinbourne, E. S., "Kinetics of the Pyrolysis of Cyclohexyl Chloride," Australian J. Chem., 11, 314-330 (1958); CA 53, 1181c

6680. Swinehart, D. F., and Vreeland, R. W., "A King-Sized Gas Chromatograph," Northwest Regional Mtg., ACS, Seattle, Wash., June 1959

6681. Swinnerton, J. W., Linnenbom, V. J., and Cheek, C. H., "Determination of Dissolved Gases in Aqueous Solutions by Gas Chromatography," Anal. Chem., 34, 483-485 (1962); 140th Natl. ACS Mtg., Chicago, Ill., September 1961, Program Abstr., p. 24B

6682. Swinnerton, J. W., Linnenbom, V. J., and Cheek, C. H., "Revised Sampling Procedure for Determination of Dissolved Gases in Solution by Gas Chromatography," Anal. Chem., 34, 1509 (1962); CA 57, 15777h

6683. Swisher, R. D., Kaelble, E. F., and Liu, S. K., "Capillary Gas Chromatography of Phenyldodecane Alkylation and Isomerization Mixtures," J. Org. Chem., 26, 4066-4069 (1961); CA 56, 8593i; 138th Natl. ACS Mtg., New York, N. Y., September 1960, Program Abstr., p. 30-O

6684. Swoboda, J., Derkosch, J., and Wessely, F., "Cyclic Acetals. I.," Monatsh. Chem., 91, 188-201 (1960); CA 54, 18412h

6685. Swoboda, P. A. T., "The Analysis of Dilute Aqueous Solutions by Gas Chromatography," Chem. & Ind. (London), 1960, 1262-1263; CA 55, 13157i

6686. Swoboda, P. A. T., "Quantitative and Qualitative Analysis of Flavour Volatiles from Edible Fats," 4th Intern. Gas Chromatography Symp., Hamburg, Germany, June 1962; pub. in "Gas Chromatography, 1962" M. van Swaay, ed., Butterworth & Co., London, 1962

6687. Sykes, A., Tatlow, J. C., and Thomas, C. R., "A New Synthesis of Fluoro Ketones," Chem. & Ind. (London), 1955, 630-631; CA 50, 166a

6688. Sykes, A., Tatlow, J. C., and Thomas, C. R., "Synthesis of Fluoro Ketones," J. Chem. Soc., 1956, 835-839; CA 50, 14625g

6689. Synge, R. L. M., "Applications of Partition Chromatography," Prix Nobel, 1952, 122-135; CA 48, 8112d

6690. Synge, R. L. M., "Principles of Chromatography," Proc. Intern. Wool Textile Research Conf., Melbourne, 1955B, 238-248; CA 51, 17329h

6691. Sze, Y. L., Borke, M. L., and Ottenstein, D. M., "Separation of Lower Aliphatic Amines by Gas Chromatography," Anal. Chem., 35, 240-242 (1963); CA 58, 7348g; 13th Pittsburgh Conf. on Anal. Chem. & Appl. Spectroscopy, Pittsburgh, Pa., March 1962, Program Abstr., p. 57

6692. Székely, G., Kormany, T., Racz, G., and Traply, G., "Gas Chromatographic Investigations," Periodica Polytechn., 2, 269-274 (1958); CA 53, 13728b

6693. Szepesy, L., and Benedek, P., "Calculation of the Continuous Action of a Gas Chromatographic Column," Magyar Kem. Lapja, 13, 369-372 (1958); CA 53, 13684g

6694. Szepesy, L., Keszthelyi, S., and Gulyas, I., "Types of Gas Chromatography Apparatus Developed at the Hungarian Oil and Gas Research Institute," Magyar Asvanyolaj es Foldgaz Kiserleti Intezet Kozlmenyei, 1961, 302-312; CA 56, 3304d

6695. Szepesy, L., and Simon, E., "Gas-Chromatographic Investigation of Petroleum Products," Acta Chim. Acad. Sci. Hung., 31, 223-233 (1962); CA 57, 2492g

6696. Szepesy, L., Simon, F., and Simon, P., "The Use of Gas Chromatography in Gas Analysis," Magyar Kem. Folyoirat, 67, 27-33 (1961); CA 55, 12149i

6697. Szepesy, L., Simon, J., and Simon, P., "Experience in the Solution of Differing Gas Analysis Problems with the Help of Specially Constructed Gas Chromatography Apparatus," Acta Chim. Acad. Sci. Hung., 27, 303-310 (1961)

6698. Szonntagh, E. L., "A New High Speed Process Gas Chromatograph," 18th Annual ISA Instrument-Automation Conf. & Exhibit, Chicago, Ill., September 1963

6699. Szonntagh, E. L., and Stewart, J. R., "Analysis of Fixed Gases by Dual-Carrier Gas Chromatography," 3rd Delaware Regional Mtg., ACS, Symposium on Vapor Phase Chromatography, Philadelphia, Pa., February 1960

6700. Szulczewski, D. H., and Higuchi, T., "Gas Chromatographic Separation of Some Permanent Gases on Silica Gel at Reduced Temperatures," Anal. Chem., 29, 1541-1543 (1957); CA 52, 975e

6701. Szymanski, H. A., (Ed.), "Progress in Industrial Gas Chromatography," Vol. 1, Plenum Press, Inc., New York, 1961. Lectures presented at the 3rd Annual Gas Chromatography Institute, Canisius College, Buffalo, N. Y., April 1961

6702. Szymanski, H. A., "New Approaches to Preparative Gas Chromatography," 4th Annual Eastern Analytical Symp., New York, N. Y., November 1962, Program Abstr., p. 25

6703. Szymanski, H. A., (Ed.), "Lectures on Gas Chromatography 1962," Plenum Press, Inc., New York, 1963. Lectures presented at the 4th Annual Gas Chromatography Institute, Canisius College, Buffalo, N. Y., April 1962, including, "A New Approach to Gas Chromatography," pp. 227-245; CA 58, 10702a

6704. Szymanski, H. A., "Gas Chromatography Using Induction Heated Columns," 14th Pittsburgh Conf. on

Anal. Chem. & Appl. Spectroscopy, Pittsburgh, Pa., March 1963, Program Abstr., p. 54

6705. Szymanski, H. A., "Column Parameters," 5th Annual Gas Chromatography Institute, Canisius College, Buffalo, N. Y., April 1963

6706. Szymanski, H. A., (Ed.), "Biomedical Applications of Gas Chromatography," Plenum Press, New York, 1964

6707. Szymanski, H. A., Kolb, J., and Kuczkowski, J., "The Influence of Column Length, Column Temperature, and Carrier Gas Flow Rate on Resolution of Components," in "Lectures on Gas Chromatography 1962," edited by H. A. Szymanski, Plenum Press, Inc., New York, 1963, pp. 53-63; CA 58, 10701g

6708. Szymanski, H. A., McMenamy, C., Kuczkowski, J., Broda, K., and May, J., "Some Experiments in Gas Chromatography," 14th Annual Mid-America Spectroscopy Symp., Chicago, Ill., May 1963; pub. in "Developments in Applied Spectroscopy," Vol. 3, Plenum Press, New York (1964) pp. 392-397

6709. Szymanski, H. A., Povinelli, R., Stamires, D., and Lynch, G., "Infrared Cell for Collecting Chromatographic Fractions," Anal. Chem., 31, 2110 (1959); CA 54, 6211c

6710. Szymanski, H. A., and Salinis, C. A., "Research in Gas Chromatography at Canisius College," 13th Annual Mid-America Spectroscopy Symp., Chicago, Ill., April-May 1962

6711. Szymanski, H. A., Salinis, C. A., and Kwitkowski, P., "Technique for Pyrolyzing or Vaporizing Samples for Gas Chromatographic Analysis," Nature, 188, 403-404 (1960); CA 55, 6057h

6712. Szymanski, H. A., Salinis, C. A., and Wagner, N., "Determination of Urea-Formaldehyde Resin on Lath Paper," in "Progress in Industrial Gas Chromatography," Vol. 1, edited by H. A. Szymanski, Plenum Press, Inc., New York, 1961, pp. 201-207; CA 56, 7967e

6713. Tabuteau, J., "Analysis of Hydrocarbon Traces in Hydrocarbons," Compt. rend. congr. intern. chim. ind., 31^e Liege, 1958; Pub. as Ind. chim. belge, Suppl., 1, 135-138 (1959); CA 54, 4271d

6714. Tadanier, J., and Cole, W., "Preparation of the Epimeric 31Aminoandrost-5-en-17-ones and 6-Amino-3α,5α-Cycloandrostan-17-ones. The mechanism of the Ammonlysis of Steroid-Δ⁵-3βp-Toluenesulfonates," J. Org. Chem., 27, 4624-4633 (1962); CA 58, 4615f

6715. Tadmor, J., "Isotopic Exchange in Gas Chromatography," J. Inorg. Nucl. Chem., 23, No. 1-2, 158-159 (1961); CA 57, 13209g

6716. Taft, E. M., and Dimick, K. P., "An Automatic Preparative Gas Chromatograph," 14th Pittsburgh Conf. on Anal. Chem. & Appl. Spectroscopy, Pittsburgh, Pa., March 1963, Program Abstr., p. 63

6717. Tagaki, W., and Mitsui, T., "Analysis of Isomeric Menthols by Gas Chromatography," Bull. Agr. Chem. Soc. Japan, 24, 217-218 (1960); CA 54, 13939d

6718. Tago, S., and Noguchi, S., "Gas Chromatographic Method for Rapid Analysis of Some Light Aromatic Hydrocarbons," J. Japan Petrol. Inst., 4, 306-310 (1961)

6719. Tahara, A., Hoshimo, O., and Ikekawa, N., "Cryptopimaric Acid and Neoisopimaric Acid," Chem. Pharm. Bull. (Tokyo), 10, 995-997 (1962); CA 59, 676g

6720. Takacs, J., Inczedy, J., and Erdey, L., "Determination of Methyl Bromide and Ethyl Bromide in Their Mixtures," J. Chromatog., 9, 247-249 (1962); CA 58, 11200b

6721. Takahashi, A., "Automatic Continuous Measurement of Minute Amounts of Gases by the Electrical Conductivity Method," Kagaku no Ryoiki, 7, 783-788 (1953); CA 48, 13536h

6722. Takahashi, Y., and Tanaka, K., "Method for Separation of Plasma Fatty Acids Using Gas Chromatography," Saishin Igaku, 15, 185-191 (1960); CA 54, 13239b

6723. Takahashi, Y., and Tanaka, K., "Gas Chromatographic Analysis of the Fatty Acid Composition of Human Fatty Liver," J. Biochem. (Tokyo), 49, 713-720 (1961); CA 55, 26195c

6724. Takahashi, Y., and Tanaka, K., "Experiments on Lipids. VIII-X. Gas Chromatography," Tampakushitsu Kakusan Koso, 7, 522-527, 638-644, 712-719 (1962); CA 58, 9396f

6725. Takamiya, N., "Gas Chromatography," Shinku-Kogyo, 5, 86-93 (1958); CA 52, 18067f

6726. Takamiya, N., "Apparatus Made in Japan," Kagaku no Ryoiki, 12, 608-613 (1958); CA 52, 19339a

6727. Takamiya, N., Kojima, J-I., and Murai, S., "A Method for Evaluating Solubility Parameters from Gas Chromatographic Data," Kogyo Kagaku Zasshi, 62, 1371-1373 (1959)

6728. Takamiya, N., and Murai, S., "Analysis of Complex Mixtures of Permanent Gases and Light Hydrocarbons by Gas Chromatography," Kogyo Kagaku Zasshi, 63, 1935-1938 (1960); CA 57, 7879i

6729. Takashima, I., Koga, A., and Kaneko, J., "Resolution of Oxygen and Nitrogen by High-Temperature Gas Chromatography with Manganese Dioxide as Stationary Phase," Kogyo Kagaku Zasshi, 65, 1223-1226 (1962); CA 58, 921h

6730. Takashima, S., and Okada, F., "Detection of Benzene in Adhesives for Rubber by Means of Gas Chromatography," Himeji Kogyo Daigaku Kenkyu Hokoku, 12, 34-38 (1960); CA 55, 15980h

6731. Takayama, Y., "Fisher-Gulf Partitioner," Kagaku no Ryoiki, 12, 599-606 (1958); CA 52, 19339a

6732. Takayama, Y., "Gas-Chromatographic Analysis of Impurities in Commercial Methyl Acrylate," Kogyo Kagaku Zasshi, 61, 682-684 (1958); CA 55, 15227c

6733. Takayama, Y., "Relationship Between Peak Area Ratio and Composition of Sample in Gas Chromatography," Kogyo Kagaku Zasshi, 61, 685-687 (1958)

6734. Takayama, Y., "Determination of 2-Methyl-5-ethyl Pyridine in Commercial 2-Methyl-5-vinylpyridine," Kogyo Kagaku Zasshi, 62, 658-661 (1959); CA 57, 7913b

6735. Takayama, Y., "A Tailingless Solid Support for Gas Chromatography," Kogyo Kagaku Zasshi, 64, 803-806 (1961); CA 57, 2827e

6736. Takayama, Y., "Stationary Support for the [Gas-Chromatographic] Column," Kagaku no Ryoiki, 15, 369-387 (1961); CA 56, 937c

6737. Takeda, I., and Mashiko, Y., "Data Presentation of Gas Chromatography by Use of a Logarithmic Scale," Kogyo Kagaku Zasshi, 62, 1798 (1959); CA 57, 7875d

6738. Takeda, K., Minato, H., and Nosaka, S., "Sesquiterpenoids. III. Derivatives of Guaiol," Tetrahedron, 13, 308-318 (1961); CA 56, 14330d

6739. Takeuchi, T., "Application of Gas Chromatography to Technoanalytical Chemistry," Kagaku no Ryoiki, 12, 614-619 (1958); CA 52, 19339a

6740. Takeuchi, T., and Hara, N., "Recent Applications of Gas Chromatography in the Petroleum Industry," J. Japan Petrol. Inst., 4, 271-279 (1961)

6741. Takeuchi, T., and Hayakawa, F., "Rapid Analysis of Aliphatic Hydrocarbons in Reagent-Grade Benzene by Gas Chromatography and a Method of Purifying Benzene," Bunseki Kagaku, 7, 712-716 (1958); CA 53, 19706b

6742. Takeuchi, T., and Ishii, D., "Determination of Traces of a Benzene, Toluene, and Xylene Mixture in Air," Kogyo Kagaku Zasshi, 64, 763-769 (1961); CA 57, 6272d

6743. Talbot, P. J., and Thomas, J. H., "The Reaction Between Hydrogen Chloride and Nitrogen Peroxide," Trans. Faraday Soc., 55, 1884-1891 (1959); CA 54, 19254i

6744. Tamaru, K., "Chromatographic Technique for Studying the Mechanism of Surface Catalysis," Nature, 183, 319-320 (1959); CA 53, 13728c

6745. Tanaka, K., and Takahashi, Y., "Gas-Chromatographic Analysis of Plasma Fatty Acid Composition in Arteriosclerotic Patients," Nippon Naika Gakkai Zasshi, 50, 790-815 (1961); CA 59, 979d

6746. Tandy, R. K., Lindgren, F. T., Martin, W. H., and Wills, R. D., "Analysis of Gas-Liquid Chromatograms by a Punched Card Technique," Anal. Chem., 33, 665-669 (1961); CA 55, 16259h

6747. Tanner, D. W., and Kipping, P. J., "Analysis of Commercially Available Fatty Acids," Nature, 193, 975 (1962); CA 57, 2357i

6748. Taramasso, M., "Vapor-Phase Chromatography of Gaseous Hydrocarbons," Ricerca sci., 26, 887-888 (1956); CA 50, 9944h

6749. Taramasso, M., "Vapor-Phase Chromatographic Analysis of Saturated Hydrocarbon Gases," Termotecnica, 10, 203-206 (1956); CA 51, 2260b

6750. Taramasso, M., and Faraci, R., "Gas Chromatography as a Means of Indicating Atmospheric Pollution by Hydrocarbons," Minerva med., 1958, 1054-1057; CA 52, 12286d

6751. Taramasso, M., and Piccinini, C., "Chromatography of Gases. Separation of Saturated and Corresponding Unsaturated Aliphatic Compounds," Rass. chim., 9, No. 6, 10-12 (1957); CA 52, 9857a

6752. Tarasov, A. I., Kudryavtseva, N. A., Ioganson, A. V., and Lulova, N. I., "Automatic Analysis of Gas in a Stream with a Kh. P.A.-1 Chromatograph," Ind. Lab., 25, 830-833 (1959); Zavodskaya Lab., 25, 803-805 (1959); CA 54, 9381i

6753. Tarasov, A. I., Kudryavtseva, N. A., Iogansen, A. V., and Lulova, N. I., "A Rapid Method for the Chromatographic Analysis of C_1-C_5 Hydrocarbons," Gazovaya Kromatografiya, Trudy 1-oi [Pervoi] Vsesoyuz. Konf., Akad. Nauk S.S.S.R., Moscow, 1959, 280-288 (Pub. 1960); CA 56, 2634d

6754. Tarasov, A. I., Lulova, N. I., Kudryavtseva, N. A., and Zemskova, E. I., "Chromatographic Gas Analyzer for Refinery Gases," Gazovaya Khromatografiya, Trudy 1-oi [Pervoi] Vsesoyuz. Konf., Akad. Nauk S.S.S.R., Moscow, 1959, 268-273 (Pub. 1960); CA 56, 1668e

6755. Tarasov, A. I., Lulova, N. I., Kudryavtseva, N. A., and Zemskova, E. I., "Laboratory Analyzer for Gas Chromatography," Izmeritel. Tekh., 1960, No. 8, 47-49; CA 55, 16028a

6756. Tarbes, H., and Bonnet, E., "Chromatographic Analysis of Hydrocarbons," J. usines gaz., 82, 261-266 (1958); CA 52, 18094i

6757. Tarman, P. B., Andreen, B. H., and Kniebes, D. V., "Comparison of Instrumental Methods of Analysis for Odorants and Other Sulfur Compounds in Natural Gas," Gas, 37, No. 9, 97-105 (1961); CA 55, 25212c

6758. Tarmy, B. L., "The Application of Frequency Response Analysis to the Measurement of Thermal Conductivity of Gases," Univ. Microfilms (Ann Arbor, Mich.), Publ. No. 20499, 151 pp.; Dissertation Abstr., 17, 834-835 (1957); CA 51, 11834c

6759. Tarmy, B. L., Bartok, W., and Lucchesi, P. J., "The Methane Dosimeter," Ind. Eng. Chem., 53, 147-150 (1961); CA 56, 2300c

6760. Tatlow, J. C., "Modern Analytical Techniques. Gas Chromatography," Chem. Age, 74, 633 (1956)

6761. Tattrie, N. H., "Positional Distribution of Saturated and Unsaturated Fatty Acids on Egg Lecithin," J. Lipid Research, 1, 60-65 (1959); CA 54, 11106g

6762. Taylor, B. W., "Fisher-Gulf Partitioner," 7th Pittsburgh Conf. on Anal. Chem. & Appl. Spectroscopy, Pittsburgh, Pa., February-March 1956, Program Abstr., p. 41

6763. Taylor, B. W., "Development of Components of an Instrument for Research and Control Applications of Vapor-Phase Chromatography," 129th Natl. ACS Mtg., Dallas, Texas, April 1956, Program Abstr., p. 17B

6764. Taylor, B. W., "An Instrument Designed for High Temperature Gas Chromatographic Analysis," Symp. on Gas Chromatography, ISA, East Lansing, Mich., August 1957

6765. Taylor, B. W., "An Instrument Designed for High Temperature Gas Chromatographic Analysis," in "Gas Chromatography," edited by V. J. Coates, H. J. Noebels, and I. S. Fagerson, Academic Press, Inc., New York, 1958, pp. 155-164

6766. Taylor, B. W., "Whole Blood Gas Analysis by Gas Chromatography," 3rd Delaware Valley Regional Mtg., ACS, Symp. on Vapor Phase Chromatography, Philadelphia, Pa., February 1960

6767. Taylor, B. W., and Poli, A. A., "Analysis of 'Fixed' Gases and Low-Boiling Hydrocarbons by Gas Chromatography," 10th Pittsburgh Conf. on Anal. Chem. & Appl. Spectroscopy, Pittsburgh, Pa., March 1959, Program Abstr., p. 58

6768. Taylor, B. W., and Pressau, J., "The Determination of Gases in Whole Blood by Gas Chromatography," Am. Physiological Soc. Mtg., Univ. of Illinois, Urbana, Ill., September 1959

6769. Taylor, F., Harkness, W., and Campbell, D. N., "Process Chromatography Utilizing Catalytic Combustion Principles," 5th Symp. on Instrumental Methods of Analysis, ISA, Houston, Texas, May 1959

6770. Taylor, G. W., and Dunlop, A. S., "The Analysis of Light Hydrocarbons by Gas-Liquid Chromatography," in "Gas Chromatography," edited by V. J. Coates, H. J. Noebels, and I. S. Fagerson, Academic Press, Inc., New York, 1958, pp. 73-85

6771. Taylor, M. P., "Possible Radiation Hazards Arising from the Use of Radioactive Detectors in Gas Chromatography," J. Chromatog., 9, 28-33 (1962); CA 58, 4793f

6772. Taylor, R. F., and Roberts, F., "The Calibration of Disk-Type Laboratory Gas-Absorption Columns," Chem. Eng. Sci., 5, 168-177 (1956); CA 50, 15135e

6773. Taylor, W. H., Mori, S., and Burton, M., "Radiolysis of Liquid Neopentane," J. Am. Chem. Soc., 82, 5817-5822 (1960); CA 55, 6113b

6774. Taylor, W. J., and Johnston, H. L., "An Improved Hot Wire Cell for Accurate Measurement of Thermal Conductivities of Gases Over a Wide Temperature Range. Results with Air Between 87° and 375°K.," J. Chem. Phys., 14, 219-233 (1946); CA 40, 3951⁴

6775. Teisseire, P., "Essential Oil of Clary Sage (Salvia Sclarea)," Recherches (France), No. 9, 10-22 (1959); CA 54, 23202b

6776. Teisseire, P., "Some Side Products of Carroll's Reaction," Recherches (France), No. 10, 18-29 (1960); CA 55, 14292b

6777. Teisseire, P., "On the Use of Gas-Liquid Chromatography in the Investigation of Chemical Reactions," Recherches (France), No. 12, 44-47 (1960)

6778. Teisseire, P., "The Essential Oil of Slavia Sclarea," Riv. ital. essenze, profumi, piante offic., oli vegetali, saponi, 42, 216-226 (1960); CA 55, 901e

6779. Teisseire, P., "Essential Oil from Clary," France et ses parfums, 4, No. 22, 243-254 (1961); CA 55, 19144d

6780. Teisseire, P., "Application of Gas Chromatography to Perfumery," Recherches (Paris), No. 12, 54-73 (1962); CA 58, 8849b

6781. Teisseire, P., "Recent Applications of Gas Chromatography to Perfumery," Bull. soc. chim. France, 1963, 384-388

6782. Teitelbaum, C. L., "New Analytical Techniques for Essential Oil and Flavor Research," Am. Perfumer Aromat., 67, No. 1, 28-30 (1956); CA 50, 5244f

6783. Teitelbaum, C. L., "Gas-Partition Chromatography. Application to Essential Oils and Other Volatile Materials," J. Soc. Cosmetic Chemists, 8, 316-327 (1957); CA 52, 662e

6784. Tenenbaum, M., and Howard, F. L., "Purification by Automatic Gas Chromatography," J. Res. Natl. Bur. Std., A66, No. 3, 255-258 (1962); CA 57, 9605f

6785. Tenney, H. M., "Selectivity of Various Liquid Substrates Used in Gas Chromatography," Anal. Chem., 30, 2-8 (1958); CA 52, 7936b; 132nd Natl. ACS Mtg., New York, N. Y., September 1957, Program Abstr., p. 32B

6786. Tenney, H. M., and Harris, R. J., "Sample Introduction System for Gas Chromatography," Anal. Chem., 29, 317-318 (1957); CA 51, 6230i

6787. Tenney, H. M., McQuaid, F. S., and McQuaid, J. W., "Some Uses of Gas Chromatography in Petroleum Process Research," Southwide Chemical Conf. on Instrumentation, ISA, Symp. on Gas Chromatography, Memphis, Tenn., December 1956

6788. Tenney, H. M., and Sturgis, F. E., "Separability of Hydrocarbons by Elution Chromatography," Anal. Chem., 26, 946-953 (1954); CA 48, 11246d

6789. Tenny, K. S., Gupta, S. C., Nystrom, R. F., and Kummerow, F. A., "The Synthesis of Tritium Labeled 9,10-Oleic Acid," J. Am. Oil Chemists' Soc., 40, 172-175 (1963)

6790. Tepe, J. B., and Wesselman, H. J., "Determination of a Series of Malonic Esters by Gas Chromatography," J. Am. Pharm. Assoc., 47, 457-458 (1958); CA 52, 13542c

6791. Terada, H., Tsuda, S., and Shono, T., "Gas Chromatography of Menthols," Kogyo Kagaku Zasshi, 65, 1569-1571 (1962); CA 58, 11404g

6792. Teranishi, R., and Buttery, R. G., "Aromagrams; Direct Vapor Analyses with Gas Chromatography," Intern. Fruchsaft-Union, Ber. Wiss.-Tech. Komm., 4, 257-266 (1962); CA 58, 13056g

6793. Teranishi, R., Buttery, R. G., and Lundin, R. E., "Gas Chromatography. Direct Vapor Analysis of Food Products with Programmed Temperature Control of Dual Columns with Dual Flame Ionization Detectors," Anal. Chem., 34, 1033-1035 (1962); CA 57, 11605f; 140th Natl. ACS Mtg., Chicago, Ill., September 1961, Program Abstr., p. 2A

6794. Teranishi, R., Corse, J. W., Day, J. C., and Jennings, W. G., "Volman Collector for Gas Chromatography," J. Chromatog., 9, 244-245 (1962); CA 58, 6166g

6795. Teranishi, R., Nimmo, C. C., and Corse, J., "Versatile Ionization Detector for Gas Chromatography," Anal. Chem., 32, 896 (1960); CA 54, 19031c

6796. Teranishi, R., Nimmo, C. C., and Corse, J., "Gas-Liquid Chromatography. Programmed Temperature Control of the Capillary Column," Anal. Chem., 32, 1384-1386 (1960); CA 55, 5050g; 137th Natl. ACS Mtg., Cleveland, Ohio, April 1960, Program Abstr., p. 25B

6797. Termini, D. J., and Glasgow, A. R., "Chromatographic Analysis of Petroleum Fractions Used in Oil-Extended Rubber," J. Res. Natl. Bur. Std., A66, 189-192 (1962); CA 57, 2385a

6798. Terres, E., and Hahn, H. H., "Use of New Procedures in an Investigation of a South African Coal Tar Obtained by Pressure Gasification. II. Investigation of Individual Fractions of the Neutral Oil," Erdol u. Kohle, 12, 823-829 (1959); CA 54, 5050a

6799. Tesarik, K., "Determination of Trace Amounts of Hydrocarbons in the Atmosphere," Gas Chromatog. Symp., Brno, Czech., June 1962; Abstr., J. Gas Chromatog., 1, No. 4, 8 (1963)

6800. Tesarik, K., Hana, K., and Janicek, M., "Manufacture of Long Capillary Tubes for Gas Chromatography," Chem. listy, 55, 1467-1469 (1961); CA 56, 5382g

6801. Testerman, M. K., and McLeod, P. C., "Ultrasonic Whistles as Detectors for Gas Chromatography," in "Gas Chromatography," edited by N. Brenner, J. E. Callen, and M. D. Weiss, Academic Press, Inc., New York, 1962, pp. 183-188

6802. Testerman, M. K., and McLeod, P. C., "Ultrasonic Whistles as Detectors for Gas Chromatography," Gas Chromatog., Intern. Symp., 3rd, Edinburgh, 1961, 183-188 (Pub. 1962); CA 58, 1895b

6803. Tharp, B. W., and Patton, S., "Coconut-Like Flavor Defect of Milk Fat. IV. Demonstration of δ-Dodecalactone in the Steam Distillate from Milk Fat," J. Dairy Sci., 43, 475-479 (1960)

6804. Thaung, U. K., Gros, A., and Feuge, R. O., "Characterization of Oils from Low-Gland and Glandless Cottonseed," J. Am. Oil Chemists' Soc., 38, 220-224 (1961); CA 55, 11885i

6805. Theer, J., "Determination of Traces of Hydrocarbons in Air and in Liquid Oxygen," Chem. Tech. (Berlin), 14, No. 3, 164-167 (1962); CA 57, 10531h

6806. Theile, F. C., Dean, D. E., and Suffis, R., "The Evaluation of Bergamot Oil," Drug & Cosmet. Ind., 86, 758-759, 837-840 (1960); CA 54, 20092b

6807. Theile, F. C., Dean, D. E., and Suffis, R., "A Modern Approach to the Evaluation of Bergamot Oil," Perfumery Essent. Oil Record, 51, 535-540 (1960)

6808. Theile, F. C., Dean, D. E., and Suffis, R., "A Modern Approach to the Evaluation of Bergamot Oil," Am. Perfumer Cosmet., 77, No. 10, Sec. 2, 43-46 (1962); CA 58, 5449e

6809. Theimer, E. T., "Recent Advances in Instrumental Examination of Fragrance Materials," Proc. Sci. Sect. Toilet Goods Assoc., No. 32, 35-40 (1959); CA 54, 11387g

6810. Theimer, E. T., "Recent Advances in Instrumental Examination of Fragrance Materials," Drug & Cosmetic Ind., 85, 745-755, 830, 831 (1959)

6811. Theimer, E. T., Somerville, W. T., Mitzner, B., and Lemberg, S., "γ-Ionones," J. Org. Chem., 27, 635-637 (1962); CA 57, 680g

6812. Theimer, E. T., Somerville, W. T., Mitzner, B., and Lemberg, S., "γ-Isomethylionone," J. Org. Chem., 27, 2934-2935 (1962); CA 57, 12335c

6813. Thielemann, H., "Application of Gas Chromatographic Methods in Technical Microbiology," Gas Chromatog. Symp., Brno, Czech., June 1962; Abstr., J. Gas Chromatog., 1, No. 4, 10 (1963)

6814. Thielemann, H., Behrens, U., and Leibnitz, E., "Quantitation of Gas Chromatography by Means of Titration. Use of an Automatic Buret," Chem. Tech. (Berlin), 13, No. 12, 737-741 (1961); CA 57, 5282f

6815. Thijsse, G. J. E., and Linden, A. C. van der, "Isoalkane Oxidation by Pseudomonas. I. Metabolism of 2-Methylhexane," Antonie van Leeuwenhoek. J. Microbiol. Serol., 27, 171-179 (1961); CA 55, 26103f

6816. Thoburn, J. M., "The Cenco Gas Chromatograph," 1st Annual Gas Chromatography Institute, Canisius College, Buffalo, N. Y., May 1959

6817. Thomas, B. W., "Gas Chromatography May Develop Into Useful Process Analyzer," Ind. Eng. Chem., 47, No. 6, 85A-88A (1955)

6818. Thomas, C. O., "Separation of Hydrogen and Deuterium − Zone Refining of Mixtures of Ordinary and

Heavy Water – Reaction of Iron with Mixtures of Ordinary and Heavy Water – Gas Chromatography with Hydrogen and Deuterium Samples," Univ. Microfilms (Ann Arbor, Mich.), L. C. Card No. Mic 58-1327, 215 pp.; Dissertation Abstr., 18, 1279 (1958); CA 52, 13364a

6819. Thomas, C. O., and Smith, H. A., "Gas Chromatography with Hydrogen and Deuterium," J. Phys. Chem., 63, 427-432 (1959); CA 53, 14817e

6820. Thomas, C. O., and Smith, H. A., "A Simple Katharometer Design," J. Chem. Educ., 36, 527 (1959); CA 54, 4070i

6821. Thomas, D. J., and Kilner, A. E., "The Synthesis of the Insecticides Aldrin and Dieldrin Labeled with Carbon-14 at High Specific Activity," Radioisotopes Phys. Sci. Ind., Proc. Conf. Use, Copenhagen, 1960, 83-90 (Pub. 1962); CA 58, 1369b

6822. Thomas, D. R., and Thomas, J. H., "The Gas Phase Reaction Between Methyl Chloride and Nitrogen Dioxide," Trans. Faraday Soc., 57, 266-278 (1961); CA 55, 18555d

6823. Thomas, R., "The Detection and Determination of Biphenyl and o-Phenylphenol in Concentrated Orange Juice by Gas Chromatography," Analyst, 85, 551-556 (1960); CA 55, 4815i

6824. Thompson, A. E., "A Flame Ionization Detector for Gas Chromatography," J. Chromatog., 2, 148-154 (1959); CA 53, 14599d

6825. Thompson, A. E., "A Fog Precipitator for Gas Chromatography," J. Chromatog., 6, 454-457 (1961); CA 56, 9899c

6826. Thompson, A. E., "A Peak Simulator for Testing Pen Recorders," J. Chromatog., 6, 539-540 (1961)

6827. Thompson, C. J., Coleman, H. J., Hopkins, R. L., and Rall, H. T., "A Microdetermination Technique for Identifying Organic Sulfur, Nitrogen, Oxygen, and Halogen Compounds," U. S. Bur. of Mines, Rept. Invest. No. 6096, 28 pp. (1962); CA 58, 4353c

6828. Thompson, C. J., Coleman, H. J., Hopkins, R. L., Ward, C. C., and Rall, H. T., "Identification of Oxygen Compounds in Gas-Liquid Chromatographic Fractions by Catalytic Deoxygenation," Anal. Chem., 32, 1762-1765 (1960); CA 55, 9147h

6829. Thompson, C. J., Coleman, H. J., Ward, C. C., and Rall, H. T., "Identification of 3-Methylthiophene in Wilmington, California Crude Oil," J. Chem. Eng. Data, 4, 347-348 (1959); CA 54, 9263f

6830. Thompson, C. J., Coleman, H. J., Ward, C. C., and Rall, H. T., "Desulfurization as a Method of Identifying Sulfur Compounds," Anal. Chem., 32, 424-430 (1960); CA 54, 10642d; 136th Natl. ACS Mtg., Atlantic City, N. J., September 1959, Program Abstr., p. 6R; Preprints, Div. Petrol. Chem., 4, No. 3, 183-194 (1959)

6831. Thompson, C. J., Coleman, H. J., Ward, C. C., and Rall, H. T., "Identification of Nitrogen Compounds by Catalytic Denitrogenation," Anal. Chem., 34, 151-154 (1962)

6832. Thompson, H. W., "Physicochemical Methods of Investigating Natural Products," Pure Appl. Chem., 2, 439-473 (1961); CA 56, 939g

6833. Thompson, J. F., Honda, S. I., Hunt, G. E., Krupka, R. M., Morris, C. J., Powell, Jr., L. E., Silberstein, O. O., Towers, G. H. N., and Zacharius, R. M., "Partition Chromatography and Its Use in the Plant Sciences," Botan. Rev., 25, 1-263 (1959); CA 53, 11089e

6834. Thompson, M. J., Louloudes, S. J., Robbins, W. E., Waters, J. A., Steele, J. A., and Mosettig, E., "Identity of the Housefly Sterol," Biochem. Biophys. Res. Communs., 9, 113-119 (1962); CA 57, 17219i

6835. Thompson, M. P., Brunner, J. R., and Stine, C. M., "Characteristics of High Melting Triglyceride Fractions from the Fat Globule Membrane and Butter Oil of Bovine Milk," J. Dairy Sci., 42, 1651-1658 (1959); CA 54, 9143e

6836. Thomson, W. A. B., "Gas-Liquid Chromatographic Identification of Volatile Substances of Biochemical Interest," Univ. Microfilms (Ann Arbor, Mich.), L. C. Card No. Mic 61-1556, 180 pp.; Dissertation Abstr., 21, 3618-3619 (1961); CA 55, 21206d

6837. Thonet, T. A., "Gas Chromatography in the Aerosol Laboratory," Aerosol Age, 4, No. 8, 20-22, 75 (1959)

6838. Thrash, C. R., "Gas Chromatographic Analysis of Various Silanic Esters," 13th Pittsburgh Conf. on Anal. Chem. & Appl. Spectroscopy, Pittsburgh, Pa., March 1962, Program Abstr., p. 58

6839. Thrune, R. I., "Gases Released When Fire Resistant Epoxy Resins are Burned," 144th Natl. ACS Mtg., Los Angeles, Calif., March-April 1963, Program Abstr., p. 1N

6840. Tiernan, T. O., and Futrell, J. H., "Inverse Temperature Programming in Gas Chromatography," Anal. Chem., 34, 1838-1839 (1962); CA 58, 2827a

6841. Tilenschi, S., "Adsorption and Separation of Hydrocarbons from Gases," Rev. Chim. (Bucharest), 5, 305-318 (1954); CA 53, 2701h

6842. Timms, D. G., Konrath, H. J., and Chirnside, R. C., "Determination of Impurities in Carbon Dioxide by Gas Chromatography, with Special Reference to Coolant Gas for Nuclear Reactors," Analyst, 83, 600-609 (1958); CA 53, 5961d

6843. Timofeev, D. P., and Erashko, I. T., "Kinetics for the Sorption of Water on Type-A Zeolites," Izv. Akad. Nauk S.S.S.R., Otd. Khim. Nauk, 1961, 1192-1197; CA 58, 1927h

6844. Timofeev, D. P., and Kabanova, O. N., "Kinetics for the Sorption of Water Vapor on Type-A Zeolites from a Stream of Carrier Gas," Izv. Akad. Nauk S.S.S.R., Otd. Khim. Nauk, 1961, 1539-1542; CA 58, 1928a

6845. Tine, G., "Gas Sampling and Chemical Analysis in Combustion Processes," Pergamon Press, Inc., New York, 1961

6846. Tinoco, J., Shannon, A., Miljanik, P., Lyman, R. L., and Okey, R., "Analysis of Fatty Acid Mixtures. Comparison of Two Absolute Methods of Determination," Anal. Biochem., 3, 514-518 (1962); CA 57, 10129c

6847. Todd, Jr., P. H., and Perun, C., "Gas-Liquid Chromatographic Analysis of Capsicum Amides," Food Technol., 15, 270-272 (1961); CA 55, 21399a

6848. Toi, B., Ota, S., and Iwata, N., "Volatile Products Obtained from Edible Oils by Open-Air Heating. II. Thermal Decomposition Products from Soybean Oil and Oxidation Products from Safflower Oil and Olive Oil," Yukagaku, 11, 504-507 (1962); CA 58, 10414c

6849. Tominaga, S., "Perkin-Elmer Model 154-C Vapor Fractometer," Kagaku no Ryoiki, 12, 590-598 (1958); CA 52, 19338i

6850. Tominaga, S., "Gas Chromatography of High Boiling Substances. I. Analysis of Bisphenol A and Its Impurities by Gas Chromatography," Bunseki Kagaku, 12, 137-143 (1963); CA 58, 11936f

6851. Tonge, B. L., and Timms, D. G., "Use of 'Molecular Sieves' as Gas Samplers," Chem. & Ind. (London), 1959, 155-156; CA 53, 20938d

6852. Torocheshnikov, N. S., and Sememova, V. A., "Chromatographic Analysis of Gas Mixtures Containing Hydrogen, Nitrogen, and Methane," Zhur. Priklad. Khim., 33, 597-602 (1960); CA 54, 18176d

6853. Torraca, G., "Gas-Liquid Chromatography," Chim. e ing. (Rome), 1957, No. 2/3, 2-19; CA 52, 9714a

6854. Toth, J., "The Basic Principles of Gas Chromatographic Apparatus Design and Analytical Techniques," Magyar Kem. Folyoirat, 64, 382-391 (1958); CA 54, 14814h

6855. Toth, J., "Vapor and Gas Adsorption on Solid Surfaces of Inhomogeneous Activity. III.," Magyar Kem. Folyoirat, 66, 431-433 (1960); CA 56, 5420g

6856. Toth, J., "Gas Adsorption on Solid Surfaces of Inhomogeneous Activity. IV.," Magyar Kem. Folyoirat, 67, 282-289 (1961); "V.," ibid., 289-293; CA 56, 5420h

6857. Toth, J., "Gas Adsorption on Solid Surfaces of Inhomogeneous Activity. VI.," Magyar Kem. Folyoirat, 67, 397-402 (1961); CA 56, 12337f

6858. Toth, J., "Gas Adsorption on Solid Surfaces with Inhomogeneous Activity. II.," Acta Chim. Acad. Sci. Hung., 31, 393-405 (1962); CA 58, 41g

6859. Toth, J., "Gas Adsorption on Solid Surfaces of Inhomogeneous Activity. IV.," Acta Chim. Acad. Sci. Hung., 33, 153-163 (1962); CA 58, 8427e

6860. Toth, J., "Gas Adsorption on Solid Surfaces of Inhomogeneous Activity. IX.," Magyar Kem. Folyoirat, 68, 377-381 (1962); CA 58, 8427f

6861. Toth, J., and Graf, L., "Adsorption Theory of Gas Chromatography," Magyar Kem. Folyoirat, 63, 71-78 (1957); CA 52, 13363h

6862. Toth, J., and Graf, L., "Determination of the Helium and Hydrogen Content of Natural Gases by Elution Chromatography," Magyar Kem. Folyoirat, 63, 216-221 (1957); CA 52, 10544c

6863. Toth, J., and Graf, L., "Combined Role of Solid Carrier and Liquid in Gas-Liquid Chromatography," Magyar Kem. Folyoirat, 64, 85-92 (1958); CA 52, 12495g

6864. Toth, J., and Graf, L., "Determination of the Helium and Hydrogen Content of Natural Gases by Elution Chromatography," Banyasz. Kutato Intezet Kozlemenyei, 1, 117-125 (1959); CA 53, 977d

6865. Toth, J., and Graf, L., "The Determination of the Gasoline Content of Natural Gases by Chromatographic Separation and Microcombustion," Magyar Kem. Folyoirat, 65, 324-328 (1959); CA 54, 11438h

6866. Toth, J., and Graf, L., "Determination of Heat of Adsorption by Elution Chromatography," Magyar Kem. Folyoirat, 66, 123-128 (1960); CA 55, 58b

6867. Toth, P., Kugler, E., and Kovats, E., "Gas Chromatographic Characterization of Organic Compounds. II. Precision Chromatograph," Helv. Chim. Acta, 42, 2519-2530 (1959); CA 54, 10409g

6868. Tous, J. G., Vioque, E., and Maza, M. P. de la, "Characteristics of Olive Oil Produced in Spain. Data About the Provinces of Andalusia. II. Methods of Determining Fatty Acids," Grasas y aceites, 10, 286-295 (1959); CA 55, 2145h

6869. Tove, S. B., "Production of Odd Numbered Carbon Fatty Acids from Propionate by Mice," Nature, 184, 1647-1648 (1959); CA 54, 13341a

6870. Tove, S. B., and Smith, F. H., "Changes in the Fatty Acid Composition of the Depot Fat of Mice Induced by Feeding Oleate and Linoleate," J. Nutrition, 71, 264-272 (1960); CA 56, 12061i

6871. Toyama, Y., and Takagi, T., "Gas-Liquid Chromatography of the Methyl Esters of Vegetable Oils Containing Punicic Acid or Eleostearic Acid," Nagoya Sangyo Kagaku Kenkyusho Kenkya Hokoku, No. 13, 29-33 (1961); CA 57, 12652f

6872. Trammel, J. A., and Janzen, J. J., "A Method for Determining the Free Fatty Acids in Milk with Gas Chromatography," 56th Annual Mtg., Am. Dairy Sci. Assoc., Madison, Wisc., June 1961; Abstr., J. Dairy Sci., 44, 1170-1171 (1961)

6873. Trams, E. G., Giuffrida, L. E., and Karmen, A., "Gas Chromatographic Analysis of Long-Chain Fatty Acids in Gangliosides," Nature, 193, 680-681 (1962); CA 56, 15997b

6874. Tranchant, J., "Recent Applications of Gas Chromatography to the Analysis of Halogen Derivatives, Amines, and Nitro Derivatives," Bull. soc. chim. France, 1963, 365-367

6875. Trapp, W. B., Pillepich, J. L., and Ruby, E. D., "Gas Chromatographic Analysis of Chloromethylated [Di-] Phenyl Ether," Anal. Chem., 32, 1737-1739 (1960); CA 55, 6264c; 137th Natl. ACS Mtg., Cleveland, Ohio, April 1960, Program Abstr., p. 35B

6876. Traynelis, V. J., and Dadura, J. G., "Pyrolysis of γ-Hydroxyalkyl Quaternary Ammonium Hydroxides and Alkoxides. Cyclohexane Derivatives," J. Org. Chem., 26, 1813-1818 (1961); CA 55, 23381h

6877. Traynham, J. G., "Benkeser Reduction of Norbornadiene and Norbornene," J. Org. Chem., 25, 833-834 (1960); CA 55, 3461i

6878. Trefny, F., "Determination of Benzene in Hydrogenated Light [Tar] Oil by Gas Chromatography," Gas- u. Wasserfach, 102, 115-118 (1961); CA 55, 11819c

6879. Trenner, N. R., "A Thermal Conductivity Method for the Determination of Isotopic Exchanges in the Simpler Gaseous Molecules," J. Chem. Phys., 5, 382-392 (1937); Correction, ibid., p. 751; CA 31, 4855²

6880. Trenwith, A. B., "Kinetics of the Oxidation of Ethylene by Nitrous Oxide," J. Chem. Soc., 1960, 3722-3726; CA 56, 6693f

6881. Trimm, D. L., and Cullis, C. F., "Radical Isomerization During the Gaseous Oxidation of 2,3-Dimethylbutane," J. Chem. Soc., 1963, 1430-1433; CA 58, 10063b

6882. Trotman-Dickenson, W. F., "Subtractive Columns for Gas Chromatography," Perkin-Elmer Instr. News, 12, No. 3, 12 (1961)

6883. Troupe, R. A., and Golner, J. J., "Process Control Methods in the Chlorination of Benzene," Anal. Chem., 30, 129-131 (1958); CA 52, 5210c

6884. Truce, W. E., and Steltenkamp, R. J., "Oxidative Rearrangement of Vinylic Sulfides," J. Org. Chem., 27, 2816-2820 (1962); CA 58, 1387h

6885. Truter, E. V., "Sorting Molecules by Size and Shape," Research (London), 6, 320-326 (1953); CA 47, 11828f

6886. Tsai, T. T., McEwen, W. E., and Kleinberg, J., "Electrolysis of Brombenzene in Pyridine Solutions," J. Org. Chem., 26, 318-323 (1961); CA 55, 12400d

6887. Tsellinskaya, T. F., Zaitseva, N. I., and Chesnakova, E. V., "Gas-Liquid Chromatography of Liquid Products of Propylene Hydroformylation," Trudy Vsesoyuz. Nauch-Issledovatel. Inst. Neftekhim. Protsessov, 1960, No. 2, 188-209; CA 56, 6665c

6888. Tsenkov, T., and Dimov, N. P., "Absorption Chromatographic Separation of Some Gas Mixtures," Khim. i Ind. (Sofia), 33, No. 2, 46-49 (1961); CA 55, 26846c

6889. Tsuchiya, A., Hashimoto, A., Tominage, H., and Masamune, S., "Dealkylation Rates of Xylene Isomers in the Presence of High Pressure Hydrogen," Bull. Japan. Petrol. Inst., 2, 85-93 (1960); CA 55, 964c

6890. Tsuda, K., Sakai, K., and Ikekawa, N., "Steroid Studies. XXXIII. Gas Chromatography of C_{27}-C_{29} Sterols," Chem. Pharm. Bull. (Tokyo), 9, 835-836 (1961); CA 57, 8631f

6891. Tsudano, S., and Kimura, M., "Application of Gas Chromatography in the Petroleum Industry Especially on the Analysis of Refinery Gas," Kagaku no Ryoiki, 12, 620-626 (1958); CA 52, 19339a

6892. Tudge, A. P., "Chromatographic Transport. I. A Simplified Theory. II. The Effect of Adsorption Isotherm Shape," Can. J. Phys., 39, 1600-1610, 1611-1618 (1961); CA 56, 2867b

6893. Tudge, A. P., "Chromatographic Transport. III. Chromathermography," Can. J. Phys., 40, 557-572 (1962); CA 57, 2826a

6894. Tuey, G. A. P., "Gas Chromatography," May & Baker Lab. Bull., 3, No. 1, 8 (1958)

6895. Tuey, G. A. P., "Gas Chromatography and Its Application to Perfumery Materials," Soap, Perfumery & Cosmetics, 31, 353-361 (1958); CA 53, 5595f

6896. Tufts, L., "An Assay Analysis by Gas Chromatography," 1st Annual Gas Chromatography Institute, Canisius College, Buffalo, N. Y., May 1959

6897. Tulloch, A. P., Craig, B. M, and Ledingham, G. A., "The Oil of Wheat Stem Rust Uredospores. II. The Isolation of cis-9,10-Epoxyoctadecanoic Acid, and the Fatty Acid Composition of the Oil," Can. J. Microbiol., 5, 485-491 (1959); CA 54, 8990a

6898. Tuna, N., Louden, M. L., and Sundeen, M. J., "Studies of Serum Fatty Acids of 'Normal' Americans and 'Normal' Japanese," Univ. Minn. Med. Bull., 31, 134-142 (1959)

6899. Tung, K-N., and Yu, W-L., "Gas Chromatography in the Analysis of Natural Gas," Shih'yu Lien Chih., 1960, No. 11, 36-37; CA 59, 355c

6900. Turina, S., Krajovan, V., and Kostomaj, T., "Simple Apparatus for the Continuous Separation of Materials from the Gas Phase. Continuous Gas Chromatography," Z. Anal. Chem., 189, 100-106 (1962); CA 57, 13166e

6901. Turk, A., and D'Angio, C. J., "Composition of Natural Fresh Air," 54th Annual Symp. of the Air Pollution Control Assoc., New York, N. Y., June 1961

6902. Turkel'taub, N. M., "Chromatographic Titrimetric Gas Analyzer," Zavodskaya Lab., 15, 653-660 (1949); CA 44, 9747d

6903. Turkel'taub, N. M., "Chromatographic Method for Separate Determinations of Microconcentration of Hydrocarbons in Air," Zhur. Anal. Khim., 5, 200-210 (1950); CA 44, 9856b

6904. Turkel'taub, N. M., "New Adsorption Methods of Analysis of Hydrocarbon Gases," Neflyanoe Khoz., 32, No. 4, 72-77 (1954); CA 48, 8517f

6905. Turkel'taub, N. M., "Chromothermographic Method of Analysis of Gases," Trudy Komissii Anal. Khim., Akad. Nauk S.S.S.R., Inst. Geokhim. i Anal. Khim., 6, 146-161 (1955); CA 50, 7663f

6906. Turkel'taub, N. M., "Partition Chromatography Method for the Separation and Determination of Hydrocarbon Gases," Zhur. Fiz. Khim., 31, 2102-2109 (1957); CA 52, 12675i

6907. Turkel'taub, N. M., "Comparative Review of the Methods of Gas Chromatography," Gazovaya Khromatografiya, Trudy 1-oi [Pervoi] Vsesoyuz. Konf., Akad. Nauk S.S.S.R., Moscow, 1959, 13-25 (Pub. 1960); CA 56, 2h

6908. Turkel'taub, N. M., "Use of Different Variants of Gas Chromatography for Analyzing Hydrocarbon Mixtures," Trudy Kom. Analit. Khim., Akad. Nauk SSSR, 13, 225-231 (1963)

6909. Turkel'taub, N. M., and Abramovich, L. Y., "Use of a Mass Spectrometer for Investigating the Separation Efficiency of Chromatographic Units," Zhur Anal. Khim., 13, 43-47 (1958); CA 52, 10795b

6910. Turkel'taub, N. M., Ainshtein, S. A., and Kuznetsov, B. V., "The Chromatographic Determination of Impurities by Use of the Flame-Ionization Detector," Khim. i Tekhnol. Topliv i Masel, 6, No. 12, 44-50 (1961); CA 57, 1553d

6911. Turkel'taub, N. M., Anvaer, B. I., Kolyubyakina, A. I., and Selenkina, M. S., "The Separation of C_2-C_5 Hydrocarbons by the Method of Gas-Liquid Partition Chromatography," Ind. Lab., 25, 159-164 (1959) (in English); Zavodskaya Lab., 25, 149-154 (1959) (in Russian); CA 54, 17857f

6912. Turkel'taub, N. M., and Ivanova, N. T., "[Gas] Chromatographic Determination of Monochlorinated C_3 Compounds," Plasticheskie Massy, 1962, No. 8, 55-59; CA 57, 14438e

6913. Turkel'taub, N. M., Kolyubyakina, A. I., and Selenkina, M. S., "Effect of Silica Gel Moisture Content on Chromatographic Separation of Gases," Zhur. Anal. Khim., 12, 302-312 (1957) (in Russian); J. Anal. Chem. U.S.S.R., 12, 302-322 (1957) (in English); CA 52, 1856c

6914. Turkel'taub, N. M., Palamarchuk, N. A., Shemyatenkova, V. T., and Syavtsillo, S. V., "Chromatographic Analysis of Organosilicon Compounds. Analysis of the Reaction Mixture After Direct Synthesis of Methylchlorosilanes," Plasticheskie Massy, 1961, No. 4, 51-56; CA 56, 2003a

6915. Turkel'taub, N. M., and Porshneva, N. V., "Adsorbents for Rapid Analysis of Low-Boiling Gases," Novosti Neft. i Gaz. Tekhn., Gaz. Delo, 1961, No. 11, 32-35; CA 58, 7348e

6916. Turkel'taub, N. M., Porshneva, N. V., and Kancheeva, O. A., "A Chromatographic, Thermochemical Gas Analyzer," Zavodskaya Lab., 22, 735-738 (1956); CA 51, 4774b

6917. Turkel'taub, N. M., and Ryabchuk, L. N., "Chromathermographic Determination of N_2O in the Presence of Ethane and Propane," Trudy Vsesoyuz. Nauch.-Issledovatel Geologorazvedoch. Neft. Inst., 1958, No. 11, 257-259; CA 55, 2355a

6918. Turkel'taub, N. M., Shemyatenkova, V. T., Palamarchuk, N. A., and Nechaeva, L. A., "Accuracy of Analysis of Mixtures by Various Methods for Interpreting [Gas] Chromatograms," Zavodskaya Lab., 26, 1075-1080 (1960); CA 56, 6631i

6919. Turkel'taub, N. M., Shemyatenkova, V. T., Palamarchuk, N. A., and Nechaeva, L. A., "Accuracy of Analysis of Mixtures by Various Methods for Interpreting [Gas] Chromatograms," Ind. Lab., 26, 1250-1256 (1960) (Pub. 1961)

6920. Turkel'taub, N. M., Shvartsman, V. P., Georgievskaya, T. V., Zolotareva, O. V., and Karymova, A. I., "Separation of Hydrocarbon Mixtures by the Chromathermographic Method," Zhur. Fiz. Khim., 27, 1827-1836 (1953); CA 48, 10476e

6921. Turkel'taub, N. M., and Zhukhovitskii, A. A., "Chromatographic Method of Analysis of Gases," Zavodskaya Lab., 22, 1032-1039 (1956); CA 51, 17565a

6922. Turkel'taub, N. M., and Zhukhovitskii, A. A., "Theory of Chromatographic Methods of Gas Analysis," Zavodskaya Lab., 23, 1023-1034 (1957); CA 53, 2922i

6923. Turkel'taub, N. M., and Zhukhovitskii, A. A., "Chromatographic Apparatus for the Analysis of Multicomponent Gas Mixtures," Zavodskaya Lab., 23, 1120-1124 (1957); CA 53, 5767d

6924. Turkel'taub, N. M., and Zhukhovitskii, A. A., "Continuous Chromatography," Geol. Nefti, 1, No. 2, 54-60 (1957); CA 52, 3317g

6925. Turkel'taub, N. M., and Zhukhovitskii, A. A., "Chromatographic Methods and Equipment for the Analysis of Mixtures of Gases and Volatile Compounds," Trudy Vsesoyuz. Nauch.-Issledovatel. Geol.-Rasvedoch. Inst., 1958, No. 10, 257-265; CA 54, 6399i

6926. Turkel'taub, N. M., and Zhukhovitskii, A. A., "Methods and Apparatus for Gas Analysis as Applied to Geochemical Prospecting for Petroleum and Gas Deposits," Izdatel Akad. Nauk S.S.S.R., 1958, 222-231 (Pub. 1959); CA 54, 11863g

6927. Turkel'taub, N. M., and Zhukhovitskii, A. A., "The Influence of Several Parameters in Gas Chromatography," Khromatog. ee Teoriya i Primenenie, Akad. Nauk S.S.S.R., Tr. Vses. Soveshch., Moscow, 1958, 291-299 (Pub. 1960); CA 56, 14900f

6928. Turkel'taub, N. M., and Zhukhovitskii, A. A., "New Chromatographic Gas Analyzers and Apparatus for the Analysis of Complex Gas Mixtures," Khromatog. ee Teoriya i Primenenie, Akad. Nauk S.S.S.R., Trudy Vsesoyuz. Soveshchaniya, Moscow, 1958, 300-307 (Pub. 1960); CA 55, 11951b

6929. Turkel'taub, N. M., Zhukhovitskii, A. A., and Porshneva, N. V., "Study of Molecular Sieves by Gas Chromatography," Zh. Prikl. Khim., 34, 1946-1953 (1961); CA 56, 13573f

6930. Turkel'taub, N. M., Zhukhovitskii, A. A., and Shvartsman, V. P., "Adsorbents in Gas Chromatography and Their Standards," Prirdnye Mineral. Sorbenty, Akad. Nauk Ukr.S.S.R., Otdel Khim. i Geol. Nauk, Trudy Soveshchaniya, Kiev, 1958, 78-87 (Pub. 1960); CA 55, 5223h

6931. Turkel'taub, N. M., Zolotareva, O. V., Latukhova, A. G., Karymova, A. I., and Kalnina, E., "Chromatographic Method for the Separation of Hydrogen, Carbon Monoxide, Methane and a Mixture of Rare Gases," Zhur. Anal. Khim., 11, 159-166 (1956); CA 50, 14446g

6932. Turner, D. A., "Lipids of Bile and Intestinal Synthesis of Glycerides," Federation Proc., 21, Suppl. No. 11, 25-27 (1962); CA 57, 14322b

6933. Turner, D. A., Jones, G. E. S., Sarlos, I. J., Barnes, A. C., and Cohen, R., "Determination of Urinary Pregnanediol by Gas Chromatography," Anal. Biochem., 5, 99-106 (1963); CA 59, 895b

6934. Turner, D. W., "A Robust But Sensitive Detector for Gas-Liquid Chromatography," Nature, 181, 1265-1266 (1958); CA 53, 1e

6935. Turner, G. S., "Process Control Systems Embodying Gas Chromatography Analyses," Gas Chromatog. Intern. Symposium, 2nd, East Lansing, Mich., 1959, 103-109 (Pub. 1960); CA 55, 20531b

6936. Turner, G. S., "Present and Future Role of Gas Chromatography in Industrial Processes," 6th Instrumental Methods of Analysis Symp., ISA, Montreal, Canada, June 1960

6937. Turner, G. S., "Developments in Process Gas Chromatography," Analyzer, 1, No. 4, 14-15 (1960)

6938. Turner, G. S., "Condensation and Evaporation Phenomena in Gas Chromatography," Analyzer, 2, No. 4, 15-16 (1961)

6939. Turner, G. S., "Process Control Systems Embodying Gas Chromatography Analyzers," in "Gas Chromatography," edited by H. J. Noebels, R. F. Wall, and N. Brenner, Academic Press, Inc., New York, 1961, pp. 103-109

6940. Turner, G. S., and Villalobos, R., "Microsampling in Process Stream Analysis," Gas Chromatog., Intern. Symp., 1961, 3, 363-369 (Pub. 1962); CA 58, 3867d

6941. Turner, G. S., and Villalobos, R., "Microsampling in Process Stream Analysis," in "Gas Chromatography," edited by N. Brenner, J. E. Callen, and M. D. Weiss, Academic Press, Inc., New York, 1962, pp. 363-370

6942. Turner, H. S., "The Synthesis of Radio-Chemically Labelled Fine Chemicals," Chem. & Ind. (London), 1955, 140

6943. Turner, N. C., "Development in Analysis of Hydrocarbon Gases by Adsorption Fractionation," Oil Gas J., 41, 48, 51, 52, 69 (1943); CA 37, 6860[6]

6944. Turner, N. C., "The Analysis of Hydrocarbon Gases by Means of Adsorption Fractionation," Natl. Petrol. News, 35, No. 18, R234-R237 (1943); CA 37, 3914[4]

6945. Turner, N. C., "The Analysis of Hydrocarbon Gases by Means of Adsorption Fractionation," Petrol. Refiner, 22, 140-144 (1943); CA 37, 3914[4]

6946. Turowska, A., and Jedrzejczyk, B., "Chromatographic Determination of Ethylene in Coal Gas," Gaz, Woda i Tech. Sanit., 31, 229-233 (1957); CA 51, 18551a

6947. Turowska, A., and Jedrzejczyk, B., "Chromatographic Analysis of Coal Gas," Gaz, Woda i Tech. Sanit., 31, 266-269 (1957); CA 51, 18551b

6948. Tuttle, W. N., and Feldstein, M., "Gas Chromatographic Analysis of Incinerator Effluents," J. Air Pollution Control Assoc., 10, 427-429, 467 (1960); CA 55, 5825g

6949. Tweet, O., and Miller, W. K., "Determination of Residual Monomer in Polymer Emulsions by Rapid Distillation and Gas Chromatography," Anal. Chem., 35, 852-853 (1963); 144th Natl. ACS Mtg., Los Angeles, Calif., March-April 1963, Program Abstr., p. 19B

6950. Twigg, G. H., "The Estimation of Hydrogen Deuteride by Means of the Micro Thermal Conductivity Gauge," Trans Faraday Soc., 33, 1329-1333 (1937); CA 31, 8264[7]

6951. Tykodi, R. J., "Thermodynamics of Adsorption," J. Chem. Phys., 22, 10 (1954); CA 49, 1400d

6952. Tyou, P., "Gas Chromatography in Determination of Hydrogen, Nitrogen and Oxygen in Steels and Cast Irons," Inst. hierro y acero, 13, 383-391 (1960); CA 54, 19290h

6953. Tyron, P., and Hans, A., "Gas Chromatography in the Determination of Hydrogen, Nitrogen, and Oxygen in Steels and Cast Irons," Rev. met., 58, 187-193 (1961); CA 55, 26832f

6954. Ucciani, E., and Naudet, M., "Halogenation of the Allylic Position of Aliphatic Unsaturated Chains. V. Place and Position Isomerization of Mono- and Dihalo Chains," Bull. soc. chim. France, 1963, 28-32; CA 59, 432a

6955. Uchida, A., and Matsuda, S., "Gas Chromatography of Phenols," Kogyo Kagaku Zasshi, 65, 574-577 (1962); CA 57, 10514e

6956. Uebayashi, A., "Application to Fermentation Products. Analysis of Fusel Oil," Kagaku no Ryoiki Zokan, No. 44, 183-198 (1961); CA 56, 3a

6957. Uezumi, N., Hasegawa, S., and Kasama, K., "Studies on Fatty Livers. I. Fatty Acid Composition of Lipids from Livers of Rats Treated with Carbon Tetrachloride," Mie Med. J., 11, 381-386 (1962); CA 58, 866h

6958. Uhlmann, L., and Prinzler, H., "Constituents of Petroleum and Related Products. Investigation of the 90-150° Fraction of a Fischer-Tropsch Medium-Pressure Synthesis Product," Chem. Tech. (Berlin), 14, 351-353 (1962); CA 58, 391g

6959. Ulbrich, V., and Dufka, O., "Partition and Identification of Glycidyl Ethers by Partition Gas Chromatography," Chem. prumysl, 10, 549-554 (1960); CA 55, 23475d

6960. Ulehla, J., "Pyrolysis and Gas Chromatography of Amino Acids," Sbornik Ceskoslov. akad. zemedel. ved, Zivocisna vyroba, 5, 567-574 (1960); CA 55, 5242g

6961. Ullman, E. R., and Fanshawe, W. J., "Unsaturated Cyclopropanes. III. Synthesis and Properties of Alkylidenecyclopropanes and Spiropentanes," J. Am. Chem. Soc., 83, 2379-2383 (1961); CA 55, 27125e

6962. Ulrich, W. F., "Chemical Analysis by Preparative Gas Chromatography," Analyzer, 1, No. 1, 8 (1960)

6963. Ulrich, W. F., Gallaway, W. S., and Johns, T., "Infrared Gas Chromatography Techniques for Trace and Micro Analyses," 11th Pittsburgh Conf. on Anal. Chem. & Appl. Spectroscopy, Pittsburgh, Pa., February-March 1960, Program Abstr., p. 54

6964. Upham, F. T., Lindgren, F. T., and Nichols, A. V., "Some Characteristics of a Strontium-90 Beta-Particle Detector for Gas-Liquid Chromatography," Anal. Chem., 33, 845-849 (1961); CA 55, 19354a

6965. Urch, D., and Wolfgang, R., "Mechanisms of Hot Hydrogen Atom Displacement Reactions with Alkanes," J. Am. Chem. Soc., 83, 2982-2991 (1961); CA 56, 3331h

6966. Urch, D., and Wolfgang, R., "Hot Hydrogen Atom Displacement Reaction at Ethylenic C-H Bonds," J. Am. Chem. Soc., 83, 2997-2998 (1961); CA 56, 3332f

6967. Urone, P., and Katnik, R. J., "A New and Efficient Liquid Phase for Gas Liquid Chromatography of Oxygenated Substances," Anal. Chem., 35, 767-768 (1963); CA 59, 1065a

6968. Urone, P., and Pecsok, R. L., "Gas Chromatographic Behavior of C_1-C_4 Saturated Alcohols and Water on Polyethylene Glycol Substrate. Effect of Solid Support Treatment," Anal. Chem., 35, 837-841 (1963)

6969. Urone, P., and Smith, J. E., "Analysis of Chlorinated Hydrocarbons with the Gas Chromatograph," Am. Ind. Hyg. Assoc. J., 1, 36-41 (1961); CA 55, 11716g

6970. Urone, P., Smith, J. E., and Katnik, R. J., "Gas Chromatographic Study of Some Chlorinated Hydrocarbons," Anal. Chem., 34, 476-480 (1962); CA 56, 14904d

6971. Usami, S., Gas-Chromatographic Determination of Trace Impurities in Vinyl Acetate," Bunseki Kagaku, 10, 141-145 (1961); CA 55, 23189f

6972. Usami, S., "Instrumental Analysis of Microcomponents in the Production of Poly(Vinyl Alcohol)," Kagaku no Ryoiki, 15, 494-502 (1961); CA 56, 8903i

6973. Vagelos, P. R., VandenHeuvel, W. J. A., and Horning, M. G., "Identification of Hydroxamic Acids by Gas Chromatography of Isocyanate Derivatives," Anal. Biochem., 2, 50-58 (1961); CA 55, 13523b

6974. Vagin, E. V., and Zhukhovitskii, A. A., "Theory of Thermal Separation of Gas Mixtures by the Adsorption Method," Doklady Akad. Nauk S.S.S.R., 94, 273-276 (1954); CA 50, 7540i

6975. Vaisberg, K. M., Kruglov, E. A., Khabibullin, M. F., and Shabalin, I. I., "A Study of the Composition of Different Grades of Naphthalene by Means of Gas-Liquid Chromatography," Koks i Khim., 1963, No. 3, 44-47; CA 59, 1416d

6976. Valstar, J. E., and Percy, L. E., "Application of Intermittent Process Analyzers to Closed Loop Control," J. Gas Chromatog., 1, No. 2, 24-27 (1963)

6977. Valussi, S., and Cofleri, G., "Variable-Flux Gas Chromatography in Butter Analysis," Boll. Lab. Chim. Provinciali (Bologna), 13, No. 1, 3-15 (1962); CA 57, 7686e

6978. Vamos, E., "Adsorption Chromatography. I and II.," Magyar Kem. Lapja, 14, 165-170, 202-207 (1959); CA 53, 18592e

6979. Van Auken, T. V., and Rinehart, Jr., K. L., "Stereochemistry of the Formation and Decomposition of 1-Pyrazolines," J. Am. Chem. Soc., 84, 3736-3743 (1962); CA 58, 10062a

6980. Vandenheuvel, F. A., "Quantitative Analysis of Submicrogram Amounts of High Boiling Compounds by Flame Ionization Gas Liquid Chromatography," Anal. Chem., 35, 1186-1192 (1963)

6981. Vandenheuvel, F. A., "Estimation and Correction of Post-Column Dead Volume Effect in Chromatography," Anal. Chem., 35, 1193-1198 (1963)

6982. Vandenheuvel, F. A., and Vatcher, D. R., "Partition Chromatography of Aliphatic Acids. Quantitative Resolution of Normal Chain Even Acids from C_{12} to C_{24}," Anal. Chem., 28, 838-845 (1956); CA 50, 11174g

6983. VandenHeuvel, W. J. A., "Medical Applications of Gas Chromatography," 5th Annual Gas Chromatography Institute, Canisius College, Buffalo, N. Y., April 1963; pub. in "Biomedical Applications of

Gas Chromatography," Plenum Press, New York, 1964 (in press)

6984. VandenHeuvel, W. J. A., "Recent Advances in Gas Chromatographic Analysis of Steroids and Related Substances," Biochemical Gas Chromatography Seminar, Wilkens Instrument & Research, Inc., New Orleans, La., July 1963

6985. VandenHeuvel, W. J. A., Creech, B. G., and Horning, E. C., "Separation and Estimation of the Principal Human Urinary 17-Keto Steroids as Trimethylsilyl Ethers," Anal. Biochem., 4, 191-197 (1962); CA 58, 4800b

6986. VandenHeuvel, W. J. A., Gardiner, W. L., and Horning, E. C., "Preparation of Thin-Film Gas Chromatographic Columns with Polyvinyl Pyrrolidone-Inactivated Supports," Anal. Chem., 35, 1745-1746 (1963)

6987. VandenHeuvel, W. J. A., Haahti, E. O. A., and Horning, E. C., "Functional Group and Stereochemical Effects in the Gas Chromatographic Separation of Steroids with Selective Phases," 140th Natl. ACS Mtg., Chicago, Ill., September 1961, Program Abstr., pp. 68Q-69Q

6988. VandenHeuvel, W. J. A., Haahti, E. O. A., and Horning, E. C., "A New Liquid Phase for Gas Chromatographic Separation of Steroids," J. Am. Chem. Soc., 83, 1513-1514 (1961); CA 55, 21226i

6989. VandenHeuvel, W. J. A., Haahti, E. O. A., and Horning, E. C., "Gas Chromatographic Separation of Drugs and Drug Metabolites," Clin. Chem., 8, 351-359 (1962); CA 57, 16745i

6990. VandenHeuvel, W. J. A., and Horning, E. C., "Gas Chromatography of Adrenal Cortical Steroid Hormones," Biochem. Biophys. Res. Communs., 3, 356-360 (1960); CA 55, 16649e

6991. VandenHeuvel, W. J. A., and Horning, E. C., "Gas Chromatographic Separations of Sugars and Related Compounds as Acetyl Derivatives," Biochem. Biophys. Res. Communs., 4, 399-403 (1961); CA 55, 27500e

6992. VandenHeuvel, W. J. A., and Horning, E. C., "Gas Chromatographic Separations of Sapogenins," J. Org. Chem., 26, 634-635 (1961); CA 55, 16594a

6993. VandenHeuvel, W. J. A., and Horning, E. C., " Retention-Time Relations in Gas Chromatography in Terms of the Structure of Steroids," Biochim. Biophys. Acta, 64, 416-429 (1962); CA 58, 6879g

6994. VandenHeuvel, W. J. A., Horning, E. C., Sato, Y., and Ikekawa, N., "Gas Chromatographic Separation of Steroidal Amines," J. Org. Chem., 26, 628-629 (1961); CA 55, 16599e

6995. VandenHeuvel, W. J. A., Sjovall, J., and Horning, E. C., "Gas-Chromatographic Behavior of Trifluoroacetoxy Steroids," Biochim. Biophys. Acta, 48, 596-599 (1961); CA 55, 24899i

6996. VandenHeuvel, W. J. A., Sweeley, C. C., and Horning, E. C., "Microanalytical Separations by Gas Chromatography in the Sex Hormone and Bile Acid Series," Biochem. Biophys, Res. Communs, 3, 33-36 (1960); CA 55, 14569c

6997. VandenHeuvel, W. J. A., Sweeley, C. C., and Horning, E. C., "Separation of Steroids by Gas Chromatography," J. Am. Chem. Soc., 82, 3481-3482 (1960); CA 54, 23198i

6998. Van der Grinten, P. M. E. M., and Dijkstra, A., "Integration with Ionization Detectors in Gas Chromatography," Nature, 191, 1195-1196 (1961); CA 56, 1309a

6999. Van der Kloot, A. P., "Application of Gas Chromatography to Brewing and Beverages," 13th Annual Mid-America Spectroscopy Symp., Chicago, Ill., April-May 1962

7000. Van der Kloot, A. P., Tenney, R. I., and Bavisotto, V., "An Approach to Flavor Definition with Gas Chromatography," Proc. Am. Soc. Brewing Chemists, 1958, 96-103; cf. Brewers Digest, p. 59 (July 1959); CA 53, 5584h

7001. Van der Kloot, A. P., and Wilcox, F. A., "An Approach to Flavor Definition with Gas Chromatography. II.," Proc. Am. Soc. Brewing Chemists, 1959, 76-80; CA 54, 7972g

7002. Van der Kloot, A. P., and Wilcox, F. A., "Studies on the Determination of Beer Volatiles by Gas Chromatography," Proc. Am. Soc. Brewing Chemists, 1960, 113-116; CA 55, 27763a

7003. Van der Kloot, A. P., and Wilcox, F. A., "The Determination of Beer Volatiles by Gas Chromatography. II.," Proc. Am. Soc. Brewing Chemists, 1961, 24-27; CA 56, 7799f

7004. Van Der Stricht, M., and Rysselberge, J. van, "Modified Bentone 34, a New Addition to Stationary Phases in Gas Chromatography to Assist in Separation of Difficult Aromatic Mixtures," J. Gas Chromatog., 1, No. 8, 29-33 (1963)

7005. Vander Wal, R. J., "The Determination of Glyceride Structure," J. Am. Oil Chemists' Soc., 40, 242-247 (1963)

7006. VanderWerf, C. A., Heasley, V., and Locateli, L., "Studies on cis- and trans-1,4-Diazido-2-butenes," 144th Natl. ACS Mtg., Los Angeles, Calif., March-April 1963, Program Abstr., p. 10M

7007. Van Duuren, B. L., and Kosak, A. L., "Isolation and Identification of Some Components of Cigarette Smoke Condensate," J. Org. Chem., 23, 473-475 (1958); CA 53, 7515g

7008. Vango, S. P., "Gas Handling Apparatus for Gas Chromatography," Chemist-Analyst, 52, No. 2, 53-54 (1963)

7009. Van Hook, W. A., and Emmett, P. H., "The Gas-Chromatographic Determination of Hydrogen, Deuterium and Hydrogen Deuteride," J. Phys. Chem., 64, 673-675 (1960); CA 54, 19262i

7010. Van Hook, W. A., and Emmett, P. H., "Trace Studies with Carbon-14. I. Some of the Secondary Reactions Occurring During the Catalytic Cracking of n-Hexadecane Over a Silica-Alumina Catalyst," J. Am. Chem. Soc., 84, 4410-4421 (1962); CA 58, 5431a

7011. Van Middelem, C. H., and Waites, R. E., "Gas Chromatographic and Colorimetric Measurement of Dimethoate Residues," 144th Natl. ACS Mtg., Los Angeles, Calif., March-April 1963, Program Abstr., p. 8A

7012. Van Swaay, M., (Ed.), "Gas Chromatography 1962," Butterworths & Co., London, 1962. Proceedings of the 4th Symp. on Gas Chromatography, Hamburg, Germany, June 1962

7013. Van Vuuren, P. J., and Serfontein, W. J., "Antibiotic Substance from a Strain of Bacillus Subtilis. III. Fatty Acids from This Antibiotic," S. African J. Agr. Sci., 5, 491-494 (1962); cf. ibid., 4, 255 (1961); CA 58, 12874f

7014. Váradi, P. F., "Quantitative and Qualitative Ionization Detector for Gas Chromatography," in "Gas Chromatography," edited by N. Brenner, J. E. Callen, and M. D. Weiss, Academic Press, Inc., New York, 1962, pp. 195-206

7015. Váradi, P. F., and Ettre, K., "Operation of the Quantitative and Qualitative Ionization Detector and Its Application for Gas Chromatographic Studies," Anal. Chem., 34, 1417-1422 (1962); CA 58, 6b

7016. Váradi, P. F., and Ettre, K., "Pyrolysis Attachment for Chromatographs Helps in Study of Polymer Degradation," Perkin-Elmer Instrument News, 14, No. 1, 1, 8-9 (Fall 1962)

7017. Váradi, P. F., and Ettre, K., "Vacuum Output Gas Chromatography," Anal. Chem., 35, 410-412 (1963); CA 58, 11934f

7018. Varma, M. C. P., "Complex Formation Between Montmorillonite and the Phenols," J. Indian Chem. Soc., Ind. & News Ed., 20, 11-16 (1957); CA 52, 2674c

7019. Vasilescu, V., "Gas-Liquid Chromatography of Synthetic Carboxylic Acids and the Corresponding Alcohols," Abhandl. Deut. Akad. Wiss. Berlin, Kl. Chem., Geol., Biol., 1959, No. 9, 136-153; CA 58, 5018d

7020. Vasilescu, V., "Gas-Liquid Chromatography of Synthetic Carboxylic Acids and of the Alcohols Produced from These Acids," in "Gas Chromatographie 1958," edited by H. P. Angele, Akademie Verlag, Berlin, 1959, pp. 136-153

7021. Vasilescu, V., "Problems of the Quantitative Evaluation of Gas Chromatography Curves of Higher Fatty Acids," in "Gas Chromatographie 1959," edited by R. E. Kaiser and H. G. Struppe, Akademie Verlag, Berlin, 1959, pp. 337-356

7022. Vasilescu, V., "The Gas-Chromatographic Separation of High-Molecular-Weight Fat Derivatives Without the Use of Vacuum Techniques," Fette, Seifen, Anstrichmittel, 63, 132-138 (1961); CA 55, 14941e

7023. Vasilescu, V., "Application of Gas-Liquid Chromatography in the Manufacture and Analysis of Detergents and Textile Auxiliaries," J. Prakt. Chem., 15, 192-205 (1962); CA 56, 14417h

7024. Vasil'eva, V. S., Kiselev, A. V., Nikitin, Y. S., Petrova, R. S., and Shcherbakova, K. D., "Graphitized Carbon as an Adsorbent in Gas Chromatography," Zhur. Fiz. Khim., 35, 1889-1891 (1961); CA 56, 5420d

7025. Vassallo, D. A., "Pyrolysis Techniques," Anal. Chem., 33, 1823-1825 (1961); CA 56, 8500f; 139th Natl. ACS Mtg., St. Louis, Mo., March 1961, Program Abstr., p. 12B

7026. Velut, M., and Jourda, J., "Chromatographic Gas Analyzer-Recorder," Gas J., 295, 270, 275-276 (1958); CA 53, 6697c

7027. Velut, M., and Jourda, J., "Recording Gas Analyzer Using Chromatography," Rev. inst. franç. pétrole et ann. combustibles liquides, 13, 1635-1647 (1958); CA 53, 8627f

7028. Ven Horst, Sister Helene, Ven Horst, H., and O'Connor, K., "A Technique for Packing Columns for Gas Chromatography," J. Chem. Educ., 37, 593 (1960); CA 55, 6057i

7029. Venugopalan, M., and Kutschke, K. O., "Gas Chromatographic Separation of Hydrogen Isotopes on Activated Alumina," Can. J. Chem., 41 548-549 (1963); CA 58, 9829a

7030. Verbeke, R., Lauryssens, M., Peeters, G., and James, A. T., "Incorporation of DL-1-C^{14}Leucine and 1-C^{14}Isovaleric Acid into Milk Constituents by the Perfused Cow's Udder," Biochem. J., 73, 24-29 (1959); CA 55, 6057i

7031. Verdier, C. H. de, and Sjoberg, C. I., "An Automatic Conductivity Bridge for Chromatographic Analyses," Acta Chem. Scand., 8, 1161-1168 (1954); CA 49, 8633g

7032. Vergnaud, J.-M., "Gas Chromatography with a Variable Flow Rate," Bull. Soc. Chim. France, 1962, 1914-1917; CA 58, 7349c

7033. Verrien, J. P., "Analysis of Natural Gas, Methods and Applications," Rev. inst. franc. pétrole et ann. combustibles liquides, 11, 641-643 (1956); CA 51, 16076h

7034. Vertalier, S., and Martin, F., "Selective Microdetermination of Alkoxy Groups by Gas-Liquid Chromatography," Chim. anal., 40, 80-86 (1958); CA 52, 8849c

7035. Verzele, M., "A Note on Preparative Scale Gas Chromatography," J. Chromatog., 9, 116-117 (1962); CA 59, 519d

7036. Veselov, V. V., "Apparatus for the Rapid Chromatographic Analysis of Hydrocarbon Mixtures," Izvest. Sibir. Otdel, Akad. Nauk S.S.S.R., 1959, No. 4, 83-89; CA 53, 20936i

7037. Vessman, J., and Schill, G., "Gas Chromatography of High Boiling Amines," Svensk Farm. Tidskr.,

<u>66</u>, 601-618 (1962); CA <u>58</u>, 9609g

7038. Viehe, H. G., Franchimont, E., Reinstein, M., and Valange, P., "Pairs of Geometrical Isomers with Preferential cis-Structure," Chem. Ber., <u>93</u>, 1697-1709 (1960); CA <u>54</u>, 24335a

7039. Vietti-Michelina, M., "Gas-Chromatography of Pyridinealdehydes," Rass. Chim., <u>13</u>, No. 6, 23-24 (1961); CA <u>57</u>, 2858i

7040. Vietti-Michelina, M., and Pilleri, R., "Gas Chromatography of Benzaldehyde and Benzyl Alcohol Mixture," Rass. Chim., <u>13</u>, No. 1, 13-14 (1961); CA <u>55</u>, 20792b

7041. Vigdergauz, M. S., "Analysis of Complex Organic Mixtures by Gas Chromatography," Usp. Khim., <u>31</u>, 73-100 (1962); CA <u>56</u>, 14897e

7042. Vigdergauz, M. S., and Gol'bert, K. A., "Selection of Conditions for the Gas-Chromatographic Separation of Complex Mixtures," Neftekhimiya, <u>1</u>, 706-715 (1961); CA <u>57</u>, 2825e

7043. Vigdergauz, M. S., and Gol'bert, K. A., "Rapid Chromatographic Analysis of Hydrocarbon Gases," Neftekhimiya, <u>2</u>, 825-830 (1962); CA <u>58</u>, 12341d

7044. Vigdergauz, M. S., and Gol'bert, K. A., "Selection of Optimal Conditions for Chromatographic Separation of Complex Organic Mixtures by Using a Series of Columns," Neftekhimiya, <u>2</u>, 852-860 (1962); CA <u>58</u>, 11933c

7045. Vigdergauz, M. S., and Gol'bert, K. A., "Tentative Nomenclature for Gas Chromatography," Neftekhimiya, <u>2</u>, 940-951 (1962); CA <u>58</u>, 7345f

7046. Vigdergauz, M. S., Gol'bert, K. A., Afanas'ev, M. I., Mashukova, G. A., and Zimin, R. A., "Gas-Chromatographic Analysis of Liquid Products from Pyrolysis and Cracking [of Petroleum]," Neftekhimiya, <u>2</u>, 405-409 (1962); CA <u>58</u>, 2304d

7047. Vigdergauz, M. S., Gol'bert, K. A., Savina, I. M., Afanas'ev, M. I., Zimin, R. A., and Bakhareva, N. I., "Gas-Chromatographic Determination of Traces of Acetylenic and Dienic Compounds in Complex Mixtures," Zavodskaya Lab., 28, 149-150 (1962); CA <u>57</u>, 2856c

7048. Vigdergauz, M. S., Gol'bert, K. A., Sidorov, V. A., and Andrianov, A. A., "Chromatographic Analysis of Gaseous and Liquid Hydrocarbon," Tr. po Khim. i Khim. Tekhnol., <u>1962,</u> No. 1, 10-15; CA <u>58,</u> 7348h

7049. Vigdergauz, M. S., Gol'bert, K. A., Zimin, R. A., and Gorshunov, O. L., "Gas-Chromatographic Analysis of the Products of Oxidation of Isobutane," Neftekhimiya, <u>2,</u> 410-414 (1962); CA <u>58,</u> 2304e

7050. Vilkas, M., and Nedumparambil, A. A., "A Case of Isomerization in Vapor Chromatography," Bull. soc. chim France, <u>1959</u>, 1651-1652; CA <u>54</u>, 10453h

7051. Villalobos, R., "Application of a Treated Solid Support in Industrial Gas Chromatography," 6th Instrumental Methods of Analysis Symp., ISA Montreal, Canada, June 1960

7052. Villalobos, R., Brace, R. O., and Johns, T., "Role of Column Backflushing in Gas Chromatography," Gas Chromatog., Intern. Symposium, 2nd, East Lansing, Mich., <u>1959,</u> 39-54 (Pub. 1961); CA <u>55</u>, 18204h

7053. Villalobos, R., Brace, R. O., and Johns, T., "The Role of Column Backflushing in Gas Chromatography," in "Gas Chromatography," edited by H. J. Noebels, R. F. Wall, and N. Brenner, Academic Press, Inc., New York, 1961, pp. 39-54

7054. Vinogradova, O. M., Zhabrova, G. M., Kadenatsi, B. M., and Yanovskii, M. I., "Radiochromatographic Study of Formation of Butylenes, in Butadiene Synthesis According to S. V. Lebedev," Zhur. Obshchei Khim., 29, 3396-3400 (1959); cf. Doklady Akad. Nauk S.S.S.R., <u>121,</u> 674 (1958); CA <u>54</u>, 17240c

7055. Vioque, E., "Column Partition Chromatography and Its Application to the Separation of Fatty Acids," Grasas y aceites, <u>6</u>, 88-93 (1955); CA <u>50</u>, 2191a

7056. Vioque, E., and Morris, L. J., "Minor Components of Olive Oils. I. Triterpenoid Acids in an Acetone-Extracted Orujo Oil," J. Am. Oil Chemists' Soc., <u>38</u>, 485-488 (1961); CA <u>55</u>, 24054f

7057. Vioque, E., Morris, L. J., and Holman, R. T., "Minor Components of Olive Oils. II. trans-9,10-Epoxystearic Acid in Orujo Oil," J. Am. Oil Chemists' Soc., 38, 489-492 (1962); CA <u>55</u>, 24054g

7058. Virus, W., "New Apparatus for Chromatographic Gas Analysis," Erdol u. Kohle, <u>11</u>, 867-868 (1958)

7059. Visser, B. F., "Gas Analysis by Thermal Conductivity Measurement," Rec. trav. chim., <u>74,</u> 507-512 (1955); CA <u>49</u>, 16034d

7060. Vitagliano, M., Leone, A. M., and Vodret, A., "Determining the Purity of Olive Oil by Gas Chromatography," Olearia, <u>14,</u> 177-190 (1960); CA <u>55,</u> 8896e

7061. Vitagliano, M., Leone, A. M., and Vodret, A., "Study of Olive Oil by Gas Chromatography and the Possibility of Detecting Its Genuinity," Riv. Ital. Sostanze Grasse, <u>38</u>, 111-120 (1961)

7062. Vitzthum, O., "The Thermal Factor in Gas Chromatography," Z. Anal. Chem., <u>189</u>, 66-77 (1962); CA <u>57</u>, 13163f

7063. Vivie, J., "Tendencies and Novelties in Instrumentation," Measures & Controle Industriel, <u>24</u>, 565-580, 645-655 (1959)

7064. Vizard, G. S., and Wynne, A., "Determination of Argon and Oxygen by Gas Chromatography," Chem. & Ind. (London), <u>1959</u>, 196-197; CA <u>53</u>, 12091f

7065. Vizard, G. S., and Wynne, A., "The Gas-Chromatographic Analysis of Mine Gases by a Zero Suppression Technique," Analyst, 87, 810-818 (1962); CA 58, 1271c

7066. Vladimirov, B. V., "Apparatus for Gas Analysis by Gas Chromatography," Neftyanoe Khoz., 34, No. 8, 61-64 (1956); CA 51, 4061c

7067. Vlastimil, H., "Gas Chromatography," Magyar Kem. Lapja, 14, 354-360 (1959); CA 54, 1023e

7068. Vlodavets, M. L., Gol'bert, K. A., Odinokov, V. N., and Sinovich, I. D., "Chromatographic Determination of Acrolein Dimer in Reaction Mixtures," Zavodskaya Lab., 28, 145-146 (1962); CA 57, 1553i

7069. Vlodavets, M. L., Gol'bert, K. A., Perovskaya, N. V., and Ternovskaya, L. P.,"Application of Gas Chromatography to the Control of Production of Phenol and Acetone Obtained by Acid Decomposition of Isopropylbenzene Hydroperoxide," Neftekhimiya, 1, 836-838 (1961); CA 57, 5303i

7070. Vnukov, A. K., and Dzedzik, R. P., "Application of Gas Chromatography to the Control of Power-Station Boilers, " Elek. Stantsii, 32, No. 5, 12-15 (1961); CA 56, 2299c

7071. Vodzinskii, Y. V., "New Instruments for Physical-Chemical Analysis of Wood Chemical Products," Gidrolizn. i Lesokhim. Prom., 15, No. 7, 8-10 (1962); CA 58, 7029f

7072. Vogel, A. M., and Quattrone, Jr., J. J., "Rapid Gas Chromatographic Method for Determination of Carbon and Hydrogen," Anal. Chem., 32, 1754-1757 (1960); CA 55, 6246d

7073. Vogel, E., Ott, K. H., and Gajek, K., "Small Carbon Rings. VI. Valence Isomerization of cis-1,2-Divinylcycloalkanes," Ann., 644, 172-188 (1961); CA 56, 1357f

7074. Vogelsang, R. F., "The Trend to Process Control by Gas Chromatography," Analyzer, 4, No. 1, 11-14 (January 1963)

7075. Voigt, J., "Gas-Chromatographic Identification of High Polymers," Kunststoffe, 51, 18-20 (1961); CA 55, 11906h

7076. Voigt, J., "Identification of Macromolecular Substances by Vapor Chromatography. II. Elastomers," Kunststoffe, 51, 314-317 (1961); CA 55, 25311h

7077. Voinov, A. P., "The Rapid Collection of Gas Samples for Analysis," Gazovaya Prom., 1958, No. 4, 48; CA 52, 12368i

7078. Volman, D. H., and Swanson, L. W., "The Photochemical Decomposition of Acetone in Aqueous Solutions of Allyl Alcohol at 2537 A.," J. Am. Chem. Soc., 82, 4141-4144 (1960); CA 55, 3166a

7079. Vorbeck, M. L., Mattick, L. R., Lee, F. A., and Pederson, C. S., "Determination of Lower-Molecular-Weight Fatty Acids by Gas Chromatography," Nature, 187, 689 (1960); CA 55, 2363f

7080. Vorbeck, M. L., Mattick, L. R., Lee, F. A., and Pederson, C. S., "Preparation of Methyl Esters of Fatty Acids for Gas-Liquid Chromatography," Anal. Chem., 33, 1512-1514 (1961); CA 56, 4076g

7081. Vorbeck, M. L., Mattick, L. R., Lee, F. A., and Pederson, C. S., "Volatile Flavor of Sauerkraut. Gas Chromatographic Identification of a Volatile Acidic Off-Odor," J. Food Sci., 26, 569-573 (1961)

7082. Voss, G., and Hessenhauer, F., "Chromatography of Gases and Vapors," Erdol u. Kohle, 10, 161-163 (1957); CA 51, 10177g

7083. Vosti, D. C., Hernandez, H. H., and Strand, J. B., "Analysis of Headspace Gases in Canned Foods by Gas Chromatography, " Food Technol., 15, 29-31 (1961); CA 56, 6420b

7084. Vranjican, D., Jurasevic, S., and Prohaska, B., "Hydrogenation of Aromatic Hydrocarbons in Oil Derivatives. II.," Nafta (Yugoslavia), 1958, 95-98; CA 53, 11807g

7085. Vrbaski, T., and Cvetanovic, R. J., "A Study of the Products of the Reactions of Ozone with Olefins in the Vapor Phase as Determined by Gas-Liquid Chromatography," Can. J. Chem., 38, 1063-1069 (1960); CA 55, 2460e

7086. Vreeland, R. W., "Thermal Decomposition of Cyclobutane," Univ. Microfilms (Ann Arbor, Mich.), L. C. Card No. Mic 61-2087, 96 pp; Dissertation Abstr., 21, 3651 (1961); CA 55, 27128i

7087. Vries, M. J. de, "Higher Fatty Acid Esters in Alcoholic Beverages. Their Extraction and Analysis by Gas-Liquid Chromatography," S. African J. Agr. Sci., 5, 395-400 (1962), CA 58, 7331g

7088. Vries, M. J. de, "Gas-Liquid Chromatographic Analysis of Lees Oil," S. African J. Agr. Sci., 5, 401-410 (1962); CA 58, 7331h

7089. Vyakhirev, D. A., and Bruk, A. I., "Effects of the Experimental Conditions of the Chromatographic Separation of Substances in the Gas and Methane, Ethane, and Propane Mixture from Silica Gel," Zhur. Fiz. Khim., 31, 1713-1719 (1957); CA 52, 5929f

7090. Vyakhirev, D. A., and Bruk, A. I., "The Influence of Various Factors on the Chromatographic Separation of Hydrocarbons in the Gaseous Phase. I. Influence of the Quantitative Composition on the Separation of Mixtures of Methane and Ethane," Uchenye Zapiski Gor'kov. Gosudarst, Univ. im. N. I. Lobachevskogo, Ser. Khim., 1958, No. 32, 43-53; CA 54, 5180d

7091. Vyakhirev, D. A., and Bruk, A. I., "Effects of the Experimental Conditions on the Chromatographic Separation of Substances in the Gas and Vapor Phases. II. The Effect of the Carrier Gas on the Separation of Mixtures of Gaseous Hydrocarbons," Zhur. Fiz. Khim., 33, 1309-1317 (1959); CA 54, 9434i

7092. Vyakhirev, D. A., and Bruk, A. I., "Influence of Carrier-Gas Composition on the Chromatographic Resolution of Gaseous Hydrocarbon Mixtures," Khromatog., ee Teoriya i Primenenie, Akad.

Nauk, S.S.S.R., Trudy Vsesoyuz. Soveshchaniya, Moscow, 1958, 260-266 (Pub. 1960); CA 55, 22991e

7093. Vyakhirev, D. A., Bruk, A. I., and Chernyaev, N. P., "The Effect of Changes in the Structure of Silica Gel on the Separation of Gaseous Hydrocarbons," Gasovaya Khromatographiya, Trudy 1-oi [Pervoi] Vsesoyuz. Konf., Akad. Nauk S.S.S.R., Moscow, 1959, 162-171 (Pub. 1960); CA 56, 10446e

7094. Vyakhirev, D. A., Bruk, A. I., and Guglina, S. A., "Volumetric-Chromatographic Method of Gas Analysis," Doklady Akad. Nauk S.S.S.R., 90, 577-579 (1953); CA 49, 13833b

7095. Vyakhirev, D. A., Bruk, A. I., and Guglina, S. A., "Volumetric-Chromatographic Method of Analysis of Mixtures of Hydrocarbon Gases," Trudy Komissii Anal. Khim., Akad. Nauk S.S.S.R., Inst. Geokhim. i Anal. Khim., 6, 137-145 (1955); CA 50, 7666a

7096. Vyakhirev, D. A., Chernyaev, N. P., and Bruk, A. I., "Effect of Experimental Factors on Gas Chromatography. III. Effect of Structure of Silica Gel on Gas Chromatography of Hydrocarbons," J. Phys. Chem. U.S.S.R., 34, 521-525 (1961); Zhur. Fiz. Khim., 34, 1096-1103 (1961); CA 57, 6654a

7097. Vyakhirev, D. A., Demina, N. D., and Aveeva, M. P., "Chromatographic Analysis of the Butylene Fraction of the Gas from Cracked Kerosene," Trudy Khim. i. Khim. Tekhnol., 2, 133-140 (1959); CA 54, 7121b

7098. Vyakhirev, D. A., Drauzin, A. Y., Reshetnikova, L. E., and Komissarov, P. F., "Separation of Mixtures of Ethyl Benzene, Isopropylbenzene, and Butylbenzene by Gas Chromatography," Tr. po Khim. i Khim. Tekhnol., 2, 600-604 (1959); CA 57, 2830c

7099. Vyakhirev, D. A., and Komissarov, P. F., "Vacuum Gas Chromatography," Doklady Akad. Nauk S.S.S.R., 129, 138-140 (1959); CA 55, 8998f

7100. Vyakhirev, D. A., Ostasheva, M. I., and Reshetnikova, L. E., "The Determination of Liquid Hydrocarbons by the Methods of Gas Chromatography," Trudy Khim. i Khim. Tekhnol., 1, 334-338 (1958); CA 54, 6409g

7101. Vyakhirev, D. A., and Ostasheva, M. I., "Determination of Benzene, Toluene, and Xylene in Their Mixtures by the Gas-Liquid Chromatographic Method," Trudy Khim. i Khim. Tekhnol., 2, 128-132 (1959); CA 54, 6409i

7102. Vyakhirev, D. A., Reshetnikova, A. E., and Sherstneva, T. V., "Determination of Ethylbenzene and Butylbenzene in Isopropylbenzene by Means of Gas-Liquid Chromatography," Trudy po Khim. i Khim. Tekhnol., 4, 342-344 (1961); CA 55, 25602f

7103. Vyakhirev, D. A., and Reshetnikova, L. E., "The Determination of Ethylbenzene Impurities in Technical Isopropylbenzene by the Method of Gas-Liquid Chromatography," Trudy Khim. i Khim. Tekhnol., 1, 339-342 (1958); CA 54, 6409g

7104. Vyakhirev, D. A., and Reshetnikova, L. E., "Separation and Analysis of Mixtures of Chlorinated Methane Derivatives by a Chromathermographic Method," Zhur Priklad. Khim., 31, 802-804 (1958) (in Russian); J. Appl. Chem. U.S.S.R., 31, 789-792 (1958) (in English); CA 52, 15344h

7105. Vyas, S. H., "Whipped Butter: Manufacture Characteristics and Gas Chromatographic Analysis of Off-Flavor Compounds," Univ. Microfilms (Ann Arbor, Mich.), Order No. 61-4994, 67 pp.; Dissertation Abstr., 22, 1946 (1961)

7106. Vykoc, J., Krepinsky, J., Herout, V., and Sorm, F., "Nature of Sesquiterpenic Hydrocarbon Calarene and Structure of β-Gurjunene," Tetrahedron Letters, 1963, 225-229; CA 59, 673f

7107. Wachs, W., "Manufacture of Modern Food Fats and Their Analysis," Z. Lebensm.-Untersuch. u. -Forsch., 113, 213-222 (1960); CA 55, 836h

7108. Wachi, F. M. L., "The Application of Gas Chromatography to the Determination of Some Physical Constants. II. High Temperature Gas Chromatography," Univ. Microfilms (Ann Arbor, Mich.), L. C. Card No. 59-2061; Dissertation Abstr., 20, 53-54 (1959); CA 53, 18592e

7109. Wagenknecht, A. C., Mattick, L. R., Lewin, L. M., Hand, D. B., and Steinkraus, K. H., "Changes in Soybean Lipids During Tempeh Fermentation," J. Food Sci., 26, 373-376 (1961); CA 56, 3863a

7110. Wagner, J., "Gas Chromatographic Separation and Determination of Amino Acids," 1960 Mtg., German Chemical Soc., Stuttgart, Germany, April 1960, Program Abstr., p. 41; Abstr., Angew. Chem., 72, 588 (1960)

7111. Wagner, J., and Winkler, G., "Gas-Chromatographic Separation and Determination of Amino Acids. Gas-Chromatographic Determination of Amino Acids as N-Trifluoroacetyl Methyl Esters," Z. Anal. Chem., 183, 1-11 (1961); CA 55, 25602a

7112. Wahlroos, O., "A High-Field Emission Ionization Detector for Gas Chromatography," Acta Chem. Scand., 15, 708-709 (1961); CA 56, 1307e

7113. Waight, E. S., and Walker, P., "The Effect of Quinones on the Gamma-Irradiation of Cyclohexane," J. Chem. Soc., 1960, 2225-2230; CA 54, 20956b

7114. Wakasugi, B., and Okhuma, Y., "Clinical Application," Kagaku no Ryoiki Zokan, No. 44, 205-222 (1961); CA 56, 3b

7115. Waksmundzki, A., "Chromatographic Analysis of Vapors and Gases," Wiadomosci Chem., 11, 617-633 (1957); CA 52, 9856g

7116. Waksmundzki, A., Soczewinski, E., and Suprynowicz, Z., "The Relation Between the Composition of the Mixed Stationary Phase and the Retention Time in Gas-Liquid Partition Chromatography," Collection Czechoslov. Chem. Commun., 27, 2001-2006 (1962); CA 58, 11977e

7117. Waksmundzki, A., Suprynowicz, Z., and Manko, R., "Zircon Concentrate as Supporting Material in Gas-Liquid Chromatography," Chem. Anal. (Warsaw), 7, 1051-1058 (1962); CA 59, 6c

7118. Walborsky, H. M., Baum, M. E., and Youssef, A. A., "Acetolysis of Bicyclo[2.2.2]octyl-2 p-Bromo-benzenesulfonate and the Absolute Configurations of Bicyclo[2.2.2]octanol-2 and cis- and trans-Bicyclo[3.2.1]octanol-2," J. Am. Chem. Soc., 83, 988-993 (1961); CA 55, 24596b

7119. Waldron, J. D., "The Mass Spectrometer in Chemical Analysis," Metropolitan Vickers Gaz., 28, No. 456, 165-175 (1957); CA 51, 16192a

7120. Walford, R. W., and Attaway, J. A., "The Application of Gas Chromatography to the Analysis of Flavor Components of Citrus Juices," 3rd Intern. Gas Chromatography Symp., ISA, East Lansing, Mich., June 1961; ISA Proc., 3, 289-295 (1961)

7121. Waligora, B., and Bylo, Z., "Potentiometric Adsorption Analysis as Applied to Determination of Adsorption Isotherms of Chromatographic Adsorbents," Zeszyty Nauk Uniw. Jagiel., Ser. Nauk Mat.-Przyrod., Mat., Fiz., Chem., No. 4, 93-126 (1958); CA 52, 17893d

7122. Walker, J. M., and Nightingale, C. F., "Simultaneous Carbon-Hydrogen-Nitrogen Determination by Gas Chromatography," 142nd Natl. ACS Mtg., Atlantic City, N. J., September 1962, Program Abstr., pp. 10B-11B

7123. Walker, J. Q., "Better Efficiency, Greater Detection Sensitivity, Speed, in Analysis," Oil Gas J., 61, No. 14, 78-80 (1963); CA 59, 354b

7124. Walker, J. Q., and Ahlberg, D. L., "Quantitative Analysis of Aromatic Hydrocarbons by Gas Chromatography," 142nd Natl. ACS Mtg., Atlantic City, N. J., September 1962, Program Abstr., p. 11B

7125. Walker, J. Q., and Ahlberg, D. L., "The Analysis of Naphthalenes by Capillary Gas Chromatography," 18th Southwest Regional Mtg., ACS, Dallas, Texas, December 1962

7126. Walker, R. E., and Westenberg, A. A., "Precision Thermal-Conductivity Gas Analyzer Using Thermistors," Rev. Sci. Instr., 28, 789-792 (1957); CA 52, 7781h

7127. Wall, R. F., " Process Infrared Analyzers - 1957," Ind. Eng. Chem., 49, No. 3, 69A-70A (1957)

7128. Wall, R. F., "Instrumentation Trends in 1957," Ind. Eng. Chem., 50, No. 1, 125A-126A (1958)

7129. Wall, R. F., "Process Control by Gas Chromatography," Ind. Eng. Chem., 50, No. 3, 51A-52A (1958)

7130. Wall, R. F., "Instrumentation for Process Development," Ind. Eng. Chem., 50, No. 4, 65A-66A (1958)

7131. Wall, R. F., "Process Analysis by Thermal Conductivity," Ind. Eng. Chem., 50, No. 5, 69A-70A (1958)

7132. Wall, R. F., "Building Your Own Analyzer," Ind. Eng. Chem., 50, No. 11, 65A-66A (1958)

7133. Wall, R. F., "Instrumentation Trends in 1958," Ind. Eng. Chem., 51, No. 1, 89A-90A (1959)

7134. Wall, R. F., "A Year's Progress in Process Control by Gas Chromatography, " Ind. Eng. Chem., 51, No. 4, 73A-74A (1959)

7135. Wall, R. F., "New Developments in Process Analysis Instrumentation," Ind. Eng. Chem., 51, No. 8, 53A-54A (1959)

7136. Wall, R. F., "Process Analyzer Response Time," Ind. Eng. Chem., 51, No. 9, 75A-76A (1959)

7137. Wall, R. F., "Trends in Instrumentation for 1960," Ind. Eng. Chem., 51, No. 12, 71A-72A (1959)

7138. Wall, R. F., Baker, W. J., Zinn, T. L., and Combs, J. F., "Process Control by Gas Chromatography in the Chemical Industry," Ann. N. Y. Acad. Sci., 72, Art. 13, 739-750 (1959);CA 53, 15423i

7139. Waller, G. R., Matlock, R. M., and Horn, M. R., "Analysis of Peanut Butter Oil by Gas-Liquid Chromatography," 16th Southwest Regional Mtg., ACS, Oklahoma City, Okla., December 1960

7140. Walling, C., and Bollyky, L., "Products of the Base-Catalyzed Reaction of Benzophenone, with Dimethyl Sulfoxide," J. Org. Chem., 28, 256-257 (1963); CA 58, 10112h

7141. Walling, C., and Helmreich, W., "Reactivity and Reversibility in the Reaction of Thiyl Radicals with Olefins," J. Am. Chem. Soc., 81, 1144-1148 (1959); CA 53, 17929h

7142. Walling, C., and Jacknow, B. B., "Positive Halogen Compounds. I. The Radical Chain Halogenation of Hydrocarbons by tert-Butyl Hypochlorite," J. Am. Chem. Soc., 82, 6108-6112 (1960); CA 55, 12268i

7143. Walling, C., and Jacknow, B. B., "Positive Halogen Compounds. II. Radical Chlorination of Substituted Hydrocarbons with tert-Butyl Hypochlorite," J. Am. Chem. Soc., 82, 6113-6115 (1960); CA 55, 12269c

7144. Walling, C., and Metzger, G., "Organic Reactions Under High Pressure. V. The Decomposition of Di-t-Butyl Peroxide," J. Am. Chem. Soc., 81, 5365-5369 (1959); CA 54, 12975i

7145. Walling, C., and Rabinowitz, R., "The Reaction of Trialkyl Phosphites with Thiyl and Alkoxy Radicals," J. Am. Chem. Soc., 81, 1243-1249 (1959); CA 53, 16932f

7146. Walling, C., and Thaler, W., " Positive Halogen Compounds. III. Allyl Chlorination with tert-Butyl Hypochlorite. The Stereochemistry of Allylic Radicals," J. Am. Chem. Soc., 83, 3877-3884 (1961); CA 56, 12718f

7147. Walsh, J. T., and Merritt, Jr., C., "Qualitative Functional Group Analysis of Gas Chromatographic

Effluents," Anal. Chem., 32, 1378-1381 (1960); CA 55, 5049g

7148. Waltz, R. H., Wisniewski, J., and Spencer, S., "Applications of a New Pyrolysis Unit to the Analysis of Non-Volatile Materials by Gas Chromatography," 11th Detroit Anachem Conf., Detroit, Mich., October 1963, Program Abstr., p. 28

7149. Warren, G. W., Haskin, J. F., Kourey, R. E., and Yarborough, V. A., "Gas Chromatography Analysis of the Reaction Products from the Hydroformylation of Isobutene," Anal. Chem., 31, 1624-1626 (1959); CA 54, 3045c

7150. Warren, G. W., Lambdin, W. J., Haskin, J. F., and Yarborough, V. A., "Gas Chromatographic Analyses of Products from Aldol Condensations," Anal. Chem., 31, 1016-1019 (1959); CA 53, 14817a

7151. Warren, G. W., Priestley, Jr., L. J., Haskin, J. F., and Yarborough, V. A., "Gas Chromatographic Analysis of Various Mixtures of Compounds Containing Chlorine," Anal. Chem., 31, 1013-1016 (1959); CA 53, 14828e

7152. Warren, G. W., Warren, R. R., and Yarborough, V. A., "Gas Chromatography: Guide to Better Extraction Processes," Ind. Eng. Chem., 51, 1475-1476 (1959); CA 54, 5209i; 135th Natl. ACS Mtg., Boston, Mass., April 1959, Program Abstr., p. 4L

7153. Warrington, Jr., H. P., "Determination of Homolog Distribution of Mixed Alkylbenzyldimethyl-ammonium Chlorides," Anal. Chem., 33, 1898-1900 (1961); CA 56, 11707a

7154. Waszeciak, P., and Nadeau, H. G., "The Use of Techniques Bases on Vapor Liquid Equilibria to Separate Molecular Species in the Range from 200 to 2500," 12th Pittsburgh Conf. on Anal. Chem. & Appl. Spectroscopy, Pittsburgh, Pa., February-March 1961, Program Abstr., p. 47

7155. Waterhouse, D. F., Forss, D. A., and Hackman, R. H., "Characteristic Odor Components of the Scent of Stink Bugs," J. Insect Physiol., 6, 113-121 (1961); CA 57, 3879e

7156. Waters, J. H., "Mechanism of Aromatic Substitution by Free Radicals," Univ. Microfilms (Ann Arbor, Mich.), L. C. Card No. Mic 61-2277, 141 pp.; Dissertation Abstr., 22, 432-433 (1961); CA 55, 25840i

7157. Watson, A. A., "The Reduction of S-Alkenyl N,N-Dimethyldithiocarbamates and -Thiocarbamates and Tetramethylthiourea with Lithium Aluminum Hydride," J. Chem. Soc., 1962, 4717-4719; CA 58, 8884g

7158. Watson, W. C., "Morphology and Lipid Composition of the Erythrocytes in Normal and Essential Fatty Acid-Deficient Rats," Brit. J. Haematol., 9, 32-38 (1963); CA 59, 940c

7159. Watson, W. C., Gordon, Jr., R. S., and Karmen, A., "Metabolism of Ricinoleic Acid in Rat and Man," 45th Annual Mtg., Federation of Am. Soc. for Exptl. Biol., Atlantic City, N. J., April 1961

7160. Watts, J. O., "Electron Capture Gas Chromatography," 12th Annual Symp. on Recent Developments in Research Methods and Instrumentation, National Institutes of Health, Bethesda, Md., October 1962

7161. Watts, J. O., and Klein, A. K., "Determination of Chlorinated Pesticide Residues by Electron-Capture Gas Chromatography," J. Assoc. Offic. Agr. Chemists, 45, 102-108 (1962); CA 56, 10635e

7162. Wawzonek, S., and Culbertson, T. P., "Studies on the Cyclization of n-Chlorodialkylamines," J. Am. Chem. Soc., 82, 441-443 (1960); CA 54, 8808g

7163. Way, R. M., "Gas Chromatographic Analysis of Oil of Lemon," Facts & Methods (F&M Scientific Corp.), 4, No. 1, 1-4 (1963)

7164. Weaver, E. R., "Gas Analysis by Methods Depending on Thermal Conductivity," in "Physical Methods in Chemical Analysis," Vol. II, edited by W. G. Berl, Academic Press, Inc., New York, 1951, pp. 387-439

7165. Weaver, N., and Law, J. H., "Heterogeneity of Fatty Acids from Royal Jelly," Nature, 188, 938-939 (1960); CA 55, 12670h

7166. Webb, A. D., and Kepner, R. E., "Fusel Oil Analysis by Means of Gas-Liquid Partition Chromatography," Am. J. Enol. Viticult., 12, 51-59 (1961); CA 55, 18004g

7167. Webb, A. D., and Kepner, R. E., "The Aroma of Flor Sherry," Am. J. Enol. Viticult., 13, 1-14 (1962); CA 57, 3873i

7168. Webb, G. A., and Black, G. S., "Determining Hydrogen in Gases with a Thermal Conductivity Apparatus," Ind. Eng. Chem., Anal. Ed., 16, 719-720 (1944); CA 39, 474[1]

7169. Webb, J. P. W., Allison, A. C., and James, A. T., "Lipid Synthesis in Fowl Blood," Biochem. J., 74, No. 3, 30P (1960)

7170. Webb, J. P. W., Allison, A. C., and James, A. T., "In Vitro Lipid Synthesis in Fowl Blood," Biochim. et Biophys. Acta, 43, 89-94 (1960); CA 55, 3688b

7171. Webb, R. D., "The Chromatograph in the Automatic Control Loop," 8th Natl. Analysis Instrumentation Symp., ISA, Charleston, W. Va., April-May 1962

7172. Webb, T. P., "The Application of Gas Chromatography to Light Hydrocarbon Analysis of Crude Oils," 7th Detroit Anachem Conf., Detroit, Mich., October 1959

7173. Weber, F., "Substitution of a Gas-Chromatographic Method for Podbielniak Total Analysis," Erdol u. Kohle, 11, 339 (1958)

7174. Webster, J. L., and Marchello, J. M., "Determination of Carrier Phase Composition in Gas Chromatography Columns," 16th Southwest Regional Mtg., ACS, Oklahoma City, Okla., December 1960

7175. Weenink, R. O., "Acetone-Soluble Lipids of Grasses and Other Forage Plants. I. Galactolipides of Red Clover (Trifolium Pratense) Leaves," J. Sci. Food Agr., 12, 34-38 (1961); CA 55, 18896f

7176. Wehe, A. H., and Mcketta, J. J., "Method for Determining Total Hydrocarbons Dissolved in Water," Anal. Chem., 33, 291-293 (1961); CA 55, 10758b

7177. Wehrli, A., and Kovats, E., "Gas Chromatographic Characterization of Organic Compounds. III. Calculation of Retention Indexes of Aliphatic, Alicyclic, and Aromatic Compounds," Helv. Chim. Acta, 42, 2709-2736 (1959); CA 54, 12722f

7178. Wehrli, A., and Kovats, E., "Preparative Gas Chromatography. A Centrifugal Cooling Trap for Separating Ice- or Cloud-forming Substances from Slow-Flowing Gases," J. Chromatog., 3, 313-316 (1960); CA 54, 17976d

7179. Wei, J., Prater, C. D., and Emery, A. R., "Analysis of Complex Hydrocarbon Mixtures. Maximum Utilization of Data from Combined Analytical Methods," 136th Natl. ACS Mtg., Atlantic City, N. J., September 1959, Program Abstr., p. 5R; Preprints, Div. Petrol. Chem., 4, No. 3, 139-154

7180. Wei-Lo, Y., "Analysis of Dissolved Hydrocarbon Gases in Gasoline by Combined Adsorption and Gas Liquid Partition Chromatography," Jan Liao Hsueh Pao, 2, 352-354 (1957); CA 52, 8519d

7181. Weingarten, H., "Electronic Effects in the Gomberg Reaction," J. Org. Chem., 25, 1066-1067 (1960); CA 54, 20977h

7182. Weingarten, H., "Steric Effects in the Gomberg Reaction," J. Org. Chem., 26, 730-733 (1961); CA 55, 22229f

7183. Weingarten, H., Ross, W. D., Schlater, J. M., and Wheeler, Jr., G., "Gas Chromatographic Analysis of Chlorinated Biphenyls," Anal. Chim. Acta, 26, 391-394 (1962); CA 57, 2829i

7184. Weinig, E., and Lautenbach, L., "Gas Chromatography as a New Technique in Forensic Toxicology and Criminalistics," Arch. Kriminol., 122, 11-17 (1958); CA 53, 963e

7185. Weinig, E., Lautenbach, L., and Geldmacher-Mallinckrodt, M., "Gas Chromatography for the Detection of Meta-Systox," Deut. Z. Ges. Gerichtl. Med., 51 565-569 (1961); CA 56, 6414h

7186. Weininger, J. L., "The Reaction of Active Nitrogen with Liquid Siloxane Heptamer, D_7," J. Am. Chem. Soc., 83, 3388-3390 (1961); CA 55, 27023a

7187. Weinstein, A., "Fraction Cutter for Gas Chromatography," Anal. Chem., 29, 1899-1900 (1957); CA 52, 3417g

7188. Weinstein, A., "Anomalous Calibration Curves in Gas Chromatography," Chem. & Ind. (London), 1959, 1347-1348; CA 54, 4238h

7189. Weinstein, A., "Analytical Accuracy in Gas Chromatography Using Thermal Conductivity Detectors," Anal. Chem., 32, 288-290 (1960); CA 54, 8419b

7190. Weinstein, A., "Adsorption Phenomena and Their Effects on Analytical Accuracy in Gas Chromatography," Anal. Chem., 33, 18-22 (1961); CA 55, 11173i

7191. Weinstein, B., and Fenselau, A. H., "Nef Reaction of 1,2,3,4,5,6,7,8,9,10,10a, 8a-Dodecahydro-9-nitrophenanthrene," J. Org. Chem., 27, 4094-4096 (1962); CA 58, 7879e

7192. Weisner, L., and Schmidt-Kuster, W. J., "Performance Characteristics of an Integrating Counter-Detector for Gas Chromatography," Intern. Symposium on Microchemical Techniques, Pennsylvania State Univ., University Park, Pa., August 1961

7193. Weiss, A. R., "Gas Chromatographic Determination of Impurities in Methyl Acrylate," 6th Annual Exhibit and Symp. on Recent Developments in Research Methods and Instrumentation, National Institutes of Health, Bethesda, Md., May 1956; Abstr., Anal. Chem., 28, 1058 (1956)

7194. Weiss, H., and Kreyenbuhl, A., "Quantitative Analysis of Phenols by Gas Chromatography," Bull. soc. chim. France, 1961, 603-606; CA 55, 16283g

7195. Weiss, J., Collins, C., Carciello, N., and Sucher, J., "Radiolytic and Pyrolytic Stability of Some Polyaromatic Compounds," 144th Natl. ACS Mtg., Los Angeles, Calif., March-April 1963, Program Abstr., p. 5J

7196. Weiss, M. D., "Analysis Instruments - Key to Process Controls," Ind. Research, 2, No. 3, 56-58 60-62 (1960)

7197. Weitkamp, A. W., "Distillation," J. Am. Oil Chemists' Soc., 32, 640-646 (1955); CA 50, 1338h

7198. Weitkamp, A. W., "Analytical Separations in the Study of Limonene Sulfides," Perfumery Essent. Oil Record, 49, 803-807 (1958); CA 53, 12590c

7199. Weitkamp, A. W., "Analytical Separations in the Study of Limonene Sulfides," in "Flavor Research and Food Acceptance," Arthur D. Little, Inc., Cambridge, Mass.; Reinhold Publishing Corp., New York, 1958, pp. 331-343; CA 53, 10276b

7200. Weitkamp, A. W., "The Action of Raney Nickel on Limonene Sulfides and Sulfones," J. Am. Chem. Soc., 81, 3434-3437 (1959); CA 53, 22057c

7201. Wellington, C. A., and Walters, W. D., "The Vapor Phase Decomposition of 2,5-Dihydrofuran," J. Am. Chem. Soc., 83, 4888-4891 (1961); CA 56, 14191i

7202. Wellman, W. E., "High Temperature Oxidation of Hydrocarbons in the Chemical Shock Tube - Synthetic Analogs of Actinomycin D," Univ. Microfilms (Ann Arbor, Mich.), L. C. Card No. Mic 60-

<u>4142</u>, 158 pp.; Dissertation Abstr., <u>21</u>, 767 (1960); CA <u>55</u>, 8414c

7203. Wells, W. W., and Makita, M., "The Quantitative Analysis of Fecal Neutral Sterols by Gas-Liquid Chromatography," Anal. Biochem., <u>4</u>, 204-212 (1962); CA <u>58</u>, 6046e

7204. Welti, D., and Wilkens, T., "Effect of an Argon-Nitrogen Carrier Gas Mixture on the Sensitivity of a Gas Chromatographic Ionization Detector," J. Chromatog., <u>3</u>, 589-591 (1960); CA <u>55</u>, 3290d

7205. Wencke, K., "Chromatographic Analysis of Mixtures of Readily Volatile Gases," Chem. Tech. (Berlin), <u>8</u>, 728-730 (1956); CA <u>51</u>, 11930f

7206. Wencke, K., "Chromatographic Method for Quantitative Gas Analysis," Chem. Tech. (Berlin), <u>9</u>, 404-406 (1957); CA <u>52</u>, 155d

7207. Wendlandt, W. W., "A New Apparatus for Simultaneous Differential Thermal Analysis and Gas Evolution Analysis," Anal. Chim. Acta, <u>27</u>, 309-314 (1962); CA <u>58</u>, 4f

7208. Werner, A. E., "A Gas Density Balance for Student Use," J. Chem. Educ., <u>33</u>, 393-395 (1956); CA <u>51</u>, 4784g

7209. Werner, Jr., E. R., "Nucleophilic Displacement of Isobutylene Dihalides," Univ. Microfilms (Ann Arbor, Mich.), L. C. Card No. <u>Mic 60-3706</u>, 161 pp., Dissertation Abstr., <u>21</u>, 767-768 (1960); CA <u>55</u>, 8277g

7210. Wesselman, H. J., "Quantitative Determination of Ethanol in Pharmaceutical Products by Gas Chromatography," 106th Conv., Am. Pharmaceutical Assoc., Cincinnati, Ohio, August 1959; J. Am. Pharm. Assoc., Sci. Ed., <u>49</u>, 320-322 (1960); CA <u>54</u>, 14583d

7211. Wesselman. H. J.. and Koons, J. R., "Insecticide Analysis. The Quantitative Determination of Heptachlor in Pesticide Formulations by Gas Chromatography," J. Agr. Food Chem., <u>11</u>, 173-174 (1963); CA <u>58</u>, 11907c

7212. Wessely, F., Swoboda, J., and Schmidt, G., "Action of Thiols and Sulfinic Acids on Quinol Acetates. II.," Monatsh. Chem., <u>91</u>, 57-78 (1960); CA <u>54</u>, 18409d

7213. West, D. L., "Determination of Deuterium, Oxygen, and Nitrogen in Helium by Gas Chromatography," U. S. At. Energy Comm., <u>TID-7568</u>, Pt. 1, 101-104 (1958); CA <u>54</u>, 9604f

7214. West, D. L., "Laboratory Evaluation of an Automatic Gas Chromatograph," U. S. At. Energy Comm., <u>DP-582</u>, 15 pp. (1961); CA <u>55</u>, 20531c

7215. West, D. L., "Composition and Stability of Ultrasene," U. S. At. Energy Comm., <u>DP-587</u>, 14 pp. (1961); CA <u>55</u>, 26735e

7216. West, P. W., Sen, B., and Gibson, N. A., "Sampling and Gas-Liquid Chromatographic Analysis of Organic Air Pollutants," 132nd Natl. ACS Mtg., New York, N. Y., September 1957, Program Abstr., p. 32B; Preprints, Div. Petrol. Chem., <u>2</u>, No. 4, D51-D65

7217. West, P. W., Sen. B., and Gibson, N. A., "Gas-Liquid Chromatographic Analysis Applied to Air Pollution," Anal. Chem., <u>30</u>, 1390-1397 (1958); CA <u>52</u>, 18971h

7218. West, P. W., Sen, B., and Sant, B. R., "Determination of Total Gaseous Pollutants in Atmosphere," Anal. Chem., <u>31</u>, 399-401 (1959); CA <u>53</u>, 11730c

7219. West, P. W., Sen, B., Sant, B. R., Mallik, K. L., and Sen Gupta, J. G., "A Catalog of Retention Times of a Number of Organic Compounds," J. Chromatog., <u>6</u>, 220-235 (1961); CA <u>56</u>, 5077d

7220. West, T. S., "Analysis of Nuclear Reactor Coolant Gases," Chem. Age (London), <u>81</u>, 15-16 (1959)

7221. West, T. S., "Time-of-Flight Mass Spectrometry for Vapor-Phase Chromatograms," Chem. Age (London), 81, 733-734 (1959)

7222. West, W. W., "Poly(Phenyl Ethers) as Liquid Phases for the Gas Chromatographic Separation of Complex Hydrocarbon Mixtures," 145th Natl. ACS Mtg., New York, N. Y., September 1963, Program Abstr., p. 12B

7223. Westaway, H., and Williams, J. F., "Composition of the Dipentene Fraction of $C_{10}H_{16}$ Terpenes," J. Appl. Chem. (London), <u>9</u>, 440-444 (1959); CA <u>54</u>, 2400b

7224. Westermark, T., "Electrical Gas Discharge - Competitor to Radiation ?," Nucleonics, <u>19</u>, No. 5, 90, 92 (1961)

7225. Westgate, M. W., "Use of the Gas-Phase Chromatograph in Solvent Problems. I.," 36th Annual Mtg., Federation of Paint and Varnish Production Clubs, Cleveland, Ohio, October 1958

7226. Westlake, W. E., "Pesticides," Anal. Chem., <u>33</u>, 88R-91R (1961); CA <u>55</u>, 12137g

7227. Westlake, W. E., "Pesticides," Anal. Chem., <u>35</u>, No. 5, 105R-110R (1963); CA <u>59</u>, 1039c

7228. Westlake, W. E., Corley, C., and Murphy, R. T., with Barthel, W. F., Bryant, H., and Schutzmann, R. L., "Chemical Residues in the Milk of Cows Grazed on Chlordan-Treated Pasture," J. Agr. Food Chem., <u>11</u>, 244-246 (1963); CA <u>59</u>, 1029c

7229. Wet, W. J. De, Haarhoff, P. C., and Pretorius, V., "Gas-Liquid Partition Chromatography. The Effect of Column Length on Efficiency," J. South African Chem. Inst., <u>13</u>, 13-18 (1960); CA <u>54,</u> 23599i

7230. Wet, W. J. de, Haarhoff, P. C., and Pretorius, V., "Granular Support in Gas-Liquid Chromatography. Effect of the Particle Size of the Granules and of the Thickness of the Liquid Film on the Column Efficiency," J. South African Chem. Inst., <u>13</u>, No. 1, 19-26 (1960); CA <u>54</u>, 21930e

7231. Wet, W. J. de, and Pretorius, V., "Some Factors Influencing the Efficiency of Gas-Liquid Partition

Chromatography Column," Anal. Chem., 30, 325-329 (1958); CA 52, 10793f

7232. Wet, W. J. de, and Pretorius, V., "Theory of Gas-Liquid Chromatography," J. South African Ind. Chemist, 13, 105 (1959)

7233. Wet, W. J. de, and Pretorius, V., "Factors Affecting the Use of Gas-Liquid Chromatography for the Separation of Large Samples. Samples Inlet System, Distribution Coefficient of Solute, and Amount of Liquid in Stationary Phase," Anal. Chem., 32, 169-174 (1960); CA 54, 7405i

7234. Wet, W. J. de, and Pretorius, V., "Factors Affecting the Use of Gas-Liquid Chromatography for the Separation of Large Sample. The Column Dimensions," Anal. Chem. 32, 1396-1399 (1960); CA 55, 2347e

7235. Wett, T. W., "Raise Sample Analysis Rate, Improve Accuracy," Chem. Processing 22, 135-136 (1959)

7236. Wetzel, W. H., and Kosters, B., "Qualitative and Quantitative Analysis of Products from Pyrolysis," Analyst, 86, 480-483 (1961); CA 56, 6632b

7237. Weurman, C., "Gas-Liquid Chromatographic Studies on the Enxymic Formation of Volatile Compounds in Raspberries," Food Technol., 15, 531-536 (1961); CA 56, 6380f

7238. Weurman, C., and Dhont, J., "Unexpected Formation of Volatile Compounds on Gas-Liquid Chromatography Columns," Nature, 184, 1480-1481 (1959); CA 54, 10842b

7239. Weyermuller, G., and Wright, N., "Speedy Mass Spectrometer," Chem. Processing, 22, No. 6, 107-108 (1959)

7240. Weygand, F., Kolb, B., and Kirchner, P., "Gas-Chromatographic Separation of the Methyl Esters of N-Trifluoroacetyl Dipeptides. II.," Z. Anal. Chem., 181, 396-399 (1961); CA 56, 7e

7241. Weygand, F., Kolb, B., Prox, A., Tilak, M. A., and Tomida, I., "N-Trifluoroacetylamino Acids. XIX. Gas Chromatographic Separation of N-Trifluoroacetyldipeptide Methyl Esters," Z. Physiol. Chem., 322, 38-51 (1960); CA 55, 8177b

7242. Weygand, F., Prox, A., Schmidhammer, L., and König, W., "Gas Chromatographic Investigation of Racemization During Peptide Synthesis," Angew. Chem. (Internl. Ed.), 2, 183-188 (1963)

7243. Whalley, C., "Changing Frontiers in the Analytical Chemistry of Paint Materials," J. Oil Colour Chemists' Assoc., 39, 193-208 (1956); CA 51, 7736e

7244. Whalley, E., "The Thermal Conductivity of Associating Gases," Disc. Faraday Soc., No. 22, 54-63 (1956); CA 51, 12583d

7245. Whalley, G. R., "The Chromatographic Separation of Fatty Acids and Their Derivatives," Soap, Perfum. Cosmet., 29, 783-789 (1956); CA 50, 13475h

7246. Whatley, T. A., "The Thermal Decomposition of Azoisopropane," Univ. Microfilms (Ann Arbor, Mich.), L. C. Card No. Mic 61-2089, 168 pp.; Dissertation Abstr., 21, 3652 (1961); CA 55, 24539d

7247. Whatmough, P., "Determination of Hydrogen in Mine Airs Using a Gas Chromatographic Technique," Nature, 179, 911-912 (1957); CA 51, 10963h

7248. Whatmough, P., "Gas Chromatograph as a Methanometer in the Coal Industry," Nature, 182, 863-864 (1958); CA 53, 3795h

7249. Wheatley, V. R., "Some Applications of Gas Chromatography," Am. Perfumer Cosmet., 78, No. 1, 27-30, 32-33 (1963); CA 58, 8848e; Addendum, ibid., No. 3, 33-34 (1963)

7250. Wheatley, V. R., Chow, D. C., and Keenan, Jr., F. C., "Lipids of Dog Skin. II. Lipid Metabolism of Perfused Surviving Dog Skin," J. Invest. Dermatol., 36, 237-239 (1961); CA 56, 798i

7251. Wheatley, V. R., and Farber, E. M., "Studies on the Chemical Composition of Psoriatic Scales," J. Invest. Dermatol., 36, 199-211 (1961); CA 55, 22556f

7252. Wheatley, V. R., and James, A. T., "Studies of Sebum. VII. The Composition of the Sebum of Some Common Rodents," Biochem. J., 65, 36-42 (1957); CA 51, 5231e

7253. Wheatley, V. R., and Sher, D. W., "Lipids of Dog Skin. I. The Chemical Composition of Dog Skin Lipids," J. Invest. Dermatol., 36, 169-170 (1961); CA 56, 798h

7254. Wheeldon, L. W., "Composition of Cabbage Leaf Phospholipids," J. Lipid Research, 1, 439-445 (1960); CA 55, 8549h

7255. Wheeler, D. J., Darby, P. W., and Kemball, C., "The Desorption of Alcohols from Metal Oxides. I. Ethanol and 2-Propanol," J. Chem. Soc., 1960, 332-340; CA 54, 10448g

7256. Wheeler, D. J., and Kemball, C., "The Desorption of Alcohols from Metal Oxides. II. The Desorption of tert-Butyl Alcohol and 1-Butanol and the Displacement of Ethanol from Anatase by 2-Propanol," J. Chem. Soc., 1960, 1840-1847; CA 54, 14866c

7257. Wheeler, O. H., Nieto, M. A., Storer, C. Bello de, Antunano, N. C., and Medina, V. J., "Analysis of Some Aldehydes and Ketones in Essential Oils by Gas Chromatography," Rev. Col. Quim. Puerto Rico, 18, 31-32 (1960); CA 57, 3573g

7258. Wheelock, T. D., Robison, C. H., and Reisch, J. C., "Permeability of Gas-Liquid Chromatography Columns," 3rd. Intern. Symp. on Gas Chromatography Columns," 3rd. Intern. Symp. on Gas Chromatography, ISA, East Lansing, Mich., June 1961

7259. Wherry, T. C., "Automatic Process-Stream Analyzers," Oil Gas J., 54, 125-126, 129 (1956); CA 51, 4772i

7260. Wherry, T. C., "Chromatography for Process Control," Chem. Eng. Progr., 56, No. 9, 49-57 (1960)

7261. Wherry, T. C., "Process Control in the Electronic Era," ISA Journal, 7, No. 3, 48-51 (1960)

7262. Whisman, M. L., "Preparation of Tritium-Labeled 1-Hexene and 1-Octene," Anal. Chem., 33, 1284-1285 (1961); CA 56, 8537g

7263. Whisman, M. L., and Eccleston, B. H., "Purification of Radioactive Organic Compounds: A Non-contaminating Gas Liquid Chromatographic System," Anal. Chem., 35, 1333-1334 (1963)

7264. Whisman, M. L., Schwartz, F. G., and Eccleston, B. H., "Susceptibility of Organic Compounds to Tritium Exchange Labeling," U. S. Bur. Mines, Rept. Invest. No. 5717, 17 pp. (1961); CA 55, 13809g

7265. White, C. S., and Lovelace, W. R., "Spectrometric Methods: Gas Analysis by Quantitative Emission Spectroscopy," AGARDograph, No. 25, 253-268 (1958); CA 53, 1980b

7266. White, D., "Use of Organic-Montmorillonite Compounds in Gas Chromatography," Nature, 179, 1075-1076 (1957); CA 51, 12598f

7267. White, D., and Cowan, C. T., "The Sorption Properties of Dimethyl-dioctadecyl Ammonium Bentonite by Gas Chromatography," Trans. Faraday Soc., 54, 557-561 (1958); CA 52, 19360g

7268. White, D., and Cowan, C. T., "Symmetrical Elution Curves in Adsorption Chromatography," in "Gas Chromatography 1958," edited by D. H. Desty, Academic Press, Inc., New York, 1958, pp. 116-124

7269. White, E. H., and Aufdermarsh, Jr., C. A., "N-Nitrosoamides. IV. N-Nitrosoamides of Primary Carbinamines," J. Am. Chem. Soc., 83, 1174-1178 (1961); CA 55, 18676b

7270. White, E. H., and Grisley, Jr., D. W., "The Preparation and Decomposition of Certain N-Nitroamides and N-Nitrocarbamates," J. Am. Chem. Soc., 83, 1191-1196 (1961); CA 55, 14358c

7271. White, J. U., and Alpert, N. L., with Ward, W. M., and Gallaway, W. S., "Microgas Cell for Infrared Spectroscopy," Anal. Chem., 31, 1267-1270 (1959); CA 54, 10b

7272. Whitham, B. T., "A Large Scale Analytical Gas-Liquid Partition Chromatographic Unit," in "Vapour-Phase Chromatography," edited by D. H. Desty, Academic Press, Inc., New York, 1957, pp. 194-212

7273. Whitham, B. T., "Application of Gas-Liquid Partition Chromatography to Solvent Analysis," in "Vapour-Phase Chromatography," edited by D. H. Desty, Academic Press, Inc., New York, 1957, pp. 395-412

7274. Whitham, B. T., "Use of Molecular Sieves in Gas Chromatography for the Determination of the Normal Paraffins in Petroleum Fractions," Nature, 182, 391-392 (1958); CA 53, 4707h

7275. Whitham, B. T., "Recent Applications of Gas Chromatographic Techniques in the Petroleum Industry," Compt. rend. congr. intern. chim., 31^e, Liege, (Pub. as Ind. chim. belge, Suppl.), 1, 563-570 (Pub. 1959); CA 54, 871a

7276. Whitney, W. K., "Toxicities, Diffusion, and Interactions of Grain Fumigants Measured by Bioassay and Gas Chromatography," Univ. Microfilms (Ann Arbor, Mich.), Order No. 62-4492, 141 pp.; Dissertation Abstr., 23, 1463-1464 (1962); CA 58, 2793c

7277. Whitney, W. K., "Inhibitors of Carbon Disulfide Decomposition During Gas Chromatography of Fumigant Vapors," J. Agr. Food Chem., 10, 470-476 (1962); CA 58, 2793e

7278. Whitson, R. L., and Fourroux, M. M., "Gas Chromatographic Control for a Sulfur Recovery Plant," ISA Journal, 7, No. 3, 40-43 (1960); 14th Annual Instrument Automation Conf. & Exhibit, ISA, Chicago, Ill., September 1959

7279. Whittier, M. B., "Quantitative Analysis of Free Sterols at Submicrogram Levels," Facts & Methods (F&M Scientific Corp.), 4, No. 1, 4-5 (1963)

7280. Whittier, M. B., Mikkelsen, L., and Armstrong, N., "Factors Affecting the Efficiency of Gas Chromatography of Steroids," 14th Annual Mid-America Spectroscopy Symp., Chicago, Ill., May 1963

7281. Whittier, M. B., Mikkelsen, L., and Spencer, S. F., "Factors Affecting the Efficiency of Gas Chromatographic Analysis of Steroids," 18th Southwest Regional Mtg., ACS, Dallas, Texas, December 1962

7282. Wiberg, K. B., and Bartley, W. J., "Cyclopropene. V. Some Reactions of Cyclopropene," J. Am. Chem. Soc., 82, 6375-6380 (1960); CA 55, 7310d

7283. Wiberg, K. B., and Foster, G., "Stereochemistry of the Chromic Acid Oxidation of Tertiary Hydrogens," J. Am. Chem. Soc., 83, 423-429 (1961); CA 56, 5811b

7284. Wiberg, K. B., and Richardson, W. H., "Chromic Acid Oxidation of Aromatic Aldehydes. Observations Concerning the Oxidation by the Chromium Species of Intermediate Valence," J. Am. Chem. Soc., 84, 2800-2807 (1962); CA 58, 1324c

7285. Wichterle, I., and Hala, E., "Semimicrodetermination of Vapor-Liquid Equilibrium," Ind. Eng. Chem. Fundamentals, 2, 155-157 (1963); CA 58, 10995e

7286. Wick, E. L., Issenberg, P., and Goldblith, S. A., "An Investigation of the Volatile Components of Ripe Banana," 140th Natl. ACS Mtg., Chicago, Ill., September 1961, Program Abstr., p. 22A

7287. Wick, E. L., Yamanishi, T., Wertheimer, L. C., Hoff, J. E., Proctor, B. E., and Goldblith, S. A., "An Investigation of Some Volatile Components of Irradiated Beef," J. Agr. Food Chem., 9, 289-

293 (1961); CA 56, 6427g; 137th Natl. ACS Mtg., Cleveland, Ohio, April 1960, Program Abstr., p. 13A

7288. Wickberg, B., "Separation of Sesquiterpenes by Partition Chromatography," J. Org. Chem., 27, 4652-4654 (1962); CA 58, 9149a

7289. Wicke, E., "Adsorption and Desorption Processes from a Gas Streaming Over a Bed of Granular Adsorbent," Kolloid-Z., 93, 129-157 (1940); CA 35, 3872[8]

7290. Wicke, E., "The Separation of Gas Mixtures by Flow Through Adsorbents," Angew. Chem., B19, 15-21 (1947); CA 41, 4349e

7291. Widmark, K., and Widmark, G., "Quantitative Gas Chromatography. Quantitative Recovery and Re-injection of a Sample," Acta Chem. Scand., 16, 575-582 (1962); CA 57, 6582e

7292. Widmer, H., and Gaumann, T., "The Combination of a Gas Chromatograph with a Mass Spectrograph," Helv. Chim. Acta, 45, 2175-2185 (1962); CA 58, 2825e

7293. Wiebe, A. K., "Elution Time and Resolution in Vapor Chromatography," J. Phys. Chem., 60, 685-688 (1956); CA 50, 12725b; 129th Natl. ACS Mtg., Dallas, Texas, April 1956, Program Abstr., p.15B

7294. Wiel, A. van der, "Two Highly Selective Solvents for Gas-Liquid Chromatography Analysis of C_2-C_6 Hydrocarbons," Nature, 187, 142-143 (1960); CA 54, 20667i

7295. Wiersma, D. S., Hoyle, R. E., and Rempis, H., "Separation and Determination of Mono-, Di-, and Tripentaerythritol by Programmed Temperature Gas Chromatography," Anal. Chem., 34, 1533-1535 (1962); CA 58, 6169d

7296. Wiersma, D. S., and Tollefson, E. L., "Column Temperature Programming in Gas-Liquid Chromatography," 2nd Alberta Gas Chromatographic Discussion, Edmonton, Alberta, February 1959

7297. Wiesner, L., Schmidt-Kuster, W. J., and Schultze, G. R., "Application of Radioisotopes to a Gas-Chromatographic Detector for Direct Indication of Concentration," Radioisotopes Phys. Sci. Ind., Proc. Conf. Use, Copenhagen, 1960, 2, 225-230 (Pub. 1962); CA 57, 5282i

7298. Wiesner, L., and Schmidt-Kuster, W. J., "Performance Characteristics of the Integrating Counter-Detector for Gas Chromatography," Microchem. J. Symp. Ser., 2, 733-747 (1962); CA 58, 6420h

7299. Wijnen, M. H. J., "Reactions of Alkoxy Radicals. VI. Photolysis of Isopropyl Propionate," J. Am. Chem. Soc., 82, 1847-1849 (1960); CA 54, 16129i

7300. Wijnen, M. H. J., "Reactions of Alkoxy Radicals. VII. The Ethoxy Radical," J. Am. Chem. Soc., 82, 3034-3040 (1960); CA 54, 23656d

7301. Wijnen, M. H. J., "Photolysis of Phosgene in the Presence of Ethylene," J. Am. Chem. Soc., 83, 3014-3017 (1961); CA 56, 3345f

7302. Wilcox, Jr., C. F., and Mesirov, M., "Preparation of 5,5-Dimethylcyclopentadiene," J. Org. Chem., 25, 1841-1844 (1960); CA 55, 12315b

7303. Wiley, R. H., Miller, W., Jarboe, Jr., C. H., Harrell, J. R., and Parish, D. J., "Gamma Radiation-Induced Isomerization of n-Propyl Chloride," Radiation Research, 13, 479-488 (1960); CA 55, 87f

7304. Wilhite, R. N., "Identification of Hydrocarbon Types in Ultrasene," U. S. At. Energy Comm., DP-571, 12 pp. (1961); CA 55, 25867i

7305. Wilkenson, L. B., "Hydrogen Analysis by Gas Chromatography," 23rd Mtg., Gulf Coast Spectroscopic Group, Houston, Texas, August 1957

7306. Wilkins, M., and Wilson, J. D., "The Measurement of Impurities in Helium. I. The Katharometer as a Continuous Analyzer for Total Impurities," U. K. At. Energy Authority, Research Group Rept. AERE C/R 2808, 21 pp. (1959); CA 53, 15858f

7307. Wilkins, M., and Wilson, J. D., "Measurement of Impurities in Helium. II. Intermittent Analyses by Gas Chromatography," At. Energy Research Estab. (Gt. Brit.), A.E.R.E. C/R 2809, 51 pp. (1959); CA 54, 12887d

7308. Wilkinson, J., and Hall, D., "A Device for Transferring Small Volumes of Gas from a Vacuum Line to a Gas Chromatograph," J. Chromatog., 10, 239-242 (1963)

7309. Wilkinson, R. W., and Winter, J. A., "Simple Apparatus for Gas Chromatography of Polyphenols," At. Energy Research Estab. (Gt. Brit.) Memo, M-820, 5 pp. (1961); CA 55, 17113b

7310. Wilks, Jr., P. A., "Multiple Internal Reflection Techniques for the Identification of Gas Chromatograph Fractions and Other Applications," 145th Natl. ACS Mtg., New York, N. Y., September 1963, Program Abstr., p. 5B

7311. Wilks, Jr., P. A., and Warren, C. W., "A Method for Collecting Vapor Phase Chromatographic Fractions Directly in Infrared Adsorption Cells," 11th Pittsburgh Conf. on Anal. Chem. & Appl. Spectroscopy, Pittsburgh, Pa., February-March 1960, Program Abstr., p. 41

7312. Wilks, Jr., P. A., and Warren, C. W., "A Method for Collecting Chromatographic Fractions in Infrared Adsorption Cells," CIC Newsletter, No. 6, 1, 3 (March 1960)

7313. Wilks, Jr., P. A., and Warren, C. W., "A Separatory Chromatograph for Automatically Collecting Fractions for Infrared Analysis," 12th Pittsburgh Conf. on Anal. Chem. & Appl. Spectroscopy, Pittsburgh, Pa., February-March 1961, Program Abstr., p. 50

7314. Williams, A. F., "Analytical Research in the Nobel Division of Imperial Chemical Industries," Proc.

Congr. Modern Anal. Chem. in Industry, June 1957

7315.   Williams, A. F., "Studies on the Accuracy of Gas-Liquid Chromatography," Discussion Group, Institute of Petroleum, Salford, Eng., September 1962; cf. Nature, 196, 818 (1962)

7316.   Williams, A. L., and Bannister, M. H., "Composition of Gum Turpentines from Twenty-Two Species of Pines Grown in New Zealand," J. Pharm. Sci., 51, 970-975 (1962); CA 58, 1634h

7317.   Williams, C. M., "Gas Chromatography of Urinary Aromatic Acids," Anal. Biochem., 4, 423-432 (1962); CA 58, 10497b

7318.   Williams, C. M., and Greer, M., "Diagnosis of Neuroblastoma by Quantitative Gas Chromatographic Analysis of Urinary Homovanillic and Vanilmandelic Acid," Clin. Chim. Acta, 7, 880-883 (1962)

7319.   Williams, C. M., and Leonard, R. H., "Microanalytical Determination of Dihydroxy Aromatic Acids by Gas Chromatography," Anal. Biochem., 5, 362-366 (1963); CA 58, 12849b

7320.   Williams, C. M., and Sweeley, C. C., "A New Method for the Determination of Urinary Aromatic Acids by Gas Chromatography," J. Clin. Endocrinol. and Metabolism. 21, 1500-1504 (1961); CA 56, 6287i

7321.   Williams, D. D., Barefort, R. D., and Miller, R. R., "Differential Thermal Analysis Apparatus for Heating and Cooling Data," Anal. Chem., 30, 492-494 (1958); CA 52, 11489h

7322.   Williams, D. D., and Miller, R. R., "Instrument for On Stream Stripping and Analysis of Dissolved Gases in Water," U. S. Dept. Com., Office Tech. Serv., AD 265,420, 12 pp. (1961); CA 58, 3197h

7323.   Williams, D. D., and Miller, R. R., "An Instrument for On-Stream Stripping and Gas Chromatographic Determination Dissolved Gases in Liquids," Anal. Chem., 34, 657-659 (1962); CA 57, 1512c

7324.   Williams, E. F., "Analysis by Vapor-Phase Chromatography Above 150°," 129th Natl. ACS Mtg., Dallas, Texas, April 1956, Program Abstr., p. 15B

7325.   Williams, E. F., "Gas Chromatography - Complementary to Spectroscopy," 11th Annual Mtg., Soc. Appl. Spectroscopy, New York, N. Y., September 1956; Abstr., Appl. Spectroscopy, 10, 221 (1956)

7326.   Williams, E. F., (Ed.), "The Analysis of Mixtures of Volatile Substances," Ann. N. Y. Acad. Sci., 72, Art. 13, 559-785 (1959); Papers Presented at the Conf. on Analysis of Mixtures of Volatile Substances, New York Academy of Sciences, New York, N. Y., April 1958

7327.   Williams, I. H., "A Rapid Method for the Preparation of Packing Material for Gas Chromatographic Columns," J. Chromatog., 5, 457 (1961); CA 56, 9379c

7328.   Williams, I. H., "Applications of the Electron Affinity Detector to the Gas Chromatographic Study of Air Pollutants," 142nd Natl. ACS Mtg., Atlantic City, N. J., September 1962, Program Abstr., pp. 19V-20V

7329.   Williams, J. F. A., Sharms, A., Morris, L. J., and Holman, R. T., "Fatty Acid Composition of [Human] Feces and Fecaliths," Proc. Soc. Exptl. Biol. Med., 105, 192-195 (1960); CA 55, 4693d

7330.   Williams, J. F. A., "Electronic Effects of Alkyl Groups. I. Adsorption Spectra of Monoaryl Carbonium Ions in Sulfuric Acid," Tetrahedron, 18, 1487-1493 (1962); CA 58, 11189d

7331.   Williams, Jr., J. H., and Wilt, R. L., "Chromatography Can Increase Alkylation Efficiency," Oil Gas J., 59, No. 20 134-137 (1961); CA 55, 18084e

7332.   Williams, Jr., J. H., and Wilt, R. L., "Chromatography Can Increase Alkylation Efficiency," Proc. Am. Petrol. Inst. Sect. III, 41, 67-71 (1961); CA 58, 1282d

7333.   Williams, M. C., and Reiser, R., "The Chemical and Biological Assay of Essential Fatty Acids," J. Am. Oil Chemists' Soc., 40, 237-241 (1963)

7334.   Williams, M. M. D., Brown, H. O., Dublin, W. B., and Boothby, W. M., "Two Physical Methods for the Determination of One Component of a Mixture of Gases," Proc. Minn. Acad. Sci., 13, 46-53 (1945); CA 45, 9310f

7335.   Williams, P. L., "The Occurrence of Linalool in Citronella Oil," Perfumery Essent. Oil Record, 50, 678 (1959); CA 54, 5021d

7336.   Williams, P. M., "Organic Acids in Pacific Ocean Waters," Nature, 189, 219-220 (1961); CA 55, 18222b

7337.   Williams, R. J. P., "Gradient Elution Analysis," Analyst, 77, 905-914 (1952); CA 47, 1530c

7338.   Williams, R. J. P., "Chromatography. General Principles," Brit. Med. Bull., 10, 165-169 (1954); CA 49, 9428b

7339.   Williams, R. L., Dunstan, I., and Blay, N. J., "Boron Hydride Derivatives. IV. Friedel-Crafts Methylation of Decarborane", J.Chem. Soc., 1960, 5006-5012; CA 55, 10303c

7340.   Williams, T. F., Wilkinson, R. W., and Rigg, T., "Radiolysis of Tributyl Phosphate," Nature, 179, 540 (1957); CA 51, 12670b

7341.   Williams, T. I., "The Elements of Chromatography," Philosophical Library, Inc., New York, 1953

7342.   Williams, V. R., and McMillan, R., "Lipids of Ankistrodesmus Braunii," Science, 133, 459-460 (1961); CA 55, 12543b

7343.   Willis, V., "Analysis by Gas Chromatography of a 'Pure' Sample with an 'Impure' Carrier," Nature, 183, 1754 (1959); CA 53, 21360b

7344.   Willis, V., "Analysis of Permanent Gases by Gas Chromatography by Using a Radioactive Ionization Type Detector," Nature, 184, 894 (1959); CA 54, 8418f

7345. Wilson, E. M., Moberg, M. L., and Pust, H. W., "Organic Functional Analysis by Gas Chromatography. I and II. Kjeldahl Nitrogen and Hydroxyl Number," 11th Pittsburgh Conf. on Anal. Chem. & Appl. Chem. & Appl. Spectroscopy, Pittsburgh, Pa, February-March 1960, Program Abstr., p. 39

7346. Wilson, E. M., Oyama, V., and Vango, S. P., "Design Features of a Lunar Gas Chromatograph," Gas Chromatog. Intern. Symp., 1961, 3, 329-338 (Pub. 1962); CA 58, 3866g

7347. Wilson, E. M., Oyama, V., and Vango, S. P., "Design Features of a Lunar Gas Chromatograph," in "Gas Chromatography," edited by N. Brenner, J. E. Callen, and M. D. Weiss, Academic Press, Inc., New York, 1962, pp. 329-338

7348. Wilson, E. M., Vango, S. P., and Oyama, V., "Design Features of a Lunar Gas Chromatograph," 3rd Intern. Symp. on Gas Chromatography, ISA, East Lansing, Mich., June 1961

7349. Wilson, J. D., "The Effect of Dietary Fatty Acids on Coprostanol Excretion by the Rat," J. Lipid Research, 2, 350-356 (1961); CA 56, 3881h

7350. Wilson, J. E., "A Sensitive Versatile Acoustic Gas Analyzer Particularly Suitable for the Analysis of Anesthetic Mixtures," Rev. Sci. Instr., 25, 927-928 (1954); CA 49, 7893f

7351. Wilson, J. N., "A Theory of Chromatography," J. Am. Chem. Soc., 62, 1583-1591 (1940); CA 34, 5369[4]

7352. Wilson, N. H., "Vapour-Phase Chromatography," Chem. & Ind. (London), 1955, 225

7353. Wilson, P. W., Kodicek, E., and Booth, V. H., "Separation of Tocopherols by Gas-Liquid Chromatography," Biochem. J., 84, 524-531 (1962); CA 57, 16984h

7354. Wilson, R. H., Jay, B. E., and Chapman, C. B., "Gas Chromatographic Method of Analyzing Body Fluids Such as Whole Blood for Mixtures of Inert and Chemically Reactive Gases," 45th Annual Mtg., Fed. of Am. Soc. for Exptl. Biol., Atlantic City, N. J., April 1961

7355. Wilson, R. H., Jay, B. E., Doty, V., Pingree, H., and Higgins, E., "Analysis of Blood Gases with Gas Adsorption Chromatographic Techniques," J. Appl. Physiol., 16, 374-377 (1961); CA 55, 22470f

7356. Wilson, R. H., Jay, B. E., and Holland, R. H., "Gas Chromatography: A Simple, Rapid, Reliable Method for Blood Gas Analysis," J. Thoracid and Cardiovascular Surgery, 42, 575-579 (1961)

7357. Wilt, J. W., and Finnerty, J. L., "Anomalous Hunsdiecker Reaction. II. Scope of the Reaction," J. Org. Chem., 26, 2173-2177 (1961); CA 55, 27208d

7358. Wilzbach, K. E., "Gas Exposure Technique for Tritium Labeling," Proc. Symposium Tritium Tracer Appl., New York, N. Y., 1957, 3-9 (Pub. 1958); CA 53, 4106c

7359. Wilzbach, K. E., "Labeling of Organic Compounds with Tritium," Oklahoma Conf.- Radioisotopes in Agriculture, Oklahoma State Univ., Stillwater, Okla., April 1959; Rept. TID-7578

7360. Wilzbach, K. E., and Riesz, P., "Isotope Effects in Gas-Liquid Chromatography," Science, 126, 748-749 (1957); Erratum, ibid., p. 1062 (1957); CA 52, 1725h

7361. Winefordner, J. D., Steinbrecher, D., and Lear, W. E., "A Vapor Detector Based on Changes in Dielectric Constant," Anal. Chem., 33, 515-521 (1961); CA 55, 19352d

7362. Wing, F. E., "Design and Performance of a New Ionization Detector System for Gas-Liquid Chromatography," 5th Natl. Symp. on Instrumental Methods of Analysis, ISA, Houston, Texas, May 1959

7363. Winkelman, J., and Karmen, A., "Use of an Ionization Chamber for Measuring Radioactivity in Gas Chromatography Effluents," Anal. Chem., 34, 1067-1071 (1962); CA 57, 9427d

7364. Winkler, W. O., "Report on the Analysis of Cacao Products," J. Assoc. Offic. Agr. Chemists, 45, 551-554 (1962); CA 57, 14252a

7365. Winstein, S., and Sonnenberg, J., "Homoconjugation and Homoaromaticity. III. The 3-Bicyclo[3.1.0] Hexyl System," J. Am. Chem. Soc., 83, 3235-3244 (1961); CA 56, 12756f

7366. Winstein, S., and Sonnenberg, J., "Homoconjugation and Homoaromaticity. IV. The Trishomocyclopropenyl Cation. A Homoaromatic Structure," J. Am. Chem. Soc., 83, 3244-3251 (1961); CA 56, 12757i

7367. Winstein, W. A., "Reversed-Phase Partition Chromatography on Microporous Polymeric Support," Anal. Chem., 34, 1334-1335 (1962); CA 57, 11834c

7368. Winters, J. C., "Chromatography Works Fast in Showing Up Differences in Crudes," Oil Gas J., 58, No. 24, 138, 140 (1960); CA 54, 25723i

7369. Winters, J. C., Jones, F. S., and Martin, R., "Gas Chromatography Guides Development of a Paraffin-Isomerization Process," World Petrol. Congr., Proc., 5th, New York, 1959, 5, 33-43 (Pub. 1960); CA 58 6627c

7370. Wirth, H., "Qualitative and Quantitative Micro-Gas-Analysis by the Desorption-Heat-Conductivity Method," Mikrochem. ver. Mikrochim. Acta, 40, 15-20 (1952); CA 47, 993g

7371. Wirth, H., "The Separation of Gases by Sorption Processes," Monatsh. Chem. 84, 156-168 (1953); CA 47, 7285a

7372. Wirth, H., "Separation of Gases by Sorption Processes. II.," Monatsh. Chem., 84, 741-750 (1953); CA 48, 2440d

7373. Wirth, H., "The Analytical Separation of Ethane, Ethylene, and Acetylene by Fractional Bromination on Activated Carbon," Monatsh. Chem., 84, 751-753 (1953); CA 48, 2522b

7374. Wirth, M. M., "The Hydrogen Microflare Detector, Using Nitrogen as a Carrier Gas," in "Vapour-Phase Chromatography," edited by D. H. Desty, Academic Press, Inc., New York, 1957, pp. 154-168

7375. Wirth, M. M., "Relative Retention Volumes of Lower Hydrocarbons on Carbitol and Dimethylformamide (Chromatographic Data Table XL)," J. Chromatog., 1, xxv (1958); cf. "Vapour-Phase Chromatography," edited by D. H. Desty, Academic Press, Inc., New York, 1957, p. 164 (n-Pentane = 1, Temperature = 0°)

7376. Wise, K. V., and Oliver, G. D., "Gas Chromatography. Effects of Some of the Variables Encountered in the Elution Technique and Some Applications of the Displacement Technique," 129th Natl. ACS Mtg., Dallas, Texas, April 1956, Program Abstr., p. 16B

7377. Wise, R. W., and Sullivan, A. B., "Determination of Antidegradants [in Rubber] by Gas Chromatography," Rubber Age (N. Y.), 91, 773-776 (1962); CA 57, 13941b; 140th Natl. ACS Mtg., Chicago, Ill., September 1961, Program Abstr., p 7V

7378. Wiseblatt, L., "Some Aromatic Components Present in Oven Gases," Cereal Chem., 37, 728-733 (1960); CA 55, 8686c

7379. Wiseblatt, L., "The Volatile Organic Acids Found in Dough, Oven Gases, and Bread," Cereal Chem., 37, 734-739 (1960); CA 55, 8686e

7380. Wiseman, W. A., "The Use of Helium Gas as the Mobile Phase in Gas Chromatography," Chem. & Ind. (London), 1956, 127-129; CA 50, 9204e

7381. Wiseman, W. A., "Behavior of Katharometers," Chem. & Ind. (London), 1957, 1356-1357; CA 52, 2461i

7382. Wiseman, W. A., "Gas Chromatography and the Perfumer," Perfumery Essent. Oil Record, 48, 380-385 (1957); CA 52, 3269i

7383. Wiseman, W. A., "Katharometer Behavior," Ann. N. Y. Acad. Sci., 72, 685-697 (1959); CA 53, 14594f

7384. Wiseman, W. A., "Effect of Carrier Gas on Katharometer Response," Nature, 183, 1321-1322 (1959); CA 53, 20925f

7385. Wiseman, W. A., "Separation Factors in Gas Chromatography," Nature, 185, 841-842 (1960); CA 54, 14815c

7386. Wiseman, W. A., "Detection by Ionization of Gases in Helium Used in Gas Chromatography," Nature, 192, 964-965 (1961); CA 56, 5384a

7387. Wiseman, W. A., and Berry, R., "Comparison of Helium and Argon in Ionization Detectors," Nature, 190, 1187-1188 (1961); CA 55, 25383b

7388. Wisniak, J., and Albright, L. F., "Hydrogenating Cottonseed Oil at Relatively High Pressure," Ind. Eng. Chem., 53, 375-380 (1961); CA 55, 15962c

7389. Wisniewski, D. F., and Stalker, G. C., "Gas Chromatography as Applied to Glycol Solution Analysis," Proc. Gas Conditioning Conf., 1960, 15-21; CA 55, 16027i

7390. Wisniewski, D. F., and Stalker, G. C., "New Use of Gas Chromatography for Glycol Analysis," Petrol. Refiner, 40, No. 2, 117-120 (1961); CA 55, 16028e

7391. Witjens, P. H., "Chemical Problems Set by Aerosol Perfuming," Parfums, cosmet., savons, 2, 431-433 (1959); CA 54, 3864a

7392. Witjens, P. H., "Chemical Problems Concerning the Perfuming of Aerosols," Perfumery Essent. Oil Record, 50, 913-915 (1959)

7393. Witjens, P. H., "Chemical Problems Concerning the Perfumery of Aerosols," Am. Perfumer & Cos., 75, No. 10, 49-50 (1960)

7394. Wittig, G., and Krebs, A., "On the Existence of Low-Membered Cycloalkynes. I.," Chem. Ber., 94, 3260-3275 (1961); CA 56, 9987f

7395. Wittig, G., and Mayer, U., "The Existence of Low-Membered Cycloalkynes. III. Formation and Behavior of Cyclohexyne," Chem. Ber., 96, 329-341 (1963); CA 58, 10141h

7396. Witting, L. A., Century, B., Harvey C. C., and Horwitt, M. K., "Fatty Acid Composition of the Lipids of Chick Brain Mitochondria," 45th Annual Mtg., Fed. Am. Soc. for Exptl. Biol., Atlantic City, N. J., April 1961

7397. Woidich, H., Gnauer, H., and Riedl, O., "Detection of Foreign Fat in Cacao Products," Z. Lebensm.-Untersuch. u. -Forsch., 112, 184-190 (1960); CA 54, 15748e

7398. Woidich, H., Gnauer, H., and Riedl, O., "Fat Migration in Filled Chocolates," Z. Lebensm.-Untersuch. u. -Forsch., 117, 478-483 (1962); CA 57, 17153e

7399. Wolf, F., and Beyer, H., "Testing of Adsorbents and Catalysts by Gas Chromatography. I. Determination of the Relative Surface of Adsorbents from Values Established by Gas Chromatography," Chem. Tech. (Berlin), 11, 142-144 (1959); CA 53, 14635f

7400. Wolf, F., and Ternow, A., "Gas Liquid Distribution Coefficient for Thin Liquid Layers on Polar Carriers," Kolloid-Z., 166, 38-43 (1959); CA 54, 39g

7401. Wolff, G., and Wolff, J. P., "Application of Gas Chromatography to the Examination of Fat Purity," Rev. franç. corps gras, 7, 73-80 (1960); CA 54, 14726i

7402. Wolff, J., "Quantitative Evaluation of Gas Chromatograms at Different Sensitivities," Wiss. Z. Tech.

Hochsh. Chem. Leuna-Merseburg, 4, No. 3/4, 275-276 (1962); CA 59, 7c

7403. Wolff, J. P., "Physico-Chemical Methods of Analysis of Fats and Their Derivatives," Ind. Ailment Agr., 75, 639-648 (1958); CA 53, 3737f; cf. J. Sci. Food Agr., 10, 4, 223 (1959)

7404. Wolff, J. P., "Vapor Chromatography to Determine the Purity of Butter," Ann. fals. et expert. chim., 53, 318-325 (1960); CA 54, 21526e

7405. Wolff, J. P., "Comparison of Different Methods for the Determination of Linoleic and Linolenic Acids in Oils," Rev. franç. corps gras, 8, 68-84 (1961); CA 55, 11884d

7406. Wolff, J. P., "Control of Lard Purity," Rev. franç. corps gras, 8, 677-685 (1961); CA 56, 9174d

7407. Wolfgang, R., and Rowland, F. S., "Radioassay by Gas Chromatography of Tritium- and Carbon-14-Labeled Compounds," Anal. Chem., 30, 903-906 (1958); CA 52, 13519g

7408. Wolford, R. W., Alberding, G. E., and Attaway, J. A., "Citrus Juice Flavor. Analysis of Recovered Natural Orange Essence by Gas Chromatography, " J. Agr. Food Chem., 10, 297-301 (1962); 140th Natl. ACS Mtg., Chicago, Ill., September 1961, Program Abstr., p. 23A

7409. Wolford, R. W., and Attaway, J. A., "The Application of Gas Chromatography to the Analysis of Flavor Components of Citrus Juices," 3rd Intern. Symp. on Gas Chromatography, ISA, East Lansing, Mich., June 1961; ISA Proc., 1961, 289-295

7410. Wolford, R. W., and Attaway, J. A., "The Application of Gas Chromatography to the Analysis of Flavor Components of Citrus Juices," in "Gas Chromatography," edited by N. Brenner, J. E. Callen, and M. D. Weiss, Academic Press, Inc., New York, 1962, pp. 457-470

7411. Wolfrom, M. L., and Arsenault, G. P., "Controlled Thermal Decomposition of Cellulose Nitrate. VII. Carbonyl Compounds," J. Am. Chem. Soc., 82, 2819-2823 (1960); CA 55, 11320b

7412. Wollast, R., and Jottrand, R., "Quantitative Gas-Chromatographic Analyses," Compt. rend. congr. intern. chim. ind., 31^e, Liege, 1958 (Pub. as Ind. chim. belge, Suppl.), 1, 184-192 (Pub. 1959); CA 54, 4237c

7413. Wollrab, V., Streibl, M., and Sorm, F., "Gas-Chromatographic Analysis of the Wax Components of Montan Wax," Chem. & Ind. (London), 1962, 1762; CA 58, 4337g

7414. Wong, M. P., and Patton, S., "Identification of Some Volatile Compounds Related to the Flavor of Milk and Cream. II. Investigations of the Volatile Components of Milk and Cream by Gas Chromatography and Mass Spectrometry," J. Dairy Sci., 45, 724-728 (1962); CA 58, 3825e

7415. Wood, Jr., F. T., and Young, J. F., "Gas Chromatographic Analysis of Gases at Douglas Aircraft Company," Perkin-Elmer Instrument News, 9, No. 3, 7-8 (Spring 1958)

7416. Wood, L. J., "Chemical Constitution of Road Tars and Coal-Tar Pitches," J. Appl. Chem. (London), 11, 130-136 (1961); CA 55, 19203e

7417. Woodford, F. P., and Gent, C. M. van, "Gas-Liquid Chromatography of Fatty Acid Methyl Esters: the Carbon Number as a Parameter for Comparison of Columns," J. Lipid Research, 1, 188-190 (1960); CA 54, 11634d

7418. Woodman, F. J., Clinton, T. G., Fletcher, W., and Welch, G. A., "Developments in the Analytical Service Given to Industrial Atomic Energy," Proc. U. N. Intern. Conf. Peaceful Uses At. Energy, 2nd, Geneva, 28, 423-435 (1958); CA 54, 5f

7419. Woods, A. E., Aurand, L. W., and Roberts, W. M., "Studies on Flavor Components from Ladino Cloyer," 56th Annual Mtg., Am. Dairy Sci. Assoc., Madison, Wisc., June 1961; Abstr., J. Dairy Sci., 44, 1152 (1961)

7420. Woolmington, K. G., "Determination of Hydrogen Cyanide by Gas Chromatography," J. Appl. Chem. (London), 11, No. 3, 114-120 (1961); CA 55, 12141c

7421. Wotiz, H. H., "Steroid Metabolism. XIII. Factors Influencing the Gas-Chromatographic Stability of Steroids," Biochim. Biophys. Acta, 63, 180-185 (1962); cf. ibid., 63, 28-32; CA 57, 17045d

7422a. Wotiz, H. H., "Steroid Metabolism. XV. Rapid Determination of Urinary Pregnanediol by Gas Chromatography," Biochim. Biophys. Acta, 69, 415-416 (1963); CA 58, 10477b

7422b. Wotiz, H. H., and Clark, S. J., Gas Chromatography in the Analysis of Steroids," Plenum Press, New York (in preparation, to be published in 1965)

7423. Wotiz, H. H., and Martin, H. F., "Steroid Metabolism. X. Gas Chromatographic Analysis of Estrogens," J. Biol. Chem., 236, 1312-1317 (1961); CA 55, 20075a; 138th Natl. ACS Mtg., New York, N. Y., September 1960, Program Abstr., p. 58C

7424. Wotiz, H. H., and Martin, H. F., "Steroid Metabolism. XI. Gas-Chromatographic Determination of Estrogens in Human Pregnancy Urine, " Anal. Biochem., 3, 97-108 (1962); CA 56, 15757c

7425. Wotiz, H. H., Naukkarinen, I., and Carr, Jr., H. E., "Gas Chromatography of Aldosterone," Biochem. Biophys. Acta, 53, 449-452 (1961); CA 56, 10478c

7426. Wrabetz, K., and Sassenberg, W., "Gas-Chromatographic Analysis of Phenol-Cresol-Xylenol Mixtures with Special Attention to the o-Cresol Content of Tritolyl Phosphate," Z. Anal. Chem. 179, 333-342 (1961); CA 55, 13157f

7427. Wreyford, D. M., "Sampling System for Multi-Stream Gas Chromatography," 5th Natl. Symp. on Instrumental Methods of Analysis, ISA, Houston, Texas, May 1959

7428. Wreyford, D. M., and Rickey, N., "Good Use of Process Analyzers," Oil Gas J., 56, 76-78, 81-82,

84-85 (1958); CA <u>52</u>, 21024a

7429. Wright, F. J., "Gas Phase Oxidation of the Xylenes. General Kinetics," J. Phys. Chem., <u>64</u>, 1944-1950 (1960); CA <u>55</u>, 18259g

7430. Wulfhekel, H., and Schaal, G., "A Gas-Discharge Detector for Gas-Chromatographic Detection of Nitrogen, Oxygen, and Other Gases," Exptl. Tech. Physik, <u>10</u>, 231-243 (1962); CA <u>58</u>, 8394d

7431. Wurst, M., "Detection Methods and Detectors Used in Gas Chromatography," Chem. listy, <u>54,</u> 1042-1058 (1960); CA <u>55</u>, 3126f

7432. Wurst, M., and Dusek, R., "Analysis of Organosilicon Compounds. II. Separation and Determination of Vinylethoxysilane by Means of Gas Chromatography," Collection Czechoslov. Chem. Commun., <u>27</u>, 2391-2397 (1962)

7433. Wurst, M., and Wurstova, E., "Use of Gas Chromatography for the Analysis of Chloro Derivatives of Benzene and Toluene," Gas Chromatog. Symp., Brno, Czech., June 1962; Abstr., J. Gas Chromatog., <u>1</u>, No. 4, 9 (1963)

7434. Wyllie, H. A., "Gas Introducer for a Vacuum System," J. Sci. Instr., <u>33</u>, 360-361 (1956); CA <u>51</u>, 3197c

7435. Wynn, J. D., Brunner, J. R., and Trout, G. M., "Gas Chromatography as a Means of Detecting Odors in Milk," Food Technol., <u>14</u>, 248-250 (1960)

7436. Yamada, M., Shitara, J., Yoneyama, S., Komoda, H., Kosaki, M., and Yoshizawa, K., "Amino Acid Fermentation by Yeasts. I. A New Pathway of Fusel Oil Formation," Hakko Kyokaishi, <u>20</u>, 134-139 (1962); cf. Tokyo Nogyo Daigaku Nogaku Shuho, <u>7</u>, 97-102 (1962); CA <u>58</u>, 9593c

7437. Yamada, S., "Gas Formed During Heat-Treatment of Coal Pitch to Which a Nitro Compound Is Added," Kogyo Kagaku Zasshi, 62, 1533-1538 (1959)

7438. Yamamoto, T., "Application [of Gas Chromatography] to Process Control in Chemical Industry," Kagaku no Ryoiki Zokan, No. <u>44</u>, 127-146 (1961); CA <u>56</u>, 3a

7439. Yamamoto, T., and Saito, K., "Analysis of Crude Methanol and By-Product Alcohols by Gas Chromatography," Yuki Gosei Kagaku Kyokaishi, <u>17</u>, 293-301 (1959); CA <u>53</u>, 21428f

7440. Yamane, M., "Quantitative Gas-Chromatographic Analysis With an Ionization Detector," Kogyo Kagaku Zasshi, <u>64</u>, 759-763 (1961); CA <u>57</u>, 2827b

7441. Yamane, M., "A New Argon Ionization Detector for Gas Chromatography," J. Chromatog., <u>9</u>, 162-172 (1962); CA <u>58</u>, 6166h

7442. Yamazaki, H., "Theoretical Consideration of Programmed-Temperature Gas Chromatography," Kogyo Kagaku Zasshi, <u>64</u>, 757-759 (1961); CA <u>57</u>, 2880h

7443. Yanak, Y., and Novak, I., "Application of Gas Chromatography to Measurement of Some Physico-Chemical Properties," Gazovaya Khromatografiya Trudy 1-oi [Pervoi] Vsesoyuz. Konf., Akad. Nauk S.S.S.R., Moscow, <u>1959</u>, 189-198 (Pub. 1960); CA <u>55</u>, 26577f

7444. Yang, J. Y., and Wolf, A. P., "Gas Phase Reactions of Recoil Carbon-14 in Anhydrous Ammonia," J. Am. Chem. Soc., <u>82</u>, 4488-4492 (1960); CA <u>55</u>, 1152f

7445. Yang, K., "Effect of Nitric Oxide in the Gamma-Radiolysis of Ethane at Very Low Concentrations," Can. J. Chem., <u>38,</u> 1234-1235 (1960); CA <u>54</u>, 20445a

7446. Yang, K., "Nitric Oxide as a Radical Scavenger in the Radiolysis of Gaseous Hydrocarbons," J. Phys. Chem., <u>65,</u> 42-45 (1961); CA <u>55</u>, 24524d

7447. Yang, K., and Gant, P. L., "Reactions Initiated by β-Decay of Tritium. II. The Tritium-Ethylene System," J. Chem. Phys., <u>31</u>, 1589-1594 (1959); CA <u>54</u>, 15014f

7448. Yang, K., and Manno, J., "The Role of Free Radical Processes in the Gamma-Radiolysis of Methane, Ethane, and Propane," J. Am. Chem. Soc., <u>81</u>, 3507-3510 (1959); CA <u>54</u>, 2900c

7449. Yanovskaya, L. A., Kucherov, V. F., and Rudenko, B. A., "Chemistry of Acetals. XII. Application of Gas-Liquid Chromatography to Analysis of Reaction Products of Ortho Esters With Vinyl Ethers," Izv. Akad. Nauk S.S.S.R., Otd. Khim. Nauk, <u>1962</u>, 2182-2189; CA <u>58</u>, 12410g

7450. Yanovskaya, L. A., Rudenko, B. A., Kucherov, V. F., Stepanova, R. N., and Kogan, G. A., "Chemistry of Acetals. XIII. A Study of Hydrolysis of Some Diacetals by the Method of Gas-Liquid Chromatography," Izv. Akad. Nauk S.S.S.R., Otd. Khim. Nauk, <u>1962</u>, 2189-2196; CA <u>58</u>, 12411a

7451. Yanovskii, M. I., and Gaziev, G. A., "The Use of Frontal Analysis in the Gas-Liquid Chromatography of Radioactive and Nonradioactive Gases," Doklady Akad. Nauk S.S.S.R., <u>120</u>, 812-814 (1958); CA <u>53</u>, 11099b

7452. Yanovskii, M. I., and Gaziev, G. A., "Gas Fluid Radiochromatograph," Vestnik Akad. S.S.S.R., <u>30</u>, No. 5, 27-31 (1960); Nuclear Sci. Abstracts, <u>14</u>, 2990 (1960); CA <u>54</u>, 20355e

7453. Yanovskii, M. I., Kapustin, D. S., and Nogotkov-Ryutin, V. A., "Method for Rapid Determination of Molar Radioactivity in Chromatography of Gases Tagged with Carbon-14," Problemy Kinetiki i Kataliza, Akad. Nauk S.S.S.R., <u>9</u>, 391-398 (1957); CA <u>53</u>, 7719b

7454. Yanovskii, M. I., Oziraner, S. N., and Lu. P-C., "Mechanism of Chromatographic Separation of Gases by Thermal Desorption," Zhur. Priklad. Khim., <u>33</u>, 1084-1091 (1960); CA <u>54</u>, 19030i

7455. Yasui, E., and Suzuki, H., "Determination of Small Amounts of Acetylene in Air with Silica Gel Treated with Liquid Oxygen as Adsorbent," Kogyo Kagaku Zasshi, <u>61</u>, 176-179 (1958); CA <u>53</u>, 19673e

7456. Yasui, H., "Constituents of Light Oil From Coal Carbonization. I. Saturated Hydrocarbons in High-Temperature Coal Carbonization Light Oil. II. Saturated Hydrocarbons Contained in Low-Temperature Coal Carbonization Light Oil," Bull. Chem. Soc. Japan, 33, 1493-1498, 1498-1503 (1960); CA 55, 12815c

7457. Yavorovskaya, S. F., "[Gas-Chromatographic] Separation and Determination of Small Amounts of Tetrachloroalkanes and Chlorinated Methanes," Gazovaya Khromatografiya, Trudy 1-oi [Pervoi] Vsesoyuz. Konf., Akad. Nauk S.S.S.R., Moscow, 1959, 302-306 (Pub. 1960); CA 56, 4105b

7458. Yavorovskaya, S. F., "Determination of Small Quantities of Chloroorganic Compounds in Air by a Gas Chromatography Method," Khim. Prom., 1961, 573-577; CA 56, 6317d

7459. Yeh, S. J., and Frisch, H. L., "Chromatographic Adsorption of Polystyrene," J. Polymer Sci., 27, 149 (1958); CA 53, 5816e

7460. Yip, G., "Cleanup in the Application of Gas and Paper Chromatography to 2,4-D Residue Analysis," 140th Natl. ACS Mtg., Chicago, Ill., September 1961, Program Abstr., p. 58B

7461. Yip, G., "Determination of 2,4-D and Other Chlorinated Phenoxy Alkyl Acids," J. Assoc. Offic. Agr. Chemists, 45, 367-376 (1962); CA 57, 2637e; 140th Natl. ACS Mtg., Chicago, Ill., September 1961, Program Abstr., pp. 21A, 63B

7462. Yokley, C. R., "Design of Discharge Tubes for Use as Elution Peak Detectors in Gas Chromatographic Analysis," 8th Annual Instrument Symp. & Research Equipment Exhibit, National Institutes of Health, Bethesda, Md., May 1958

7463. Yokley, C. R., and Ferguson, R. E., "Separation of the Products of Cool Flame Oxidation of Propane," Combustion and Flame, 2, 117-128 (1958); CA 52, 14131c

7464. Yoshida, F., and Koyanagi, T., "Liquid Phase Transfer Rates and Effective Interfacial Area in Packed Adsorption Columns," Ind. Eng. Chem., 50, 365-374 (1958); CA 52, 10654h

7465. Yoshimi, N., Yamauchi, T., Yamao, M., and Tanaka, S., "Hardening Reactions of Urea-Formaldehyde, Followed by Means of Gas Chromatography and Infrared Absorption Spectroscopy," Kogyo Kagaku Zasshi, 65, 1131-1135 (1962); CA 58, 626f

7466. Yoshimi, N., Yamauchi, T., Yamao, M., and Tanaka, S., "Analysis of Urea Resins. IX. Study of Hardening Reactions of Dimethylolurea Dialkyl Ethers by Gas Chromatography and Infrared Absorption Spectroscopy," Kogyo Kagaku Zasshi, 65, 1484-1487 (1962); CA 58, 10349h

7467. Yoshizawa, K., "The Formation of Higher Alcohols in the Fermentation of Amino Acids by Yeast. The Formation of [Optically] Active Amyl Alcohol from α-Ketobutyric Acid by Cell-Free Extract of Yeast," Agr. Biol. Chem. (Tokyo), 27, 162-164 (1963); CA 59, 2129h

7468. Yoshizawa, K., Furukawa, T., Tadenuma, M., and Yamada, M., "The Formation of Higher Alcohols in the Fermentation of Amino Acids by Yeast. The Determination of Alcohols with Gas Chromatography," Agr. Biol. Chem. (Tokyo), 25, 326-332 (1961); CA 55, 13758i

7469. Young, D. M., "A Doser for Admitting Measured Amounts of Vapor," Rev. Sci. Instr., 24, 77-78 (1953)

7470. Young, I. G., "The Sensitivity of Detectors for Gas Chromatography," Gas Chromatog., Intern. Symposium, 2nd, East Lansing, Mich., 1959, 75-83 (Pub. 1961); CA 55, 18203f

7471. Young, I. G., "The Sensitivity of Detectors for Gas Chromatography," in "Gas Chromatography," edited by H. J. Noebels, R. F. Wall, and N. Brenner, Academic Press, Inc., New York, 1961, pp. 75-83

7472. Young, J. A., Durrell, W. S., and Dresdner, R. D., "Fluorocarbon Nitrogen Compounds. IV. The Reaction of Metallic Fluorides with Carbon-Nitrogen Unsaturation in Perfluoro-2-azapropene," J. Am. Chem. Soc., 81, 1587-1589 (1959); CA 53, 16935g

7473. Young, J. F., "Theoretical Prediction of Resolution in a Gas Chromatography Column," 133rd Natl. ACS Mtg., San Francisco, Calif., April 1958, Program Abstr., p. 48B

7474. Young, J. F., "A Derivation of the Equation for Elution Chromatography Assuming Linear Rate Constants," in "Gas Chromatography," edited by V. J. Coates, H. J. Noebels, and I. S. Fagerson, Academic Press, Inc., New York, 1958, pp. 15-23

7475. Young, J. R., "The 'Family Plot' of Retention Volumes for Alkyl Ketones on Dinonyl Phthalate," Chem. & Ind. (London), 1958, 594-595; CA 53, 810e

7476. Young, J. R., "The Thermal Decomposition of Ketene," J. Chem. Soc., 1958, 2909-2921; CA 52, 19362f

7477. Young, W. G., Sharman, S. H., and Winstein, S., "Allylic Rearrangements. XLVII. The Silver Ion-Assisted Hydrolysis of α- and γ-Methylallyl Chlorides. Preservation of Configuration in Allylic Cations," J. Am. Chem. Soc., 82, 1376-1382 (1960); CA 54, 20834f

7478. Youngs, C. G., "Analysis of Mixtures of Amino Acids by Gas Phase Chromatography," Anal. Chem., 31, 1019-1021 (1959); CA 53, 14818h

7479. Youngs, C. G., "Glyceride Structure of Fats," J. Am. Oil Chemists' Soc., 36, 664-667 (1959); CA 54, 2785g

7480. Youngs, C. G., "Determination of the Glyceride Structure of Fats," J. Am. Oil Chemists' Soc., 38, 62-67 (1961); CA 55, 7869g

274

7481. Yu, C.-C., Yang, H.-C., Su, F.-C., Hsieh, C.-C., and Liu, C-H., "Gas-Liquid Chromatographic Separation and Identification of Methylchlorosilanes," Hua Hsueh Hsueh Pao, 25, No. 6, 420-423 (1959); CA 54, 20668g

7482. Yu, C-Y., Balandin, A. A., and Slovokhotova, T. A., "Gas Chromatography for Analysis of Catalytic Conversion Products of Isomeric Cresols," Vestnik Moskov. Univ., Khim., Ser. II, 16, No. 6, 62-66 (1961); CA 56, 10885a

7483. Yu, W-L., "Analysis of Dissolved Hydrocarbon Gases in Gasoline by Combined Adsorption and Gas-Liquid Partition Chromatography," Jan Liao Hsueh Pao, 2, 352-354 (1957); CA 52, 8519d

7484. Yu, W-L., and Tung, K-N., "Analysis of cis-2-Butene and Isopentane in Gasoline by Gas-Liquid Chromatography," Jan Liao Hsueh Pao, 4, 411-413 (1959); CA 54, 18944g

7485. Yuan, C., and Bloch, K., "Conversion of Oleic Acid to Linoleic Acid," J. Biol. Chem., 236, 1277-1279 (1961); CA 55, 17762g

7486. Yueh, M. H., and Strong, F. M., "Some Volatile Constituents of Cooked Beef," J. Agr. Food Chem., 8, 491-493 (1960); CA 55, 19058g; 135th Natl. ACS Mtg., Boston, Mass., April 1959, Program Abstr., p. 13A

7487. Yurovskii, Y. M., "The Separation Problem in the Analysis of Natural Gases," Razved. i Promysl. Geofiz., Vses. Nauchn.-Issled. Inst. Geofiz. Metodov Razvedki, 1959, No. 30, 71-73; CA 57, 3694i

7488. Yurovskii, Y. M., Vladimirov, B. V., and Galkin, L. A., "Chromathermographic Analysis of Hydrocarbon Gases in Gas Core-Sampling," Razved. i Promysl. Geoliz., 1959, No. 30, 60-70; CA 55, 10984f

7489. Zabor, R. C., and Emmett, P. H., "The Adsorption of Normal Paraffins on Cracking Catalysts," J. Am. Chem. Soc., 73, 5639-5643 (1951); CA 46, 2786i

7490. Zahn, C., and Langer, S. H., "Bibliography on Gas Chromatography," U. S. Dept. Interior, Bureau of Mines Information Circular 7856, 40 pp. (1958); CA 53, 809f

7491. Zakaib, D. D., "Direct Determination of C₂ to C₅ Hydrocarbons in Olefinic and Nonolefinic Gasolines by Gas-Liquid Chromatography," Anal. Chem., 32, 1107-1110 (1960); CA 54, 25726d; 11th Pittsburgh Conf. on Anal. Chem. & Appl. Spectroscopy, Pittsburgh, Pa., February-March 1960, Program Abstr., p. 38

7492. Zanini, C., Dal Pozzo, A., and Dansi, A., "Principal Components of Oil from Dwarf Pine Needles," Boll. chim. farm., 100, 83-92 (1961); CA 55, 20335g

7493. Zarembo, J. E., and Lysyj, I., "Use of a New Stationary Liquid Phase in Gas Chromatography Determination of Alcohols in the Presence of Large Amounts of Water," Anal. Chem., 31, 1833-1834 (1959); CA 54, 24130c

7494. Zawisza, A., "The Chromatographic Method of Gas Analysis and Its Industrial Uses," Koka, Smola, Gaz, 2, 157-165 (1957); CA 52, 2652c

7495. Zbiral, E., "The Effect of Bromine on Benzoquinol Acetate," Monatsh. Chem., 91, 280-288 (1960); CA 54, 24489g

7496. Zderic, J. A., Bonner, W. A., and Greenlee, T. W., "The Stereochemistry of Raney Nickel Action. VIII. Carbon-Carbon Bond Hydrogenlyses Catalyzed by Raney Nickel," J. Am. Chem. Soc., 79, 1696-1698 (1957); CA 51, 11296g

7497. Zeelenberg, A. P., "Slow Oxidation of Hydrocarbons in the Gas Phase. II. Neopentane," Rec. Trav. Chim., 81, 720-728 (1962); CA 58, 5487c

7498. Zehner, R. C., and Bain, J. C., "Glidden Aromatics," Drug & Cosmet. Ind., 86, 332-333 (1960); CA 54, 15831a

7499. Zeman, I., "Quantitative Interpretation of Fatty Acids Analysis by Means of Gas Chromatograph - Chrom I," Gas Chromatog. Symp., Brno, Czech., June 1962; Abstr., J. Gas Chromatog., 1, No. 4, 10 (1963)

7500. Zeman, I., and Pokorny, J., "Gas Chromatographic Analysis of Cyclopentenyl Fatty Acids," J. Chromatog., 10, 15-20 (1963)

7501. Zhdanov, S. P., Kalmanovskii, V. I., Kiselev, A. V., Fiks, M. M., and Yashin, Y. I., "Porous Glasses as Adsorbents in Gas Chromatography," Zh. Fiz. Khim., 36, 1118-1120 (1962); CA 57, 6581h

7502. Zhdanov, S. P., Kiselev, A. V., and Pavlova, L. F., "Adsorption of Benzene and n-Hexane Vapors and Their Liquid Solutions by Zeolites Types 10X and 13X," Kinetika i Kataliz, 3, 445-448 (1962); CA 58, 2868f

7503. Zhdanov, S. P., Yastrebova, L. S., and Koromal'di, E. V., "Porous Glasses as Molecular Sieves," Sintetich. Tseolity, Poluchenie, Issled. i Primenenie, Akad. Nauk S.S.S.R., Otd. Khim. Nauk, 1962, 68-74; CA 58, 8743c

7504. Zhukhovitskii, A. A., "Theory of Gas Chromatography," Uspekhi Khim., 28, 1201-1215 (1959); CA 54, 8215h

7505. Zhukhovitskii, A. A., "Some Developments in Gas Chromatography in the U.S.S.R.," in "Gas Chromatography 1960," edited by R. P. W. Scott, Butterworths, London, 1960, pp. 293-300

7506. Zhukhovitskii, A. A., "New Developments in Gas Chromatography," Gas Chromatography Symp., Brno, Czech., June 1962; Abstr., J. Gas Chromatog., 1, No. 4, 6 (1963)

7507. Zhukhovitskii, A. A., Kazanskii, B. A., Karimova, A. I., Pavlova, P. S., Sterligov, O. D., and Turkel'taub, N. M., "Chromatography of C_5 Hydrocarbon Mixtures," Zhur. Anal. Khim., 14, 721-728 (1959); CA 54, 12901f

7508. Zhukhovitskii, A. A., Kazanskii, B. A., Sterligov, O. D., and Turkel'taub, N. M., "Chromatographic Analysis of Hydrocarbon Mixtures of C_5-Composition," Doklady Akad. Nauk S.S.S.R., 123, 1037-1040 (1958); CA 53, 7718f

7509. Zhukhovitskii, A. A., Malyasova, L. A., and Turkel'taub, N. M., "Analysis of Unresolved Peaks with Similar Retention Times (Iteration Chromatography)," Neftekhimiya, 2, 831-836 (1962); CA 59, 1067g

7510. Zhukhovitskii, A. A., Selenkina, M. S., and Turkel'taub, N. M., "A Chromatographic Method of Identification of the Components of Hydrocarbon Mixtures," Khim. i Tekhnol. Topliv i Masel, 5, No. 11, 57-64 (1960); CA 55, 18084f

7511. Zhukhovitskii, A. A., Selenkina, M. S., and Turkel'taub, N. M., "Consecutive Columns in Gas Chromatography," Zh. Fiz. Khim., 36, 993-998 (1962); CA 57, 6581i

7512. Zhukhovitskii, A. A., and Turkel'taub, N. M., "Chromathermographic Method of Separation and Analysis of Gases," Uspekhi Khim., 25, 859-871 (1956); CA 50, 15324g

7513. Zhukhovitskii, A. A., and Turkel'taub, N. M., "Apparatus for the Continuous Analysis of Gases," Zavodskaya Lab., 22, 1252-1255 (1956); CA 51, 13471d

7514. Zhukhovitskii, A. A., and Turkel'taub, N. M., "Use of the Thermal Factor in Gas Chromatography," Doklady Akad. Nauk S.S.S.R., 116, 986-989 (1957); CA 52, 17892i

7515. Zhukhovitskii, A. A., and Turkel'taub, N. M., "Thermal Factors in Adsorption Methods of Separation of Mixtures," Uspekhi Khim., 26, 992-1006 (1957); CA 52, 4383c

7516. Zhukhovitskii, A. A., and Turkel'taub, N. M., "Errors in Chromatographic Analysis Due to Incomplete Separation," Zavodskaya Lab., 24, 796-798 (1958); CA 54, 24088i

7517. Zhukhovitskii, A. A., and Turkel'taub, N. M., "Chromatographic Methods and Instruments for Gas Analysis," Khim. Nauka i Prom., 4, 207-215 (1959); CA 53, 15858g

7518. Zhukhovitskii, A. A., and Turkel'taub, N. M., "The Thermal Factor in Chromatography," Gazovaya Khromatografiya, Trudy 1-oi [Pervoi] Vsesoyuz. Konf., Akad. Nauk S.S.S.R., Moscow, 1959, 107-117 (Pub. 1960); CA 55, 26823d

7519. Zhukhovitskii, A. A., and Turkel'taub, N. M., "Effectiveness Criteria in Gas Chromatography," Uspekhi Khim., 30, 877-894 (1961); CA 56, 36i

7520. Zhukhovitskii, A. A., and Turkel'taub, N. M., "Vacantochromatography," Doklady Akad. Nauk S.S.S.R., 143, 646-648 (1962); CA 57, 5283b

7521. Zhukhovitskii, A. A., and Turkel'taub, N. M., "Stepped Chromatography," Doklady Akad. Nauk S.S.S.R., 144, 829-832 (1962); CA 57, 9189e

7522. Zhukhovitskii, A. A., and Turkel'taub, N. M., "New Variants of Gas Chromatography for Automatic Regulation of Petrochemical Processes," Neftekhimiya, 2, 818-824 (1962); CA 58, 12351b

7523. Zhukhovitskii, A. A., and Turkel'taub, N. M., "Gazovaya Khromatografiya (Gas Chromatography)," Moscow: Gos. Nauchn.-Tekhn. Izd. Neft. i Gorn.-Toplivn. Prom., 1962, 442 pp.

7524. Zhukhovitskii, A. A., and Turkel'taub, N. M., "Increasing the Efficiency of Gas Chromatography," Zavodskaya Lab., 28, 133-136 (1962); CA 57, 1511c

7525. Zhukhovitskii, A. A., Turkel 'taub, N. M., and Georgievskaya, T. V., "Continuous Chromathermography," Doklady Akad. Nauk S.S.S.R., 92, 987-990 (1953)

7526. Zhukhovitskii, A. A., Turkel'taub, N. M., and Shvartsman, V. P., "The Theory of Chromatography," Zhur. Fiz. Khim., 28, 1901-1909 (1954); CA 50, 7541b

7527. Zhukhovitskii, A. A., Turkel'taub, N. M., and Sokolov, V. A., "Theory of Chromathermography," Doklady Akad. Nauk S.S.S.R., 88, 859-862 (1953); CA 47, 11882d

7528. Zhukhovitskii, A. A., Turkel'taub, N. M., Vagin, E. V., and Shvartsman, V. P., "The Spreading of Bands During Chromathermographic and Thermal Separation," Doklady Akad. Nauk S.S.S.R., 96, 303-306 (1954); CA 51, 7223i

7529. Zhukhovitskii, A. A., Zolotareva, O. V., Sokolov, V. A., and Turkel'taub, N. M., "New Method of Chromatographic Analysis," Doklady Akad. Nauk S.S.S.R., 77, 435-438 (1951); CA 46, 11011b

7530. Ziebland, H., "The Thermal Conductivity of Nitrogen, Oxygen, and Argon in the Liquid and Gaseous State," Dechema Monograph, 32, 74-82 (1959); CA 54, 1002d

7531. Ziebland, H., and Burton, J. T. A., "The Thermal Conductivity of Nitrogen and Argon in the Liquid and Gaseous States," Brit. J. Appl. Phys., 9, No. 2, 52-59 (1958); CA 52, 10704g

7532. Ziegenhain, W. C., "Molecular Sieves," Petrol. Engr., 29, No. 9, C6-C12 (1957); CA 51, 15103f

7533. Zielinski, E., "Determination of Helium and Neon in Gases by Chromatography," Chem. Anal. (Warsaw), 5, 297-307 (1960)

7534. Zielinski, E., "Neon as Eluent in the Gas Chromatographic Determination of Helium in Gases," Chem. Anal. (Warsaw), 5 605-610 (1960); CA 57, 4021e

7535. Zielinski, Jr., W. L., Moseley, Jr., W. V., and Bricker, R. C., "Gas Chromatography for the Qualitative Determination of Oil Content in Organic Coatings," Offic. Dig. Federation Soc. Paint Technol., 33, 622-634 (1961); CA 55, 17039i; 138th Natl. ACS Mtg., New York, N. Y., September 1960, Program Abstr., pp. 7Q-8Q

7536. Zientara, F., and Owades, J. L., "Analysis of American and Foreign Beers by Gas Chromatography," Am. Brewer, 93, 37-39 (1960); CA 55, 2011d

7537. Ziffer, H., VandenHeuvel, W. J. A., Haahti, E. O. A., and Horning, E. C., "Gas Chromatographic Behavior of Vitamins D_2 and D_3," J. Am. Chem. Soc., 82, 6411-6412 (1960); CA 55, 13522b

7538. Zike, J., "Process Chromatograph Controls Deethanizer Operation," Oil Gas J., 59, No. 7, 92-94 (1961); CA 56, 10856h

7539. Zilversmit, D. B., Sweeley, C. C., and Newman, H. A., "Fatty Acid Composition of Serum and Aortic Intimal Lipids in Rabbits Fed Low- and High-Cholesterol Diets," Circulation Research, 9, 235-241 (1961); CA 55, 12572b

7540. Zinn, T. L., Baker, W. J., Norlin, H. L., and Wall, R. F., "Exploratory Process Gas Chromatography," in "Gas Chromatography," edited by V. J. Coates, H. J. Noebels, and L. S. Fagerson, Academic Press, Inc., New York, 1958, pp. 281-285

7541. Zinn, T. L., Baker, W. J., and Wall, R. F., "The Use of Hydrogen as a Carrier Gas in Process Gas Chromatography," 1958 Natl. ISA Mtg., Symp. on Instrumental Methods of Analysis, Houston, Texas, May 1958, Proc., pp. 251-254

7542. Zlatkis, A., "Resolution of Isomeric Hexanes by Gas-Liquid Chromatography," Anal. Chem., 30, 332-333 (1958); CA 52, 10808e; 132nd Natl. ACS Mtg., New York, N. Y., September 1957, Program Abstr., p. 38B; Preprints, Div. Petrol. Chem., 2, No. 4, D127-D129 (1957)

7543. Zlatkis, A., "The Use of Capillary Columns in Gas Chromatography," Symp. on Gas Chromatography, CIC, Toronto, Ontario, February 1960

7544. Zlatkis, A., "Currents in Gas Chromatography," 11th Pittsburgh Conf. on Anal. Chem. & Appl. Spectroscopy, Pittsburgh, Pa., February-March 1960, Program Abstr., p. 38

7545. Zlatkis, A., "Ionization Detectors and Capillary Columns," in "Lectures on Gas Chromatography 1962," edited by H. A. Szymanski, Plenum Press, Inc., New York, 1963, pp. 87-104; 4th Annual Gas Chromatography Institute, Canisius College, Buffalo, N. Y., April 1962

7546. Zlatkis, A., "Electron Capture Detection Systems," 10th Detroit Anachem Conf., Detroit, Mich., October 1962, Program Abstr., p. 31

7547. Zlatkis, A., "New Developments in Detection Systems for Gas Chromatography," 5th Annual Gas Chromatography Institute, Canisius College, Buffalo, N. Y., April 1963

7548. Zlatkis, A., and Kaufman, H. R., "Use of Coated Tubing as Columns for Gas Chromatography," Nature, 184, 2010 (1959); CA 54, 14814c

7549. Zlatkis, A., and Kaufman, H. R., "New Approaches to the Resolution of Olefins," Gas Chromatog., Intern. Symposium, 2nd, East Lansing, Mich., 1959, 339-342 (Pub. 1960); CA 55, 23309f

7550. Zlatkis, A., and Kaufman, H. R., "New Approaches to the Resolution of Olefins," in "Gas Chromatography," edited by H. J. Noebels, R. F. Wall, and N. Brenner, Academic Press, Inc., New York, 1961, pp. 339-342

7551. Zlatkis, A., Ling, S., and Kaufman, H. R., "Resolution of Isomeric Xylenes by Gas-Liquid Chromatography," Anal. Chem., 31, 945-947 (1959); CA 53, 18727f; 134th Natl. ACS Mtg., Chicago, Ill., September 1958, Program Abstr., p. 23B

7552. Zlatkis, A., and Lovelock, J. E., "Gas Chromatography of Hydrocarbons Using Capillary Columns and Ionization Detectors," Anal. Chem., 31, 620-621 (1959); CA 53, 11902c

7553. Zlatkis, A., O'Brien, L., and Scholly, P. R., "Gas Chromatographic Separation of Meta- and Para-Xylenes in Aromatic Mixtures," Nature, 181, 1794 (1958); CA 52, 20995e

7554. Zlatkis, A., and Oro, J. F., "Amino Acid Analysis by Reactor-Gas Chromatography," Anal. Chem., 30, 1156 (1958); CA 52, 15345a; 133rd Natl. ACS Mtg., San Francisco, Calif., April 1958, Program Abstr., p. 13C

7555. Zlatkis, A., Oro, J. F., and Kimball, A. P., "Direct Amino Acid Analysis by Gas Chromatography," Anal. Chem., 32, 162-164 (1960); CA 54, 9612h

7556. Zlatkis, A., and Ridgway, J. A., "Methane Conversion Detector for Gas Chromatography," Nature, 182, 130-131 (1958); CA 55, 19353d

7557. Zlatkis, A., and Sivetz, M., "Analysis of Coffee Volatiles by Gas Chromatography," Food Res., 25, 395-398 (1960); 135th Natl. ACS Mtg., Boston, Mass., April 1959, Program Abstr., p. 13A

7558. Zlatkis, A., and Walker, J. Q., "Capillary Columns for Gas-Solid Chromatography," 13th Pittsburgh Conf. on Anal. Chem. & Appl. Spectroscopy, Pittsburgh, Pa., March 1962, Program Abstr., p. 50

7559. Zlatkis, A., and Walker, J. Q., "Direct Sample Introduction for Large Bore Capillary Columns in Gas Chromatography," J. Gas Chromatog., 1, No. 5, 9-11 (1963)

7560. Zlatkis, A., and Walker, J. Q., "Surface Modification of Capillary Columns for Use in Gas Chromatography," Anal. Chem., 35, 1359-1362 (1963)

7561. Zmitko, J., Brodsky, J., and Biza, V., "Automatic Recording of Gas Chromatography Analyses,"

Chem. Tech. (Berlin), 9, 458-459 (1957); CA 52, 7788e

7562. Zmitko, J., Brodsky, J., and Biza, V., "Automatic Recorder of Gas-Chromatographic Analysis," Chem. prùmysl, 7, 414-416 (1957)

7563. Znamenskaya, N. B., and Korol, A. N., "Decreasing the Analysis Time for C_{1-4} Hydrocarbon Mixtures. [High-Speed Gas Chromatography with Packed Columns]," Tr., Nauchn.-Issled. Inst. Avtomatiz. Proizv. Khim. Prom. i Tsvetn. Met., 1962, No. 10, 70-74; CA 59, 1064c

7564. Zolotareva, O. V., "Chromathermographic Analysis of Natural Gases," Trudy Vsesoyuz. Nauch.-Issledovatel. Geologorazvedoch. Neft. Inst., 1958, No. 11, 245-256; CA 55, 3954e

7565. Zomzely, C., Marco, G., and Emery, E., "Gas Chromatography of the n-Butyl-N-trifluoroacetyl Derivatives of Amino Acids," Anal. Chem., 34, 1414-1417 (1962); CA 57, 15781e; 140th Natl. ACS Mtg., Chicago, Ill., September 1961, Program Abstr., p. 48C

7566. Zorin, A. D., Ezheleva, A. E., and Devyatykh, G. G., "Analysis of Butane-Butylene Fractions by Gas-Liquid Partition Chromatography," Trudy Khim. in Khim. Tekhnol., 1, 605-610 (1958); CA 54, 6385f

7567. Zubyk, W. J., and Conner, A. Z., "Analysis of Terpene Hydrocarbons and Related Compounds by Gas Chromatography," Anal. Chem., 912-917 (1960); CA 54, 19313e

7568. Zulaica, J., and Guiochon, G., "Analysis of Plasticizers by Gas Chromatography," Compt. Rend., 225, 524-526 (1962); CA 57, 13981a

7569. Zulaica, J., and Guiochon, G., "Fast Qualitative and Quantitative Microanalysis of Plasticizers in Plastics by Gas Liquid Chromatography," Anal. Chem., 35, 1724-1728 (1963)

7570. Zulaica, J., Landault, C., and Guiochon, G., "Application of Gas Phase Chromatography to the Study of Plasticizers," Bull. soc. chim France, 1962, 1294-1301; CA 57, 15350a

7571. Zweig, G., "Pesticide Residue Analysis by Combination Gas Chromatography and Colorimetry," 140th Natl. ACS Mtg., Chicago, Ill., September 1961, Program Abstr., p. 61B

7572. Zweig, G., and Archer, T. E., "Determination of Thiodan by Gas Chromatography," J. Agr. Food Chem., 8, 190-192 (1960); CA 55, 14791h; 136th Natl. ACS Mtg., Atlantic City, N. J., September 1959, Program Abstr., p. 37A

7573. Zweig, G., and Archer, T. E., "Trace Analysis of Pesticide Chemicals by a Combination of Gas Chromatography and Infrared Spectrophotometry," 138th Natl. ACS Mtg., New York, N. Y., September 1960, Program Abstr., p. 12A

7574. Zweig, G., Archer, T. E., and Beckman, H. F., "Pesticide Residue Determination by a Combination of Gas Chromatography and Spectrophotometry," 18th Intern. Congr. of Pure & Appl. Chemistry, Montreal, Canada, August 1961

7575. Zweig, G., Archer, T. E., and Raz, D., "Sprout Inhibitor Residues. Residue Determination of Naphthaleneacetic Acid and Its Methyl Ester in Potatoes by a Combination of Gas Chromatography and Ultraviolet Spectrophotometry," J. Agr. Food Chem., 10, 199-203 (1962); CA 57, 10298i

7576. Zweig, G., Archer, T. E., and Rubenstein, D., "Residue Analysis of a Chlorinated Insecticide (Thiodan) by a Combination of Gas Chromatography and Infrared Spectroscopy," J. Agr. Food Chem., 8, 403-405 (1960); CA 55, 17997f

7577. Zwierzak, A., and Pines, H., "Base Catalyzed Reactions. XXIV. Equilibration of cis- and trans-Methylstilbene in the Presence of Potassium tert-Butoxide as Catalyst," J. Org. Chem., 27, 4084-4085 (1962); CA 58, 8866d

Analytical Chemistry, 188-194
Analyst, 195, 196
Analyzer, 197
British Chemical Engineering, 198
Bunseki Kagaku, 199
Burrell Corporation, 200
Canadian Chemical Processing, 201
Chemical Age, 202, 203
Chemical Engineering, 204 206
Chemical & Engineering News, 207-271
Chemical Engineering Progress, 272
Chemistry in Canada, 273
Chemistry & Industry (London), 274, 275
Chemical Processing, 276-278
Chemical & Processing Engineering, 279
Chemical Week, 280-282
Chimie Analytique, 283, 284
Chimica e Industria, 285
Control Engineering, 286, 287
Cryogenic Engineering Laboratory, Natl. Bureau Standards, 288
Erdol u. Kohle, 289
European Technical Digest, 290
Federation Paint and Varnish Production Club, Official Digest, 291
Gas-Chrom Newsletter, 292
Gas Pipe, 293, 294
Ideas in Development, 295
Industrial & Engineering Chemistry, 296
Industrial Laboratories, 297
International Perfumer, 298
ISA Journal, 299-303
Journal of Agricultural and Food Chemistry, 304
J-M Celite Div., Johns-Manville, 305-308
Inst. Sewage Purification, J. Proc., 309
Manufacturing Chemist, 310
Nature, 311
Nutrition Reviews, 312
Oil & Gas Journal, 313-316
Perfumery Essential Oil Record, 317-320
Perkin-Elmer Instrument News, 321-328
Pesticide Research Bulletin, 329
Petroleum Week, 330-332

Reichstoffe u. Aromen, 333
Research for Industry, 334
Seifen, Ole, Fette, Wachse, 335
South African Industrial Chemist, 336
What's New, 337
Anselmi, S., 338-342
Anson, F. C., 5963
Anson, P. C., 343, 344
Antenucci, F. M., 4741
Anthony, D. S., 345
Anthony, K. V., 3244
Antila, M., 4325
Antonaccio, L. D., 1872
Antonis, A., 346
Antunano, N. C., 7257
Anvaer, B. I., 6911
Aoyagi, M., 6038
Aoyanagi, M., 347
Aplin, R. J., 2455
Apostolakis, M., 348
Applequist, D. E., 349-353
Applewhite, T. H., 829
April, A., 354
Arad-Talmi, Y., 355
Arai, S., 356, 357
Araki, N., 2708
Araki, S., 358-362, 4716
Araki, T., 363-366
Araki, Y., 367
Aranda, V. G., 368
Aratani, T., 369, 4734-4737
Archer, D. A., 370, 371
Archer, E. D., 372, 2348
Archer, M., 1387
Archer, T. E., 373, 7572-7576
Arcus, A. C., 374
Arcus, C. L., 375-377
Ardizio, P., 5154
Argabright, P. A., 4894
Argyle, P., 6640
Aris, R., 378, 379
Aristov, B. G., 380
Aristova, V., 381
Arkell, A., 382, 5216
Arkima, V., 6323
Armbruster, C. W., 2838
Armitage, F., 383
Armstrong, N. W., 384
Armstrong, W. H., 5277
Arndt, R. R., 385
Arnett, E. M., 386-390
Arnold, J. H., 391
Arnold, L. K., 392, 393
Arnold, R., 394
Arnold, R. T., 395
Arpino, A., 396
Arsenault, G. P., 7411
Arthur, H. R., 397
Artsybasheva, Y. P., 398
Arumeel, E., 399, 2042

Arvidson, G., 400
Asahara, T., 401-403
Asai, K., 3011
Asatoor, A., 404
Ascoli, F., 405, 406
Ashbury, G. K., 407, 408
Ashikawa, J. K., 5767, 5768
Ashley, Jr., J. W., 409, 5771, 5772
Asinger, F. A., 410
Askew, W. B., 5859
Askins, J. W., 411
Aspinall, G. O., 412-417
Asscher, M., 418
Asselineau, C., 419
Asselineau, J., 419-422, 1684
Ast, H. J., 423
Atherley, J. F., 95
Atkinson, E. P., 424, 425
Atkinson, E. R., 6548
Atkinson, R. d'E., 426
Attaway, J. A., 427, 428, 7120, 7408-7410
Attrill, J. E., 429
Atwood, J. G., 430
Aubeau, R., 431-435
Audette, R. C. S., 436
Audinot, M., 5514
Audran, R., 437
Aufdermarsh, Jr., C. A., 7269
Augustine, R. L., 438
Aul, F., 2814, 2815, 2818, 2819
Aunstrup, K., 439
Aurand, L. W., 7419
Ausloos, P., 440, 441, 950
Aust, R., 442
Austin, F. L., 443
Avancini, D., 1740-1743
Avdeeva, A. A., 444
Aveeva, M. P., 7097
Avens, A. W., 4741
Averill, W., 445-458, 1400-1402, 1493, 1494, 2124-2127, 2132, 2597
Aves, E. K., 1345
Avgul, V. T., 459, 460, 1371
Avodonina, E. N., 5210
Axelrod, L. R., 3122
Ayers, B. O., 461-464, 1737-1739, 3810, 3812, 4513, 4514
Aylward, F., 465-467
Ayrapaa, T., 468
Ayres, S. M., 4534
Aznavourian, W., 469, 470, 4807

Baba, T., 471
Babayan, V. K., 3867
Babini, B., 472, 4625, 5375, 6015
Baboshin, B. K., 6317
Bacchetta, V. L., 4173
Bachman, G. B., 473

280

Bachmann, L., 474
Bachmann, O., 1925
Bächtold, H., see Baechtold, H.
Baddiel, C. B., 475
Baddour, R. F., 2599, 2600
Bade, M. L., 476
Bader, R. F. W., 6658
Badger, G. M., 477-486
Badings, H. T., 487, 488
Baechtold, H., 2207, 2208
Baeder, D. L., 4524
Bagby, M. O., 489, 6392
Bagwell, E. E., 2446
Bahr, G., 490
Baikowitz, H., 3021
Bailey, G. F., 5726
Bailey, S. D., 491
Bailey, W. J., 492-496
Baillie, J., 413, 414
Bailly, F. H., 497
Bain, J. C., 7498
Baines, C. B., 498
Bair, L. R., 4721
Baird, J. H., 4820
Baitinger, W. E., 135-137
Baker, A. R., 499
Baker, B. E., 500
Baker, B. G., 153
Baker, B. R., 5780, 5781
Baker, C. D., 501, 502
Baker, D., 503
Baker, E. B., 1164
Baker, J. M., 504
Baker, R. A., 505, 506
Baker, R. H., 6059
Baker, R. W. R., 507, 508
Baker, W. J., 510-515, 3019,
   6252, 7138, 7540, 7541
Balandin, A. A., 516, 7482
Balarev, K., 4953
Balasubrahmanyam, S. N., 517,
   5733
Baldwin, R. A., 518
Baldwin, W. H., 3143
Ball, D. H., 519, 4803
Ball, J. S., 3094, 4289
Ballance, P. E., 520
Ballard, H. S., 4642, 4643
Ballinger, P., 521, 560
Ballod, A. P., 522
Bambara, P., 4212
Bambrick, W. E., 3381, 3382
Banerjee, S. K., 523
Bang, H. O., 524, 3616, 3617
Banister, P. G., 1560
Banks, R. E., 525
Bannerman, M. A., 17-19
Bannister, D. W., 526
Bannister, M. H., 527, 528, 7316
Baque, C., 529, 530
Barabanov, N. L., 531

Baragli, S., 5911
Baraud, J., 532-537, 2509, 4246
Barber, D. W., 538
Barber, H. J., 539
Barbier, M., 540, 5367
Barbour, W., 541
Barbul, M., 5576-5578
Barcellini, A., 3989
Barclay, L. R. C., 542
Bardwell, J., 543
Barefoot, R. R., 544, 545
Barefort, R. D., 7321
Baret, C., 5514
Barford, R. A., 546, 3114, 3115,
   4531, 4595, 4873
Barger, B. D., 2406
Barkalov, I. M., 2004
Barker, E., 547
Barker, H., 548
Barker, K. H., 2638
Barker, P. E., 549
Barlow, A., 550
Barlow, D. O., 3207
Barlow, J. S., 551
Barnard, D., 552
Barnard, J. A., 553-557
Barner, R., 2874
Barnes, A. C., 3698, 4211, 6933
Barnes, C. S., 558
Barnes, Jr., I. L., 5133
Barnes, M. M., 2832
Baron, C., 559
Barr, J. K., 6067-6069
Barr, J. T., 3857
Barrall, II, E. M., 560, 561
Barras, R. C., 562
Barrer, R. M., 38, 563-605
Barrera, J. B., 2693
Barrett, F. C., 1057
Barrionuevo, M., 6168, 6187
Barron, E. J., 2941
Barrow, J. G., 606
Barrow, R. E., 5008
Barry, D. L., 4741
Barry, J. A., 607
Barry, R., 608
Bars, A., 5820, 5821
Bars, E. A., 6454
Barsky, M. H., 609
Bartels-Keith, J. R., 610
Barter, Jr., C. J., 4036
Barthel, W. F., 7228
Bartkus, E. A., 6465
Bartle, J., 611
Bartlet, J. C., 612, 5228, 6399
Bartley, W., 613, 831, 2532
Bartley, W. J., 1061, 7282
Bartok, W., 614, 615, 4522,
   6759
Barton, D. H. R., 616
Bartsch, R. C., 617, 1677
Bashkirov, A. N., 618

Basmadjian, D., 619, 620
Bassemir, R. W., 621
Bassery, L., 2520, 4748
Bassett, D. W., 622, 623
Bassette, R., 624-627, 5361
Bassler, L., 656
Basson, R. A., 628
Bastar, L., 5739
Bastin-Merkeman, M. J., 629
Bate, R., 3902
Bates, R. B., 630-634
Bates, T. H., 635
Bathory, J., 2375
Baum, M. E., 7118
Baumann, F., 636, 2466, 2467,
   3659,6473
Baumann, F. B., 637, 638
Baumann, G. P., 6538
Baumgarten, H. E., 639
Bavisotto, V., 7000
Bavisotto, V. S., 640-644
Bavley, A., 2776
Baxter, R. A., 645-647, 3889
Baxter, R. M., 436, 648, 649,
   3852
Bayer, E., 650-665
Bayes, K., 666
Bayles, G. G., 667
Baylouny, R. A, 492, 493, 668
Baynham, J. W., 576-578
Bazant, V., 3688, 4102, 4118,
   4119, 4121, 4122
Bazinet, M. L., 491, 669, 4884,
   4885
Beach, Jl Y., 3659
Beal, R. E., 2386
Beale, P. A. A., 1758
Beard, W. E., 6566
Beare, J. L., 670-672, 1573
Beason, L. R., 5758
Beaufils, J. P., 1313
Beaven, G. H., 673, 3650
Beaver, G. H., 1460
Becher, P., 674
Bechtold, E., 474, 1594, 4161
Beck, E., 675, 1914
Beck, E. C., 676, 3724, 3725
Beck, J. E., 4680
Beck, M. G., 677-679, 4927,
   4930,6103
Beck, M. S., 384
Beck, P. E., 2601
Becker, E. I., 4477, 5909
Beckert, W., 5812
Beckman, H., 680-683
Beckman, H. F., 684, 7574
Beckman Instruments, Inc., 685
Beckstrom, R. D., 3858
Becktenwald, G. W., 5546
Beckwith, A. L. J., 686
Bednas, M. E., 687, 5965
Beebe, R. A., 688, 3213

Beech, J. A., 6025
Beersum, W. van, 3363, 3364
Beerthuis, R. K., 689-693
Beggs, W. S., 4365
Behmann, F. W., 694
Behr, O. M., 695
Behrendt, S., 696-698
Behrens, U., 6814
Beilina, L. I., 4171
Bekesy, M., 699
Bekkum, H. van, 700
Belchetz, L., 579
Bell, J. H., 6471
Bell, K. M., 701
Bellar, T. A., 105, 106, 702-704
Bellis, H. E., 705
Beltran, E. G., 5177
Beltz, P. B., 3162
Belyakova, L. D., 706
Benard, C., 4951
Bendel, E., 707
Bender, S. R., 708, 4180, 4181
Bendoraitis, J. G., 709, 3112
Bendz, G., 710
Benedek, P., 711-722, 2376, 2377, 6693
Benediktova, V., 723
Ben-Efraim, D. A., 724
Benford, C. L., 1608
Benjamin, W., 725
Benkeser, R A., 726, 727
Bennett, C E., 384, 678, 728-732, 1654, 1655, 4657-4661
Bennett, R. D., 3647
Benneville, P. L. de, 3366
Benoy, G., 5530
Bens, E. M., 733-737, 4761
Bensadoun, A., 738
Benson, G. W., 739
Benson, K. A., 844
Benson, R. E., 740
Benson, S. W., 741
Bentley, R., 742
Berces, T., 743
Bercev, B., 744
Berchtold, G. A., 1513
Berck, B., 745
Berdick, M., 746, 4578, 4581-4583, 4778
Berezkin, V. G., 747-753
Berg, C., 754-758
Berg, E. W., 979
Berge, P. C. van, 760
Berger, A., 979
Berger, C. R. A., 5281
Berger, R., 761
Bergmann, G., 762, 763, 3610, 3611
Bergmann, J. G., 764
Bergmann, W., 4259
Berka, I., 765
Berkenkotter, P., 684

Berkowitz-Mattuck, J. B., 766
Berl, W. G., 767
Bernal, P. G., 6326
Bernard, A. H., 768
Bernhard, R. A , 769-777, 1418, 1419
Bernick, S., 5030
Bernsohn, J., 778
Beroes, C. S., 779
Beroza, M., 26, 27, 780-782, 3419
Berridge, N. J., 783
Berry, J. F., 4175
Berry, R., 784, 785, 7387
Bersohn, I., 346
Berson, J. A., 786-790
Berton, A., 791, 792, 2611
Bertrand, D., 6004
Beschea, C., 5576, 5577
Bestougeff, M. A., 793
Bethea, R. M., 794-801
Bethge, P. O., 802
Bethune, J. L., 803, 804, 5818, 5819
Beuerman, D. R., 805, 806
Bevenue, A., 373, 680-683, 807
Beveridge, J. M. R., 4208, 4766
Bevilacqua, E. M., 808, 809
Bey, K., 810
Beyer, H., 7399
Beynon, J. H., 811-814
Beynon, K. I., 815
Bezard, J., 1438
Bhalerao, V. R., 816, 817
Bhati, A., 818
Bhatnagar, V. M., 819
Bhattacharyya, S. C., 3422
Biddiscombe, D. P., 820
Bidmead, D. S., 821
Biegler, R., 898, 6157-6163
Biemann, K., 822, 823, 6239
Bier, M., 824
Biermann, W. J., 825
Bierhacki, W., 826
Biggers, R. E., 827
Bignardi, G., 828
Billek, G., 4148
Biller, W. F., 5710
Bills, D. D., 4380
Binder, R. G., 829
Bindernagel, H., 830
Bingham, S. A., 2813-2815
Biran, L. A., 831, 5784
Birch, S. F., 832
Birchall, J. M., 833
Bird, G. R., 834
Birdwell, B. F., 835
Birkmeier, R. L., 674
Bischoff, C., 836
Bishop, C. A., 1761
Bishop, C. T., 34, 837-841, 4803, 5444

Bishop, D. G., 842
Bishop, J. R., 843
Bissot, T. C., 844
Biza, V., 7561, 7562
Black, G. S., 7168
Black, L. T., 845
Blackie, A., 846, 847
Blackmore, R. L., 848
Blackwell, J., 849
Blades, A. T., 850
Blades, J., 4292, 5042
Blaha, J., 2339
Blair, J. W., 851
Blake, A. R., 852, 853
Blackmore, G., 854
Blakeway, J. M., 855
Blanc, C., 2841
Blanchard, Jr., E. P., 856
Blanchard, H. S., 857
Blanchard, K. R., 858
Blanchard, M., 2520-2523
Bland, D. E., 4148
Blandenet, G., 5851
Blank, F., 838, 859
Blank, M. L., 5665
Blankenhorn, D. H., 860, 3581
Blankenship, L. C., 76
Blaustein, B. D., 861
Blay, N. J., 862, 863, 1973, 7339
Blecharczyk, S. S., 864
Blinn, R. C., 2832-2835
Bloch, K., 7485
Bloch, M. G., 865, 866
Blom, L., 867-869
Blomquist, A. T., 870
Blomstrand, R., 871-878, 1829
Bloomfield, D. K., 879-881
Blount, W. P., 1530
Blum, H. A., 882
Blum, M. S., 883
Blumenthal, J. L., 884
Blumer, M., 5669
Blundell, R. V., 885, 886
Bly, R. K., 1514
Blyholder, G., 887-889
Boatman, C., 890, 2931
Bobak, A., 6356-6358, 6361
Bober, M., 2710
Bobrova, V. P., 4124
Boch, R., 891
Bochinski, J., 6550
Bochinski, J. H., 892-894, 3707
Bock, H., 895
Boddy, P. J., 896
Bodnar S. J., 897
Boehle, E., see Böhle, E.
Boehm, E. E., 899, 900
Boeke, J., 901-904
Boelsma-van Houte, E., 967, 970-973
Boer, H., 905, 906

Boettcher, C. J. F., see
  Böttcher, C. J. F.
Boettger, H. G., 907, 908
Bogen, P., 3612
Boggs, M. M., 1183
Boggus, J. D., 883, 909
Bogue, D. C., 910
Bohemen, J., 911-916
Böhle, E., 898, 6157-6163
Bohm, Z., 917, 918
Boileau, J., 3395
Boisselle, A. P., 3109
Bokhoven, C., 919
Bolland, J. L., 920
Bolling, J. M., 921
Bollyky, L., 7140
Boltz, D. F., 1691
Bombaugh, K. J., 922-927
Bomstein, J., 3170, 3171
Bond, G. C., 928-930
Bond, J., 2041
Bonelli, E., 931
Boniforti, L., 339-342, 932, 933
Bonner, W. A., 934, 935, 7496
Bonnet, E., 6756
Bonnet, J., 936
Bonnichsen, R., 937
Bonnier, J. M., 938, 939
Boord, C. E., 940, 941, 4231-
  4233, 4880
Booth, H., 370, 371, 942
Booth, V. H., 7353
Boothby, W. M., 7334
Borchert, A. E., 5356
Borecky, J., 2484
Boreham, G. R., 943-945
Boren, H. G., 946
Borer, K., 947, 948
Borfitz, H., 949
Borke, M. L., 6691
Borker, E., 6378
Borkowski, R., 950
Born, F., 662, 1927
Bornemann, P., 951
Borsch, R. J., 2790
Bortner, T. E., 4878
Boruff, C. S., 443
Borup, R. E., 4471
Bosanquet, C. H., 952-954
Bosch, J. van den, 2168, 3717
Boschan, R., 955
Bosin, W. A., 956, 957
Boskin, M. J., 1752
Bossart, C. J., 958, 4604
Bosshard, E., 959
Bota, T., 960
Boteiu, A., 2825
Bothe, H. K., 961-963
Bothner-By, A. A., 4146
Botquin, G., 1979
Böttcher, C. J. F., 964-973
Botter, F., 974

Boufford, C. E., 975
Boughton, B., 976, 977
Boulet, G. A., 4471
Bouthilet, R. J., 978
Bovijn, L., 979
Bowman, E. R., 4816
Bowman, R. E., 980
Bowman, R. L., 981, 3821-3828,
  3830-3832
Boyce, C. B., 6592
Boyce, E. A., 6609
Boyd, C. M., 429
Boyd, G. S., 982
Boyle, J. F., 562
Boyle, Jr., J. J., 983, 984
Boys, S. L., 985
Bozak, R. E., 1682
Brace, R. O., 986, 6569, 7052,
  7053
Bracht, G., 987
Bradford, B. W., 988, 989
Bradley, D. C., 990
Bradley, J. K., 991
Bradley, J. N., 992
Bradley, Jr., R. L., 993
Bradley, R. M., 995
Bradley, W. E., 757
Bradow, R. L., 4945
Brady, A. P., 994
Brady, R. O., 995
Bragdon, J. H., 996
Braid, M., 3034, 3044
Brako, F., 6269
Braman, R. S., 4207
Branch, R. F., 997
Brand, J. C. D., 998
Brand, L., 136, 137
Brandenberger, H., 999
Brandli, J., 5719
Brandsch, J., 2825
Brandt, L. W., 1000
Brandt, W. W., 1001-1006, 1988,
  3297-3299, 6149, 6150
Brasseaux, D. J., 6183-6185
Brauer, G. M., 1007, 4322, 6605
Brauer, H., 1008
Brauner, K., 2207
Braverman, J. B. S., 1009
Brazhnikov, V. V., 2498
Brealey, L., 1010
Breck, D. W., 1011, 1012, 5762
Bredel, H., 1013, 1014, 5670
Breed, L. W., 1015, 2881
Brennan, D., 1016-1018
Brennan, H. M., 2406
Brenner, H. H., 1625
Brenner, N.. 1019-1044, 1454,
  1455, 2128-2131, 2133-2135,
  2177, 3048-3051, 4607, 5244,
  6143, 6144
Brenner, R. R., 1045
Bresky, D. R., 1042, 1043, 1430

Bresler, S. E., 1046-1050
Breslow, D. S., 3069
Bresnick, S. R., 4884, 4885
Breton, C., 1051
Breton, J. L., 1052
Bretting, C., 5078
Brewer, J. E., 4319
Brewerton, H. V., 527
Brhacek, L., 4168, 4169
Bricker, R. C., 7535
Bricteux, J., 1524-1528
Brida, J., 6356-6361
Brieskorn, C. H., 1053
Brill, W. F., 1054, 1055
Brillyantov, N. A., 1056
Brimley, R. C., 1057
Brinckman, F. E., 1058
Brinton, R. K., 1059
Briscoe, F. J., 4608
Brister, T. B., 1060
Broadbent, H. S., 1061
Brobst, K. M., 1332
Brochere-Ferreol, G., 1062
Brochmann-Hanssen, E., 1063-
  1068
Brockerhoff, H., 2941
Broda, K., 6708
Brodel, H., 1069
Brodie, B. B., 3253
Brodskii, A. M., 1070-1077
Brodsky, J., 1078, 1079, 7561,
  7562
Brokaw, R. S., 1179
Bromley, L. A., 1080
Brook, B. M., 1081
Brook, D. W., 580, 581
Brook, J. H. T., 1082
Brooks, C. J. W., 1083, 1084
Brooks, F. R., 2360
Brooks, G. T., 1085
Brooks, J., 1086, 1087
Brooks, V. T., 1088-1092
Broom, A. D., 438
Broome, J., 1093
Broughton, D. B., 1094
Broughton, M. E., 2272
Broussard, L., 1095
Brown, A. L., 1096
Brown, A. W. A., 2199
Brown, B. L., 709
Brown, B. R., 1093
Brown, G. M., 1097
Brown, G. R., 1098
Brown, H. C., 1099-1109, 4652,
  6582, 6583
Brown, H. O., 7334
Brown, I., 1110, 1111
Brown, J. B., 2277, 3261, 4249,
  6492
Brown, J. F., 1112
Brown, L., 5586, 5587
Brown, M. F., 702

Brown, P., 1113
Brown, P. M., 3840-3842
Brown, R. A., 4360, 4361
Brown, S. A., 1114
Brown, W. G., 5282, 5283
Brown, W. H., 6620
Brownell, W. B., 1115
Browning, L. C., 1116, 1117
Brucker-Voigt, L., 4039
Bruenner, R. S., 1118
Bruk, A. I., 1119, 7089-7096
Brunauer, S., 1120
Bruner, F., 1121
Bruni, G., 5959
Brunnee, C., 1122
Brunner, J. R., 4412, 6835, 7435
Bruno, G., 4132
Bruno, J. J., 2515
Bruno, S., 1123
Bruns, K., 6022
Brus, G., 1124
Bruyn, J. de, 1125
Bryan, F. A., 1126
Bryan, F. R., 1127
Bryanskaya, E. K., 5559
Bryant, H., 7228
Bryant, L. M., 2534
Bryce, W. A., 1128, 1129, 3886, 5976-5981
Bua, E., 1130
Bubenikova, V., 3535, 3536
Buchanan, A. S., 4840
Buck, K. R., 1096
Buck, K. W., 1131
Buckler, S. A., 2211
Buckles, R. E., 1132, 1133
Buckley, K. B., 5043
Bucur, C., 960
Bucur, R., 1134, 1135
Budgen, D. E., 1499
Budzikiewicz, H., 1872, 1876
Buechel, K. H., 1136
Buechi, G., 856
Buechi, J., 1137, 3370, 3371
Buehler, A. A., 2232
Bukata, S. W., 1138
Bukhari, M. A., 1139
Bull, III, H., 6309
Bull, L. B., 1834
Bull, W. C., 925-927
Bultitude, F. W., 582
Bumb, F. C., 1140
Bumgardner, C. L., 1141-1143, 1515
Bunina, N. N., 6235
Bunyan, P. J., 1165
Buoncristiani, D., 1144, 1145
Burchfield, H. P., 1146-1149, 4811, 6593
Burdon, J., 1113, 1150
Burdy, L. M. J., 4949

Burford, R. R., 1151
Burg, E. A., 1152
Burg, S. P., 1152-1155
Burgess, A. R., 1156, 1157
Burgher, R. D., 20-23, 3561
Burgos, J. L., 5798, 5799
Burgt, M. J. van der, 1158
Burham, R. L., 1133
Burk, M. C., 1159, 3811, 3812
Burkat, T. M., 1879
Burke, J., 1160, 1161
Burkhalter, T. S., 4554
Burkin, A. R., 1162
Burks, Jr., R. E., 1163, 1164
Burn, A. J., 1165
Burnell, M., 1249
Burnell, M. R., 1166-1169, 2453, 3639, 3640
Burnett, M. C., 1170
Burnett, M. G., 1171
Burnham, H. D., 4318
Burov, A. N., 1172
Burow, F. H., 4395, 4396, 4739, 5290
Burr, J. G., 65
Burrell Corp., 1173
Burrous, M. L., 726
Burrows, E. P., 1514
Burrows, G., 1174
Burt, R., 1175
Burton, J. T. A., 7531
Burton, M., 6773
Burton, P. E., 1516
Burton, R., 5440
Burwasser, H., 4406
Busch, H., 6524
Busey, H. M., 4852
Bush, I. E., 1176
Bush, S. J., 1177
Bushong, P. A., 1178
Butler, D., 3144
Butler, J. N., 1179-1181
Butler, R. A., 1182
Butterfield, R. O., 6136
Buttery, R. G., 477, 478, 1183-1185, 1881, 6792, 6793
Butz, W. H., 1186, 1187
Buzon, J., 1188-1191
Buzzelli, G., 5048
Bywater, S., 2719

Cabeza, F., 5647-5650
Cabiddu, S., 1192
Cacace, F., 80-83, 1193-1207
Cadman, J., 6377
Cadman, W. J., 1208-1212, 3641, 4473
Cadogan, J. I. G., 1165, 1213, 1214
Cadogan, W. P., 4379
Cady, G. H., 1215

Cady, P., 1216
Caffrey, Jr., J. M., 1217
Caggiano, G., 4680
Cain, E. F. C., 1218, 1219, 2749
Cairncross, I. M., 415
Caldarera, C. M., 1220, 5046
Calfee, R. K., 5786
Call, F., 1221
Callahan, J. A., 1687
Callear, A. B., 1222-1224
Callen, J. E., 1033, 5333, 5334
Callisen, F. I., 1225
Calvarano, I., 1226
Calvarano, M., 1227-1230
Calvert, J. G., 3921, 4152, 6430
Calvin, M., 67, 4335
Calzolari, C., 1231
Cameron, D. W., 1232, 1233
Camier, J., 4679
Camin, D. L., 1234
Campbell, D. N., 6769
Campbell, G. C., 1061
Campbell, J. A., 671, 672, 1573
Campbell, J. K., 1235
Campbell, R. G., 6400
Campbell, R. H., 1236, 2795, 2796
Campi, E., 5129
Cancel, L. E., 5174
Cannings, F. R., 1237
Cannon, H. J., 1238
Cannon, P., 1239
Canter, R., 1508
Capony, J., 5055
Caprioli, G., 1240, 1809
Caran, J. G., 1241
Caran, S. H., 1241
Carazzola, G. A., 4321
Carberry, J. J., 1242, 1243
Carel, A. B., 1244, 5920, 6259, 6260
Carew, J. R., 1245
Carges, G. L., 6057
Cargill, R. L., 1683
Carle, D. W., 1246-1251, 3639, 3640, 4818, 4819
Carlisle, Jr., L. J., 4794-4796
Carlsson, O., 6383
Carlstrom, A. A., 1252, 3608
Carman, E. H., 3801
Carman, P. C., 1253
Carnahan, C. L., 1254
Carnes, M. A., 2170
Caroti, G., 1255
Carpenter, F. G., 1256
Carr, A., 6004
Carr, Jr., H. E., 7425
Carr, Jr., H. F., 1257
Carr, J. B., 3333
Carroll, K. K., 1258-1261
Carroll, R. B., 1262
Carruthers, W., 1263-1265

Clift, T. L., 1445
Clinton, T. G., 7418
Closs, G. L., 1446
Clough, K. H., 1447-1452
Clough, S., 813
Coates, J. I., 1453, 2639
Coates, V. J., 1035-1037, 1431, 1454-1456, 1536, 2137, 2138
Coats, F. H., 6531
Cobb, W. Y., 1457
Cobler, J. G., 1458-1460, 4940
Cochrane, C. C., 1461
Cochrane, C. N., 5962
Cocker, W., 1462-1464
Codegone, C., 1465, 1466
Coe, F. R., 1467
Coe, P. L., 1468
Coetzee, J. F., 1469
Coffman, D. D., 1944
Coffman, J. R., 1470-1473, 6395
Cofleri, G., 6977
Coggeshall, N. D., 1474, 4396
Cogswell, H. W., 5277
Cohen, I. R., 108, 109
Cohen, R., 3698, 6933
Coke, J. L., 89-91
Colard, P., 1475
Colburn, C. B., 1476
Cole, B. T., 1477
Cole, D. D., 1478
Cole, E. W., 3344
Cole, W., 6714
Coleman, H. J., 1479-1481, 6827-6831
Coleman, M. H., 1482
Coley, J. R., 2406
Collaud, C., 6649
Collerson, R. R., 120, 121, 820
Collier, H. G., 52
Collins, C., 7195
Collins, C. H., 1483, 2402, 2933
Collins, G. A., 1089-1092
Collins, R. P., 1484
Colson, E. R., 1485
Colthup, E. C., 4882
Comberiati, J. R., 3844, 3845
Combes, R., 4759
Combs, J. F., 3159, 3160, 7138
Combs, R. L., 2823
Comendant, F., 4892
Comings, E. W., 1486, 4336, 4337
Condit, P. C., 6240
Condon, R. D., 1487-1494
Conia, J., 2492
Conia, J. M., 92, 1495, 1496
Coniglio, J. G., 4009
Conklin, D. B., 5857
Conner, A. Z., 1497, 7567
Connolly, D. J., 870
Connor, J., 4370, 4579
Constantin, M. J., 6060

Contaxis, C. C., 4952
Conti, L., 4391-4393
Convent, L., 4646
Cook, C. D., 1498
Cook, J. G. H., 1499
Cook, J. W., 1500
Cook, S., 6522
Cook, W. A., 2584
Cooke, N. J., 1501, 2958-2960
Cooke, W. D., 1502-1505, 2356-2359, 3816, 3817
Cooney, P. M., 2145
Cooper, A. D., 4418
Cooper, C. V., 4775
Cooper, F. P., 839-841, 4803, 5031
Cooper, G. D., 1506
Cooper, J. A., 1507-1510, 4799
Cope, A. C., 1511-1523
Coppens, L., 1524-1528
Coppock, J. B. M., 1529-1532
Corbin, J. R., 1533-1536
Corcoran, W. H., 6006
Cordon, J. L. M., 368, 1537
Corey, E. J., 1283, 1346
Corkery, A., 1953
Corley, C., 7228
Corman, B. G., 6391
Cornatzer, W. E., 4798
Cornforth, J. W., 1538, 5582, 5584
Cornforth, R. H., 5582-5584
Cornu, A., 4278
Cornwell, D. G., 3260, 3261, 4249
Corse, J., 1539, 1540, 1857-1861, 1863, 6795, 6796
Corse, J. W., 3345
Corsini, F., 472, 4625, 5375, 6105
Corson, B. B., 4248
Cort, L. A., 375
Corwin, A. H., 4045
Cossitt, R. E., 2789
Costa, B., 3902
Costa, G., 4394
Coste, J., 1955
Cotabish, H. N., 1541
Cotrupe, D. P., 2209
Cotter, J. L., 1542
Cotter, R. J., 1543, 6051
Coull, J., 1544
Coulson, D. M., 1316, 1317, 1545-1553, 1810, 1811
Courteix, J., 4677
Courtenay, S. G. P., 1554
Cowan, C. B., 739, 1555
Cowan C. T., 1556, 7267, 7268
Cowan, J. C., 1995, 2485, 6135
Cowan, P. J., 1557
Cowley, B. R., 1558
Cox, J. S. G., 1559

Cox, J. T., 1085
Coxon, R. V., 1560
Coyle, C. F., 3090
Craats, F. van de, 1561, 1562, 3227
Craft, C. C., 4397
Craggs, J. D., 4861
Craig, B. M., 671, 1563-1573, 6897
Craig, D., 1574, 1575, 1577
Craig, L. C., 1576, 1577
Craig, N. C., 1578
Cram, W. W., 1579, 1580
Cramers, K. A. M. G., 4828
Craven, E. C., 1581
Crawford, G. W., 835
Crawford, H. M., 1582
Crawford, R. V., 1583
Creech, B. G., 1509, 1510, 3243, 3246, 4799, 6985
Creech, W., 1584
Cremer, E., 474, 1585-1603, 3942
Crescenzi, V., 405, 4392, 4393
Crespi, V., 1604
Crews, O. P., 5781
Crippen, R. C., 1605-1607
Crisler, R. O., 1608
Crisp, P. C., 371
Critcher, D., 549
Croitoru, P. P., 1609
Crombie, L., 1610
Cronan, C. S., 1611
Crooks, D. A., 813
Crooks, W. M., 6485
Cropper, F. R., 1612-1615
Cross, C. K., 500
Crosser, O. K., 5258
Crossley, S. P., 3984
Crouse, R. H., 1616
Crowe, D. F., 2248
Crowe, P. F., 77, 3254-3258
Crowther, S., 1617
Crum, W. M., 1618, 1619
Crumb, J. W., 1620
Crump, G. B., 1621, 2407
Crupi, F., 1845
Csicsery, S. M., 1622, 1623
Cudmore, J. W., 5245
Cuffel, R. F., 6006
Cugudda, E., 4640
Culberson, C. F., 1624
Culbertson, T. P., 5828, 7162
Cull, N. L., 1625
Cullis, C. F., 475, 1157, 1626-1628, 6881
Cullum, T. V., 832
Culvenor, C. C. J., 1834
Cumming, C. A., 6224, 6225
Cundall, R. B., 1629
Cunningham, G. P., 1469
Curby, R. J., 93
Currah, J. E., 544, 545

Curran, T. D., 3589
Curren, W. J., 1630, 4807
Currie, L. A., 4526
Curry, A. S., 1631
Curtin, D. Y., 1632
Curugno, N., 1633
Cuthbertson, F., 1634
Cvejanovich, G. J., 1635, 1636
Cvetanovic, R. J., 1222, 1223, 1637-1639, 2180, 2318, 6044, 7085
Cvrkal, H., 1640, 1641
Cymerman-Craig, C., 1642
Czaran, L., 2212

Dabney, III, W. T., 1643
Dadura, J. G., 6876
Daghetta, A., 2285
Dahl, T., 1462
Dahlback, O., 875, 876
Dahmen, E. A. M. F., 1644, 1645, 1849
Dailey, R. E., 1646
Dal, V. I., 1647-1649
Dalgleish, C. E., 404
Dal Nogare, S., 728, 729, 1650-1665, 3712, 5990
Dal Pozzo, A., 7492
Daly, J. W., 1666
Dam, H., 3568
Damen, H., 3107
Dan, T., 1667, 5337
Danatos, S., 5317
Danby, C. J., 1668
Dandiya, P. C., 648
Daneman, H. L., 1669-1671
D'Angio, C. J., 6901
Daniel, J. C., 2841
Daniels, N. W. R., 1529-1531, 1672, 1673
Dann, A. T., 1834
Dansi, A., 7492
Danyi, M. D., 6486, 6487
Darby, P. W., 1674, 1675, 7255
Darley, E. F., 1676, 6541
Darling, D. J., 1677
Dart, M. C., 1678
Datskevich, A. A., 1679, 4133
Datta, P. R., 1680, 1681
Dauben, W. G., 1682, 1683
Dauchy, S., 1684
Davenport, J. B., 1685
David, D. J., 6538
Davidson, J. A., 2163
Davidson, J. M., 1686
Davidson, L. V., 1687
Davies, A. G., 7
Davies, A. J., 407, 408, 1688
Davies, D. I., 1689
Davies, G. R., 1690
Davies, L., 4358

Davies, V., 1691
Davis, A., 1692-1695, 6010, 6011
Davis, A. D., 1696, 1697
Davis, C. E., 1698
Davis, D. K., 5370
Davis, D. S., 1699
Davis, G. A., 3725
Davis, J. J., 1431, 1700, 1701
Davis, R. C., 2401
Davis, R. E., 1702, 1703
Davis, T. C., 3352
Davis, T. W. M., 1704-1706, 2190, 5228
Davison, V. L., 6492
Davison, W. H. T., 1707
Dawson, Jr., H. J., 1708-1710
Day, E. A., 1711-1716, 3961, 4380, 4403, 5422
Day, J. C., 6794
Day, P., 1717
Daynes, H. A., 1718
Dayton, S., 1719
Deady, L. W., 1720
Deal, C. H., 1721, 1722, 5518, 5519, 5599
Dean, D. E., 6625, 6806-6808
Dean, R. A., 832
DeAngelis, G., 1723
Deans, H. A., 6511
Deavours, M. F., 1724
Debbrecht, F. J., 1725
DeBoer, F. E., 1726
Debuch, H., 1727
Decora, A. W., 1728-1732
Decoteau, A. E., 890
Deeds, M. L., 1132
Deemter, J. J. van, 1733, 1734
Deenan, L. L. M. van, 4104
DeFord, D. D., 1735-1739, 5305, 5546
DeFrancesco, F., 1740-1743
De Goey, J., 1847
Deinema, M. H., 1744
Deisler, P. F., 1745
De la Mare, P. B. D., 521
De la Mere, H. E., 1746
DeMan, J. M., 1747
Dement'eva, M. I., 1748
Demina, N. D., 7097
Demole, E., 1749
Denekas, E., 1750
Denis, J., 1751
Denney, D. B., 1752, 1753
Denning, Jr., G. S., 1754
Denny, B., 1755-1757
Densham, A. B., 1758, 1759
Dent, L. S., 1760
De Puy, C. H., 1761-1768, 2402
Derbesy, M., 5131
Derby, J. V., 1769

Derfer, J. M., 940, 941
Derkosch, J., 6684
DeRose, A., 1770
Derr, E. L., 5518, 5519
DeSimone, G. J., 5049
Desnuelle, P., 6060
De Somer, P., 2168
Despa, S., 1771
Desty, D. H., 1772-1803, 3712
Deszyck, E. J., 2776
DeTar, D. F., 1804
DeTomaso, P., 1151
Deuel, H., 1805
Deutsch, I., 1806
Deutschman, J. E., 1051
Dev, S., 5179
DeVault, D., 1807
DeVay, J. E., 4954
Devienne, F. M., 1808
DeVita, M., 1240, 1809
DeVries, J. E., 1317, 1550, 1551, 1810, 1811
DeVries, L., 1812
DeVries, M., 1813, 4855-4857
DeVries, T., 4829, 4830
Devyatykh, G. G., 1814, 1815, 7566
Dewar, M. J. S., 1816
Dewar, R. A., 1817, 4845-4847
DeWet, C. R., 628
de Wet, W. J., see Wet, W. J. de
Dewhurst, H. A., 1818-1825, 6003
Dewitt, H., 2170
Dhont, J., 7238
Dhont, J. H., 819, 1826-1828
Dhopeshwarker, G. A., 1829-1831
Diaper, D. G. M., 1832, 6510
DiCenzo, R. J., 1833
Dick, A. T., 1834
Dick, J. R., 5380
Dickenson, J. D., 1835
Dickholtz, R., 4653
Dickman, J. T., 1836
Diemair, W., 1837, 1838
Dieringer, L. F., 3380
Dietrich, P., 1839
Dietz, H. G., 629
Dietz, R., 1816
Dietz, W. A., 1840-1843
Dietze, S., 1844
Di Giacomo, A., 1845
DiGiaimo, M. P., 5399
Dijkstra, A., 919, 1846, 6998
Dijkstra, F., 1158
Dijkstra, G., 690, 1847, 1848, 3916, 3917
Dijkstra, R., 1849
DiLeone, R., 1753
Dille, R. M., 1850
Dillon, M., 1851
Dimbat, M., 1852, 1853, 2361

Dimick, K. P., 138, 931, 1539, 1854-1865, 2999, 3000, 3345, 3590, 3591, 6716
Dimov, N., 5236
Dimov, N. P., 6888
Dinelli, D., 1866
Dinerstein, R. A., 1867, 6114, 6115
Dinneen, G. U., 1728-1732
Dintenfass, H. T., 1868
D'Iorio, B. I., 6528
Di Prima, A., 1869, 6594
DiStefano, F., 933
Dixon, H. B. F., 1870
Dixon, J. A., 5128
Dixon, W. S., 1871
Djerassi, C., 1872-1876, 5122
Djurtoft, R., 439
Dobbins, W. E., 4485
Dobbs, H. E., 1877
Dobiasova, M., 1878, 3521
Dobychin, D. P., 1748, 1879
Dodd, H. C., 6571
Doering, C. E., 1880
Doering, W. v. E., 1881-1888, 4007
Doerr, R. C., 506
Doerrscheidt, W., 1889
Dohi, K., 2963
Dole, V. P., 1890, 4868
Dolejsek, Z., 1891
Dollear, F. G., 2303
Dolphin, J. L., 1892
Domange, L., 1893-1896
Dominguez, A. M., 1897
Domsch, K. H., 5082
Domsky, I. I., 4400-4402
Donaldson, M. M., 6112
Donegan, L., 1898
Donner, W., 1250, 1899-1901
Doolen, O. K., 1902, 1903
Dora, R. A., 1904, 1905
Dorfman, L. M., 1906
Dorfner, K., 1907, 1908
Dorsey, J. A., 1909, 1910
Doscher, T., 5890
Doty, V., 3576, 7355
Doubek, M., 5353
Douglas, A. G., 3684
Douglass, C. D., 6282
Doumani, T. F., 2703
Dowden, D. A., 928
Downing, D. T., 1911-1913
Downs, S., 2906
Doyle, G. J., 2077
Doyle, L. C., 1638
Doyle, R. R., 4360, 4361
Dragel, D. T., 1914
Drake, B., 1915
Draper, J. W., 5677
Drauzin, A. Y., 7098
Drawert, F., 1916-1927

Drehman, L. E., 3113
Drekopf, K., 1928
Dresdner, R. D., 1929-1931, 5765, 7472
Dressler, D. P., 1932
Dreux, J., 3161
Drew, C. M., 1933-1939, 4833, 4834
Drew, E. H., 4683
Drews, B., 1940
Drews, H., 1941
Dreyer, H., 1942
Driatskaya, Z. V., 5426
Drienovsky, P., 1943
Driesbach, L., 6186, 6187
Drimus, I., 960
Drinker, P. H., 3284
Drinkwater, J. W., 407, 408
Driscoll, J. L., 97, 491, 2796-2799, 4682
Droegemeir, G., 5988
Drouhin, N., 2821
Drushel, H. V., 6251
Drysdale, J. J., 1944, 4738
Dubinin, M. M., 1945-1952
Dublin, W. B., 7334
Dubois, L., 1953, 1954, 5004-5006
Dubosc, J-P., 1955
Dubovikova, A. P., 6035
Dubsky, H. E., 1956-1958, 3532-3534
Ducros, M., 1382
Dudenbostel, Jr., B. F., 1843, 1959-1962
Duffey, D. C., 4372
Duffey, J. G., 1963
Duffield, J. J., 1964-1966
Dufka, O., 6959
Dugan, P. R., 5114
Duggleby, P. M., 387
Duke, J. R. C., 1967
DuMay, H., 1968
Dumazert, C., 1969, 1970
Duncan, J. L., 150
Duncan, W. R. H., 1971
Dunckley, G. G., 374
Dunkley, W. L., 3059, 3595
Dunlop, A., 744
Dunlop, A. K., 5759
Dunlop, A. S., 6770
Dunn, F. J., 1972
Dunstan, G. H., 3154-3156
Dunstan, I., 862, 1973, 7339
Dunston, M. L., 1974
Dunstone, E. A., 2307-2310
Dupaigne, P., 1975
Dupire, F., 1976-1979
Du Plessis, L. A., see Plessis, L. A. du
Durr, R. A., 4818, 4819
Durrell, W., 1980
Durrell, W. S., 7472

Durrett, L. R., 1981-1985, 2000
Dusek, R., 7432
Duskova, L., 1986
Duswalt, Jr., A. A., 1987, 1988
Dutch, P. H., 1989
Dutton, H. J., 1990-1996, 3695, 4721, 5284, 6136-6138, 6140
Duuren, B. L. van, 1997
DuVall, A. H., 1998
Dux, J. P., 6661
Duyckaerts, G., 5057
Dvoretzky, I., 1983-1985, 1999, 2000, 5486, 5801, 5802, 6340, 6341
Dwyer, R., 2001
D'yachenko, A. I., 4921
Dyer, E., 2002
Dykstra, S., 2003
Dymicky, M., 6527, 6528
Dyr, J., 4184
Dzantiev, B. G., 2004, 5631
Dzedzik, R. P., 7070

Eaborn, C., 1835
Eanes, R. D., 1687, 2005, 6440
Earle, F. R., 4976-4978
Earle, M. J., 3250
Eastham, J. F., 2006
Eastman, R. H., 1666, 2007, 2008
Easton, C. B., 2009
Ebeid, F. M., 1175, 2010, 2011
Eberhagen, D., 4041
Eberhardt, M., 2404
Eberle, D., 3091
Eberly, P. E., 2012, 2013
Ebert, Jr., A. A., 2014
Eccleston, B. H., 7263, 7264
Echigoya, E., 116
Echizen, A., 4717
Eckerson, B. A., 2015
Eckert, J. W., 5082
Eckhardt, F., 2016-2018
Edds, D. L., 168
Ede, P., 1051, 2019
Edelhausen, L., 868, 869
Eden, M., 2020
Edgecombe, F. H. C., 1289, 2021-2023
Edmands, R. E., 606
Edwards, A. W. T., 2189
Edwards, D. G., 3589
Edwards, F. G., 3324
Edwards, G. J., 427, 428
Edwards, Jr., H. M., 2024
Edwards, J. A., 2025
Edwards, R. T., 4471
Edwards, W. G. H., 2026
Effenberger, M., 2027
Egan, C. J., 6631
Egerer, W., 3307
Egger, K., 2028

Felgenhauer, R., 1920, 1921
Felix, D., 6239
Fellows, C. G., 2219
Fellows, E. G., 2220
Fells, I., 2221
Felter, R. E., 4526
Feltkamp, H., 2222, 3308
Felton, H. R., 2223-2232
Fenaroli, G., 2233
Feng, P. Y., 2234, 2235
Fenning, C. M., 5318
Fenselau, A. H., 7191
Fenske, M. R., 3702
Ferguson, R. E., 7463
Ferguson, W. C., 3283
Ferrand, M. R., 2236
Ferrand, R., 1189, 2237, 2238
Ferraro, J. R., 5436b, 5458
Ferrer, P., 2239
Ferrero, C., 2240
Ferrers, P., 2241
Ferrier, R. J., 2242, 2243
Ferrin, C. R., 2244-2246, 3094,
    4289
Fessenden, R., 1292, 1293
Fessenden, R. J., 2247-2250
Fett, E. R., 2251
Fettis, G. C., 2252, 2253
Feuerberg, H., 2254
Feuge, R. O., 6804
Fiedler, G., 5060
Fiehman, J., 2255
Field, D. C., 2256
Field, Jr., H., 1646, 6671-6676
Fields, E. K., 1941
Fikentscher, L., 2353
Fiks, M. M., 3782, 3783, 7501
Filatova, E. D., 1074-1076
Filatova, N. V., 1372
Filipic, V. J., 1681
Filippov, G. G., 2257
Findeis, A. F., 2258, 2259
Finke, M., 951
Finkelstein, M., 2260, 2261
Finlayson, A. J., 2262
Finnegan, R. A., 1873, 2263
Finnegan, W. G., 2264
Finnerty, J. L., 7357
Firestone, D., 2041, 2265-2267
Fischer, J., 2268-2271, 4110
Fish, A., 1626, 1627
Fisher, A., 1237
Fisher, G. S., 3912
Fisher, N., 1532, 2272
Fisher Scientific Co., 2273-2276
Fishman, J., 2277
Fitch, G. R., 2278
Fitzgerald, J. S., 2279-2281
Flanders, R. L., 2282
Flaquer, J. O., 368
Fleck, S. A., 4395, 4396
Fleetwood, C. W., 2283

Fleischmann, L., 2284, 2285
Fletcher, W., 5178, 7418
Flett, M. St. C., 2286
Flinn, R. A., 3003
Flipse, R. J., 159
Flook, W. A., 2287
Floss, E., 3805
Flowers, M. C., 2288-2292
Fluck, A. A. J., 2293
Flumerfelt, G. C., 2294
Fodderie, C., 5883
Foerderreuther, M., 6190, 6191
Foldiak, G., 6098
Follain, G., 1190, 2295
Folmer, O. F., 5916, 5917, 5920
Folmer, Jr., O. F., 2296-2298
Fontan, C. R., 2299, 5383-5387
Fontell, K., 2300, 5036-5038
Ford, H. W., 2301, 6170
Ford, J. F., 1237, 2302
Ford, T., 4651
Forde, M. B., 528
Fordham, W. D., 848
Fore, S. P., 2303
Forest, C. A., 2485
Forman, P., 3902
Forrest, C. W., 2050-2052
Forrestal, L. J., 2304
Forrester, F., 5296
Forrester, J. L., 1133
Forrester, J. S., 2305
Forss, D. A., 1711, 1712, 2306-
    2314, 5422, 7155
Forster, T. L., 3607
Fort, A. W., 2315
Fortune, W. B., 2316
Forziati, A. F., 6605
Foster, A. B., 1131, 1139, 2317
Foster, D. W., 4047
Foster, G., 7283
Foster, G. E., 1319
Foster, N. F., 2318
Foster, N. G., 1479
Foster, R. A., 2319, 2320
Fouassin, A., 2321
Foulletier, L., 2322
Fourroux, M. M., 2323-2325,
    7278
Fowler, R. B., 1574, 1575
Fox, F. T., 2326
Fox, J. E., 2327
Fox, S., 1530
Fraade, D. J., 35, 2328-2333
Fradkov, A. B., 1056
Franc, J., 2334-2347
Francesco, F. de, see
    DeFrancesco, F.
Franchimont, E., 7038
Francis, S. A., 372, 2348
Frank, A. J., 6123
Frank, Yu. A., 2349
Frankel, E. N., 2350, 2351

Franz, W. F., 2352
Franzen, V., 2353
Fraser, R. R., 1632, 2354
Frederick, D. H., 2355-2359
Fredericks, D. E., 3173
Fredericks, E. M., 2360, 2361
Fredericks, P. S., 344, 2362-
    2364
Freed, V. H., 3330
Freedman, R. W., 1609
Freeguard, G. F., 2365, 2366
Freeman, G. R., 1668, 2367,
    2368, 5727
Freeman, M. P., 2949
Freeman, N. K., 4410, 5199
Freeman, R. B., 3112
Freeman, S. K., 2369, 2370, 4968
Freenor, F. J., 2249, 2250
Freidlin, L. K., 5181
Freimuth, H., 1607
Freiser, H., 2371, 3903, 3904
French, R. B., 2372
Frenzel, J., 2373, 2374, 2647,
    2648
Freund, H., 6127
Freund, M., 2375-2377
Freund, W., 5061
Frey, H. M., 2288-2292, 2378-
    2382
Friedlander, M., 4969
Friedman, H. L., 2383, 2384
Friedman, L., 2266, 2267
Friedman, R. L., 2385
Friedrich, J. P., 2386
Friedrich, K., 1889, 2387, 2388,
    3612, 3613, 6242
Friel, D. D., 2389, 2390
Frisch, H. L., 7459
Frisone, G. J., 2391-2393
Fritz, G., 2394-2400
Fritz, I. T., 2401
Froemsdorf, D. H., 1762, 1764,
    2402
Frolovskii, P. A., 142, 2403
Fromm-Czaran, E., 2214
Fry, A., 2404, 5328
Frye, C. G., 2405, 2406
Fryer, F. H., 2407
Fryer, J. F., 2408
Fuchs, W., 2409, 2410
Fuerst, H., see Fürst, H.
Fujii, A., 2962
Fujii, M., 4592
Fujiishi, K., 6241
Fujino, Z., 6045
Fujishima, I., 2411
Fujita, E., 2412
Fujita, K., 2431
Fujita, M., 2414, 3876
Fujita, Y., 5301-5303
Fujiwara, F., 2415, 2416
Fukayama, G., 4909

Fuks, N. A., 2417
Fukuda, T., 2418-2425
Fukui, K., 2426, 4199
Fukunaga, S., 2427
Fukushima, M., 2428
Fukuzumi, K., 3397
Fulco, A. J., 2429
Fuller, D. H., 2430
Fuller, E. N., 2576
Fuller, G., 2431
Fulton, J. D., 1617
Fulton, W. C., 1482
Funasaka, W., 2432-2434
Funch, J. P., 3568
Funk, J. E., 2435-2437
Funke, P. T., 4706
Furlani, A., 1231, 2438
Fürst, H., 2439-2441
Furukawa, J., 2692
Furukawa, T., 7468
Furuta, S., 4769
Furuyama, S., 2442, 2443
Futrell, J. H., 2444, 6840

Gable, C. M., 5758, 5759
Gabor, T., 583
Gadsden, R. H., 2445, 2446
Gaertner, K., see Gärtner, K.
Gaeumann, T., see Gäumann, T.
Gage, J. C., 2447
Gager, Jr., F., 2448, 6246, 6247
Gagliardi, P., 5630
Gagnaire, D., 2449
Gajan, R. J., 4035, 5720
Gajek, K., 7073
Galanina, N. L., 522
Galbraith, A. R., 695
Galbraith, F. J., 4362, 4364
Gale, D. M., 630-632
Galkin, L. A., 7488
Gall, E. J., 4710
Gall, J. S., 808
Galla, S. J., 2450, 2451
Gallagher, M. J., 2452
Gallaway, W. S., 1901, 2453-2457,
    6552, 6553, 6570, 6610-6612,
    6963, 7271
Gallay, L. R., 24
Galwey, A. K., 2458-2460
Gamson, R. M., 2461
Gander, G. W., 2462, 2463, 3599-
    3602, 3605, 3606
Gandini, C., 1743
Ganin, Y. V., 2464, 6275
Gannon, W. F., 3272
Gant, P. L., 2465, 7447
Garattini, S., 5372
Garbini, L. J., 5007
Garbuglio, C., 2499
Garcia, F. G., 4904

Gardiner, K. W., 892, 2466, 2467,
    2865, 2866
Gardiner, Jr., W. C., 4229
Gardiner, W. L., 6986
Gardner, H. M., 6327
Gardner, J. N., 2468
Gardner, K., 2469
Gardner, P. D., 3040
Gardner, P. E., 838
Garilli, F., 2470
Garn, P. D., 2471
Garner, A. Y., 2472
Garner, F. H., 2473, 5360
Garner, J. W., 1616
Garnett, J. L., 2474
Garnova, T. G., 2475
Garoglio, P. G., 2476, 2477
Garratt, D. C., 2478
Garratt, P. G., 2204, 2479
Gärtner, K., 2480
Garton, G. A., 1971, 2481
Garzo, G., 2482, 2483
Gasparic, J., 2484
Gass, J. P., 6301
Gast, J. H., 1508
Gast, L. E., 2485
Gaston, L. K., 2486
Gatrell, R. L., 2487-2489
Gatzke, A. L., 6510
Gaudemaris, G. de, 938, 939
Gaulin, C. A., 2490
Gault, F. G., 2491, 2492
Gäumann, T., 2493, 2494, 3208,
    7292
Gaylor, V. F., 2495
Gaziev, G. A., 2496-2498, 5362-
    5364, 7451, 7452
Geach, C. J., 1784, 1785
Gechele, G. B., 2499
Gee, M., 2500, 2501
Gehrke, C. W., 2502-2504, 5800
Geiger, S., 3308
Gelchsheimer, E., 3309
Geldenhuis, P., 2505
Geldmacher-Mallinckrodt, M.,
    7185
Gellerman, J. L., 2882, 6106-
    6109
Gellhorn, A., 725, 2506
Gemmill, A. V., 2507
Genas, M., 2508
Genevois, L., 534-536, 2509
Genge, C. A., 2510
Genin, G., 2511
Genkin, A. N., 2512-2514
Gensler, W. J., 2515
Gent, C. M. van, 968, 970-973,
    1437, 7417
Gent, P. L., 5045
Gentili, M., 5911
Geodekyan, A. A., 6454
George, W., 5704

Georgieff, K. K., 2516
Georgievskaya, T. V., 6920, 7525
Gerberich, H. R., 2517, 2518
Gerdes, W. F., 2519
Gerhardt, G. E., 4647
Germain, J. E., 1313, 2491, 2492,
    2520-2525, 4748
Germanos, G. A., 4901
Gerrard, W., 1770, 2526, 2527
Gerritsma, K. W., 3790
Gershbein, L. L., 5323
Gerson, T., 2528-2530
Gesser, H., 825
Getoff, N., 2531
Getz, G. S., 613, 2532
Getzendaner, M. E., 2533
Gevantman, L. H., 2534
Ghanayem, I., 2535
Gherman, I., 2945
Gherman, M., 2091
Ghiglione, C., 1969, 1970
Giacomello, G., 1199
Giammaria, J. J., 2536
Gianetto, A., 2537
Giannardi, G. B., 2476, 2477
Gibbs, D. S., 2538
Gibson, G. W., 2006
Gibson, N. A., 7216, 7217
Gibson, S. P., 95
Giddings, J. C., 2539-2583, 3907,
    6568
Gier, J. de, 4104
Gierschner, K., 4863
Giesecke, W., 5952
Giever, P. M., 2584
Gifford, A. P., 2585
Giladi, J., 2586
Gil-Av, E., 2587-2595, 3121,
    6263-6265
Gilbert, B., 1872, 1873
Gilbert, R. E., 3040, 3041
Gilbertson, J. R., 2596
Gilde, H-O., 6433, 6434
Gilfillan, J. H., 6614
Gill, H. A., 2597
Gilladi, J., 6314-6316
Gillespie, C. K., 6372
Gilliland, E. R., 2598-2600, 4379
Gillis, B. T., 1514, 2601
Gillis, R. G., 2602, 2603
Gilman, C. J., 1150
Gilmore, W. F., 3270, 3271
Gilmour, H., 2604
Gimblett, F. G. R., 2605
Ginnings, D. C., 5280
Ginsburg, A. E., 525
Ginsburg, L., 2606
Giovannelli, G., 4640
Giovannozzi-Sermanni, G., 1279-
    1281, 2607
Giovetti, G. L., 5829
Girard, C. A., 2315

291

Girling, G. W., 2608, 2609, 6226
Giroux, J. W., 2610
Gitoneas, P., 4970
Giuffrida, L., 3826-3829
Giuffrida, L. E., 6873
Giullot, M., 2611
Gjaldbaek, J. C., 2612, 2613, 4334
Gjerstad, G., 2614-2616
Gjertsen, P., 2617, 2618
Glaser, A., 348, 2619
Glasgow, A. R., 6797
Glasson, W. A., 2234
Glatz, A., 5207
Glazunov, P. Y., 6484
Glew, D. N., 2620
Glick, C. F., 2621
Gloesener, E., 2623, 6509
Gloesener, R., 2624
Glogoczowski, J., 2625
Glueckauf, E., 1453, 2626-2642
Gnauck, G., 2643-2648
Gnauer, H., 7397, 7398
Goble, A. G., 2649, 5545
Godet, M., 2650
Godet, M. R., 4431
Godfrey, F. M., 1786, 1787
Godin, P. J., 1898
Godsell, J. A., 2651
Goeckner, N. A., 959, 2652
Goeders, C. N., 1761, 1763
Goedkoop, W., 2653, 2654
Goering, H. L., 2655-2657
Goerlitz, D. F., 2502, 2503, 5800
Goetz, R. W., 2658
Goggin, P., 1328
Gohlke, R. S., 2659-2663
Golay, M. J. E., 2139, 2664-2676, 3712
Gol'bert, K. A., 71, 2677, 2678, 7042-7049, 7068, 7069
Gold, H. J., 2679-2682
Gold, V., 2683, 2684
Goldbaum, L. R., 1897
Goldberg, G., 2685
Goldberg, S. I., 5828
Goldblatt, L. A., 829
Goldblith, S. A., 7286, 7287
Goldfen, A., 4770
Golding, W. E., 2686
Gol'dinov, A. L., 2687
Goldman, L., 6000
Goldschmidt, S., 2688
Goldsmith, D. J., 2689
Goldup, A., 1784, 1785, 1788-1795, 2690, 2691
Golner, J. J., 6883
Gomi, S., 2692
Goncz, I., 2715
Gonzalez, A. G., 1052, 2693
Good, C. D., 2694, 2695
Goodloe, M. H. R., 606
Goodman, D. S., 2696, 5582, 5584

Goodman, L., 1291, 5781
Goodno, J. A., 4211
Goodwin, E. S., 2697-2699
Gordillo, A. L., 2700
Gordon, A. S., 1938, 1939, 2701, 4833-4838, 6428, 6429
Gordon, R. J., 2702
Gordon, R. S., 4639
Gordon, S., 2703
Gordon, S. M., 2704
Gordus, A. A., 2705-2707
Gorin, P. A. J., 172
Gorshunov, O. L., 7049
Gorvaev, M. I., 3369
Gosnell, A. B., 6407, 6408
Goto, R., 363-366, 2708
Gotsis, A., 3072
Gottlieb, O. R., 5155
Gotz, A., 2709, 2710
Gough, G., 1759
Gould, I. A., 2979
Goulden, R., 2697-2699
Gouze, M. L., 4848
Gover, T. A., 2711
Gower, D. B., 508, 2712
Graciantous, J., 2713
Graf, L., 2714, 2715, 6861-6866
Gragnoli, G., 4640
Graham, J. E., 4207
Graham, W. H., 1288
Grampoloff, V. A., 5156-5160
Gramstad, T., 2716
Grandy, G. L., 2717, 4099, 4100
Grant, D. H., 2718, 2719
Grant, D. W., 2720-2728
Gras, B., 1313
Grasselli, J. G., 2729
Grassie, N., 2718
Grassmann, H., 2730
Graven, W. M., 2731, 2732
Gray, Jr., F. B., 2733-2735
Gray, G. M., 2736-2739, 4568, 4569
Graziano, V., 2740
Green, F. C., 1666
Green, G. E., 2741
Green, R. M., 3284
Green, S. W., 2742, 2743
Greene, S. A., 2744-2755
Greenlee, T. W., 5533, 7496
Greenwood, C. T., 2756
Greenwood, F. L., 2757
Greenwood, N. N., 2758
Greer, M., 7318
Gregg, S. J., 2759, 2760
Gregoriou, G., 4941
Gregory, E. R. W., 672
Gregory, N. L., 2761, 2762, 4500, 4501
Greiner, A., 2763
Greiner, R. W., 2655
Gribben, T., 2770

Grieco, D., 2764, 2765
Griesbaum, K., 5342
Griessbach, R., 2480
Griffin, B. F., 1610
Griffiths, J., 2766
Griffiths, J. H., 2767
Griffiths, S. T., 885, 886
Grigoryan, K. A., 2768
Grilly, E. R., 3671
Grimes, M. D., 5203
Grimmer, G., 348, 2619
Grisley, Jr., D. W., 7270
Grob, C. A., 2769
Grob, R. L., 2770, 5902
Grobe, J., 2395, 2396
Groennings, S., 2029-2031, 2033, 4072, 5193, 5563, 5564
Groringer, H. S., 2771
Gros, A., 6804
Gross, A. L., 1235
Grossi, E., 2772-2774, 5438
Grosskopf, K., 2775
Grossman, J. D., 2776, 3376
Groves, J. D., 5381
Grubbs, E. J., 3272
Gruber, H., 4350
Gruber, K., 2777, 4148-4150
Grubner, O., 1891, 2778-2783, 4103
Gruener, Ch., 6265
Gruger, Jr., E. H., 5805
Grund, H., 6164
Grundon, M. F., 2784
Grune, W. N., 2785-2790
Gruner, B. J., 630, 631
Grushetskaya, E. V., 4225
Gryzlova, L. V., 516
Guarino, A., 1200-1203
Gube, M., 2844
Gubser, H., 2791
Gudzinowicz, B. J., 97, 1236, 2792-2803, 4682, 6435, 6436
Guenther, E., 2804-2806
Guermont, J. P., 5514
Guertin, D. L., 2807
Guglina, S. A., 7094, 7095
Guidry, C. L., 5331
Guild, L. V., 2808-2820, 3885, 4455, 4525
Guillaumin, R., 2821
Guillemin, C. L., 2822
Guillet, J. E., 2823
Guiochon, G., 1189, 2196, 2824, 4256, 7568-7570
Gulyas, I., 6694
Gunesch, H., 2825, 2826
Gunn, W. H., 2827
Gunner, S. W., 2828, 2829
Gunning, H. E., 4064
Gunstone, F. D., 501, 502, 2830
Gunther, F. A., 2831-2835, 5343
Gunzler, H., 2836

Gupta, S. C., 6789
Gurtler, J., 878
Gur'yakov, I. I., 2837
Gutberlet, L. C., 2406
Gutsche, C. D., 2838, 2839
Guyer, A., 2840
Guyer, P., 2840
Guyot, A., 2841
Guzman, A., 2842
Gygi, R., 2843

Haack, E., 2844
Haag, W. O., 2845, 4258, 5534
Haahti, E., 2846-2852
Haahti, E. O. A., 2184, 2853-2859
   3240-3242, 3247-3249, 4550,
   6987-6989, 7537
Haarhoff, P. C., 760, 2704, 2860-
   2863, 7229, 7230
Habboush, A. E., 2864
Haber, H. S., 2865, 2866
Habgood, H. W., 622, 623, 2408,
   2867-2873
Habich, A., 2874
Hachenberg, H., 2875, 2876, 3233,
   3234, 6188-6192
Hackman, R. H., 7155
Hadaway, H., 5205
Haddock, L. A., 2877, 2878
Hadley, E. H., 2879
Haertle, W. R., 1118
Hagdahl, L., 2880
Hager, W., 4327
Haggerty, Jr., W. J., 1015, 2881
Hahn, H. H., 6798
Hahto, M. P., 2609
Haines, T. H., 2882
Haines, W. E., 2246, 4471, 5308
Hajra, A. K., 2883, 5718
Hake, C. L., 2884, 4816
Hala, E., 7285
Halasz, I., 2885-2902, 6088, 6089
Halcour, K., 410
Halden, W., 2903
Hale, W. F., 494, 495
Halek, G. W., 5357
Hall, D., 7308
Hall, E., 744
Hall, Jr., H. K., 2904
Hall, H. L., 2905
Hall, K. D., 2906
Hall, L. G., 4653
Hall, N. D., 542
Hall, R. J., 5465
Hall, W. K., 2907-2909
Hallgren, B., 2910-2915
Halsey, G., 1162
Halter, R. C., 2916
Hamada, I., 401
Hambling, J. K., 3959
Hamelin, R., 2917, 2918
Hamence, J. H., 2919

Hamer, J. C., 2920
Hamill, W. H., 2304
Hamilton, C. H., 2921
Hamilton, J. B., 3028
Hamilton, L. H., 2922-2926, 6422
Hamilton, P. B., 2927
Hamilton, R. J., 2035
Hamilton, W. C., 1244, 2928,
   5449, 5450, 5916-5919, 6260
Hamlin, A. G., 3401-3404
Hammar, C. G. B., 2929
Hammer, H., 4112
Hammond, E. G., 890, 2055-2057,
   2930-2932, 3699
Hammond, G. S., 1483, 2402,
   2933
Hampton, B. L., 2934
Hampton, M. C., 5307
Hampton, M. G., 584
Hampton, W., 4032
Han, A., 6203
Han, Y. W., 5846
Hana, K., 6800
Hanack, M., 2935-2938
Hanada, Y., 2939
Hanahan, D. J., 2940-2942
Hanaineh, L., 1083
Hancart, J., 2943
Hand, D. B., 7109
Handa, K. L., 2944
Handley, R., 820
Handy, C. T., 3228
Hanel, R., 5660
Hanes, A., 2945-2947, 6033
Hankinson, C. L., 1478, 2948,
   2979
Hanlan, J. F., 2868, 2949
Hanna, C., 2950
Hanneman, W. W., 2951, 2952
Hanni, H., 2953, 2954, 5836
Hans, A., 6953
Hansen, A. E., 5406
Hansen, N. R., 2955
Hansen, R. P., 1501, 2956-2960,
   3060, 4804
Hanson, D. N., 2961
Hanus, V., 1891
Haq, I. U., 1200, 1204, 1205
Hara, A., 2962
Hara, N., 2963, 2964, 6740
Haraldson, L., 2965-2967
Harborne, J. B., 2968
Harbourn, C. L. A., 1786, 1787,
   1796, 1797
Harden, J. C., 1655, 1658
Hardiman, J., 2969
Hardy, C. J., 2970-2973, 5568-
   5573
Hardy, F. R. F., 1626, 1628
Haresnape, J. N., 1798
Hargis, B., 3576
Hargreaves, M. K., 2974

Harkness, W., 6769
Harlan, J. W., 3858, 4304a, 4305
Harley, J., 2975, 2976
Harlow, E. S., 3387
Harold, F. V., 2977, 2978
Harper, W. J., 1373, 1478, 2948,
   2979, 5849
Harrell, J. R., 7303
Harrington, R. E., 2538
Harris, B. L., 2980-2983
Harris, E. R., 1294
Harris, Jr., J. F., 2984, 6283,
   6498
Harris, J. J., 2985
Harris, P., 1367
Harris, R. J., 6786
Harris, W. E., 2408, 2869-2873,
   2986-2989, 4547, 4791
Harrison, A., 1085
Harrison, G. F., 2990-2992
Harrison, R. B., 2993
Harrocks, J. A., 2994, 5429
Harteck, P., 2995
Hartley, C. B., 980
Hartman, J. S., 169
Hartman, L., 2996, 2997
Hartmann, H., 931, 2998-3000
Hartree, E. F., 3001
Hartung, G. K., 3002, 3003
Hartwell, F. J., 499
Hartwell, J. M., 1556
Hartzler, H. D., 3004
Haruki, T., 3005-3012
Harva, O., 3013-3015
Harvey, C. C., 3263, 7396
Harvey, D., 988, 3016-3018
Harvey, F. H., 3019
Harvey, H. E., 3020
Harvey, W. E., 3020
Harwood, H. J., 1461, 3021
Harwood, J. F., 4357
Hasegawa, S., 6957
Hasegawa, T., 3022
Hashim, S., 1387
Hashimoto, A., 6889
Hashimoto, S., 1719
Haskin, J. F., 3023, 3024, 7149-
   7151
Haskin, L. A., 3025
Haslam, J., 3026-3035
Haslewood, G. A. D., 2712
Hasselstrom, T., 3036, 3037
Haszeldine, R. N., 525, 833
Hatch, L. F., 3038-3042
Hatten, M. J., 3156
Haubenstack, H., 2045
Haupt, R., 1595
Hauptschein, M., 3043, 3044
Hausdorff, H. H., 1040, 1596,
   3045-3051
Hause, J. A., 3052
Hauser, C. R., 4107

293

Hauthal, H. G., 1880
Haverkamp-Begemann, P., 6134
Havlik, A. J., 4295
Hawes, R. C., 5195
Hawke, J. C., 2529, 2996, 2997, 3053-3060
Hawkes, S. J., 2526, 2527, 3061
Hawkins, E. G. E., 29
Hawkins, J. M., 2788
Hawkins, L. H. C., 3062
Haworth, R. D., 3063
Hawthorne, J. N., 3306
Hay, D. G., 3064
Hayakawa, A., 3392
Hayakawa, F., 6741
Hayashi, S., 4734-4737
Hayes, H., 5035, 6109
Hayes, W. V., 3211
Hazdra, J. J., 727
Hazeldean, G. S. F., 3065, 6227, 6228
Hazen, G. G., 3052
Heald, E. F., 3066, 5133
Healy, J. W., 6037
Hearfield, R. C., 3067
Heasley, V., 7006
Heath, C. E., 4523, 4524
Heath, M. T., 2992
Heaton, W. B., 3068
Heck, R. F., 3069
Hecke, F. van, 3070
Hedgley, E. J., 3071
Hedsted, D. M., 3072
Hefendehl, F. W., 3073
Heft, C. H., 3074, 3075
Heft, K. H., 442
Heftmann, E., 3076-3078, 3647
Heide, R. ter, 3079, 4053-4057, 5973
Heigl, J., 3080
Heilbronner, E., 3778, 4138, 4139
Heilbronner, I. E., 3081
Heine, E., 2887
Heine, Jr., K. S., 3704
Heinemann, W., 3082, 3083
Heines, Sister Virginia, 3084
Heinisch, B., 643
Heinrich, B. J., 4471
Heinze, G., 951, 4328
Heinze, H. O., 2016-2018, 3085, 3086
Heinzel, M., 3311
Heinzig, E., 2439
Heitz, J., 2825
Heitzman, R. J., 3087
Hele, P., 3088
Heller, H., 958, 4604
Hellman, H. M., 3089
Hellman, M., 3090
Hellmann, H., 3091
Hellstrom, K., 3092

Hellyer, R. O., 3093
Helm, R. V., 3094
Helmreich, W., 7141
Helms, C. C., 1430, 3095-3097
Helwig, H. L., 3618
Helzhauser, H., 3098
Henbest, H. B., 1678, 2784, 3099
Henchman, M., 3100
Hendel, C. E., 1183
Henderson, J. F., 3101
Henderson, J. I., 3102
Henderson, J. N., 6279
Henderson, L., 2474
Henderson, R., 4816
Hendricks, W. J., 3103
Hendrickson, R., 3928-3930
Hendry, R. A., 1270, 1271
Henly, R. S., 3104
Henneberg, D., 3105-3108, 6154, 6155
Henner, E. B., 1359
Henniker, J., 2196
Hennion, G. F., 3109, 3110
Henrikson, B. W., 349, 350
Hepfinger, N., 388
Hepler, L. G., 3111
Hepner, L. S., 709, 3112
Hepp, H. J., 3113
Herb, S. F., 546, 3114-3117, 3368, 4595, 4596, 4873, 6233
Heredy, L. A., 3118
Herington, E. F. G., 820, 3119
Herk, L., 3120
Herk, Jr., L. P., 4363
Herlan, A., 5515
Herling, J., 2587-2590, 3121, 6264, 6265
Hernandez, H. H., 7083
Hernandez, Jr., R., 3122
Herndon, W. C., 4371, 4372
Herout, V., 4162, 4535-4537, 5052, 5273, 7106
Herr, W., 12, 3123, 3124, 6122, 6586
Herrin, C. B., 3125
Herrmann, E., 354
Hersch, P., 3126
Hersh, C. K., 3127
Hertog, H. J. Den, 4654, 5520
Herzberg-Minzly, Y., 2591, 2592
Heseltine, H. K., 3128
Hess, H. V., 2352
Hesse, G., 3129-3131
Hessenhauer, F., 7082
Heston, W. B., 3413
Heuschkel, G., 3132
Heveran, J. E., 1006
Hewett, D. R., 3133
Hewitt, E. J., 3037, 3134, 3135, 4580
Hewitt, G. C., 3136

Hey, D. H., 1213, 1214, 3137, 3138
Heydtmann, H., 3139
Heyes, T. D., 3140
Heywood, A., 1612-1614
Hickerson, J. F., 3141, 3142
Hickinbottom, W. J., 849
Hickling, J. W., 5437
Hickman, H. M., 4415, 4416
Hield, H. Z., 2084
Higgins, C. E., 3143
Higgins, E., 7355
Higgins, G. M. C., 552
High, L. B., 1559
Highet, R. J., 4452, 4453
Higman, H. C., 1681
Higson, H. G., 3144
Higuchi, T., 6700
Hildebrand, G. P., 5771-5773
Hildebrand, R. P., 2977, 2978, 3145, 3146
Hill, D., 5698
Hill, D. A. W., 990
Hill, D. W., 41, 1182, 3147-3152
Hill, H. I., 2676
Hill, K. J., 178
Hill, K. R., 473
Hill, R., 2877
Hill, R. V., 4239, 4242
Hillen, L. W., 3153
Hills, G. L., 2311
Hills, P. R., 2204, 2479
Himoe, A., 6584
Hindin, E., 3154-3156
Hinds, L., 585
Hines, W. J., 2200, 2201, 5203, 6396
Hines, W. L., 3157
Hinkins, R. L., 5601
Hinkle, E. A., 3158-3160, 5393
Hinnen, A., 3161
Hinsvark, O. N., 3162, 3163
Hirano, S., 3164
Hirose, Y., 3165-3167, 5298, 6005
Hirsch, J., 56, 3168, 3385, 5470
Hirst, E. L., 416
Hirt, T. J., 3169
Hisatsune, I. C., 3577
Hishta, C., 3170-3175
Hissel, J., 3176
Hiu, D. N., 3177
Hively, R. A., 3178, 3179
Ho, H., 6576-6580
Hoag, L. E., 3180
Hoare, M. R., 3181-3184
Hobbs, A. P., 3185
Hobbs, M. E., 5705, 5708
Hobden, F. W., 3186
Hobson, A., 3187
Hodges, R., 2035
Hoek, A. van den, 3188, 6379
Hoelscher, H. E., 4812

Hoenes, Jr., H. J., 3189
Hofer, L. J. E., 156, 6533, 6534
Hoff, J. E., 3190, 7287
Hoffman, A. J., 3191
Hoffman, A. S., 5602
Hoffman, L. L., 1569
Hoffmann, D., 3192
Hoffmann, E. G., 3193, 3194
Hoffmann, G., 3195-3198
Hoffmann, N. L., 4319
Hoffmann, R. W., 3199
Hofmann, J. E., 3200
Hofmann, M., 3201-3203
Hofstader, R. A., 3204, 4647
Hofstee, T., 3205
Hoftyzer, P. J., 3206
Hoh, G., 3207
Hohr, L., 1352
Hoigne, J., 3208
Holaday, D. A., 3209
Holland, P. D., 5396
Holland, R. H., 7356
Hollingsworth, C. A., 2816
Hollis, B., 1409
Hollis, O. L., 3210, 3211
Holman, R. L., 5205
Holman, R. T., 2300, 4253, 5035-
    5038, 6109, 7057, 7329
Holmberg, J., 468
Holmes, J. C., 3212, 6247
Holmes, J. M., 3213
Holmes, P. D., 1237
Holmes, W., 3214
Holmes, W. L., 3215, 4174, 4405
Holmgren, A., 5296
Holness, D., 3216-3220
Holst, J. J., 2031, 5192, 5563,
    5564
Holt, A. S., 5031, 5032
Holt, C., 3221
Holzhauser, H., 3222-3224, 3767,
    3775
Homma, T., 2189
Honda, S. I., 6833
Honegger, C. G., 3225
Honegger, R., 3225
Hoogschagen, J., 3226
Hooimeijer, J., 3227
Hook, J. R., 3151, 3152
Hooker, C. N., 3059
Hoover, F. W., 3228
Hopkins, C. Y., 1362-1364, 3229-
    3231
Hopkins, R. L., 1479, 1480, 6827,
    6828
Horak, M., 4102
Horak, W., 3232
Horn, M. R., 7139
Horn, O., 3233, 3234
Horn, R. C., 4288
Horner, L., 3235
Horner, P. J., 3236

Horning, E. C., 1642, 2184, 2853-
    2855, 2857-2859, 3237-3249,
    4550, 4551, 5543, 6665, 6985-
    6997, 7537
Horning, M. G., 3250-3253, 4676,
    6973
Hornstein, I., 77, 3254-3258
Hornung, W., 3310
Horrii, T., 5027
Horrocks, B. J., 3619
Horrocks, L. A., 3259-3261
Horton, A. D., 827, 3262
Horvath, C., 2888, 2889
Horwitt, M. K., 3263, 7396
Horwitz, W., 2266, 2267
Horwood, J. F., 2307
Hoshimo, O., 6719
Hosoi, T., 2692
Hotchkiss, D. K., 4563
Hougen, L. R., 3264
Hough-Grossby, A. W., 3265
Houghton, A. A., 3266
Houghton, G., 2436, 2437, 3267,
    5315
Houle, C. R., 5806
Houlihan, W. J., 3268, 3269
House, H. O., 3270-3272
House, R., 4076
Houser, E. A., 3273, 3274
Houston, R. H., 3275
Houze, N., 4075
Hovermann, W., 3276, 3614
Howard, F. L., 6784
Howard, G. A., 1696, 1897, 3277-
    3282
Howard, H. E., 3283
Howard, J. F., 2040
Howard, R., 6281
Howard, R. F., 3721
Howard, T. J., 375, 376
Howe, D. D., 2052
Howe, W. H., 3284
Howell, C. F., 1512, 1514, 1517
Howells, T. J., 2221
Howsam, S. G., 4800, 4801
Howton, D. R., 3285
Hoyle, R. E., 7295
Hrapia, H., 3286-3290, 4329
Hrivnac, M., 750, 3291-3293,
    3522, 3523
Hrusovsky, M., 6444
Hrutfiord, B. F., 3294
Hsieh, C-C., 7481
Hsu, Y-S., 1351
Huang, H-F., 5436a
Hubbard, G. L., 4852
Hubbert, H. E., 4945
Hubele, A., 3311
Huber, H., 3759, 3760, 5060
Huber, H. F., 1597-1599
Huber, J., 5212
Huber, J. F. K., 3295

Hubicki, J. A., 3052
Hubis, W., 1474
Huck, G., 5425
Huck, H. W., 3296
Hückel, W., see Hueckel, W.
Hudson, Jr., B. E., 3297-3299,
    5342
Hudson, J. R., 3300
Hudson, R. L., 5390, 5391
Hudy, J. A., 2510, 3301
Huebner, V. R., 3302-3305
Huebscher, G., 3306
Hueckel, W., 3307-3316
Huelin, F. E., 3317
Huestis, L. D., 3318
Huff, H., 994
Huffman, J. W., 3319
Hugel, M. F., 5367
Hughes, D., 1360, 3320
Hughes, H. W. D., 556, 557
Hughes, K. J., 3321-3324, 3351,
    3353, 3354, 4218
Hughes, M. A., 3325, 3326
Hughes, R. B., 3327-3329
Hughes, Jr., R. E., 3330
Hughes, R. W., 3331
Huguet, M., 1189, 3332
Huitric, A. C., 3333
Hull, W. Q., 3334, 3335
Hultschig, M., 3336, 4110
Hummel, D., 3337
Hummel, R. W., 3338
Humphrey, E. F., 4525
Humphrey, G. L., 3841, 3842
Humphrey, M., 843
Hunsmann, W., 3339
Hunt, G. E., 6833
Hunt, M. H., 6429
Hunt, P. P., 3340, 3341, 6416
Hunt, R. H., 1910
Hunter, G. L. K., 3342
Hunter, I. R., 3343-3349
Hunter, T. F., 2181
Hupe, K. P., 657-659
Hurn, R. W., 2244, 2245, 3321-
    3324, 3350-3354
Hurrell, R. A., 131, 3355
Hurst, G., 1631
Hurst, G. S., 4878
Hurwitz, H., 6353
Hurwitz, P., 409
Hussey, A. S., 3356, 6059
Hussong, R. V., 3410
Hutchinson, R. B., 3357
Huttig, E., 4206, 4600
Hutton, D. G., 3702
Huyser, E. S., 3358-3361
Huyten, F. G., 3362-3365
Hwa, J. C. H., 3366
Hybl, C., 3367
Hyden, S., 3368
Hyder, G., 5643

Hyer, H., 4944

Ibbitson, D. A., 586, 587
Ibraev, G. Z., 3369
Ichizuka, I., 5238
Iconomou, N., 3370
Iconomou-Petrovich, N., 1137, 3371
Iden, R. B., 3372
Idleman, J. A., 4813-4815
Iftimescu, C., 1771
Iguchi, M., 3373, 3374
Ikeda, M., 4215
Ikeda, N., 6294
Ikeda, R. M., 2776, 3375-3377, 6522. 6523
Ikegami, A., 6045
Ikegawa, N., 3378
Ikekawa, N., 3379, 6719, 6890, 6994
Ikram, M., 1206
Imhoff, D. H., 758
Ince, J. H., 5659
Inczedy, J., 6720
Ineichen, M., 2840
Ingram, W. T., 3380
Innes, W. B., 3381, 3382
Inoue, H., 5116
Institute of Petroleum, 3383
Insull, Jr., W., 56, 2195, 3384-3386, 6591
Iogansen, A. V., 4124, 4217, 6752
Ionescu, A. G., 6034
Ippoliti, P., 1723
Irby, R. M., 3387
Ireland, C. E., 1332, 1333
Irvine, L., 3388, 3389
Irving, H. M., 3390
Irwin, G. A., 5389
Isaac, P. C. G., 1368-1370
Isbell, A. F., 4438, 4439
Isbell, R. E., 3391
Ishi, Y., 3392
Ishii, A., 3874-3877
Ishii, D., 6742
Ishikawa, A., 2963
Ishikawa, M., 367, 3393
Ishizuka, I., 6635-6637
Isler, O., 6047
Issenberg, P., 3394, 7286
Issoire, J., 3395, 3396
Istrate, E., 2091
Itaya, M., 3012
Ito, S., 3397
Itsek, S. E., 5065
Ivanova, L. N., 5180, 5940
Ivanova, N. T., 6912
Ivanova, R. V., 3398
Ives, I. G. C., 3399
Iveson, G., 2053, 3400-3404
Iwakiri, Y., 5027
Iwata, N., 6848

Iwaya, K., 5086
Iyer, R. M., 3405
Jack, E. L., 3406
Jacknow, B. B., 3407, 7142, 7143
Jackson, A., 815
Jackson, H. W., 3408-3410
Jackson, K. L., 3411, 3412
Jackson, M. W., 3413
Jackson, R. E., 3414
Jackson, R. G., 3415
Jacobs, M., 4147, 4971-4973
Jacobs, M. B., 3416, 3417
Jacobs, T. L., 3418, 5848
Jacobson, M., 3419
Jacura, Z., 1512
Jaffe, F., 3420
Jahnsen, V. J., 3421
Jain, K., 3651
Jain, T. C., 3422
James, A. T., 122-124, 673, 1538, 1890, 3385, 3386, 3423-3475, 4502, 4674, 5205, 7030, 7169, 7170, 7252
James, D., 2766
James, D. H., 3476-3478
Jamieson, G. R., 3479-3484
Janak, J., 723, 749, 750, 1395-1397, 1641, 1957, 3291-3293, 3485-3559, 4154, 4155, 4158, 4159, 5206, 5268-5270, 5272, 6426
Janda, J., 3560
Jandorek, R. D., 5898
Jangaard, P. M., 3561
Janicek, M., 6800
Janitzki, U., 3562
Janssen, R., 707
Jantzen, E., 2619
Janzen, J. J., 6872
Jarboe, C. H., 3563, 7303
Jardine, D. A., 1289
Jarreau, C. L., 2198
Jart, A., 3564-3568, 4030, 5078
Jaulmes, P., 3569-3571
Jaureguiberry, G., 3572
Jaworski, M., 3573, 3574
Jay, B. E., 3575, 3576, 7354-7356
Jayadevappa, E. S., 3577
Jedrzejczyk, B., 6947
Jeffery, P. G., 3133, 3578-3580, 3992
Jeffs, A. R., 3028-3033
Jelliffe, R. W., 3581
Jellinek, J. S., 3582
Jenard, H., 3583
Jenckel, L., 1122
Jenkins, G. I., 4173
Jenkins, J. W., 3584
Jenkins, N., 1467
Jenkins, P., 3585, 3586
Jenkins, R. E., 3587
Jennings, A. P. H., 3588

Jennings, Jr., E. C., 3589-3591
Jennings, Jr., F. C., 6548
Jennings, W. G., 3592-3596, 6794
Jensen, P. W., 3597
Jensen, R., 3598, 4346
Jensen, R. G., 2462, 2463, 3599-3607
Jensen, R. M., 5397
Jentoft, R. E., 3608
Jentzsch, D., 762, 763, 3276, 3609-3614
Jessamy, J., 1719
Jesse, W. P., 3615
Jessop, A. S., 4804
Jesting, E., 524, 3616, 3617
Jeung, E., 3618
Jewett, P., 3619
Jikilee, B. M., 3620
Johansson, G., 3621
Johncock, P., 3622
Johns, T., 893, 894, 1187, 1210-1212, 1251, 1901, 2454-2456, 3623-3644, 3707, 5999, 6377, 6569, 6963, 7052, 7053
Johnsen, S. E. J., 3158
Johnson, A. W., 942
Johnson, C. H., 3838
Johnson, D. E., 3645, 3646, 5081
Johnson, D. F., 3647
Johnson, E. A., 673, 3648-3650
Johnson, F., 942
Johnson, H. D., 2503, 5800
Johnson, H. E., 1514
Johnson, H. S., 3651
Johnson, Jr., H. W., 3652-3658
Johnson, J. A., 4937, 4419
Johnson, J. F., 561, 637, 638, 1252, 2466, 2467, 2951, 2952, 3608, 3659, 5601-5609, 6473-6475
Johnson, J. H., 1061, 1935
Johnson, J. K., 1688
Johnson, J. L., 6409
Johnson, J. W., 4471
Johnson, L., 1161
Johnson, L. F., 4769
Johnson, R. E., 3660-3670, 3866-3868
Johnson, Jr., R. L., 4296
Johnson, W. H., 2259
Johnson, W. S., 5186
Johnston, H. L., 3671, 6774
Johnston, Jr., J. W., 3672
Johnston, P. V., 3673, 3674
Johnston, V. D., 3675-3683, 5590-5593
Johnstone, R. A. W., 1263-1265, 3063, 3684-3686
Jokl, J., 2341, 2342
Joklik, J., 3687, 3688
Joly, D., 4278
Jones, A. G., 2877, 3689

Kelley, R. P., 2992
Kelley, T. R., 6336, 6337
Kellner, S. M. E., 3908
Kellock, T. D., 3473
Keltakallio, A., 3013-3015
Kemball, C., 1016-1018, 1674, 1675, 3909, 4590, 7255, 7256
Kemp, P., 3306
Kennedy, C. D., 3910, 3911
Kennerly, G. W., 4882
Kennett, B., 4916, 4917
Kennett, B. H., 3317
Kenney, R. L., 3912
Kent, J. A., 3913
Kent, N. R., 1631
Kent, T. B., 3914
Kepner, R. E., 7166, 7167
Keppler, J. G., 690, 691, 1848, 3198, 3915-3917
Kerckoff, W. G., 694
Kerenyi, E., 3918, 3919
Kergomard, A., 2188, 3920
Kern, M., 707
Kern, W., 1352
Kerr, J. A., 3921-3927
Kessler, J. E., 2471
Kessler, T., 2621
Kesterson, J. W., 3928-3930
Keszler, I., 3931
Keszthelyi, S., 3918, 3919, 6694
Kettner, K. A., 1676
Keulemans, A. I. M., 122, 125, 3295, 3712, 3932-3952, 4828
Keutmann, E. H., 6467, 6468
Keyes, F. G., 3953
Keyzer, H., 3093
Khabibullin, M. F., 6975
Khan, M. A., 42, 43, 3954-3958
Khan, N. A., 500
Khan, T., 4517
Kharasch, M. S., 3959
Kharin, A. N., 3960
Kharlamov, N. R., 1950
Khatri, L. L., 3961
Khotimskaya, M. I., 4113
Khrapova, E. V., 5480a
Kielar, E. A., 4453
Kienitz, R., 3768, 3962, 3963
Kieselbach, R., 3713, 3964-3971
Kieser, M. E., 3972, 3973
Kikuchi, Y., 2412, 3974
Kilheffer, J. V., 3975
Kilner, A. E., 6821
Kim, G., 3976
Kimball, A. P., 7555
Kimber, R. W. L., 478-480
Kimel, W., 5296
Kimura, H., 4220
Kimura, K., 3977
Kimura, M., 3978, 3979, 6891
Kindley, L. M., 5468, 5469
King, B. D., 504

King, G. G., 6437
King, L. C., 3980
King, R. W., 1234, 1764, 2173, 3981, 3982, 4471
King, Jr., W. H., 3297-3299
King, W. J., 3983
Kingsbury, K. J., 3984
Kingsley, G. R., 3985, 3986
Kingston, B. H., 3987
Kingston, C. R., 3988
Kinoshita, S., 4730
Kinsell, L. W., 3989, 4909
Kipling, J. J., 3990
Kipping, P. J., 3133, 3578-3580, 3991, 3992, 6747
Kiraly, J., 6088, 6089
Kirby, F. B., 4107
Kirby, G., 1041
Kirch, E. R., 3849
Kircher, H. W., 3993-3996
Kirchner, P., 7240
Kirihara, T., 6293
Kirk, A. D., 3997
Kirk, P. L., 2299, 3988, 5196, 5383-5388
Kirkland, J. J., 3998-4006
Kirmse, W., 1882, 4007
Kirsch, F. W., 4008
Kirschman, J. C., 4009
Kirschner, M. A., 4010
Kirsten, W. J., 4011
Kiselev, A. V., 380, 706, 3783, 4012-4017, 7024, 7501, 7502
Kiseleva, N. N., 1879
Kiser, R. W., 4877
Kishimoto, K., 361, 4018, 4019
Kishimoto, Y., 4020
Kisic, A., 4021
Kislinskii, A. N., 5033
Kistiakowsky, G. B., 1180, 1181, 2382, 4022
Kitade, K., 4220
Kitagawa, I., 4023
Kitahara, K., 4024-4028
Kitahara, M., 4029
Kitajima, M., 2939
Kitao, T., 5285
Kitaoka, Y., 5285
Kitt, G. P., 2638, 2640-2642
Kivalo, P., 3014, 3015
Kjaer, A., 4030
Klaas, P. J., 4031
Klasse, J. M., 4032
Klauser, H. E., 4440
Klaver, R. F., 637, 2466, 2467, 3659, 4033
Kleber, W., 4034
Klein, A. K., 4035, 7161
Klein, E., 4036
Kleinberg, J., 6886
Kleis, A. A. B., 700
Klemm, H., 5274

Klemm, L. H., 4037, 4038
Klenk, E., 4039-4044
Kleschick, A., 6186, 6187
Klesper, E., 4045
Klima, J., 4046
Kliman, B., 4047
Klingman, C. L., 4048, 4049
Klinkenberg, A., 667, 4050-4052
Kliss, R. M., 4101
Kloss, A., 4546
Klouwen, M. H., 4053-4057
Knapman, C. E. H., 4058-4062
Knapp, R. E., 168
Knick, H., 2141
Kniebes, D. V., 162, 4063, 6757
Knight, A. R., 4064
Knight, H. S., 2032, 2033, 4065-4073
Knight, J. A., 4074, 4075, 6309-6311
Knight, J. D., 4076
Knight, P., 2992
Knights, B. A., 4077-4079
Knowles, J. R., 4080
Knox, J. H., 2179, 2182, 2252, 2253, 2756, 3102, 4081-4093
Knox, L. H., 1883, 1884
Knox, W. R., 4094
Kobashi, Y., 4095, 4096
Kobayashi, M., 3854, 4097
Kobayashi, R., 4872, 5890, 6512
Kobayashi, T., 2692
Koberstein, E., 4098
Koch, R. C., 2717, 4099, 4100
Koch, S. D., 4101
Kochloefl, K., 4102, 4120-4122
Kocirik, K., 4103
Kodicek, E., 7353
Koegl, F., 4104
Koegler, H., 4105
Koehler, J., see Köhler, J.
Koennecke, H. G., 3202
Koerner, W. E., 2069-2071
Kofler, W., 4106
Kofron, W. G., 4107
Koga, A., 6729
Kogan, G. A., 7450
Kogan, V. B., 2513
Kögl, F., see Koegl, F.
Kogler, H., 4108-4110
Kohler, G. O., 829, 3349
Köhler, J., 4044
Kohlik, A. J., 4038
Kohlmann, P., 6253
Kohn, G. K., 2833
Kohn, R., 4528-4530
Kojima, J-I., 6727
Kojima, T., 2433, 2434
Kokes, R. J., 4111
Kolb, B., 7240, 7241
Kolb, J., 6708
Kolbanovskii, Y. A., 1074, 1075

Kolbel, H., 4112
Kolesnikova, L. P., 4113, 6456
Kolloff, R. H., 4114
Kolobikhin, V. A., 4115, 4116
Kolyubyakina, A. I., 6911, 6913
Komae, H., 369, 4734-4737
Komers, R., 3524-3529, 4117-
  4122, 4537
Komissarov, P. F., 7098, 7099
Komoda, H., 7436
Kondo, T., 401
Kondrat'ev, D. A., 4123
König, W., 7242
Konigsbacher, K., 4370
Konigsbacker, K. S., 3037
Konishi, T., 4029
Konnecke, H. G., 3290, 4328, 4329
Konrath, H. J., 6842
Kontorovich, L. M., 4124
Kooiman, P., 4125
Koons, J. R., 7211
Koons, P. D., 6338, 6339
Koops, J., 488
Kopaczyk, K. C., 3673
Koppe, R. K., 31-33
Kormany, T., 6692
Korn, E. D., 4126
Kornyakov, V. S., 2496, 2497,
  5362-5364
Koro, S., 2427
Korol, A. N., 4127, 4128, 7563
Koroleva, N. M., 4539
Koromal'di, E. V., 7503
Korotov, S. Y., 4129
Korte, F., 1136, 4130
Korvzee, A. E., 4131
Kory, R. C., 2926
Korytnyk, W., 1099
Kosak, A. I., 1997
Kosak, A. L., 7007
Kosaki, M., 7436
Koshinen, O., 2850
Kosikowski, F. V., 6074, 6075
Koski, W. S., 3862, 5451, 6126
Kossler, I., 1891
Kossoy, A. D., 5012
Köster, R., 3107, 4132, 6154,
  6155
Kosters, B., 6412, 6413, 7236
Kostomaj, T., 6900
Kostyk, I., 6186, 6187
Kostyo, A. E., 3118
Kotel'nikov, B. P., 2464, 4133,
  6275
Kourey, R. E., 3023, 7149
Kovacs, A. S., 4134
Kovats, E., 122-124, 3081, 4135-
  4139, 6613, 6867, 7177, 7178
Kowanko, N., 481, 482
Koyanagi, T., 7464
Kracke, F. L., 946
Krajkeman, A. J., 4140, 4141

Krajovan, V., 6900
Kramer, D. N., 2461
Kramer, F. R., 1486
Kramer, G. M., 4142
Kramer, K., 4143
Kramlich, W. E., 4144
Kranz, Z. H., 1912, 1913, 4145
Krapcho, A. P., 4146
Krasheninnikov, S. K., 751
Kratz, P., 4147
Kratzl, K., 4148-4150
Kraus, H., 4151
Kraus, J. W., 4152
Kraus, K. W., 1295-1298
Kraus, T., 4161
Kraus, U., 4327
Krause, L. A., 4633
Krebs, A., 7394
Krebs, K. G., 2104
Krejci, M., 3530-3534, 4153-4159,
  5960
Krell, E., 4160
Kremer, E., 4161
Krepinsky, J., 4162, 7106
Kress, A., 4468
Kretschmer, F., 4163
Kreuchunas, A., 4164
Kreyenbuhl, A., 4165-4167, 7194
Krhut, A., 4168, 4169
Krichmar, S. I., 4170, 4171
Krige, G. J., 2704
Krimmel, J. A., 6121
Kring, E. V., 4173
Kritchevsky, D., 4174, 4175
Kritz, W. R., 4176-4178
Krivoruchko, F. D., 4179
Kroman, H. S., 708, 4180, 4181
Kronemberger, K., 1122
Kronmueller, G., 4182, 4183
Kronskop, N. C., 4609
Krotoszynski, B. K., 2235
Kruger, F. A., 4249
Kruglov, E. A., 6975
Krumphanzl, V., 4184
Krupka, R. M., 6833
Krupp, H., 4186
Krylov, B. K., 4187
Krzeminski, Z. S., 4188
Ksinsik, D., 2396, 2397
Ku, C-W., 5097
Kuan, T-S., 4216
Kuang, Y-D., 4517
Kubiczkova, H., 4189, 5792
Kubinova, M., 4190-4196
Kucera, E., 2781
Kucharczyk, N., 4189, 5792
Kucherov, V. F., 5938-5940, 7449,
  7450
Kuck, J. A., 4197
Kuczkowski, J., 6708
Kudo, S., 4198, 4199
Kudryavtseva, N. A., 6752-6755

Kuehn, P. R., 4946
Kuffner, F., 3781, 4200
Kugler, E., 6867
Kuhl, M., 3098, 3224, 3775, 4201-
  4203
Kuhn, A., 2095, 2096
Kuhn, H.-J., 1922
Kuhn, S. J., 5309, 5310
Kuhn, W., 4204, 4205
Kuhnhanss, G., 4206
Kuhns, L. J., 4207
Kuksis, A., 4208, 4766
Kulanovic, D., 116
Kulcsar, G. J., 4209
Kulcsar-Novakova, M., 4209
Kuley, C. J., 4210
Kuliev, A. M., 2768
Kulka, K., 2804-2806
Kulonen, E., 2851
Kumar, D., 4211
Kummer, D., 2398, 2399
Kummerow, F. A., 816, 817,
  3673, 3674, 4023, 5446, 5447,
  6789
Kundel, H., 725
Kung, J. T., 4212, 4213
Kunst, E. D., 4214
Kunugi, T., 4215
Kuo, H-F., 4216
Kupfer, G., 660, 661, 1920, 1921,
  1923, 1924
Kurkchi, G. A., 4217
Kurlyama, K., 5116
Kurn, R. W., 4218
Kuroda, T., 4219
Kurono, G., 4220
Kurz, J., 3311
Kusama, T., 2425
Kuschel, J., 4112
Kusuda, Y., 3167
Kusy, V., 4221, 4222
Kutschke, K. O., 852, 853, 1639,
  5652, 7029
Kuwada, D. M., 4223
Kuzdzal-Savoic, S., 4224
Kuz'mina, A. V., 4539
Kuz'mina, L. P., 165, 6455, 6457
Kuznetsov, B. V., 6910
Kvitkovskii, L. N., 4225
Kwan, T., 2413, 2442, 2443, 4226,
  4227, 5301-5303
Kwan, Y., 4518
Kwantes, A., 3205, 3227, 3943-
  3946, 4228
Kwiatkowski, G. T., 6053
Kwie, W. W., 4229
Kwitkowski, P., 6711
Kyryacos, G., 941, 4230-4233,
  4880

Laaksonen, A. L., 2852
Lacey, Jr., J. C., 1164

Lewis, K. G., 4377
Lewis, L. L., 4378
Lewis, W. K., 4379
Leyton, G. R., 6326
Lhotsky, G., 3367
Li, H. C., 4520
Libbey, L. M., 4380
Libers, R., 4094
Liberti, A., 1275, 4381-4394
Lichtenberg, J. J., 5895
Lichtenfels, D. H., 4395, 4396, 5480
Lieberman, M., 4397
Liebmann, H., 843
Liebster, J., 1878
Liepina, R., 6234
Liggett, R. W., 4398
Lightfoot, E. N., 4399
Lijinsky, W., 4400-4402
Lillard, D. A., 4403
Lille, U., 4404
Limoncelli, E. A., 4617, 4618
Lin, T. H., 4405
Lincoln, R. M., 4406
Linde, H. W., 4407
Lindeman, L. P., 4408, 4409
Linden, A. C. van der, 6815
Lindgren, F. T., 4410, 4411, 5218, 6746, 6964
Lindlar, H., 6047
Lindquist, K., 4412
Lindsay, R. C., 1713, 1714
Lindsey, Jr., R. V., 740
Lineweaver, H., 4413
Linfield, W. M., 676, 3725
Ling, S., 7551
Lingren, B. O., 802
Link, H., 2769
Link, W. E., 4414-4418, 6109
Linko, Y. Y., 4419
Linnenbom, V. J., 6681, 6682
Linnig, F. J., 5392
Lins, G., 4420
Linturi, M., 937
Lipman, C., 1463
Lipowicz, J., 388
Lippold, G., 3807
Lipsky, S. R., 4260-4265, 4421-4435, 4503, 6266, 6267
Lires, O. A., 4469
Lis, E. W., 4436
Liska, B. J., 4279, 6539
Liska, D., 5793
Lister, F., 1055
Litchfield, C., 4437-4439
Little, L. H., 4440
Littlewood, A. B., 947, 4441-4449
Litvin, E. F., 5181
Lityaeva, Z. A., 4450
Liu, C-H., 7481
Liu, S. K., 6683

Lively, L. D., 1244, 4285-4287, 5448-5450, 5919
Llewellyn, D. R., 7
Lloyd, D. I., 4451
Lloyd, H. A., 4452, 4453
Lloyd, M. I., 2817-2819, 4454, 4455
Loc, L., 375
Locateli, L., 7006
Lochte, H. L., 4456, 4457
Locke, L. N., 4458, 4459
Lockhart, E. E., 4460, 4461
Lodge, Jr., J. P., 4462
Loffler, A., 2825
Loffler, W., 4463
Loft, J. T., 4464
Logan, T. J., 473
Lohman, F., 4465
Lohmar, R. L., 489, 1170, 6392, 6394
Lohr, L. J., 4466, 4467
Lombard, R., 4468
Lombardo, G., 2470
Lombardo, J., 4469
Long, C. N. H., 4470
Long, D. R., 4471
Long, J. F., 2186
Long, R., 5360
Longenecker, W. H., 4472
Longhetti, A., 4473
Longmire, C. L., 4474
Longuevalle, S., 1893-1896
Looney, R. W., 4475
Lopez, G. Z., 368, 1537
Lopiekes, D., 4101
Lorant, M., 4476
Lord, R. C., 2162
Lorenz, D. H., 4477
Lorz, W., 4478
Losse, A., 4479, 4480
Losse, G., 4480
Lotz, J. R., 4481, 4482, 6629, 6630
Louden, M. L., 6898
Louedec, A., 4483, 4484
Louloudes, S. J., 4485, 6834
Loury, M., 4486
Lovelace, A. M., 1980
Lovelace, W. R., 7265
Loveland, J. W., 2173, 5463
Lovelock, J. E., 2762, 3450, 3451, 4432-4434, 4487-4506, 6331, 7552
Low, M. J. D., 4507
Lowe, A. E., 4508, 5585-5587
Lowe, H. J., 4509-4511
Lowrey, W., 977
Lowrie, R. S., 1276, 1277
Lowry, R. R., 4512, 6424
Loyd, R. J., 1737-1739, 4513, 4514
Lu, P-C., 4515-4519, 7454

Lu, P-T., 4520
Lu, T-F., 4519, 4520
Lucas, D. M., 4521
Lucchesi, C. A., 4793
Lucchesi, P. J., 614, 4522-4524, 6759
Luce, C. C., 4525
Luce, R. L., 4526
Luchsinger, W., 4527
Lück, H., 4528-4530
Luck, J. M., 6564
Luckhurst, G. R., 1792, 2691
Luddy, F. E., 546, 3116, 4531, 4596, 4873
Luders, R. C., 5766
Ludwig, E. H., 984
Luebbe, Jr., R. H., 4532
Lueck, H., see Lück, H.
Luh, B. S., 4533
Lukacovic, L., 6358, 6362
Lukas, D. S., 4534
Lukes, V., 4535-4537
Lukhovitskii, V. I., 2687
Luk'yanovich, V. M., 1950
Lulova, N. I., 4538, 4539, 6654, 6752-6755
Lunaas, T., 4530
Lund, N. A., 3266, 4541
Lund, P. G., 4542
Lundberg, G. W., 5466
Lundberg, W. O., 4543
Lundeen, A., 4544
Lundin, R. E., 5726, 6793
Lundquist, R. T., 1118
Lunt, E., 539
Lupu, C., 2947
Lusena, C. F., 75
Luskina, B. M., 4545
Luther, H., 4546
Lutwick, G. D., 4547
Lutz, C. A., 4548
Lutz, E. F., 4549
Luukkainen, T., 2184, 3243, 4550, 4551
Lykken, L., 6177
Lyman, R. L., 6846
Lynch, C. A., 3110
Lynch, C. T., 4552
Lynch, E. R., 4553
Lynch, G., 6709
Lynch, J., 4554
Lynn, T. R., 4555, 5424
Lyons, J. M., 4556
Lyons, M. J., 4557
Lyssy, G. H., 6350
Lysyj, I., 4558-4562, 7493

Ma, A. P., 4563
Mable, S. E. R., 3152
Maccioni, A., 1192
Maccoll, A., 4564, 4565

301

MacDonald, W. B., 3474
MacDonell, H. L., 4566
Macfarlane, M. G., 4567-4569
MacGregor, D. R., 6626
Mach, E. E., 5834
Machacek, Z., 4570
Machata, G., 4571, 4572
Machiroux, R., 4573
Machlin, L. J., 4639
MacIver, D. S., 2908
Maciw, B. O., 4574
Mack, H., 657
Macka, M., 1078
Mackay, D. A. M., 3135, 4575-
    4583, 4778
Mackay, G., 4584
Mackenna, R. M. B., 976
Mackenzie, N., 588, 589, 928
Mackie, A., 4585
Mackle, H., 4586
MacLachlan, A., 4587
MacLeod, D. M., 589, 590
MacRitchie, A. L., 3080
Maczek, A. O. S., 4588
Madan, M. P., 4589
Madden, W. F., 4590
Maddock, K. C., 3244
Mader, P. P., 5190
Mader, W. J., 5376
Madison, J. J., 4591
Maeda, T., 4592
Magalhaes, M. T., 5155
Magee, E. M., 4593
Maggs, R. J., 6229
Magidman, P., 546, 3114-3117,
    4594-4596, 6233
Maglitto, C., 1743
Magnolet, J. C. P., 4597
Magritte, H., 4598
Maher, T. P., 4599
Mahoney, L. R., 1765
Mahrwald, R., 4600, 6129
Maier, H. J., 1042, 1043, 4601-
    4608
Maimoni, A., 2961
Main, R. K., 2534
Mair, B. J., 4609, 4610
Mairanovskii, V. G., 4611
Maizus, Z. K., 3839
Majer, J. R., 4612
Majhofer-Orescanin, B., 4613
Makita, M., 742, 4614, 7203
Makower, B. B., 1862
Malafeev, N. A., 4615, 4616
Maley, L. E., 1542
Malgiolio, J., 4617, 4618
Malin, L., 4619-4621
Maling, H. M., 3250, 3253
Malinowska, K., 4622
Malins, D. C., 4623, 4624
Mallard, T. M., 1569
Mallik, K. L., 2578, 7219

Malyasova, L. A., 7509
Manaresi, P., 1130
Manfredi, G., 472, 4625, 5375,
    6105
Mangan, Jr., G. F., 4626
Mangold, G. B., 4627
Mangold, H. K., 4624, 4628, 4629,
    6109
Manjarrez, A., 111
Manjarrez, M., 2842
Manka, D. P., 4630, 4631
Mankel, G., 3864, 3865
Manko, R., 7117
Mann, J. B., 1972
Mann, T., 3001
Mannering, G. J., 140, 141
Manning, R. J., 4632
Manno, J., 7448
Manolopoulo, M., 523, 5440
Mansur, R. H., 4633
Mantell, C. L., 4634, 4635
Manuel, T. A., 4636
Mapelson, W. W., 4637
Marantette, J. C., 4471
Marbet, R., 6047
Marcali, K., 4638
Marchello, J. M., 7174
Marco, G., 7565
Marco, G. J., 2089, 4639
Marcolongo, F., 4640
Marcus, A. J., 4641-4643
Marcus, N., 1455
Mare, H. E. de la, 4644
Marechal, J., 4645, 4646
Maresh, C., 4647, 6641
Marhoff, F. A., 943-945
Maricq, L., 4648-4650
Marinetti, G. V., 4651
Marino, G., 1100, 1101, 4652
Marion, J. E., 2024
Markert, F., 6483
Markevich, A. M., 1324
Markosov, P. I., 4450
Markov, M. A., 4123
Markova, Z. A., 5877
Markowitz, J., 4175
Marks, M. L., 1140, 4653
Marks, P. A., 2506
Marktscheffel, F., 496
Marot, J., 2943
Marple, W. L., 5620
Marr, A. G., 776
Marsh, R. F., 73, 74
Marshall, S. A., 2234
Marten, A., 4204, 4205
Martens, R. J., 4654
Martin, A. E., 4655
Martin, A. J., 730-732, 2009, 2143,
    2144, 4656-4661
Martin, A. J. P., 3452-3467, 3712,
    4662-4675
Martin, D. B., 3251, 3252, 4676

Martin, F., 4677-4679, 7034
Martin, G. E., 4680
Martin, H. F., 2796, 2800, 3681,
    4682, 7423, 7424
Martin, J. C., 4683
Martin, J. F., 820
Martin, J. L., 4684
Martin, L., 1705, 4685
Martin, L. H., 3802
Martin, R., 7369
Martin, R. L., 764, 4686-4693
Martin, S. B., 4694-4697
Martin, W. H., 6746
Martinelli, M., 4698-4701, 5046
Martinez, C. J. L., 4702, 4703
Martinez, F. W., 730, 731, 4657-
    4661
Martire, D. E., 4704-4706
Martsinkevich, L. E., 1649
Martynova, E. N., 2464
Marukyan, G. M., 516
Maruyama, K., 3878
Maruyama, M., 4707-4709, 4716
Marvel, C. S., 4710, 4711
Marvell, E. N., 4712
Marvillet, L., 4713-4715
Marwald, R., 4600
Masamune, S., 6889
Mashiko, Y., 362, 4716-4718, 6737
Mashukova, G. A., 7046
Mason, E. A., 4719
Mason, G., 4401, 4402
Mason, K. G., 4720
Mason, L. H., 1993, 3695, 4721,
    5284
Mason, M., 4722
Mason, W. B., 6467, 6468
Massart, R., 4723
Massingham, W. E., 2146
Mastio, G. J., 1932
Masuko, Y., 4724
Matas, M., 6444
Mathews, W. S., 4725
Matlock, R. M., 7139
Matousek, S., 4726-4729
Matsubara, I., 4730
Matsuda, H., 4731
Matsuda, S., 4731, 6955
Matsuda, T., 66, 4732
Matsumoto, M., 5119
Matsumura, Y., 4733
Matsuura, T., 4734-4737
Matteson, D. S., 4738
Matthes, K. J., 9
Matthews, J. S., 4739, 6626
Matthews, R. F., 4740
Mattick, J. F., 2044
Mattick, L. R., 4304, 4741-4743,
    7079-7081, 7109
Mattson, F. H., 4744
Mattsson, S., 4745
Matyska, B., 1891, 1986

302

Maucher, D., 3311
Maume, B., 559
Maurel, A., 4746
Maurel, R., 4747, 4748
Maurice, A., 537
Maury, M. J., 5887, 5888
Mavrov, M. V., 4955
Mawson, J., 4749, 4750
May, D. S., 3155, 3156
May, J. E., 2352, 6708
Mayer, S. W., 4751
Mayer, T. J., 4609
Mayer, U., 7395
Mayeux, S. J., 897
Mayrick, R. G., 4586
Maza, M. P. de la, 6868
Mazitova, F. N., 4752, 4753
Mazliak, P., 4754-4759
Mazzetti, F., 4335
McAdams, W. H., 4760
McAlpine, S. G., 4292, 4293
McBain, J. W., 994
McBride, H. D., 4471
McBride, W. R., 735, 4761
McCaffrey, I., 3830-3832
McCaleb, S., 6318
McCall, E. B., 4553
McCallum, J. D., 4762-4765
McCallum, N., 578, 591
McCarthy, A. I., 491
McCarthy, E. M., 1552
McCarthy, M. J., 4208, 4766
McCarthy, R. D., 2155, 4767, 5421, 5424
McCarthy, W. C., 4768
McCasland, G. E., 4769
McClenachan, E. C., 2043
McCloud, G. T., 4374
McCollum, J. D., 4908
McComas, D. B., 4770
McConhaughey, P. W., 1541
McConnell, D. G., 2350
McCord, W. M., 2445, 2446
McCoy, L. L., 4771-4773
McCrae, W., 2036
McCrea, J. M., 1702
McCreadie, S. W. S., 4774
McDaniel, D. H., 2816
McDaniel, J. C., 1766
McDaniel, R. L., 4074
McDermott, P. S., 4775
McDonald, H. J., 4776
McDonald, I. R. C., 527, 528
McDonald, R., 3155
McDonnell, H., 4777
McDonough, E. G., 4778
McDowell, C. A., 701, 4779
McDowell, P. E., 6100
McEwen, D. J., 4780-4785
McEwen, J. I., 4786
McEwen, W. E., 6886

McFadden, J. L., 4787-4789
McFadden, W. H., 2988, 2989, 4790, 4791
McFarland, C. H., 4792
McGee, T. W., 1133
McGinness, J. D., 4793
McGlasson, W. B., 4556
McGovern, L. J., 4794-4796
McGreer, D. E., 351, 4797
McGregor, R. F., 4798, 4799
McGugan, W. A., 4800, 4801, 6277
McGuire, D., 388
McGuire, D. K., 1469
McHenry, Jr., K. W., 1745
McInnes, A. G., 2957, 4802-4806
McIntosh, R. G., 4791
McIntyre, E. A., 469, 470, 1630, 4807
McIntyre, H. C., 4808
McKay, D. G., 3866
McKay, D. K., 4809
McKay, J. B., 935
McKee, H. C., 4810, 4811
McKelvey, J. M., 4812
McKenna, Jr., T. A., 4813-4815
McKennis, Jr., H., 4816
McKetta, J. J., 5845, 7176
McKinley, Jr., J. J., 4817
McKinney, C. B., 4818, 4819, 5854
McKinney, C. R., 3721
McKinney, R. W., 3420, 4820
McLafferty, F. W., 2663
McLaren, L., 4089
McLaughlin, M. F., 1138
McLeod, P. C., 6801, 6802
McLeod, Jr., W. D., 1298
McMahon, W. A., 4810
McMenamy, C., 6708
McMichael, H. E., 4821
McMillan, G., 4822
McMillan, G. R., 4823, 4824
McMillan, R., 7342
McMillan, Jr., W. G., 4825
McMurry, T. B. H., 1462, 1463
McNair, H., 4826
McNair, H. M., 3947, 4827-4830
McNamara, L. S., 4831
McNees, R. S., 2263
McNeil, D., 4832
McNesby, J. R., 1936-1939, 2708, 4833-4838
McNulty, J. A., 1326
McQuaid, F. S., 6787
McQuaid, J. W., 6787
McReynolds, W. O., 4839
McSloan, D., 6581
McTigue, P. H., 4840
McWilliam, I. G., 4841-4847
Mead, J. F., 1830, 1831, 2429, 3285, 4848-4851, 5030
Meador, G. R., 4471
Meadows, G. E., 4852

Meakins, G. D., 4853
Mechelynck-David, C., 4854
Mecke, R., 1813, 4855-4858
Medema, D., 700
Meder, V., 6161
Meder, W. A., 6098
Medina, V. J., 7257
Medlock, R. S., 4859
Medvedeva, N. I., 4860
Meek, J. M., 4861
Meeker, R. L., 2197
Meerssche, M. van, 5079
Mehlenbacher, V. C., 4862
Mehlitz, A., 4863
Meier, G., 490
Meier, W. M., 592
Meigh, D. F., 4864-4866
Meijer, W. A., 969
Meilgaard, M., 4867
Meinertz, H., 4868
Meinstein, S., 5010
Meister, A., 3645, 3646
Melamed, N., 4869
Melhuish, W. H., 2529
Melikadze, L. D., 4870
Melkonian, G. A., 4871
Mellado, G. L., 4872
Mellon, E. F., 4873
Mellon, M. G., 4874
Mellor, N., 4876
Melnick, L. M., 4378
Meloan, C. E., 805, 806, 1314, 4877
Meloni, C. R., 6366
Melpolder, F. W., 4360-4362, 4364
Melton, G. E., 4878
Melville, H. W., 920
Melvin, E. H., 5275, 5276
Menapace, H. R., 940, 941, 4233, 4879, 4880
Mendeloff, A. I., 2048
Mendelsohn, J. M., 4886
Mendoza, M. P., 2187
Menetee, B. S., 5704, 5706
Menke, K. H., 6085
Menna, M. E., 2997
Mercea, I., 1135
Mercea, V., 1135
Mercer, D., 2770
Mercier, D., 1839
Mercuri, O. F., 1045
Meresz, O., 3071
Merits, I., 4881
Meriwether, L. S., 4882
Merlini, L., 5539
Merritt, Jr., C., 4626, 4883-4891, 6628, 7147
Merritt, J., 4892, 6431
Merten, F., 4893
Mesehke, R. W., 4281
Mesirov, M., 7302
Messerly, J. P., 3172, 3173

Messner, A. E., 4894
Mestres, R., 3569-3571
Metcalf, R. L., 4895
Metcalfe, L. D., 4896-4902
Metzger, G., 7144
Metzner, K., 6167
Meyer, Jr., A. S., 429, 4903
Meyer, H., 4904
Meyer, R. A., 2921, 4905, 4906
Meyer zu Reckendorf, W., 4907
Meyerson, S., 1941, 4908
Michael, T. C., 4478
Michaels, G. D., 3989, 4909
Michaelsen, E. R., 2490
Michajlova, S., 2343
Michalek, W., 4910
Michalovic, J. G., 1695
Michel, K. W., 5711
Michelina, M. V., 4911
Mickel, J. P., 4912-4915
Middlehurst, J., 4916, 4917
Miedlich, H., 4918
Mieras, D. G., 4585
Miettinen, J. K., 4919
Mignolet, J. C. P., 4920
Mikhailova, E. A., 4921
Miki, M., 3785-3789
Mikkelsen, L., 679, 4922-4930,
    6103, 6476, 6481, 7280, 7281
Mikl, O., 1078, 4196
Mikolajcik, E., 2948
Mikolajczak, K. L., 4978, 6393
Mile, B., 4612
Miler, B. S., 4419
Miles, D. M., 2056
Miles, H. T., 4931
Miljanich, P., 5307
Miljanik, P., 6846
Millar, R. W., 5758, 5759
Miller, A. J., 4932-4935
Miller, B., 4936
Miller, B. S., 4937
Miller, C. T., 4938
Miller, D. B., 4939
Miller, D. L., 4940
Miller, E. D., 4365
Miller, F. D., 617, 1677
Miller, F. M., 2461
Miller, J., 1544, 4941
Miller, J. E., 5693
Miller, J. J., 4942
Miller, L. D., 4943
Miller, L. J., 3889
Miller, P. H., 1715
Miller, R., 4944
Miller, R. L., 6116
Miller, R. R., 7321-7323
Miller, T., 56
Miller, W., 7303
Miller, W. K., 6949
Miller, W. T., 1299
Milligan, C. E., 542

Milligan, I. B., 4945
Million, J. G., 4946
Milloy, A. D., 393
Mills, A. L., 4341
Mills, G. A., 4478
Mills, P. A., 5325
Milne, E., 2756
Milton, R. M., 1011, 1012
Milun, A., 5201, 5202
Mil'vitskaya, E. M., 4947
Minachev, K. M., 4123
Minato, H., 6738
Minkoff, G. J., 1175, 2011, 4948
Minter, C. C., 4949, 4950
Miquel, R., 4951
Miranda, B. T., 2357-2359
Miras, C. J., 4952
Mirev, D., 4953
Mirocha, C. J., 4954
Mironov, V. A., 4955
Miropol'skaya, M. A., 4956
Mirov, N. T., 6518
Mirskii, Y. V., 380
Misic, D., 4957
Miskalis, A. J., 2621, 4958
Missa, L., 4723
Mistretta, A. G., 4959, 4960
Mistrik, I. E. J., 4961
Mitchell, H. I., 3191
Mitchell, Jr., J., 4962, 4963
Mitchell, R., 4964
Mitchell, T., 4965
Mitchell, T. J., 3388, 3389
Mitchell, W., 2293
Mitchen, J. H., 5081
Mitsui, T., 6717
Mittal, J. P., 3405
Mitzner, B., 6811, 6812
Mitzner, B. M., 4147, 4966-4974
Mitzner, S., 921
Miville, M. E., 5775
Miwa, G. C., 4976, 4977
Miwa, T. K., 489, 4975-4978,
    6392, 6394
Miyahara, S., 4979, 4980
Miyake, H., 4981, 4982
Miyazaki, H., 4215
Mizuoka, T., 347
Mkhchiyan, M. A., 5426
Mladenov, I., 5236
Mlejnek, O., 4983-4986
Moberg, M. L., 2750, 5694, 7345
Moerikofer, A. W., 1102
Moessner, F., 3307
Moffat, A. J., 4987
Moger, G. G., 4988
Moghadame, P. E., 1191, 4989,
    4990
Mogilevskii, G. A., 6454
Mohnke, M., 4331, 4991-4994
Mohr, W., 4995
Mohrhauer, H., 4042

Moir, R., 744
Mokrushin, S. G., 4996
Mold, J. D., 3780, 6561
Mole, T., 1886
Molinari, M. A., 4469
Moll, F., 4997
Molle, L., 4648-4650, 4998
Molnar, W. S., 4999
Momigny, J., 5000
Momyer, Jr., F. F., 5001
Monacelli, R., 339-342
Monet, G. P., 5002
Monkman, J. L., 1953, 1954, 5003-
    5006, 5790
Monotani, M., 6046
Monroe, R. E., 4485
Montefinale, G., 1199, 1201, 1202
Montes, A. L., 2700
Montgomery, M. E., 4403
Mooney, E. F., 1770, 2526, 2527
Mooney, J. B., 5007
Mooney, T. F., 5008
Moore, A. D., 5009
Moore, C. E., 5010
Moore, D., 4508, 5585-5587
Moore, D. R., 5011, 5012
Moore, P. T. 1521, 1522
Moore, R. J., 2702
Moore, W. R., 1520-1522, 5013-
    5018
Morgan, C. R., 5621
Morgan, D. J., 5019
Morgan, D. M., 3984
Morgan, G. O., 954, 3017
Morgan, H., 1409
Morgan, M. E., 1484, 5020, 5021
Morgan, T. O., 2489, 5022
Morgantini, M., 5023-5025
Morgareidge, K., 5026
Mori, M., 5027
Mori, S., 6773
Morice, I. M., 5028, 5029
Morice, M., 2996
Morieson, A. S., 2977, 2978
Morin, R. J., 5030
Moritz, A. G., 478
Mork, H. M., 6172
Morley, H. V., 5031, 5032
Morman, J. F., 998
Morozova, O. E., 5033
Morrell, F. A., 3212, 6247
Morris, C. J., 6833
Morris, J. C., 5308
Morris, L. J., 5034-5038, 7056,
    7057, 7329
Morris, M., 5050
Morris, R., 3643
Morris, R. A., 5039
Morrison, R. L., 5040, 5041
Morrison, W. R., 5042
Morrissette, R. A., 4415-4418,
    6109

Morrow, H. N., 5043
Mortensen, E. M., 5044
Mortimer, J., 5189
Mortimer, J. V., 5045
Moruzzi, G., 4699, 4700, 5046
Moryashchev, A. K., 5047
Moscatelli, E. A., 3245, 6666
Moseley, Jr., W. V., 7535
Mosen, A. W., 5048
Moser, H. C., 6297
Mosettig, E., 6834
Mosher, H. S., 2003, 4941
Mosher, W. A., 5049
Moshier, R. W., 5050
Moshinskaya, M. B., 2475
Mosley, J. R., 1972
Moss, T. E., 5264
Mosser, H. C., 1541
Mossini, F., 5051
Motl, O., 5052
Motta, L., 1130
Mottlau, A. Y., 5053
Moudry, V., 2347
Mount, S. F., 4341
Mounts, T. L., 1996
Moureu, H., 1384
Mourgues, L. de, 5054-5056
Moussebois, C., 5057
Mousseron, M., 1389, 5058
Mousseron-Canet, M., 5058
Moyer, J. C., 4743
Moyles, A. F., 3693, 3694
Muccini, G. A., 5059
Muelder, W. W., 4816
Mueller, E., 5060, 5812
Mueller, E. P., 5061
Mueller, U., 2763
Mueller, W. A., 786
Muenster, H. K., 5062
Muhlstadt, M., 5063
Muhs, M. A., 5064
Mukhina, T. N., 531, 3813, 5065
Mulder, I., 4104
Mullen, R. T., 5066
Muller, G., 5067
Muller, R., 1600, 1601
Müller, R. H., 5068-5073
Müller, S., 999
Multer, H. J., 758
Mumpower, R. C., 5074
Munari, S., 828, 5907
Munch, R. H., 1321, 5075-5077
Munch-Petersen, J., 5078
Mund, W., 5079
Munday, C. W., 5080
Munemiya, S., 364
Mungall, T. G., 5081
Munnecke, D. E., 5082
Munsell, M. W., 6465
Munteanu, I., 5579
Murad, E., 440, 5083, 5084
Murai, S., 6727, 6728

Murakami, Y., 5085
Muraosa, K., 5086
Murata, S., 2426
Murata, Y., 5087
Murdmaa, K. O., 1951, 1952
Murdoch, I. A., 5088
Murfee, Jr., J. A., 5089
Murie, R. A., 2827
Murphy, R. T., 7228
Murray, G. S., 4627
Murray, K. E., 1912, 1913, 4145, 5090-5092
Murray, L. J., 1103
Murray, P. J., 2977, 2978
Murray, R. W., 5093, 6056
Murray, W., 1086, 1087
Murray, W. J., 5094
Murrill, D. P., 6247
Murto, J., 5095
Murty, N. L., 1570-1572, 5096, 5779
Musaev, I. A., 5097
Musgrave, B., 4307-4310
Musgrave, L. R., 5895
Musgrave, W. K. R., 1328, 1634, 3622, 5098
Music, J. F., 1686
Musker, W. K., 4338
Musso, G., 5371
Musso, H., 5099
Muysers, K., 5100
Muzyczuk, J., 5101
Myddleton, W. W., 5102
Myers, H. W., 5103, 5104, 5696
Myers, L. J., 5105
Mylroi, M. G., 3018

Nabivach, V. M., 1647-1649, 5106, 5107
Nachbauer, E., 5108
Nadeau, H. G., 4816, 5109-5111, 5213, 6477, 7154
Nadenicek, J. D., 5666
Nag, D. S., 3312
Nagase, Y., 3374
Nagatani, M., 6241
Nagatomi, H., 2426
Nager, M., 6343
Nagy, Z., 716, 717, 3189
Naimushin, N. N., 1076, 3777
Nair, G. V., 5112, 5113
Nair, III, J. H., 5114
Nair, P. P., 5115
Nakae, T., 5120, 5121
Nakagawa, T., 5116-5118
Nakahara, Y., 5119
Nakanishi, T., 5120, 5121
Nakano, T., 5122
Nakata, I., 5117
Nakayama, F. S., 6398
Nakazaki, M., 94

Napier, D. H., 5123
Napier, Jr., E. A., 5124
Napier, I. M., 478, 483, 5125
Napier, K. H., 154, 155
Narasimhachari, N., 5126
Narayanaswami, K., 1816
Naro, P. A., 5127, 5128
Narr, B., 5060
Naryshkina, T. I., 6304
Nasini, A. G., 5129
Natelson, S., 5130
Naudet, M., 5131, 5132, 6954
Naughton, J. J., 5133
Naukkarinen, I., 7425
Nausch, E., 4463
Naves, Y. R., 5134-5173, 6649
Nawar, W. W., 5174-5178
Nayak, U. R., 5179
Nazarov, I. N., 5180
Nazarova, N. M., 5181
Nebbia, L., 5182
Nechaeva, L. A., 6918, 6919
Necsoiu, I., 5207
Nedorost, M., 3535, 3536, 5183
Nedumparambil, A. A., 7050
Neely, E. E., 5184
Neely, W. Brock, 5185
Neeman, M., 5186
Neerman, J. C., 1127
Negishi, K., 4198, 4199
Negri, R. G., 5187
Nehring, D., 1942
Neidhardt, M. P., 4816
Neiman, M. B., 5674
Neimark, I. E., 5235
Nel, W., 385, 628, 2505, 2975, 5188, 5189
Neligan, R. E., 5190
Nelsen, F. M., 2034, 5191-5193
Nelson, D. C., 5194, 5195
Nelson, D. F., 5196
Nelson, G. J., 5197-5199
Nelson, J. P., 5200-5202
Nelson, K. H., 5203
Nelson, L. E., 726
Nelson, N. A., 5204
Nelson, R. L., 4090
Nelson, W. R., 5205
Nemec, L., 5485
Nemeckova, A., 5206
Nemtrsov, M. S., 2513, 2514
Nenitzescu, C. D., 5207
Nenz, A., 2499
Nerdel, F., 5482-5484
Nerheim, A. G., 5208, 5209
Nesmeyanov, A. N., 5210, 5631
Nettesheim, G., 2410
Neubauer, N. R., 5211
Neudert, W., 5212
Neumann, C. L., 92
Neumann, E. W., 5213
Neuray, M., 1527

Neureiter, N. P., 5214
Neuworth, M. B., 3118
Nevenzel, J. C., 4851
Neville, G., 5215
Nevskaya, E. M., 4615
Newburger, S. H., 3705
Newham, J., 929, 930
Newitt, E. J., 1157
Newland, J., 4798
Newman, H. A., 7539
Newman, M. S., 382, 5216
Newsome, J. R., 5257
Newton, A. S., 2444
Newton, P. R., 4562
Neylan, D. N., 5217
Ng, Hawkins, 3346, 3347
Ng, Y. L., 397
Nichiyama, A., 3374
Nicholas, P. P., 632
Nichols, A. V., 4410, 4411, 5218, 5767, 5768, 6964
Nichols, B. W., 465
Nicholson, D. L., 989
Nickell, E. C., 5667, 5668
Nickless, G., 5
Nicksic, S. W., 3720
Nicolaides, N., 5219, 5220, 6536
Niegisch, W. D., 5221
Nield, E., 5222-5224
Nieto, M. A., 7257
Niezabitowski, W., 4304a, 4305
Nigam, I. C., 4358, 5225-5230
Nightingale, C. F., 5231, 7122
Nikelly, J. G., 5232, 5233
Nikitin, Y. S., 7024
Nikkari, T., 2848-2852
Niklasch, F., 5234
Nikolicova, L., 1397
Nikolina, V. Y., 5235
Nikolinski, P., 5236
Nimmo, C. C., 6795, 6796
Niolle, G., 1124
Nishi, S., 5237
Nishibayashi, M., 5694
Nishida, Y., 6299
Nishimura, K., 3166, 6005
Nishimura, S., 5238
Nixon, A. C., 5239
Nobe, K., 884
Noble, F. W., 5240
Nodop, G., 5241
Noebels, H. J., 1187, 1456, 5242-5244
Noehren, T. H., 5245
Noel, C. J., 5246
Nogotkov-Ryutin, V. A., 7453
Noguchi, S., 5247, 6718
Noguchi, T., 766
Nojima, S., 1270-1272
Noland, J. S., 5248
Noller, C. R., 6104
Nomura, T., 5249

Nonaka, M., 4413, 5250, 5251
Noonan, J. M., 4566
Norem, S., 3096, 3097
Norem, S. D., 2139, 2676, 5252, 5253
Norikov, Y. D., 5254
Norlin, H. L., 511, 7540
Norman, R. O. C., 1558, 2864, 3913, 4080, 5255, 5256
Norman, V., 5257
Norris, F., 2906
Norris, H. D., 2536
Norris, M. S., 5290, 5291
Norris, T. A., 6298
Norris, T. G., 5258
Norris, W. P., 5259
Norrish, H. H., 4525
Norrish, R. G. W., 3181, 3182, 5260, 5261
Norteman, Jr., W. E., 5773
Northington, M., 5262
Norton, C. J., 5263
Norton, J., 5264
Nosaka, S., 6738
Notari, B., 5265
Notation, A. D., 10
Nott, J., 5185
Notton, B. M., 613
Nouse, D. C., 1329
Novak, I., 7443
Novak, J., 3537-3542, 5266-5272
Novotny, J., 484, 485
Novotny, L., 5273
Novotny, Z., 5885
Nowak, P., 5274
Nowakowska, J., 2351, 5275, 5276, 6138-6140, 6492
Noyes, Jr., W. A., 4824, 5084, 5600
Nukada, K., 4718
Nukada, M., 5249
Nunez, L. J., 5277
Nunn, R. F., 1499
Nunziata, L., 5278
Nurok, D., 2593
Nuss, G., 5279
Nuttall, R. L., 5280
Nykanen, L., 6324
Nystrom, R. F., 1993, 3695, 5281-5284, 6789

Oae, S., 5285
Oaks, D., 3000
Oaks, Jr., D. M., 5109-5111
Obata, S., 6299
Ober, S., 3712
Ober, S. S., 5286
Oberhansli, P., 6501
Obolentsev, R. D., 5287
O'Brien, J. P., 6172

O'Brien, L., 1044, 1262, 1455, 5288, 7553
Ochsner, P., 5161-5164
Ocker, H. D., 5289
O'Connor, J. G., 5290, 5291
O'Connor, K., 7028
Ode, W. H., 5292, 5293
Odense, P. H., 23
Odermatt, A., 5165-5168
Odinokov, V. N., 7068
O'Donnell, A. R., 3981, 3982
Oe, M., 2964
Oehlmann, G., 5294
Oertel, H., 2619
Oette, K., 4043, 4044, 5295
O'Farrell, S., 2354
Offer, G., 1940
Ofner, A., 5296
Oganesyan, G. A., 5297
Ogawa, I. A., 1766
Ogawa, M., 3167, 5298
Ogilvie, J. L., 5299, 5486
Ognyanov, I., 5300
Ogorodnikov, S. K., 2513, 2514
Ohkoshi, S., 5301-5303
Ohkusa, T., 6295
Ohline, R. W., 5304, 5305
Ohloff, G., 5306, 5483, 5484
Ohlson, R., 6386-6389
Ohno, K., 1271, 1272
Okada, F., 6730
Okada, Y., 6295
Okany, A., 648
O'Keefe, A. E., 6247
Okey, R., 5307, 6846
Okey, R. A., 4436
Okhuma, Y., 7114
Okunishi, T., 367
Okuno, I., 5308
Olah, G. A., 5309-5311
Olah, K., 5312
Olander, C. J., 2966, 5313
Oldenkamp, R. D., 5314, 5315
Oldfield, J. F. T., 5316
Oldham, G. F., 29, 5657, 5658
O'Leary, Sister Mary Adeline, 3084
Olive, T. R., 5317
Olivecrona, T., 400
Oliver, G. D., 4303, 7376
Oliver, J. P., 6236
Oliver, T., 5318
Olivier, K. L., 5319
Olsanski, V. L., 6438
Olsen, C. J., 787, 788
Olsen, R. S., 6469
Olson, A. C., 5320, 5321
Olson, A. M., 6389
Olson, T. J., 2197
Olund, S. A., 636
Omori, H., 2424, 2425
O'Neal, M. J., 1910

Perovskaya, N. V., 7069
Perrett, R. H., 911, 5452-5454
Perriere, G. de la, 974
Perrine, W. L., 5455, 5456
Perry, E., 5457
Perry, H. M., 2293
Perry, J., 5458
Perry, J. A., 5459, 5460
Perry, M. B., 2828, 2829, 3700, 3701
Perry, M. V., 5461
Perry, S. G., 3355, 3948, 3949
Perry, W. H., 5462
Perun, C., 6847
Pete, O., 6033
Peterkin, M. E., 5463
Peters, G., 3468
Peters, J., 3869
Peters, K., 5464
Peters, R. A., 5465
Petersen, R. C., 2260, 2261
Peterson, A. H., 352, 353
Peterson, D. L., 5466, 5467
Peterson, J. I., 5468, 5469
Peterson, M. L., 3168, 5470
Peterson, P. E., 5471, 5472
Peterson, W. R., 6515
Petho, A., 5473, 6091
Petitjean, D. L., 5474-5477
Petranek, J., 2484, 5478, 5479
Petrocelli, J. A., 5480
Petrov, A. A., 5033
Petrova, R. S., 380, 1386, 5480a, 7024
Petrowitz, H. J., 5481-5484
Petty, W. L., 3418
Petzelt, B., 5485
Peurifoy, P. V., 5486
Peyron, L., 5487
Peyrot, P., 5488, 5489
Pfenninger, H., 5490, 5491, 5634
Phelps, F. T., 5492
Philip, Jr., R. H., 2789, 2790
Philipp, E. E., 809
Philippe, G., 5058
Phillips, C., 2766
Phillips, C. S. G., 163, 526, 538, 947, 948, 1276, 1277, 2767, 3476-3478, 3712, 4449, 4588, 5493-5506
Phillips, D. D., 5507, 6572
Phillips, G., 5508
Phillips, J. R., 3756, 5509
Phillips, L. G., 2878
Phillips, R. E., 5510
Phillips, T. R., 3404, 5217, 5511-5513
Piccinini, C., 6751
Pichat, L., 5514
Pichler, H., 5515-5517
Pickering, G. B., 4725
Piehl, F. J., 5626
Pierotti, G. J., 5518, 5519

Pieterse, M. J., 5520
Pietra, S., 2499
Pietsch, H., 5521, 5522
Piez, K. A., 5523
Pigeret, F., 3920
Piguzova, L. I., 4538
Pilar de la Maza, M., 2713
Pillai, C. N., 5524, 5535, 5536
Pillepich, J. L., 6875
Pilleri, R., 4911, 5525, 7040
Pilo, C., 5526
Pinchas, S., 6265
Pinchin, F. J., 5527
Pinder, A. R., 5528, 5529
Pines, H., 1623, 2845, 5265, 5524, 5530-5538, 7577
Pingree, H., 3576, 7355
Pinkava, J., 3776
Piontkovskaya, M. A., 5235
Piozzi, F., 5539, 5540
Piper, E. A., 3469, 3470, 4502
Piper, J. U., 5204
Pippen, E. L., 4413, 5250, 5251
Piringer, O., 5541
Pirotte, J., 979
Pirrie, R., 4292
Pisano, J. J., 2185, 5542, 5543
Pispisa, B., 406
Pitkethly, R. C., 2302, 5544, 5545
Pitman, M., 172
Pittman, A. G., 4456, 4457
Pitts, Jr., J. N., 3779, 5546, 5547
Plate, A. F., 4947
Platek, J., 5548
Plenat, F., 1389, 1390
Plessis, L. A. du, 5549, 5550
Plimmer, J. R., 1265, 3685
Podall, H. E., 5468, 5469
Podbielniak, W. J., 5551-5557
Podkletnov, N. E., 5558, 5559
Poe, R. W., 5560
Pohler, L. W., 2916
Pokorny, J., 7500
Polak, L. S., 752, 753, 5561
Polezzo, S., 1866
Polgar, A., 5562
Polgar, A. G., 5563, 5564
Polgar, N., 5565
Poli, A. A., 6767
Policastro, S. G., 921
Polievka, M., 4961
Pollak, P., 2610
Pollara, L. Z., 4706
Pollard, A., 3972
Pollard, F. H., 5, 2166, 5566-5573
Pollard, G. E., 5507
Pollard, G. R., 6572
Polonsky, J., 1062
Polyakova, T. A., 5574
Ponomarev, A. S., 5575
Pont, E. G., 2312, 2313
Pop, A., 5576-5579

Pope, C. G., 5580
Pope, Jr., H. L., 6472
Popescu, R., 5578
Popjak, G., 1391, 2696, 3088, 5581-5587
Porath, J., 5588
Porcaro, P. J., 5589-5593
Porsch, F., 2191, 2192, 5594, 5595
Porshneva, N. V., 2497, 6915, 6916, 6929
Porter, J. A., 894
Porter, J. W., 5596
Porter, K., 5260, 5597
Porter, L. K., 6566
Porter, M., 5598
Porter, P. E., 1853, 5519, 5599
Porter, R. F., 1514
Porter, R. P., 5600
Porter, R. S., 561, 5601-5609
Possagno, E., 1196, 1198, 1202, 1203, 1207
Potter, R. A., 5610, 5611
Potterat, M., 2843, 5612
Poukka, R., 5613, 5614
Poulos, T. J., 5615, 5616
Poulson, R. E., 6554, 6555
Pourchez, A., 4313, 4314, 5378, 5617
Poutsma, M., 5618
Povinelli, R. J., 5619, 6709
Powe, W. C., 5620
Powell, A. L., 5621
Powell, H., 1631, 5622
Powell, H. M., 4297
Powell, J. W., 5623-5625
Powell, Jr., L. E., 6833
Powell, P., 5504
Powers, G. W., 5626
Poy, F., 5627-5630
Pozdeev, V. V., 5631
Prabucki, A. L., 5632-5634
Praill, P. F. G., 5635
Prater, C. D., 7179
Pratt, G. L., 5636, 5637
Pratt, H. K., 4556
Pratt, T. H., 5638
Prelog, V., 5639
Prenkschas, W., 6168
Prescott, W. B., 5398
Presman, B. I., 2513
Pressau, J., 6768
Preston, S. T., 5551-5557, 5640-5643
Pretorius, V., 385, 628, 760, 2704, 2860-2863, 2975, 2976, 5188, 5189, 5644, 7229-7234
Prevot, A., 5645-5650
Price, A. R., 5651
Price, D. T., 4449
Price, M. J., 4306
Price, S. J. W., 5652-5654
Prichard, E., 5527

Richardson, C. O., 2503
Richardson, C. W., 5800
Richardson, D. B., 2000, 5801, 5802, 6340, 6341
Richardson, D. M., 1479
Richardson, R. D., 5803, 5804
Richardson, T., 5805, 5806
Richardson, W. H., 7284
Richey, N., 7428
Richmond, A. B., 5807
Richmond, J. W., 1672, 1673
Richmond, R. S., 4928
Rickborn, B., 1579
Riddick, E. B., 6411
Riddle, V. M., 5808
Ridgeway, Jr., J. A., 5809, 5810, 7556
Riedel, O., 5811
Riedl, O., 7397, 7398
Rieker, A., 5812
Riemenschneider, R. W., 546, 3115-3117, 4531, 4595, 4596, 5813, 5814, 6233
Riesz, P., 5815, 7360
Rifkind, B. M., 4292, 4293
Riganesis, M. D., 5816
Rigby, C. A., 5817
Rigby, F. C., 803, 804
Rigby, F. L., 5818-5821
Rigg, T., 7340
Riggs, W. A., 1698
Righi, H., 5822
Rightmire, R. A., 1098
Rijnders, G. W. A., 3205, 3363-3365, 3945, 3950, 4228, 5823, 5824
Riley, B., 5825, 5826
Riley, C., 1499
Riley, D. W., 599
Riley, F. W., 5831
Rillaers, G., 5827
Rinck, G., 3139
Rinehart, Jr., K. L., 5828, 6979
Rinetti, M., 5829
Ring, R. D., 975, 5830, 5831
Rinker, R. G., 6006
Rinzler, S. H., 1387
Rio, A., 5832
Ripa, D., 5832
Ripoll, J. L., 92
Rippere, R. E., 5833
Risinger, G. E., 5834
Rispoli, G., 1845
Ritchie, C. D., 3703-3705
Ritchie, M. L., 1532
Ritter, D. M., 2694, 2695, 4548, 5402, 6236
Ritter, H., 5835
Ritter, J. J., 3037
Ritter, W., 2953, 2954, 5836
Rizack, M. A., 1890
Roaldi, 1695, 5837

Robb, J. C., 550, 896, 1224, 4324, 4612, 5838
Robbins, W. E., 6834
Roberts, A., 1093
Roberts, A. I. L., 3326
Roberts, A. L., 5839
Roberts, C. B., 5185, 5840, 6571
Roberts, F., 6772
Roberts, F. M., 6298
Roberts, G. A. H., 4251
Roberts, H. L., 1284, 1285, 5841
Roberts, J. B., 5842-5844
Roberts, J. D., 1288, 5186, 6329
Roberts, L. R., 5845
Roberts, R., 2204, 2479
Roberts, R. M., 2874, 5846, 5847
Roberts, W. M., 7419
Robertson, D. N., 2884
Robertson, G. R., 5848
Robertson, J. A., 5849
Robertson, R. H. S., 5850
Robin, J., 5851
Robins, A. B., 600, 601
Robinson, B., 5852
Robinson, J. F., 4293
Robinson, J. W., 5853
Robinson, M. A., 5854, 5855
Robinson, R. H., 5856, 5857
Robinson, R. J., 4937, 5858
Robison, C. H., 5777, 7258
Robison, R. A., 2579
Robson, J. W., 5859
Robson, P., 5860
Roch, L. A., 640-644
Rochina, V., 5056
Rock, H., 122-124, 663, 5861-5863
Rock, S. M., 6281
Rockoff, C. V., 5864, 5865
Rockwood, B. N., 3860
Rodda, H. J., 5125
Rodger, M. N., 2037
Rodig, O. R., 5866
Rodrigo, A. R., 5278
Roe, A., 4945
Roe, E. T., 5867
Roesler, H., 5915
Rogers, F. E., 5382
Rogers, J. A., 2804-2806, 4268, 5868-5870
Rogers, L. B., 99, 100, 409, 1964-1966, 4407, 5871-5874
Rogers, L. H., 2301, 5875
Rogers, R. L., 4406
Roginskii, S. Z., 5876, 5877
Rogozinski, M., 5878-5880
Rohleder, H., 3898, 5881
Rohrer, J. C., 6354
Rohrschneider, L., 5882
Rolle, L. A., 3375, 3377, 6523
Rombaut, J., 5883
Romeike, A., 5884

Romeny-Wachter, C. C. ter H., 967, 970, 972, 973
Romovacek, J., 5885
Ronca, G., 1220
Ronning, M., 6425
Rooney, J. J., 3909, 4586
Rooney, T. B., 4807
Root, J. W., 4307
Root, M. J., 5886-5888
Roper, Jr., J. N., 5889
Roper, W. A., 5890
Rose, A., 3104, 4555, 5434, 5891-5893
Rose, B. A., 2993, 5894
Rose, J. E., 4945
Rosegay, A., 3089
Roselius, L., 1603
Rosen, A. A., 5895
Rosen, P., 2195
Rosenberg, A. M., 6439
Rosene, C. J., 3563
Rosenfeld, R. S., 5896-5899
Rosenfeld, W. D., 5900
Rosengreen, B. K., 1507
Rosengren, K., 5901
Rosie, D. M., 4894, 5902
Rosner, H., 4206
Ross, D. E., 1670
Ross, D. L., 1513, 1523
Ross, G. N., 5903
Ross, J. M., 5904
Ross, K. M., 415
Ross, R. A., 5905
Ross, S. D., 2261
Ross, W. A., 2685
Ross, W. D., 7183
Rossi, B. B., 5906
Rossi, C., 5907
Rost, H. E., 768
Roth, J. F., 5908
Roth, W. R., 1888
Rotheram, M. A., 3889
Rothman, L. A., 5909
Rotini, O. T., 5910, 5911
Rotzsche, H., 5912-5915
Rouayheb, G. M., 5449, 5450, 5916-5920
Roubal, W. T., 5921, 5922
Rouessac, F., 1496
Rouit, C., 122-124, 5923, 5924
Roussel, J., 5925
Rovesti, G., 5926
Rovesti, P., 5926, 5927
Rowan, Jr., R., 5928-5930
Rowe, C. E., 5931, 5932
Rowe, V. K., 2884
Rowell, D. A., 6202
Rowland, F. S., 3620, 4307-4310, 5933, 7407
Rowlinson, J. S., 4251, 5934
Roxburgh, J. M., 5935
Roy, H. E., 2754

Royals, E. E., 5936
Rubenstein, D., 7576
Rubin, I. B., 4903
Rubinstein, R., 4405
Ruby, A., 1118
Ruby, E. D., 5937, 6875
Rudakov, G. A., 6317
Rudenko, B. A., 5180, 5938-5940,
    7449, 7450
Rudloff, E. von, 5112, 5113, 5126,
    5941-5950
Rudy, T. P., 3959
Ruggieri, S., 5951
Ruhlmann, K., 5952
Rull, T., 2508
Runeberg, J., 5526, 5953-5957
Runge, H., 5958
Runti, C., 5959
Rusek, M., 3545-3553, 5271, 5272,
    5960
Rushing, D. E., 5961
Rushneck, D. R., 541, 4282
Russell, A. S., 5962
Russell, C. D., 5963
Russell, D. S., 5964, 5965
Russell, G. A., 5966-5972
Russell, J. L., 2600
Rust, F. F., 1746, 4644
Rusterholz, W. E., 621
Ruys, A. H., 5973
Ruzicka, D. J., 1129, 5974
Ryabchikova, G. G., 6484
Ryabchuk, L. N., 6917
Ryan, P. W., 727
Rybicka, S. M., 5324, 5975
Ryce, S. A., 5976-5981
Ryden, J., 6385
Ryder, Jr., W. S., 5982
Rye, T., 2061
Ryhage, R., 2913, 5582, 5584
Ryhage, R. R., 419, 5983, 5984
Rylander, P. N., 4908
Rysselberge, J. van, 4646, 5985-
    5987, 7004

Sabel, C., 6162
Sabelashvili, S. D., 166, 167
Sachs, V., 5988
Sadauskis, J., 3615
Sadowski, F., 5989
Safavi, T., 4402
Saffert, W., 4993, 4994
Safier, L. B., 4643
Safranski, L. W., 728, 729, 1662-
    1665, 5990
Sagert, N. H., 5991
Saha, J. G., 11
Sahasrabudhe, M., 5228, 5229
Said, A. S., 1167, 1168, 2320,
    5855, 5992-5999
Saifer, A., 6000, 6001

Saini, G., 5129
Saint-James, D., 6002
St. Pierre, L. E., 1824, 6003
Saint-Rat, L. de, 6004
Saito, K., 7439
Sakai, K., 6890
Sakai, T., 367, 3166, 4220, 6005
Sakaida, R. R., 6006
Salas, L. J., 1552
Salinis, C. A., 6007-6011, 6710-
    6712
Salmi, A. M., 2852
Salo, T., 6012
Salomaa, P., 6013
Saltzman, B. E., 6014
Salumaa, T., 5898
Salvadorini, R., 1144, 1145
Salvioli, Jr., G. P., 472, 4625,
    5375, 6015
Sambasivarao, K., 6016
Samm, J., 6017
Sammons, H. G., 6018
Samokhvalov, G. I., 4956
Sampugna, J., 2463, 3600-3607,
    5397
Samsel, E. P., 1458-1460, 4940,
    6019
Sancetta, S. M., 2451
Sand, D. M., 6020, 6107, 6108,
    6110, 6111
Sand, P. T., 6021
Sandermann, W., 6022-6024
Sanderson, W. A., 1213, 1214
Sandler, S., 6025, 6026
Sandulescu, D., 960, 2945-2947,
    6027-6034
Sanin, P. I., 5097
Sant, B. R., 7218, 7219
Santavy, F., 5206
Saori, H., 2412
Sarakhov, A. I., 1949
Sarancha, E. T., 6035
Sarlos, I. J., 6933
Sarlos, L. J., 3698
Saroff, H. A., 5523, 6036, 6037
Sasaki, K., 113
Sasaki, N., 6038
Sass, J., 2211
Sass, S., 6039
Sasse, W. H. F., 481, 482, 6040
Sassenberg, W., 6041, 7426
Sassiver, M. L., 6042
Satchell, R. S., 2683, 2684
Sato, R., 3795-3798
Sato, S., 356, 357, 6043, 6044
Sato, T., 6045
Sato, T. R., 6600
Sato, Y., 6046, 6994
Satsmadjis, A., 1097
Sattler-Dornbacher, E., 2531
Saucy, G., 6047
Sauer, K., 4022

Sauer, Jr., M. C., 57, 2705, 2706,
    6048
Sauer, R. W., 2490
Sauerland, H. D., 6049, 6050
Sauers, C. K., 1543, 6051
Sauers, R. R., 6052-6055
Saunders, F. C., 3138
Saunders, M., 6056
Saunders, R. A., 814
Saunders, W. H., 6057
Sauvage, J. F., 3356, 6058, 6059
Savary, P., 6060
Savidan, L., 6061, 6062
Saville, B., 5598
Saville, D. A., 2282
Savina, I. M., 7047
Savinov, I. M., 3783
Savitsky, A., 6063, 6064
Sawicki, E., 2326, 6065
Sawyer, D. T., 5774, 6066-6069
Sawyer, F. M., 2177, 5177, 5178
Sax, K. J., 6070
Saxena, C. S., 4719
Saxena, S. C., 6071, 6072, 6497
Scanlon, P. M., 2472
Scarborough, J. M., 65
Scarpellino, R., 6073-6075
Schaal, G., 7430
Schaap, L., 5537
Schaefer, C. A., 6076
Schaefer, F. C., 6077
Schaeppi, W. H., 6078
Schamp, N., 6079, 6080
Schams, E., 1837
Schams, F., 1838, 6081
Scharfe, G., 6082, 6083
Scharpenseel, H. W., 6084, 6085
Schauenstein, E., 2903
Schay, G., 2212, 2214-2216, 5312,
    6086-6095
Schechter, M. S., 6096
Scheidegger, A. E., 6097
Schenck, G. O., 6098, 6099
Schepartz, A. I., 6100
Scheuer, P. J., 3177
Schiek, R. C., 127
Schiess, P. W., 2769
Schill, G., 7037
Schiller, R., 6101
Schimelpfenig, C. W., 6102
Schindler, R., 4856, 4857
Schirmer, Jr., W., 4374
Schlafer, L., 3235
Schlater, J. M., 6103, 7183
Schlegel, W., 6104
Schlemmer, W., 4699, 4700
Schlenk, H., 2882, 6020, 6105-
    6111
Schleyer, P. v. R., 858, 6112
Schlotzhauer, W. S., 6117
Schmauch, L. J., 1710, 6113-6115
Schmeltz, I., 6116, 6117

Schmid, C. E., 6118
Schmid, H., 2874
Schmid, P., 4034
Schmidbaur, H., 6119
Schmidhammer, L., 7242
Schmidt, A., 6120
Schmidt, F., 3123, 3124
Schmidt, G., 7212
Schmidt, J. J. E., 6121
Schmidt, M., 6119
Schmidt-Bleek, F., 6122, 6586
Schmidt-Collerus, J. J., 6123
Schmidt-Kuster, W. J., 6124, 6125, 6178, 7192, 7297, 7298
Schmied, H., 6126
Schmitz, A. A., 4901, 4902
Schmitz, F. J., 1300, 1301
Schmuller, J., 6156
Schneck, E., 6192
Schneider, C. R., 6127
Schneider, I. A., 6128
Schneider, P., 4102
Schneider, W., 2890-2896, 6129
Schneider, W. J., 2485
Schneyder, J., 6130
Schnier, H., 5835
Schoedler, C., 6131, 6132
Schoell, H., see Schöll, H.
Schoen, W., 5809, 5810
Schoenfelder, C. W., 6133
Schogt, J. C. M., 1125, 6134
Scholfield, C. R., 1994-1996, 6135-6140
Scholl, F., 6141, 6142
Schöll, H., 4044
Schollenberger, R. S., 4743
Scholly, P., 458, 1044
Scholly, P. R., 1430, 1492-1494, 5288, 6143, 6144, 7553
Schols, J., 6145
Schols, J. A., 1848, 3916, 3917, 6146, 6147
Scholz, R. G., 6148-6150
Schomburg, G., 3108, 6151-6155
Schools, Jr., P. E., 6671, 6672, 6677
Schrade, W., 898, 6157-6163
Schrader, R., 6164
Schramm, G., 3084
Schramm, R. M., 6165
Schreiber, H. P., 6166
Schreiber, R. A., 1703
Schreyer, G., 2897-2899
Schriesheim, A., 3200, 4142
Schroder, E., 5294
Schrodt, V. N., 5892
Schroter, M., 6167, 6168
Schuberth, H., 6169
Schuck, E. A., 2077, 6170, 6171, 6542
Schuetz, R. D., 6172
Schuhknecht, W., 2208, 6173, 6174

Schuldiner, S., 4950
Schulek, E., 6175, 6176
Schuler, R. H., 2494, 5059
Schultz, D. R., 6177
Schultz, J. F., 6533, 6534
Schultze, G. R., 6178, 7298
Schulz, G., 6179
Schulz, H., 5515, 5516, 6180
Schumacher, H., 1137, 3370, 3371
Schutzmann, R. L., 7228
Schwarberg, J. E., 5050
Schwartz, C. E., 6181
Schwartz, D. P., 5395
Schwartz, F. G., 7264
Schwartz, G. M., 1446
Schwartz, R. D., 6182-6185
Schwarz, H. P., 6186, 6187
Schwecke, W. M., 1470, 1471
Schweizer, E. E., 1515, 1517
Schwenk, U., 3233, 3234, 6188-6193
Schwerdt, R., 4600
Scott, B. A., 6194
Scott, C. G., 4059-4062, 6195-6202
Scott, D. S., 6203
Scott, G. P., 6204
Scott, M. D., 2784
Scott, P. G. W., 2165, 6205
Scott, R. P. W., 1355, 3065, 5330, 6206-6229
Scott, R. R., 6230, 6231
Scott, S. J., 3646
Scott, T. W., 6232
Scott, W. B., 1365
Scott, W. E., 6233, 6541
Scrubis, B., 777
Scully, N. J., 1994
Seager, S. L., 2580-2583, 6568
Searle, H. T., 1340
Searles, Jr., S., 6234
Secci, M., 1192
Sechenov, G. P., 6235
Sechi, A. M., 4701
Seckarova, H., 4986
Seely, G. R., 6236
Segal, H. S., 6237
Seher, A., 3864, 6238
Seidel, C. F., 6239
Seifert, W. K., 6240
Seimovich, R. G., 516
Seiyama, T., 6241
Seiz, W., 2688
Sekerka, B., 6242
Sekiguchi, S., 3392
Sekiya, Y., 3165
Selenkina, M. S., 6911, 6913, 7510, 7511
Self, R., 6243
Seligman, P. C., 6244
Seligman, R. B., 6245-6247
Seligson, D., 4809

Selin, T. G., 6248
Selke, E., 6139
Sell, G., 5729
Sellmann-Persson, G., 468
Seltzer, J., 5899
Selwitz, C. M., 6249
Sememova, V. A., 6852
Semenova, V. N., 4014
Semenovskaya, T. D., 459
Semenovskii, A. V., 5939
Semina, G. N., 4124
Semlyen, J. A., 163, 5504
Sen, B., 6250, 7216-7219
Seng, G. C., 10
Sen Gupta, J. G., 7219
Senn, Jr., W. L., 6251
Seno, S., 4707-4709
Serfass, E. J., 6252
Serfontein, W. J., 7013
Sergeev, G. B., 4988
Seris, G., 6253
Serpinet, J., 1189, 6254-6257
Servello, F., 406
Setinek, K., 6258
Setzkorn, E. A., 6259, 6260
Sevenster, P. G., 6261, 6262
Shabalin, I. I., 6975
Shabtai, J., 2589, 2590, 2594, 2595, 3121, 5538, 6263-6265
Shafran, R. N., 5181
Shahin, M. M., 4435, 6266, 6267
Shahrai, V. A., 753
Shalit, H., 4478
Shamaiengar, M., 4610
Shank, R. S., 6268
Shannon, A., 6846
Shapiro, H., 6269
Shaposhnikov, Y. K., 6270
Shapras, P., 6271
Sharkey, W. H., 4738
Sharman, S. H., 7477
Sharms, A., 7329
Shaw, G. C., 1498
Shaw, P. D., 6272
Shaw, R., 6273, 6274
Shaw, S. J., 1464
Shawhan, S. D., 1234
Shchegrova, K. A., 1353
Shcherbakov, P. M., 6275
Shcherbakova, K. D., 3783, 4015, 5480a, 7024
Shea, J. L., 6172
Shearer, D. A., 891, 6276, 6277
Shechter, H., 6268
Sheeran, S. R., 3207
Shefter, V. E., 1748
Shelton, E. J., 607
Shelton, J. R., 6278, 6279
Shemyakin, F. M., 6280
Shemyatenkova, V. T., 6914, 6918, 6919
Shepard, R. C., 3889

Tai, H., 1332, 1333
Tai, L., 4216
Takacs, J., 6720
Takagi, T., 6871
Takagi, Y., 402
Takahashi, A., 6721
Takahashi, Y., 6722-6724, 6745
Takamiya, N., 6725-6728
Takanishi, T., 4716
Takashima, I., 6729
Takashima, S., 6730
Takayama, Y., 6731-6736
Takeda, I., 4718, 6737
Takeda, K., 6738
Takenishi, T., 4724, 5087
Takeuchi, S., 3872
Takeuchi, T., 2411, 3022, 6739-
    6742
Takeuchi, Y., 6460
Talbot, G. S., 1671
Talbot, P. J., 6743
Tam, S. W., 397
Tamaru, K., 6744
Tanabe, H., 3877
Tanaka, K., 6722-6724, 6745
Tanaka, S., 7465, 7466
Tanaka, Y., 4028
Tandy, R. K., 6746
T'ang, H-Y., 4216
Tang, Y-N., 4307
Tanner, D. W., 6747
Taponeco, G., 1144, 1145
Tappel, A. L., 5805, 5806
Taramasso, M., 1866, 6748-6751
Tarasov, A. I., 4538, 4539, 6654,
    6752-6755
Tarbes, H., 6756
Tarimu, C. L., 5828
Tarman, P. B., 6757
Tarmy, B. L., 6758, 6759
Tarrago, X., 4317
Tarrant, P., 5804
Tatchell, A. R., 3282
Tatlow, J. C., 1113, 1150, 1468,
    2146-2148, 2431, 2651, 3087,
    5222-5224, 5860, 6427, 6543,
    6687, 6688, 6760
Tattrie, N. H., 6761
Tattrie, W. H., 4806
Taufer, M., 2903
Tavs, P., 1302
Taylor, B. W., 2202, 4809, 6762-
    6768
Taylor, F., 6769
Taylor, G. W., 6770
Taylor, L. M., 1984, 1985, 6343
Taylor, M. P., 6771
Taylor, O. C., 6541
Taylor, R. F., 6772
Taylor, T. E., 1929
Taylor, W. H., 6773
Taylor, W. J., 6774

Teach, E. G., 3418
Tealdo, G. F., 5907
Techman, T., 5006
Tedder, J. M., 343, 344, 2362-
    2364
Teicke, R., 6161, 6163
Teisseire, P., 6775-6781
Teitelbaum, C. L., 6782, 6783
Teitelbaum, P., 824
Teller, E., 4825
Tenenbaum, M., 6784
Tenma, S., 5302, 5303
Tenney, H. M., 6785-6788
Tenney, R. I., 7000
Tenny, K. S., 6789
Tepe, J. B., 6790
Terada, H., 6791
Teranishi, R., 1185, 1540, 6792-
    6796
Terent'ev, A. P., 4545
Termini, D. J., 6797
Ternovskaya, L. P., 7069
Ternow, A., 7400
Terres, E., 6798
Tesarik, K., 3554-3558, 4156-
    4159, 6330, 6799, 6800
Testerman, M. K., 6801, 6802
Teszler, A. P., 6545
Teubel, J., 6129
Thacker, R., 5934
Thackray, M., 3153
Thain, E. M., 7, 1898
Thaler, W., 7146
Thaler, W. A., 5342
Thaller, V., 899
Tharp, B. W., 6803
Thaung, U. K., 6804
Thaysen, E. H., 524
Theall, G., 3163
Theer, J., 6805
Theile, F. C., 6806-6808
Theimer, E. T., 4974, 6809-6812
Theivagt, J. G., 1115
Thevelin, M., 2169
Thiebaut, R., 1955
Thiele, K., 3314
Thielemann, H., 6813, 6814
Thielking, H., 2400
Thiemann, A., 12
Thijsse, G. J. E., 6815
Thimann, K. V., 1154, 1155
Thirion, B., 1385
Thoburn, J., 3712
Thoburn, J. M., 6816
Thodos, G., 4957, 5358, 5359,
    6076
Thomas, B. W., 6817
Thomas, C. O., 6818-6820
Thomas, C. R., 6687, 6688
Thomas, D. B., 855
Thomas, D. J., 6821
Thomas, D. R., 6822

Thomas, G. H., 1327, 4077-4079
Thomas, J. H., 6743, 6822
Thomas, K. D., 2222, 3315
Thomas, M. E. de, 1045
Thomas, R., 6823
Thomas, T. L., 1012
Thomas, W. H., 5622
Thomasson, H. J., 56
Thompson, A. E., 6824-6826
Thompson, B., 3644
Thompson, C. J., 1479-1481,
    6827-6831
Thompson, D. W., 2473
Thompson, G. P., 6533, 6534
Thompson, H. W., 6832
Thompson, J. F., 6833
Thompson, M. J., 4485, 6834
Thompson, M. P., 6835
Thompson, R. D., 29
Thompson, R. J., 5657, 5658
Thomson, W. A. B., 2054, 6836
Thonet, T. A., 6837
Thorburn, S., 6370
Thorneman, T., 2967
Thorpe, R. E., 5239
Thrash, C. R., 6838
Thrune, R. I., 6839
Thuerkauf, M., 4204, 4205
Thürkauf, M., see Thuerkauf, M.
Tieman, C. H., 6572
Tiernan, T. O., 6840
Tilak, M. A., 7241
Tilenschi, S., 6841
Tiley, P. F., 2278
Till, F., 2482, 2483
Till, I., 2483
Timmerhaus, K. D., 5374, 5677
Timms, D. G., 6842, 6851
Timms, P. L., 5505, 5506
Timofeev, D. P., 6843, 6844
Tine, G., 6845
Tinoco, J., 4436, 6846
Tipotsch, D. G., 2455, 2456
Tipson, C. L., 1272
Tischendorf, G., 4206
Tishchenko, S., 974
Titov, V. B., 1073, 1076, 1077
Tivin, F., 3746, 3747
Tobin, Jr., H., 4111
Todd, J. E., 3862
Todd, Jr., P. H., 6847
Toi, B., 6848
Tokairin, H., 2962
Tokumaru, S., 471
Tolgyesi, W. S., 5311
Tollefson, E. L., 7296
Tomida, I., 7241
Tominaga, K., 6038
Tominaga, S., 6849, 6850
Tominage, H., 6889
Tomita, H., 4718
Tomlinson, R. H., 6325

317

Varma, K. R., 3422
Varma, M. C. P., 7018
Vasenius, L., 5614
Vasilescu, V., 7019-7023
Vasil'eva, V. S., 7024
Vasishth, R. C., 607
Vassallo, D. A., 7025
Vatcher, D. R., 6982
Vaughan, G. A., 2723-2728
Vaughan, J., 1720
Vecsler, F., 6001
Vedel, M., 1390
Vedeneev, K. P., 6270
Veinberg, A. Y., 4956
Velut, M., 7026, 7027
Venanzi, L. M., 1277
Venema, A., 6608
Ven Horst, H., 7028
Ven Horst, Sister Helene, 7028
Venter, J., 1526, 1528
Venugopalan, M., 7029
Verbeke, R., 7030
Vercillo, A., 933
Verdier, C. H. de, 7031
Verdin, A., 538
Veres, K., 3521
Vereshchinskii, I. V., 6484
Vergnaud, J.-H., 7032
Verkade, P. E., 700
Vernotte, P., 6253
Verrien, J. P., 7033
Verrier, M., 1475
Versie, J., 2623
Vertalier, S., 4677-4679, 7034
Verver, C. G., 3951
Verzele, M., 5827, 7035
Veselov, V. V., 7036
Vessman, J., 7037
Vetter, W., 823
Vezinet, P., 5132
Vial, M. H., 4277
Vickers, J. H., 6307
Vidavskii, N. N., 68
Viehe, H. G., 7038
Vietti-Michelina, M., 5525, 7039, 7040
Vigdergauz, M. S., 2678, 7041-7049
Viljhalmsson, S., 3595
Vilkas, M., 7050
Villalobos, R., 6940, 6941, 7051-7053
Vinogradova, L. M., 1119
Vinogradova, O. M., 5876, 5877, 7054
Viola, S. J., 4873
Vioque, E., 2713, 6868, 7055-7057
Virobyants, R. A., 4752, 4753
Virus, W., 7058
Vishnyakova, M. M., 1948, 1949
Visser, B. F., 7059
Vita-Finizi, P., 5540

Vitagliano, M., 7060, 7061
Vitali, T., 5051
Vitzthum, O., 7062
Vivie, J., 7063
Vizard, G. S., 7064, 7065
Vladimirov, B. V., 7066, 7488
Vlakhov, R., 5300
Vlastimil, H., 7067
Vlodavets, M. L., 7068, 7069
Vnukov, A. K., 7070
Vodehnal, J., 1891
Vodret, A., 7060, 7061
Vodzinskii, Y. V., 6270, 7071
Voelker, M. W., 2203
Voelker, W. A., 6508
Vofsi, D., 5838
Voge, H. H., 3952
Vogel, A. M., 7072
Vogel, E., 7073
Vogelenzang, E. H., 4998
Vogelsang, R. F., 7074
Voigt, J., 7075, 7076
Voigt, K. D., 348
Voinov, A. P., 7077
Voitko, L. M., 3960
Vojtovic, K. V., 3559
Volman, D. H., 5597, 7078
Volpenhein, R. A., 4744
Volsi, D., 355, 418
Vorbeck, M. L., 7079-7081
Voronin, V. G., 5047
Voss, G., 7082
Vosti, D. C., 7083
Vranjican, D., 7084
Vrbaski, T., 7085
Vreeland, R. W., 6680, 7086
Vries, J. X. de, 616
Vries, M. J. de, 7087, 7088
Vyakhirev, D. A., 60, 1119, 5713, 7089-7104
Vyas, S. H., 7105
Vykoc, J., 7106
Vyrodov, V. A., 4129

Wachi, F. M., 2755, 3748, 3749, 7108
Wachs, W., 7107
Waclawik, J., 4243
Wagenknecht, A. C., 7109
Waggoner, T. B., 2884
Wagner, H., 5989
Wagner, J., 7110, 7111
Wagner, M., 725, 4206, 4600
Wagner, N., 6712
Wagner, W., 2062
Wahl, G., 664
Wahlroos, O., 7112
Waight, E. S., 7113
Wain, M. E., 5781
Wainman, H., 3128
Waites, R. E., 7011

Wakasugi, B., 7114
Waksmundzki, A., 5336, 7115-7117
Walborsky, H. M., 7118
Waldman, M. H., 6166
Waldron, J. D., 7119
Walford, R. W., 7120
Walia, J. S., 787, 788, 790
Waligora, B., 7121
Walker, G., 3989
Walker, G. C., 436, 648
Walker, Jr., H. G., 2500, 2501
Walker, J., 4471
Walker, J. M., 5231, 5400, 5401, 7122
Walker, J. Q., 1865, 7123-7125, 7558-7560
Walker, P., 7113
Walker, R. E., 7126
Wall, R. F., 510-512, 515, 3159, 5244, 7127-7138, 7540, 7541
Waller, G. R., 4722, 7139
Walling, C., 7140-7146
Walsh, J. T., 669, 4626, 4884, 4885, 4887-4891, 6628, 7147
Walter, J. F., 5765
Walters, W. D., 2517, 2518, 3908, 7201
Walther, B., 1550, 1551
Walton, W. H., 5663
Waltz, R. H., 7148
Wang, D. T., 3361
Wansink, E. J., 3790
Wantland, C. F., 1984, 1985
Warawa, E. J., 1875
Ward, C. C., 1481, 6828-6831
Ward, C. P., 5839
Ward, D. N., 4799
Ward, H. R., 5015-5018
Ward, Jr., L. F., 6572
Ward, P. F. V., 5465
Ward, T. L., 2303
Ward, W. M., 7271
Ward, W. R., 1581
Warham, T. J., 1800
Warner, W. C., 3843, 3845
Warren, C. W., 7311-7313
Warren, G. W., 3023, 3024, 7149-7152
Warren, R. R., 7152
Warren, R. W., 4467
Warrington, Jr., H. P., 7153
Warshawsky, A., 5878-5880
Warwicker, E. A., 1626
Washburn, R. M., 518
Waszak, S., 4243
Waszeciak, P., 7154
Watanabe, M., 4096
Waterhouse, D. F., 7155
Waterman, H. I., 1158, 3103, 6443
Waters, J. A., 6834
Waters, J. H., 7156
Waters, O. J., 6650

Waters, W. A., 686, 1558
Watson, A. A., 5598, 7157
Watson, E. J., 6573
Watson, E. S., 1431
Watson, W. C., 7158, 7159
Watts, J. D., 783
Watts, J. O., 1116, 1117, 7160, 7161
Watts, R. M., 2942
Wawzonek, S., 7162
Way, R. M., 7163
Weaver, E. R., 5369, 7164
Weaver, N., 7165
Webb, A. D., 7166, 7167
Webb, G. A., 7168
Webb, J., 3471, 3472
Webb, J. P. W., 1890, 3451, 3473, 3474, 7169, 7170
Webb, R. D., 7171
Webb, T. P., 7172
Webber, J. M., 1139
Weber, C. W., 4946
Weber, E., 6193
Weber, F., 7173
Weber, H. P., 2908
Weber, O., 3898
Weber, R. P., 5666, 5668
Webster, J. L., 7174
Webster, O. W., 3228
Weenik, R. O., 7175
Weerdt, W. J. van de, 1158
Wegner, E. E., 2900-2902
Wehe, A. H., 7176
Wehrli, A., 7177, 7178
Wei, J., 7179
Weidenbach, G., 4110
Weigel, H., 2254
Weil, K., 5464
Wei-Lo, Y., 7180
Weingarten, H., 7181-7183
Weinig, E., 7184, 7185
Weininger, J. L., 7186
Weinstein, A., 7187-7190
Weinstein, B., 7191
Weisner, L., 7192
Weiss, A. R., 7193
Weiss, F. T., 4073, 5064
Weiss, H., 4167, 7194
Weiss, J., 7195
Weiss, K., 5329
Weiss, M. D., 1033, 7196
Weiss, T. J., 3860
Weissmann, G., 6024
Weitkamp, A. W., 7197-7200
Weitzel, D. H., 5677
Welch, G. A., 7418
Wellington, C. A., 7201
Wellman, W. E., 7202
Wells, C. H. J., 4093
Wells, D. V., 1804
Wells, J., 2770
Wells, W. W., 742, 4614, 7203

Welti, D., 821, 7204
Wencke, K., 7205, 7206
Wender, E., 4272
Wender, I., 861
Wendlandt, W. W., 7207
Wenger, E., 1053
Wentworth, J. T., 3068
Wepster, B. M., 700
Werner, A. E., 7208
Werner, Jr., E. R., 7209
Wertheimer, L. C., 7287
Werthessen, N. T., 5205
Wesselman, H. J., 6790, 7210, 7211
Wessely, F., 6684, 7212
West, D. L., 7213-7215
West, P. W., 7216-7219
West, T. S., 7220, 7221
West, W. W., 7222
Westaway, H., 7223
Westenberg, A. A., 7126
Westermark, T., 7224
Westgate, M. W., 7225
Westlake, W. E., 6096, 7226-7228
Weston, W. J., 1266
Westwood, J. H., 1139
Wet, W. J. de, 5188, 7229-7234
Wett, T. W., 7235
Wetzel, W. H., 6413, 7236
Weurman, C., 1828, 7237, 7238
Wexler, A., 6269
Weyermuller, G., 7239
Weygand, F., 7240-7242
Whalley, C., 7243
Whalley, E., 7244
Whalley, G. R., 7245
Whatley, T. A., 7246
Whatmough, P., 7247, 7248
Wheatley, V. R., 976, 977, 3475, 5776, 7249-7253
Wheeldon, L. W., 4568, 4569, 7254
Wheeler, B. M., 1463
Wheeler, D. J., 7255, 7256
Wheeler, Jr., G., 7183
Wheeler, O. H., 7257
Wheeler, P., 3989, 4909
Wheeler, R. J., 4811
Wheelock, T. D., 799-801, 5777, 7258
Whelan, J. M., 1543
Wherry, T. C., 7259-7261
Whetstone, R. R., 6572
Whisman, M. L., 1479, 7262-7264
White, A. M. S., 1093
White, C. S., 7265
White, D., 3326, 7266-7268
White, E. A. D., 585, 604, 605
White, E. H., 7269, 7270
White, F. A., 638
White, G. B., 1323
White, J. C., 4903
White, J. U., 7271

White, L. G., 6232
White, R. G., 5211
White, R. M., 5933
Whitfield, F. B., 1318
Whiting, D. H., 3018
Whiting, M. C., 899, 900, 3696, 3697, 5623-5625
Whitman, B. T., 42-48, 1081, 3958, 7272-7275
Whitnah, C. H., 624-627
Whitner, V. S., 606
Whitney, J. E., 4213
Whitney, W. K., 7276, 7277
Whitson, R. L., 7278
Whittenbaugh, J. A., 4793
Whittier, M. B., 7279-7281
Whittingham, G., 3181, 3182
Whittle, E., 1343
Whittle, J. E., 3126
Whyman, B. H. F., 1795, 1798, 1800-1803
Whyman, C., 3072
Wiberg, K. B., 7282-7284
Wichterle, I., 7285
Wick, E. L., 3394, 7286, 7287
Wickberg, B., 7288
Wicke, E., 7289, 7290
Wicks, Jr., G. E., 5716
Widmark, G., 7291
Widmark, K., 7291
Widmer, H., 7292
Wiebe, A. K., 7293
Wiebe, R., 5275, 5276
Wiel, A. van der, 7294
Wiersma, D. S., 7295, 7296
Wiesner, L., 7297, 7298
Wiggs, S. M., 6018
Wightman, R. E., 2324, 2325
Wijnen, M. H. J., 4822, 7299-7301
Wilcox, Jr., C. F., 7302
Wilcox, F. A., 7001-7003
Wilder, Jr., P., 1624, 5708
Wilding, M. D., 3858
Wildman, W. C., 4453
Wiles, R. A., 6658
Wiley, R. H., 7303
Wiley, V. G., 940
Wilhelm, R. H., 1745
Wilhite, R. N., 7304
Wilkens, T., 7204
Wilkenson, L. B., 7305
Wilkins, M., 7306, 7307
Wilkinson, J., 7308
Wilkinson, R. W., 7309, 7340
Wilkinson, V. J., 4251
Wilks, Jr., P. A., 7310-7313
Willard, J. E., 57, 2152-2154, 2705-2707, 2711, 4532, 6048
Williams, A. E., 814
Williams, A. F., 1086, 1087, 4774, 5094, 7314, 7315
Williams, A. L., 528, 7316

Williams, C. M., 6667, 7317-7320
Williams, D. D., 7321-7323
Williams, D. L. H., 521
Williams, E., 3712
Williams, E. A., 3253
Williams, E. F., 7324-7326
Williams, G. H., 3137, 3138, 6284
Williams, H. D., 6414
Williams, I. H., 7327, 7328
Williams, J., 863
Williams, J. A., 7329
Williams, J. F., 7223
Williams, J. F. A., 7330
Williams, Jr., J. H., 7331, 7332
Williams, J. P., 4566
Williams, M. C., 5096, 7333
Williams, M. M. D., 7334
Williams, P. L., 7335
Williams, P. M., 7336
Williams, R. J. P., 526, 3390,
    7337, 7338
Williams, R. L., 862, 863, 1973,
    7339
Williams, T. F., 635, 7340
Williams, T. I., 7341
Williams, V. R., 7342
Williamson, A. G., 2026, 6194
Willingham, C. B., 4481, 4482,
    6629, 6630
Willis, H. A., 3033
Willis, V., 7343, 7344
Willner, D., 790
Wills, P., 467
Wills, R. D., 4411, 6746
Wilshire, J. F. K., 6669, 6670
Wilson, C. W., 2681, 2682
Wilson, D. W., 1278
Wilson, E. M., 2750, 7345-7348
Wilson, H., 4859
Wilson, H. N., 6614
Wilson, J. D., 7306, 7307, 7349
Wilson, J. E., 7350
Wilson, J. M., 1872, 1876
Wilson, J. N., 7351
Wilson, K., 3983
Wilson, N. H., 7352
Wilson, P. W., 7353
Wilson, R. H., 3575, 3576, 7354-
    7356
Wilson, R. R., 885, 886
Wilson, T. L., 6393, 6394
Wilt, J. W., 7357
Wilt, R. L., 7331, 7332
Wilton, V. B., 1552
Wilzbach, K. E., 1906, 5815, 7358-
    7360
Wimer, D. C., 1115
Wincel, H., 3887
Windey, J. P., 5436b
Windsor, D. A., 6371
Winefordner, J. D., 7361
Wineman, R. J., 4101

Wing, Jr., F. E., 2151, 7362
Winkelman, J., 7363
Winkelman, J. W., 3832
Winkler, G., 7111
Winkler, W. O., 7364
Winn, A. V., 2008
Winslow, E. H., 1825
Winstein, S., 96, 145, 7365, 7366,
    7477
Winsten, W. A., 7367
Winter, J. A., 7309
Winter, M., 6500
Winters, J. C., 4692, 4693, 7368,
    7369
Winzen, W., 1928
Wirth, H., 7370-7373
Wirth, M. M., 7374, 7375
Wise, K. V., 515, 7376
Wise, R. W., 7377
Wiseblatt, L., 7378, 7379
Wiseman, E. H., 6544
Wiseman, W. A., 7380-7387
Wisniak, J., 7388
Wisniewski, D. F., 7389, 7390
Wisniewski, J., 732, 7148
Wisniewski, J. V., 4930
Witham, B. T., 3136
Witjens, P. H., 7391-7393
Witsch, H. G., 658, 659, 664, 665
Wittig, G., 7394, 7395
Witting, L. A., 7396
Witts, K., 5060
Wnuk, R. J., 1118
Woelbrock, F., 5655
Woidich, H., 7397, 7398
Wolf, A. P., 7444
Wolf, F., 7399, 7400
Wolf, H. O., 4134
Wolfe, C. L., 2909
Wolff, G., 7401
Wolff, I. A., 489, 4976-4978, 6392,
    6394
Wolff, J., 7402
Wolff, J. P., 7401, 7403-7406
Wolff, R., 3572
Wolff, R. E., 1875
Wolfgang, R., 3100, 5638, 6966,
    7407
Wolford, R. W., 427, 428, 7408-
    7410
Wolfrom, M. L., 7411
Wollast, R., 7412
Wollner, E., 6482
Wollrab, V., 7413
Wolny, J., 3132
Wolthuis, E., 5618
Wong, F. F., 1267, 1268
Wong, M. P., 7414
Wong, N. P., 2041
Wood, Jr., F. T., 7415
Wood, L. J., 7416
Wood, T. M., 417

Woodford, F. P., 971-973, 7417
Woodman, F. J., 7418
Woods, A. E., 7419
Woods, E. F., 2446
Woolford, R. G., 4711
Woolmington, K. G., 4240-4242,
    6244, 7420
Wooten, Jr., W. C., 2823
Wormall, A., 976
Worth, G., 5100
Wotiz, J. H., 2816
Wotiz, H. H., 1257, 7421-7425
Wrabetz, K., 6041, 7426
Wragg, A. L., 1707
Wreyford, D. M., 7427, 7428
Wright, A. N., 4143
Wright, F. J., 7429
Wright, N., 7239
Wrolstad, R. E., 3596
Wronka, J. A., 4471
Wu, C. Y., 389, 390
Wu, H-F., 4519
Wulfhekel, H., 7430
Wurst, M., 2344-2347, 7431-7433
Wurstova, E., 7433
Wylde, J., 5058
Wyllie, H. A., 7434
Wynder, E. L., 3192
Wynn, J. D., 7435
Wynne, A., 7064, 7065

Yamada, M., 7436, 7468
Yamada, S., 7437
Yamamoto, T., 7438, 7439
Yamane, M., 7440, 7441
Yamanishi, T., 7287
Yamao, M., 3164, 7465, 7466
Yamartino, R. L., 2040
Yamashita, K., 403
Yamauchi, T., 7465, 7466
Yamazaki, H., 7442
Yamazaki, T., 3847, 3848
Yanak, Y., 7443
Yang, H-C., 7481
Yang, H-J., 4216
Yang, J. Y., 7444
Yang, K., 2298, 2465, 7445-7448
Yanotovskii, M. T., 4611, 4956
Yanovskaya, L. A., 7449
Yanovskii, M. I., 2349, 2496-2498,
    5362-5364, 5876, 5877, 7054,
    7451-7454
Yanyukova, A. M., 522
Yarborough, V. A., 3023, 3024,
    4999, 7149-7152
Yasenev, B. P., 6454
Yashin, Y. I., 1172, 3782, 7501
Yastrebova, L. S., 7503
Yasui, E., 7455
Yasui, H., 7456
Yasumori, Y., 361, 4019

Yats, L. D., 86, 87
Yatsugi, H., 4732
Yavorovskaya, S. F., 7457, 7458
Yee, J. L., 5387
Yeh, S. J., 7459
Yip, G., 7460, 7461
Yirovskii, Y. M., 6454
Yokley, C. R., 7462, 7463
Yoneyama, S., 7436
York, W. B., 4464
Yoshida, F., 7464
Yoshigi, H., 4097
Yoshimi, N., 7465, 7466
Yoshizawa, K., 7436, 7467, 7468
Young, D. M., 2620, 6147, 7469
Young, I. G., 7470, 7471
Young, J. A., 1929-1931, 7472
Young, J. F., 7415, 7473, 7474
Young, J. O, 4380
Young, J. R., 7475, 7476
Young, J. S., 1084
Young, R., 416
Young, V. O., 2302
Young, W. G., 5319, 7477
Youngs, C. G., 1573, 7478-7480
Youssef, A. A., 7118
Yu, C-C., 7481
Yu, C-Y., 7482
Yu, H-F., 5436a
Yu, W-L., 6899, 7483, 7484
Yuan, C., 7485
Yudilevich, M. D., 460
Yudina, I. P., 4616
Yueh, M. H., 7486
Yufit, S. S., 5940
Yurovskii, Y. M., 7487, 7488

Zaal, P., 3946
Zabor, R. C., 7489
Zabrocki, L. L., 1138
Zacharius, R. M., 6833
Zagalo, A., 4773
Zahn, C., 4274-4277, 7490
Zaichenko, V. N., 4450
Zaitseva, N. I., 6887
Zakaib, D. D., 7491
Zalkow, V., 94
Zalman, M., 5207
Zander, M., 4267
Zanini, C., 7492
Zanten, B. van, 5354
Zarembo, J. E., 7493
Zavazal, J., 6452
Zaverina, E. D., 1948-1950
Zavodsky, R., 6632
Zawisza, A., 7494
Zazurca, L. G., 4702, 4703
Zbiral, E., 7495
Zderic, J. A., 7496
Zeelenberg, A. P., 7497
Zehner, R. C., 7498
Zeisberger, R., 3312
Zeman, I., 7499, 7500
Zemskova, E. I., 6654, 6754, 6755
Zemskova, Z. K., 5033
Zhabrova, G. M., 5876, 5877, 7054
Zhavoronkov, N. M., 4615, 4616
Zhdanov, S. P., 7501-7503
Zhukhovitskii, A. A., 72, 1679,
    3883, 6547, 6921-6930, 6974,
    7504-7529

Zhukovskaya, E. G., 1948, 1949,
    1951, 1952
Ziebland, H., 7530, 7531
Ziegenhain, W. C., 7532
Zielinski, E., 7533, 7534
Zielinski, Jr., W. L., 7535
Zientara, F., 7536
Ziffer, H., 7537
Zike, J., 7538
Zilversmit, D. B., 7539
Zimin, R. A., 7046, 7049
Zinn, T. L., 511, 513-515, 7138,
    7540, 7541
Ziomek, J. S., 5436b, 5458
Zirker, K., 4858
Zlatkis, A., 3861, 4504, 4506,
    5331. 7542-7560
Zmitko, J., 1079, 7561, 7562
Znamenskaya, N. B., 7563
Zobel, H., 6394
Zoebelein, H., 2688
Zollinger, H., 6627
Zollner, G., 3541, 3542
Zolotareva, O. V., 6920, 6931,
    7529, 7564
Zomzely, C., 7565
Zorin, A. D., 1814, 1815, 7566
Zubyk, W. J., 7567
Zucco, P. S., 1722
Zulaica, J., 7568-7570
Zutphen, H. J. van, 2066
Zvonov, N. V., 1077
Zweifel, G., 1103, 1105-1109
Zweig, G., 373, 7571-7576
Zwicker, J. D., 1051, 2019
Zwierzak, A., 7577

A

Absorption; a. columns; a. rates
  367, 588-9, 711-2, 1386, 1757, 1950, 4129, 4498-9, 4953, 6696, 6772
Accuracy, analytical
  3050, 3629, 4747, 5338-9, 7315, 7189-90
Acetals
  519, 1311, 1366, 3059, 3483, 3595, 3745, 3787, 3894-5, 4236, 5041, 5387, 5637, 5734, 6684, 7391-3, 7450
Acetates; acetic acids; acetolysis
  668, 891, 950, 1605, 1764, 2037, 2402, 2757, 2829, 3024, 3888, 4072, 4173, 4313, 4342-3, 4552, 4565, 4683, 4894, 5387, 5489, 5734, 5797, 5963, 6045, 6584, 6991, 7118
Acetone and acetone mixtures, derivatives
  736-7, 897, 1059, 1605, 1943, 1987, 2663, 3024, 3101, 3595, 4072, 4127, 4730, 5050, 5387, 6692, 7069, 7078
Acetonitriles
  1235, 1469, 3024, 5387, 5963
Acetophenones
  5089, 5618
Acetylene and derivatives
  69, 483, 614, 695, 713, 715, 726, 740, 928, 1100-1, 1105, 1109, 1282, 1886, 2036, 2090, 2374, 2376, 2422, 2468, 2516, 2826, 3109, 3132, 3228, 3332, 3339, 3696-7, 3701, 3861, 3914, 4124, 4450, 4549, 4725, 4882, 4989, 5050, 5182, 5387, 5427, 5747, 6190, 6530-1, 7047, 7110-1, 7373, 7455
Acetylsalicylic acid - see Pharmaceutical
Acidification reactions
  3346
Acids, organic (see also Lipids and specific types, e.g., Aliphatic)
  427, 610, 617, 1062, 1297, 1301, 1685, 3155, 3346-8, 4119, 4773, 4853, 4977, 5449, 5450, 5617, 5704, 5724, 5808, 6046, 6156, 6288, 6525, 7319, 7336, 7379
Acrolein; acrylates; acrylics
  745, 789, 1055, 3120, 4250, 4321, 4940, 5387, 5715, 6204, 7068
Activity coefficients
  1382, 2166-7, 2513, 2971, 3205, 3853, 4704-6, 4872, 5518, 6968, 6970
Acyl compounds
  955, 4238, 5309, 6020, 6108-10, 6684
Adhesives - see Rubber
Adipose tissue - see Biochemical; Biomedical; Lipids
Adsorbents; adsorption (see also Column packing; Glass; Stationary; Teflon; Zeolites)
  38, 61-3, 99, 100, 404, 563-4, 602, 619-20, 667, 688, 699, 706, 714, 733-4, 738, 884, 1036, 1162, 1274, 1380, 1412-3, 1544, 1588, 1591-3, 1597-9, 1602, 1749, 1828, 1868, 1879, 1945-7, 1949-52. 2012-3, 2033, 2109, 2212, 2214, 2216, 2377, 2418, 2480, 2499, 2552, 2631, 2638, 2714, 2751-2, 2759, 2768, 2778, 2783, 2868, 2929, 2980-3, 3111, 3129, 3131, 3188, 3206, 3213, 3226, 3478, 3498, 3533, 3576, 3753-4, 3922, 3990, 4006, 4012, 4017, 4379, 4440, 4507, 4516, 4552, 4566, 4597, 4610, 4634-5, 4689, 4691, 4752, 4825, 4871, 4920, 4953, 4991, 5002, 5303, 5315, 5336, 5409, 5464, 5466, 5480a, 5588, 5855, 5872, 6088-94, 6166, 6182-4, 6200, 6231, 6367, 6379, 6408, 6455, 6693, 6855-63, 6892, 6915, 6930, 6951, 7024, 7122, 7190, 7267-8, 7289-90, 7371-2, 7399, 7501, 7505, 7515
Aerosols
  278, 685, 3584, 4330, 5887-8, 6561, 6837, 7391-3
Agricultural applications (see also Food; Nitrogen-containing; Phosphorus; Urea; specific names)
  26-7, 34, 258, 269, 329, 334, 392, 413-6, 466, 540, 558, 681, 931, 956-7, 991, 1085, 1139, 1306, 1362-4, 1458-60, 1500, 1546-7, 1898, 2039, 2163, 2266, 2372, 2533, 2614, 2698-9, 2797, 2831-2, 2994, 2999, 3020, 3114, 3419, 3589, 3643, 3647, 3796, 3860, 3870, 4220, 4279, 4741, 4793, 4954, 5009, 5021, 5122, 5325, 5343, 5367, 5465, 5507, 5899, 5953-7, 6096, 6237, 6287-9, 6300, 6393, 6539, 6572, 6593, 6719, 6804, 6821, 6833, 6897, 6991, 7175, 7185, 7211, 7228, 7276, 7316, 7388, 7575-6
Air, analysis of; pollution of; toxic gases
  31, 104-9, 160-1, 256, 359, 446, 506, 702-4, 765, 1221, 1342, 1400, 1525, 1615, 1640, 1751, 1892, 1989, 2034, 2077, 2091, 2115, 2197, 2209, 2244-5, 2287, 2301, 2326, 2331, 2584, 2747, 3068, 3192, 3321-4, 3350-3, 3380-2, 3413-7, 3485-91, 3618, 3690, 3826, 3879, 3983, 4179, 4218, 4226, 4298, 4355-7, 4557,

4633, 4733, 4784, 4810, 4949, 4989, 5008, 5039, 5190, 5414, 5462, 5508, 5662, 5710, 5875, 5961, 5988, 6014, 6065, 6170-1, 6281, 6293, 6419, 6506, 6533-4, 6541-2, 6571, 6742, 6750, 6799, 6805, 6901-3, 6948, 7216-8, 7247, 7328, 7455, 7458

Air, peaks - see Peak

Alcohols (see also specific listings)

20, 43, 52, 96, 172, 240, 427, 533, 553-7, 661, 729, 736-7, 820, 897, 921, 934, 1139, 1156-7, 1210, 1297, 1920, 1925, 1963, 2078, 2284-5, 2321, 2327, 2617-8, 2653-4, 2843, 2977-8, 3167, 3180, 3232, 3277-80, 2770, 2843, 2854, 2937, 2954, 3015, 3024, 3312, 3344, 3376, 3748, 3787-9, 3853, 3870, 3894-5, 4072, 4113, 4127, 4135, 4146, 4246, 4271, 4313, 4351, 4377, 4415-7, 4552, 4555, 4649-50, 4730, 4746, 4756, 4830-1, 4888-9, 4894, 4989, 5023, 5027, 5040, 5049, 5105, 5116, 5232, 5296, 5378, 5385-7, 5448-51, 5489, 5581-3, 5590, 5635, 5853, 5866, 5878-9, 6012, 6035, 6042, 6052, 6069, 6257, 6270, 6344-5, 6358-61, 6429, 6456, 6476, 6481, 6486, 6525, 6558, 6625, 6632, 6658, 6727, 6868, 6968, 6991, 7040, 7078, 7108, 7151, 7210, 7255-6, 7467-8, 7493

Alcoholic beverages; blood alcohol

223, 439, 443, 468, 533-7, 640-3, 656, 685, 803-4, 978, 1206, 1262, 1308, 1395-7, 1442, 1605, 1813, 1916, 1920, 1925, 1963, 2078, 2284-5, 2321, 2327, 2617-8, 2653-4, 2843, 2977-8, 3167, 3180, 3232, 3277-80, 3300, 3344, 3421, 3583, 3641-2, 3752, 3785-9, 4184, 4420, 4572, 4581-3, 4648, 4746, 4855-7, 4867, 5040-1, 5387, 5734, 5789, 5797, 5818-21, 5842-3, 6323-4, 6558-60, 6608, 6652, 6956, 6999, 7000-3, 7087-8, 7166-7, 7536

Aldehydes

856, 959, 1105, 1109, 1408, 2307, 2340, 2424, 2736-8, 2825, 2918, 3371, 3375, 3595, 3856, 3922-3, 4135, 4779, 4888-9, 5041, 5282, 5335, 5387, 5449-50, 5621, 5633, 5724-5, 5734, 6134, 6234, 6345, 6358, 6361, 6523-5, 6558, 7040, 7150, 7257, 7284, 7391-3, 7556

Algae - see Oceanographic

Alinear ideal chromatography

3295

Aliphatic compounds (see also specific names)

127, 344, 2346, 2725, 3423, 3482, 4135, 4313, 4629, 5617, 5727, 5966-8, 6062, 6410, 6473, 6691, 6741, 6751, 7177

Alkaloids - see Pharmaceutical

Alkanes; alkenes; alkenyls

539, 748, 1076, 1102, 1107, 1522, 1771, 1819-22, 2166, 2252-3, 2839, 4888-9, 5321, 5449-50, 5598, 5725-6, 6058, 6274, 6471, 6670, 6965, 7073, 7157, 7457

Alkoxy compounds

1926, 2881, 2910, 4148-50, 4623, 4678, 6468, 7034

Alkyl compounds; alkylation (see also Trialkyl)

5, 32, 370, 490, 560, 584, 614-5, 673, 849, 975, 992, 1100, 1300, 1446, 1458-60, 1485, 1495-6, 1770, 1796-7, 1849, 1984-5, 2166, 2347, 2370, 2682, 2798, 2966, 3123, 3200, 3368, 3649, 3724-5, 3894-5, 3981-2, 3997, 4007, 4038, 4120-2, 4286, 4371, 4506, 4548, 4565, 4720, 4761, 4791, 4888-9, 5033, 5118, 5264-5, 5320-1, 5310, 5524, 5712, 5725, 5847, 5901, 5971, 6119-21, 6152, 6234, 6259-60, 6264, 6327, 6459, 6486, 6961, 7153, 7331-2, 7394-5, 7475

Allenes

143-4, 740, 1885, 5016-8

Allyl compounds

1015, 1054, 1129-31, 1391, 1678, 1888, 2003, 2658, 2696, 2701, 2757, 3039-41, 3318, 4066, 4711-2, 5281, 5387, 5581-3, 5672, 5974, 6042, 6329, 7146, 7477

Alpha particles - see Radiochemistry

Aluminum and compounds; uses

496, 591, 604-5, 858, 884, 1849, 2046-7, 2520, 2751, 2845, 2900-2, 3039-41, 3847, 3909, 4037, 4098, 4132, 4752, 5017, 5044, 5050, 5055, 5282-3, 5315, 5477, 5530-4, 5598, 5962, 6195, 6307, 6455, 7029, 7266

Amides

27, 1199, 1405, 2028, 2260, 2393, 4237, 4901, 6847, 7269

Amines; diamines; etc.

127, 639, 727, 870, 1065-67, 1164, 1310, 1348, 1405, 1433, 1511, 1515-20, 1523, 1799, 2183-5, 2222, 2338, 2346, 2415, 2424, 3024, 3099, 3171, 3207, 3225, 3327, 3423, 3429, 3440, 3542, 3579, 4066, 4418, 5201, 5210, 5387, 5449-50, 5691, 5831, 5851, 6056, 6257, 6288, 6329, 6410, 6563, 6609, 6691, 6814, 6874, 6989, 7037, 7162, 7269

Amino acids; a. compounds

249, 532, 639, 650, 662, 824, 2043, 2064, 2249, 2953, 3345-9, 3645-6, 3700, 3759-60, 4675, 4768, 4869, 5523, 5539, 5543, 5952, 6036-7, 6257, 6411, 6524, 6714, 6960, 7110-1, 7478, 7554-5, 7565

Ammonia; ammonium compounds

67, 470, 584, 761, 1141-3, 1478, 2759, 3213, 3396, 3467, 4146, 4242, 4989, 5549-50, 6014, 6037, 6691, 6876, 7153, 7267

Amplifier, electron voltage
917
Ampoule technique
285, 1352
Anaerobic techniques
2789, 3155, 5130
Analcite
578, 585-6
Analog computers and integrators
5071, 6611-2
Analysis - see Apparatus; Chromatograms; Gas a.; Separations
Analyzers
161, 288, 316, 332, 394, 1745, 3382, 5039, 5662, 6063, 7026-7, 7132, 7350
Androsterones - see Hormones; Steroids
Anesthesiology (see also specific compounds)
41, 274, 504, 1182, 1319, 1893, 2170, 2445-6, 2451, 2622-3, 2906, 2950, 3147-8, 3547, 3550, 3576, 3671, 4334, 4509-11, 4637, 5025, 5245, 5262, 5679, 6639, 7151, 7350
Anetholes
1192, 5136, 5151, 5154
Anhydrides
5282, 5635
Anilines - see Dyes; Nitrogen-containing
Anisoles
818, 1298, 6583
Anthracene, anthracene oil
1281, 5527, 6458
Antimony
4312, 6014
Antioxidants
6238
Apparatus - design, construction, and operation (see also Automatic; Chromatographs; Detectors; Gas sampling; Injection; Integrators; Thermal)
3-4, 19, 98, 133, 163, 222-9, 257, 265-6, 300, 382, 437-9, 460, 611, 646-7, 659, 729-31, 747, 751, 854, 865-6, 918, 944-5, 1013, 1023, 1039-40, 1112, 1116, 1178, 1246, 1250, 1290, 1313-6, 1365, 1384-5, 1417, 1425-9, 1455, 1487-8, 1508, 1536, 1557, 1658, 1663, 1677-9, 1700-1, 1745, 1750, 1833, 1853, 1899, 1905, 1968-70, 2091, 2195-7, 2202, 2224, 2241, 2255, 2330, 2377, 2389, 2403, 2420, 2440, 2490, 2505, 2561, 2625, 2650, 2659, 2929, 2955, 2961, 2976, 3056, 3095, 3234, 3351, 3401-3, 3543-4, 3557, 3560, 3609, 3624, 3657, 3688, 3776, 3850-1, 3885, 3898, 3998, 4160, 4269, 4367, 4454, 4466, 4535, 4597, 4610, 4812, 4888-93, 4912, 4966, 5068, 5100, 5130, 5195, 5252, 5258, 5353, 5371, 5486, 5500, 5505, 5546, 5676, 5760, 5856, 5929, 6002, 6014, 6025-7, 6120, 6253, 6262, 6343, 6359, 6575-7, 6602, 6694, 6763-7, 6793, 6814, 6854, 6900, 6916, 6923-5, 6928, 7008, 7016, 7026-7, 7036, 7066, 7135, 7196, 7207-9, 7272, 7309, 7321-2, 7334, 7390, 7452, 7462, 7469, 7513, 7517
Aqueous - see Water
Arachidic and arachidonic acids and compounds
1144-5, 2764, 2882, 6673
Areas - elution, peak, surface
609, 612, 1588, 1769, 1846, 2052, 3158, 4858, 5332
Argon and a. mixtures
2746-7, 2943, 2964, 3013, 3485, 3491, 3543-4, 3802, 3818, 3827, 4155, 4158-9, 4226, 4239, 4242, 4288, 5088, 5609, 5826, 6072, 6267, 6915, 7064, 7204, 7387, 7530-1
Argon (ionization) detectors
961, 1493-4, 2165, 2319-20, 2460, 2483, 3405, 3414, 3661-4, 3668, 3827, 4282, 4421, 4487, 4492-6, 4502, 4684, 4729, 4828, 4878, 5009, 6153, 6183, 6216, 6221, 6230, 6267, 6664, 7204, 7387, 7441, 7545
Aromas - see Odors; also Dairy; Essential oils; Food; Perfumes
Aromatic compounds   (see also Essential oils; Perfumes; specific names)
12, 70, 107, 154, 198, 404, 477-80, 483-6, 646, 849, 906, 1081, 1281, 1287, 1328, 1334-5, 1353, 1771, 1799, 1919, 2236, 2244-5, 2338-9, 2405, 2425, 2470, 2595, 2624, 2724-6, 2939, 3118, 3312, 3332, 3394, 3429, 3441, 3459, 3523, 3610, 3702, 3843, 3888, 4273, 4276, 4453, 4892-4, 4964, 5054, 5106, 5139, 5264, 5320-1, 5449-50, 5479, 5530-1, 5537, 5909, 6229, 6263, 6333-4, 6382, 6446, 6451, 6530, 6631, 6663, 6667-9, 6718, 7004, 7084, 7124, 7156, 7177, 7284-6, 7317, 7365, 7378,
Aroxyls
5812

Arsenic and compounds
    2796, 2798, 2800, 3373-4, 6014
Aryl compounds; arylation
    1015, 1265, 1720, 2795-6, 3138, 5311, 5742, 6412
Asphalt - see Refinery
Aspirin - see Pharmaceutical
Atherosclerosis - see Lipids
Atmosphere - see Air
Atomizer
    2447
Attenuators
    638, 4969
Automatic and continuous analyses; process chromatographs and controls
    3, 142, 161, 218, 246, 264-7, 273, 461, 611, 712-22, 909, 918, 1112, 1159, 1195-6, 1755-6, 1899, 1909-10,
    1967-8, 1972, 2025, 2219-20, 2268-9, 2330, 2519, 2561, 2584, 2717, 2786, 2789, 3007, 3011, 3095, 3212,
    3227, 3265, 3469, 3556-7, 3624, 3653-7, 3812, 3834, 3869, 4033, 4049-50, 4176-8, 4204, 4451, 4539, 4786,
    4874, 4891-2, 4969, 5069, 5183, 5389, 5669, 5771-2, 5803, 5856, 6002, 6063, 6132, 6210, 6440, 6461,
    6489-90, 6507, 6524, 6537, 6603, 6611, 6698, 6716, 6721, 6752, 6900, 6924, 6935-9, 6976, 7031, 7127, 7135,
    7171, 7196, 7214, 7513, 7525, 7538, 7561-2
Azo compounds - see Dyes; Nitrogen-containing

B

Backflushing
    1428, 2251, 4785, 7052-3
Baker-Nathan effect
    6584
Bakery products - see Food
Band areas; b. shapes
    915-6, 1594, 3267, 3864, 7528
Barbiturates - see Pharmaceutical
Base-line control; b.-l. drift
    2070-1, 2814-5
Basicity measurements
    390
Beer - see Alcoholic
Benzene (see also Benzene derivatives and specific types of derivatives)
    40-2, 85, 367, 484, 544, 549, 706, 729, 819, 868, 1101, 1121, 1343, 1349, 1826, 1982, 2173, 2427, 2494,
    2621, 2703, 2843, 2864, 3292-3, 3559, 3577, 3815, 4016, 4072, 4131, 4335, 4566, 4600, 4677, 4689, 4830,
    4890-2, 4911, 4958, 4989, 5066, 5110-1, 5360, 5387, 5489, 5793, 5885, 5916-7, 5971, 5976, 6182-4, 6345,
    6461, 6487, 6584, 6633-4, 6730, 6741-2, 6878, 6883, 6910, 6955, 6970, 7032, 7101, 7266, 7433, 7502
Benzene derivatives (see also specific types of derivatives)
    40, 480, 542, 549, 681, 686, 695, 819, 906, 922, 936, 1100, 1199, 1213-4, 1281, 1468, 1479, 1558, 1648,
    1689, 1799, 1803, 1826, 2086-7, 2264, 2726, 2864, 2945, 3082, 3103, 3137-8, 3170-1, 3481, 3559, 3875,
    4002, 4117-9, 4146, 4170, 4179, 4451, 4831, 4942, 5011, 5106, 5110-1, 5387, 5439, 5489, 5576, 5739, 6182-4,
    6412, 6441, 6465, 6582, 6886, 7032, 7069, 7098, 7102-3, 7118, 7140, 7495
Beryllium and compounds
    1987, 3153, 4903
Beta-ray emitters and ionization detectors
    905, 1437, 4348-9, 5202, 6085, 6351, 6795, 7470-1
Beverages - see Alcoholic; Dairy; Food
Bibliographic data; books; data retrieval (see also Data)
    119, 170-1, 652-3, 728, 759, 767, 807, 837, 1033, 1057, 1120, 1149, 1176, 1187, 1303-5, 1388, 1456, 1585,
    1659, 1718, 1751, 1774-6, 2114-6, 2273, 2276, 2286, 2296, 2940, 3077-8, 3127, 3624, 3667, 3689, 3719,
    3723, 3764-5, 3771-2, 3934, 3942, 3951, 4059-62, 4087, 4247, 4274, 4300-2, 4374, 4446, 4634, 4716, 4724, 4820,
    4861, 4877, 4895, 5244, 5430, 5496, 5542, 5641, 5685, 5774, 5783, 5848, 5906, 6061, 6087, 6097, 6167,
    6214, 6322, 6502, 6614, 6701-3, 6706, 6845, 7012, 7341, 7490, 7523
Bile - see Biochemical; Biomedical
Biochemical applications (see also Biomedical; Pharmaceutical)
    9, 55-6, 141, 176-9, 213, 247, 263, 345, 348, 472, 501, 507, 519, 613, 692-4, 725, 768, 778, 831, 837-9,
    842, 860, 871-80, 937, 946, 983-4, 1123, 1144-5, 1149, 1152-3, 1233, 1261, 1302, 1329-30, 1368-70, 1472,
    1584, 1605, 1617, 1646, 1829-30, 1865, 1890, 1932, 1971, 2048, 2063-4, 2168, 2184-5, 2189, 2193, 2217,

2243, 2277, 2317, 2429, 2445-6, 2450-1, 2468, 2500-2, 2506, 2529-32, 2619, 2736, 2756, 2764, 2772-4, 2828-9, 2849-59, 2922-6, 2940-2, 3052-4, 3071, 3088, 3147-8, 3168, 3215, 3239, 3244, 3263, 3285, 3331, 3450-1, 3468, 3473-6, 3591, 3673-4, 3700, 3717, 3866, 3884, 3984, 3989, 3994-6, 4020, 4040, 4047, 4104, 4175, 4188, 4249, 4261-3, 4397, 4405, 4423-5, 4429, 4470, 4483-6, 4505, 4534, 4568-9, 4572, 4625, 4639, 4676, 4681, 4698-9, 4700-1, 4798-9, 4803-5, 4809, 4816, 4881, 4909, 4931, 5030, 5046, 5130, 5197-9, 5205-7, 5323, 5372, 5375, 5406, 5435, 5446-7, 5539, 5542-3, 5565, 5581, 5613, 5639, 5717, 5728-9, 5776, 5805-6, 5813-4, 5896-9, 5931-2, 5988, 6000, 6015, 6036-7, 6060, 6157-63, 6186, 6232, 6352, 6365-6, 6422, 6467-8, 6478, 6571, 6590, 6644, 6662-7, 6671-7, 6706, 6722-4, 6745-6, 6766-8, 6815, 6834-6, 6846, 6898, 6932-3, 6957, 6973, 6980, 6984-97, 7013, 7059, 7114, 7158-9, 7160-70, 7203, 7250-2, 7279-81, 7317-20, 7329, 7353-6, 7396, 7421-5, 7485

Biological applications - see Biochemical; Biomedical; Pharmaceutical

Biomedical applications (see also Air; Alcoholic; Anesthesiology; Biochemical; Enzymes; Lipids; Pharmaceutical; Steroids)
233, 263, 337, 345, 504, 507, 694, 708, 736-7, 838, 860, 871-3, 877-80, 891-3, 946, 959, 964-73, 976-7, 982-4, 1052, 1065-7, 1329, 1367, 1387, 1499, 1509-10, 1605-7, 1631, 1643, 1684, 1719, 1834, 1865, 1897, 1911, 1932, 2058, 2155, 2170, 2185, 2189, 2299, 2445-51, 2506, 2596, 2739, 2794, 2799, 2856-7, 2914-5, 2922-3, 2925-7, 2950, 3092, 3147-8, 3168, 3191, 3209, 3214-5, 3222, 3239-44, 3331, 3435, 3447, 3474, 3562, 3575-6, 3581, 3626, 3644, 3667, 3707, 3884, 3977, 3985-6, 4058, 4148, 4175, 4261, 4292-6, 4326, 4423-5, 4470, 4509-11, 4534, 4557, 4582, 4614, 4623, 4637-43, 4651, 4848, 4850, 4952, 5100, 5130, 5206, 5245, 5262, 5375, 5385-7, 5581, 5639, 5679, 5728-9, 5766, 5780, 5786, 5829, 5951, 5988, 6014, 6109, 6157-63, 6171, 6365, 6422, 6468, 6478, 6532, 6571, 6663-7, 6671-5, 6706, 6745, 6766, 6836, 6983-5, 6996, 7059, 7114, 7250-3, 7318, 7350, 7354-6, 7424

Blast furnaces - see Furnaces

Blood analyses - see Alcoholic; Biochemical; Biomedical; Lipids

Boilers - see Furnaces

Books - see Bibliographic

Boron; compounds and reactions
2, 96, 862-3, 947, 1058, 1099, 1102-9, 1973, 2689, 2985, 3862, 4132, 4207, 4299, 4441, 4468, 4548, 5103-4, 5311, 5402, 5509, 5696, 6152-4, 6236, 7339

Botanicals - see Agricultural

Branching - see Chain

Brewery products - see Alcoholic

Bromine; compounds and reactions
260, 373, 745, 764, 818, 858, 1133, 1390, 1521, 1875, 1980, 2036, 2253, 2427, 2656, 2688, 2796, 2800, 2987-11, 3040-2, 3091, 3199, 3404, 3535-6, 3795-7, 3959, 4264, 4299, 4320, 4506, 4790-1, 4830, 4988, 5000, 5016, 5089, 5116, 5281, 5311, 5738, 5749, 5909, 5970, 6014, 6076, 6369, 6498, 6562-4, 6583, 6658, 6886, 7032, 7373, 7495

Butadienes
1575, 1814, 2512, 2595, 3040, 3537-8, 3861, 4008, 4116, 4179, 6263, 6495, 7054

Butanes; butenes; butyl compounds
29, 92, 181, 343-4, 351, 687, 701, 797, 844, 1128, 1252, 1292, 1438, 1472, 1629, 1635, 1814, 1886, 2021, 2033, 2077, 2318, 2348, 2364, 2512-3, 2517-8, 2656, 2663, 2988-9, 3023, 3139, 3143, 3200, 3336, 3870, 3886, 3908, 4115, 4440, 4554, 4791, 4813, 4830-3, 4837, 4928, 5260, 5319, 5466, 5538, 5689, 5711-3, 5966, 5972, 5980, 5991, 6284, 6329, 6369, 6473, 6563, 6755, 6881, 7006, 7049-54, 7097, 7149-51, 7209, 7340, 7484, 7508, 7518, 7566

Butter - see Dairy

Butyl compounds
542, 553-4, 820, 853, 950, 1100, 1320, 1605, 1761, 1987, 2363, 3361, 3368, 3407, 3589, 3793, 3875, 4072, 4152, 4264, 4377, 4564, 4822, 4830, 4834, 4890, 5049, 5105, 5387, 5851, 5897, 6035, 6069, 6270, 6329, 6345, 6358, 6361, 6429, 6582, 6592, 7098, 7142-4, 7151, 7256, 7493

Butyric compounds
1873, 3923, 4540, 5387, 6358, 6361

C

Cadmium and compounds
1292, 1296, 1301, 4552, 5653

Caffeine - see Pharmaceutical

Calcium compounds
7266

Calibrations

513, 1171, 1508, 1541, 2024, 2082, 3051, 3624, 3945, 4192, 4765, 4812, 4944, 5192-4, 5266, 5393, 5474, 5864-5, 6007, 6014, 6772, 7188, 7417

Camphor and related compounds

559, 649, 1283, 2233, 2415, 2624, 2936-9, 3312, 3716, 5173, 5592

Capillary columns and traps

430, 449-54, 458, 512, 654-7, 775, 907-8, 985, 1030-2, 1039, 1234, 1332, 1380, 1424, 1487-90, 1779-80, 1788-95, 1798, 1847, 1909-10, 1983-5, 2110-2, 2122-4, 2133-5, 2546-8, 2563, 2662, 2667-8, 2691, 2860-1, 2872, 2885-9, 2894-9, 3065, 3108, 3276, 3586, 3614, 3688, 3765-7, 3774, 3783, 3955, 4088-9, 4103, 4166, 4262, 4331, 4489, 4693, 4783-5, 4968, 4972-3, 4991, 4994, 4999, 5123, 5253, 5477, 5485, 5512, 5589, 5683-4, 5709, 5916-8, 6182-5, 6210-2, 6218, 6227-8, 6243, 6333, 6338-9, 6795-6, 6800, 6927, 7123, 7125, 7543-8, 7558-60

Caranes, carenes

5142, 5160, 5306

Carbenes

992, 1446, 1883-7, 3004, 6056

Carbethoxyl group

2044-5

Carbohydrates - see Biochemical; Food

Carbon; determination of; uses of

266, 1195-6, 1366, 1987-8, 2865-6, 2976, 3162, 3213, 3733, 3837, 4545, 4647, 5007, 5193, 5231, 5580, 6641, 7024, 7072, 7122

Carbon dioxide

41, 67, 431, 715, 867, 975, 1034, 1051, 1525, 1589, 1897, 1963, 2019, 2489, 2905, 3147-8, 3213, 3578-9, 3671, 3707, 3726-9, 3802, 3831, 4171, 4242, 4315, 4533, 4556, 4559-60, 4713-4, 4809, 4949, 4989, 5085, 5100, 5662, 5883, 6014, 6094, 6191, 6293, 6343, 6364, 6426, 6681, 6692, 6696, 6700, 6755, 6842, 7205-6

Carbon disulfide

1210, 2014, 3559, 4911, 4989, 6537, 6656, 7277

Carbon monoxide

463, 475, 1210, 1524, 1589, 3485, 3491, 3532-4, 3671, 4112, 4713-4, 4989, 5074, 5085, 5521, 5597, 5609, 5747, 5867, 5904, 5960, 5988, 6014, 6076, 6191, 6293, 6455, 6700, 6915-6, 6931, 7188, 7205

Carbon suboxide

666, 3169

Carbon tetrachloride

418, 745, 3888, 4989, 5108, 5387, 5807, 5951, 6957, 7151

Carbonates

1256, 2796-9, 6412

Carbonium ions

7330

Carbonyl compounds

258, 428, 1183, 1711-2, 1838, 1862, 2044, 2905, 3342, 4029, 4281, 4403, 4419, 4529, 4636, 4937, 5237, 6146-7, 6545

Carboxylic acids and compounds

17-8, 81, 700, 1203, 1644, 2099, 2103, 2386, 2502, 3089, 3170, 4978, 5275, 5283, 5295, 5789, 5867, 6465

Carotenoids

5596

Carrier gases

262, 293, 913, 919, 1467, 1792, 2361, 2691, 2754, 2964, 3049, 3530-1, 3624, 3814, 3831, 4222, 4513, 5267, 5470, 5549, 5657-8, 5752-7, 6114-5, 6189, 6194, 6219-20, 6252, 6707, 6844, 7091-2, 7174, 7204, 7231, 7374, 7376, 7384, 7386, 7541

Carroll reaction

6776

Carvo- compounds

3912, 5226

Catalysis (see also specific materials)

116, 156, 438, 481-2, 516, 583, 618, 622-3, 929-30, 960, 1055, 1061, 1158, 1365, 1591, 1649, 1710, 1955, 2022, 2074-5, 2128-30, 2297-8, 2318, 2474, 2499, 2520-3, 2649, 2845, 2907-8, 3003, 3226, 3381, 3888, 3909, 3952, 3976, 3981-2, 4098, 4111, 4173, 4237-8, 4377, 4408, 4524, 4549, 4646, 4912, 5054-5, 5089, 5328-9, 5517, 5530-4, 5545, 5635, 5846, 5876, 5905, 5969-71, 6034, 6058, 6095, 6129, 6140, 6165, 6319-20, 6501, 6533-4, 6769, 6828, 6931, 7010, 7140, 7284, 7482, 7489

Cellulose and compounds

523, 766, 1458-60, 3294, 3993, 4695, 5185, 5440, 7411

Ceramic applications
6142

Cesium compounds
591

Chabazites - see Zeolites

Chain compounds (see also specific types)
1293, 4978, 5867, 7417

Charcoal - see Adsorbents; Carbon

Cheese - see Dairy

Chelates
259, 825, 1987, 6321

Chlorine and compounds (see also Carbon tetrachloride; Chloroform; Ethyl chloride; Halogens; Trichloro)
2, 29, 43, 260, 269, 343, 373, 418, 516-8, 521, 544, 607, 680-6, 745, 764, 818, 828, 909, 922, 1015, 1051, 1054, 1078, 1133, 1239, 1285, 1297, 1339, 1446, 1556, 1689-91, 1980, 2014, 2019-21, 2049, 2084, 2091, 2250, 2322, 2347, 2369-70, 2395, 2444, 2469, 2611, 2692, 2698, 2743, 2794-9, 2827, 2878, 2945, 2990-1, 3110, 3139, 3144, 3147-8, 3170, 3211, 3235, 3400-7, 3535-6, 3591, 3622, 3627, 3687, 3797-8, 3870-1,3888, 4002-7, 4081, 4240, 4342-3, 4377, 4477, 4564, 4718, 4830, 5000, 5032, 5103-4, 5108-11, 5184, 5214, 5282, 5311, 5319, 5325, 5343, 5387, 5489, 5572, 5671, 5713, 5807, 5847, 5886, 5966, 6014, 6045, 6052, 6056, 6076, 6176, 6234, 6254-5, 6283, 6293, 6419, 6465, 6488-9, 6539, 6583, 6593, 6875, 6883, 6910, 6969-70, 7032, 7104, 7142-3, 7146, 7151-3, 7162, 7183, 7433, 7457-8, 7461, 7576

Chloroform and derivatives
1010, 1131, 1319, 2622, 3024, 3174-5, 4334, 4371-2, 4566, 4649, 5387, 6487, 7151

Chocolate - see Cocoa; Food

Cholesterol and related compounds
247, 348, 476, 860, 874, 966, 982, 1261, 1391, 1646, 1719, 2168, 2530, 2696, 2857, 3214, 3647, 4175, 4484, 5030, 5046, 5323, 5372, 6103, 6352, 6366, 6671, 6674-6, 6991

Chromathermography
3960, 5208, 5304-5, 6654-60, 6893, 6905, 6920, 6974, 7505, 7512, 7515, 7525-9

Chromatograms
272, 362, 424-5, 827, 1587, 1669, 1724, 1870, 2005, 2137-8, 2145, 2314, 2475, 2639, 2837, 3150-2, 3403-4, 3624, 3836, 3885, 3910, 3945, 4222, 4368, 4923-5, 5312, 5470, 5548, 5553, 6213, 6224-5, 6313, 6348-9, 6364, 6449, 6575, 6696-7, 6726, 6746, 6918-9, 7272, 7402, 7412

Chromatographs (see also Automatic and Preparative)
68, 73-4, 129, 161, 188, 191, 240, 244, 253, 261, 269, 281-3, 317-9, 326, 335, 407-8, 865, 1140, 1169, 1250, 1365, 1693, 1899, 1900, 2009, 2025, 2038, 2070, 2143-4, 2223, 2244-5, 2268-9, 2390, 2597, 2731, 2786, 2802, 2899, 3010, 3152, 3234, 3310, 3367, 3405, 3428, 3449, 3470, 3497, 3556, 3678, 3837, 3917, 4032, 4045, 4 049, 4176-8, 4481, 4658, 4696-7, 4890, 4970, 4989, 5069-70, 5270-2, 5299, 5340, 5344-6, 5353, 5459-60, 5485, 5604, 5610, 5615, 6032, 6038, 6083, 6147, 6244, 6309, 6325, 6342, 6346, 6457, 6466, 6538, 6680, 6716, 6731, 6751-4, 6762-7, 6816, 6849, 6867, 6902, 6935, 6939, 7022, 7058, 7063, 7214, 7248, 7272, 7313, 7346-8, 7452, 7499

Chromium and compounds
1006, 3976, 5050, 5531, 6069, 6510, 7284

Cigarettes; cigars - see Tobacco

Clathrates
819

Clinical applications - see Biochemical; Biomedical

Coal; c. gas; c. tar; coke; coking gas; and products (see also specific chemicals)
6, 11, 925, 987, 1110, 2018, 2187, 2238, 2409, 2608-9, 2724, 3118, 3389, 3722, 3844-5, 4102, 4599, 4958, 5515, 5791, 5835, 6049-50, 6226, 6696, 6947, 7248, 7437, 7456

Coatings - see Column packing; Paints; Resins; Stationary phase

Cobalt and compounds
538, 2658, 3069, 3959

Cocoa; cocoa butter; chocolate
1996, 2066, 3067, 3699, 4530, 4541, 4995, 7364, 7397-8

Coffee - see Foods

Collection apparatus and techniques (see also Fractionation)
669, 814, 1333, 1569, 1935, 2679, 2883, 3033, 3107, 3221, 3257, 3654, 3703, 4147, 4182-3, 4347, 4466, 5125, 5175, 5228, 5436b, 5619, 6111, 6277, 6315, 6480, 6561, 6709, 6794, 7077, 7311-3

Colorimetric analyses
3726-30, 3733, 5875

Columns, general aspects (see also Capillary; Dual; Efficiency; Golay)
16, 323, 447-52, 464, 512-4, 683-5, 718-22, 794, 1004, 1372, 1410, 1505, 1508, 1673, 1735, 1815, 1964,

D

Decanes; decalin
    502, 1309, 2508, 2769, 2838, 5448, 6441
Decomposition methods
    955, 2732, 4083, 4694, 7431
Demonstration apparatus
    132-3, 854, 1016, 1116, 1536, 1557, 5014, 6102, 6309, 6708, 6854, 7208
Derivative preparation
    141, 4924, 4929
Design - see Apparatus; Chromatograph; Specific components
Detectors (see also Argon; Beta-ray; Electron capture; Flame; Gas; Infrared; Ionization; Thermal)
    21, 131, 138, 161, 189, 242, 252, 373, 475, 560, 628, 655, 697-8, 775, 905, 944-5, 956-7, 1160, 1171, 1193,
    1326, 1392, 1422-3, 1432, 1493-4, 1547-9, 1552, 1696, 1722, 1775, 1810-1, 1901, 1909-10, 1934, 2011,
    2164, 2169, 2186, 2251, 2258-9, 2297-8, 2337, 2349, 2361, 2458-9, 2674, 2741, 2767, 2775, 2792, 2803,
    2835, 3049, 3106, 3212, 3287-9, 3296-9, 3382, 3481-3, 3506, 3558, 3621-4, 3637, 3658, 3662-9, 3684,
    3721, 3731, 3821-5, 3864, 3880-2, 3966, 4161, 4187, 4315, 4329, 4360-2, 4409, 4435, 4488, 4495-9, 4504,
    4573, 4611, 4653, 4818-9, 4841, 4892, 4905, 4918, 4962-3, 5098, 5189, 5215, 5240, 5297, 5449-50, 5515,
    5541-4, 5675, 5681, 5825-6, 6014, 6031, 6113, 6131, 6153, 6163, 6221, 6241-2, 6246, 6251-6, 6261, 6266-7,
    6331, 6355, 6407, 6550, 6554, 6771, 6964, 6793, 6801-2, 6910, 6934, 7052-3, 7112, 7297-8, 7328, 7334,
    7344, 7350, 7361, 7430-1, 7470-1, 7545-7, 7556
Detergents; surfactants; etc. (see also Soap)
    855, 1728-31, 1796-7, 3724-5, 5116-8, 7023
Deuterium and compounds
    386-8, 619-20, 934, 974, 1121, 1575, 1989, 5303, 5609, 5677-8, 5811, 5966, 6369, 7009
Developments - see Review
Di- - see simple compound listings
Diels-Adler reaction
    786, 789, 960, 1132, 2592, 4038, 4947
Dienes
    668, 695, 1107, 1519, 1624, 1632, 1683, 2162, 2263, 2769, 2937, 3132, 4195, 4908, 4955, 5721, 6055, 6099,
    6596, 6877, 7047, 7302
Differential curves, d. methods
    3513, 4959-60, 6598, 7207, 7321, 7431
Difficulties, errors, in G.C.
    21, 23, 717, 2569, 3045, 3050, 3465, 3672, 3680, 5332, 5645, 6151, 6291, 6421, 6981, 7402, 7516
Diffusion; d. cells; d. coefficients
    28, 108, 379, 391, 581, 597, 602, 910, 915-6, 952, 1048, 1243, 1734, 2212-4, 2540-3, 2545-7, 2550-4,
    2573-9, 2580-2, 2598, 2634, 3226, 3967, 4052, 4283, 4399, 4812, 5258, 5454, 5681, 5686, 6113, 6428
Dispersion
    378-9, 2539-43, 3267
Displacement reactions and technique
    1449, 2811
Distillation columns and curves
    1215, 2270-1, 2781, 4075, 5065, 5606, 6193, 7152, 7197
Distribution coefficient; d. ratio
    1965, 7233, 7400
Double bonding, location of
    836, 1137, 1767, 3443, 3855
Drugs - see Pharmaceuticals
Drying oils - see Paints
Dual columns (see also Multiple; Partition)
    1404, 1491, 1703, 1763, 2060, 2070, 2569, 2880, 3351, 4376, 4559-61, 4888-90, 4930, 5349, 6262, 6633-4,
    6699, 6767, 6793
Dyes (see also Nitrogen-containing)
    2246, 5126, 6649

E

Educational apparatus - see Demonstration
Efficiency and sensitivity of apparatus
    131, 323, 464, 509, 771, 864, 881, 914, 919, 1144-5, 1355, 1376, 1380, 1406, 1652, 1656-7, 1673, 1736-9,
    1787, 1852, 2121, 2356-8, 2361, 2665, 2674, 2735, 2824, 2861-2, 2898, 3006, 3009, 3363-4, 3390, 3624,

3656-7, 3944, 3955, 4129, 4222, 4518-20, 4547, 4876, 4929, 5286, 5404, 5454, 5549, 5681-4, 5686, 5752-5, 5998, 6066-7, 6208-11, 6218-20, 6223, 6227-31, 6310-1, 6615-6, 6736, 6909, 7204, 7229-34, 7376, 7402, 7519, 7524

Effluents in G.C.
1195, 2471, 3212, 3721, 3880, 4632, 4887, 5006, 7147, 7363

Electrochemical applications
862, 1579, 1870, 2260-1, 2511, 4080, 6068, 6433-4, 6326, 6600, 6721

Electron capture; e. acceptance; detectors
329, 560, 682, 703, 1422, 1676, 1708-9, 2698, 2761-2, 2797, 2999, 3668, 4264, 4286, 4311, 4425, 4496-9, 4500-3, 4506, 4577, 5009, 5215, 5812, 6153, 6221, 6237, 6539, 6571, 6600, 7160, 7224, 7328, 7545-6

Elimination reactions
492-5, 1142-3, 1523, 1792

Elution chromatography; e. curves
609, 915, 1915, 2338, 2631, 2811, 2847, 3158, 3183-4, 3469, 3510, 4313, 4316, 4466, 4775, 5258, 5473, 5580, 5844, 6086, 6788, 6862-4, 6925, 7089, 7268, 7293, 7337, 7376, 7462, 7474, 7511

Enamels - see Paints

Enzymes (see also Biomedical; Food)
76, 228, 995, 1391, 2696, 3134, 3602-4, 4126, 4580, 5421, 5581

Epoxy compounds
2046, 2188, 2689, 3114, 3270, 4976, 5035, 5038, 5164, 6839, 6897

Equilibrium; e. constants; non-e.
1046-7, 1050, 2543, 2553, 3748, 3954-5, 4748, 4872, 5064, 5640, 7108, 7285

Equivalent chain lengths
4978, 6108

Erythritols
736-7, 7295

Essential oils and constituents (see also Aromatics; Perfumes; Terpenes)
139, 196, 339, 396-7, 427-8, 546, 633-5, 648, 660, 690, 770-7, 828, 848, 857, 899, 1009, 1053, 1123, 1192, 1226-33, 1308, 1401, 1417-9, 1442, 1463-4, 1640-1, 1666, 1706, 1844-5, 1862, 1894-5, 2085, 2190-1, 2204, 2233, 2240, 2293, 2479, 2507, 2616, 2689, 2804-6, 2842, 2944, 3000, 3037, 3073, 3145, 3166, 3217-8, 3230, 3278-81, 3375-7, 3393, 3422, 3540, 3572, 3849, 3854, 3928-30, 3954, 3987, 4097, 4140-1, 4172, 4212, 4219, 4244, 4254, 4268, 4304a-5, 4358, 4385-6, 4391, 4536, 4541, 4580-3, 4685, 4725, 4736, 4793, 4972-3, 5052, 5134-73, 5225-30, 5300, 5445, 5483-7, 5526, 5593-5, 5736, 5789, 5798-9, 5816, 5819, 5868-70, 5927, 5945-7, 5953-7, 6005, 6047, 6238-9, 6294, 6299, 6306, 6373-5, 6394, 6401-3, 6499, 6500-1, 6519-23, 6556, 6595, 6638, 6649, 6738, 6775-83, 6806-12, 7056-7, 7106, 7163, 7249, 7257, 7335, 7408

Esters; diesters; etc.
93, 251, 492-5, 533, 537, 656, 661, 668, 685, 856, 1265, 1301, 1436, 1484, 1564, 1570, 1642, 1646, 1829, 2461, 2764, 3015, 3059, 3215, 3256, 3343, 3361, 3482, 3539-40, 3691-2, 3748, 3853, 3874, 3894-5, 4002, 4117-9, 4276, 4313, 4480, 4555, 4888-9, 4940, 4978, 5037, 5131, 5282, 5439-41, 5449-50, 5617, 5628, 5724, 5739, 5936, 6036-7, 6051, 6106-10, 6204, 6345, 6439, 6501, 6528, 6558, 6647, 6790, 7087, 7110-1, 7417

Estrogens - see Steroids

Ethanes
161, 531, 1072-3, 2182, 2379, 2524, 3398, 3870, 4255, 4517, 4830, 4989, 5084, 5086, 5466, 5609, 5656, 5747, 6473, 6713, 6748, 6755, 6916-7, 7038, 7089-90, 7151, 7373, 7445, 7448

Ethers
41, 389, 504, 741, 878, 897, 1458-60, 1668, 1884, 2353, 2622, 2906, 2911, 3091, 3147-8, 3481-3, 3894, 4072, 4188, 4195, 4623, 4649-50, 4712, 4793, 4830, 4888-9, 4894, 5011, 5083, 5116, 5245, 5262, 5387, 5489, 5605, 5672, 5742, 6303, 6435-6, 6487, 6655, 6875, 6959, 7350, 7466

Ethoxy compounds
990, 5380, 7300

Ethyl alcohol (see also Alcoholic beverages; Alcohols)
43, 557, 897, 921, 1210, 1395-7, 1605, 3344, 3870, 4072, 4351, 4830, 4989, 5023, 5040, 5387, 5489, 6012, 7210, 7255, 7493

Ethylenes; ethylation; ethyl compounds (see also Triethyl)
41, 71, 87, 116, 161, 339, 463, 504, 516, 615, 745, 861, 896, 909, 1152-5, 1172, 1310, 1330, 1410, 1578, 1620, 1689, 1873, 1986, 2022, 2091, 2378, 2444, 3147-8, 3330, 3392, 3524, 3626, 3687, 3795, 3893, 4072, 4143, 4342-3, 4397, 4450, 4506, 4517, 4532, 4539, 4556, 4559-60, 4570, 4612, 4830, 4865-6, 4940, 4989, 5083, 5109, 5116, 5213, 5241-5, 5262, 5359, 5387, 5427, 5521, 5537, 5573, 5605-9, 5656, 5691, 5747-9, 5880, 5969, 6014, 6069, 6094, 6190-1, 6255, 6263-5, 6636-7, 6655, 6713, 6755, 6785, 6880, 6915, 6946, 6966, 7151, 7373

Ethynyl compounds
998, 3110, 4942

Fats; fatty acids, alcohols, esters, oils, etc. - see references at Lipids

Firebrick - see Solid supports

Fischer-Tropsch synthesis
    861, 888-9, 1674-5, 6958

Fixed-bed operation - see Moving

Flame ionization, emission, photometry detectors (see also Ionization)
    50, 160-1, 287, 317, 385, 548, 696, 1013-4, 1166, 1392, 1401, 1491-3, 1626, 1784-5, 1817, 2069, 2110-2,
    2118-9, 2133-6, 2287, 2411, 2453-7, 2885, 2894-9, 2975, 3012, 3098, 3102, 3224, 3612-3, 3662-8, 3782,
    3972, 4018, 4496, 4828, 4842, 4845-7, 4916-7, 5004-5, 5019, 5039, 5268-72, 5326-7, 5448-50, 5597, 5611,
    5657-8, 5790, 5919, 6153, 6206-7, 6237, 6338-9, 6552-3, 6824, 7123, 7374, 7470-1, 7545

Flash exchange chromotography
    3343, 4250, 5607, 5722-4

Flavones and related compounds
    2028, 5126, 6104

Flavors (see also Odors; Alcoholic; Dairy; Food)

Flotation collectors
    6640

Flowmeters; flow reactors; f. rate; f. reversal
    142, 570, 771, 800-1, 846-7, 901-4, 1253, 2013, 2082, 2534, 2599, 2600, 2731, 2744, 2778, 3050, 3157,
    3476, 3624, 3809, 3892, 3907, 4084, 4209, 4601, 5258, 5732, 6014, 6025, 6092, 6097, 6181, 6203, 6350,
    7376

Fluorine and compounds (see also Halogens; Polyfluoro)
    41, 343, 374, 525, 833, 955, 1113, 1213, 1236, 1239, 1284-5, 1343, 1468, 1578, 1929-30, 1944, 1980,
    2147-8, 2188, 2225-8, 2252, 2322, 2431, 2445-6, 2651, 2687-9, 2743, 2755, 2761, 2798, 2879, 2936-8,
    2984, 3043, 3087, 3091, 3400, 3404, 3584, 3622, 3756, 3870, 3888, 4004, 4083, 4289, 4468, 4562, 4720,
    5050, 5108, 5222-4, 5255, 5309, 5381, 5442, 5465, 5471, 5509, 5572, 5642, 5763, 5807, 5860, 5886, 5912-3,
    6036, 6076, 6127, 6133, 6183, 6254, 6283, 6427, 6458, 6498, 6543-4, 6562-3, 6687-8, 6785, 7032, 7038,
    7240-1, 7472, 7565

Fog precipitator
    6825

Food chemistry (see also Agricultural; Alcoholic; Cocoa; Dairy; Lipids; Odors; Oils; Sugars)
    76, 233, 275, 333-4, 340-2, 346, 393, 427-8, 445-6, 465-8, 491, 500, 503, 519-20, 547, 624, 660, 670, 681,
    746, 807, 821, 843, 932-3, 993, 1114, 1154-5, 1163-4, 1172, 1183-7, 1230-1, 1270-2, 1315, 1375, 1395,
    1470-3, 1529-32, 1539, 1545, 1571, 1621, 1681, 1685, 1837-8, 1857-63, 1889, 1919, 1925, 1975, 2041, 2066,
    2084, 2163, 2171-2, 2174-7, 2205, 2283, 2308, 2372, 2438, 2476-7, 2507, 2530, 2652, 2681-2, 2713, 2771,
    2903, 2919, 3036, 3057, 3063, 3117, 3195-6, 3257, 3266, 3317, 3327-8, 3347, 3375, 3394, 3564, 3592-6, 3605,
    3630, 3755, 3798, 3858-60, 3865, 3874-7, 3930, 3940-1, 3972, 4035, 4134, 4144, 4245, 4304-5, 4346, 4385,
    4397, 4413, 4460-1, 4533, 4541, 4556, 4575-84, 4675, 4696, 4740-4, 4754-9, 4767, 4804, 4853, 4863-6,
    4884-6, 4891, 4937, 4995-7, 5026, 5175-6, 5221, 5248-51, 5289, 5298, 5307, 5316, 5325, 5331, 5361, 5373,
    5397, 5447, 5612, 5723-4, 5736, 5794-6, 5910-1, 5921-2, 5982, 6004, 6073-5, 6156, 6177, 6233, 6359,
    6371-8, 6395, 6492, 6507-8, 6519-23, 6597, 6626, 6653, 6678, 6686, 6761, 6792-3, 6823, 6848, 6868,
    7060-1, 7081-3, 7107, 7120, 7139, 7166, 7237-8, 7254, 7286-7, 7364, 7378-9, 7397-8, 7401-10, 7486, 7557,
    7572, 7575

Forensic science
    192, 675, 1208-11, 1395-9, 1704-6, 2218, 2327, 3150, 3641, 3988, 4340-1, 4351, 4521, 4572, 4581-3, 4648,
    5388, 5785, 5884, 5988, 6509, 7184

Formic acid and compounds (see also Chloroform)
    177-9, 685, 852, 926-7, 1345, 1605, 2260, 2393, 3023, 3170, 3897, 4480, 4839, 5319, 5489, 5605, 5734,
    5925, 6026, 6100, 6483, 6558, 6785, 7149, 7375

Fractionation (see also Collection; Separations)
    276, 321, 326, 413, 669, 1215, 1845, 1935-9, 2995, 3095, 3097, 3107, 3557, 3703, 3787, 3813, 4075, 4147,
    4182-3, 4436, 4610, 4966, 5125, 5414, 5436b, 5900, 5929, 6111, 6276-7, 6315, 6569-70, 6648, 6709, 6904,
    6943-5, 7187, 7310

Free electrons
    4497

Free radicals
    216, 787-8, 1082, 2234, 2362, 3358, 6432-3, 7156

Freidel-Crafts reaction
    5847, 5971, 7339

Fragrances - see Essential oils; Odors; Perfume

Frontal analysis
    1411-5, 2215, 4072, 6088-9, 6093, 6379, 6404, 7451
Fuels - see Coal; Natural gas; Petroleum; Refinery
Fumigants
    3128, 4718, 7276-7
Functional group analysis, response
    1312, 3190, 4883, 4898, 4924, 4929, 5449-50, 5919, 7147
Fundamental principles of G.C. (see also Review; Theory)
    1304-5, 1586, 2174-5, 3933, 3956, 3962, 5412, 5433
Furan derivatives
    496, 1291, 1763, 1884, 2601, 6465, 7201
Furnaces; f. atmospheres; coke ovens; etc.
    246, 394, 444, 987, 1042-3, 2274, 2621, 3003, 3838, 4821, 4958, 5348, 5515, 6132, 6696, 7070, 7379
Fusel oils
    535, 978, 2509, 3165, 3167, 3300, 3785-6, 5298, 5491, 6323, 6956, 7166, 7436

G

Gallium and compounds
    5050
Gas analysis (see also Hydrocarbons; Natural g.; Permanent g.; and specific names)
    263, 309, 470, 603, 694, 711-2, 754-8, 762, 882, 893, 987, 1153, 1173, 1365, 1402, 1407, 1447, 1536,
    1561-2, 1596-602, 1636, 1649, 1703, 1718, 1745, 1800, 1806, 1850, 1932, 1968, 2050, 2206-8, 2255-7,
    2274, 2294, 2610, 2635, 2750-1, 2929, 3131, 3176, 3180, 3339, 3381, 3477, 3492-7, 3517, 3543-51, 3587,
    3615, 3726-9, 3813, 3857, 3870, 4046, 4063, 4110, 4151, 4163, 4177, 4205, 4243, 4319, 4329, 4348-9, 4404,
    4439, 4463, 4591, 4630, 4752, 4788, 4821, 4894, 4948-9, 5079, 5183, 5217, 5297, 5349, 5369, 5415, 5464,
    5493, 5522, 5546, 5677-8, 5728, 5747, 5864, 5923-5, 6002, 6021, 6120-3, 6164, 6175, 6235, 6253, 6266-7,
    6431, 6452-3, 6473, 6601-2, 6681, 6693, 6699, 6905, 6922-3, 6928, 7026-7, 7094-5, 7126, 7164, 7265,
    7290, 7370-2, 7415, 7451-4, 7512
Gas density detectors, balances, and meters
    209, 1867, 1912, 3650, 3662, 3666, 4151, 4163, 4393, 4674, 5080, 5090, 5209, 6153, 6550, 7208, 7430
Gas phase reactions
    1936, 2707, 3924
Gas-solid chromatography
    803, 1597, 1897, 1987, 2050, 2465, 2551, 2574, 2746, 2783, 2905, 3262, 3326, 3509, 4006, 4994, 5017, 5314,
    5580, 6199, 6200-2, 6296, 6927, 7558
General principles - see Fundamental
Geological applications (see also Metals; Oil; specific elements; etc.)
    499, 1241, 2733-4, 3066, 3487, 5670, 6455, 6926, 7065
Germanium and compounds
    5, 250, 948, 2693, 5503-6, 6119
Glass; g. beads; g. columns
    864, 1127, 1748, 1798, 1879, 2355, 2552, 2563, 2578, 3171, 3256, 4331, 4440-3, 4566, 4968, 5232, 5253,
    6069, 7501
Glucose and related compounds (see also Sugars)
    406, 838, 840, 1052, 2844, 3995, 4125, 5444, 6159, 6991
Glycerides (see also Triglycerides)
    2407, 3599, 4806, 6932
Glycerides, glycerols, etc. (see also Lipids)
    159, 466, 856, 878, 890, 1444, 1573, 2829, 2911, 3260, 3302-4, 3604-5, 4040, 4744, 5324, 5423, 6238, 6438,
    6558, 6932, 6959, 7005, 7479-80
Glycols, glycose compounds
    40, 181, 339, 559, 841, 1272, 1444, 2028, 2535, 2606, 2829, 3259, 4758, 4816, 5116, 5131, 5441, 5449-50,
    5830, 6465, 6476-7, 6481, 7389-90
Golay columns; g. equation (see also Capillary columns)
    1401, 1487-9, 1700-1, 2113-4, 2117, 2120, 2123, 2126, 2132, 2140, 2597, 5413
Gomberg reaction
    7181-2
Grignard reagent
    1295-8, 2816, 2917-8, 3959, 4941, 5078, 5812, 6053

Hydrogen; hydrogenation; hydrogenolyses

57, 69, 230, 266, 344, 434, 438, 474, 515, 619-20, 843, 850, 865-6, 896, 928-9, 974, 979, 1056, 1195-6, 1221, 1269, 1394, 1467, 1483, 1588, 1928, 1987-8, 1992, 2080-1, 2090, 2237, 2274, 2405, 2413, 2422, 2442-3, 2458, 2465, 2641-2, 2683-4, 2711, 2741, 2865-6, 2976, 2992, 3003, 3162, 3290, 3340, 3370, 3485, 3488, 3491, 3532-4, 3554-5, 3664, 3671, 3794, 3802, 3846-8, 3853, 3860, 3887, 3991-2, 4157-9, 4221, 4227, 4231, 4242, 4309, 4408, 4477, 4516, 4533, 4545, 4591, 4647, 4813, 4910, 4950-1, 4961, 4989, 4994, 5004-5, 5015-7, 5044, 5053, 5103, 5231, 5301-2, 5329, 5360, 5416, 5513, 5530, 5570, 5609, 5652, 5656, 5677-8, 5747, 5803, 5811, 5856, 5883, 5909, 5950, 6013, 6048, 6058-9, 6083, 6091, 6128, 6137-40, 6183, 6189, 6252, 6284, 6296, 6319-20, 6385-7, 6415-6, 6443, 6455-7, 6565, 6571, 6641, 6728, 6755, 6759, 6767, 6818-9, 6852, 6862-4, 6915-6, 6931, 6950-3, 6965, 7009, 7029, 7072, 7084, 7122, 7168, 7206, 7247, 7283, 7305, 7374, 7388, 7478, 7496, 7507, 7541

Hydrogen halides

70, 3236, 3400, 3404, 5103, 6014, 6076, 6743

Hydroxamic acid

6973

Hydroxy acids

502, 558, 842, 1610, 2350, 2688, 4020, 4928, 5036, 6973

Hydroxy(1) compounds, miscellaneous

395, 502, 523, 1184, 1475, 1610, 2688, 3751, 4006, 4010, 4928, 5126-8, 5306, 5716, 5852, 6467, 6504, 6876, 7345

Hygiene, industrial (see also Air)

160, 792, 3890, 4179, 4633, 4733, 5961, 6742

Hypersorption

716-7, 754-8, 2092, 2377, 6537

I

Identifications - see Separations; also individual listings

Indium and compounds

5050

Induction heating

6703-4

Industrial hygiene; i. toxicology

792, 3890

Inert gases - see Rare gases and specific names

600, 980, 2450, 3896, 5358

Infrared applications, detectors, gas cells

146-50, 322, 405, 705, 792, 834, 1332-3, 1392, 2016-8, 2218, 2236, 2455-6, 2729, 2836, 2847, 3033, 3068, 3086, 3094, 3193, 3201, 3577, 3803-4, 4023, 4320, 4394, 4530, 4632, 4999, 5228, 5875, 5964, 6153, 6242, 6251, 6260, 6265, 6277, 6295, 6448, 6569-70, 6709, 6963, 7127, 7271, 7310-3, 7325, 7466, 7576

Injection systems and devices (see also Sample; Syringes)

4, 1417, 1424, 1908, 1957, 2061, 2749, 2824, 3126, 4512, 4597, 4770, 4920, 5130, 5177, 5436b, 5602, 5771-2, 5784, 6019, 6028, 6285, 6314, 6360, 6703

Ink, printing

621, 3030

Inlet systems; i. volume; i. pressures (see also Sample)

1702, 1983, 2863, 3465

Inorganic applications (see also specific elements and compounds)

99, 100, 1636, 3747, 3766, 4188, 4348-9, 4894, 5217, 5503, 5506, 5864, 5958, 6619

Insecticides - see Agricultural

Instruments - see Apparatus and specific types

Integrators; integral detectors, counters

37, 324, 887, 918, 1431, 1655, 1687, 1958, 2466-7, 2890-3, 2909, 3287-9, 3411-2, 3588, 3657, 3662, 3898, 4201, 4329, 4764, 4992, 5455-6, 6068, 6192, 6376, 6472, 6610-2, 6998, 7192, 7297-8, 7431

Internal standards

645, 2062, 3051, 3372, 4303, 4342-3, 5124, 6467, 6593

Iodine and compounds (see also Halogens)

2711, 2983, 3535-6, 4405, 4532, 4824, 4830, 5387, 6076, 6268, 6486, 6501, 6563

Ion sieves - see Molecular sieves

Ionization detectors; i. chambers; i. gauges; i. potentials (see also Flame)

286, 336, 541, 608, 784, 962-3, 1126, 1193, 1420, 1423, 1477, 1491, 1700-1, 2050, 2095-6, 2151, 2195,

2218, 2349, 2459, 2496-7, 2646, 2680, 2801-3, 2817-20, 2848-51, 3098, 3159-60, 3660-3, 3670, 3826-7, 4422, 4489-96, 4504, 4575, 4581-3, 4653, 4721, 4726-8, 4826, 5000, 5194, 5262-3, 5355, 5362-3, 5379, 5760, 5977-8, 6125, 6209, 6266-7, 6331, 6549-50, 6578, 6998, 7014-5, 7204, 7344, 7362-3, 7384-7, 7440-1, 7470-1, 7545

Iridium compounds
1136
Iron; alloys and compounds
1467, 2460, 2888, 3732-3, 3837, 3848, 4168-9, 4636, 6592-3, 6952-3
Iso- - see normal compound
Isomerism - see Stereochemistry
Isoprene and related compounds
383, 709, 1575, 1891, 1986, 3686, 4191, 5939, 6500, 6632, 7507
Isotherms
1413, 1593, 1597-9, 2212-4, 2365, 2630-2, 2760, 2870, 3213, 3478, 5315, 5855, 6088-9, 6093, 6200, 6379, 6892, 7121, 7505
Isotope analysis; i. exchange
974, 1049, 1135, 1347, 2637-8, 4871, 6715, 6879, 7360

K

Katharometers - see Thermal
Ketones; ketenes
438, 639, 856, 899, 1093, 1282, 1295, 1300, 1327, 1475, 1485, 1509-10, 1605, 1628, 1875, 2043, 2047, 2340, 2404, 2424, 2857, 2904, 2917, 3024, 3089, 3172, 3272, 3371, 3803, 3853, 3870, 3894-5, 4135, 4236, 4833, 4888-9, 4894, 4961, 5060, 5063, 5216, 5226, 5282, 5328, 5547, 5706, 5724, 6099, 6159, 6294, 6363, 6428, 6467-8, 6493, 6499, 6525, 6562, 6647, 6687-8, 6985, 7257, 7475-6
Kinetics
86-7, 521, 1046-7, 1222-3, 1366, 1476, 1937, 1992, 2459, 2544, 2547, 2551, 2683-4, 2732, 2841, 2845, 3038, 3139, 3854, 3745, 4050, 4083, 6485, 7504
Krypton
433, 2636, 2640, 2747, 3543-4, 4153, 4158-9

L

Labeled compounds - see Radiochemistry
Laboratory set-up and operation
231, 368, 6102
Lactic acid; lactones
616, 1114, 1136, 1230, 3751, 4773, 6159
Lacquers; l. solvents - see Paints
Lead and compounds
5, 560, 1689, 1708-9, 3756, 4286, 4500-1, 4506, 4773, 5390-1
Leather industry (see also alphabetical listings)
1357-8, 4873
Lecithin
2736, 5665
Lignin - see Cellulose
Linoleic, linolenic acids and compounds
690, 966, 1144-5, 1992, 1995, 2057, 2515, 2764-5, 2932, 3197-8, 3867, 4709, 5096, 5628, 5779, 6135-40, 6492, 7405
Linseed oil
2485, 3357, 3860
Lipids; fatty acids and compounds (see also Biochemical, -medical; Cholesterol; Cocoa; Dairy; Glycerides and Tri-; Linoleic; Olefins; Oleic; Steroids)
9, 13-16, 20-5, 56, 61, 76-7, 84, 159, 176-9, 213-7, 238, 251, 283-4, 295, 312, 338-42, 348, 374, 381, 392-3, 400-3, 420-2, 436, 455, 465-8, 472-6, 487-9, 500-7, 524, 536, 546, 551, 558, 606, 613, 670-4, 685, 708, 725, 730, 738, 768, 778, 816, 829-31, 842-5, 859-60, 871-7, 898, 906, 932, 964-7, 970-6, 982-4, 995-9, 1044-5, 1125, 1144-5, 1170, 1183, 1216, 1220, 1259-61, 1270-2, 1302, 1322, 1331, 1338, 1345, 1361-7, 1387, 1404, 1437-8, 1462, 1482, 1500, 1529-32, 1563-5, 1571-3, 1583, 1614, 1643-6, 1672, 1684, 1705, 1727, 1740-7, 1829-31, 1836-7, 1878, 1890, 1922, 1971, 1995-6, 2024, 2033, 2041, 2054, 2058, 2063, 2069, 2089, 2127, 2150, 2155, 2163, 2168-72, 2193-5, 2199, 2217, 2267, 2272-7, 2300-3, 2350-1, 2372,

2426-8, 2464, 2481, 2502-6, 2510, 2516, 2528-32, 2596, 2619, 2713, 2736-40, 2772-4, 2821, 2846, 2850-4, 2903, 2911-5, 2931-4, 2940-2, 2953-60, 2969, 2979, 2996-7, 3022, 3053-60, 3067, 3072, 3088, 3115-7, 3133, 3140, 3196-8, 3230, 3244, 3250-66, 3285, 3372, 3384-5, 3397, 3406, 3433-7, 3443-4, 3450-4, 3464, 3471-4, 3561-8, 3599-607, 3616-9, 3673-4, 3695-9, 3717, 3799, 3819, 3824, 3828, 3852-68, 3877, 3958-61, 3977, 3984, 3989, 4009, 4020-8, 4039-41, 4104, 4126, 4133, 4174-5, 4180, 4220, 4249, 4260-3, 4290-3, 4304, 4314, 4326-7, 4382-90, 4410-1, 4416, 4424-6, 4432-9, 4483-6, 4528-31, 4541-3, 4551, 4567-9, 4594, 4619-29, 4639-43, 4651, 4676, 4681, 4696-701, 4709, 4722, 4742-5, 4754-5, 4767, 4798-9, 4848-51, 4862, 4868, 4873, 4896-902, 4909, 4943, 4977-9, 5024-30, 5034, 5042, 5046, 5091, 5120-4, 5131, 5168, 5198, 5205, 5218, 5233, 5276, 5284, 5295, 5307, 5333-4, 5368, 5375-7, 5397, 5403-6, 5423-4, 5435-8, 5441, 5557, 5565, 5613-4, 5620, 5628-32, 5646, 5650, 5665-7, 5673, 5717, 5735, 5767-8, 5776, 5784, 5800, 5805-6, 5813-4, 5829, 5836, 5911, 5921-2, 5931-2, 5951, 5975, 6000-1, 6015-8, 6023, 6046, 6078, 6105-9, 6116, 6134-6, 6157-63, 6186-7, 6232-3, 6238, 6282, 6300-1, 6352, 6370-1, 6393, 6407, 6424-5, 6438, 6464-7, 6532, 6536, 6540, 6587-91, 6626, 6653-7, 6671-7, 6686, 6722-4, 6745-7, 6761, 6803-4, 6814, 6835, 6846, 6868-73, 6932, 6957, 7013, 7022, 7055-7, 7079-80, 7087, 7109, 7158-9, 7165, 7169-70, 7175, 7245, 7249-54, 7329, 7333, 7342, 7349, 7396-8, 7401-3, 7417, 7423-4, 7479-80, 7499-500, 7539

Lipoproteins; lipolysis
75-77, 4126, 5218
Liquid gases - see Cryogenic
Liquid phase - see Stationary
Liquids, analysis of
667, 1116, 1352, 1432, 4407, 5105, 5452, 6346
Literature - see Bibliographical
Lithium and compounds
496, 539, 604, 727, 1446, 1682, 2046-7, 2395, 3039-41, 5016-8, 5282-3, 5598, 7157
Lubricating compounds
951, 2059, 5603, 5626
Lunar chromatographs - see Space

M

Magnesium compounds
1604, 2395, 3091, 7266
Manganese compounds
538, 6729
Manometric analyses
5546, 6120
Manufactured gases - see Natural
Marijuana - see Narcotics
Mass spectrometry
189, 210, 627, 811-4, 822, 1122, 1392, 1667, 1858, 1872, 1901, 1909-10, 1938-9, 2454, 2660-2, 2836, 2913, 3105-8, 3201, 3212, 3283, 3721, 3808, 4033, 4187, 4360-4, 4408-9, 5337, 5515, 5833, 5903, 6154-5, 6254, 6336-7, 6505, 6531, 6909, 7119, 7221, 7239, 7292
Mass transfer
28, 2565, 3065, 3967, 5452-4, 5686, 5769
Mathematical models
2435-7
Measurements; measuring devices
912, 1597-9, 2287, 3187, 3713, 3556, 3776, 3892, 4491, 5123, 5462, 6014, 6409, 7443
Medical applications - see Biomedical; etc.
Menthanes; menthenes; menthols; menthones
559, 1895, 2614, 3121, 3268, 3308, 3751, 4057, 5012, 5144, 5159, 5212, 5482, 5591, 5949, 6717, 6791
Mercaptans
32, 1266, 2178, 3815, 4344, 4384, 4888-9, 5342, 5976, 6204, 6473, 6487, 6628, 6643, 7212
Mercury; m. compounds; uses of
356-7, 490, 701, 2036, 2758, 4064, 4612, 5261, 5653, 6648
Metallurgical applications (see also Metals and specific listings)
270, 529-30, 830, 1467, 1726, 1928, 2208, 2460, 2941, 3732-3, 3837, 3847-8, 4168-9, 4350, 4378, 4913-5, 5007, 6123, 6173, 6651, 6725, 6952-3, 7247
Metals; minerals; rocks (general); gases in (see also specific names)
145, 259, 270, 481-2, 529-30, 588-9, 825, 830, 929-30, 1277, 1300, 1726, 2049, 2207-8, 2538, 2943, 3163, 3578, 3580, 3746, 3749, 3901-4, 4146, 4168-9, 4350, 4378, 4913-5, 5007, 5011, 5050, 5081, 5969-71, 6123, 6188, 6321, 6651, 6725, 7255

Methanes; methoates; methoxy compounds

67, 156, 370-1, 743, 761, 1055, 1239, 1368-70, 1526, 1809, 2090, 2376, 2445, 2458, 2524-5, 2687, 2743, 2984, 3272, 3358-9, 3487, 3532-4, 3671, 3714, 3870, 4064, 4107, 4153, 4158-9, 4231, 4239, 4291, 4310, 4450, 4453, 4475-7, 4910, 5085-6, 5129, 5331, 5521, 5571, 5638, 5652, 5662, 5713, 5732, 5747, 5838, 5886, 5900, 6128, 6234, 6254, 6273-4, 6283, 6457, 6681, 6728, 6759, 6767, 6852, 6931, 7011, 7089-90, 7104, 7205-6, 7248, 7448, 7457, 7556

Methanol

339, 729, 921, 1674, 3863, 3870, 4064, 4173, 4258, 4680, 4708, 4989, 5387, 5489, 5600, 5734, 6012, 6691, 7439

Methyl compounds; methylene, etc. (see also Trimethyl)

14, 40, 91, 143-4, 353, 375-7, 412, 419, 441, 493, 522, 534, 546, 594, 668, 686, 741, 745, 783, 809, 841, 852, 870, 900, 929-30, 959, 1102, 1107, 1118, 1180-1, 1237, 1265, 1286, 1301, 1339, 1341, 1343, 1346, 1356, 1479, 1483, 1511, 1518, 1559, 1605, 1622, 1627, 1682, 1709, 1714, 1746, 1754, 1831, 1881-7, 1999, 2000, 2033, 2040, 2248-50, 2253, 2260-4, 2288-90, 2351-3, 2378, 2380-2, 2387, 2396, 2469, 2483, 2492, 2499, 2501-2, 2517-8, 2587-91, 2694-5, 2718-9, 2764-5, 2823, 2832, 2933, 2958, 2989, 3028, 3004, 3060, 3120, 3132, 3161, 3181-2, 3201, 3215, 3225, 3256, 3269, 3280-1, 3308, 3309, 3315, 3328, 3338, 3358, 3360, 3395-6, 3458, 3460-7, 3524, 3617, 3648, 3695, 3701, 3804, 3840, 3878, 3886, 3913, 3926, 3975, 3981-2, 3993, 3999, 4002, 4010, 4022, 4025, 4036, 4064, 4092-3, 4117-9, 4121, 4188, 4229, 4248, 4255, 4262, 4314, 4322, 4390, 4410, 4438, 4453, 4456-7, 4478, 4480, 4531, 4535, 4614, 4644, 4652, 4679, 4709-11, 4722, 4738, 4748, 4836, 4840, 4910, 4928, 4940, 4955, 4956, 4978, 4980, 5011, 5042, 5063, 5093-5, 5116, 5120, 5124, 5132, 5180-1, 5185-6, 5211, 5246, 5248, 5256, 5264, 5306, 5333-4, 5365, 5422, 5453, 5489, 5531-3, 5614, 5617, 5628-9, 5653-4, 5668, 5694-5, 5705-6, 5715, 5718, 5721, 5733, 5784, 5801-2, 5830, 5834, 5852, 5915, 5934, 5966, 5971, 5976, 5980, 5984, 6013, 6036, 6137, 6108, 6182-4, 6234, 6236, 6268, 6297, 6304-5, 6312, 6318, 6340-1, 6353-4, 6501, 6510, 6535-6, 6573, 6590-1, 6605, 6609, 6635-6, 6691, 6720, 6732-4, 6809-10, 6815, 6822, 6829, 6871, 6875, 6881, 6914, 7080, 7110, 7111, 7140, 7153, 7157, 7193, 7240-1, 7267, 7294, 7302, 7339, 7375, 7417, 7466, 7481, 7575, 7577

Microanalyses - see Trace

Microbiological applications - see Alcoholic; Biochemical; Biomedical

2882, 3856

Microcoulometry - see Coulometry

Microwaves, use of in G. C.

3621

Migration methods

2541, 4959-60

Milk - see Dairy

Mixed-bed columns

5773

Mixtures - see Separations

Molecular sieves; ion sieves (see also Zeolites)

45, 99, 100, 167, 315, 380, 474, 564-6, 570-9, 592, 599, 600, 924, 1035-7, 1094-5, 098, 1239, 1760, 1947-52, 2001, 2012, 2030, 2129, 2352, 2731, 2782, 2840, 3127, 3533, 3706, 4014-6, 4255, 4317, 4538, 4610, 4994, 5235, 5249, 5263, 5290-1, 5426, 5467, 5904, 5929, 6286, 6851, 6929, 7274, 7503, 7532

Molecular weight determination

727, 2183, 3074-5, 4392-3, 4927, 5201, 5290, 5503-6, 7443

Molybdenum and compounds

2759, 5530

Monoalkyl benzenes

849, 4553, 5535

Monoethanolamine

3579, 6014

Monitoring - see Automatic

Monomers

6269, 6271, 6949

Moving and fixed-bed operation

28, 714, 2278

Multiple columns (see also Dual)

514, 1619, 3624, 4970, 6008, 6083, 7427

Naphthas; naphthalenes; and derivatives
89, 482, 863, 1353, 1409, 1799, 1841, 2026, 2415, 2470, 2724, 3291-3, 3356, 3420, 3459, 3913, 3952, 4267, 4342-3, 4464, 4894, 5058, 5264, 5624, 5791, 6129, 6179, 6345, 6441-3, 6975, 7125, 7575

Narcotics (see also Pharmaceutical)
1704-5, 3136, 3884, 3988, 4340-1, 5785

Natural gases; fuel g.; manufactured g.
6, 78, 162, 943-5, 1454, 1806, 2083, 2200-1, 2221, 2425, 2613, 2730, 3185, 3487, 3628, 3992, 4905, 4932, 4935, 5061, 5086, 5510, 5670, 6146, 6642, 6696, 6749, 6757, 6862-4, 6899, 7033, 7487, 7564

Nef reaction
7191

Neon; n. lamps
2626, 2638, 2746-7, 3485, 3491, 3543-4, 3554-5, 3802, 4157-9, 5374, 5544, 7533-4

Neutron irradiation - see Radiochemistry

Nickel and compounds
538, 1276-7, 3400, 6040, 6069, 7200, 7496

Nicotine - see Tobacco

Niobium and compounds
3901-4

Nitric acid; nitrous acid
410, 639, 955, 2249, 3316, 4928, 6329

Nitrogen and gaseous compounds
38, 41, 109, 266, 473, 522, 884, 1034, 1952, 2077, 2189, 2216, 2283, 2451, 2732, 2747, 2753, 2943, 2983, 3147-8, 3213, 3532-4, 3547-50, 3576, 3579, 3671, 3760-1, 3802, 3818, 4153, 4158-9, 4237, 4242, 4288, 4317, 4378, 4533, 4647, 4713-4, 4928, 4989, 5025, 5084-8, 5108, 5130, 5231, 5280, 5374, 5400-1, 5451, 5462, 5479, 5521, 5677-8, 5713, 5775, 5782, 5807, 5845, 5883, 6006, 6014, 6076, 6083, 6126, 6171, 6261, 6330, 6398, 6426, 6566, 6681, 6700, 6729, 6743, 6755, 6827, 6852, 6880, 6915-7, 6952-3, 7122, 7204-6, 7345, 7430, 7445, 7530-1

Nitrogen-containing compounds (see also Agricultural; Dyes; Heterocyclic; and specific compounds)
12, 67, 351, 355, 370-1, 473, 539, 548, 680, 688, 736-7, 750, 795-9, 826, 955, 991, 1087, 1235, 1311, 1339, 1405, 1408, 1469, 1476, 1543, 1730-2, 1816, 1923, 2166, 2246, 2338, 2433, 2461, 3002, 3024, 3086, 3094, 3120, 3137, 3171, 3199, 3272, 3442, 3522, 3591, 3704, 3714, 3750, 3759, 3921, 4415, 4456-7, 4467, 4558, 4715, 4761, 4797, 5042, 5106, 5110-1, 5186, 5216, 5237, 5265, 5310, 5341, 5387, 5425, 5573, 5625, 5725, 5780-1, 5801-2, 5830, 5963, 6077, 6106, 6133, 6240, 6430, 6509, 6545, 6562-4, 6609, 6627, 6636, 6696, 6785, 6874, 6931, 7006, 7032, 7191, 7246, 7269-70, 7411, 7437, 7472

Nomenclature in G.C.
122-4, 432, 3712, 3773, 7045

Nuclear (proton) magnetic resonance applications
2243, 3333, 4718

Nuclear processes - see Radiochemistry

Oceanographic and limnological applications
6109, 7336

Octadec- compounds
501, 1363, 2429, 2956, 3173, 3231, 3855, 3859, 4325, 4939, 6016, 6137

Octanes, octenes, and compounds
410, 707, 724, 1286, 1349, 2441, 3197, 4566, 5457, 6473

Odors and flavors (see also Essential oils; Food)
111, 190, 228, 445-6, 520, 636, 656, 660, 774, 821, 1146, 1163, 1183-5, 1267, 1331, 1373, 1442, 1457, 1470-3, 1539, 1711-2, 1857-63, 1919, 1924, 1975, 2066, 2177, 2205, 2306-12, 2953, 3036, 3134-5, 3177, 3255-7, 3582, 3592-6, 3858, 3941, 3972, 4346, 4363, 4461, 4581-4, 4626, 4740, 4863, 4884-6, 5175-6, 5221, 5250, 5417-22, 5445, 5463, 5612, 5659, 5723, 6081, 6324, 6473, 6505-7, 6519-20, 6559, 6597, 6608, 6626, 6686, 6782, 6792-3, 6809-10, 7000-1, 7081, 7155, 7167, 7557

Oil prospecting; oil field uses
497, 1241, 1365, 2042, 2733-4, 3499, 4404, 4627, 5670, 5890, 5895, 6454

Oils - see specific types, also Agricultural; Essential oils; Foods; Lipids; Petroleum; Refinery

Oils, animal
690, 1123, 2372, 3397, 3561, 4039-41, 4253, 4623, 5029, 6587

Oils, seed (see also Agricultural; Food; Linseed)
392, 546, 671-2, 1362-4, 2163, 2266, 2372, 2476-7, 2713, 2920, 3117, 3229, 3231, 3564, 3755, 3860, 5028, 5036, 5373, 5666, 5910, 6370, 6393, 6463, 6492, 6804, 6848, 6868, 7060-1, 7388, 7401

Peptides (see also Biochemical)
823, 3585, 7242

Perfluoro-, perhalo- compounds (see also Fluorine; Halogens)
1931, 2799, 3044, 3756-7, 4638, 5382, 5509, 5652, 5804, 5934, 6103, 6390, 6785

Performance index for columns; p.-number scale
941, 2665

Perfume industry (see also Aromatics; Essential oils)
111, 298, 317-8, 369, 630-2, 848, 1377, 1608, 1869, 2085, 3216-20, 3582, 3675, 3920, 3987, 4140-1, 4172, 4234-5, 4465, 4615, 4620, 5139, 5153, 5869, 5926, 6299, 6594, 6780-1, 6806-11, 6895, 7249, 7335, 7382, 7391-3, 7498

Periodic table
2159

Permanent gases (see also Gas; Natural)
41, 432, 600, 603, 608, 784-5, 2050, 2169, 2459, 2647-8, 2731, 3367, 4261, 4435, 5379, 6700, 6728, 7344

Permeability
5536, 5777

Peroxides; peroxy compounds
7, 853, 857, 1138, 1156-7, 1165, 1213-4, 1676, 2003, 3207, 3361, 3368, 3868, 3997, 4683, 4822-3, 5193, 5536, 5674, 6743, 7069, 7144

Perturbation velocities
6511

Pesticides (see also Agricultural)
1, 4, 183, 269, 329, 334, 680, 684, 807, 956-7, 1160-1, 1307, 1317, 1326, 1422, 1545-8, 1550-3, 1811, 1873, 2039, 2698, 2831, 2846, 2998-9, 3154-6, 3638, 5114, 5325, 5788, 6096, 6177, 6372, 7167, 7185, 7226-7, 7571-4

Petroleum industry (see also Hydrocarbons; Lead; Natural gas; Oil; Refinery; and names of specific products)
44-8, 114-5, 130, 161, 197, 231, 245, 277, 282, 313-6, 330-2, 411, 497, 562, 709, 719, 793, 885-6, 943, 951, 1017, 1077, 1081, 1094, 1188, 1353, 1407, 1451, 1479-81, 1533-5, 1635, 1667, 1709, 1755-6, 1795, 1801-2, 1840-3, 1960-3, 1981-3, 2025, 2029-34, 2042, 2059, 2150, 2282, 2297-8, 2323-5, 2328, 2401, 2410-4, 2604, 2733-4, 2807, 2916, 2962-3, 3002-3, 3082-3, 3112, 3136, 3141-2, 3265, 3283, 3382-3, 3499, 3574, 3691-2, 3706, 3775, 3835, 3918-9, 3943, 3978-82, 4065, 4105, 4109, 4205, 4214, 4225, 4278, 4286, 4289, 4352-5, 4360-6, 4396, 4402-6, 4456-9, 4464, 4471-3, 4506, 4521, 4546, 4627, 4690-3, 4753, 4794-6, 4814-5, 4870, 4892, 4906, 4932-9, 4965, 4989-90, 5061, 5203, 5258, 5277, 5290-1, 5366, 5389-94, 5437, 5461-7, 5480, 5501, 5551-9, 5579, 5603, 5606-8, 5622, 5626, 5660, 5666, 5891, 5924, 5985, 6182-5, 6290, 6298, 6345, 6367-8, 6454, 6623, 6694-5, 6740, 6749, 6754-5, 6787, 6797, 6829, 6865, 6891, 6924, 6943-5, 6958, 7026-7, 7046, 7052-3, 7123, 7172-3, 7179-80, 7259, 7274-5, 7304, 7331-2, 7368-9, 7428, 7489, 7491, 7519, 7522, 7532, 7538

Pharmaceutical applications (see also Biomedical; Narcotics; Toxicology)
140, 258, 607, 649, 762-3, 986, 1088, 1264, 1328, 1506, 1616, 1970, 1987, 1998, 2185, 2279-80, 2334-6, 1607, 1704, 1834, 1872, 1895-6, 2065, 2111, 2184-5, 2190, 2299, 2313-6, 2613-5, 2793-4, 2799, 2844, 3136, 3171-5, 3191, 3370-4, 3379, 3430, 3539-40, 3581, 3585, 3793, 3870-2, 3884, 4188, 4219, 4340-1, 4452-3, 4581, 4592, 4682, 4997-8, 5196, 5376, 5383-8, 5465, 5627, 5700-3, 5707, 5766, 5785, 5884, 6039, 6104, 6272, 6442, 6509, 6594, 6833, 6989, 7202, 7210, 7537

Phase equilibrium
6512

Phenanthrenes
5540, 6458-9

Phenols and derivatives (see also specific compounds)
140, 258, 607, 649, 762-3, 986, 1088, 1264, 1328, 1506, 1616, 1970, 1987, 1998, 2185, 2279-80, 2334-6, 2416, 2484, 2535, 2728, 2794, 2799, 2968, 3090, 3171, 3481, 3524-8, 3840-2, 3889, 4127-8, 4167, 4682, 4793, 5428, 5661, 5742, 5860, 6024, 6041, 6121, 6436, 6446-7, 6470, 6560, 6823, 6850, 6955, 6989, 7018, 7069, 7194, 7426

Phenyls; biphenyls; phenylation
10, 64-5, 376, 486, 645, 934-5, 1300, 1320, 1408, 1513, 2315, 2344-8, 3103, 3137, 3171, 3199, 3319, 3539, 3648, 4143, 4477, 4710, 4799, 4987, 5093, 5126, 5207, 5543, 5909, 6103, 6441, 6475, 6536, 6559, 6670, 6683, 6823, 6989, 7183, 7222, 7309

Phosphorus and compounds (see also Agricultural; Biomedical)
2, 465-6, 487-8, 968, 1165, 1340-1, 1391, 1695, 1752, 1770, 2193, 2211, 2274, 2472, 2521-3, 2605, 2695-6, 2736, 2795-6, 2827, 2941, 3001, 3053-8, 3143, 3189, 3306, 4040, 4043, 4104, 4259, 4318, 4345, 4449, 4567-9, 4641-3, 4900, 4936, 5197-9, 5476, 5665, 5741, 5881, 5931-2, 6014, 6052-4, 6297, 6424, 6514, 6536, 6572, 6785, 7145, 7254, 7301, 7340

Photochemistry
151, 440, 701, 1059, 2077, 2377, 2663, 2838, 3101, 3651, 3731, 3750, 3779, 3923, 3925, 4024, 4027, 4495, 4496, 4611, 4822-4, 4838, 4945, 5084, 5261, 5600, 5967, 6171, 6221, 6494-6, 6541, 6550, 6814, 7078, 7299

Phthalic compounds
42-3, 1564, 1799, 2166-7, 2366, 2682, 3337, 4117, 4276-7, 4940, 5010, 5441, 6527, 6629-30, 7475
Phthioic acid derivatives
5983-4
Physical, physico-chemical methods
1587-8, 1592, 1759, 2906, 3710, 5480a, 5528-9, 5681, 6021, 6832, 7443
Physiological measurements - see Biomedical
Pigments - see Paints
Pilot plants, G. C. in
297, 3809
Piperic acid derivatives
1310, 3791, 5229
Plant sciences - see Agriculture
Plastics; polymers; polyphenyls (see also Resins; Rubber; Vinyls)
143-4, 172, 237, 242, 302, 442, 526, 550, 561, 635, 768, 1007, 1078, 1352, 1416, 1459, 1498, 1707, 1723, 1980, 2017, 2022, 2023, 2091, 2106-7, 2265, 2383, 2472, 2719, 2823, 3004, 3021, 3026-31, 3110, 3136, 3201, 3399, 3507, 3597, 3691-3, 3861, 3865, 3974, 4250, 4303, 4321-2, 4324, 4375, 4476, 4711, 4717, 4808, 4882, 4924-9, 4940-2, 4983, 5083, 5122, 5213-4, 5246, 5274, 5396, 5439-41, 5509, 5538, 5602, 5607, 5661, 5715, 5720, 5817, 6034, 6269, 6271, 6515, 6605, 6719, 6732, 6949, 7016, 7025, 7073-5, 7151, 7193, 7309, 7459, 7568-70
Plate height; p. number; p. temperature; HETP
798-801, 912, 1699, 1870, 1965, 2252, 2543, 2547-71, 2579, 2583, 2633, 2898, 3711, 3907, 3964-7, 3971, 4103, 4129, 4331, 4814, 5258, 5454, 5686, 5855, 5992-5, 6219-20, 6291, 6567, 7231, 7504
Platformate
4592
Platinum and compounds
1276-7, 2474, 3952, 5533, 6059, 6319
Podbielniak analysis
129, 6193, 7173
Polarography; polar compounds
63, 262, 563, 566, 596, 808, 1868, 1986, 2071, 2514, 3171-2, 3305, 4038, 4072, 4287, 4479, 4606, 4611, 5183, 5330, 5345, 5882, 5914, 6110, 6644, 7400, 7443
Pollution - see Air
Polycyclic compounds
1953, 4101
Polyenes; polyenoics; polyester, polyether, polyethylene compounds
40, 66, 187, 561, 1322, 1337, 1540, 1855, 2528, 2765, 3245, 3264, 4042-4, 4525, 4907, 4984, 5117, 5129, 5213, 5441, 5692, 6968
Polyfluoro- compounds (see also Trifluoro-)
525, 833, 5223
Polyglycols
39, 6238, 6785
Polyhydric, polyhydroxy compounds
742, 3015, 4984, 6238, 6383
Polymers - see Plastics; Rubber
Polyols
2098, 4555, 7295
Polyphenols - see Phenols
Polyphenyls - see Phenyls
Polysaccharides - see Sugars
Polyvinyls
3258, 6972, 6986
Pore radius; p. volume
5851, 6097, 6455
Porphyrins
4045, 5031
Potassium and compounds
576, 2936, 4107, 4320
Prediction of data
248, 2156-7, 2869, 4606, 5893, 7473
Pregnanediol; pregnenes; etc.
1507, 3698, 5200, 6933, 7422

Preparative chromatography
   10, 19, 72, 121, 132, 197, 264, 382, 424-5, 654-9, 664, 735, 894, 1020-1, 1178, 1186, 1244, 1249-51, 1313,
   1320, 1406, 1443, 1750, 1854, 1864-6, 2142-6, 2205, 2227-31, 2392, 2548, 2610, 2643-4, 2669,
   2675-6, 2863, 2875, 2955, 3081, 3149, 3204, 3280, 3362, 3608, 3636, 3639-42, 3683, 3869, 3910, 3998,
   4019, 4102, 4139, 4269, 4328, 4520, 4535, 4632, 4762, 4812, 4854, 4921, 5436b, 5458, 5486, 5514, 5569,
   5616, 5699, 5736-7, 5771-2, 6066, 6119, 6210, 6222, 6450, 6480, 6548, 6645, 6694, 6702-3, 6716, 6784,
   6812, 6962, 7035, 7178, 7327
Pressure effects
   1008, 2555, 3006, 3050, 3093, 3850, 5267, 5777, 5930, 6555, 6568, 7376
Process chromatography; p. analysis; p. control
   35, 58, 101-2, 142, 189, 210-1, 238, 243, 267, 301, 330, 367, 461, 511, 713-22, 958, 989, 1060, 1112,
   1159, 1323, 1427-30, 1445, 1535, 1611, 1619, 1842-3, 1850, 1904-5, 1959-62, 2005, 2093-4, 2219-20,
   2270-1, 2282, 2323-5, 2328-33, 2373, 2390, 2401, 2661, 2788, 2916, 3007-11, 3018, 3095, 3265, 3284,
   3291, 3415, 3436, 3624, 3762, 3768, 3836, 4110, 4170, 4458, 4539, 4574, 4602, 4605, 4749, 4760, 4777,
   4794-6, 4808, 4892, 4933-4, 5242, 5249, 5279, 5317, 5389, 5393, 5461, 5822, 5924, 6064, 6118, 6210,
   6217, 6221, 6238, 6368, 6440, 6489-90, 6574, 6752, 6769, 6883, 6935-9, 6967, 7069, 7074, 7127-38, 7196,
   7260-1, 7331-2, 7428, 7438, 7522, 7529, 7538-41
Programming - see Automatic; Temperature p.
Promethium
   2349, 2496, 5262, 5362-3
Propanes; propenes; propyl compounds
   373, 401, 521-2, 555, 614-5, 700, 729, 743, 820, 835, 896-7, 1087, 1100, 1130, 1139, 1142, 1156-7, 1198,
   1390, 1478, 1581, 1605, 1720, 1766, 1882, 2067, 2086, 2179-82, 2198, 2288-90, 2379, 2692, 2989, 3042,
   3118, 3124, 3314, 3797, 3870, 3908, 3914, 3921, 4072, 4085, 4517, 4554, 4718, 4824, 4830, 5016-8, 5086,
   5116, 5211, 5238, 5259, 5318, 5387, 5425, 5466, 5656, 5672, 5689, 5732, 5738, 5740, 5747, 5847, 6012,
   6165, 6191, 6263, 6329, 6356-7, 6473, 6484, 6504, 6755, 6916-7, 6961, 7069, 7098, 7102-3, 7151, 7246,
   7299, 7303, 7366, 7448, 7463, 7493, 7521
Propionic acid and compounds
   1216, 1605, 2688, 3468, 3927, 5387, 7299
Proportional counters; p. detectors
   1193, 3620, 4307
Propylene
   1180, 2379, 2677, 4523, 5109, 5466, 5656, 5689, 6263, 6887
Proteins - see Biochemical
Purification of materials
   1726, 1835, 3636, 6784, 7263
Pyrans; pyrazines; derivatives
   1720, 1839, 3161, 3358, 3714-5, 4797, 6979
Pyridines
   525, 1086, 1090-2, 1728-9, 2281, 2686, 3086, 3423, 3442, 3522, 3791-2, 4011, 4095-6, 4654, 5094, 5210,
   5265, 5285, 5387, 5520, 5525, 5792, 6052, 6886, 7039
Pyrocatechol derivatives
   6024
Pyrolysis
   184-6, 395, 479, 480, 483-5, 492, 494-5, 517, 531, 553-4, 555, 557, 561, 668, 819, 850, 939, 947, 1007,
   1761, 1763-4, 1826, 1827, 2043, 2088, 2106, 2383, 2402, 2663, 2692, 2757, 3136, 3368, 3507, 3511-6,
   3563, 3591, 3693-4, 3844, 3997, 4250, 4267, 4322, 4401, 4564, 4565, 4696-7, 4710, 4838, 4924-9, 4939,
   5065, 5196, 5213-4, 5236, 5306, 5539, 5602, 5637, 5654, 5689, 5936, 5974, 5991, 6413, 6515, 6661, 6670,
   6679, 6711, 6876, 6960, 7016, 7025, 7076, 7143, 7148, 7195, 7236
Pyrroles and derivatives
   942, 1834, 2246, 6986

                                                         Q

Qualitative analyses
   3035, 3051, 3735, 3744, 3764-5, 4136, 4376, 4498-99, 4883, 4887-90, 5302, 5560, 7014, 7147, 7370
Quantitative methods (see also Trace analyses)
   533, 869, 1039, 1117, 1312, 1421, 1846, 1853, 2110-2, 2133-5, 2140, 2277, 2425, 2885, 2894-6, 3035, 3046,
   3050-1, 3193, 3224, 3239, 3244, 3510, 3513, 3598, 3624, 3629, 3658, 3735, 3744, 3764-6, 4145, 4149, 4150,
   4154, 4197, 4208, 4222, 4417, 4498-9, 4747, 4765, 4774, 4894, 4923-5, 5232, 5302, 5305, 5448, 5475, 5704,
   5707, 5837-8, 5865, 5902, 5916-7, 5975, 6146, 6153, 6331, 6336-9, 6359, 6365, 6385, 6467, 6569-70, 6980,
   7014-5, 7124, 7265, 7211, 7318, 7370, 7440

Quaternary and diquaternary compounds
370-1, 542, 1141, 1143, 1515-20, 2261, 3024, 4899, 6876

Quinols; quinolines; and quinones
371, 1096, 1799, 2246, 2281, 2434, 3522-3, 5716, 6040, 6989, 7113, 7212, 7495

R

Radiochemistry (see also Tritium and specific elements)
12, 49-51, 57, 64-5, 70, 79-83, 95, 614-5, 635, 697, 747, 753, 761, 779, 850, 888-9, 962-3, 1056, 1074-5, 1163-4, 1193-4, 1198-1204, 1202-5, 1217, 1237, 1392, 1745, 1764, 1818-25, 1878, 1906, 1943, 1991-3, 2074, 2152-4, 2180, 2204, 2234, 2259, 2301, 2304, 2367-8, 2444, 2458-9, 2479, 2494, 2635-6, 2649, 2683, 2703-7, 2717, 2720-1, 2731, 2792, 2801-3, 2874, 2879, 2950, 3100-1, 3123, 3153, 3236, 3469, 3523, 3615, 3620, 3748, 3813, 3820-2, 3813, 3828-33, 3887, 4074, 4100, 4177, 4255, 4266, 4308-9, 4335-9, 4398, 4475, 4488, 4508, 4516, 4522, 4526, 4529, 4587, 4786, 4791, 4868, 4878, 4991, 5001, 5057-9, 5066, 5195, 5239, 5403, 5514, 5547, 5561, 5586-7, 5727, 5760, 5933, 6048, 6084-5, 6098, 6101, 6122, 6170, 6221, 6266-7, 6296, 6325, 6345, 6391, 6407, 6507, 6542, 6586, 6604, 6627, 6771-3, 6821, 6842, 6942, 6965-6, 6981, 7029, 7054, 7113, 7220, 7262-3, 7298, 7340, 7344, 7358-9, 7363, 7407, 7418, 7445, 7448, 7451-3

Radium, uses
2801-3, 6407

Raman spectroscopy
2836

Rare gases (see also specific elements)
1254, 2538-40, 2643-8, 2746-7, 3490, 3494, 3543-4, 3801, 4156, 4488, 5001, 6497

Reaction area; r.-gas chromatography; r. mechanisms; r. products
347, 1918-22, 2729, 3748, 4365, 4593, 5053, 6166, 6777

Reactor-gas chromatography
2130, 7220, 7554-5

Rearrangements
8, 787-8, 950, 1888, 2404, 2602, 2874, 3089, 3270-1, 3318, 3358, 3716, 4238, 4712, 5328, 5520, 5672, 5846, 5847, 6432

Recorders; recordings
917, 1669-71, 1902-3, 2014, 2722, 3469, 3513, 3588, 4969, 5073, 5332, 5664, 6261, 6355, 6551, 6610, 6648, 6826, 7561-2

Recovery process, use of G. C. as
5858

Reduction reactions
727, 1282, 2046-7, 3039, 3041, 3716, 4941, 5060, 5282, 5598, 5618, 5639, 5742, 6138, 6877

Refinery products (see also Hydrocarbons; Natural gases; Petroleum; and specific names)
94, 108, 154, 198, 231, 265, 399, 477-80, 483-6, 588-9, 646, 750, 832, 863, 868, 885-6, 906, 943-5, 1072-3, 1081, 1087, 1094-8, 1110, 1334, 1365, 1407, 1429, 1758, 1799, 1990, 2015, 2042, 2173, 2236-7, 2241-5, 2339, 2352, 2425, 2490, 2655-7, 2702, 2726, 3064, 3132, 3192, 3332, 3398, 3559, 3610, 3702, 3843, 3888, 3959, 3981-2, 4001, 4065, 4072, 4151, 4269, 4404, 4414, 4600, 4609, 4752, 4973, 5045, 5065, 5105, 5130, 5360, 5448, 5479, 5537, 5861, 5975, 6025, 6050, 6082, 6146-7, 6229, 6427, 6713, 6718, 6745, 6845, 6878, 6946, 7004, 7066, 7084, 7097, 7118, 7124, 7152, 7365, 7483-91, 7518

Refrigerants - see Cryogenic

Resins; gums; heartwood extractives
413, 416, 1501, 2098-103, 2100, 2102, 2934, 3301, 3844-5, 4325, 5112-3, 5441, 5947, 6712, 6839, 7316, 7466

Resolution in G. C.
276, 286, 409, 779, 827, 1018, 1652, 2408, 2548, 3355, 3816-7, 4514, 5773, 6008, 6011, 6568, 6707, 7293, 7473

Response time
3347, 5772, 6113-5, 7136

Restrictions in columns
136-7

Retention data
13-15, 40-2, 125-6, 661, 687, 748, 779, 798-801, 931, 1100, 1189, 1286, 1383, 1435-6, 1792, 1799, 1803, 2156-60, 2166, 2180, 2233, 2237, 2295, 2234-6, 2421, 2424, 2502, 2549, 2666, 2693, 2725-6, 2745, 2754, 2779, 2783, 2857, 2869-70, 2897, 2949, 2991, 3051, 3110, 3121, 3183-4, 3301, 3332, 3440-2, 3524-7, 3567, 3610, 3624, 3737, 3741-3, 3840, 3851-3, 3864, 3884, 3894-5, 3906, 4051, 4078, 4117, 4135-7, 4295, 4316, 4447, 4452, 4470, 4552, 4558-60, 4566, 4588, 4607, 4689, 4704-5, 4709, 4827, 4830, 4843, 4883-9, 5103, 5110-1, 5116, 5219, 5226, 5385-8, 5441, 5525, 5572, 5609, 5617, 5656, 5724-5, 5745, 5773, 5824, 5928-30.

6069, 6100, 6144-50, 6198, 6236, 6327, 6361, 6369, 6406, 6409, 6420, 6458, 6473, 6487, 6558, 6727, 6736, 6746, 6785, 6868, 6955, 6967-70, 6991-3, 7052-3, 7116, 7151, 7177, 7190, 7219, 7417, 7443, 7475, 7493, 7509-11, 7565

Reviews
6, 10, 88, 117-8, 128, 158, 191-208, 214-20, 222-9, 234-5, 254-7, 279-80, 289, 296-9, 303-8, 310-18, 402, 411, 420, 432, 544-5, 689, 711, 744, 901-2, 905, 997, 1001-5, 1024-8, 1040, 1097, 1174, 1191, 1255, 1304, 1378, 1453, 1497, 1533-4, 1554, 1565, 1660, 1665, 1690-4, 1721, 1772-83, 1791, 1807-8, 2010, 2016, 2035, 2073, 2079, 2105, 2137-8, 2187, 2205, 2224, 2239, 2417-9, 2423, 2507, 2519, 2574, 2612, 2625, 2650, 2666, 2672, 2690, 2716, 2742, 2791, 2804-12, 2867, 2927-8, 2953, 2970-3, 2986, 3016, 3034-5, 3046-8, 3061, 3079, 3085, 3096, 3130, 3164, 3221, 3233, 3424, 3431, 3439, 3445-6, 3455-7, 3477, 3497, 3500-5, 3508-9, 3519-20, 3609, 3635-7, 3675-82, 3718, 3738-42, 3758, 3766, 3770, 3784, 3835, 3865, 3890, 3900, 3907, 3915, 3931, 3936-9, 3947-50, 3963, 3978, 4082, 4215, 4284, 4332-3, 4370, 4395, 4459, 4481-2, 4573, 4622, 4645, 4655, 4664-72, 4703, 4776-7, 4802, 4844, 4867, 4919, 4924-9, 4938, 4981, 4996, 5003, 5013-4, 5062, 5072-7, 5119, 5187, 5234, 5247, 5292-3, 5314, 5394, 5410-1, 5415, 5431-2, 5480, 5488, 5494-9, 5501-2, 5551-9, 5566, 5613, 5627-30, 5644-5, 5676, 5681, 5698, 5714, 5750-3, 5823, 5894, 5973, 5982, 5990, 6014, 6043, 6079-80, 6131, 6153, 6173, 6196, 6201, 6205, 6215, 6221, 6246, 6280, 6322, 6347-9, 6380, 6417, 6442, 6449, 6546, 6574-7, 6580, 6598-9, 6606-7, 6659-60, 6690, 6701, 6710, 6760, 6817, 6832-3, 6853-4, 6894-6, 6907, 6918-21, 6925-7, 6936, 6962, 6978, 7041, 7067, 7082, 7128, 7133-7, 7249, 7275, 7282, 7314, 7338, 7351-2, 7431, 7494, 7505-6, 7517-9, 7544, 7547

Rhenium compounds
1061

Rubber; adhesive compounds
383, 471, 809, 1575, 1814, 1891, 2254, 4008, 4195-6, 4250, 4717, 4813-5, 5179, 5232, 5236, 5392, 6730, 6797, 7054, 7076, 7377

Rubidium and compounds
591

Rumen - see Dairy; Lipids

S

Salts, general
1942, 2951-2, 6164

Sampling, aspects of (see also Collection; Injection; Syringes)
4, 33, 134-5, 180, 320, 358, 506, 513, 659, 669, 749, 771, 798, 800-1, 834, 846-7, 981, 993, 1000, 1221, 1238, 1248, 1289, 1371, 1355, 1485, 1542, 1702, 1800, 1871, 1908, 1957, 1967, 1983, 2014, 2061, 2094, 2256, 2282, 2305, 2384, 2528, 2610, 2620, 2749, 2802, 2813, 2921, 2967, 3019, 3147, 3151, 3180, 3273-4, 3380, 3382, 3543-4, 3586, 3634, 3767, 3776, 3779, 3810, 3836, 3892, 4455, 4512, 4734, 4770, 4774, 4782, 4810, 4859, 4920, 4924, 4927, 5022, 5067, 5105, 5123, 5130, 5188, 5258, 5370, 5385, 5436b, 5437, 5443, 5466, 5509, 5636, 5663, 5770, 5784, 6014, 6019, 6028, 6281, 6285, 6292, 6310-5, 6334, 6355, 6359, 6377, 6419, 6480, 6485, 6513, 6579, 6682, 6711, 6786, 6851, 6940-1, 6980, 7077, 7187, 7216-7, 7233-5, 7291, 7424, 7427, 7469, 7559

Sanitary science applications (see also Air; Biomedical)
309, 1369, 2027, 2785-7, 2789-90

Saponins and derivatives
2693, 6992

Scandium and compounds
1987

Schleich's mixture (see also Anesthesiology)
1893, 2623

Schiemann reaction
5311

Scintillation counters (see also Radiochemistry)
1193, 1991, 3828-30, 3833, 4508, 5585-7, 6604

Sebum (see also Lipids)
976-7, 2855, 3475, 7252

Seed oils - see Oils, Seed

Semiconductor applications (see also specific uses and materials)
892, 3746, 5468-9, 6241

Sensitivity of equipment (see also Accuracy; Efficiency)
286, 385, 739, 1392, 2011, 2361, 3782, 4444, 4845, 4847, 5678, 5755, 6153, 7204, 7470-1

Separations (see also Fractionation; Partition)
5, 18, 21, 32, 41, 59, 60, 78, 111-5, 127, 138, 143-4, 230, 242, 249, 270, 363-8, 399, 432-4, 474, 544-52,

565-6, 579, 626, 648, 650, 659, 687, 690, 711-13, 754, 756-8, 769, 772, 775, 865-6, 878, 882, 948, 974, 988, 1009, 1049, 1064, 1065-7, 1070-1, 1086-9, 1118, 1121, 1144-5, 1173, 1196, 1254, 1267-9, 1275-8, 1281, 1286-7, 1325, 1347, 1353, 1357, 1380, 1384, 1408, 1418, 1439, 1453-5, 1472, 1474, 1478, 1502, 1536-7, 1556, 1576-7, 1588, 1596, 1600, 1602, 1626, 1647, 1664, 1680, 1699, 1705, 1738, 1771, 1793, 1799, 1873, 1889, 1911, 1919, 1940, 1964, 1970, 1986, 2014, 2060, 2090, 2108-9, 2141, 2147, 2154, 2166, 2180, 2185, 2244-6, 2257, 2322, 2340, 2347, 2369-70, 2375, 2379, 2388, 2397, 2407, 2413-5, 2431, 2450, 2465, 2473, 2490, 2546, 2555, 2586, 2630, 2636-42, 2660, 2723, 2731, 2737, 2746, 2750-51, 2765, 2789, 2793, 2794, 2795, 2798, 2799-800, 2822, 2827, 2829, 2858, 2887, 2897, 2900-2, 2910, 2943, 2953, 2957, 2995, 3030, 3050, 3083, 3090, 3093, 3103, 3130, 3147-8, 3171, 3172-3, 3195, 3215, 3239, 3242, 3244, 3258, 3290, 3293, 3301, 3317, 3322, 3324, 3326, 3327, 3340-9, 3373-4, 3390, 3396, 3400, 3404, 3421, 3423, 3425-7, 3429, 3459, 3467, 3485-91, 3522-6, 3527, 3532-4, 3535-6, 3541, 3546, 3553, 3569-71, 3584, 3600, 3606, 3611, 3687, 3701, 3720, 3736, 3746, 3756, 3766, 3794, 3805-7, 3815, 3862, 3870, 3874, 3884, 3903-4, 3960, 3992-3, 4020, 4030, 4045, 4066, 4095-9, 4103-6, 4118-9, 4120, 4122, 4144-5, 4150, 4153-62, 4180, 4189, 4205-6, 4221, 4231, 4263, 4264, 4272-6, 4288, 4317, 4347, 4387, 4415, 4426, 4432, 4452, 4469, 4506, 4513-6, 4520, 4535, 4555-60, 4591, 4606, 4614-5, 4681, 4702-5, 4709, 4718, 4751, 4758, 4781-3, 4806, 4813-5, 4860, 4885, 4890, 4898, 4910-1, 4921, 4950-1, 4973, 4990, 5015-7, 5031, 5043, 5050, 5086, 5099, 5106-8, 5210, 5219, 5232, 5249, 5281, 5287-8, 5301-2, 5337, 5345, 5374, 5383, 5386-7, 5388, 5390, 5415, 5434, 5436a, 5448, 5464, 5478-9, 5489, 5504, 5525, 5567-8, 5571, 5605, 5628, 5633, 5651, 5678, 5694, 5700, 5709, 5713, 5732, 5739, 5746-7, 5775, 5798-9, 5807, 5830, 5838-40, 5845, 5851, 5854, 5873, 5891, 5914-7, 5940, 5952, 5976, 5986-7, 5998, 6012, 6062, 6083, 6128, 6178-80, 6182-4, 6202, 6229, 6262, 6264, 6313, 6316-7, 6321, 6344, 6358, 6362-3, 6369, 6388, 6395, 6396, 6410, 6415-6, 6435-6, 6441, 6446-7, 6456, 6457-8, 6462, 6467, 6482, 6486, 6515, 6525, 6536, 6565, 6628, 6643, 6669, 6691, 6700, 6728, 6748, 6751, 6767, 6785, 6788, 6796, 6806-9, 6814, 6818-9, 6841, 6885, 6892, 6900, 6906-9, 6910, 6913-6, 6927, 6931, 6959, 6967, 6989, 6991-7, 7004, 7009, 7038, 7041-4, 7089-93, 7096, 7098, 7110-11, 7154, 7179, 7198-9, 7206, 7222, 7233-4, 7240-1, 7266, 7286-7, 7288-90, 7294-5, 7317, 7367, 7371-3, 7385, 7390, 7423, 7457, 7487, 7508-10, 7512, 7518-21, 7528, 7542, 7549-50, 7553

Serum analyses - see Biochemical; Biomedical; specific substances

Sewage treatment - see Sanitary science

Silicon; s. compounds; reactions; uses (see also Glass)
2, 5, 30, 38, 373, 576, 579, 583-91, 604-5, 706, 726, 851, 948, 990, 1015, 1127, 1299, 1506, 1556, 1604, 1805, 2243, 2247-50, 2388, 2394-9, 2400, 2480, 2482, 2521-3, 2728, 2751, 2881, 3687, 3976, 3909, 4017, 4143, 4271-2, 4436, 4441, 4449, 4554, 4711, 4720, 4871, 5055, 5219, 5232, 5381, 5453, 5503-6, 5517, 5850, 5907, 5915, 5952, 5969-70, 6003, 6017, 6119, 6124, 6178, 6248, 6258, 6272, 6383, 6406, 6418, 6455, 6644, 6692, 6700, 6785, 6838, 6910, 6913-4, 6985, 7010, 7089, 7093, 7096, 7186, 7266, 7432, 7481

Silver and compounds, uses of
167, 687, 1964, 3121, 5064, 5550, 6388

Sludge - see Sanitary science

Smoke, smog (see also Air; Tobacco)
212, 236, 1235, 1267, 1279, 1440-1, 1616, 1997, 2197, 2607, 3381, 3387, 4147, 5416, 5701-8, 6117, 6281, 6470-1, 7007

Soap industry (see also Detergents)
676, 2415, 3863, 5648, 5973

Sodium and compounds
29, 605, 2263, 2415, 2478, 5011, 5081, 5938, 6234

Soil chemistry - see Agriculture

Solanesol
271, 2776

Solubility; solutes
113, 378-9, 760, 1453, 2629-30, 2632, 3863, 4217, 4827, 5118, 5690, 6328, 6727, 7233

Solvents in G. C.; analysis of commercial s. (see also Paint)
291, 654, 786, 1298, 1358-9, 1606, 2006, 2032, 2072, 2435, 2488, 2513, 2611, 2632, 3030, 3307, 3705, 4266, 4357, 4633, 4704-6, 4827, 4829, 4929, 5003, 5330, 5489, 5519, 5577, 5680-1, 5697, 5966-8, 5972, 6230, 6231, 6284, 6382, 6584, 6594, 7225, 7273, 7294

Sonic waves; s. and ultrasonic methods
5240, 5675, 6163, 6801-2, 7334, 7350

Sorbic acid; sorbitol
933, 3052, 3876, 5078, 6624

Sorption data; s. processes; etc. (see also Adsorbents)
571, 602, 1679, 1948, 2215, 2760, 2780, 3130, 5002, 7371-2

Space applications
253, 1169, 1900, 4811, 5130, 6529, 7346-8, 7415

Spectroscopy, G. C. and (see also Infrared; Mass s.; NMR; Raman; Ultraviolet)
4264, 4425, 4503, 5875, 6145, 7265, 7325

Support materials - see Adsorbents; etc.

Surface area; s. chemistry

    1588, 1769, 2052, 2108-9, 2131, 2498, 2574, 2767, 2778, 2783, 3111, 3305, 3530-1, 4012, 4076, 4440, 4825, 5118, 5191, 5908, 5962, 6142, 6166, 6585, 6744

Syringes; microsyringes

    290, 2610, 2921, 4270, 6668

T

Tailing

    1399, 2033, 2686, 3014, 4066, 5904

Tall oil - see Resins

Tallow - see Waxes

Tantalum and compounds

    3901-4

Tars and constituents

    11, 750, 936, 969, 1041, 1334, 1335, 1527, 1580, 1976-9, 2237, 2241, 2434, 3388, 4216, 4401, 4793, 6798, 6878, 7416

Teflon

    2274, 4256, 6526

Temperature, aspects of in G. C. (see also High-temperature; T. programming; Thermal)

    200, 409, 477-80, 483-6, 646, 688, 820, 1366, 1452, 1639, 1662, 1904, 1933, 2416, 2526-7, 2599, 2752, 2897, 3050, 3056, 3183-4, 3208, 3766, 4316, 4978, 5258, 5568, 5642, 5732, 5756-7, 5844, 5854, 6115, 6219-20, 6408, 6703, 6951, 7062, 7231-3, 7293, 7515

Temperature programming (see also Automatic)

    206, 328, 361, 384, 460, 494, 495, 637, 677-9, 731, 795-6, 907-8, 949, 956-7, 985, 1167-8, 1381, 1549, 1653-61, 1679, 1725, 2031, 2070, 2097, 2101, 2126, 2139-44, 2408, 2549, 2556-9, 2564-7, 2597, 2869-73, 3633-4, 3907, 4084, 4108, 4401, 4454, 4656-63, 4783, 4843, 4885, 4891, 4985, 5303, 5617, 5844, 5928-30, 5996-9, 6049, 6067, 6103, 6567, 6628, 6703, 6793, 6796, 6840, 6916, 7131, 7231, 7296, 7376, 7442

Terms; terminology

    2666, 3624, 3713, 3655, 4369

Terpenes and derivatives

    8, 369, 405, 527, 559, 630-5, 650, 660-1, 769, 775-6, 788, 899, 1052, 1275, 1418, 1439, 1608, 1641, 1680, 1715, 1845, 1874, 2008, 2085, 2192, 2233, 2451, 2693, 3032, 3145-6, 3216, 3377, 3422, 3920, 3980, 4053-5, 4056, 4162, 4306, 4387, 4468, 4736-7, 4972-3, 5052, 5122, 5148, 5173, 5179, 5300, 5318, 5481-4, 5596, 5942-50, 6022, 6317, 6504, 6516-8, 6650, 6678, 6811, 7056, 7198-9, 7200, 7223, 7335, 7567

Tests

    790, 1954

Tetra- compounds

    5, 250, 371, 479, 496, 501, 590, 709, 735, 990, 1257, 1309, 1884, 1931, 2067, 2242, 2449, 2601, 3099, 3161, 3873, 4107, 4276, 4277, 4500-1, 4506, 4720, 4761, 5110-1, 5116, 5144, 5161, 5380, 5448, 6133, 6441, 7151, 7294, 7390, 7457

Textile industry

    4717, 7023

Thallium and compounds

    145, 4306, 5050

Theory, general, of G. C. (see also Fundamental)

    173-5, 813, 1027-8, 1378, 1453, 1733, 1807, 1851, 2539-83, 2562, 2627-35, 2639, 2953, 3076, 3188, 3569-71, 3907, 3957, 5314, 5413, 5732, 7232, 7351, 7504, 7526

Thermal conductivity; T. C. detectors and cells; T. analysis; T. decomposition; T. response; T. stability

    23, 36, 97, 120, 373, 543, 548, 609, 739, 849, 895, 912-20, 938-9, 1080, 1128, 1135, 1153, 1179, 1224, 1234, 1321, 1324, 1465-6, 1486, 1506, 1555, 1560-2, 1686-8, 1718, 1808, 1926, 2002, 2020, 2181, 2232, 2289-94, 2439, 2471, 2497, 2517-8, 2595, 2718, 2770, 3017, 3074-5, 3128, 3139, 3143, 3194, 3479-84, 3612, 3662-8, 3671, 3763, 3769, 3801-2, 3908, 3932, 3945-9, 3953, 3965, 3969, 4154, 4165, 4171, 4202-3, 4229, 4257, 4287, 4294, 4323, 4336-7, 4444-5, 4474, 4497, 4589-90, 4653, 4719, 4775, 4788-9, 4876, 4894, 4949-50, 4957, 4986, 4989, 4993, 5056, 5101, 5280, 5358-9, 5369, 5398, 5553, 5575, 5677-8, 5752-7, 5857, 5902, 5981, 6071-2, 6076, 6114-5, 6153, 6185, 6194, 6203, 6258, 6297, 6322, 6440, 6460, 6497, 6530-1, 6581, 6602, 6621, 6758, 6774, 6820, 6839, 6879, 6950, 7059, 7062, 7126, 7129-31, 7164, 7189, 7207, 7231, 7244, 7321, 7381-4, 7470-1, 7476, 7518, 7530-1

Thermistors; thermionic devices

    1392, 1555, 1696-7, 3297-9, 5056, 5098, 6407, 7014-5, 7470-1

Thermodynamics of G. C.
43, 85, 155, 1578, 1588, 2167, 3119, 3957, 4273, 6951

Thin-layer chromatography (see also Capillary)
2028, 2888, 4358, 4626, 5038, 6238, 7400

Thinners - see Paints

Thio- compounds
140, 482, 520, 552, 832, 1306, 1479, 1609, 1803, 1931, 2246, 3091, 3292, 3293, 3330, 3418, 3815, 4014, 4397, 4682, 4768, 4888-9, 4911, 5331, 5342, 5476, 5543, 5758, 5860, 5976, 6179, 6473, 6632, 6785, 6989, 7572

Tin and compounds
5, 2347, 2371, 3756-7, 3901, 3903-4, 4477, 4731, 5509, 5909

Titanium and compounds
990, 2021-2, 3901-4, 3749, 6266-7

Titrations
867, 3857, 6814, 6902

Tobacco and products
212, 271, 517, 1235, 1263-4, 1279-81, 1440-1, 1616, 1633, 1997, 2385, 2448, 2607, 3381, 3387, 3563, 3685-6, 4095-6, 4189, 4581-3, 5074, 5257, 5416, 5525, 5692-3, 5700-8, 6116-7, 6245-7, 6470-1, 6476, 6527-8, 6545, 6657, 7007

Toluenes and derivatives
27, 83, 87, 544, 729, 1101, 1207, 1286, 1343, 1766, 1982, 2087-8, 2173, 2335, 2354, 2369, 2726, 2983, 3208, 3293, 3307, 3310-2, 3326, 3793, 4566, 4677, 4830, 4890-2, 4904, 4911, 4958, 5204, 5211, 5387, 5399, 5916-7, 6165, 6345, 6358, 6584, 6627, 6633-4, 6742, 6955, 7032, 7101, 7266, 7433

Toxicology (see also Biomedical; Forensic; Narcotics; Pharmaceutical)
792, 1212, 1398, 1613, 1631, 1834, 3884, 4571, 5385, 5387-8, 5465, 6014, 6509, 7184, 7276

Trace analyses
71, 161, 193, 232-3, 242, 269, 331, 453, 458, 463, 729, 780-2, 791, 909, 1003, 1025, 1038, 1051, 1252, 1385, 1392-4, 1576, 1593-5, 1601, 1639, 1651, 1749, 1974, 2026, 2069, 2087, 2125, 2141, 2173, 2184, 2197, 2203, 2307, 2374, 2455-6, 2490, 2646, 2761, 2905, 2947, 3126, 3162, 3168, 3365, 3382, 3429, 3554-5, 3624, 3709, 3727-8, 3789, 3914, 4073, 4270, 4291, 4315-7, 4348-9, 4370, 4400-2, 4436, 4450, 4500-1, 4545, 4570, 4576, 4638, 4647, 4673, 4708, 4770, 4812, 4905, 4922-4, 4929, 4948, 4993, 5025, 5081, 5094, 5230, 5266, 5414, 5427, 5521, 5611, 5647, 5710, 5803, 5808, 5826, 6014, 6038, 6095, 6127, 6190-1, 6255-6, 6477, 6483, 6569-70, 6641, 6713, 6799, 6805, 6827, 6903, 6963, 6971, 6980, 7047, 7034, 7343, 7455, 7457, 7569

Transfer devices and systems
1935, 2758, 4164, 4774, 5965, 7308, 7464

Transport, chromatographic
5965, 6892-3

Traps (see also Capillary)
1332-3, 2129, 2236, 2391, 2883, 4632, 4967, 4971, 6243, 6507, 6561, 6794, 6825, 7178, 7187

Trialkyl compounds
1104, 2221, 3107, 4544, 5712, 6155, 7145

Trichloro compounds
2, 1398, 2021, 2884, 3147-8, 3359-60, 3584, 4034, 4554, 5103, 5110-1, 5387, 6248, 6258, 6910, 7151

Triethyl compounds
1165, 1348, 2606, 4816, 5830, 5851, 6477

Triglycerides (see also Glycerides)
346, 423, 2407, 2462, 3253, 3607, 3891, 4208, 4766, 4806, 5434, 5446, 5665, 6060, 6163, 6438, 6835

Trimethyl compounds
1286, 1389, 1444, 2248, 3328, 4074, 4248, 4271, 4689, 5116, 5970, 6041, 6057, 6112, 6236, 6272

Tritium
12, 70, 79-82, 1194, 1197, 1201-2, 1207, 1877, 1906, 1993, 2149, 2474, 2683, 2706, 2950, 3820, 3832, 4307-10, 5284, 5585, 5631, 5638, 5815, 5933, 6048, 6586, 6789, 7262, 7264, 7358-9, 7407, 7447

U

Ultrasound - see Sonic

Ultraviolet spectroscopy
405, 629, 792, 2372, 2516, 2836, 3094, 3684, 3860, 3880, 3882, 4892, 6649, 7575

Uranium and compounds (see also Radiochemistry)
3025, 3400, 3404, 7367

Urea and compounds (see also Agricultural; Biomedical)
490, 2150, 2375, 5759, 6712, 6885, 7157, 7465-6

Urinalyses (see also Biochemical)
937, 1605, 2857, 5896, 5898, 6467, 6663, 6667, 7317-8, 7320

V

Vacuum G. C.
98, 529-30, 7099
Valves
1618, 3625, 3810
van Deemter equation
659, 1734, 2669, 3711, 3968, 4547, 4607, 5413
Van Slyke method
3707, 3759
Vapor pressures; v. fractometer; v.-liquid equilibria; v. phase chromatography; etc.
85, 326, 820, 901-4, 916, 994, 1465-6, 2537, 3149, 3183-4, 3477, 3610, 3702, 3719, 3778, 3853, 3899, 4251, 4488, 4491, 4872, 4906, 4966, 5264, 5371, 5493, 5640, 5892, 5923, 6169, 6618-9, 6793, 6849, 6855-7, 7085, 7115, 7154, 7201, 7285, 7324, 7469
Ventilization
946
Vinyl compounds (see also Plastics; Polyvinyls)
69, 398, 485, 1058, 1980, 2091, 2291, 2825, 3004, 3691-2, 3757, 3861, 4922, 5083, 5357, 5402, 5439, 5509, 5730, 5838, 6278-9, 6884, 6971, 7073, 7376, 7432, 7449
Vitamins (see also Pharmaceutical)
1260, 1834, 3793, 5115, 5219, 5596, 7353, 7367, 7537
Volatile components (see also Agricultural; Dairy; Essential oils; Foods; Perfume)
1403, 2218, 3331, 3425-7, 3452, 3973, 4383, 4665, 5387, 5567, 6525, 7326
Volumetric analyses
6235, 7094-5

W

Water, analysis of; in substances; substances in
22, 325, 389, 469, 505, 519, 729, 761, 764, 897, 979, 1010, 1218-9, 1252, 1952, 1969-70, 1974, 2065, 2069, 2072, 2210, 2489, 2983, 3176, 3348, 3541, 3542, 3576, 3870, 4073, 4112, 4213, 4223, 4344, 4631, 4723, 4989, 5023, 5087, 5109, 5361, 5387, 5567, 5709, 5878-9, 5920, 6012, 6176, 6190, 6381, 6609, 6681, 6685, 6803, 6868, 6968, 7176, 7322, 7336, 7390, 7493, 7556
Waxes, tallow
245, 400, 1912-4, 1981-2, 2535, 2853, 3133, 3382, 3860, 3894-5, 4145, 4252, 4360-6, 4756-9, 5092, 5220, 5232, 5290, 5463, 5941a, 6103, 6345, 6482, 7413

X

Xenon
433, 2640, 2746, 3543-4, 4099, 6072, 7505
Xylene, xylols, xylose
86, 363-6, 415, 629, 986, 1287, 1343, 1793, 1799, 2087-8, 2243, 2341-2, 2726-7, 3070, 3326, 3995, 4275-6, 4958, 5107, 5288, 5524, 5661, 5736, 5916-7, 5986, 6069, 6345, 6479, 6487, 6633-4, 6742, 6796, 6889, 6955, 7101, 7385, 7426, 7429, 7502, 7551, 7553

Z

Zeolites, chabazites
166, 563-9, 577-81, 586-7, 596-9, 601, 1012, 1101-2, 1760, 1947-52, 2840, 3495-6, 3504, 3532-4, 4015, 4153-7, 4538, 5762, 6843-4, 7502
Zinc, compounds and uses
1987, 5654, 6268